TECHNOLOGY FORECAST: 1998

PRICE WATERHOUSE

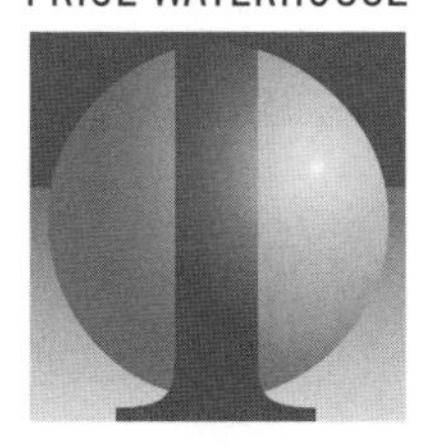

TECHNOLOGY CENTRE

Price Waterhouse Global Technology Centre
Menlo Park, California 94025, U.S.A.

Technology Forecast

Version 8
January 1998

Price Waterhouse Global Technology Centre
68 Willow Road
Menlo Park, California 94025, U.S.A.

Phone: +1-650-322-0606
Fax: +1-650-321-5543

To order copies of the *Technology Forecast*, order #PWTC-01-08, please see page v.

Printing 10 9 8 7 6 5 4 3 2 1

Table of Contents

Level 1 – Introduction **1**

Level 2 – Interviews **5**

2.1 – Interview with Commissioner Bangemann 7
2.2 – Interview with Donald Gips 15
2.3 – Interview with Thomas Malone 23
2.4 – Interview with Dr. Othman Yeop Abdullah 33

Level 3 – Components **43**

3.1 – I/O Technologies 45
3.2 – Storage and Batteries 69
3.3 – Smart Cards 107

Level 4 – Communications **135**

4.1 – Transmission Media 137
4.2 – Networking Systems and Protocols 165
4.3 – Telecommunications Services 203
4.4 – The Internet and Intranets 231

Level 5 – Systems and Architectures **283**

5.1 – Operating Systems Environments 285
5.2 – Software Component Architectures 317
5.3 – Computing Platforms 341
5.4 – Security 373

Level 6 – Software **427**

6.1 – Application Development Environments 429
6.2 – Databases 469
6.3 – Groupware, Workflow, Document and Knowledge Management 491
6.4 – Systems and Network Management 521

Level 7 – Applications **551**

7.1 – Electronic Commerce 553
7.2 – Data Warehousing 599
7.3 – Corporate Applications 643

Level 8 – Back of Envelope Numbers **685**

Index **701**

Acknowledgments

Editorial Director	Andrew Binstock
Associate Editors	Corinne DeBra, Barbara Jurin
Contributing Editors	Eric Berg, Michael Katz
Partner in Charge	Paul Turner
Contributors	Steve Baker, Laura Bayer, Geoffrey Bock, Ralf Burget, Steve Farmer, Mark Ferelli, Joan Garrett, David Giannini, Mark Hall, Emily Kay, Ronni Marshak, Sandy Metz, Patrick O'Neil, Joe Panettieri, Suzanne Purnell, Gail Graziano Roberts, John van den Hoven, Cynthia Weaver, Alan Zeichick
Advisory Review Board	Donald Almeida, Mark Austen, Fred Balboni, Joseph Bellissimo, Mike Boberschmidt, Tom Chambers, Jagdish Dalal, Stephen Darcy, Andrew Davies, Peter Davis, Allen Edwards, Steve Edwards, Joël Garlot, Jim Goodwin, Patrick Gray, Scott Hartz, Steve Higgins, William Hoffman, Kerry Horan, Peter Horowitz, Doug Kalish, Lew Krulwich, Hideki Kurashige, Mark Lutchen, Donald McGovern, Nigel Montgomery, Bob Muir, Howard Niden, Michael Olszewski, Roger Pavitt, Paul Pederson, Mike Schroeck, John Singel, Arnie Segal, Noel Taylor, Dick Vincent, Kirk Walden, Paul Weaver, Dale Young
Reviewers	Faried Abrahams, Ken Amann, Utpal Amin, Roger Auinger, Doug Baird, John Ball, Alastair Cran, Ajay Dhingra, Jack Dooley, Adolf Dorig, Lionel Drew, Clarke Ferber, Giri Giridharan, Richard Golding, Larry Gosselin, Lynnette Groves, Alan Guscott, Walter Hamscher, Jim Holloway, John Hunt, Mark Huppert, Jeffrey Keller, Chris Knowles, Peter Kranz, George Kurtz, Priscilla Kwok, Michael Littlejohn, Matthew Mancuso, Pete Marcotty, Chris Martin, James McGrath, Stephen Mongulla, Tom Morgan, David Morris, Bernard Morrow, Phu Ngo, Dave Oshman, Irv Paskowitz, Deepan Patel, Todd Pearsall, Ned Pendergast, Bernard Plagman, Geoffery Price-Gunter, Terry Retter, John Riccio, Alex Rosenbaum, Galen Schreck, Eric Schultze, Edmond Sheehy, Jude Sylvestre, David Thiele, Rick Turner
Business Development	Gayle Rocklage
Research Services	Ann Mueller, Bo Parker, Jim Reed, Glorianne Wong
Lotus Notes Version	Jennifer Usher
Graphics and Layout	Loretta Au, Susie Campbell, Arlene Kilkenny
Indexer	Lesley Schneider
Copy Editors	Lisa Ramos, Robyn Frendberg
Cover	Andrew Hathaway
Distribution	Benny Beronda, Steve Johnson, Joyce Uriwal

Special thanks to Dana Thorat.

Preface

Welcome to the 1998 version of the *Technology Forecast.*

The *Technology Forecast* is a publication of the Price Waterhouse Global Technology Centre. It is prepared with the assistance of a variety of subject experts. Working with authors' drafts, Price Waterhouse practice units worldwide collaborate in an extensive editorial review process to produce the final text. Our direct contributors are recognized in the Acknowledgments section. In addition, this book is a result of the knowledge available within Price Waterhouse.

The Lotus Notes edition of the *Technology Forecast* is in a database, called `PWTC Technology Forecast`, on Price Waterhouse's network of Lotus Notes servers. This on-line edition provides a structured view of the document.

Obtaining Copies

The *Technology Forecast* (order #PWTC-01-08) is available for US$450 by calling +1-800-654-3387 (U.S. calls only) or +1-314-997-2540, or by sending a fax to +1-314-997-1351. To place an order over the Internet, refer to *http://www.pw.com/tc/tforder*. The price includes shipping and handling. American Express, MasterCard, and Visa are accepted. Payment by check can also be arranged.

The *Technology Forecast* is available both in print and in electronic formats. Price Waterhouse clients should direct requests for printed copies to their engagement contact.

Clients of Price Waterhouse may license an electronic version of the *Technology Forecast* in conjunction with a subscription to the Lotus Notes `PWTC Technology Update Service` database. For details, contact Gayle Rocklage at the Technology Centre, +1-650-688-6611 or via e-mail to `Gayle_Rocklage@notes.pw.com`.

Printed copies of the *Technology Forecast* are available for internal use. Partners and staff should direct requests for document #PWTC-01-08 by Notes to `NAC Distribution@Price Waterhouse-US`. For more information, PW partners and staff may refer to either the `PWTC Tech Forecast Update` or `PWTC Technology Forecast` Lotus Notes databases.

Commentary and Questions

The editors and authors of the *Technology Forecast* welcome your comments. We also want to hear how you are using the *Technology Forecast* in your work, and how you might use it in the future. A self-addressed reader response card is included at the back of the *Forecast*.

An electronic mail address has been established to submit comments, questions, or suggestions about the *Technology Forecast*.
Price Waterhouse users of Lotus Notes can reach the editors at:

`Technology Forecast Editors@Price Waterhouse-US`

Internet users can send their comments to:

`Technology_Forecast_Editors@notes.pw.com`

Readers without access to electronic mail can contact the Technology Centre by phone, fax, or mail. We look forward to hearing from you.

Organization

The *Technology Forecast* is organized into levels, each covering a general technology subject. We begin with an introduction and interviews with public policymakers about future technology directions. We then examine components and transmission media – the level of technology that is least visible to the user. Subsequent levels progress up the hierarchy through systems, architectures, software, and applications.

Each level of technology covered in the *Technology Forecast* builds on the information contained in the previous one. It is not assumed that the reader will read the entire *Technology Forecast* sequentially. Each chapter is designed to serve as an individual reference resource.

As an additional reference aid, the Index makes it easy to locate discussions of specific topics, products, and companies as well as cross-references to acronyms used in the text.

The content of each chapter is organized into the following sections:

- *Executive Summary* – A brief overview of the chapter's contents.
- *Technology* – Basic background information. This section includes how the technology evolved, how it works, and basic terminology. Recent major events and trends are also discussed.
- *Market Overview* – A summary of how the technology is used in products by major vendors in the marketplace.
- *Forecast* – Forecasts for the progress of the technology over the next 1 year to 3 years (referred to as the "forecast period").
- *References* – A list of references cited in the chapter, plus a bibliography of related reference materials and URLs of selected mentioned companies.

Please note: We have chosen the following usage for the *Forecast*:

- All monetary amounts are expressed in U.S. dollars ($).
- Units of measure are expressed in either metric or U.S. Customary system, depending upon common industry usage.
- A billion is 1,000 million (1,000,000,000 or 10^9).

About Price Waterhouse

For nearly 150 years, Price Waterhouse has been helping the world's leading companies solve complex business problems. Today, through a worldwide network comprising 56,000 professionals, Price Waterhouse assists clients in effecting organizational and strategic change; using information technology for competitive advantage; complying with statutory audit and tax requirements; and implementing strategies to improve business performance and increase shareholder value.

To find out more about Price Waterhouse, please call your local office or visit our World Wide Web home page at `http://www.pw.com`.

About the Technology Centre

The Price Waterhouse Global Technology Centre, located in Menlo Park, California, provides Price Waterhouse engagement teams and their clients with analysis, evaluation, and implementation of emerging information technologies.

The Centre has a staff of more than 40 researchers, software developers, technology analysts, information specialists, and administrators. These professionals have extensive backgrounds in advanced uses of existing technology as well as in the practical application of emerging technologies.

Technology Centre staff deliver on-site presentations on technology trends to Price Waterhouse clients and staff around the world, using the *Technology Forecast* and other proprietary thought leadership publications as references. These presentations supplement our general consulting services, which focus on information technology and business strategy issues.

The Centre also provides fee-based custom research and competitive analysis on information technologies and industry trends. This research is used to give the firm and its clients a competitive edge in the marketplace.

To keep Price Waterhouse professionals and clients abreast of new findings and tools, the Centre also produces the PWTC Technology Forecast Update Service, which abstracts recent developments in key technology industries. Subscribers can access this service by using the Web, e-mail, or fax delivery options. The service is also available in Adobe Acrobat format on CD-ROM.

The Technology Centre engages in research and software development designed to maintain Price Waterhouse's leadership in delivering professional services. For example, Technology Centre knowledge management initiatives are incorporated in a wide range of audit, business advisory, tax, and consulting services.

All of these activities are the result of a strong research mission coupled with a problem-solving approach that puts state-of-the-art technology theories into practice.

List of Figures

How Capacitive-Sensing Touchscreens Work 49
How CCDs Create an Image 51
How FEDs Work 57
Inside a Light-Valve LCD Projector 58
The Light-Reflection Process in DMD Projectors 59
Lithium-Ion Battery 96
Lithium-Polymer Battery 97
Zinc-Air Battery 98
Worldwide Disk Drive Shipments by Vendor 99
Worldwide Hard Disk Units Shipped by Capacity 100
Mainframe Storage Market Share 101
Worldwide Forecast of Portables by Battery Type 102
A Smart Card Family Tree 108
German GeldKarte Multi-Application Architecture 110
More Memory, Less Cost per Card 111
Smart Card Transaction Authorizations 115
Contactless Card and Combicard 116
French Offline Debit Transactions 120
French Smart Cards Cut Fraud 121
Danmønt Cards and Transactions on the Rise 122
Reloadable Electronic Purse System 123
Worldwide Smart Card Usage: 1997 130
Worldwide Smart Card Issuer IT Spending: 1997 130
Cards, Sectors of Activity, and Applications 131
Repeaters Required for Fiber versus Copper Cabling 139
Fiber Cable and Copper Cable 140
How SONET Works 142
How WDM Works 143
How Hybrid Fiber-Coax Systems Work 152
Networking Devices Overview 174
Worldwide Network Market Revenue – 1996 196
Switched Lines Can Be Cost-Effective 221
Packet/Cell-Based Service Revenue by Segment 227
ISDN BRI Usage Growth 227
Push Technology 247
Internet Telephony 262
Lotus Domino Architecture 268
Microsoft's Commercial Internet System 269
Netscape's Open Network Environment Architecture 270
Oracle's Web Request Broker Architecture 271
Oracle's Network Computing Architecture 271
Growth in Web Hosts 272
Internet Users 273

Internet and Intranet Server Shipments....277
The Three Waves of Client/Server Computing....320
Traditional Program Functions versus Objects....325
Two Ways of Invoking Objects Using a CORBA ORB....327
The Architecture of Enterprise JavaBeans....330
Microsoft's DCOM Architecture....333
Architecture of the Microsoft Transaction Server....334
Worldwide Distributed Transaction Middleware Revenue....337
1997 Computing Product Segments....348
The PC Models Defined in the PC 98 Specification....357
Worldwide Computer Factory Revenue: 1996-2000....361
Revenue Market Share for Midrange Vendors – 1996....362
Worldwide Workstation Sales by Type....363
Worldwide Revenue from PC Sales....364
Worldwide Shipments of Mobile Computers....365
Worldwide Unit Sales of PDAs and Hand-Held Computers....365
Realignment of Computing Platforms as Projected for 2002....367
Kerberos Security....381
DCE Security....383
Government Regulation of Cryptography and Related Technologies....397
Projected Growth in the Security Market....419
Anti-Virus Vendor Worldwide Market Share – 1996....421
Worldwide Development Tools Revenue by Category....462
Worldwide Development Tools Revenue by Platform....463
Java Integrated Development Environments Revenue....464
The Workflow Continuum....500
Routing Models for Workflow Processes....502
Document Management as an Integration Environment....506
The Knowledge Continuum....510
SNMP Components....531
WBEM Architecture....534
CARP Caching....536
Worldwide Systems Management Software Revenues....542
CA's Unicenter TNG Architecture....545
Internet Commerce....556
EDI and the Web....558
Electronic Commerce Architecture....565
Pandesic Process....569
SET Payment Authorization....575
Cisco Connection Online....587
Worldwide Internet Commerce Market....591
Creating the Data Warehouse....609
Technical and Business Metadata....614
OLAP as a Multidimensional Data Cube....617
A Typical Dimensional Model....617
Web-Accessible Architecture....631

Web-Enabled Architecture 631
Web-Exploited Architecture 632
SAP Business Information Warehouse 633
Worldwide Data Warehouse Market 636
SAP R/3 Architecture 654
Worldwide Corporate Application Vendor Revenues – 1996 680
Rate Ranges for Service Types 689
Radio Frequency Spectrum 692
Radio and Television Radio Frequencies (Part 1) 692
Radio and Television Radio Frequencies (Part 2) 692
U.S. Allocated Personal Communications Services (PCS) 692
Computer and Telecommunications Hardware Spending 694

List of Tables

Smart Card Chip and Card Maker Alliances....129
Transmission Media Alternatives....138
EIA/TIA-568-A-1 Optical Fiber Specifications....142
Network Installation Costs....153
Fiber Savings – Operational Costs per Node per Year....154
Three Basic Types of Telecommunications Satellites....156
Share of Premises Wiring Types in Europe....161
Comparison of IPsec and PPTP....185
Worldwide Hub Port Shipments by Technology....198
Worldwide LAN Switch Market....198
DSL Technologies....212
Examples of Wireless Products....218
Circuit Switching versus Packet Switching....220
U.S. BRI and PRI ISDN Revenue Model and Forecast....228
New Generic Top-Level Domain Names....238
URLs for Internet Search Sites....250
Characteristics of Major Operating Systems....287
Windows Operating-System Features....298
Worldwide Desktop Operating System Sales....308
Worldwide Customer Revenue by Server Operating System....310
Representative Workstation Processors....345
Representative Hardware Defined in PC 98 Specification....358
Features of Network Computers versus NetPCs....360
Managed Internet Security Service Providers....375
Representative Biometric Technologies and Vendors....377
TCSEC and ITSEC Product Evaluation Criteria....384
Estimated Time Needed to Crack DES Encryption Key....386
Recent Mergers and Acquisitions in the Security Sector....418
Popular Older Programming Languages....431
Object Technology Definitions....434
Popular Object-Oriented Programming Languages....435
Analysis and Design Models and Diagram Types....444
Software Configuration Management Capabilities....445
Software Testing and Quality Assurance Tools....448
Worldwide Server Platform DBMS Revenue – 1996....484
Worldwide Software Revenue for Database Vendors – 1996....485
Worldwide Systems Management Software Revenue for ISVs....542
Management Platforms Software Revenue....543
Worldwide Storage Management Software Revenues....543
EC Applications by Industry Groups....561
Different Types of Electronic Payment Systems....572
Alternative Electronic Payment Standards....577
Early Adopters of Internet Commerce....581

PCWeek's Top Ten Web sites 582
Key Electronic Commerce Infrastructure Providers 593
Contrasts Between OLAP and OLTP 602
Data Mart versus Data Warehouse 607
Representative Warehouse Generation Vendors 612
Representative OLAP Data Management Vendors 620
Representative Data Access Tool Vendors 622
Representative Data Mining Tool Vendors 626
Historical Component Price/Performance Improvement 685
Some Microprocessor Architectures 686
Document Storage Requirements 687
Common Storage Media 687
Audio Storage Requirements 688
Video Storage Requirements 688
Memory Requirements for Video Resolution Levels 688
Telecommunication Line Rates 690
Approximate Transmission Times for Typical Documents 690
Approximate Transmission Times for Files 691
Common Contemporary Satellite Uses 691
Information Services Budget as Percent of Total Revenue 693
Prefix Reference 695
Common Paper and Envelope Sizes 696
Significant Technology Introductions on IBM Computers 697
Historical Events 698

1 Introduction

Because of the growing complexity of information technology (IT), the reach of this edition of the *Technology Forecast* exceeds all previous versions. In the following chapters, you will find detailed explanations of dozens of technologies of interest to corporate IT organizations as well as predictions as to how these technologies will develop during the forecast period (roughly 1998 to 2000). As we studied these technologies, we frequently saw how the development of technology discussed in one chapter depends significantly on technology discussed in another. These interdependencies are difficult to define in the individual sections of the *Forecast*, and so they are presented here for your thoughtful consideration. Moreover, to see how key technologists in the public sector view these interrelations, we have included four interviews with key advisors and policymakers from around the globe.

In 1997, business came to recognize and respect the Internet as a significant, strategic, commercial opportunity. In previous years, the Internet was viewed as a new entry point for existing customers into the sales cycle, although seen internally as a useful carrier for intra-company communication. As a result, companies that viewed themselves as Internet-aware generally had corporate-wide e-mail, an intranet, and a Web site. By the end of 1997, however, these offerings were *de rigueur*, and companies that had not done anything more were in sudden danger of falling behind. The reason for this discrepancy is electronic commerce.

Electronic commerce may be the "killer application" of the Internet. Corporations have come to recognize that Web sites are used by more than existing customers and that electronic commerce is more than handling existing transactions via a new medium. The new realization is that good Web sites are powerful magnets for new customers, and – more important – that electronic commerce can be far more cost-effective than traditional sales processes. This advantage has been particularly evident in high-volume, retail businesses. For example, in 1997, almost all discount investment houses promoted Web trading of stocks and bonds, while charging record low commission levels. Also, following the model popularized by Amazon.com, traditional bookstore chains moved to establish their own Web presence.

This trend is likely to accelerate, fueled in part by an increasing confidence in Internet security. Although consumers still feel skittish about relaying confidential information over the Internet, the dire predictions of frequent interception of customer passwords and payment information failed to come true in 1997. Most security breaches on the Internet were concerted, malicious attacks by crackers against specific Web sites or Internet service providers. The biggest concerns for consumers were the need for faster access and relief from junk e-mail, unaffectionately termed "spam."

The next few years should also see electronic commerce move squarely into the area of high-value sales transactions, especially in business-to-business transactions. Indeed, although consumer-oriented sites gain most of the publicity, the real story for the next few years will be in business-to-business electronic commerce. Extranets, and to some extent new forms of EDI, will drive this push.

The Internet left its imprint throughout the entire IT market, however, and its effects are likely to be felt for years.

For example, software development saw the continued upswing in the use of Java, Sun's portable programming language, as it evolved from its Internet orientation. Now, the language increasingly is used in traditional application programming, where downloading freestanding executable portions of the program is not a necessity. The frequent extensions such as JavaBeans, with which Sun has endowed Java, have moved it from a pure development language into an execution platform, validating one of Microsoft's major concerns about the upstart language: that it would become a viable alternative to the Windows family of operating systems as the standard to which applications are written.

The Internet also has driven and expanded the corporate view of distributed objects. This computing model, once seen as a cutting-edge technology of interest to large companies with special needs (and advanced technical skills), has become more mainstream and is likely to be the standard programming paradigm within the next few years.

Desktop products also bore the imprint of the Internet and the Web. For example, all new releases of word processors, desktop publishing packages, and document-management tools saw added emphasis on Web-oriented documents. The seemingly stale issue of competition between Web browsers took on new life as Microsoft faced U.S. government scrutiny over the bundling of its browser with Windows 95. As the *Technology Forecast* went to press, Netscape announced its browser would be distributed at no cost and that the source code for it would be available to developers.

The deciding factor for Internet growth is access speed. In response to this challenge, a host of new technologies has arisen, and major vendors in the telephone, cable television, satellite communications, and utilities markets have formed a variety of alliances to compete actively to provide Internet access. Cable modems, xDSL, set-top boxes, and satellite communications are technologies competing to provide greater bandwidth to different market segments. Soon, 56Kbps modems and ISDN will look like slow technologies, indeed. In terms of vendors, consolidation seems to be a common theme: The hotly fought battle over who would acquire MCI is one example, as is AT&T's acquisition of Teleport.

Within corporations, the bandwidth issue has been a major driver for upgrades of network infrastructure. Fast Ethernet became firmly established in 1997, just as new technologies such as Gigabit Ethernet and ATM prepared to dislodge it. Only a year ago, the adoption of both these new technologies seemed much further in the future.

One Internet-derived technology that has not yet proven its viability is the network computer. Vendors of these devices did not attain nearly the success they were hoping for last year. Whether they can convince users to adopt the thin-client approach in the long term is a question of significant debate. Few people expect a definitive answer in 1998, except perhaps in very limited niches where network computers have an immediate, compelling advantage. However, the advent of these devices has resulted in a new focus by Intel and Microsoft on total cost of ownership for corporate PCs and the emergence of sub-$1,000 consumer PCs.

By far the most striking development in computing systems is the growing convergence between RISC and Intel segments of the server and workstation markets. Most major vendors of RISC-based servers have announced plans to sell Intel-based systems. Because of the increasing overlap of platform functionality, consolidation among vendors is a likely outcome. Evidence of this consolidation is already apparent: Compaq acquired Tandem and, at press time, it had just announced its intention to acquire Digital Equipment. Earlier, Gateway 2000 purchased ALR, a company known for innovation in its Intel servers. The changes in this market have been driven by the growing role of Microsoft's Windows NT as a viable workstation and server product – although not yet for the most highly demanding enterprise applications. Apple, too, felt the pressure brought about by this development, going so far as to announce that the next version of the Macintosh operating system would ship on Intel platforms as well.

The most significant issue for in-house applications lay away from all these concerns: 1997 was the year business began responding in earnest to the Year 2000 problem. The denial and apathy of previous years disappeared, and serious, intense work was seen throughout most corporations. It is difficult, though, to see what the effects of this newfound commitment will be. Analysts are divided as to whether funds poured into remediation will serve mainly to fix existing programs or to port the programs to new platforms or to canned third-party applications. In either case, the Year 2000 problem will be a major distraction for all IT sites during the next 3 years. Europeans face the additional problem of converting applications to comply with the European Monetary Union and to handle transactions denominated in Euros, the newly designated coin of the realm.

Although IT managers will be distracted by these two issues for the next few years, they also will have to consider the greater embrace of electronic commerce, the demand for greater bandwidth, the consolidation of RISC and Intel server platforms, and the continued ascendance of Windows NT.

You will find greater detail on these issues and technologies in the pages that follow. Enjoy. If you have suggestions to help improve the *Technology Forecast*, please contact us at `Technology_Forecast_Editors@notes.pw.com.` After all, this book is written for you.

– Andrew Binstock

Editorial Director

2 Interviews

2.1 Interview with Commissioner Bangemann

Martin Bangemann is the European Union (EU) Commissioner for Industrial Affairs, Information Technologies, and Telecommunications. Trained as a lawyer, he has spent more than 20 years in German and European politics. Bangemann became an EU commissioner in 1989, and since 1993 has headed DG-XIII, the EU's directorate-general for information technology and communications. Since 1989, Bangemann also has been the head of DG-III, the EU's directorate-general for industrial affairs.

In 1996, he established an Information Society project office at EU headquarters in Brussels to promote the potential of digital technologies in the public and private sectors.

Bangemann is well known in Europe as the author of the 1994 *Bangemann Report,* which described the need for accelerated telecommunications deregulation and called upon the private sector to fund technology initiatives.

In direct response to this report, the city of Stockholm

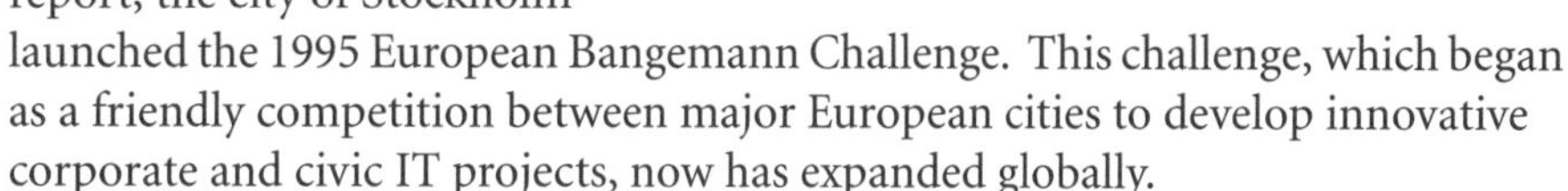

launched the 1995 European Bangemann Challenge. This challenge, which began as a friendly competition between major European cities to develop innovative corporate and civic IT projects, now has expanded globally.

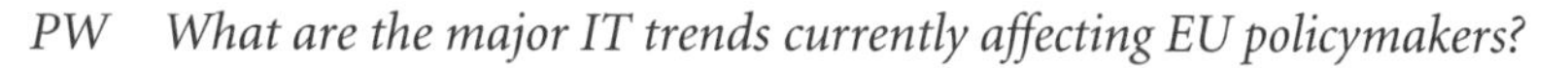

PW *What are the major IT trends currently affecting EU policymakers?*

Bangemann The major issue I believe the industry is facing is convergence. The convergence effort will constitute the greatest challenge to policymakers in the years to come. In fact, it will pose a challenge more complex than the liberalization of the telecommunications market. I say this today with certainty, but this was not always the case.

When we invited experts 2 years ago to help us begin our internal discussion and think about convergence, one of the experts strongly believed that convergence was nonsense and that everything would remain the same. This illustrates how difficult prophecy is in this field – and how challenging it is to set polices and regulations in a field that is constantly changing.

PW *Convergence is a word that can be used in many different ways. How are you thinking about it?*

Bangemann Convergence represents the blurring of frontiers between telecommunications, computer, audiovisual, and publishing technologies, services, and industries. Although in the past telecommunications services were linked to a specific infrastructure, today many types of services (voice, data, video) can be transmitted over any kind of network (fixed, wireless, satellite) in a network-independent way.

Obviously, because convergence facilitates global communications, this trend will change the way we develop technology standards and laws. In my work at the EU, we also are studying how convergence will change the way we do business.

To give a very down-to-earth example, I recently learned about a farmer in the U.K. who used to drive his sheep down from the hills in which he was living to an auction hall. Today, he does not do that any longer. Now, he enters each sheep's weight into the Internet and then he gets purchase offers electronically. So, using technology, the farmer doesn't have to move his sheep until they are sold.

Meanwhile, at the high end of manufacturing, the Airbus aircraft enterprise is not based in one place the way the U.S. automobile industry used to be based in Detroit, Michigan. Airbus parts are produced around the world, put together by Airbus Industries. To have direct contact with their customers and their suppliers, these bit companies built an intranet. This intranet now is used by all the companies to exchange information among themselves. They have built an electronic community.

Geographic borders and distances become irrelevant as a result of global communications networks. In the digital environment, data knows no frontier and information increasingly is becoming a traded good. The prospect of global electronic commerce will reinforce that trend.

In addition, convergence means that telecommunications, media, and IT companies now are able to offer new services outside their traditional business sectors. We certainly have seen this development in the consumer market, where we increasingly are using integrated consumer devices for purposes such as telephony, television, and personal computing.

There also has been an expansion of the services offered by banks as they build their Internet and intranet capabilities. Soon, banks will offer not only their traditional banking services but also new services, including things such as travel arrangements and insurance services. People linked to the bank will use the intranet to exchange information among themselves and develop their own electronic communities.

PW *How is the EU managing the new policy issues that have been brought about by the convergence of telecommunications, media, and IT?*

Bangemann The European Commission recently issued a "green paper" inviting public debate and discussion on convergence policy issues. One important issue we will be exploring is what approach should be taken to complete the transition from analog to digital services. We also will be looking at how convergence affects the relationship between regional and international standardization.

Our task will be challenging, however, because convergence, by its very nature, will blur the issues dealt with by different organizations. For example, present regulation is based on the sectorial difference and limits between industries. The broadcasting industry has one set of technology rules and standards, and the telecommunications industry has another. Now, we may have to make new rules because the old ones may not be able to be extensible.

For us to be successful, it is extremely important that the future regulatory framework for communications and media in Europe is open, flexible, and adaptable.

PW *What are some of the economic factors that will affect how quickly new technologies are adopted by European companies and consumers?*

Bangemann Digital technology allows a substantially higher capacity of traditional and new services to converge toward the same transporting networks, both on a national and on a global scale. Now, flexible digital technologies are enabling telecommunications, media, and IT companies to offer new services outside their traditional business sectors.

PW *Can you give examples of the effect of information technology and the information society?*

Bangemann Information technology is a powerful tool, but it is only a means to achieve our economic and social ends. In other words, information technology will have powerful implications not only for our economies, but also for our societies as a whole. Although the term "information highway" once was ubiquitous, today more people are speaking in broader terms about an "Information Society."

Finland is one European country that has embraced the Information Society with open arms and it is now a world leader in mobile communications. Although mobile communications initially were marketed as high-end business tools, Finland's mobile communications industry is based on a thriving home market that has the world record for mobile phone penetration. This strong domestic base has been a crucial launch pad for world success.

In the wake of this success, thousands of new jobs have been created – new jobs that are highly skilled and well paid. Higher productivity leads to even higher wages and more purchasing power. Today, every Finnish citizen benefits from the new wealth brought into the country and from the cheaper and more mobile access to telecommunications services now available.

Now, as we continue to build the European Information Society, we must deal with a wide range of economic issues. One critical success factor thus far has been the EU's liberalization of telecommunications markets. Going forward, growing international competition has forced us to realize that quality is no longer a purely European trademark. With our new competitors continually improving their quality and reliability, Europe must look for new, value-added enhancements. This task is especially important if we are to compete with the low-wage countries in the production of goods and services.

PW *What other strategies will help ensure that Europe becomes a strong player in the global marketplace?*

Bangemann First, telecommunications must be affordable. High tariffs will prevent users from using their new purchasing power. Liberalization of the telecommunications market in Europe will be completed in 1998, exposing European monopoly suppliers to competition. This situation promises to inject a new dynamism into the European market.

Second, new information products and services must be useful, usable, and affordable. Consumers will buy only things they want, things they find easy to use, and things they find cheap. An essential role of pilot projects is to test technology in new markets and adapt it to the needs of users.

Third, European governments can play a role by sponsoring research and development projects designed to close our technology gap with the U.S. and Japan. For example, a project sponsored by the European Commission's Esprit program is bringing together major European manufacturers to design the next generation of ultra-compact CMOS semiconductor chips. We also are starting to shift our R&D efforts into the field of software production, which has not been a traditional European strength. As part of our effort to strengthen this industry, we are trying to foster collaborative rather than protectionist attitudes in the European software industry.

And finally, in addition to using private and public funds to build the physical infrastructure of the Information Society, money must be directed at education. We must promote universities, research centers, and science parks as centers of excellence in which the Information Society can coalesce. EU activities such as project WOLF (World Wide Web Opportunities in Less Favored regions) are aimed at exploring technology opportunities for less favored regions in the wired world.

PW *What are some of the social and cultural factors that will affect how quickly new technologies are adopted by European companies and consumers?*

Bangemann The type of technology we are developing now carries with it a deep sociological change, a deep change of morals and ethics, and a deep change of political systems. That poses a threat to some people. Europeans obviously are concerned about job loss, but they also are concerned about the loss of cultural identity. Part of our work is to eliminate such fears.

I have made several speeches with titles such as "Will modern technology kill the book?" And I have had to stress that new technologies aren't the end of publishing but the beginning of a whole new era because it is much easier to publish with the new technologies.

Going beyond the new technologies that are transforming the media and communications industries, we need to examine content as well. Although the technological availability of content is now already possible on a global scale, we need to explore ways to develop content that will interest people around the globe. Is there something in which everybody is interested? Do you have to produce different content for different cultures? Are you blocked by the different moral standards that are part of a particular culture?

Once you begin to explore these questions, you can see that rules are needed on privacy, on the security of information, and on defining how people technologically can choose the content they want. I often hear Americans say that European technology policy is driven heavily by regulations and that Americans rely more on market forces. However, we all have come to realize that if you do not have rules for secure banking systems, for example, no one will do banking.

Nevertheless, we will face challenges in producing these rules. Because technology is being developed so rapidly, everything we have produced in the last 9 years will be thrown out the window within the next 5 years. So new forms of regulation must be more general and flexible. We also should be working to develop global rules. In fact, I recently proposed that we try to develop an International Charter.

PW *What would an International Charter include?*

Bangemann An International Charter would cover issues such as global standards to ensure global interoperability; mutual recognition of authorizations and licenses; digital signatures; encryption; different aspects of content regulation, including protection against illegal and harmful content; customs; and data privacy.

There is evident tension between, on the one hand, market-led interests and the rapid development of technology, and, on the other hand, the need to preserve public interests and to have a predictable framework. Therefore, this Charter must foster new commercial opportunities while protecting people from potential abuse. Above all, for it to work, such a framework needs to be industry-led and globally agreed upon.

Progress already has been made in the introduction of global rules, notably with regard to intellectual property rights in the World Intellectual Property Organization (WIPO) and market access in the World Trade Organization (WTO). Now, we need to build on and recognize these current agreements and develop new ones to combat the emergence of cyber-crime, for example.

PW *Will this type of Charter erase individual national cultures?*

Bangemann Such global cooperation does not signify an end to individual cultures. A culture is not a closed system.

Actually, it's quite the reverse. If a culture tries to close itself and to shut down the curtain to other cultures, it will die. Culture is not something you can preserve, like a herring or a tomato in a tin. In business, we now have large global corporations in which employees in all parts of the world are working together without losing their identities. Our own organization, the European Union, provides an exchange of different European national cultures. At the same time, English, French, German, and Italian cultures continue to thrive.

The real challenge now is to ensure that the implementation of global technology policies is done democratically. Because a democracy involves the participation of citizens and parliaments, we must guarantee that each citizen has a vote. We have to find ways to manage democratic control of technical and political international organizations involved in developing technology policy.

PW *On January 1, 1999, the European Monetary Union (EMU) will be introducing a single, unified Euro currency. How are plans proceeding, and how can companies prepare for the change?*

Bangemann Companies really are going to have to expedite their efforts to meet the 1999 deadline. I'm concerned that EC surveys conducted in October 1997 showed only 20 percent of companies are aware that there may be technology implications to the switch. These same surveys indicate that just one company in ten has a clear action plan.

Adapting information technology, customer-supplier functions, accounting and billing systems, and treasury operations for a unified currency presents both a problem and an opportunity for companies of all sizes. It is clear that those companies that adopt new technologies to solve the problem will have a huge competitive advantage.

Thus far, our surveys show that larger companies are better prepared for the Euro, but even the smallest companies will need computerization if they are to survive the switch. However, this requirement may be challenging for the smallest enterprises because many small European companies still do not even have a fax machine.

PW *How does the introduction of the Euro interact with Year 2000 conversion efforts?*

Bangemann It's another major conversion effort. The information technology problem of adapting computers to the Euro is complicated further by the fact that companies also are reprogramming or converting their computer systems to handle dates accurately starting in 2000. A limited number of consultants can help companies prepare for each of these challenges, and demand for their services is skyrocketing.

PW *This year, several global telecommunications manufacturers competing in the European market are proposing standards for a third-generation of mobile telephony known as the Universal Mobile Telecommunications System (UMTS). Why is a new standard needed?*

Bangemann New, third-generation mobile telephony will support data, multimedia, and packet-switched services for both corporate and consumer markets. For example, within the next decade, we will see traditional telephone handsets evolve to offer many features of a television set, a personal computer, and a radio. We also will see service providers combining UMTS services with low-earth-orbit (LEO) systems to provide complete coverage of the whole world, without having to go into fixed networks. That is really the next important step in the technology.

PW *How will the UMTS benefit businesses and consumers?*

Bangemann To answer that question, let's look back at the history of cellular telephony. Initially, mobile communications were marketed as a high-end business service. Today, especially in the Scandinavian countries, we are seeing up to 50 percent penetration in the consumer market.

Now that consumers and executives have experience with mobile systems, they are beginning to envision the next-generation systems that will provide communication to everyone, everywhere, at any time – whether the communication involves speech, data, or even video. The new UMTS service is intended to help cordless users who require excellent communications with high data rates when in the home or office, as well as satellite users who need truly worldwide coverage. UTMS is extremely flexible: It provides coverage from LEO satellites at one extreme and in pico-cells such as airplanes or trains at the other extreme. This development opens up new possibilities for greater system availability and remote computer network access.

One of the most important factors in ensuring that the system works successfully is integrating UMTS seamlessly with the fixed network so that users receive nearly identical services whether they are using fixed or mobile phones.

PW *What challenges are involved in setting new global standards for the UMTS?*

Bangemann Currently, the world standard for digital mobile telephony is the European-derived Global System for Mobile communications (GSM). Right now, GSM has more than 60 percent of the world market. Early in 1998, however, the European Telecommunications Standards Institute (ETSI) will decide on the new third-generation standards.

Thus far, Ericsson and Nokia appear to want a wideband Code Division Multiple Access (CDMA) standard, which also has Japanese support. The Japanese position is particularly important because these two companies believe that as global players, they should cooperate with the Asian market, which has the highest growth potential. Meanwhile, Siemens AG and Alcatel Data Networks are working together with North Americans, including Motorola and Nortel, on a separate standard that would be a combination of CDMA and Time Division Multiple Access (TDMA). These two standards are very different: You cannot combine them. Of course, when the first GSM standards were being developed 12 years ago, the EU was smaller. We are much larger now, and that may make it more difficult to get agreement on a single standard.

Ultimately, industry will decide the new third-generation standard or standards. The EU would like only one industry standard because we want to renew the success of GSM. However, we cannot and we do not want to force this issue. No official authority or government agency should be charged with bringing out binding standards. Our philosophy is let industry do that work itself.

PW *There is global consensus about the success of the Bangemann Challenge. What are the goals and accomplishments of this international IT competition?*

Bangemann The Bangemann Challenge originated in Sweden in 1995, when the city of Stockholm offered a friendly challenge to other European cities to show how innovative they could be in developing IT projects. The mayor of Barcelona, Spain, also took a leading role in developing this program in which over a period of 2 years, cities would exchange their experiences and invite their citizens to take part. We wanted to generate interest and show people ways that technology could be used to benefit their communities.

In the beginning, 25 European cities participated with 110 committed projects, and they worked out a method of handing out awards. Because that was a success, a global Bangemann Challenge is now underway. This time, Stockholm and its fellow European cities and regions are challenging cities in Asia-Pacific, Africa, and the Americas to provide the most benefit for their citizens through easy-to-use applications of information technology in 11 categories. The categories include the use of IT in the modern workplace and virtual organization, the use of IT in education and lifelong learning, and the use of IT tools to foster participation by citizens and promote public service by governments.

PW *What project awards in 1997 most impressed you?*

Bangemann The city of Rotterdam, Holland, for example, won an award for applying technology to create even flows of highway traffic. A health care award was shared by the cities of Stockholm, Sweden and Edinburgh, Scotland: In Stockholm, doctors were trained to use hand-held computers to enter data while conducting hospital ward rounds, and in the city of Edinburgh, midwives learned to link portable computers to hospital databases. Other programs focused on advances in distance education and studies of how worker productivity is affected by telecommuting.

The real winners were all the participants, who formed the biggest contact network in Europe for sharing experiences and know-how in IT. In many ways, these local champions accelerated the development and practical use of IT in Europe. It is this spirit of networking and sharing that I'm hoping will develop on a worldwide basis.

2.2 Interview with Donald Gips

Donald Gips is the Chief Domestic Policy Advisor in the Office of the Vice President of the United States. He advises the Vice President on a broad range of economic and social issues, with a focus on communications and technology matters in both domestic and international arenas.

Prior to his appointment, Gips was Chief of the Federal Communication Commission's International Bureau, beginning in May 1996. As Bureau Chief, he oversaw many important international initiatives, including the World Trade Organization talks on basic telecommunications, the international accounting rates proposal, numerous policies in the satellite area, and the successful DTH/DBS and paging agreements signed with Mexico. Gips joined the FCC in January 1994 and served until February 1996 as Deputy Chief of the Office of Plans and Policy, where he led the team that designed the Personal Communications Service (PCS) bandwidth plan and auction rules for the FCC's first-ever spectrum auctions. Before coming to the FCC, Gips worked at McKinsey & Company.

Gips was graduated magna cum laude from Harvard University in 1982, where he received the Ames Award for leadership and character. He also received the Paul Revere Frothingham prize for scholarship and contribution to the university. He went on to receive a master's degree in Public and Private Management from the Yale School of Organization and Management in 1989. He was awarded a Rockefeller fellowship in 1983 to work on development issues in Asia and a CORO Foundation Public Affairs fellowship in 1984.

PW *As someone who has been involved with technology in the government arena for some time, what are your expectations for the future?*

Gips If you look at the evolution of technology and examine which technologies have come to fruition and which haven't, the result is not what you would expect. The industry's short-term projections generally have been missed, and the longer-term projections frequently have been off-base and often missed the winning technology. For example, one often-quoted study by a consulting organization said there would be 900,000 cellular customers by the year 2000. Looking back, we know this number was incredibly low. Today, more than 50 million wireless subscribers are in the U.S.

PW *It sounds similar to the initial worldwide market projection for five computers.*

Gips Yes. Tom Watson, IBM's founder, made that statement in 1945. Now, the interesting thing about the cellular study is that it was far more optimistic than anything else out there, and thus, Craig McCaw was able to use it to get his original financing. So, the lesson is that we have been surprised by wireless technologies in general, whether it's DBS [Direct Broadcast Satellite] or many other technologies. Wireless, in general, has exceeded our expectations.

On the other hand, we've had more difficulties upgrading land-line networks than we expected – whether in the cable boxes or in some of the early trials pushing video over the phone network. This process has proved much harder to implement at acceptable prices than anybody thought it would be, so when we look at the whole picture, we are ahead in some areas and behind in others. It is so tough to predict. You know, if you would have said 5 years ago that the Internet was going to be where it is today, people would have thought you were crazy. Now, you can't even watch a television commercial without seeing an Internet address.

PW *They would have asked, "What is the Internet?"*

Gips Yes, it really is remarkable. As we sit here and try to predict what the transforming trend is going to be, our track record as a society shows us that we are pretty unlikely to pick the right answer, but that technology will continue to transform our lives.

Are we on the information highway? The Internet is the information highway right now. It was 2 years ago when Al Gore first coined the term information highway. He saw the Internet potential back then, and he funded the early research networks that helped get us to where we are today.

PW *The Vice President has a pretty strong track record in technology.*

Gips Right. It's a little scary to work for somebody who's got such good vision. Forecasting is very, very difficult. Let me give you the classic example of spectrum for wireless. I think PCS services are going to be powerful and wireless will continue to develop because we purposefully created very few rules. You can use the spectrum virtually as you please as long as you are not interfering with your neighbor. If you want to use it for fixed wireless, go ahead. If you want to use it for mobile telephony, go ahead. Fifteen years from now, if somebody figures out how to compress video enough to shoot it down the spectrum, you will be able to do that, too. That's an example of letting the market work.

That is not to say there isn't a role for government. At the same time, you need to put in place rules of competition so no one company can own all the spectrum slots. They have to be shared. There has to be a cap on how much spectrum one company can have. That's how the goal of preserving competition is accomplished while still allowing the market to work.

PW *Will it all work out on the wireless side with PCS vendors, and will it merge together without becoming another duopoly, as we have on the cellular side?*

Gips The rules are in place to guarantee that four or five different licensees will share the mobile spectrum. I don't see how it could consolidate unless they had a very compelling argument that the market could only support two providers. Today, we're not seeing any evidence of that.

PW *What about a little further out? What are your more optimistic predictions for five or 10 years from now?*

Gips In the future, computing will be ubiquitous – information technology will reach the point where you won't have to wear five different things on your body. The Xerox PARC people use an analogy I really like. When you're driving a car, you don't really listen to the sound of the engine. But as soon as there's a clunk in the engine, you know there's a problem. To me, computers don't do that for you right now.

We are in the age of information overload. Technology doesn't utilize all our processing capabilities. If you walk through a park, your brain is processing millions of bits of different types of information, whether it's the scent of the flowers or the breeze blowing through the trees. Technology today basically uses your eyes only. It doesn't use your peripheral vision. Part of information overload is that we haven't found good ways of engaging all our senses. We need to be able to process information in ways that make it a more positive experience.

PW *What are some of the other bigger challenges?*

Gips I'm very concerned about the residential bandwidth issue. With the advent of the Telecommunications Act of 1996, we've seen a lot of positive developments: Many new wireless services are being offered that businesses are utilizing, and satellites seem poised to help deliver these services eventually to residential areas as well. The real challenge is continuing to provide the bandwidth to the home and making sure that it's available for most, if not all, people. Eventually, these networks will all converge, and we must ensure that all citizens can access the products that flow over the networks.

PW *Haven't we almost started to see the cable and phone companies move in the opposite direction?*

Gips When I say convergence, I don't mean those companies become one or use the same technology. I mean they are all trying to deliver the same product, which is broadband bandwidth. They are taking different paths to deliver broadband data, such as cable's modem architecture, the phone companies providing xDSL technologies, or wireless and satellite technologies.

PW *So it's ubiquitous service available in the home?*

Gips Yes. Our national goal is to figure out a way to have the private sector get there. We need to base it on a competitive model because that's the only way we're going to continue to get the improvements we need. This is a very important national policy issue, one I think will become a critical global policy issue as well.

PW *What is the hardest part about achieving these goals?*

Gips Information technology is constantly transforming how we live, work, and play. The role of government is to try to make sure that as we go through these transformations, we let the market work and then figure out where government can and should intersect with the market.

Information technology creates a power shift. It globalizes our decision-making. The recent stock crash in Asia and the U.S. stock market's reaction showed how intertwined all the financial markets have become. Government's role as financial controller is judged by how the world markets are doing, and that is largely the result of advances in information technology that speed information around the world.

PW *What about controlling information on these networks, such as pornography?*

Gips Pornography and obscene material on the Internet are really a global issue. To protect our children, it is no longer adequate to control only what happens in our own country because people in other countries can reach our children, and vice versa. Ultimately, we want to give parents the tools they need to control the material to which their children have access.

We are in a global world that is increasingly shaped by information technology. Getting more information out to people must become a positive experience. We need to be careful, though: As we get more information out to more people, issues of privacy become more important. As information gets easier to copy and duplicate, we need to determine government's role in ensuring intellectual property is protected.

PW *Do you view privacy and intellectual property as being extremely important?*

Gips Yes, and they lead us to difficult issues such as encryption. Many of the old norms we established in the analog world do not translate well to the digital world. We must develop a whole new set of societal norms regarding intellectual property, privacy, encryption, pornography, and obscenity.

The Vice President believes strongly that our first role should be "do no harm." However, it's a very different role for the industry. We're calling on industry to take the lead.

PW *Has the Telecommunications Act of 1996 fulfilled its mission?*

Gips It's too early to tell. Certain market sectors are vibrantly competitive, but other sectors have been slower to develop. The true judgment won't be available for several years, and right now we are pushing hard. The FCC faces serious challenges in continuing to open up local phone, video, and data service to competition.

In addition, we are challenged to make sure that broadband deployment is incentivized as we open it up to competition. Protecting and promoting the universal service that has connected everyone in our country is also essential.

We have some social objectives, too, such as ensuring this technology is a productive part of our children's classrooms. We need to train the work force of the future properly as well as ensure that health care providers and libraries all have access. We want all people to be able to use these technologies. That is the bundle of social issues that needs to become part of this discussion.

PW *Are you happy with the Act as it stands right now?*

Gips Now is the time to let the Act work and for the FCC and the Department of Justice to promote competition and stay on track. We also need to work with the states, which of course play a very important role in implementing provisions of the Act.

PW *Are there areas of the government that exemplify best uses of technology?*

Gips A lot of the "reinventing government" initiative has centered around ways to flatten organizations, to make them more responsive and based on the concept of distributed intelligence rather than the old centralized mainframe approach.

Look at what NASA has done. Dan Golden worked on a reinvention modeled after the Vice President's challenge and came up with the mantra "faster, better, cheaper," which is what the Mars Rover mission was all about.

PW *And it was a huge success.*

Gips Yes. He's reduced costs significantly, and NASA is doing a tremendous number of new things. So that's really an example of an organization that went out and reinvented itself. Is the job done? No. Things still can be improved? Yes. But we're making huge strides.

PW *Part of the change seems to be a new way of thinking that encourages working with off-the-shelf solutions instead of doing everything the "custom" way.*

Gips Right. You have to break the not-invented-here syndrome. That is one of the first challenges that faces everyone who tries to reinvent how they do things. It's something we're trying to do throughout government procurement now.

PW *Talking about this new way of doing things in government brings to mind some difficult issues, such as whether to legislate obscenity or pornography.*

Gips We have to enforce laws vigorously against pedophiles who are pursuing kids on the Internet. We can't just ignore the problem. Industry must develop tools for parents to make controlling access as simple as using a remote control for your TV. It has to be simpler than programming the VCR.

Providing parents with tools to help ensure their kids have a safe place on the Internet will be critical to its success as a broad-based medium. Now that doesn't mean there is always blocking. It doesn't mean there is one rating system. Society's job is to give parents real choices about what they want their children exposed to and what they don't.

PW *Putting the control in the hands of the parent rather than the government seems much closer to the telephone model of regulation, which says that a phone company isn't responsible for the content of a call – than to the television model of regulation.*

Gips That's what we're pushing for. Whether Washington and the public will be willing to accept that model is going to depend on whether the industry delivers effective tools.

PW *What roles will the industry and the general public have in this process?*

Gips I spent a lot of time doing development work in India, and one thing that really struck me was that no one ever stands in a line in India. They crowd around and fight to get onto a bus or a train.

No law states you have to stand in line in the U.S. We've just built a value, a cultural norm that says it's the right way to behave. We need to develop more of these norms to deal with this world. And society can do it. We've done it with litter, which is another example where it used to be the norm to throw trash out the car window, and we changed that norm. We've said, "No, that's not O.K." Are we perfect? No. But we've made a huge improvement.

We as a society need to establish norms for proper behavior on the Internet. And the Internet itself has built a set of norms. Some of these norms, such as spamming, are being challenged now. But we must continue to develop norms that work for everyone in the country.

People in the industry often forget how intimidating technology can be for the average person. That will change. Kids growing up with technology are not intimidated, and they will establish their own norms.

Still, we have to strike a balance to encourage growth. Just as we continue to let technology innovate, we must make sure that society is protected.

PW *There has been a lot of anticipation about the effect the information highway will have. What do you think the effect will be?*

Gips I used to work at the FCC, and the only way you could find out what was going on at the FCC was by paying lobbyists to provide you with the information. With the introduction of the information highway, that information now is available publicly on the Internet. So, now we're debating everything from kids' television to connecting the classroom. It's a tangible example of how the Internet empowers people to get involved.

Basic government services increasingly will be delivered over the Internet. People will begin paying their taxes over the Internet, and they will access information to be more involved in the democratic process.

Also, the speed and community nature of the Internet will enable virtual communities to have an effect on specific issues.

PW *Will it change the political process?*

Gips Yes. Power will shift from the center to the periphery. The information highway will reshape our politics because it will empower citizens' groups. Now, the downside may be that the veracity of information is going to be in question. Editors and commentators must serve as the quality control, so the public will have confidence that information is accurate and reliable.

PW *Are you optimistic about these changes?*

Gips Our country always has thrived on individualism and entrepreneurialism within the context of common values. That is the model for the Internet and for the Information Age, and that is why we should do very well in it.

The President and Vice President challenged the information technology industry to regulate itself. The biggest challenge in self-regulation is not for government to hold back – it is for industry to step up.

Look at issues such as privacy and child protection. My fear is that industry may view these issues as Washington problems, as government relations problems.

Yet these are probably the two biggest areas where industry must step up and take the leadership role. If industry just views these as government relations problems and hires lobbyists to avoid them, it's missing the bigger issue. If these technologies are ever to become mass market, people have to feel safe: They must feel their children are protected, and they must feel their private information is protected on the Internet.

The only way this will happen is if industry invests as earnestly in providing that safety as it invests in marketing a new software release. I don't see enough of that happening yet. I'm hopeful the industry's mindset will change and executives will see this issue as something their strategic planners need to worry about, not just their Washington lobbyists.

This subject goes right to the core of whether the self-regulation model works and whether the Internet will really ever take off a mass-market medium.

PW *Are there any other challenges for industry?*

Gips Everyone wants to make money on the Internet. Direct marketing, using that information, and selling it is one easy revenue source. It is in the long-term financial interest of companies to handle this exchange correctly, but short-term interests may differ.

The hardest thing about a self-regulatory model is that without rules, any player can come in and lower the standard, and others may be forced to respond even though they don't want to. That's one of the biggest risks. We hope the major players will take the higher road and do the right thing.

2.3 Interview with Thomas Malone

Professor Thomas W. Malone is the Patrick J. McGovern Professor of Information Systems at the MIT Sloan School of Management. He is also the founder and director of the MIT Center for Coordination Science and is one of two founding co-directors of the MIT Initiative on Inventing the Organizations of the 21st Century. His research focuses on how computer and communications technologies can change the ways people work together in groups and organizations. Among other things, Professor Malone is well-known for having led the team at MIT that developed the Information Lens system, a pioneering groupware tool in which intelligent agents help users find, filter, and sort large volumes of electronic information.

Professor Malone has been a cofounder of three software companies and has consulted for a number of other organizations. He has published more than 50 research papers and book chapters and frequently has been quoted in publications such as *Fortune*, *Scientific American*, and *The Wall Street Journal.*

Before joining the MIT faculty, Professor Malone was a research scientist at the Xerox Palo Alto Research Center (PARC), where his research involved designing educational software and office information systems. His background includes a Ph.D. from Stanford University and degrees in applied mathematics, engineering, and psychology.

PW *For this year's book, we're taking a comprehensive look at where IT has been the last few years and where it is going. In particular, we're focusing on what some of the changes are going to mean not only for the industry itself, but also for economies and governments. You've been doing a lot of work on collaborative computing, on the next step in terms of people working together over computer networks. Are collaborative computing and groupware applications going to be the next big thing?*

Malone Yes, I think you could argue that they will. In fact, in the long run, I think the very word "computer" may be a misnomer. The word suggests machines that compute, that do arithmetic computations. I think if you look at how computers are likely to be used in the future, we will probably remember this technology not as *computing* technology, but as *coordinating* technology.

That is, computers and communications technologies increasingly will be used to help connect the work and activities of people – to *coordinate*, not just to *compute.*

PW *Could you give some examples of this trend?*

Malone The obvious examples are things such as e-mail, groupware, and the World Wide Web. Most traditional transaction processing systems – accounting, materials planning, and so forth – also are about coordinating the work of different people.

In fact, if you look back far enough, even telephones and television are technologies for connecting people.

PW *But isn't the telephone a one-to-one, spontaneous technology, whereas television is a one-to-many, largely scheduled technology?*

Malone In the way they are used today, yes, but it is interesting to look historically at how people predicted the uses of different technologies. In the early days of the telephone, many people anticipated it would be a broadcast medium. The first telephone visionaries predicted you would use the telephone to broadcast symphonies and dramas to your living room, so that you could listen to them at home.

PW *We may not be too far away from that prospect.*

Malone It is interesting if you lay this scenario out. People initially thought telephones would be used for broadcasting, and in the early days of radio, they believed it would be used as radio telegraph, for point-to-point communications. In fact, the actual uses of these technologies have been just the opposite of what was originally predicted. What may happen in the future, however, is that those uses will switch again.

For example, radio through cellular telephones is being used increasingly for point-to-point communication, whereas wires, through networks such as the Internet, are being used increasingly to broadcast communication.

A century or two from now when historians look back on the time in which we're living, the period beginning with the earliest electronic communication technologies – such as the telegraph and telephone – and the later development of computational technologies – such as computers – may blur into a historical second. Then, in the late 20th century, those technologies suddenly came together and rapidly thrust us into a new era.

If you want to be grandiose, it may not be too farfetched to say that future generations may refer to the distant past as pre-electronic communication time. Today, therefore, we're living on the cusp of the new electronic age.

PW *What makes that change happen? What's the big bang?*

Malone Network phenomena have strong critical mass effects. In other words, the value of the technology to any one user is proportional to the number of users in the world. That's certainly been true of telephones, faxes, e-mail, and the World Wide Web.

For most technologies, growth is affected by how many people have access to the technology because your likelihood of even hearing about it is dependent on its popularity. With network technologies, however, this effect is even stronger because the actual value of the innovation to you depends on the number of other people who already have adopted it.

In part, this is the "increasing returns" phenomenon that Brian Arthur and others have written about, which helps explain the success of Microsoft and Intel. In other words, the usefulness of your PC as opposed to your Macintosh is indirectly affected by how many other people have PCs, even if you are not connected to a network, because the availability of software and so forth is affected by how many other people have PCs.

This effect is probably more important with the information goods of a knowledge economy than it is with physical goods. Therefore, we should expect that there will be this relative concentration within certain knowledge economy industries, probably even more than we've seen in the past.

PW *Are there forces that could slow this concentration?*

Malone The other force that may work against increasing returns or network effects is the increasing pace of innovation. Each time you move to a new generation of technology or product, you essentially start a new increasing returns game – a new race to see who can get the most first.

You can imagine an economy in which you keep opening up new games, and each game quickly comes to be dominated by one or a few players. Before too long, that game becomes obsolete, and you are off running in another game where, again, the outcome is uncertain for a time. How those two forces will work against each other is very hard to predict.

PW *What about predicting when a new technology will take off?*

Malone The tricky thing about making predictions is that you often can predict qualitative changes fairly confidently, but the actual timing of them is inherently unpredictable. For example, when fax machines first were invented around the turn of the century, you could pretty confidently predict that electronic transmission of document images would someday be an important element of human communication. Yet we went for 70 or 80 years before that possibility came to any degree of widespread commercial realization. Then, seemingly overnight, somewhere in the late 1980s, it went from, "Oh, yes, I think I know somebody who has a fax machine" to, "Oh, you're in business, then of course you have a fax machine."

PW *The same thing is proving true with e-mail today and probably will extend to collaborative or group computing.*

Malone To me, one of the biggest surprises of the last 5 years has been the incredibly rapid diffusion of the Internet and the World Wide Web. Again, I think it was easy to predict 10 or maybe even 20 or 50 years ago that something like the Internet would happen, though it would have been hard to predict the exact technical form.

Now, the Internet, the World Wide Web, and the amazing speed with which they have been growing have become symbols for what is happening in business. People have varying opinions about whether the Internet as we know it will continue as the primary infrastructure for communication in the future. However, it is obvious that if it's not the Internet, it is going to be something very much like the Internet.

PW *Why is the Internet the right model? We had plenty of networks before the Internet.*

Malone I think the decentralized nature of the Internet has had a lot to do with its success. Think about the alternatives. If AT&T had been running the Internet, could the Internet possibly have grown as fast as it has, doubling in size every year since 1988? Would AT&T have been capable of keeping up with the demand?

PW *It probably would have been difficult for any centralized authority to keep up.*

Malone Right. So many of us have a centralized mindset, however, that it is hard to see what is possible. I find that the Internet is a really useful example to beat people over the head with the possibility that it is not always necessary to have centralized control to do good things.

Think about this: What if you viewed the Internet not only as a technical infrastructure for enabling business, but also as an organizational model for how to manage a business? You could take the technical architecture and the governing architecture of the Internet as a model for what other organizations could be like.

PW *You've written a lot about models for organizations. How has technology influenced organizations and how we work?*

Malone I believe organizations go through three distinct historical stages: a progression that is enabled by decreasing communication costs.

The first stage is what we call "the cowboys," the unconnected, decentralized decision-makers. It is an organizational form that was almost universal until the late 19th century. Much of human history has been a story of small towns and villages filled with independent farmers and shopkeepers, people who made decisions based only on local information.

Communication technologies have altered this reliance on local information. Mass print media, and to a greater degree electronic communications technologies, now make it possible to share information cheaply and easily around the world.

One way of interpreting the history of business in the 20th century is by working through the implications of these decreasing communication costs. We call the second stage "the commanders," and it represents the time when centralized decision-making became possible.

With this development it became economically feasible to move information around cheaply to many distant places; those with access to that distant information could make better decisions. People brought the best information available in the world to some centralized place because they found they could make better decisions than the local decision-makers scattered throughout their organizations. The big multinational corporations are clearly examples of this centralized decision-making structure.

What we're beginning to see now in the latter part of this century, and what is likely to become much more common in the 21st century, is the third stage in this evolution. Decreasing costs of communication will soon make it possible to bring the relevant information not just to one central point, but to all points in the organization. These decentralized decision-makers can be as well informed as the centralized decision-makers were in the past. We call these decentralized, connected decision-makers "cyber-cowboys."

Cyber-cowboys are independent decision-makers acting on global information. In fact, many business trends of the latter part of the 20th century fit into this model.

PW *That is an interesting model for looking at how business has evolved. Doesn't the third stage require a lot of teamwork and communication?*

Malone Right. To make effective team decisions, more people have to have more information, so more communication needs to occur. That is the point where the technology comes in – it helps make the extra communications possible.

PW *What are the advantages of working as a team?*

Malone Team-based and networked organizations often are more flexible than a traditional hierarchy. They can respond more quickly to changes in the environment or to innovative ideas because they don't have to convince everyone all the way up the management chain. They can just do it at the team level.

This type of structure often is more desirable from a human point of view. Team-based organizations provide an outlet for a very natural kind of human behavior, which is the opportunity for personal growth that comes from joining together with one's colleagues to achieve a common goal.

Take empowerment, for example, which involves pushing decisions lower and lower down in an organization. That change is possible, in part, because it is economically feasible to have people lower in the organization well enough informed to make good decisions. We also are seeing a lot more outsourcing, a lot more networked organizations, and a lot more virtual corporations. Market-like interactions among people are increasing. The free market is itself a very decentralized form of decision-making because all buyers and sellers make their own decisions about whether to engage in a given transaction.

PW *How will these trends affect the size of corporations?*

Malone I think we will see more small organizations. Information technology makes it easier to coordinate with outside suppliers and therefore makes it more feasible for companies to outsource the production of goods and services. In our research, we have found that industries that increase their use of information technology tend to have a decrease in the average size of companies with a lag of a couple of years. My personal prediction is that in 20 years or so, 10- to 30-person companies may become the most important economic form of organization.

Small companies have fewer of the disadvantages they had in the past. In addition, they have the advantages of intimate teamwork and motivation. It is often easier to motivate people when their individual effort contributes to a visible bottom line for a company in which they play a major role, instead of being just one of tens of thousands of people scattered all over the world.

PW *Where are we likely to see more decentralized decision-making?*

Malone Decentralized decision-making works best when there are specific motivational factors. For example, creative or knowledge work often is done better by people who enjoy some autonomy rather than by people who are just following orders.

Everything we know about the economy of the future suggests that knowledge work and innovation are likely to become more important. I think that entrepreneurial activity will become more wide spread throughout the population because the technology makes it more economically effective and possible.

PW *Organizations becoming less hierarchical might have some serious ramifications. How do you see globalization and cheap communications affecting government?*

Malone There are at least three levels at which these implications are likely to play out in government:

- First, as information and communication costs decrease, it will be easier for government to perform many of its functions in a decentralized way.
- Second, government organizations will become more decentralized, and the need for centralized government will be diminished.
- The third implication of these pressures is more far-reaching. Governments are based on a geographical region of people who live near each other. Communication technologies decrease the effect of distance. The same technological and economic forces that make distance less important in organizing companies and social networks also diminish the importance of distance in organizing governments. So, today's governments, which are based primarily on distance and geography, are going to become less important in the world of the future.

PW *If that happens, what will take their place?*

Malone No one really knows the answer to that question. However, here is one example of what is possible: One function that government provides today is a forum for legal mediation of contractual disputes. So, geographically based governments offer a legal and court system for dealing with conflicts that have not been resolved by independent contractors in the market.

Today, a legal contract can specify that disputes will be resolved via a certain form of binding arbitration. I can imagine a world where it would be common for most business contracts to specify a particular court system for the resolution of disputes that are not part of a geographical country.

These could be arbitration services that would be transnational in nature and that could compete with each other for who earned the reputation as the fairest and most efficient system. This is an example of how functions currently fulfilled by the geographical jurisdiction of a government could be fulfilled in a more decentralized manner.

PW *We don't yet have a way to resolve issues related to things such as taxation, intellectual property, and privacy without resorting to geographic governments. These are not well-established principles in cyberspace. How will this situation be resolved?*

Malone One possibility is that in cyberspace, the governing legal systems may depend upon the region of cyberspace within which the action occurs.

Just as you pass a physical border from one geographical country to another, you often pass borders from one region of cyberspace to another. For example, when you log onto a particular Web site, you have entered that organization's region. You can have larger groupings of these regions in cyberspace, just as you have larger grouping of neighborhoods in geographical space.

Laws that govern a particular transaction would be dependent upon where you were in cyberspace when the transaction occurred, not upon where you were geographically.

It wouldn't matter whether you were in the Dutch Antilles or Tokyo or Singapore. What would matter, for example, would be whether you were in the `.com` or the `.edu` domain. There might be one set of laws for `.com` and another set of laws for `.edu`.

PW *Does this mean that although you are in cyberspace, information is not going to be free?*

Malone Right. In cyberspace, just as in physical space, some information will be free and some won't. One way to think about this situation is by conducting a thought experiment. Imagine you could have access to any information, anywhere in the world, for free – any piece of paper could be available for free electronically. Imagine also that in any place in the world you could see and hear what is happening for free.

Now what would this world be like? The first observation is that access to all of that information probably won't be free because some of the information has great value. Even if the transporting of it were free, the creation of the information certainly is not free, and so we are going to need much better ways of compensating the creators of valuable information.

Today in physical space, compensation for valuable information depends on patents and copyright law. However, the new technology is making it increasingly obvious that those laws are not sufficient in their current form. We need much better legal and economic mechanisms for valuing and exchanging information.

PW *Those laws are all based on a paper model.*

Malone Right. The old laws, the copyright laws, for instance, are based on a paper model. Patent laws are based on a model of mechanical devices, of which informational devices such as computer software usually are not very good examples. So we'd better sort out new legal, contractual, and economic mechanisms for valuing and exchanging information.

PW *What else do you expect to see?*

Malone A lot more ubiquitous computing and computational intelligence will be embedded in all sorts of devices in our physical environment. This movement will be somewhat independent of the trend of ubiquitous communication. We could have a lot of intelligent devices that don't talk to each other as well as those that talk to each other and to us.

Perhaps 10 or 20 years from now, personal cellular telephones will be widespread among the middle class. For instance, it is pretty obvious that by the time my children, who are now 4 and 6 years old, are in high school or college, they and their friends will have cellular telephones with many applications built into them.

All this intelligence and connectivity in our physical environment will lead to some surprising changes that we do not really understand. Just imagine if your plates knew when they'd last been washed, if your refrigerator knew what was inside it and what was about to rot, and if your clothes knew when they'd last been worn and who wore them.

It also becomes very clear that better tools for finding, filtering, and safeguarding information are needed. And we need better ways of dealing with issues of privacy. When it is conceptually possible to be able to see into anyone's bedroom anywhere in the world, it becomes obvious that it's not desirable for that to be easy. A whole new set of mechanisms need to be invented not for just finding information but for filtering it and protecting the individual rights and privacy of people in the process.

PW *What are some of the tools you envision?*

Malone We're on the verge of seeing project collaboration tools being widely adopted. The readable Web is step one. It has eliminated most of the infrastructure problems for access, for letting people see it. However, the Web doesn't yet have easy capabilities for writing or modifying Web pages.

PW *Do you think electronic media will surpass paper in most relevant dimensions?*

Malone For that to happen, electronic media will have to get much more portable, have somewhat higher resolution, and become cheaper, more robust, and easier to handle.

Imagine you had a piece of film the size and shape of a newspaper, but it was electronic film with good resolution that was portable, flexible, and not easily breakable. You've also got versions of it that fit in your pocket. At some point, devices such as this will surpass paper, but obviously we're still not there.

PW *You've done a lot of work in the area of coordination. Could you explain some of your theories?*

Malone We have found it useful to create some definitions. We start by defining coordination as managing dependencies between activities. That definition may sound obvious, but it's actually the result of years of refinement.

Now, if coordination is managing dependencies between activities, you can analyze it more deeply by saying, "What kinds of dependencies can there be? What kinds of processes can you use to manage different dependencies?"

For example, we see a lot of what we call a shared-resource dependency. This situation occurs whenever two activities use the same resource, whether it is a machine on a factory floor or the budget from which it is are drawn. In either case, there is a dependency to be managed involving the resource, and there is a coordination problem.

Another kind of coordination problem arises whenever one activity produces something that's used by another, say when one machine on the assembly line produces something that gets used by the next machine. We call this a flow dependency, and it actually implies three other dependencies that need to be managed for people to work together.

First, there is the prerequisite dependency: The item has to be produced before it can be used. Second, there is the transfer dependency: It has to be transferred from where it's produced to the place where it will be used. And third is what we call a usability dependency: Whatever gets produced should be usable by whoever is using it. Loosely speaking, this all amounts to having the *right things* in the *right place* at the *right time*.

PW *What's the first step for a company in putting these theories into practice?*

Malone When you are designing things such as cars or computers, each has a kind of "design space." With cars, for instance, the design space has dimensions such as these: What color is the car? What kind of engine does it have? How many wheels does it have? If you think of human organizations as being designed, then you can think of there being design spaces for them, too.

At a certain cost for communicating and computing, some regions in that design space are economically feasible and therefore desirable to explore. Information technology is greatly decreasing the constraints on what kinds of communicating and coordinating are possible in organizing human activity and is greatly increasing the parts of the design space that are feasible to explore. For example, I think many organizations should be exploring more decentralized ways of organizing their activities.

PW *Then why is the centralized approach often the first way people or groups approach creating an organization or taking on a task?*

Malone That's an interesting question. If your goal is to coordinate some set of activities to achieve some goal, you can use a variety of coordination strategies. One strategy is to set up a hierarchical structure in which some actors tell other actors what to do. Another strategy is to create a set of "rules of engagement" through which independent actors interact with each other in such a way that the same overall goals are achieved.

Most of us today have what Mitch Resnick at the MIT Media Lab calls the "centralized mindset." For instance, when we see a flock of birds flying in formation, we assume that the bird at the front is the leader and that the leader is somehow organizing the flock to fly. In fact, we now know from biologists that each bird is simply following a set of little rules that results in the emergence of this pattern.

In principle, almost anything that can be coordinated in a hierarchical way also can be coordinated in a nonhierarchical or decentralized way.

PW *How do all these technical developments challenge us?*

Malone It is important to recognize that these new technologies do not predetermine our future. They aren't some inexorable force that makes us do certain things and not others. For the most part, technology will be giving us many more choices. We therefore need to think much more clearly about getting what we really want in the first place.

For example, new technologies will make it possible for many more people to change the physical location of their work. Many of us, for example, will be able to work at home much of the time. However, I'm not sure we will want to do so. Those of us who have tried telecommuting and home office work know that it has advantages and disadvantages.

PW *It seems as if we've only dabbled at the investment and other changes necessary for that experiment.*

Malone Right. In fact, there are many other areas where we, as individuals and as organizations, need to think much more clearly about what we really want. For example, many of us get a sense of financial security from being employed by a stable, large organization. What if, in the future, many people no longer work for these large organizations, but work instead in temporary networks as individual contractors? Where will we get our sense of financial security then?

Another potential problem is where do we go for learning, for sharing war stories, or for having our reputations established and our credentials verified? Where do we go for a sense of community, a sense of identity that many of us get from the organizations for which we work?

The new technologies make it possible to create organizations that might be more efficient at completing tasks but that may provide fewer opportunities for satisfying socialization. In the future, that doesn't have to be the case. It is also possible for us to create organizations that may give us more satisfying opportunities to interact with our fellow human beings through a variety of media.

PW *Is it a question of rethinking how we work?*

Malone Absolutely. If we want to be more economically efficient, we're going to have to pay more attention to a wider range of human needs of our employees, of our customers, our suppliers, and of our society in general.

References

Brynjolfsson, E., T.W. Malone, V. Gurbaxani, and A. Kambil. 1994. Does information technology lead to smaller firms? *Management Science.* Vol. 40: 1628-1644.

Interview with Tom Malone: free on the range. 1997. *IEEE Internet Computing*, Vol. 1, No. 3, May/June: 8-20.

Laubacher, Robert J., Thomas W. Malone, and the MIT Scenario Working Group. 1997. *Two scenarios for 21st century organizations: shifting networks of small firms or all-encompassing "virtual countries"?* MIT Initiative on Inventing the Organizations of the 21st Century. Working paper 21C WP #001. January. `http://ccs.mit.edu/21c/21CWP001.html`

Malone, T. W. 1997. Is "empowerment" just a fad? Control, decision-making, and information technology. *Sloan Management Review.* Vol. 38: 23-35.

Malone, T. W., J. Yates, and R.I. Benjamin. 1987. Electronic markets and electronic hierarchies. *Communications of the ACM.* Vol. 30: 484-497.

Malone, T. W. and J.F. Rockart. 1991. Computers, networks, and the corporation. *Scientific American.* September: 128-136.

Malone, T. W., K.R. Grant, F.A. Turbak, S.A. Brobst, and M.D. Cohen. 1987. Intelligent information sharing systems. *Communications of the ACM.* Vol. 30: 390-402.

Malone, Thomas W., Kevin Crowston, Jintae Lee, et. al. 1997. *Tools for inventing organizations: toward a handbook of organizational processes.* Technical Report CCS WP No. 198. February. `http://ccs.mit.edu/ccswp198/`

Resnick, Mitchel. 1997. *Turtles, termites, and traffic jams: explorations in massively parallel microworlds.* Cambridge, Mass.: MIT Press. Complex Adaptive Systems series.

URLs of Selected Mentioned Companies

MIT Center for Coordination Science `http://ccs.mit.edu`

2.4 Interview with Dr. Othman Yeop Abdullah

Dr. Othman Yeop Abdullah is Executive Chairman of the Multimedia Development Corporation (MDC) in Malaysia. The MDC is a company fully owned by the Ministry of Finance of Malaysia, established to lead the development and management of Malaysia's Multimedia Super Corridor (MSC). The MDC functions as a client-focused agency to companies operating in the MSC.

The MDC provides information and advice on the MSC, assists in expediting permit and license approvals, and connects companies to potential local partners and financiers. The MDC is an organization wholly focused on ensuring the success of Malaysia's MSC and the companies operating it.

PW *What is the Multimedia Super Corridor?*

Othman The MSC's goal is to provide a comprehensive package for the development of the first world-class environment to help companies of the world test the limits of technology and propel themselves for the future. The MSC also will accelerate Malaysia's entry into the Information Age and help to realize Malaysia's vision to achieve a fully developed, mature, and knowledge-rich society by the year 2020.

The MSC is an area 15 kilometers wide and 50 kilometers long that starts from the Kuala Lumpur City Centre and extends south to the site of the region's largest airport, the Kuala Lumpur International Airport, which is scheduled to open in early 1998. Two of the world's first "smart cities" are being developed in the MSC. One is Putrajaya, the new seat of government and administrative capital of Malaysia, where the concept of electronic government will be introduced. The other is Cyberjaya, an intelligent city with multimedia industries, R&D centers, the Multimedia University, and the operational or regional headquarters for multinationals who wish to direct their worldwide manufacturing and trading activities using state-of-the-art multimedia technology.

PW *What time frame does the Multimedia Development Corporation envision for the Multimedia Super Corridor?*

Othman We are planning on a 20-year time frame for the full implementation and execution of the MSC, when Malaysia will have achieved a global competitive position in the Information Age. That will happen in three phases:

- In the first phase, which is where we are now, the MDC will develop the infrastructure of the MSC, attract a core group of world-class companies, launch seven flagship applications, put in place a world-leading framework of cyber laws, and establish Cyberjaya and Putrajaya as first-class intelligent cities. Of the seven flagship applications identified as the shapers of the MSC, the immediate focus is on electronic government, smart schools, telemedicine, and the national multipurpose smart card.
- In the second phase, the MSC will roll out and link to other cyber cities in Malaysia and the world. It will create a "web" of corridors and establish a second cluster of world-class companies. We also will set global standards in flagship applications, harmonize a global framework of cyber laws, and establish a number of intelligent, globally linked cities.

- During the final phase, we expect Malaysia will be transformed into a knowledge-based society. We will be a true global test-bed for new IT and multimedia applications and a cradle for technology companies. We will have a cluster of intelligent cities linked to the global information highway and become the platform for the International Cybercourt of Justice.

PW *What effect do you expect the MSC to have on your economy?*

Othman The aim of the Multimedia Super Corridor is to make the economy more resilient with more depth and value in all the sectors. We have identified that the way to do this is by positioning the economy from our current "input-driven" mode to a "productivity-driven" strategy. The enabling factor to affect the change is information technology and multimedia.

Input-driven mode is perceived as a mindset in the planning and managing of the economy. It also involves the methodology used in measuring success and performance in economic activities. Growth in the past has been directed mainly from creating tax incentives to attract foreign direct investments [FDI]. Cheap labor and incentives to draw inexpensive land were other critical input factors in the growth chemistry. Success often was measured by the quantum of FDI, the dollar value of export, and the growth of GDP [gross domestic product].

The relationship between the factors of production and the value derived was not adequately emphasized. The types of technology, the desired diffusion of skills, and the relative growth of productivity in the factors of production were measured but not addressed as part of the planning cycle until recently. However, the economy progressed in the last 10 years. As we achieve a level of sophistication, the country begins to lose some of the previous comparative advantages, particularly cheap labor and availability of inexpensive real estate. Other emerging economies such as China, Indonesia, and Thailand with cheap labor and a similar set of tax incentives are able to attract foreign capital inflow.

The MSC as an economic strategy is expected to focus on the new technologies and investment that can raise the total factor productivity of manufacturing, service, and other sectors of the economy.

At the same time, new investments from international leading-edge multimedia and IT companies with the participation of local IT companies is envisaged to generate new applications that will blossom into new products and new industries. Some of the applications are expected to feed into existing manufacturing and service sectors and transform them into global players.

PW *How do trends in manufacturing industries influence the development of the MSC?*

Othman Trends in manufacturing have been moving from labor-intensive to technology-intensive applications. Investments in R&D have been a crucial factor in ensuring new product lines and sustainable growth of manufacturing companies. The clustering approach in planning and the complexity of managing the entire business integration has reinforced the MSC objectives.

Take Malaysia's national car project as an example. Let me give you a little background. Proton (Perusahaan Otomobil National Bhd) was set up in 1983 to undertake the manufacture of the Proton Saga car under the national car project. Proton project's main objective is to spearhead Malaysia's industrialization drive and to act as a catalyst for the development of local component sub-sectors. We started Proton as a joint-venture between Malaysian Governments through HICOM Bhd (70 percent); Mitsubishi Corp., Japan (15 percent); and Mitsubishi Motors Corp., Japan (15 percent). Construction of the manufacturing plant started in 1983. The plant originally had a designed production capacity of 80,000 units per year working a two-shift operation. This capacity has since been increased to 100,000 units per year. The plant officially was opened in 1985. Today, Proton Saga cars now are exported to Indonesia, Singapore, Brunei, and the United Kingdom.

For Proton to develop an up-to-date design, the conventional method of designing is no longer effective and relevant. Proton needs to use the latest methods, employing multimedia technology with three-dimensional modeling and computer-aided design. IT is utilized not only for the design phase, but also for planning, distribution, marketing, and so on to efficiently follow through the entire value chain from design to distribution.

So, we have projected the Multimedia Super Corridor to help influence the manufacturing industries in three areas. First, it can help raise the competitiveness of the manufacturing sector. This means we can elevate from pure assembly operations to something more sophisticated and eventually derive higher value from the investment. Second, it can help reduce the dependence on cheap labor, and third, it can help integrate all the activities on the manufacturing value chain, from R&D, manufacturing design, marketing, distribution, and so on.

PW *Do you have some specific new technologies in the back of your mind that if you see them deployed in the MSC, you are going to stand up and clap and say, "Yes, that's great!" Something that you really think is important?*

Othman No specific technology or product can be singled out as a sole success indicator for MSC.

It is almost the whole range of multimedia value chain. If you look at the value chain of multimedia, one exciting component may be microelectronics, research, design, and industry applications. Another component may be computer animation and graphics.

Multimedia contents have tremendous prospects for the emergence of new industries. Education can emerge as a new profit-making industry. If a typical subject in biology today takes 5 years to study, multimedia is able to collapse the learning time to one-half or one-quarter of that period because learning has become simplified. Students do not have to laboriously go through text or a single source only. They can look at the graphics, listen to sound, and interact and access hypertext, depending on your their level and capacity.

Another exciting product from multimedia application relates to advertising and entertainment. For example, video-on-demand is now possible through multimedia and IT developments. The other area that is most exciting is electronic commerce.

PW *Can you give us some examples of how multimedia can assist electronic commerce?*

Othman Electronic directory and procurement are good examples. Let's say you want to buy a house. You cannot make that decision looking at a catalog. You have to visit the site, which may be some distance away. You may need to travel an hour or two and laboriously move from one house to another, physically examining each property. However, through multimedia, you can conduct a virtual visit of the site. With a multimedia catalog, you can go through the kitchen, examine the texture of materials used in the kitchen, inspect the bedroom, the lounge, and then move on to another site. You are able to narrow your choice and finally make a decision without taking so much time visiting all physical sites.

Using these new technologies, advertising and marketing become more convenient to both parties – seller and buyer. Particularly in the service sector, the middlemen can be removed completely. The discount is tremendous. It is a very efficient way of doing business.

These are some of the exciting approaches and applications that technologies developed in MSC can contribute to make electronic commerce flourish.

Electronic commerce, however, is a global phenomenon. Opportunities are in the global marketplace. With the increasing use of electronics and the Internet to conduct business, time and physical limitations become less problematic. However, many issues need to be addressed before electronic commerce across boundaries can be expanded. One is to raise the current level of security in the transactions as well as improve the infrastructure.

PW *Smart cards are viewed as a way to improve electronic commerce security. In many parts of the world, multipurpose smart cards probably wouldn't be enabled without government support. One thing that is interesting about the Multimedia Super Corridor is the cooperation between government and private industry. Do you view that as an advantage?*

Othman Yes, definitely. The collaboration between government and private industry is critical for the success of electronic commerce. The top-down support at the initial stage is essential to jump-start and ensure that the efforts being undertaken have some degree of direction. For the groundswell to occur, it needs time and a great deal of diffusion, promotion, and so on. The National IT Agenda is important to create awareness and spearhead national transformation and priorities. MSC is a shared vision with the Prime Minister as the champion. People need to identify with this vision and feel that this is not the Prime Minister's vision, but the national vision.

Market forces have their own agendas in determining what is beneficial to society and to stakeholders. We always quote President Kennedy's space programs. If the Americans had allowed market forces to take their course, there would not have been any significant investment in space projects. This is because of the space program's long gestation period and that there is no bottom line for support. The commercial benefits were clear only many years later: the spinoffs are very difficult to anticipate.

So, the U.S. government played the leading role. At the same time, private industry cooperation is key for the program's ultimate commercial success. The dominant role of the government is only during the initial stages. As the project matures, the role of the government becomes less and less prominent.

The MDC, which was established by the Malaysian government, was created to facilitate this type of development. The MDC can interact with the industries as well as the government. It is free from bureaucratic red tape, and it has the latitude to make investments. The presence of MDC is vital in the initial stages – in the creation of the Multimedia Super Corridor.

PW *Tell us about the multipurpose smart card and electronic government. What's that going to mean to the people in Malaysia 5 years from now?*

Othman We are embarking on a multipurpose smart card project for both financial and non-financial applications. It will serve as an electronic identity card with multiple functions. Many smart cards are being used throughout the world, but not a single country or company has developed a smart card solution that has financial as well as non-financial applications. We are offering a green playing field by assuring companies that by the year 2000, about 3 million Malaysian citizens will be using this technology with a single card for financial and non-financial applications. And, working with consortiums, the companies took up the challenge.

The multipurpose smart card will bring the benefits of IT to the public. For example, rather than queuing in front of government offices to renew their driver's licenses or to make take payments or other transactions with the government, citizens can go to electronic kiosks, use the electronic payment system, and complete the necessary transactions with minimum effort. This capability would significantly affect government and society.

For the Multimedia Super Corridor to have meaning to the society at large, it must have tangible and visible benefits on people's lives. The Electronic Government project is a step in that direction: it will improve access to government services for our citizens. Electronic Government is a noneconomic application; it is socioeconomic in nature.

Electronic Government applications also will enhance governmental efficiency. For example, the electronic procurement system will enable the government to access digital catalogs, procure the best products of supplies, and acquire them in an even more efficient and cost-effective manner. Another advantage would be a move toward a paperless government. Instead of sending letters through post, the government, its agencies, and the public can communicate through e-mail.

At the same time, government leaders quickly can retrieve any information that is relevant to their decision-making process. Here, the government must demonstrate leadership and technology competence.

PW *Tell us about the cyber laws.*

Othman To support the development of the Multimedia Super Corridor and the growth of new applications, relevant cyber laws are needed. At the same time, for the flagship applications to be successful, there is also the need for enabling legislation. The legislation that has been passed by the Malaysian Parliament includes the Digital Signature Act, the Amendment to the Copyright Act, the Computer Crime Act, and the Telemedicine Act. The Multimedia Convergence Bill and Electronic Government Bill currently are being drafted. Some of this legislation needs to be harmonized across borders with other countries.

In addition to a set of laws, we need an arbitration center, either under the WTO or other international forum, that can settle disputes and defaults speedily. I raised this matter recently during the International Telecommunications Union meeting in Geneva. The response has been positive.

PW *What about intellectual property?*

Othman Clearly, intellectual property protection is prerequisite to investment in R&D and multimedia content development. Malaysia has put the legislation in place and extended the definition of intellectual property to cover various contents in multimedia. At the same time, the enforcement mechanisms have been strengthened to give teeth to the legislation.

PW *How is the MSC being funded?*

Othman The MSC is funded primarily by industry. The government's contribution has been to provide the initial loan to the MDC as start-up capital.

PW *Do you think the monetary challenges facing Asian economies will have an effect on the MDC?*

Othman The effect will be very marginal. The current economic downturn in the region reinforces the need for greater value and depth in the economy.

In fact, it points to the need for the MSC to speed up and move forward. It also underscores the rational behind MSC to be internalized within all sectors of the Malaysian economy. We must be innovative and very global in our outlook and link up with new partners to enable entry into new markets. The use of new technology to increase productivity and competitiveness is critical.

An integral part if the MSC plan is the involvement of foreign companies, especially leading high-technology firms. When these companies come to invest in the MSC, they will attract other companies. We believe that the presence of these leading edge firms can help benefit the small and medium-scale indigenous companies by way of technology transfer. I see a bright future for Malaysia and the region in the MSC.

PW *How do you envision government and business working together? How do you attract companies to invest in MSC?*

Othman The government has always regarded the private sector as the engine of growth. Over the years, we have developed an effective working relationship with industries at all levels of government, allowing the private sector to participate and make constructive contribution in the formulation of policies that are amenable to their business operations.

This arrangement has also been extended to the MSC with the setting up of the International Advisory Panel. The members are specially selected from chief executives of leading edge global companies as well as independent intellectuals in information technology. The Prime Minister [Dato' Seri Dr. Mahathir Mohamad] acts as Chairman, and the Panel advises the Prime Minister on critical issues that need to be addressed to ensure success of the MSC and to attract foreign investments.

The formulation of the Concept Request for Proposals [CRFPs] for the flagship applications is something unique and a departure from current bureaucratic practice in government contracts. Leading-edge companies, working together with other "web" shapers and the lead government agencies, contribute expertise and time to define the applications. The CRFPs provide the guidelines for companies to bid for the applications and propose the total solutions. We believe that these arrangements and the inherent attractiveness of the MSC concept – coupled with financial and non-financial incentives – provide a comprehensive package to support investment and growth of our multimedia industry.

PW *What is the status of the MSC?*

Othman As I mentioned earlier, the physical elements of the MSC – the KLCC Twin Towers, the new international airport, Putrajaya, and Cyberjaya will be fully completed in October and December 1998, respectively. For Cyberjaya, the first phase encompasses approximately 1,400 hectares.

In fact, the MDC moved into Cyberjaya in September 1997, one year ahead of schedule. We currently are occupying a 28-acre site that also includes a resort and a business complex for about 20 companies. The 2.5- to 10-Gbps broadband telecommunications network using ATM switches will be fully functional by the end of 1998. MDC has set an aggressive target of locating 50 companies in Cyberjaya by the end of 1998. We recognize the value and importance of creating a community in the MSC that encourages a free flow of ideas, spurs technological innovation, and unleashes the creative talents of its members. Hence, our emphasis is on a fast-track development for Cyberjaya to quickly catalyze a critical mass of companies and knowledge workers living and working in a world-class environment.

As for companies investing in the MSC, as of January 1998, we have about 180 companies applying for MSC status; that is an average of one company per day since we began accepting applications in June 1997. About 35 percent of these applicants are local companies, 26 percent are foreign-local joint ventures, and the remainder are from Europe, North America, Japan, and Southeast Asia. Of the 103 companies approved, 78 are already in operations and will move into Cyberjaya once Phase One is completed by the end of 1998. From their business plans and projections, the 180 companies propose capital investments in excess of 4 billion Malaysian Ringgit over the next 5 years, with expected revenues of more than 13 billion Malaysian Ringgit. The forecasts reflect strong sentiments on the value proposition of the MSC and the business prospects for investors. Two independent studies have indicated that the MSC will generate US$25 billion in value to the Malaysian economy after 5 years of operations.

Regarding the progress of the flagship applications, the lead agencies have completed evaluation of the responses from companies to the CRFPs for electronic government, smart schools, telemedicine, and the multipurpose smart card, and are currently in multi-track negotiations with the various partners. We expect that the successful companies will be announced between February and June 1998, according to the different flagships, and implementation will commence immediately thereafter.

The CRFP process is a departure from previous government practice in the awarding of contracts as it welcomes responding companies to submit their own proposed technology solutions and business model for consideration. The flagship applications seek solutions that are not readily available anywhere in the world. This is a clear example of the "test bed" nature of the MSC initiative, and successful companies have the opportunity to replicate their solutions outside of Malaysia once they have successfully tested them here.

PW *What about human resources for the project?*

Othman The shortage of knowledge workers is a global problem. In the U.S. alone, it is reported that some 190,000 high technology jobs remained unfilled. Worldwide, the figure is around 400,000, and the shortfall is expected to continue for at least another decade. MSC companies will require about 17,000 knowledge workers over the next 5 years. The country has a comprehensive strategy to deal with the growing demand for knowledge workers as the MSC develops.

In the short term, MSC status companies have been given full flexibility to recruit knowledge workers from abroad under the MSC's Bill of Guarantees. The long-term solution is to meet the demand for knowledge workers from within the country. The overall objective is to raise the group enrollment ratio in tertiary educational institutions from the present 13 percent to 20 percent by the year 2000. The higher education policy of the country has been adjusted to encourage the growth of private education, and we will see the total number of universities doubling by the year 2000 with the establishment of new private universities.

In addition, Multimedia University has been established as an international university to lead the way in producing the next generation of knowledge workers for the MSC – already there are some 1,300 students in a temporary campus and they will move to Cyberjaya when the permanent campus is completed in October 1998. Scholarships also will be provided to attract top students from abroad. We are also discussing a novel idea by Kenichi Ohmae to create an "MSC virtual degree" that will allow students anywhere to study various components of a relevant degree from multiple institutions – via the electronic media. This would go a long way to create a web of potential knowledge workers spanning the globe, many of whom we hope eventually will make their homes in the MSC.

The government also has extended the benefits under the MSC Bill of Guarantees to qualifying institutions of higher education. This offer is meant to encourage investments in training and educational institutions that focus on curricula relevant to producing skills for the MSC – in multimedia, information technology, computer animation, and engineering sciences.

At the end of the day, the MSC will be enriched by the diversity of nationalities, cultural heritage, and intellectual capital. We believe that our efforts to create this world-class environment for knowledge workers of the Information Age will attract the best and the brightest from around the world, as an ideal place to live, work, and play. The government is fully committed to the transformation of Malaysia into a knowledge-based society by the year 2020, and the MSC will be the catalyst for this transformation.

PW *How will you measure the success of the project?*

Othman We have set a number of measurable indicators to gauge the success of the MSC. The corporate indicators include the number of leading-edge companies making the MSC their home in Asia, the depth of the R&D and value-added activities of MSC companies, and the quality of the partnerships and joint ventures formed by these companies in expanding globally. We have set ourselves what I believe to be a very achievable target of attracting at least 50 world-class companies into the MSC by the turn of the century. Already, we have about 30 companies, both local and foreign, that have the technical and marketing competence to compete in the global marketplace.

On the infrastructure side, the quality and pricing of telecommunications services, the lifeblood of tomorrow's knowledge-based industries, will remain a key performance indicator. The MSC's success also will be measured by our sustained ability to pioneer a robust framework of cyberlaws that is conducive to technological innovation, creativity, and the growth of electronic commerce. Our cyberlaws also must continue to balance societal interests with the valid concerns for privacy and freedom of expression of individuals. Our ability to bring about a harmonization of these cyberlaws among our closest trading partners will be critical in the MSC's goal of building global bridges to the world. There is a similar objective to ensure a common, open platform across the MSC flagship applications as they are rolled out, not only for domestic interoperability but also to ensure that eventually we can connect to and communicate with the rest of the world, and vice versa. Only then can the full potential of the global information infrastructure be harnessed by the world community.

A number of macro indicators also will be used to benchmark the MSC. The growth of high-value skills in Malaysia is vital to our success, and its foundation lies in our ability to provide universal access to education and training with advances in multimedia and telecommunications. In the age of globalization and global competition, the MSC is critical in transforming all sectors of our economy, from agriculture to manufacturing to services, via a renewed emphasis on innovation and increased productivity of human capital and investment. This goal will ensure that the country remains on its growth trajectory toward the year 2020.

PW *That's quite a challenge.*

Othman These are ambitious but realistic targets. Remember, the MSC is a national priority that has the full and undivided commitment of the Malaysian government to ensure its success. We have little choice but to move forward – the Information Age is already upon us.

Our society is fortunate to be in a state of readiness to take full advantage of the opportunities created by the MSC. There is no substitute for perseverance and hard work. I am confident that we shall rise to the challenge.

3 Components

3.1 I/O Technologies

3.1.1 Executive Summary

As de facto standards become outdated, basic input devices, such as keyboards and mice, are becoming more complex. Currently, few true standards exist, because a device must be compatible with the most popular operating systems: Apple's MacOS 8, and Microsoft's Windows 95 and Windows NT 4.0.

Specialized input and control devices, namely, touchscreens and speech recognition, are remaining just that – specialized. Although touchscreens continue to evolve, the current generation is more than adequate for today's key application, the free-standing kiosk; there appears to be little user need or interest in desktop touchscreens. Speech recognition has shown considerable improvement in the past 2 years, spurred by higher-powered Pentium-family processors and better mathematical algorithms. However, despite speech recognition's ready availability (it is built into IBM's OS/2 Warp 4.0) and low cost (continuous-recognition software is available for only a few hundred dollars), it has not yet appealed to the mainstream.

Image capture devices, such as scanners and video and digital cameras, are moving from expensive, specialized uses (such as graphics arts and video conferencing systems) into the mainstream. To that end, improvements in charge-coupled device (CCD) technology are making possible imaging devices with higher resolution, better ability to show color, and faster operating speeds (translating into either faster per-page scanning or better quality video that captures a higher number of frames per second) at ever-decreasing costs.

In the world of display output devices, the cathode-ray tube (CRT) display – the kind traditionally used for desktop monitors – has reached a level of maturity. New functionality is limited to "bells and whistles," not drastic improvements in image quality or reductions in pricing. Liquid-crystal displays (LCDs), however, are showing significant improvements in brightness, resolution, reduced power consumption, and sheer size. Given the improvements in LCD technology, it is very likely that they will make inroads into the desktop display market, particularly displacing 17-inch CRTs, which require a great deal of electricity and are heavy, bulky, and expensive.

Print output devices have become stratified into four levels. In the personal-printer market, sub-$500 color ink-jet printers and low-cost laser printers vie for market share; consumers often value the color output, and business users prefer the crispness of laser output. In the networked business market, laser printers rule supreme; current advances are increasing speed and resolution as well as sophisticated paper-handling features, such as multiple output trays and collating/stapling. Graphics-arts and marketing professionals have several color-printer technologies to choose from, each optimized for different purposes; despite improvements in technology, prices for the devices and consumables (ink and media) remain high. Finally, high-volume printers faster than 100 pages per minute (ppm) remain centrally controlled devices attached to mainframes, mini-computers, or dedicated print servers. Although impact printers are still used to produce multi-part forms, the market is shifting to laser printers as current advances yield somewhat higher speeds, higher quality output, and recently, the introduction of high-speed, color laser printing.

3.1.2 Input Devices

Nothing may seem more common than computer input devices, such as the familiar mice and keyboards, but this is merely a reflection of how far computing has evolved since the batch-processing, punched-card era. Personal computer (PC) input devices have now entered their third phase.

In the first phase of the late 1970s and early 1980s, diverse companies such as Apple, Atari, Commodore, IBM, Kaypro, and Tandy-Radio Shack developed their own keyboard layouts, each attempting to define a standard.

The second phase, from the mid-1980s to the early 1990s, saw the emergence of de facto standards, such as the IBM PC keyboard, the Apple Macintosh keyboard, and the one-button Macintosh mouse.

The third phase, of the late 1990s, is characterized by rampant experimentation designed to increase general productivity and enable new classes of applications. For example, consider the keyboard and mouse. A traditional software maker, Microsoft, has developed innovative ergonomic "split" keyboards designed to combat repetitive-motion injury (such as carpal tunnel syndrome), and mice with so-called scrolling wheels meant to ease a Web browser's scrolling functions. Other companies are experimenting with input devices such as keyboards with built-in scanners, magnetic-stripe readers, or smart card interfaces; cordless mice; and mouse alternatives (such as touchpads) for both laptops and desktop systems.

Innovation is not limited to keyboards and pointing-devices. Personal flat-bed scanners, once a luxury reserved for artists and desktop-publishing workers, are affordable and can be used to fax hard-copy originals from computerized fax boards. Voice recognition has improved, with continuous speech dictation possible on PCs. Even CCD-based video cameras are appearing for desktop videoconferencing. Digital still-frame cameras, also known as digital cameras, also are based on CCD technology. These digital cameras, priced anywhere between $200 and $1,000, have emerged as a hot consumer category in 1997. This trend is expected to continue as CCD resolution and sensitivity increase, and as manufacturers add additional storage to these devices. Although their output quality does not match that of ordinary photographic film and paper, digital cameras are used in applications where the lower quality is acceptable, such as in creation of Web pages or in newspaper photography, or where quick data transmission of the photograph is essential. The third phase of input devices clearly has arrived, and if anything, the rate of experimentation is increasing, with new productivity gains sure to follow.

■ Keyboards

The computer keyboard remains the primary way to control a computer or terminal as well as enter data into applications. Despite the technological improvements in other input technologies, such as voice recognition, it is unlikely that the importance of the keyboard on traditional PCs will fade. In fact, the choice of keyboard has recently become more difficult due to increasing innovation and competition.

In 1983, when the IBM Personal Computer was introduced, there was one type of keyboard. Clone makers mimicked the IBM PC keyboard design (which was different from IBM's well-known standard, the Selectric typewriter); their products differed only in the switch technology placed under the keys and thus, in the tactile feel of the keyboard. Today, a significant difference in feel still exists between various manufacturers' keyboards. However, that is the least of the differences. New keyboard designs are aimed at preventing repetitive-motion injuries, easing the use of particular operating systems or applications, overcoming the inconvenience of separate pointing devices, or catering to other special needs.

The most visually obvious innovation is the ergonomic keyboard, popularized by Microsoft's Natural Keyboard, but also offered by Adesso, DynaPoint, Lexmark, PC Concepts, and others. The new keyboards offer one or more of three classes of change: split, where a normal flat keyboard is split down the middle with both halves bent back at an angle; molded, where the surface of the keyboard is curved in three dimensions and each key is adjusted to the size of the finger that should press it; and tilted, where keys are offset to fit the length of the fingers. A common feature in these new keyboards is a large wrist rest, designed to allow the hands to be held in a more natural, less-stressful position. Some users swear by ergonomic keyboards; others cannot adapt easily to the non-traditional feel. These inexpensive (less than $200) keyboards are nearly always after-market purchases, typically purchased on request for employees who suffer from repetitive-motion ailments or who otherwise indicate a preference for the new design.

Another change is the addition of one or more application-specific keys. Actually, this is not truly an innovation, because application-specific keys were found on many terminals and early computers; however, such keys were not part of the original IBM PC standard. Microsoft broke this decade-long tradition by introducing the "Windows" key on the Natural Keyboard. This key acts the same as the Ctrl-Esc combination under Windows 95 and Windows NT 4.0: it pulls up the Start menu. Because the action of the Windows key can be duplicated with a single mouse click or the Ctrl-Esc key combination, these keyboards do not actually enable new functionality.

Other keyboards integrate pointing devices. This trend began with laptop and notebook computers; the earliest successes were Apple's trackball, IBM's TrackPoint pointing device, and now the equally common touchpad. Manufacturers, including Alps, IBM, and Kensington, now offer similar pointing devices for desktop keyboards; these are targeted toward portable computer users who wish to maintain a consistent keyboard interface between their notebook and desktop computers.

An area of slow, but steady, evolution is the so-called chord keyboard. In conventional computer keyboards, modeled after the typewriter, most keys correspond to a single letter or number. In one-handed chord keyboards, patterned after a stenographer's keyboard, a small number of keys (fewer than a dozen, occasionally with as few as five keys) are pressed in simultaneous combinations to create all desired numbers, letters, punctuation marks, and control sequences. Chord keyboards have never made a significant impact on the general market, due to popular familiarity with the QWERTY keyboard, but they were used in specialized applications; for example, the U.S. Post Office used chord keyboards in the 1960s to enter numbers for mail sorting. Chord keyboards, such as WorkLink's BAT Personal Keyboard, appeal to individuals who have physical disabilities, or specific requirements for a keyboard that can be operated completely with one hand (perhaps so the other hand can use a computer-aided-design digitizer or operate other equipment). Wearable "chord keyboard gloves," which allow the user to type simply by wiggling finders in the air, are an interesting development, pioneered by research at the MIT Media Lab. It will be years before such devices are generally available, however, because these applications are highly specialized.

■ Mice/Trackballs

In the world of graphical user interfaces (GUIs), such as Apple's MacOS or Microsoft's Windows, a pointing device plays a vital role. It moves a cursor around the screen, highlights text or spreadsheet cells, pulls down and selects menu options, and pushes buttons on Web browsers. Although it is possible to operate most GUIs without a pointing device, it is neither easy nor intuitive. The situation is more acute with the World Wide Web: Many Web pages are not designed with keyboard shortcuts, and thus can be accessed only via a pointing device.

New PC systems typically ship with a basic two-button mouse for Windows-based PCs and a one-button mouse for the Macintosh. Mice are bundled with nearly all new computers; better-quality mice, offering improved ergonomics and greater sensitivity, are manufactured by Kensington, Logitech, Microsoft, and others, and can be purchased for less than $50. (UNIX workstations use three-button mice that are shipped from the workstation manufacturer.)

Mouse alternatives exist for users of portable computers and those performing special applications. Laptop and notebook users usually have two main choices: TrackPoint and touchpad, neither of which have moving parts. The TrackPoint, popularized by IBM's ThinkPads but available on systems from Texas Instruments, Toshiba, and other makers, is a small, pressure-sensitive button, shaped like a pencil eraser, located between the B, G, and H keys on the notebook keyboard. The touchpad is a small, pressure-sensitive pad (about 2.5 by 1.5 inches) upon which the user rub a finger to move the cursor; these are often found on Acer, Compaq, Dell, Hitachi, and other laptops. Consumers choose between TrackPoint or touchpad based on personal preference; within corporations, hardware-standardization policies generally preclude such a choice for the corporate end user. A third option is the trackball, the oldest graphical pointing device; it is basically an inverted mouse, consisting of small ball located either on the leading edge of the keyboard or manually latched to one side. Because the moving ball and its motion sensors are moving parts, they require frequent cleaning, and thus trackballs are rarely encountered on new portable computers.

Mouse alternatives exist for desktop PCs, too; touchpads, TrackPoints, and trackballs are available integrated into standard PC keyboards. External devices are available as well; typical touchpads by Cirque and trackballs by Kensington sell for less than $100. In many cases, use of these devices is a matter of personal preference, although many graphic artists find that a 3-inch or larger trackball is more comfortable for drawing, photograph editing, and desktop publishing.

A new variation is the Intellimouse, a two-button mouse with a small wheel located vertically between the left and right buttons. This wheel is exploited by newer Microsoft applications, such as the Internet Explorer 3.x Web browser or the Word 97 word processor, and permits vertical scrolling without moving the mouse to the vertical scroll bar or pressing the PgUp or PgDn key. It remains to be seen whether other software makers will support the Intellimouse wheel or whether other mouse manufacturers will adopt the concept.

Digitizing tablets, used for two decades by computer-aided designers and engineers, have remained a relatively static technology. Such tablets vary in size from little larger than a standard sheet of paper to big, freestanding devices large enough to hold full-sized blueprints. To use a digitizing tablet, the user places a sheet of paper on top of the tablet, accurately places a pointing device (called a puck) on a specific point on the sheet of paper, and presses a button. The coordinates of this point are then relayed to the CAD or engineering software. Innovations such as wireless pucks have replaced less-convenient corded pucks. These tablets remain a specialized technology. The only exception to this model is a class of smaller digitizers used by computer artists. These devices use a pressure-sensitive "pen" to allow the artist to draw freehand on the tablet and thus create artwork directly in a drawing program. Three-dimensional digitizers also have begun to appear at reasonable prices.

■ Touchscreens

A convenient computer control device without moving parts is the touchscreen. Around for more than two decades, the basic touchscreen has found a home in kiosks and other areas where user input is limited to "pushing a button," selecting an area from a map, or doing very limited typing.

Touchscreens appear to be ordinary CRTs or LCDs. However, a data-acquisition device between the user and the display senses the touch of a finger or special probe, and relays *x-y* coordinates back to the application. Numerous technologies enable modern touchscreens; the three most common are capacitive sensing, infrared sensing, and resistance sensing.

In capacitive sensing, sensors embedded between the screen and a sheet of outer glass react with the body's electrical field. (See Figure 3-1.) Infrared sensing relies upon transmitter/receivers located around the screen's perimeter; when a finger breaks the horizontal and vertical beams, this information is relayed to the software. Resistance sensing consists of two layers of conductive material, separated by flexible spacers. When a finger or stylus presses on the outer layer, it makes electrical contact with the inner layer, and the position of the contact can then be determined.

Figure 3-1: How Capacitive-Sensing Touchscreens Work

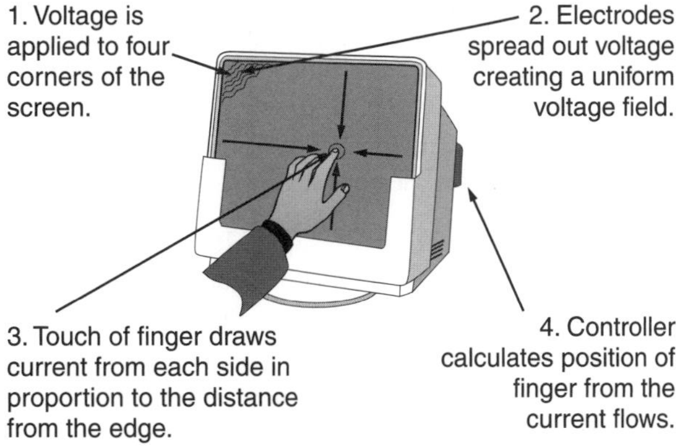

Source: MicroTouch Systems, 1997

An emerging touchscreen technology is guided acoustic wave (GAW), which exploits the fact that sound waves travel at a constant rate through glass. Sound waves are transmitted through a glass overlay; when the glass is touched, the surface is deflected and the distance the sound waves must travel is increased. By calculating the point at which the sound waves suffer greatest delay, the point of contact can be determined.

Another innovation, from MicroTouch Systems, is called ThruGlass. Using projected capacitive technology, ThruGlass can sense touch through as much as 1 inch of ordinary window glass, making it attractive for store displays – the touchscreen can remain inside the store and can be accessed from outside through a plate-glass window.

The other major manufactures of touchscreens are Elo TouchSystems, IBM, and Mitsubishi America, each of which makes products using a variety of technologies. Many other companies integrate these four companies' touchscreens into kiosks.

■ Desktop Scanners

A veritable explosion in the number and variety of desktop scanners has occurred recently, along with a commensurate increase in functionality and decrease in cost. Desktop business-quality scanners come in two types: high-resolution flatbed scanners that look like photocopiers and are used occasionally with automatic document feeders, and sheet-fed "convenience" scanners similar in appearance to fax machines. Both types are being used to scan photographs and artwork for Web pages or desktop publishing, or to scan text pages for optical character recognition (OCR) or entry into document databases. Convenience scanners, which often trade lower resolution for higher speed and built-in paper handling, are also used with personal or networked fax software to allow faxing of hard-copy originals.

Flatbed scanners, designed for photographs or artwork, are now low-priced (quoted prices as of late 1997). For example, the Microtek ScanMaker V300 (street price approximately $150) offers actual 24-bit (millions of colors) 300 by 600 dots per inch (dpi) scanning using a CCD scan head; software interpolates the image dots to provide an image that appears to be 4,800 by 4,800 dpi resolution. At the high end, Microtek's $1,400 ScanMaker III or Umax's $1,400 PowerLook II each scan 36-bit color images (tens of billions of colors) at an actual 600 by 1,200 dpi, interpolated to 9,600 by 9,600 dpi. Other scanner manufacturers, such as HP, make comparable flat-bed scanners at similar prices. Higher-end models are distinguished by faster scanning time, larger scanning surfaces (8.5-inch by 14-inch or larger), and high-speed interfaces using SCSI rather than the relatively slow parallel port. Some models also have options for scanning transparencies, such as 35 mm slides or negatives. Another high-end option is color calibration, so that scanned images more closely match the original.

Sheet-fed scanners are designed for higher speed, but lower resolution. For example, the $230 Logitech PageScan Color offers 24-bit color, but only at a low 100 to 200 dpi (adequate for creating Web pages, but not for desktop publishing), or 256 shades of gray at 400 dpi – comparable to that of a crisp fax image. This output is appropriate because scanned images are often fed directly into fax software. These printers are often accompanied by OCR software; the leading desktop products are Caere's OmniPage and Xerox's TextBridge.

Specialized scanners have started emerging during the past 2 years. OCR-enhanced business-card scanners, supported by popular contact-management software, allow name and address data to be captured without typing.

■ Video Cameras

Video cameras have found new usefulness in the digital domain. Small cameras, from companies such as Connectix and Intel, typically connect to a PC's parallel port; some models connect via the new Universal Serial Bus (USB) port built into late-model PCs (or via an expansion card for pre-1997 machines). These cameras are used for desktop videoconferencing or Internet video telephony. Larger cameras, such as ParkerVision's CameraMan series, attach to a computer or communications device (such as an ISDN terminal adapter or standard dial-up modem) via a serial port.

In both cases, the main ingredient is a CCD – the same technology used in digital still cameras and scanners. The CCD is a single silicon chip that combines a rectangular array of light-sensitive cells with circuitry to process and digitize the image the cells record. As the camera's shutter exposes the CCD's imaging substrate, each cell builds a charge proportional to the amount of light it receives. The resulting image is read cell by cell. The cells in the bottom row pass their charges to a serial shift register below the array. Cells in the rows above pass their charges down one row. Cell by cell, the CCD chip reads the contents of the shift register, converting the number of electrons to an amplified analog signal. Row by row, the analog signal is processed into a digital image that is stored in the camera's memory. (See Figure 3-2.) Capturing color images requires color filters (red, green, and blue) because the individual CCD cells measure light intensity, not frequency.

Figure 3-2: How CCDs Create an Image

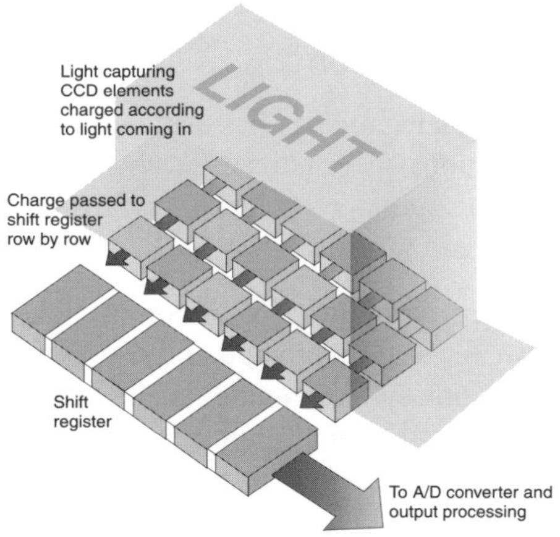

Source: *Byte*, 1997

Speed of image capture differentiates low-cost cameras, such as the Connectix QuickCam (around $200), from the $2,000+ solutions from ParkerVision and other professional-camera manufacturers. CCD cells take a measurable amount of time to record the intensity of the light; the faster the recording, the more quickly the CCD can be reset for another frame. Currently, low-end CCD video cameras can capture low-resolution images (as high as 320 x 240 pixels) at 15 to 24 frames per second (fps) in bright light; higher-end systems can capture 1,024 x 1,024 pixels or larger images at 30+ fps in moderate light. As CCD technology improves, the CCD will have increased ability to operate in dimmer light. To provide context, a standard North American television image is broadcast in a format called NTSC (National Television Standard Committee). The analog NTSC standard, supported by televisions, is nominally 768 x 576 pixels, refreshed at 30 fps. (By comparison, the PAL, or Phased Alternate Line, system used in

Germany, the U.K., and other countries offers higher resolution, at 625 x 830 pixels, but at an effective rate of only 25 fps.) Other areas of development are resolution, reducing the size and increasing the number of cells, and widening the field of view.

When the video camera is examined in the context of videoconferencing, however, additional technologies are needed. Cameras such as the Connectix QuickCam must be aimed and focused manually, and they offer only a single focal length. Professional digital video cameras offer automatic focusing, manual- or computer-controlled image zoom, and tilting (vertical movement) and panning (horizontal movement). Location presets allow the camera operator to configure the video system for each chair in a conference room, allowing the camera to quickly tilt/pan, zoom, and focus on the appropriate speaker.

A new technology is dynamic speaker location. For several years, vendors such as PictureTel have offered a wireless "token," which can be passed manually from speaker to speaker within a room; sensors in the room track the token and aim the video camera to the closest preset location. The new technology, which ParkerVision calls autoTRACK and which PictureTel calls LimeLight, uses an array of special directional microphones (often on a single stand) positioned near the video camera to triangulate the location of the loudest sound source in the room. The camera can be instructed either to move to the nearest preset location or to focus automatically on the speaker's exact location, adjusting zoom based on distance to present a pleasing image composition. The PictureTel LimeLight offers an advanced feature: If two people are conversing in one room, it aims the camera between them and adjusts the zoom to cover both individuals.

■ Speech Recognition

Speech recognition technology has made considerable progress in the late 1990s. Accurate continuous-speech dictation systems to help word processing applications and discrete-speech recognition systems to control basic PC applications verbally have begun to appear.

In practical terms, computer recognition of spoken words falls into two main application areas: verbal command and speech recognition. Verbal command or discrete-recognition systems are designed to enable spoken selection of a small number of options; they are found in so-called interactive voice response (IVR) systems, in which a telephone caller can either press a keypad number or say the number to select an item from a menu. Recent advances in software, sound processing, and increased processing power have also led to reliable verbal-command software for PCs. This task is easier than continuous recognition because the list of words to be interpreted is relatively small (typically, only a few dozen), the context is well known so little ambiguity occurs, and the command words are naturally spoken one at a time.

By comparison, true continuous speech recognition is a difficult problem. Individual speakers have accents and may speak differently at different times due to context, excitement, or even health. The lexicon of words is large (tens of thousands of words) and homonyms may be disambiguated only by context. For this reason, continuous speech systems are speaker-dependent – that is, they require training to an individual speaker's voice. This training is accomplished by having the speaker read a predetermined script containing a wide variety of phonemes several times until the software has mastered the pattern. The most successful commercial products have been in specific technical fields, such as law and medicine, where the vocabularies are well-defined and much smaller. Leaders in speech recognition for basic multimedia PCs are Dragon Systems' NaturallySpeaking, and IBM's ViaVoice and ViaVoice Gold. Kurzweil and Verbex also make and sell continuous-speech recognition systems.

Speech recognition systems follow a four-step process:

1. The sound is captured and digitized. In standard PC-based systems, this process typically is accomplished by a Sound Blaster-compatible sound board and a low-fidelity microphone. In professional speech systems, sound is captured using high-fidelity microphones and then fed into digital-signal processors (DSPs) that are optimized to capture voice at a high sampling rate while rejecting spurious background noise.
2. A technique called feature extraction breaks the digitized signal into individual sound components. Various mathematical algorithms are used in this step, and the feature-extraction stage is the basis for many recent improvements in speech recognition.
3. The sound components are interpreted, and decisions are made about them. Depending on the software, the components may be reassembled at this stage into segments ranging from individual vowel or consonant sounds to complete words or phrases.
4. The output from the preceding stage is compared against a pattern database to determine whether the sound is a word and whether it makes sense in context. If so, a correct spelling may be determined; if not, it may be flagged on-screen for approval or repeated back to the user using computer-generated speech ("Did you say 'cancel order'? If so, say 'yes,' if not, say 'no.'").

Developments in speech recognition focus on the algorithms for feature extraction. Areas under development include Hidden Markov Models, a mathematical technique that can be used to model sound patterns, and neural networks, an artificial intelligence technique that allows for easy training and tuning of a speech system. Along with the improvements in mathematical models, increased computer power – both within the main processor itself and in specially designed speech-recognition DSPs – is making the task of speech recognition easier, if only by throwing more computing power at it.

3.1.3 Display Output Devices

Two primary technologies dominate computer displays: the familiar CRT, also found in most television sets, and LCDs, found in laptop computers, personal organizers, and other electronics equipment.

CRTs are a mature technology. The benefits of CRTs are bright images, ruggedness, and low prices due to the number of CRT factories around the world. LCDs offer different advantages, namely thinness (a complete LCD panel, no matter its image size, is rarely more than a couple of inches thick) and low weight. Until now, those advantages were offset by small sizes due to limitations in LCD fabrication technology, low-resolution images, and very dim displays. Recent advances are changing the landscape: bright LCD panels as large as 16.1 inches (comparable in viewing area to a 19-inch CRT) are commercially available. The biggest drawback now is price: Large LCDs cost at least five times as much as CRT equivalents. As production technology for LCD screens improves and prices decline, the appeal of the LCD will begin to overcome the familiarity of the CRT.

The size and resolution of the display device itself are only two of several factors in overall display usability and image quality. Factors such as color depth, refresh rate, and smoothness of image movement are controlled by the computer's graphics subsystem, also known as a graphics controller. In low-cost business desktop PCs and in notebooks, the graphics controller is built onto the PC's motherboard; in higher-end workstations, the controller is an easily upgradable expansion card.

A revolution is also underway in the display-projection systems market. Room-sized devices have been based on CRT technology; new technologies based on LCDs and on a new innovation, the digital light processor, are challenging CRTs here as well. (See "Digital Light Processing Projectors" on page 59 for an explanation of this technology.) New technologies for wall-mounted displays also are appearing – in the form of field-effect displays and gas-plasma displays.

■ Cathode-Ray Tube Displays

Bigger is better – at least, that is the saying. When it comes to computer displays, that saying translates into additional display "real estate." The assortment of desktop displays, based on CRT technology, has remained largely unchanged for several years. The basic desktop systems use a 15-inch (13.8- to 14-inch viewable) CRT; office workers with higher-priced systems are increasingly using 17-inch (15.7- to 16.1-inch viewable) and 19-inch (17.5-inch viewable) screens. Larger CRTs, normally 20-inch (18.4- to 18.9-inch viewable) or 21-inch (19.6- to 20.0-inch viewable), are reserved for graphics professionals, such as those running CAD, desktop publishing, or other high-end applications. In the UNIX environment, 20- or 21-inch displays are common for many workstation users.

The 14- to 15-inch screen is acceptable for normal office work and data entry, with usable resolution up to 1,024 x 768 pixels. The 17-inch display, with usable resolution up to 1,280 x 1,024 pixels, can show a normal sheet of 8.5- by 11-inch paper sideways; it is useful for showing a great deal of document or spreadsheet real estate, or for users who wish to have several windows on-screen simultaneously. The 20- or 21-inch display, with useful resolution up to 1,600 x 1,200 pixels, can show two 8.5- by 11-inch pages side-by-side; these displays are ideal for full-sized document imaging, CAD, and desktop publishing as well as for use by programmers who need to view several windows at once. Special larger CRTs are designed for boardrooms or classrooms. These devices typically have pixel resolutions similar to those of 21-inch CRTs; for example, Viewsonic's 29GA (29-inch tube, 27.0-inch viewing area) can display 1,280 x 1,024 pixel digital images as well as normal analog television/videocassette output from NTSC/PAL sources.

Due to confusing terminology, it is often difficult to compare CRTs. One area of confusion, that of the CRT's actual size versus the maximum viewable size, has been resolved via a class-action lawsuit in the U.S. Now, CRT sellers must advertise both numbers, making comparison easier.

Another area of confusion regards the actual CRT technology used. The most common type is known as a shadow mask, where the red, green, and blue phosphor "dots" are arranged in small triangles, called triads. The smaller the diagonal distance between two phosphor dots of the same color – called the "dot pitch" – the sharper the image. In the other leading technology, aperture grille, the red, green, and blue phosphors are arranged in thin horizontal lines. The smaller the vertical distance between two lines of the same color – the "stripe pitch" – the sharper the display. For 14-inch through 21-inch CRTs, a dot pitch of 0.25 mm is considered excellent, 0.28 mm is good, and 0.31 mm is "fuzzy" and to be avoided.

The term "interlaced" also can cause confusion. Inexpensive CRTs are non-interlaced at lower resolutions (such as 640 x 480 pixels or 800 x 600 pixels) and interlaced at high resolutions; more expensive models are non-interlaced at all resolutions. In an interlaced CRT, the technology is akin to a standard television, where the electronic beam first draws all odd-numbered lines starting from the top of the screen, then comes back to the top to start drawing the even-numbered lines. Flickering occurs because the unrefreshed lines fade before they are redrawn. In the non-interlaced mode, all lines on the screen are drawn in one pass, producing less flicker and less eye strain for the user. Note that standard NTSC televisions are interlaced, which is why they appear to flicker more than non-interlaced CRT monitors.

Recent innovation in CRTs is in two main areas of development: lower energy consumption and easier setup and configuration.

CRTs complying with the U.S. Department of Energy's EnergyStar program are now common. These CRTs include features that reduce the current draw to extremely low levels – less than 1/10 of an ampere, compared to approximately 2 to 3 amperes in normal operation – if the graphics controller puts the CRT into "sleep" mode.

Increasingly sophisticated setup programs also allow the user to move or resize the screen quickly and easily for optimum viewing. Depending on the model, the setup program also may allow the user to tune the shape of the image, compensating for slight manufacturing variations on the CRT assembly line or for the local magnetic field. Larger CRTs targeted at the graphics market also include the ability to adjust the color mix; the most advanced models include sensors to adjust on-screen colors to match real-world printed color samples. A primary benefit of such menu-based setup programs is that they allow the user to make major modifications to the screen image, color, and clarity electronically, instead of requiring manual adjustments to the CRT hardware by a technician. Note that large CRTs (20-inch or larger) are very sensitive to local magnetic fields. It is not uncommon for firms to hire technicians to adjust or calibrate monitors used for CAD or high-end imaging after receiving them for installation and readjust them again later as the devices age.

Liquid Crystal Displays

From their humble beginnings as dim, barely legible black-and-white displays, the LCDs used in laptops and notebooks have shown amazing evolution over the past decade. In the mid-1980s, a laptop screen was low-contrast and hard to read. Today, huge 13.3-inch and 14.1-inch active-matrix color displays rival CRTs for size, brightness, and clarity.

Nearly all laptops use one of two types of thin-film-transistor (TFT) LCD. Dual-scan, the less-expensive type, uses tiny diodes that either absorb (dark pixel) or reflect (bright pixel) ambient light. The brightness of a dual-scan LCD is determined by the amount of light in the room or the brightness of a back-light behind the LCD panel. Dual-screen displays typically can be read clearly only by a user facing the panel straight on. The more expensive type is active-matrix, where the individual diodes emit red, green, or blue light. Active-matrix panels offer a much wider viewing angle, and they can be read easily in both bright and dim environments.

Both types of screens have suffered from limitations compared to CRTs, such as slow response rate (often leading to "ghosts" as the LCD elements change from transmission to reflection more slowly than images change on the screen). Active-matrix panels also suffer from low manufacturing yields – a single defective element can ruin a display panel – which accounts for their higher cost. Where the smallest acceptable desktop CRT measures 13- to 13.5-inches (viewable), LCDs made slow progress from 9-inch diagonal to 10.1-inch, to 11.4-inch, to 12.1-inch diagonal. Only now are manufacturers selling larger screens, such as IBM's top-of-the-line ThinkPad 770, which sports a 14.1-inch, active-matrix display – comparable to a 15-inch CRT. At lower cost, and greater portability, are the 13.3-inch displays in Hitachi's VisionBook Pro 7560, IBM's ThinkPad 765 models, and Toshiba's Tecra 740 and 750 series. Early 1998 should see other models announced in the 13.3- and 14.1-inch form factor, arguably the largest desirable size in a notebook computer.

New on the scene are stand-alone flat-panel displays; they use the same technology as laptop/notebook active-matrix screens and are only 3 or 4 inches thick. Such displays, which cost several thousand dollars for the equivalent viewing area of a 15-inch CRT, are currently used only in situations such as securities-trading workstations where the large physical depth or high heat dissipation of CRTs make the high cost of flat-panel display justifiable. The very low manufacturing yields drive the high cost of these displays. For example, IBM's 16.1-inch panel (with viewable area comparable to a 19-inch CRT) costs $7,200; Toshiba's 15.1-inch panel (equivalent to a 17-inch CRT) costs $4,000. As they become less expensive, LCDs' smaller size (especially compared to 17-inch or larger CRTs), low power consumption, and ability to be hung on a wall will drive increasing acceptance in the office and kiosk environment.

A major method for improving LCD manufacturing technologies is similar to one for increasing microprocessor fabrication yields: fabrication of more or larger panels in one operation. Initially, only four 8.4-inch LCD panels could be cut out of the so-called first-generation 300 mm by 420 mm substrate. Currently, the 400 mm by 500 mm second-generation substrate can yield four 10.4-inch or two 12.1-inch panels. This was followed by the 550 mm by 650 mm third-generation substrate, which yielded six 12.1-inch panels. The newest 650 mm by 850 mm three point five generation substrates allow the manufacture of six 13.3-inch panels, or a smaller number of 14.1-inch panels.

Alternative Display Technologies

Color gas-plasma and field emission displays (FEDs) offer promise, initially for specialized applications, but perhaps soon in more mainstream products.

Gas-plasma displays, which work with gas-filled tubes similar to those found in fluorescent lights, are not new – they have been around for more than a decade, and were well known in orange monochrome panels, such as those used by IBM terminals and the AC-powered Toshiba T3100-series and T5200-series laptops from the late 1980s. Today's gas-plasma screens, such as those from Fujitsu and Toshiba, contain gas-filled tubes coated with red, green, and blue phosphors. Electrons emitted from the tubes strike the phosphors, resulting in a crisp, bright image. Due to the requirements of the gas panels, color gas-plasma screens are large – in the 40-inch to 60-inch diagonal range; prices begin at $10,000 for the display panel itself. This price, and the wide viewing angle, make the screen most useful for public areas such as meeting rooms, where it can be used as computer or videoconferencing displays. An existing limitation of gas-plasma displays is that current-generation phosphors degrade rapidly; they are rated for about 10,000 hours of use. Also, although they can be used outdoors, they are very difficult to read in direct sunlight.

FED panels are a cross between CRTs and active-matrix LCDs. Like a CRT, the FED's image comes from electrons striking red, green, and blue phosphors. Unlike the CRT, the electronics are emitted not from a magnetically aimed cathode, but from tiny electron emitters embedded behind each pixel – wired up using a matrix, like an LCD panel.

According to PixTech, the pioneer of this technology, the advantages of the FED over the CRT are a full 160-degree view angle, quick response at any temperature (LCDs become sluggish at low temperatures), only 2.5 mm thickness, high brightness, and low power consumption. Currently, PixTech has developed 5.2-inch panels, with 8.5-inch panels announced for 1998. Another vendor in the FED market is Candescent Technologies, formerly known as Silicon Valley Video. Its FED panel, which the company calls ThinCRT, has been built in engineering prototypes measuring 2.3 inches diagonally. Screens of this type are aimed at specialized installations where even an active-matrix LCD does not offer sufficiently wide viewing angles, or in low temperatures where all conventional LCDs become too cold to operate.

Figure 3-3: How FEDs Work

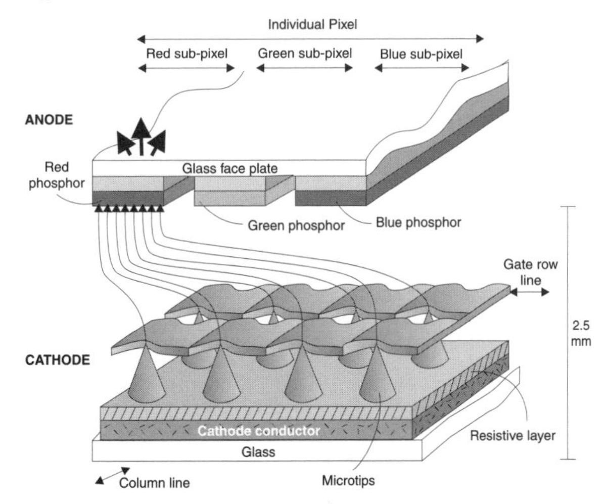

Source: PixTech, 1997

As show in Figure 3-3, FEDs are composed of thousands of very small microtips deposited on metal strips placed on the bottom plate of glass. These strips form the columns of a matrix-addressed display. Metal strips with holes forming rows of "gates" run crosswise above the microtips. A current is applied to one row of microtips, causing cold cathode emission of electrons in proportion to how much voltage also is present in each column strip. The emitted electrons are attracted to the phosphor screen on the other glass plate because of its higher charge. As in a CRT, when electrons strike the phosphors, colored light is produced.

Projection Systems

Video projection systems have come a long way from the dim, washed-out LCD panels – essentially, laptop screens without back lighting or opaque case –that a presenter plopped atop an overhead projector.

Beyond those LCD panels – still in widespread use because of their portability and reasonable price tag – are three additional technologies for projecting a computer image onto a screen. (The quality of a computer image is measured by its brightness in lumens. A lumen is a measure of the amount of light exiting from the projector; the greater the number of lumens, the brighter the light.)

■ Light-Valve Projectors

Light-valve LCD projectors, such as Barco's BarcoVision 9200, are designed to project in a large hall, such as an auditorium. They create a very bright light that is separated into its red, green, and blue components using coated mirrors. (See Figure 3-4.) Those three colored light streams are fed through three LCD panels, which either absorb or transmit the light on a pixel-by-pixel basis. The beams are then recombined into a single stream and projected onto the screen for viewing. Advances in light-valve projectors have led to increased resolutions (currently as high as 1,280 x 1,024 pixels) and brighter and whiter lights (high-end systems use 1,800-watt metal-halide bulbs). Light-valve projection systems are priced from $25,000 to $100,000, with image brightness up to 6,000 lumens.

Figure 3-4: Inside a Light-Valve LCD Projector

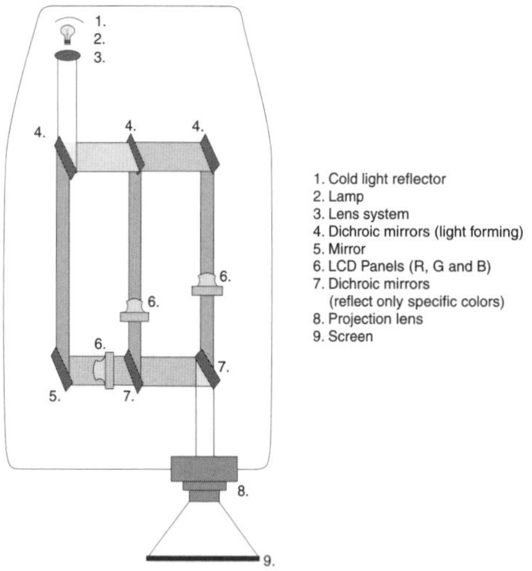

Source: Barco, 1997

■ Three-Beam CRT Projectors

Three-beam CRT projectors, based on the same technology as projection television systems, are priced in the $15,000 to $25,000 range for brightness up to 1,000 lumens, and they can project images suitable for large meeting rooms. These systems use three very bright monochrome CRTs, usually 5 to 7 inches diagonally, corresponding to the red, green, and blue picture elements. Light output from the three CRTs is projected through the appropriate color lens; great care is required in setting up the projector so the three images align perfectly. Beyond increasing resolution and brightness, and reducing the size and weight of the projector, relatively little can be done to improve this technology.

■ Digital Light Processing Projectors

Digital light processing (DLP), pioneered by Texas Instruments, is the newest technology for self-contained projectors. The DLP projector's unique component, the Digital Micromirror Device (DMD), functions like a matrix of mirrors, one for each pixel. A bright light source is aimed at the DMD, and when a pixel is on, the light is reflected out through a projection lens. When the pixel is to appear dark, the mirror is rotated so that the light is reflected to a light-absorbing surface instead. (In Figure 3-5, for example, the two outer mirrors at the bottom of the figure are turned on and reflect the light through the projection lens and onto the screen. The center mirror is tilted to the off position. This mirror reflects light away from the projection lens onto a light absorber, so no light reaches the screen at that particular pixel.) By using a color filter system and by varying the amount of time each DMD pixel mirror is on, a full-color, digital picture is projected onto the screen.

The DMD is a monochrome device; to achieve red, green, and blue dots, data for each of the three colors is sent to the DMD in sequence, timed to correspond with a rotating wheel that contains the three appropriately colored optical filters. First-generation DLP projectors offer brightness from 300 to 3,000 lumens.

Figure 3-5: The Light-Reflection Process in DMD Projectors

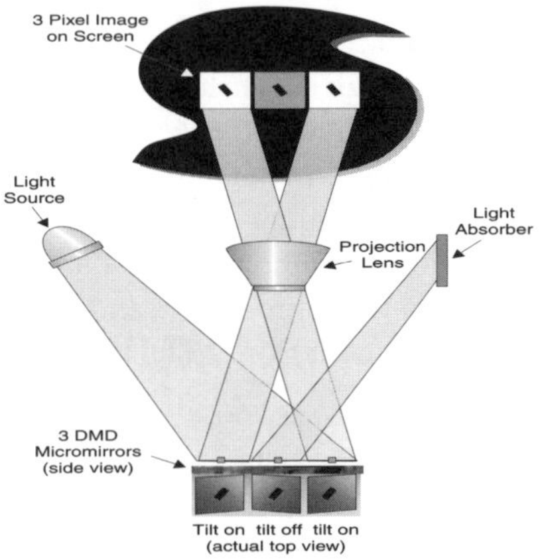

Source: Texas Instruments, 1997

Graphics Controllers

The largest CRT or LCD panel is worthless without the digital circuitry to drive it; this is the function of the graphics controller. A graphics controller, which is found either on a PC's main system board (the motherboard) or installed in an expansion slot, contains three key components: a graphics coprocessor, which creates the images; a clock crystal, which synchronizes the controller's output to match the frequency of an attached display device; and video memory, which stores the bit-by-bit data corresponding to pixels on the display device.

Before the advent of the dedicated graphics coprocessor, the computer's main microprocessor calculated the position and color of every pixel on the screen. This was a time-consuming process that resulted in tradeoffs between graphics quality and the computer's ability to perform other work. The graphics coprocessor, sometimes called a graphics accelerator, relieves the microprocessor of the graphics calculations. The operating system can communicate directly to the graphics coprocessor, issuing complex instructions, such as "draw a 50 percent translucent blue circle 20 pixels in diameter centered on coordinates 120 horizontal, 200 vertical" or "move that circle up 10 pixels, uncovering what it had covered before." The graphics processor performs the appropriate calculations to indicate whether each pixel should be on or off, and it writes the appropriate patterns into the video memory. These changes are synchronized with the display device's refresh frequency by a clock crystal; at the appropriate time, data stored in the video memory is sent to the display device. The higher the clock/display frequency, the less flicker is visible.

Most low-cost graphics controllers today, and those built into system motherboards, are so-called two-dimensional (2-D) controllers, that perform image calculations assuming all images are on a flat plane. 2-D graphics controllers, pioneered by companies such as ATI Technologies and Tseng Labs, are a relatively mature technology; vendors continue accelerating the chips to handle higher resolutions more quickly. However, the biggest consideration is to reduce the size and cost of the graphics coprocessor and other components.

The area of intensive development is 3-D coprocessors. The software application can specify dimensions, texture, and movement of objects in three dimensions, and the 3-D graphics coprocessor will calculate the appropriate parameters for showing the graphics on a display device. The market for 3-D controllers is rooted in CAD and scientific visualization, but is being driven to higher volume and hence lower cost by computer game enthusiasts.

3-D graphics processors contain new algorithms for alpha blending, allowing one object to show through a translucent object; anti-aliasing, using intermediate-colored pixels to soften jagged edges of linear objects; texture mapping, twisting a color pattern, such as that of a brick wall, to fit realistically onto a 3-D shape; and Z-buffering, deciding which objects are visible and which are temporarily hidden behind foreground objects. Emerging 3-D graphics controllers require high-performance coprocessors and a great deal of video memory; products under development from ATI, Hercules, and Videologic use as much as 16 MB of video memory.

Today's leading graphics controllers use application-specific integrated circuits (ASICs). These chips offer much lower cost of design than traditional microprocessors, can be optimized for the specific tasks needed in graphics applications, and can be manufactured quickly and affordably in relatively small quantities. This feature shortens the design cycle and allows for rapid innovation and increased competition.

One final emerging trend is the addition of special circuitry to 2-D and 3-D controllers designed to accelerate the playback of digital video; such devices are often labeled multimedia accelerators.

3.1.4 Print Output Devices

Computer printers certainly have evolved tremendously since the type-chain printers and line printers used by early mainframe systems, and the slow, loud, low-resolution dot-matrix impact printers embraced by early PC adopters. In many ways, printers have evolved in parallel with PCs and their applications; the release of the first laser printers coincided with the release of desktop publishing systems. The demand for fast printers even led some manufacturers to put more power into their printers than into their computers: Apple's first LaserWriter, for example, boasted a Motorola 68020 processor, although the then state-of-the-art Macintosh was based on the earlier and slower Motorola 68000.

Despite the frequent announcement of the paperless office, the number and variety of printers is staggering, as is the number of technologies in use. Impact dot-matrix, ink-jet, laser, thermal transfer, die-sublimation, and others vie for general-purpose, mass-market sales as well as high-margin, specialized applications. Printers are also becoming available in larger sizes, with B- and C-sized printers now common. (See Table 8-14 on page 696.)

This section examines several rapidly evolving areas of printer technology: professional-quality color printers, mid- and high-volume devices, and enterprise printers.

Professional-Quality Color Printers

Professional color printers are often noisy and slow. They are also expensive – both as to the initial purchase and the subsequent consumables: ribbons, inks, cleaning materials, and with some technologies, special papers.

These devices typically are targeted at artists, desktop publishing staff, marketing designers, and service bureaus. Marketing designers, especially, use these color prints for one-to-one marketing; for example, printing the color illustration of a unique item (such as a house or an objet d'art) to a specific individual who fits a defined marketing profile. Professional-quality color printers are all available as network-connected devices, shared within an organization or department and managed by a print server.

Thermal transfer printers, also known as thermal wax printers, apply colored wax to paper to produce graphics and text. The image consists of rows and columns of dots, which are applied by passing the paper under a fixed array of heating elements. A layer of transfer film is interposed between the paper and the heating array; when heat is applied, the ink on the transfer film is melted onto the paper. Thermal transfer printers available today have a maximum resolution of 600 by 600 dpi; to the eye, the image has thousands of colors. Typically, these printers produce deep, saturated color images on a variety of media, including overhead transparencies, but not including standard office paper. The ink costs $0.50 to $1 per letter-sized page.

A recent innovation in thermal transfer is the variable-dot-size printer, pioneered by Mitsubishi. The new Mitsubishi print engine, found in Tally Printers' $11,400 SpectraStar T8050, uses variable dot sizes. Although still limited to 300 by 300 dpi, the printer can create up to 64 different dot sizes and can also place dots in different patterns. This capability allows for sharper edges and image quality that approaches that of the thermal dye printers (discussed below).

Color laser printers use the same laser electrophotography method as the more familiar monochrome printers, except that four colored toners – cyan, magenta, yellow, and black – are transferred to the printer's drum, and from the drum to the paper. The toner is then heat-bonded to the paper. Color laser printers have the highest resolution of any mainstream color technology, with many models operating at 1,200 by 1,200 dpi (with full-color speeds upward of 3 ppm). Toner for an A-sized page with 15 percent coverage, equivalent to a non-saturated image or color-enhanced text, would cost approximately $0.05 for HP's Color LaserJet 5M.

Inexpensive (less than $500) color ink-jet printers spray water-soluble ink from a moving print head. These printers can print on a variety of paper sources, including standard office paper. Specially designed paper and transparency film will produce higher-quality images, albeit at higher cost. The lowest-cost ink-jet printers use a single ink cartridge to hold the cyan, magenta, yellow, and black ink needed for color images; a drawback is that when a single color is exhausted, the entire cartridge must be replaced. Higher-priced models use four separate ink cartridges; one model with four large cartridges is the Epson Stylus Color 3000. Resolution on ink-jet printers is between 300 by 300 and 600 by 600 dpi. Approximate ink cost per page varies between $0.05 and $0.20 on plain, letter-sized paper.

A recent innovation in ink-jet printing is HP's Photo REt II (Resolution Enhancement technology) used in the DeskJet 720C, 722C, and 890C printers. The Photo REt system uses as many as 16 small ink droplets, with a volume of 10 picoliters, to make up individual dots; the original HP DeskJet, for comparison, used droplets with a volume of 86 picoliters. These smaller droplets allow for greater color control and placement of single pixels as well as smoother edges on irregular objects.

Solid-ink printers come in two versions. They either spray melted wax on the page or spray wax on a rubber drum, which is then rolled against the paper. With solid ink, the colors do not run, and they can be printed on virtually any kind of paper, transparency film, plastic, or even cloth. Solid-ink printers are often used for the same purposes as thermal-transfer printers, thanks to their greater media flexibility and low ink cost. Ink cost per page ranges from $0.25 to $0.57 for a letter-sized page.

Thermal dye printers, also known as dye-sublimation printers, press an ink-permeated ribbon against the paper and apply heat so that dyes in the ribbon diffuse into a polyester coating on the paper, which has the look and feel of a glossy photo. Because the colored dots melt together and form a single color, thermal dye printers, with their 300 by 300 dpi resolution, are often used for digital photographic output as well as for full-color proofs of graphics images. However, the ink-and-paper combination of thermal dye printers is the most expensive of the foregoing types, at $2 to $3 for an A-sized page.

Mid-Volume Printers

Personal printers, such as under-$500 color ink-jet or monochrome laser printers, can produce between 3 and 12 ppm, with 1 or 2 small paper trays (maximum capacity 250 sheets). Laser printers for small work groups or departments typically can handle higher volumes, at 12 to 30 ppm. These printers are often available with multiple large paper trays, capable of handling 250 to 500 sheets, or optional external trays with 1,000-sheet-plus capability. Some have the option of two-sided printing (known as duplexing).

Mid-volume printers, in the 15 to 25 ppm range, can support greater volumes of work. Printers in the 16 to 17 ppm range can handle small work groups or larger departments with relatively constant printer usage throughout the day. Faster 20 to 24 ppm models are more appropriate where a large number of printed pages are produced hourly or where smaller work groups create demand sufficient to cause workflow problems with slower printers.

The technology of choice in mid-volume printers is the monochrome laser printer. Laser printers work via a process analogous to the electrophotographic technology found in the typical office photocopier. The image to be printed is received in digital form from a computer and is converted and mapped into an array of bits, corresponding to the size of the printer image and its resolution. For example, a 300 by 300 dpi printer will require 90,000 dpi. If the printable area of 8.5- by 11-inch paper is 8.0 by 10.5 inches, or 84 square inches, 945,000 bytes of bit-addressable memory would be required to form the image. Because translating the incoming image into an array of bits suitable for printing is a slow process, high-volume printers maintain many pages in memory at once, updating only those portions of the page that change. This feature is especially useful when printing forms, and it is what distinguishes these printers from lower-volume printers, which re-create each page from scratch.

Once the bits in memory have been calculated to match the dots to be printed on the page, an ion or magnetic charge, corresponding to each "on" bit, is written onto a rotating drum; these changes attract toner. The toner, in turn, is pressed onto the paper or other medium; the paper and toner are fused together by heat and pressure as the paper exits the printer.

In the print engine itself, little recent technological innovation has occurred. The biggest changes involve increases in resolution, moving from the 300 by 300 dpi resolution of the early 1990s to a near-universal 600 by 600 dpi today. Higher resolutions are readily available, such as Lexmark's Optra S series, at 1,200 by 1,200 dpi, or the Xerox Docuprint 4517's 1,200 by 600 dpi. (Note that other technologies for mid-volume printers, including impact dot-matrix, still exist and are required for printing multi-part forms or on continuous fan-folded paper. Due to the decreasing cost, better output appearance, and growing speed of laser prints, the demand for impact printers increasingly will be relegated to these specialized operations.)

Beyond higher resolution and enhanced print-engine speed, the main areas of current and anticipated technology evolution are in software manageability and paper handling, including various document-publishing options.

Pioneered by companies such as HP, software-based management of networked printers has come to life. No more waiting for "printer out of paper" messages to flash on the screen: software such as HP's JetAdmin can monitor a properly equipped printer's paper or toner level remotely, detect jams, and log usage patterns. These management products, which also include tools such as Cabletron's Spectrum, Computer Associate's Unicenter, HP's OpenView, and IBM/Tivoli's TME+, can alert administrators via pop-up windows, e-mail, or even pagers. These are all large management software packages that report on many devices in the enterprise, of which printers are just one example.

Media handling has progressed substantially from the early days of tractor-fed fan-folded paper, or even the single-tray 250-sheet feeders of small laser printers. Many networked mid-range printers can now handle duplex (two-sided) output. Paper comes from 3 or 4 large-capacity bins, with 250- to 500-sheet bins standard, and 750- to 2,500-sheet bins available as options. Larger paper sizes also are increasingly supported; nearly all printers in the 15- to 20-ppm range can accept both 8.5- by 11-inch letter and 8.5- by 14-inch legal paper; a few models in that range, such as HP's LaserJet 4V/MV, can accept 11- by 17-inch ledger paper. Ledger paper support is nearly universal in the 20+-ppm printer range. On many printers, the paper bin is software-selectable.

Output handling technology, borrowed from the photocopier world, is another area of innovation for networked printers. Many models from HP, IBM, Lexmark, and Xerox offer multiple 250-page bins that can be assigned to individual users or groups of users, thereby eliminating the need for job header pages (also known as cover sheets). Locked output bins for confidential printouts are also increasingly common.

High-Volume Printers

Although high-volume laser printers, such as the IBM 3800, have been around for more than 20 years, recent innovations in paper handling and dot-addressable graphics output have moved laser printers into a leadership role in the 100+-ppm space.

It is essential to keep the point engine of high-volume printers busy, not only to maintain maximum throughput but also to achieve the greatest return on investment from the printer. Thus, high-volume printers typically are located centrally and fed from mid-range or mainframe computers via dedicated 10 Mbps or faster Ethernet or Token Ring links. Another successful alternative is to attach all printers to a dedicated print server, which not only can keep appropriate printers busy, but can manage the print stream, performing such tasks as converting PostScript and other file formats into the appropriate data for the given printer.

Another important consideration is paper supply. No vendor currently offers a sheet-fed printer that can work at 200 ppm. At those speeds, paper is fed in rolls, in a way similar to a web-fed printing press. Due to the inertia involved in starting and stopping large rolls of paper, an additional design constraint is that the speed of the roll must remain constant – and thus, the wait time between pages must be constant, even if adjacent pages vary greatly in complexity. Again, this is a departure from slower sheet-fed printers, which can print on an irregular schedule.

Other technological issues are unique to high-volume printers, including the cutting and collating of output. In some cases, output is rolled onto another roll and cut, folded, and otherwise processed elsewhere. Another issue is replenishment of toner; high-volume printers use large bottles of toner that can be refilled on the fly, rather than preloaded cartridges that must be changed manually.

Other approaches used in high-volume printing include printer clusters and two-up printing.

Printer clusters were pioneered by T/R Systems and embraced by several high-volume printer manufacturers. Printer clusters employ multiple lower-throughput printers, controlled by a single printer controller, to offer a scalable printing solution. For example, Xerox's DocuPrint 184 offers 194 ppm, using two 300 by 300-dpi, 92-ppm Xerox color print engines and an IBM RS/6000 print server. T/R Systems' Multiple PrintStation 024 Performance similarly groups as many as eight 24-ppm laser printers for a throughput of 192 ppm. Document reassembly can be a bit of a challenge, if not managed attentively.

Two-up printing barely warrants the term technology, but at a stroke, it doubled the volume of high-end printers. The trick: Print two pages side-by-side in one operation, and then slice them apart. This trick is used by Océ Technologies' PageStream 440, a 300 by 300-dpi printer that can handle 308 ppm of A4 paper in landscape (horizontal) or 440 ppm of two-up A4 paper in portrait (vertical). Two PageStream 440 printers can work together to produce two-sided output.

Similarly, IBM's inforPrint 4000 offers continuous forms printing with either 15- or 17-inch wide print line, with all-points-addressable duplex (two-sided) printing at speeds of up to 1000 impressions per minute for two-up printing at 600 dpi resolution. The printer can output as many as eight logical pages between the two sides of a duplexed sheet and can produce standard bitmapped graphics, vector (line) graphics, text, and bar codes.

Emerging Print Trends

Four printing trends for the future of personal and mid-volume printers do not revolve around the print engine or ink-on-paper methodology, but on paper handling, Internet connectivity, and multi-function devices.

- **Paper handling** – HP popularized the word Mopier for an enhanced version of its mid-volume (24 letter-sized ppm) LaserJet 5Si. The word refers to the concept of multiple original copier – that is, the increased use of a laser printer to print multiple copies of a multi-page document, rather than printing a single original that is then replicated by a conventional photocopier. To meet this goal, the LaserJet 5Si Mopier is equipped with four input sources for different media (from 8.5 by 11 inches to 11 by 17 inches) with a 3,100-page total capacity; five output bins for collating letter- or legal-sized paper; a sixth bin for stapling letter-sized documents; and a seventh bin for transparencies, labels, envelopes, and other special media. The printer also supports duplexing. Although the Mopier will not replace the high-volume and flexible capabilities of the traditional office photocopier, it can reduce the need for the copier in many situations.

- **Internet connectivity** – Today, the most common way to access a networked printer is to work through the network operating system. The network operating system "owns" the printer and provides access control, logging and management, and queuing of print jobs. To the user, the printer appears as a local area network (LAN) resource. An emerging trend is to use the Internet instead of a LAN's own utilities to locate, map, and manage these printers as well as send them a document for printing. An immediate application of Internet-based printing is in service bureaus, where end users can access those printers remotely, arranging for separate delivery of the print documents (akin to a return of the centralized mainframe printer). However, as intranets become common, the use of Web-based utilities, such as HP's Web JetAdmin, to manage and interact with networked printers is likely to become easy and common.

- **Multi-function devices** – The product category of multi-function printers is a marketing and packaging distinction, not a technological evolution. Devices, such as the Brother MFC-7550MC, the HP OfficeJet series, and the Xerox WorkCenter 250, are designed for the executive or small-office/home-office (SO/HO) worker who needs a personal printer and fax send/receive capability. These multi-function devices marry a basic monochrome or color ink-jet engine or monochrome laser engine, a low-resolution monochrome scanner, and a fax send/receive modem in one package; the benefit to the consumer is a single, low-footprint device that does not take up much space. More advanced models, such as those from Brother, also include multiple-mailbox digital answering machines. Future models from all manufacturers will include higher-resolution printers and scanners, both with better color capability and higher throughput.

3.1.5 Forecast

Input Devices

- Ergonomic computer keyboards will become more common. Despite the proliferation of additional input devices, the keyboard will remain the primary input device (except on hand-held computers) throughout the forecast period.
- Mouse and pointing-device technology will remain relatively static. The use of touchscreens in kiosks will increase.
- Scanners will improve in resolution and color depth; as sales volumes rise, prices will continue to fall.
- Desktop digital video cameras will increase to handle large images (up to 640 x 480 pixels) at a frame rate up to 30 fps. Professional video cameras will decrease in price.
- As prices drop and resolution improves, digital still cameras will become more popular for capturing images for use in printed and electronic documents.
- Compared to the major improvements of the past 2 years, continuous speech recognition technology will see minor improvements in algorithms. However, the increasing power of the PC microprocessor will provide the raw power to improve speech recognition even further.

Display Output Devices

- Because the technology for CRTs is relatively mature, few changes will occur in price or performance.
- Rapid advances in LCD technology will yield larger, higher-resolution panels at lower costs. By the end of the forecast period, these will begin to appear not only on portable computers, but also on desktops as a CRT replacement.
- Bright LCD and DLP projectors with built-in light sources will become smaller and more affordable, and they will displace LCD projection panels and CRT-based projectors.
- Initially, 3-D graphics controllers will be commonly available on consumer PCs and high-end workstations. Due to the lack of 3-D features in productivity applications, they will not be standard on business-class PCs until volume production drives costs to the point where it is cheaper to increase 3-D volumes by discontinuing production of 2-D controllers.

Print Output Devices

- Professional-quality color printers will not see much of a reduction in capital or consumable costs.
- Laser printers for general office use will see increasing resolution up to 1,200 by 1,200 dpi, as well as increased engine speeds and lower costs; some devices will hit 2,400 by 2,400 dpi. Whenever possible, users will choose laser over other technologies, except at the very low end of the price spectrum.
- Printers for small business and home use will remain at approximately the same price point ($300 to $400). Laser printers will improve resolution and paper-handling options but will remain black-and-white only. Color printers will improve resolution and color quality.
- High-volume, production-oriented laser printers will concentrate on improving speed rather than resolution.

- Manufacturers will continue to add advanced paper-handling functionality to networked laser printers, reducing – but not eliminating – the demand for conventional photocopiers.

3.1.6 References

■ Articles

Stone, David. 1997. Color for the pros. *PC Magazine.* October 21.

■ Periodicals, Industry and Technical Reports

Weilerstein, Kenny. 1995. *Competitive outlook: color thermal transfer printers.* Datapro Information Services Group. August 9.

Weilerstein, Kenny. 1996. *Competitive outlook: 100-plus page-per-minute printers.* Brian Dooley and Datapro Information Services Group. November 4.

Weilerstein, Kenny. 1997. *Competitive outlook: 15-to-25 page-per-minute printers.* Datapro Information Services Group. April 2.

■ URLs of Selected Mentioned Companies

Acer `http://www.acer.com`
Adesso `http://www.adessoinc.com`
Alps `http://www.alpsusa.com`
Apple `http://www.apple.com`
Atari `http://www.atari.com`
ATI Technologies `http://www.atitech.com`
Barco `http://www.barco.com`
Brother `http://www.brother.com`
Cabletron `http://www.cabletron.com`
Caere `http://www.caere.com`
Candescent Technologies `http://www.candescent.com`
Cirque `http://www.cirque.com`
Compaq `http://www.compaq.com`
Computer Associates `http://www.cai.com`
Connectix `http://www.connectix.com`
Dell `http://www.dell.com`
Dragon Systems `http://www.dragonsys.com`
DynaPoint `http://www.spec-research.com`
(DynaPoint is a manufacturing division of Spec Research)
Elo TouchSystems `http:www.elo.com`
Epson `http://www.epson.com`
Fujitsu `http://www.fujitsu.com`
Hercules `http://www.hercules.com`
Hewlett-Packard `http://www.hp.com`
Hitachi `http://www.hitachi.com`
IBM Pinter Division `http://www.printers.ibm.com`
Intel `http://www.intel.com`
Kensington `http://www.kensington.com`
Kurzweil `http://www.kurzweil.com`
Lexmark `http://www.lexmark.com`
Logitech `http://www.logitech.com`
Microsoft `http://www.microsoft.com`
Microtek `http://www.microtek.com`
MicroTouch Systems `http://www.microtouch.com`
Mitsubishi America `http://www.mitsubishi.com`
Mitsubishi `http://www.mitsubishi.com`
Oce Technology `http://www.oce.com`
ParkerVision `http://www.parkervision.com`
PC Concepts `http://www.pcconcepts.com`
PictureTel `http://www.picturetel.com`
PixTech `http://www.pixtech.com`
Sound Blaster `http://www.soundblaster.com` or
`http://www.creativelabs.com` (Sound Blaster is produced by Creative Labs)
T/R Systems `http://www.trsystems.com`
Tally Printers `http://www.tally.com`
Tandy-Radio Shack `http://www.tandy.com`
Texas Instruments `http://www.ti.com`
Toshiba http://www.toshiba.com
Tseng Labs `http://www.tseng.com`
Umax `http://www.umax.com`
Verbex `http://www.verbex.com`
Videologic `http://www.videologic.com`

ViewSonic *http://www.viewsonic.com*
WorkLink *http://www.worklink.com*
Xerox *http://www.xerox.com*

3.2 Storage and Batteries

3.2.1 Executive Summary

In the last few years, the demand for large amounts of reliable storage has increased dramatically. Graphical user interfaces (GUIs), the size of their applications, more sophisticated operating system software, and a persistent resistance to managing storage resources have caused disk drive capacities on typical personal computer (PC) systems to rise from 10 MB in 1983 to 2 GB or more in 1997. Even the use of data compression software has not slowed the move to higher-capacity disk drives on PCs.

On servers and mainframe systems, disk storage has moved away from large form factor disk drives to 3.5-inch high-capacity (9 GB, 18 GB, and 23 GB), high-performance magnetic disk drives. Tape is moving toward higher-capacity, faster transfer rates, greater reliability, and multiple form factors for high-performance input/output. Tape systems for mainframes and midrange servers are used for backup and restore, archiving, and data interchange. Software and data files are still commonly archived and moved between mainframe and midrange or networked systems.

Document and image systems, data warehousing, and other large-scale corporate applications continue to drive the need for storage. For example, many document management systems have requirements that are best met by optical disc drive jukeboxes. Jukebox storage subsystems can provide terabytes (millions of megabytes) of near-on-line storage that can be accessed in a few seconds.

The cost-per-megabyte and physical size of disk and tape drives are shrinking dramatically. Improvements in technology will sustain this trend through the forecast period. Built-in hardware- or software-based data compression also helps to increase capacities and transfer rates at little cost.

Platform choices also influence storage selection criteria. Selection criteria include capacity, transfer rate, and cost of ownership. The Santa Clara Consulting Group estimates that capacity requirements for servers will increase 20 to 30 percent annually. The transfer rate of a given technology is an important consideration, with users trying to accommodate limited backup windows (the amount of time per day or per week that the site can take the system offline for backups).

Cost of ownership is also gaining importance, taking into account the quantity of media required per year. The cost of subsequent media purchases is as important, if not more important, than the initial cost of the drive.

Battery life has become an important criteria for buyers of portable electronic products. The battery industry is traditionally an area where innovation comes slowly. Improvements continue in lithium-ion, lithium-polymer, and zinc-air battery technologies. Each technology is at a different stage of development; however, all promise greater energy storage, lighter weight, lower costs, and a safer environment. Lithium-ion batteries are becoming the battery of choice for notebook computers as well as cellular phones because they are smaller and lighter than other alternatives. Nickel-metal hydride batteries are expected to replace nickel-cadmium rechargeable batteries until more efficient technologies become available.

Battery form factor is another important issue. The best hope for the future of portable electronics is a systems approach that would create standardized products with internal batteries capable of operating for a full day. To date, however, increases in power consumption from brighter displays, faster processors, and high-speed modems seem to consume any improvements in battery technology as soon as they appear, so that useful battery life remains constant.

3.2.2 Magnetic Tape Storage

Computer storage devices serve many useful purposes: storing data and software on-line, archiving and backing up data, and distributing data and software. According to most market researchers, storage makes up 40 to 60 percent of the cost of a computer system. For this reason, storage – including permanent, secondary devices such as tape, disk, and optical drives – is increasingly recognized as equal to processors and communications as a key user concern. Tape storage capacity is measured in bits per inch (bpi), tracks per inch (tpi), and total tape area (length times width); tape drive speed is measured in inches per second (ips).

Magnetic tape, which accesses data sequentially, was the computer industry's first magnetic storage medium. Despite the growth of more sophisticated, random-access storage techniques such as magnetic and optical disk storage, tape remains in high demand because of its low cost and reliability. For example, the digital audio tape (DAT) used on PCs and servers stores data for only pennies per megabyte.

Historically, magnetic tape has served three related functions: software distribution, data backup, and data archiving. These functions have become less entwined, with software increasingly distributed on CD-ROM, data backup remaining on tape, and archiving done on tape, optical disc, or CD-ROM.

Today, tape is used most frequently for archiving files, systems back up and restoration, and off-site data storage for disaster recovery in mainframes and servers. There are three tape drive market segments:

- High-end mainframe or large host devices
- Midrange servers or network-based devices
- Low-end desktop devices (PCs)

Mainframe Tape Formats

The first mainframe magnetic storage devices were created in 1952, when IBM developed its 726 Tape Unit for use with the IBM 701 Defense Calculator. By the early 1980s, half-inch reel-to-reel tape formats were common.

■ Half-Inch Cartridge Tape Drives

In 1985, IBM introduced its half-inch 18-track 3480 drive, which protected the tape by using a square cartridge instead of an open reel. IBM's half-inch 3480 and 3490 tape cartridges store 2 to 4 bytes each on 18 or 36 parallel tracks at 38,000 bps. Today, half-inch cartridge tape drives are produced by Hitachi Data Systems (HDS), IBM, Quantum, Sony, StorageTek, and others.

■ Automated Tape Silos

Automated tape retrieval systems use robots to retrieve and replace the tens of thousands of tape cartridges used by information-intensive industries such as banking, insurance, and health care. The cartridges themselves are stored in structures called silos. For example, StorageTek's Powderhorn robotic cartridge library system holds up to 6,000 tape cartridges in a silo that is 8 feet tall and 12 feet in diameter.

■ Multipurpose Systems

In September 1995, IBM released the 128-track Magstar MP (multipurpose) system. The 3590 Magstar Tape Subsystem offers transfer rates up to 20 MBps. Each half-inch cartridge can hold more than 30 GB of compressed data or 10 GB of uncompressed data. The tape cartridge uses metal-particle, half-inch tape specifically formulated to support higher areal densities. This process yields a 50 times improvement over 3480 cartridge capacity and represents a 12 times improvement over 3490 cartridge capacity. Magstar's media handling places the beginning of tape (BOT) at the midpoint of the tape, thus reducing the amount of tape to run (because the tape does not have to stream from the physical beginning) and improving seek times (the time it takes to find the requested data).

■ D2 Tape Systems

D2 tape technology uses cartridges with the same form factor as video tapes and with a storage capacity of up to 12 TB per cartridge (without compression). Cartridges are stored in a robotic silo, giving total capacities of up to a petabyte (PB). A system normally contains 4 to 16 tape drives, each of which is connected to the host computer via a Fast-Wide SCSI bus. Data transfer rates are typically about 9 MBps per drive, but it is often possible to stripe data across multiple drives to increase the effective transfer rate.

Major manufacturers of D2 tape systems include Ampex in the U.S., which markets a range of systems designated the DST 81x series, and the German company, Grau.

D2 tape systems are gaining popularity for high-end applications requiring near-line storage of terabytes to petabytes of data. Applications with these characteristics include data warehouses, geophysical data such as seismic surveys, and increasingly, large digital libraries such as near-video-on-demand (NVOD) systems and large catalogs of items such as museum specimens or artworks.

In these types of applications, single data objects are typically large binary objects (BLOBs) ranging in size from a few to several tens of gigabytes. For example, a 90-minute movie encoded as an MPEG-2 file at the broadcast data rate of 8 MBps is approximately an 8-GB file.

Typically, the requirement is for access to a small section of the data only at a given time rather than retrieval of a whole file or data set. Examples of such access requests include the next few minutes of video to be broadcast (which can retrieved in advance) or a data warehouse query that concerns a single product out of thousands for which data are held.

In this environment, D2 tapes have a significant advantage because the data stored on the tape is addressable by block, not by file; depending on the application itself, single data blocks can be retrieved. Typically, retrieval of any given block of data held on a tape in the silo can be performed in approximately 15 seconds. Given the vast volume of data that can be stored in these systems, this is an extremely fast access time. Performance characteristics of D2 systems fall somewhat closer to those of on-line hard disk drives rather than conventional tape or optical jukebox systems.

■ High-Density Data Tape Systems

High-density data tapes (HDDT) systems serve a niche market that requires high-speed access to medium data volumes. Typical HDDT systems have input/output (I/O rates) ranging from 100 to 256 MBps and cartridge capacities of about 1 TB. Cartridges have manufacturer-dependent form factors and specialized host computer connections.

HDDT tapes are extensively used in the broadcast arena for nonlinear editing of real-time video data (which requires a data rate of 270 MBps per video stream) and in high-speed, burst-mode communications links such as satellite systems.

Major manufacturers of HDDT systems include Sony and Thorn-EMI.

Server and PC Tape Formats

A single tape can easily accommodate a backup copy of most hard drives, a task that is difficult to perform using floppy disks. Today, the main tape format for servers and PCs is 8 mm (originally developed for camcorders and then adopted for data use), the first commonly available high-capacity tape.

■ 8-mm Tape

For example, Exabyte's 8-mm cassettes can store 20 GB of uncompressed data (40 GB of compressed data) on a single tape cartridge the size of a deck of playing cards. The 8-mm tape backup market was dominated by Exabyte until 1997, when Sony introduced its Advanced Intelligent Tape (AIT) architecture to meet the data storage needs of the midrange market, including workstations, PC networks, and video server environments.

Sony's AIT architecture uses a new 8-mm helical scan recording format as well as new head and media technology in a 3.5-inch form factor. A primary difference between Sony AIT and Exabyte Mammoth helical-scan drive mechanisms is the manner in which they control the speed of the tape past the read/write heads. Sony AIT uses the traditional servo-driven capstan/pinch roller combination to achieve a constant tape speed. In contrast, Mammoth drives are capstanless, meaning the servo motors driving the take-up and supply hubs must constantly and precisely vary in speed as the diameter of the tape spool changes.

Announced in May 1997, Sony's SDX series of tape drives and media offer a basic capacity of 25 GB (50 GB with compression), with a data transfer rate of 3 MBps (6 MBps with compression). These products compete with two existing midrange offerings: Quantum's DLT-4000 and Exabyte's Mammoth drive. Sony's SDX drives feature 5 GB more basic capacity than either of the two competitors; they also offer the same data transfer rate as the Mammoth and twice that of the DLT-4000. Sony AIT also uses a more efficient compression algorithm, producing a maximum compression ratio of 2.6:1, compared with the ratio of 2:1 achieved in other technologies.

Sony's AIT products are not compatible with other products, although the cartridge is the same size as existing 8-mm tape cartridges. SDX drives will accept only special cartridges designed using Sony's Advanced Metal Evaporated (AME) tape formulation. (Sony also markets AME tapes for the Mammoth drive.) The drives include a cartridge-sensing system that prevents the use of other tapes. Although the cartridges contain a special identification, the drive also includes a write-protect feature that is activated if an AIT cartridge is inserted by mistake into a non-AIT 8-mm drive.

Sony says its AIT drives offer a head life of 30,000 hours and a mean time between failure (MTBF) of 200,000 hours; each cartridge can be used 20,000 times, and media life is projected at 30 years. The SDX-300c drive, first shipped in late 1997, uses a data compression chip co-developed with IBM that incorporates IBM's latest Adaptive Lossless Data Compression (ALDC) technology. Previously available only in mainframe-class tape, ALDC offers an average data compression ratio across multiple data types of 2.6:1, compared with 2.0:1 or less with other compression technologies. (The data compression ratio is a figure expressing the difference between the basic capacity of a tape and its compressed capacity.)

An innovative feature of AIT (in both the media and the drives) is the Memory In Cassette (MIC) system, which utilizes a 16-KB memory chip built into the data cartridge. This EEPROM flash memory chip holds the tape's log as well as user-defined information, and helps improve data access time and provide data management capability. For example, average file-access times for the Sony drives with MIC are 27 seconds, compared with 55 seconds for Mammoth drives and 68 seconds for DLT-4000 drives. However, this feature is new only for 8-mm tape, which cannot use the tape as storage (because it can be removed without rewinding); any tape that does not have a take-up reel in the cartridge, such as DLT, already includes this feature.

■ 4-mm Digital Audio Tape

4-mm tape, also known as digital audio tape (DAT), was initially a CD-quality audio format designed for recording music. In 1988, HP and Sony defined the Digital Data Storage (DDS) format standard and quality level so that DAT tapes could be used by computers. DDS is a sequential format – that is, all data recorded to the tape falls after the previous block of data.

DDS technology was commercialized in 1989 with the introduction of a 1.3-GB system. Although lower in capacity and performance than the 8-mm alternatives, DDS was accepted rapidly in the PC server market because of its attractive price, multiple drive sources, and clear migration path. DDS-2 products were introduced in 1994 offering 4 GB of storage, and in 1996, the industry started shipping DDS-3 products featuring 12 GB of storage and a 1-MBps data rate.

Key advantages of DDS technology include backward compatibility with a large installed base, the 3.5-inch form factor, low cost of ownership, multiple drive suppliers, and perceived reliability. Disadvantages include a slow data rate and concern about whether the next-generation, higher-capacity version can be achieved, given the current generation's high packing density (the number of bits or tracks per inch of recording surface) and media availability.

The popularity of DAT continues today, especially in automated tape libraries. In addition, DDS is an attractive product in the 10-GB to 15-GB application markets, where performance is not the most important criteria.

■ Digital Linear Tape

Digital linear tape (DLT) technology was originally developed by Digital for use with its midrange VAX systems and is now used primarily for backing up and archiving data on servers in medium- to large-scale local area networks (LANs). The tape is half-inch wide, and cartridges provide up to 35 GB of storage uncompressed (70 GB compressed), with a data transfer rate of 5 MBps. Quantum subsequently bought the DLT technology from Digital and now licenses it to other manufacturers. DLT drives are included as standard equipment in high-capacity servers sold by many leading PC vendors.

Effective use of a tape drive involves keeping the drive streaming so that it keeps up with the host. The newest DLT iteration, the DLT-7000, features an adjustable block size, which allows more data to be put into each block and thus requires fewer stopping points. In addition, the DLT-7000's compaction capability makes sure that the drive will be able to feed data into the cache buffer nearer to the rate that the buffer empties on the other side.

The DLT-7000 uses the same half-inch metal-particle tape media as the DLT-4000 drives. Preformatting the tape is unnecessary; the drive writes a burst at the beginning of a data set and another at the end.

Quantum remains the only source for DLT drives and has established two families of DLT products: the Flagship line and the Valueline. The DLT-7000 represents the high-end Flagship line, which offers higher capacities and better performance due, in part, to Quantum's new Symmetric Phase Recording (SPR) technology. (The SPR format enhances traditional DLT recording by writing adjacent tracks with an alternating, slightly offset head angle, eliminating cross-track interference and the need for guard bands. This process makes smaller track widths possible, allowing the DLT-7000 to utilize a track density of 416 tpi.) Quantum's Valueline will be targeted toward the sub-15-GB market. Using single-channel metal-in-gap (MIG) read/write heads, the company is developing a new product (for release by the end of 1997) to challenge the widespread use of 4 mm DAT and high-end quarter-inch cartridge (QIC) solutions in the PC server backup market.

Limited availability of DLT products for the library manufacturers that package them into finished robotic subsystems has been the sole drawback to Quantum's approach. However, the company's new facility in Colorado Springs should help address previous production problems.

One reason given for the widespread popularity of DLT technology was Exabyte's delay in releasing its newest competitive offering, the Mammoth drive. Originally expected to be released in 1994 or 1995, Mammoth finally shipped in March 1996. Mammoth provides a 20-GB capacity with a 3-MBps transfer rate.

■ Quarter-Inch Cartridges

QIC tapes come in two form factors: 3.5-inch minicartridges and 5.25-inch data cartridges. (That is, quarter-inch is the width of the tape itself, and 3.5-inch or 5.25-inch is the size of the box it comes in.) QIC was used primarily with UNIX workstations in the 1980s and as a desktop PC backup medium until the mid-1990s when the advent of the Travan architecture brought QIC into use with file servers. In 1994, five leading QIC companies – 3M (later Imation), HP's Colorado Memory Systems Division, Conner Peripherals (purchased by Seagate),

Iomega, and Sony Recording Media – announced plans to develop new drive and minicartridge products that would more than double existing QIC minicartridge capacity by utilizing Travan technology.

Travan technology optimizes available space in a 3.5-inch drive housing. Mechanical changes allow the drive to use existing minicartridges, QIC-Wide, and Travan cartridges interchangeably. (QIC-Wide is an extension to the QIC tape from Sony that provides more storage capacity. It increases recording density, uses .315-inch-wide tape, and uses the standard QIC minicartridge with a redesigned housing.) Initial Travan cartridge offerings require no changes in media formulation and use existing drive electronics and available head technology.

Working with the QIC Committee (the QIC standards organization), the five companies developed a product roadmap for the Travan cartridge and the resulting drive recording formats. Initial points along the roadmap included modified versions of the QIC-80, 3010, and 3020 recording formats using the new cartridge. Basic storage capacities for the new Travan products are 400 MB for the modified QIC-80 drive/cartridge (current minicartridge capacity is 125 MB), 800 MB for the new 3010 drive/cartridge (current minicartridge capacity is 340 MB), and 1.6 GB for the new 3020 drive/cartridge (current minicartridge capacity is 670 MB).

Although the minicartridge is the most common QIC form factor, the 5.25-inch data cartridge is also available. This market is represented by multilinear recording (MLR) technology from Tandberg Data (Oslo, Norway), offering up to 16 GB of storage and a 1.5-MBps data rate.

Key advantages of MLR technology are high reliability, a large installed base of users with libraries of compatible media, and a low cost of ownership. Perceived disadvantages include lower capacity and performance than some competing technologies, a single source for drives, and the perception that quarter-inch technology does not represent the future of tape backup.

MLR products are valuable for maintaining backward compatibility with libraries of older QIC media. In addition, MLR products compare favorably with the current generation of DAT technology, offering faster data rates at similar drive prices. Unless MLR can amass an installed base of users large enough to drive down media costs, however, the technology will not win market share from DAT products based on cost of ownership.

3.2.3 Hard Disk Drives

Hard disks are the most commonly used mass storage devices because of their low cost, high speed, and large storage capacity. Hard disk drives are sealed storage units that read from and write to rotating magnetic disk platters. Hard disk drives are also sometimes referred to as Winchester drives, from a code name IBM used for an early removable disk that put the heads and platters in a sealed unit for greater speed. It was this disk's dual 30-MB modules, or 30-30 design, that prompted developers to adopt the Winchester rifle nickname. The term later referred to any fixed hard disk where the heads and platters were not separable.

Most disk drives (including fixed and removable hard disks, floppy disk drives, and optical drives) store data on platters divided into concentric tracks. (However, CD and DVD use a single, spiral track.) Each track is further divided into segments called sectors. In a two-step process, a read/write head moves radially across the rotating disks to locate the right track, and the head then waits as the disk rotates until the right sector is underneath it. Today, most hard disk drives are housed in hermetically sealed containers to avoid contamination.

Hard drives come in a variety of sizes or form factors, ranging from the 14-inch, 10-inch, and 8-inch units previously used for traditional mainframe and high-end server storage systems to 5.25-inch, 3.5-inch, and 3-inch drives used today for high-end servers, workstations, and PCs, to the 2.5-inch and 1.8-inch drives currently used in laptops and hand-held computers. Over time, the disk drives used in various types of computers have become identical; today's mainframe systems often use the same devices as PCs.

At the same time that disks and disk drives have gotten smaller, they also have become faster, denser, and less expensive – factors that continue to influence the market for drives.

Size

Today, the 3.5-inch drive accounts for more than 90 percent of the market. The growth of laptop and portable computing has increased the demand for even smaller drives. For example, the new 3-inch form factor from Western Digital and JTS is challenging the 2.5-inch drives used in portable computers for a share of this market. Drive size alone will not be the determining factor, however, unless there is a cost advantage to be gained from more space as well.

The demand for larger storage capacities on mainframes, servers, and workstations has caused some manufacturers (primarily Seagate) to begin supplying 5.25-inch drives with multiple gigabyte capacities. In addition, demand for larger-capacity drives at lower costs has led to the recent reintroduction of 5.25-inch hard drives, especially in consumer PCs. Increasingly, however, mainframe storage is being served by high-end disk array systems included in storage subsystems by Digital, EMC, IBM, and others.

Speed

Disk drives are made up of one or more rigid disks attached to a motor assembly that rotates the disks at a constant speed. The speed at which hard disks revolve is significant because it determines how long it takes for the requested data to rotate into position, which is referred to as rotational latency time. The transfer rate (how long it takes for the data to pass beneath the read/write heads that magnetically record and retrieve data on the rotating disks) is a function of speed and the density of bpi. In practice, because the technology limits the speed (that is, the transfer rate) at any given time, changing the speed only changes the bpi to adjust accordingly.

The number of heads on disk drives also has increased because more heads lower latency and provide faster access to a given track and sector.

In 1992, the speed of high-performance disks was 5,400 revolutions per minute (rpm). Current high-performance disks spin at up to 7,200 rpm. For example, the 7,200-rpm rate is used for AV disks to guarantee a minimum constant data rate matched to multimedia video requirements. Seagate's Cheetah 4LP 4-GB hard disk, announced in early 1997, offers 10,000-rpm performance and a rotational latency of 3 milliseconds (ms).

Density/Capacity

Over the past 2 years, disk drive areal density (the number of bits that can be stored in a given area) has increased by 60 percent or more. In early 1996, state-of-the-art storage capacity for 3.5-inch disks was 1 GB. By late 1997, however, new hard drives featured 9 GB to 18 GB of data in standard 3.5-inch packages and 33 GB to 46 GB in 5.25-inch packages. In contrast, in 1993, state-of-the-art drives for PCs had a storage capacity of 500 MB.

In December 1997, IBM introduced two new high-capacity drives that use the company's giant magneto-resistive (GMR) head technology, especially ceramic heads. The Deskstar 16GP has a storage capacity of 3.2 GB to 16.8 GB; the Deskstar 14GXP will provide up to 14.4 GB of storage. PCs with the new drives will be available in early 1998. IBM plans eventually to make the GMR technology available in devices ranging from notebooks to IBM subsystems.

Drive capacity has increased, in part, due to improvements in thin-film technology, and newer MR head technologies. Other significant factors include the introduction of the partial response, maximum likelihood (PRML), read channel electronics, and internally built diagnostics such as on-the-fly self-calibration. PRML is a digital signal processing technique used to differentiate a valid signal from noise by measuring the rate of change at various intervals of the rising waveform. The magnetic field strength generated by stored data bits on a hard disk platter has uniform characteristics, whereas random noise does not. Therefore, PRML allows valid data to be separated more easily from noise. PRML thus increases the number of bpi that can be recorded, compared with earlier methods. Improvements such as PRML have allowed vendors of read/write heads to achieve smaller physical dimensions. New coating compositions allow smaller micro-domains and, therefore, higher storage density.

Cost

According to Dataquest, the cost per megabyte of hard disk storage fell from $75 in 1982 to $0.92 in 1993. In 1996, the average cost of 3.5-inch disks was $0.15 per megabyte, and larger-capacity 5.25-inch desktop drives averaged $0.10 per megabyte. However, as the cost per megabyte falls, the number of megabytes per drive has increased. Thus, the price for the typical disk drive installed in a given type of system has remained relatively constant over the past 4 to 5 years. Cheaper electronics such as Enhanced Integrated Drive Electronics (EIDE) have decreased the cost of disks for PCs even further.

New Technologies

A significant technology development in hard disk drives is just beginning to emerge: near-field recording (NFR). Developed by TeraStor, NFR is essentially a magneto-optical (MO) technique, but it uses a read/write head that is radically different and borrows an important design feature from traditional Winchester hard drive technology. The result is a drive that potentially could deliver ten times the storage of today's drives at about the same performance level and price (and at lower data rates).

NFR uses a new kind of lens, called a solid immersion lens, that produces a smaller footprint than a traditional MO lens, enabling more bits to fit in each unit area. It also makes use of a flying optical head that automatically maintains the right distance from the disk and places the magnetic coils closer to the disk, further increasing performance compared with a standard MO drive.

TeraStor's first NFR product, expected in mid-1998, will be a single-sided 20-GB disk drive, with double-sided 40-GB disks to follow in 6 months. Pricing will be comparable to Jaz and SyJet. Performance should be close to SyJet speeds, with an expected media cost of $0.05 per 100 MB – almost 100 times cheaper than CD-R or tape. Therefore, NFR could pose a threat to DVD and compact disc technologies.

In September 1997, TeraStor and Quantum announced a new strategic partnership. Under the terms of the agreement, Quantum will license TeraStor's initial NFR removable disk drive, and the companies will jointly develop a version optimized for use in robotic libraries. Quantum will also have the right to develop, manufacture, and market a variety of future products based on NFR technology.

Although the computer industry is driven primarily by Moore's Law, which says that the number of transistors on a chip doubles every 18 months, it also has been driven by the hard disk drive industry's ability to increase storage density by about 60 percent per year since 1991. This increase is the result of improvements to magnetic heads, disk surfaces, and the mathematical codes that govern the drives. According to Disk/Trend, IBM has been the company that has instituted most of these gains.

At the end of 1997, IBM announced that it had broken the magnetic disk drive storage "barrier" of 10 billion bits of data per square inch using a new technology, which will appear first in products in the year 2001. The technology will be used in the 2.5-inch non-removable disk drives used in portable computers. At this size, a drive will be able to hold 6.5 GB of data, making ultra-slim laptop computers with vast storage capacities possible. A 3.5-inch drive will hold up to 13 GB of data.

3.2.4 Disk Drive Interfaces

Today, most PCs and workstations use one of two high-performance disk interface standards: the EIDE interface or the Small Computer Systems Interface (SCSI). Both EIDE and SCSI use intelligent devices with built-in controllers; these controllers are then connected to a host adapter that plugs into the system bus.

EIDE

Hard disk drives featuring EIDE interfaces are standard in most new PCs. Because they offer good performance and are inexpensive, EIDE drives are expected to remain the most popular choice over the next 2 to 3 years. EIDE also supports tape drives and CD-ROMs. Beginning in 1994, most PCs were shipped with EIDE host adapters and drives.

In 1996, Quantum and Intel announced a new protocol for the EIDE interface called Ultra DMA (or Ultra DMA-33), which doubles the standard EIDE hard disk burst data transfer rate to 33 MBps. Quantum is licensing the technology at no charge to peripheral and PC chip set vendors, and the protocol has been endorsed by leading hard drive manufacturers such as IBM, Maxtor, Seagate, and Western Digital.

SCSI

Hard disk drives featuring SCSI interfaces are generally found in most PC-class and UNIX servers, high-end workstations, and some desktop computers. SCSI drives are more expensive than EIDE drives, but they offer greater reliability and lower error rates, and they support more devices. As a result, they are often used for graphics work, server-based storage, large databases, and desktop publishing. SCSI technology has been a standard feature in Apple's Macintosh, PCs, and UNIX workstations.

Ultra SCSI, which began shipping in products in 1996, transfers information at 20 MBps to 40 MBps. Ultra SCSI is now a mature technology, and newer generations such as Ultra-2 and Ultra-3 promise to extend the life of parallel SCSI technology and provide speeds close to that of Fibre Channel.

Ultra-2 SCSI doubles the interface bandwidth to 40 MBps to 80 MBps, and a variety of products are already available. By the end of 1997, for example, Adaptec had introduced a new disk drive controller for the Ultra-2 SCSI interface (the AIC-8420); Quantum had announced two new Ultra-2 SCSI hard disk drives (the Atlas III and Viking II drives), as had Seagate (the Barracuda 18 and Barracuda 9LP drives); and Symbios had introduced an Ultra-2 SCSI host adapter (the SYM8951U).

Serial Storage Architecture

In 1992, a team of manufacturers led by IBM developed a new architecture called Serial Storage Architecture (SSA). SSA is a powerful, high-speed serial interface designed to connect high-capacity SCSI storage devices and subsystems to network servers and workstations. SSA products began to appear on the market in early 1995.

SSA expands on SCSI capabilities by enabling a greater number of storage devices to be attached to an interface. For example, where SCSI allows 7 devices (or 15 for Wide SCSI) to be attached on a string, SSA attaches 127 devices on a loop and gives users the ability to connect 96 disk drives to a single adapter. SSA total interface bandwidth is 80 MBps. Each port in an SSA interface can carry two 20-MBps conversations simultaneously (one inbound and one outbound). In contrast, SCSI is limited to one conversation at a time. Dual- or quad-headed connections provide fault tolerance and reliability, particularly in high-availability cluster computing environments.

SSA offers other benefits over SCSI systems in terms of performance, affordability, compactness of design, flexible configurations, and fault tolerance.

At first, the industry moved quickly to support SSA through organizations such as the SSA Industry Association. Formed in January 1995, the Association boasted approximately 40 members and represented all segments of the computer industry at the height of SSA's popularity.

In February 1996, however, Seagate announced that it was discontinuing development on SSA. Faced with a lack of support from a key drive manufacturer, other players such as Adaptec and Micropolis also backed away from SSA, which began to emerge as an IBM-only technology.

Today, IBM still remains solidly behind SSA and is planning to release a next generation in 1998 that will double link speeds for a total throughput of 160 MBps. IBM is also planning to capitalize on the performance strengths of SSA by merging them with the connectivity fabric of Fibre Channel in a new product offering in mid-1998.

Fibre Channel Arbitrated Loop

Fibre Channel Arbitrated Loop (FC-AL) is a high-speed serial storage interface technology developed by a team of manufacturers led by Seagate and including HP and Quantum. FC-AL was originally designed to compete with SSA by offering a transmission rate of slightly more than 100 MBps in half-duplex mode and 200 MBps in a full-duplex configuration. Among the applications that require this kind of data rate are video storage and retrieval, supercomputer modeling, and image processing. In addition, FC-AL was designed to be compatible at the device driver (software) level with the installed base of SCSI devices.

FC-AL offers dual path redundant access to each device, self-configuring hot-plug connectivity, RAID performance enhancements (see below), and the ability to locate systems and devices up to 30 meters apart via coaxial cable, or up to 10 kilometers apart with fiber optic cable.

The first FC-AL products started to appear in mid-1997. Seagate's Barracuda drives were the first hard disk drives with an FC-AL option. Several controller and switch firms are offering products as of late 1997.

In addition to FC-AL, there is also switched FC technology, and vendors such as Brocade currently are producing FC switches. In addition, Sequent and others already have announced plans to use switched FC technology in their new servers because switching offers better aggregate throughput than FC-AL (much like switched Ethernet).

Fibre Channel Loop

In September 1996, IBM announced that it would work with Adaptec and Seagate to merge SSA and FC-AL into a specification called Fibre Channel Loop (FCL; initially called Fibre Channel Enhanced Loop, or FC-EL). FCL will combine the best aspects of both technologies – SSA's fault tolerance and FC-AL's high speed over long distances.

FCL may hold great theoretical promise. However, in October 1997, the Technical Committee for Device Level Interfaces (part of the National Committee for IT standards, also known as the T11 Committee), which is in charge of Fibre Channel standards, tabled consideration of the FCL specification due to concerns over complexity and backward compatibility. T11 is trying to restart the standardization process but failed to come up with a statement of future work on FCL at its December 1997 meeting.

Even so, this merger of technologies is indicative of a larger trend in which the distinction between storage technologies and network channels is starting to blur. Historically, storage interconnects worked over distances measured in meters and networks covered much larger distances. In fact, the term switching fabric (once used to describe an intelligent switching system that connected two complex networks) is now being broadened to include devices that enable different network and storage channels and protocols to interconnect seamlessly.

3.2.5 Storage Architectures

For computer systems larger than a PC, disks are not just attached to a computer but are part of a storage subsystem. The subsystem may include features such as increased reliability, hierarchical storage management, backup and recovery, automated tape libraries, and so on. In addition, storage subsystems may be part of a storage architecture that provides a blueprint for how computer systems and storage subsystems connect throughout the enterprise.

RAID

Hard drives in all computer systems are susceptible to failures caused by temperature variations, head crashes, motor failure, controller failure, and changing voltage conditions. In addition, operating system malfunctions, viruses, and heavy I/O traffic also affect disk reliability.

To improve reliability and protect the data in their mass storage systems, many companies use Redundant Array of Independent Disks (RAID) storage products. RAID is a technology that links groups of standard hard drives with a specialized microcontroller. The microcontroller coordinates the drives so that they appear as a single logical drive but can take advantage of the multiple physical drives by storing data redundantly, thus protecting against data loss due to the failure of any single drive.

The term disk array has been used for many years to represent most multi-drive configurations. In 1987, David Patterson and other researchers at the University of California, Berkeley, published a paper entitled, "A Case for Redundant Arrays of Inexpensive Disks: RAID." The study sought methods to increase storage, bandwidth, and fault tolerance of disk subsystems using arrays of inexpensive small drives. According to this paper, the capacity of single hard disks had improved, but the performance and reliability of these single large disks had not grown at an analogous rate.

There are various forms or levels of RAID, which differ on exactly how the data is distributed redundantly across the multiple disks in an array. The most misunderstood element of RAID technology is the way level numbers are assigned. RAID levels are defined as 0 through 7, but a higher number does not imply an increase to the disk subsystem's performance, reliability, or scalability. Instead, the levels refer to different approaches to taking advantage of redundant storage to improve reliability, speed, bandwidth, and availability.

RAID storage subsystems typically are sold with several disk drives and include their own power supplies, cooling fans, and cabling. A typical disk array system may have five disk drives that provide a fault-tolerant capacity equivalent to four stand-alone disk drives. Using the extra disk strategically, a RAID system can continue to operate (ensuring that all the information stored on the system remains available) even if one disk fails. Once the bad disk is replaced, the redundancy is automatically reconstructed.

Disk mirroring (keeping an exact copy of one disk on another) is the simplest way to protect data, but it requires twice the disk capacity and associated cost. Mirroring, known as RAID 1, is used by many information-intensive industries. RAID 1 is a common RAID configuration because it is easy to implement and provides a performance advantage on systems that have more read accesses to data than write accesses. Because there are two copies of the data, either disk drive can be used to respond to read requests. However, RAID 1 also requires the greatest amount of extra disk capacity in the array.

Other encoding schemes are available so that the data on disks can be replicated at less than 100 percent overhead. RAID 5 is also a common RAID implementation, in part because it is less expensive. In RAID 5 systems, each block of data is spread across multiple disk drives for storage. In addition, the RAID controller calculates a summary of the data block, known as parity information, and stores it on a different drive from any of the original data. If any one of the disk drives holding the original data block fails, the remaining parts of the data block, plus the parity data, are sufficient to reconstruct the original data in its entirety with no downtime. The amount of overhead is dependent on the number of drives in the disk array, but typically would be about 25 percent of the original data. Over the past 2 years, RAID systems have begun to be used on PC servers as well as in high-end servers, mainframes, and network storage systems.

■ Failure-Tolerance Classification

Changing definitions of RAID technology have led to the formation of the RAID Advisory Board (RAB), an organization of some 60 storage-related companies. The goal of the organization is to promote the understanding and use of RAID and related storage technologies. The group has three key programs: education, standardization, and certification.

In September 1996, the RAB acknowledged the growing importance of disk system resiliency by abandoning its original definitions of RAID levels and the product certification program based on them in favor of classifying disk systems based on failure tolerance. In time, buyers can expect products to be classified by vendors more in terms of extended data availability and protection (EDAP) properties than RAID levels. The new definitions also include requirements for duplicate power supplies, fans, and controllers.

The RAB's goal is to encourage vendors and users to focus on all the determinants of data and not just the RAID level. Many RAID products, such as those sold by EMC, offer multiple RAID options simultaneously.

The RAB's three new storage classifications are as follows:

- **Failure-Resistant Data Systems Plus** (FRDS+) – Designed to protect against data loss due to any single component failure with the system and against loss of data due to single disk failure. FRDS+ adds automatic hot disk swap and protection against data loss due to cache failure or external power failure.
- **Failure-Tolerant Disk Systems Plus** (FTDS+) – Offer continuous data availability in the event of any single system component failure. FTDS+ adds data access protection against host and host I/O bus failure and external power failure, requires hot swapping of any major component, and requires the ability to connect at least two hosts using separate I/O buses.
- **Disaster-Tolerant Disk Systems Plus** (DTDS+) – Must be divisible into two or more zones that cooperate to protect against loss of access to data in the event of one system's complete failure. Protection also is provided against massive power outages, cooling system failures, and disconnection of power or bus cables. DTDS+ adds a distance between zones of at least 1 kilometer.

■ New RAID Solutions

The increasing use of distributed work-group servers for corporate applications is driving the development of RAID disk arrays that can support data storage from servers of diverse platforms. For example, HDS's 5700 line of RAID array systems can support increasing data storage demands in a mixed UNIX and Windows NT environment.

EMC, IBM, and Sun are each developing inter-platform support capabilities into new and existing RAID disk product lines. In mid-1997, EMC announced that it had optimized its disk arrays to handle different data types created by each brand of the UNIX operating system. IBM is offering special adapters that allow its 7133 arrays to store data from non-IBM platforms, such as HP and Windows NT. Sun also opened its disk arrays to HP's UNIX and to Windows NT.

Enterprise Storage Solutions

Data-intensive technologies such as the Internet, data mining, and data warehousing are driving companies to seek new and more effective storage solutions that can be deployed throughout the enterprise. Storage needs are growing at 70 percent per year. For example, the Standish Group estimates a 200-GB corporate data warehouse will grow to more than 1 TB of storage by the year 2000.

One primary characteristic that distinguishes an enterprise storage solution from a simple collection of high-capacity disks is the "intelligence" built into the storage subsystem. That intelligence enables vendors to address some of the most pressing issues in data storage by offering the ability to share data across diverse platforms, perform offline data management (copying and backup), and use commodity technologies as needed.

Users are also putting pressure on vendors to blend the disparate storage subsystem market into an interoperable, enterprise architecture. In fact, many corporations are attempting to recentralize management of decentralized storage. The justification for consolidated enterprise storage stems from the improvement in storage management efficiency and the avoidance of costly losses.

Notable vendors in the enterprise storage marketplace – EMC, IBM, Hitachi, Amdahl (now owned by Fujitsu), and others – are discussed below in order of market share.

■ EMC

EMC has been the leading vendor in the storage subsystem marketplace for the last several years. Following a price collapse in its primary mainframe storage business in 1995, EMC rebounded in 1997 to regain its former position and market share. Its products also are resold by HP and NCR.

All EMC Symmetrix systems support simultaneous storage for heterogeneous networks composed of mainframes, all versions of UNIX, Windows NT, AS/400, RS/6000, and Digital Alpha systems through EMC's Symmetrix Enterprise Storage Platform (ESP) software. In addition, all systems offer the highest level of data protection available – DTDS+ – as classified by the RAB. EMC's Symmetrix systems feature high performance and capacity; for example, the 3700 product line provides up to 3 TB of storage, with support for up to 128 23-GB disks.

Symmetrix products also work in tandem with EMC's Intelligent Storage Architecture (ISA). EMC's modular Mosaic:2000 architecture forms the foundation of ISA, allowing the company to deliver the most enhanced software and hardware components in each generation of Symmetrix products.

At the beginning of 1997, EMC began to phase out its AS/400 Harmonix storage systems in favor of its new Symmetrix line. (The Harmonix HX3SR line competed with IBM's 9337 external RAID storage subsystem.)

The EMC Symmetrix Enterprise Storage system includes a Remote Data Facility (SRDF) and EMC TimeFinder software. Through SRDF and EMC TimeFinder, residing on Symmetrix, users create real-time, host-independent, mirrored volumes of production-site information at a physically separate site. A single Symmetrix system is capable of supporting up to 32 heterogeneous servers simultaneously (including mainframes) with up to 3 TB of capacity and provides a variety of software-based information-sharing, information management, and information-protection capabilities.

■ IBM

In June 1997, IBM introduced a new storage enterprise architecture called Seascape, which uses independent building block components that can stand alone or be combined into a total storage solution. One key goal of Seascape is to make storage devices accessible to multiple platforms so that data is accessible independent of operating systems and regardless of the type of storage media used. The connectivity options available under Seascape let users access storage from PCs, midrange servers, mainframes, and third-party hardware; these options also enable data sharing through management software.

Seascape's snap-in, modular storage approach lets companies take advantage of technology advances in any one of the building blocks and leverage existing storage products. For example, customers using IBM's SSA disk subsystem can snap these into future Seascape offerings. The "future-ready" element sets Seascape apart from earlier storage systems that required the user to remove the entire storage system to accommodate technology upgrades.

Earlier storage systems relied on custom-designed hardware controllers with microprocessors to perform core functions. The Seascape architecture, in contrast, uses portable software that is less expensive to manage and can be brought to market faster than hardware technologies. In Seascape, IBM also uses the same PowerPC RISC processors found in RS/6000s and AS/400s as storage controllers, which makes them less expensive and more available than custom-made controllers.

New products based on the Seascape architecture, introduced in 1997, include IBM's Network Storage Manager, ADSTAR Distributed Storage Manager (ADSM) backup software, and Magstar MP 3575 Tape Library DataServers. The cornerstone of the new architecture is Network Storage Manager, which provides 1 TB to 9 TB of data storage. Network Storage Manager is a self-contained, fully configured combination of hardware and software building blocks that includes ADSM, SSA disk storage system, tape libraries, and storage platform adapters. When combined, ADSM, the IBM 7133 serial storage disk subsystem, and the IBM Magstar tape family become a Virtual Tape Server.

IBM has announced that in 1998, under the Seascape architecture, it will provide the capability for both UNIX-based servers and S/390 enterprise servers to access storage space in the RAMAC 3 Array system. IBM's future plans include the Versatile Storage Server, a disk-array subsystem for on-line data storage that will let customers re-allocate storage space dynamically among AS/400, Windows NT, and UNIX systems. The new server will support systems from HP and Sun in addition to IBM.

■ Hitachi

HDS integrates enterprise-wide storage solutions in a family of enhanced RAID offerings. HDS's storage strategy enables companies to share resources and information across multiple platforms or dedicate portions of a subsystem to AIX, HP-UX, Solaris, and Windows NT platforms. HDS also offers a storage subsystem specifically designed for Very Large Databases (VLDBs), which are used for applications such as data warehousing where terabytes of data must be stored and accessed.

In early 1997, HDS announced a series of storage advances, including a shared subsystem solution for heterogeneous data center environments, that builds on its flagship HDS 7700 Scalable Array. The HDS 7700 is a mainframe-class storage system that also supports distributed systems environments. It offers multiple levels of RAID and features remote copy capabilities for additional data protection and recovery. In November 1997, the HDS 7700 received the highest DTDS+ protection rating from the RAB.

Storing and retrieving data for disparate processing platforms requires an architecture that can be adapted to different environments. HDS's Multiplatform Resource Sharing (HMRS) is the first phase of Hitachi's approach, which enables UNIX-based platforms to store and retrieve data from HDS 7700 subsystems, which are also connected to mainframe systems. With Hitachi Multiplatform Data Exchange (HMDE), MVS-based OLTP systems can transfer data quickly to UNIX-based data warehousing systems.

■ Amdahl

Amdahl's Spectris disk storage subsystems provide S/390 users with scalable RAID storage. In September 1997, Spectris received the RAB's highest FTDS+ rating for data availability and protection in a single-site storage facility. Amdahl's Logical Volume Series (LVS 4500, LVS 4100, and LVS 4000) RAID storage systems offer connectivity and accessibility for high-performance UNIX and Windows NT environments. In July 1997, Amdahl integrated Quantum Corp.'s solid-state disk technology in the LVS 4500 storage system, extending mission-critical data center values across the enterprise.

Amdahl's Transparent Data Migration Facility (TDMF) software allows customers to move data transparently within their storage subsystems. With TDMF, customers can transfer data between any two like-geometry storage devices of appropriate capacity, regardless of vendor or model, thus providing uninterrupted access to data and business applications.

■ Other Solutions

In January 1997, Sun introduced a data warehousing enterprise storage device to answer the growing need for terabytes of storage capacity and higher computing power. Hoping to build its reputation as a storage provider, Sun also is licensing Symbios Logic Inc.'s intelligent RAID storage hub with hot-swappable components, up to 80 4-GB drives, and FDDI connections. The new RAID subsystem supports UNIX variants other than Sun's Solaris for the first time.

Sun's acquisition of the storage business and technology of Encore Computer Corp. in mid-1997 is intended to bolster its plans to become a major storage provider. Encore's Infinity SP arrays let users store mainframe and open systems data in the same box, and is one of the few products that competes directly with EMC's ESP technology.

Like Sun, Data General is extending the capabilities of its products: In June 1997, it announced plans to incorporate Fibre Channel technology in its new Clariion disk arrays. Similarly, MTI Technology is adding solid-state disk capability to its Gladiator line of RAID disk arrays for use with mission-critical databases. MTI also is competing directly with EMC in niche situations where users need high-speed database storage.

3.2.6 Floppy Disks

A floppy disk, also called a diskette, is a removable magnetic storage medium on which data can be recorded, erased, and recorded again. The floppy disk itself is made up of a flexible plastic platter coated with a magnetic medium used to record and store data.

The floppy disk was introduced by IBM in an 8-inch format in 1971, and was first used by IBM engineers and salespeople to carry programs and program updates to customers. The disks, which were lighter and less cumbersome than magnetic tape, ultimately provided valuable storage capacity for a variety of home and business applications, especially spreadsheets and word processing.

Today, the disks continue to be used for transferring data from one computer to another. However, floppy disks hold far less data than hard disk drives and transfer information more slowly.

Currently, floppy disks are available in two formats: a 3.5-inch rigid case with a capacity of 800 KB to 2.88 MB (standard 3.5-inch drives are currently 1.44 MB), and a 5.25-inch flexible envelope with a capacity of 100 KB to 1.2 MB.

Sales of floppy disks are expected to decline over the next 5 years for several reasons. In the late 1980s and early 1990s, PCs were generally equipped with two floppy disk drives: one 5.25-inch drive and one 3.5-inch drive. Today, however, as the market has shifted away from the older 5.25-inch format, most PCs are shipped with only a single 3.5-inch drive.

High-Density Floppy Disk Drives

In the 10 years following the introduction of the 3.5-inch diskette, there were few innovations in floppy disk technology. The first alternative diskette technology was introduced in 1995: high-density floppy drives. High-density floppy disk drives are faster than standard 1.44-MB floppy drives and use disks with much higher capacities, ranging from 25 MB to more than 200 MB. Examples of high-capacity floppy drives include Iomega's 100-MB Zip drive and SyQuest's 135-MB EZ and 230-MB EZFlyer models. These drives are not backward-compatible; that is, they cannot read or write traditional 3.5-inch floppy disks.

The Zip drive was developed by Iomega and currently is also sold by Fujitsu and Epson. The Zip drive rotates its flexible disk about 8 times faster than a 1.44-MB standard floppy disk drive. At this faster spin rate, the flexible disk is stretched until it is rigid. This process allows a denser, more precise placement of recording tracks, resulting in greater storage capacity; the 100-MB Zip disk offers storage equal to 70 floppy disks. The Zip drive also weighs only 1 pound, making it highly portable.

The Iomega Zip drive began shipping in March 1995. By October 1996, 3 million units had been sold, according to market research firm, Dataquest. Dataquest projects that 8 to 10 million Zip drives will be sold by the end of 1997. The price of the unit has been dropping steadily: In 1996, Zip drives were selling for around $200, with individual disks costing about $20; in 1997, Zip drives were often sale-priced at around $100, with disks costing less than $12 each (in bulk).

In December 1997, SyQuest introduced the SparQ, which combines the speed of a fixed hard disk with the flexibility of 1-GB removable cartridges. It provides an average seek time of 12 ms and an average sustained read/write data rate of 5.63 MBps. The SparQ is available in parallel-port and internal EIDE configurations; SyQuest plans to release a SCSI version in early 1998.

Sony and Fuji Photo Film are offering to replace the floppy disk drive with another high-capacity alternative. The High-Capacity Floppy Disk (HiFD) will have a dual-sided formatted capacity of 200 MB, and will read and write to the older 1.44-MB and 720-KB floppies. Although O.R. Technology's LS-120 a:drive also reads and writes to both formats, Iomega's proprietary Zip format cannot read standard floppy disks. The HiFD will have a data transfer rate of 3.6 MBps. Compared to a standard floppy drive transfer rate of .06 MBps, the Zip drive's rate of 1.4 MBps, and the a:drive's 0.6 MBps, the HiFD is the fastest performer among removable drives. Sony and Fuji expect to ship HiFD units by mid-1998.

Multi-Function Floppy Drives

A consortium headed by Compaq, Imation, Matsushita, and O.R. Technology recently introduced the LS-120 drive (soon to be renamed the SuperDrive). Initially available for IDE, the LS-120 should be available for SCSI by the end of 1997. The LS-120 can replace a floppy drive with a unit that reads both 1.44-MB and new 120-MB floppies, thus providing both backward compatibility and almost a one hundred-fold increase in capacity, and indicating that it might become the floppy drive of choice in new computers. Unfortunately, the first LS-120 units tested were slower than an Iomega Zip drive attached to a SCSI card, although speed is nearly equal when the LS-120 is compared with a parallel port Zip drive (the way most such units are sold). A speedier floppy format from Mitsumi that also would be backward compatible is reportedly in development, but commercial shipments may arrive too late to have much impact on the direction of the market.

Removable Disk Drives

Removable disk drives were first used on the mainframes of the 1950s and 1960s. At that time, the drive mechanisms were extremely expensive, so different applications used dedicated disk packs that were mounted only while the programs were executing. As disk drive prices fell, sealed, non-removable drives were incorporated into mainframes.

Smaller, more portable, removable disk drives became valuable again in the 1980s, this time as a data transfer medium for the printing industry. Both Iomega and SyQuest began producing 5.25-inch 44-MB drives that backed up graphics, artwork, and print. It took another decade before small, removable drives were developed for more widespread home and commercial use. Their development was driven, in part, by the larger storage requirements resulting from the increased multimedia content of business and consumer software. Increased business travel also spurred a need for reliable, portable data transfer media that could hold larger amounts of data.

Removable hard drives are faster than removable floppy disk drives because their disks are composed of compact, rigid platters that can spin at a faster rate. The 1-GB Iomega Jaz cartridge, for example, spins at a rate fast enough for multimedia playback as well as storage. In 1996, external Jaz drives sold for less than $600; in 1997, however, prices had dropped below $400.

In November 1997, Iomega revealed another variation on its popular Zip and Jaz drives: the Clik!, which holds 40 MB in a cartridge smaller than a matchbook. The company says the Clik! will show up in hand-held PCs, digital cameras, and other pocket-sized products by mid- to late 1998. Clik! is expected to compete against flash memory/PC Cards. Cartridges are initially expected to cost $10 each. One analyst has already predicted that Clik! will boost Iomega's revenues by as much as $1 billion by the year 2000.

Also in November, SyQuest introduced a new high-performance platform: The Quest 4.7-GB system combines the speed and throughput of high-end fixed hard disk systems with the flexibility of removable storage. A member of SyQuest's Rocket platform, announced earlier in 1997, Quest is the first removable cartridge drive to integrate dual-stripe magneto-resistive recording heads. A single 4.7-GB cartridge can hold the digital equivalent of a full-length feature film, making it the first rewritable removable storage option with enough capacity for DVD mastering.

Optical Storage Media

Optical storage media use lasers that write to and read from discs. In general, the discs are pre-formatted with grooves and lands (tracks) that enable an optical pickup and recording head to locate specific information.

To record a bit, a laser generates a small spot on the disc. These spots vary in phase, intensity polarization, and reflectivity, depending on the type of information being recorded. The differences then can be read and interpreted by a readout beam, which is emitted from a detector in the optical head of a disc player or reader.

The storage capacity of an optical storage system is a direct function of spot size (minimum dimensions of a stored bit) and the dimensions of the media. Data transfer rates are determined by the linear density and the rotational speed of the optical storage system drives. Optical discs are slower than magnetic disks, and optical disc drive mechanisms are more expensive than magnetic drives.

Automated optical disc changers, known as optical jukeboxes, are popular with companies that use hierarchical storage management (HSM) to transfer data between primary and secondary storage, based on when the information was last accessed, among other priorities specified to the HSM system. Optical storage is frequently used with HSM systems because it is a reliable medium that falls between the less-expensive but slower tape drive options and the more-expensive but faster magnetic disk drive options. Jukeboxes typically hold up to several hundred discs and function with one or more drives to provide access to multi-volume libraries of recorded data. Jukebox vendors include Cygnet Storage Solutions, HP, Micro Design International, Sony Electronics, and Philips.

Magneto-Optical Drives

Magneto-optical (MO) drives are the next step up from floppy drives. Currently, the SyQuest EZFlyer 230 is the fastest offering in this category, and at $0.12 cents per MB, media cost is less than that of the Zip drive. At about $0.04 cents per MB, the 3.5-inch, 230-MB MO cartridges (such as those used in Olympus' SYS.230) cost about the same as Zip disks and can be read and written by 640-MB MOs. Both 230-MB contenders offer good media value: the EZFlyer offers excellent speed, and the MO offers archival reliability. However, neither MO drive can claim Iomega's reported 7 million installed base, making Zip the drive of choice if data interchange is a priority.

3.2.7 Optical Storage Devices

The first optical discs came in large (12- or 14-inch) form factors and were used primarily as a data archiving mechanism. CD-ROM discs and drives were commercially introduced in the early 1980s. Today, they are the most widely used optical storage devices. Market research firm Disk/Trend estimates that more than 54 million CD-ROM drives were sold in 1996, up from 2.5 million drives in 1992. Most CD-ROM disc drives play audio and data discs interchangeably.

CD-ROM discs are created using essentially the same laser technology that is used to manufacture CD audio discs. To read the discs, a red laser beam passes over pits that are stamped into the reflective surface of the disc. As the beam passes over the disc, photodetectors and other electronics translate variations in the amount of reflected light into the 0s and 1s of digital data.

Once a master disc has been produced, copies can be manufactured inexpensively: the unit cost is about $1 for quantities of 2,000 discs. Because of this inexpensive duplication, CD-ROM discs – with a capacity of about 650 MB – are ideal for publishing large amounts of text or data or large numbers of computer programs or images in an electronically readable form.

Recordable CD-ROM discs and drives hold 650 MB, transfer information more quickly than floppy disks, and offer users read and write capabilities. Standard (non-recordable) CD-ROM drives have already become popular as floppy disk drives on newer systems, especially those targeted toward home entertainment and educational uses. With faster drives available at prices that continue to decrease steadily, CD-ROMs have begun to supplant floppy disk drives for software distribution in most desktop systems. In addition, CD-ROM discs have almost completely replaced half-inch and quarter-inch magnetic tape for the distribution of software and documentation for workstations.

CD-Recordable Technology

CD-Recordable (CD-R) technology creates CDs via a recording system that uses write-once-type optical discs standardized by Philips and Sony. Individual discs are long-lasting and relatively inexpensive (about $10 apiece). Each disc can store about 650 MB. It takes from 20 to 60 minutes to record a full CD-R disc, depending on the speed of the drive.

Write-once recordings, often referred to as WORM (write-once read-many) have an extremely long read life. As a result, it is often used for archiving and audit trail applications. Updating requires destroying the existing data (all 0s made 1s), and writing new data to an unused part of the disc.

There are two kinds of WORM technologies: ablative and continuous composite. Ablative WORM is the original WORM technology that makes a permanent change in the optical material. Continuous composite write WORM is a mode in multi-function 5.25-inch MO drives that emulates a WORM drive. The media is not permanently changed, but the drive contains firmware that ensures recorded areas are not rewritten.

The larger 12-inch and 14-inch WORM drives are still manufactured, but due to multi-function MO drives and other optical technologies on the horizon, ablative WORM systems are expected to be a declining market.

CD-Rewritable Technology

One of the main advances in CD-ROM technology in recent years has been CD-Rewritable (CD-RW) drives. CD-RW drives, which began to appear on the market in late 1997, let users create CDs that can be rewritten 1,000 times or more, overcoming the traditional limitation of CD-R drives' WORM technology. CD-RW disks store up to 500 MB of data. The new CD-RW drives read CD-RW disks as well as regular CD-ROM discs.

HP announced its first rewritable drive, the SureStore CD-Writer Plus, in October 1997; internal and external versions are available. Other CD-RW drives are available from Philips Electronics and Yamaha Systems. For example, Yamaha began shipping three new CD-RW drives in October 1997. Its internal CRW4260t and the external CRW4260tx are 4X-record, 6X-read, 2X-rewrite devices; the CRW2260 is a 2X-read and 2X-write design. (A 2X CD-ROM drive spins twice as fast as the first CD-ROM, a 4X CD-ROM drive spins four times as fast, and so on.)

CD-RW technical standards are being developed by a coalition of five technology companies: HP, Philips, Ricoh, Sony, and Verbatim (a subsidiary of Mitsubishi). These companies have announced that CD-RW discs will sell for about $25 each, and initially, CD-RW drives will sell for less than $1,000. However, most analysts believe that prices will need to drop well below the $500 range before CD-RW devices will be widely adopted, particularly in the consumer market. Analysts expect prices for CD-RW drives to drop by about 20 percent by the end of 1998.

Several issues may limit the ultimate acceptance of CD-RW technology. First, although CD-RW will theoretically coexist with DVD (see below), the first DVD drives had problems reading both CD-R and CD-RW media; second-generation DVD drives, due out by early 1998, should address this problem. Second, compatibility and media cost may hinder widespread adoption: existing CD-ROM drives cannot read CD-RW discs, and CD-RW media currently cost many times more than CD-ROM discs, although prices are expected to fall slowly.

DVD

Advances in digital data storage and laser technology led to the development of DVD discs (and players), which began appearing on the market in early 1997. Some companies refer to these discs as Digital Video Discs, and others refer to them as Digital Versatile Discs. Because the entertainment and computing industries cannot agree on a definition, the devices now are simply called DVD. DVD discs physically resemble CDs, but they can hold far more data (the equivalent of up to eight CD-ROM discs, or enough space to hold a 133-minute movie accompanied by a surround-sound audio track).

DVD discs are available in two formats: basic DVD, suitable for consumer electronics home entertainment systems, and DVD-ROM, which is specifically designed for use in a PC. Both DVD formats store text, graphics, video, and audio in a digital format that computers can read.

DVD standards are the result of a compromise between ten leading entertainment and consumer electronics companies known as the DVD Consortium (now called the DVD Forum). In early 1995, two competing proposals for DVD-ROMs were put forth: one by Sony and Philips; and the other by Toshiba Corp. and Time Warner. In September 1995, a compromise was reached so that the technology could be quickly adopted.

The DVD standard specifies that players will be able to read current CDs and CD-ROM discs as well as DVDs. The companies agreed that single DVDs would be 12 cm (5 inches) in diameter and be able to hold up to 4.7 GB on a single recording layer on a single side. Double-layer discs feature capacities of 8.5 GB per side. Both single- and double-layer discs will also be bonded into two-sided discs holding up to 17 GB. Like CD-ROM discs, DVDs will be used to store music, movies, and multimedia packages. However, DVD released titles will also be enhanced with features that, for example, might offer movie viewers a choice of camera angles or several different sound-track languages.

DVD-ROM drives are able to read CD-ROM discs and DVDs. Rewritable DVD drives are expected to appear by the end of 1998. IDC projects that worldwide DVD-ROM unit shipments will surpass CD-ROMs by the year 2000, climbing to 117 million in the year 2001.

■ Divx

In September 1997, a partnership called Digital Video Express backed by electronics retailer Circuit City and a Los Angeles-based entertainment law firm unveiled plans to produce a competing version of DVD called Divx (for Digital Video Express) with discs whose contents would only be readable for a limited period of time. Compared with the existing DVDs, which are aimed at consumers who want to buy and keep movies, the Divx systems is aimed at those users who might want just a single viewing, much like those who now rent video cassettes.

Divx is not a new format; instead, it builds on the DVD format. All standard DVDs will play on a Divx player, but because of additional encryption, Divx discs will not play on standard DVD players. Divx players will support multi-channel digital sound, including Dolby's AC-3 surround sound audio format. Divx players also will support digital audio output for use with an external digital-to-analog converter as well as other sound components.

All movies available through Divx will be fully encrypted using Triple-DES (Data Encryption Standard). Discs and players will be individually serialized, allowing the system to pass disc- and player-specific encryption "keys" to players and generate usage reports from players on a disc-by-disc basis. (See Chapter 5.4, *Security,* for more information on encryption methods.) As a result, Divx will be able to grant individual customers limited access to individual discs and track each time the disc is played.

The Divx system will allow viewers to pay a rental price of less than $5 for a disc that does not need to be returned because the movie stored on it will be locked and unavailable 48 hours after it is first played. If viewers want another viewing at any time in the future, they will be able to complete an electronic transaction to pay for it. Another option would charge a fee, probably less than $20, for unlimited viewing of a disc.

Divx can be incorporated into DVD players, allowing users to take advantage of its features. The first Divx players and accompanying software should appear by mid-1998. Divx players will be manufactured initially by Matsushita/Panasonic, SG-Thomson, and Zenith, the same companies that currently manufacture standard DVD players. Four Hollywood studios – DreamWorks, Paramount, Universal, and Walt Disney – have all agreed to provide titles for Divx release.

Sony and Toshiba have stated they will not support Divx or incorporate the technology into their DVD players, however. Instead, these companies are pushing for open DVD standards. In October 1997, the National Association of Video Distributors (NAVD) also took a stand against Divx, claiming that it forces consumers to purchase different and more expensive hardware, have a modem or other phone connection, and pay a rental rate higher than the current average video or DVD.

■ DVD-RAM

A DVD-RAM is a rewritable optical disc that uses DVD technology. DVD-RAM discs that can be erased and recorded multiple times will use phase-change technology to hold the recorded data. (Phase-change technology uses a short, high-intensity laser pulse to create a bit by altering the crystalline structure of the material. The bit either reflects or absorbs light when read. A medium-intensity pulse is used to restore the crystalline structure.) DVD-RAM discs will be suitable for users with very high storage needs.

The first DVD-RAM drives from Hitachi, Matsushita, and Toshiba, which will ship in early 1998, will probably cost between $700 and $900 (rewritable discs will run about $50 to $100 each), at least 50 percent more than a DVD-ROM setup, although analysts expect prices to fall to less than $300 by the end of 1999. In addition, the first DVD-RAM discs will have a capacity of only 2.6 GB per side; next-generation, DVD-RAM drives, due to ship in 1999, will feature 4.7 GB (the equivalent of DVD-ROM).

The DVD standards area is complicated, however, because two rivals to the existing DVD-RAM specification have emerged. Unlike the DVD Forum's standard that supports 2.6 GB per side (already used by Hitachi's pioneering drives), six drive makers, including HP, Philips, and Sony, announced in August 1997 that they would pursue their own rewritable DVD format with discs that store 3 GB per side. Dubbed DVD Rewritable or DVD+RW, the technology will not be compatible with DVD-RAM. Both the DVD-RAM and proposed DVD+RW standards have been submitted to the European Computer Manufacturers Association (ECMA), a European standards body. The original DVD-RAM format 1.0, announced in September 1997, is also being submitted to a Japanese standards body.

NEC has announced that it, too, will release a proprietary rewritable drive. The company hopes to leapfrog both the foregoing formats with a disc capable of storing 5.2 GB per side.

Because of its recordability, DVD-RAM may find an early market as a backup storage device for graphic artists and others who share large files. Like other DVD technologies, these machines will play most variations of DVD formats as well as CD audio and CD-ROM discs.

■ Other DVD Derivatives

Beyond the forecast period, *Scientific American* reports that derivatives of the basic DVD technology might ultimately produce a product that could hold more than 50 GB on a 1.2-mm-thick platter – the equivalent of a small library on a single disc.

Magneto-Optical Technology

Until DVD-RAM drives become available, the only removable storage drives with greater-than-1.5-GB capacity are the high-end 5.25-inch MO drives. With high spin rates and 4 MB of onboard cache on some drives, MO performance approaches that of fast hard disks. Once data is written, it is re-recordable and easily transportable. Like hard drives, most MO drives do not require special drivers but are supported by the operating system. Currently, the most attractive MO drives are 2.6-GB units that are a bit slower than ordinary Winchester drives.

The 5.25-inch MO disks are double-sided and hold a total of 650 MB, 1.3 GB, or 2.6 GB, but must be manually flipped for each side. A 5.2-GB version is expected by early 1998, and analysts predict that a 10.4-GB version will be available by the year 2000. MO media prices are already dropping. For example, in July 1997, Pinnacle's 4.6-GB Apex MO disks, which previously cost $169 each, had dropped to $99 apiece; the cost of Vertex 2.6-GB disks had been lowered from $89 to $69 each.

Magneto-optic technology uses a combination of magnetic disk and optical methods. Data is written by a laser and a magnet. The laser heats the bit to the Curie point (the temperature at which molecules can be realigned when subjected to a magnetic field). Then, a magnet changes the bit's polarity.

Reading is accomplished with a lower-power laser that reflects light from the bits. The light is rotated differently, depending on the polarity of the bit, and the difference in rotation is sensed. Writing takes two passes. The existing bits are set to zero in one pass, and data is written on the second pass.

An alternative MO technology is represented by the new Light Intensity Modulation Direct Overwrite (LIMDOW) drives. Instead of writing in two steps like a standard MO, LIMDOW writes in one step, resulting in an average increase in speed of 50 percent. LIMDOW units can also read and write to older, non-LIMDOW MO cartridges.

In 1997, Sony Electronics introduced two new 2.6-GB MO drives (the SMO-F541-DW and SMO-F544-DW) incorporating LIMDOW technology, doubling write speeds of previous generation drives. The increase in write performance, up to 4.0 MBps sustained transfer rate, and the low average seek time of 25 ms opens up several new applications to MO technology where performance and expandability are key issues.

The next-generation MO drive standard, originally dubbed MO7 (because of its planned 7-GB capacity), will be based on 120-mm cartridges and will be able to read CD-ROM and DVD discs as well as rewritable DVD-RAM discs. Compared with DVD's 4.7-GB capacity and 200-ms to 300-ms seek times, MO7's 7 GB capacity and 30-ms seek time present an attractive alternative.

In late 1997, however, the consortium backing MO7's development scaled back the capacity of MO7 from 7 GB to 6 GB to speed time to market. The consortium – which includes Fujitsu, Hitachi, Hitachi Maxell, Matsushita, Olympus Optical, Philips, Sanyo Electric, Sharp, Sony, and LG Electronics – also changed the technical name from MO7 to ASMO (for advanced storage magneto-optical) technology. According to the consortium, commercial shipments of 6-GB ASMO drive systems are scheduled for spring 1998. In the meantime, 5.25-inch MO drives may offer the best solution for users with large amounts of data that need to be securely, but rapidly, stored and accessed.

3.2.8 Batteries

The battery industry is traditionally an area where innovation comes slowly. The commercial emergence of rechargeable, nickel-cadmium batteries in the mid-1960s made possible the portable tools, camcorders, cellular phones, and computing devices that consumers now take for granted. Today, although portable electronics technologies have expanded and improved, battery technology is just starting to respond.

Because battery life is an important factor for consumers of portable electronic products, many computer notebook and laptop makers are redesigning their products to use less power. At the same time, manufacturers are rushing to enhance their portable systems with larger memories, faster processors, better color displays, and additional peripherals that consume more power. The tradeoff between useful features and battery life is having a profound effect on which products succeed in the marketplace. Users of hand-held computers need even greater battery performance than traditional notebook users; hand-held computers typically require all-day or all-week functioning on a single set of batteries.

Battery Basics

All batteries work by coupling two chemical reactions – an oxidation reaction (which produces electrons) with a reduction reaction (which consumes electrons) – and providing a path for the emitted electrons to flow from the negative battery terminal to the positive. The total amount of electrical energy the battery produces is limited by the amount of chemical energy stored by the materials involved in the reactions.

In disposable or primary batteries, the charge/discharge process works in one direction. These batteries are sold fully charged and are disposed of after the battery has delivered electrical energy. Rechargeable or secondary batteries, on the other hand, have a two-way process of charge and discharge that makes use of an external power supply. Each time this power supply reverses the flow of current, the battery is recharged.

One of the most common difficulties with battery-powered equipment is the gradual deterioration in battery performance after the first year of service. In fact, many computer service centers report that up to half of equipment failures are actually battery-related.

Battery Technology Alternatives

Many established and startup companies are focusing their efforts on improving the efficiency of today's widely used nickel-cadmium (NiCd) and nickel-metal hydride (NiMH) batteries. Other manufacturing and development efforts have focused on emerging battery technologies that involve lithium-ion, lithium-polymer, and zinc-air chemistries. The need to engineer longer battery life and a choice of form factors to accommodate new consumer electronics is driving battery development.

The following issues are key to the viability of rechargeable battery technologies:

- **Energy storage** – Refers to the amount of energy that can be stored in a battery safely and practically. Energy storage determines the length of time a battery will operate between charges.
- **Energy density** – The amount of energy stored per unit of battery weight or volume
- **Weight** – A critical factor in the actual portability of an electronic product
- **Size and shape** – Cylindrical cells take up more space when packed together in a battery compartment. Flat, rectangular cells are more space-efficient.
- **Self-discharge** – Refers to the rate at which a battery loses energy when not in use. Self-discharge determines the battery's shelf life.
- **Cycle life** – Refers to the number of times a battery can be charged and discharged without losing significant rechargeability (also called service life)
- **Environmental concerns** – Fears about the toxicity of the cadmium used in nickel-cadmium batteries contributes to the increased use and development of other rechargeable technologies
- **Cost** – Low price contributes to the widespread use of nickel-cadmium batteries, and has created a standard with which newer technologies must compete.

■ Nickel-Cadmium

Most portable power-consuming devices, such as power tools, video cameras, notebook computers, and cellular telephones, would never have been commercialized in portable form without NiCd batteries.

NiCd batteries are much less bulky than previous batteries, and thus suitable for portable use. The advantages of NiCd chemistry are that it provides a high level of power with a low cost per watt-hour of energy. NiCds provide a cycle life of 300 to 700 charges, giving a typical user 18 months to 3 years of battery usage. NiCd chemistry also has a proven track record and is an established technology.

The drawbacks to NiCd compared with newer technologies is that these batteries have relatively low energy density, which limits battery use in a typical notebook computer to an average of about 2 to 3 hours before recharging. These batteries also have a characteristic known as the memory effect: If the battery is not fully discharged before recharging it, over time it develops a "memory" of the level of charge it held before being recharged and cannot be recharged beyond that level. In addition, the battery has a discharge rate of about 1 percent per day when not in use. Finally, the cadmium used in NiCd batteries is a toxic heavy metal that is harmful to the environment. NiCd cells account for half of all toxic cadmium in U.S. landfills.

■ Nickel-Metal Hydride

NiMH batteries, developed in the early 1990s, are now in common use. NiMH can hold at least 1.25 times as much power per unit as a standard NiCd battery. Compared with a NiCd battery of equal size, NiMH batteries run for 30 percent longer on each charge. Furthermore, NiMH batteries do not have the "memory" effect; they can be recharged to their full potential even if they are not fully discharged. NiMH is also more environmentally friendly because it contains no toxic metals.

However, the price of the NiMH battery is about 30 to 50 percent higher than that of the NiCd battery. The downside to NiMH technology is overall battery life: batteries last for 400 charge and discharge cycles. In addition, although a NiCd battery can be safely charged in 90 minutes, the NiMH battery needs about 3 hours to charge under the same conditions.

■ Lithium-Ion

Lithium-ion (Li-ion) batteries, already used in cellular phones and camcorders, are smaller and lighter than both NiCd and NiMH batteries. The Li-ion battery weighs about half of the equivalent NiCd battery but costs about the same. Li-ion technology also has the potential for extremely long life: the total number of charge/discharge cycles can exceed 1,000 cycles.

One serious drawback is that Li-ion batteries have a tendency to ignite. In response to this problem, manufacturers now make them in a safer but weaker liquid electrolyte form. Li-ion batteries are also very sensitive to overcharging. A typical Li-ion battery is illustrated in Figure 3-6.

Figure 3-6: Lithium-Ion Battery

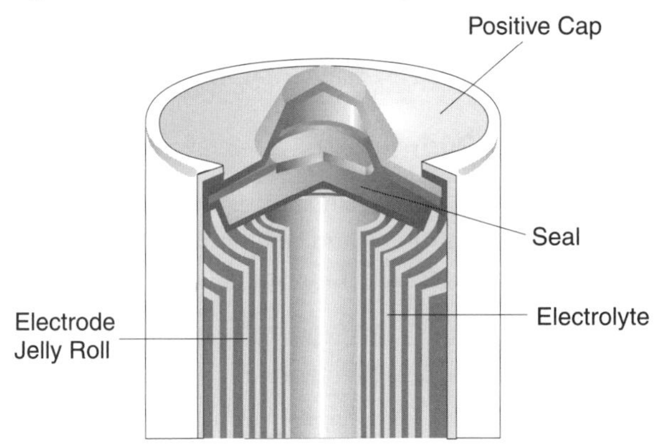

Source: *Mobile Computing & Communications*, 1997

Because of their light weight and high energy density, lithium-ion batteries are becoming favorites in the high-performance, rechargeable-battery market. Rechargeable lithium-ion batteries also have a high unit-cell voltage – about 3 V to 4.2 V, compared with 1.5 V for NiCd and NiMH cells.

In May 1997, Hitachi Maxell and its Battery Engineering subsidiary announced a rechargeable Li-ion battery the size of a credit card, with an energy potential of 3.6 V, that can be charged and discharged up to 800 times. The battery also can be formed into shapes, including a 90-degree curve, and has a thickness of 4 mm, making it suitable for portable phones. A postcard-sized version contains the equivalent power of two of the cylindrical batteries commonly used for notebook computers and other portable equipment.

Lithium-ion polymer has the same properties and performance characteristics as Li-ion battery technology, but polymer is a solid, flexible material that can be stretched thin. As a result, notebook makers can wrap a lithium-ion polymer battery behind the display or under the keyboard, saving space and weight. Such batteries would be a boon to the creation of significantly thinner portable computing devices, including notebook PCs, personal digital assistants, and cellular phones. In response, new companies such as Lithium Technology are springing up that plan to focus on developing next-generation lithium-ion polymer batteries; Lithium Technology's first shipments are scheduled for mid-1998.

■ Lithium-Polymer

Solid-state lithium-polymer rechargeable batteries are safer than liquid lithium-ion, offer greater design flexibility, cost considerably less, and are environmentally friendly. Polymer technology has been in development for years by such companies as Duracell and Ultralife Batteries, but its complex structure and manufacturing problems have delayed release. HP has previewed an ultra-thin notebook computer with a polymer battery jointly developed by Mitsubishi and Ultralife that is scheduled to ship in early 1998. However in late 1997, serious quality problems forced HP to postpone plans to release notebooks equipped with lithium-polymer batteries; an early 1998 rollout is scheduled with a variation of the product. Mitsubishi, which was planning to use the same battery from Ultralife in its forthcoming 0.7-inch-thick Panther notebooks, has decided to retrofit the units with more standard lithium-ion batteries.

In-Charge rechargeable lithium-polymer AA-size batteries produced by Tadiran Electronic Industries are already being used in cellular phones and other portable electronics applications, according to *Electronic Engineering Times*. A typical lithium-polymer battery is shown in Figure 3-7.

Figure 3-7: Lithium-Polymer Battery

Lithium Foil (Anode)
Insulator
100 microns (μ)
Metal Foil (Current Collector)
Electrolyte
Cathode

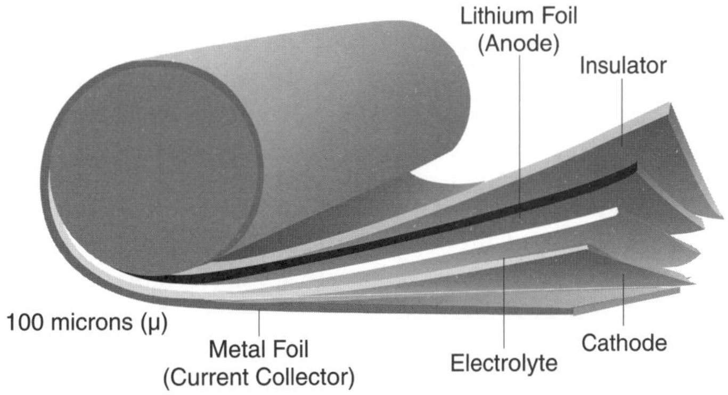

Source: *Mobile Computing & Communications*, 1997

■ Zinc-Air

The greatest performance potential so far for a new rechargeable technology comes from zinc-air batteries. Long used for primary batteries, this chemistry combines lightweight zinc with oxygen from the air to fuel a chemical reaction. Zinc-air offers the advantage of being one-third lighter than nickel-cadmium and half the weight of lithium. In addition, with an energy density of 150 watt hours per kilogram, zinc-air batteries are the first to promise 8 hours or more runtime on a single charge. Figure 3-8 illustrates a typical zinc-air battery.

Although often considered a niche product, zinc-air technology nevertheless compares well with Li-ion when it comes to energy density per unit weight. A typical zinc-air battery provides 10 to 25 watts of power; 120-watt per hour zinc-air battery packs are being used in the HP Omnibook 800 and weigh about the same as the computer, approximately 3.4 pounds. Lighter-weight batteries also are in development and are expected to debut in 1998.

Zinc-air rechargeable batteries are well-suited for long runtime applications. Until recently, that market was largely unserved, but AER Energy Resources is developing products in this area and already offers a 10-cell zinc-air battery for a satellite telephone that permits 8 hours of talk time and 50 hours of standby. (Previous batteries typically provided 1 hour of talk time and 3 hours of standby.)

Figure 3-8: Zinc-Air Battery

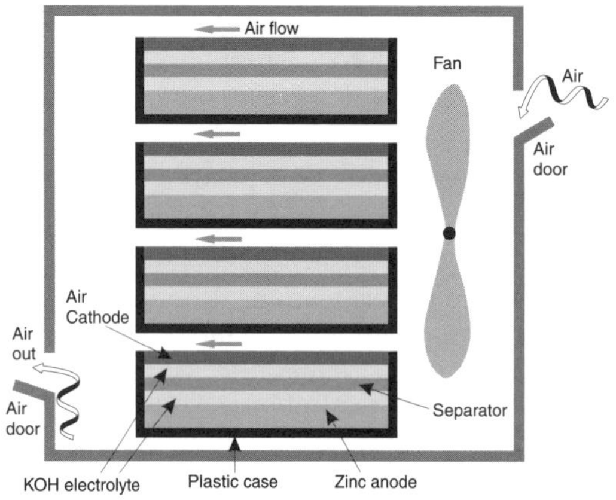

Source: *IEEE Spectrum*, 1997

Smart Batteries

Battery technology has evolved slowly in relation to semiconductor technology – with battery innovations centered on chemical rather than electronic technologies. Over the past few years, however, battery manufacturers have begun to apply electronics technology to their products, leading to the development of "smart" batteries that adjust to changing environments.

Smart batteries shift design attention to how the batteries interact with the devices in which they are used by creating circuits built into battery shells that manage their use. These circuits perform two functions: they optimize the battery-charging functions and report back to the user on battery status. When a management circuit is packed with the battery in an integrated energy pack, the combination becomes an intelligent battery that can communicate with the host computer and provide fuel-gauging and charge-control capability.

Smart battery technologies promise to improve power management capabilities for portable devices by minimizing power drain on the battery and accurately reporting on the battery's remaining life. Smart rechargeable batteries contain microprocessors that can measure, calculate, update, and communicate accurate information about battery status and usage. Users can then disable various power-consuming devices such as CD-ROMs and screen lighting to use remaining battery capacity most efficiently.

Industry trends pushing smart battery development include the emergence of dual-battery notebooks, docking stations that include charging bays, and flexible battery form factors to fit the requirements of thinner, more compact notebooks.

Thus far, the notebook computer segment of the computer industry is making the greatest investment in smart batteries because the cost of the notebooks is high enough to make it cost-effective to spend more on state-of-the-art batteries.

A consortium of chip makers, battery manufacturers, and other companies are developing standard specifications for smart batteries in hand-held devices. The Smart Battery Systems Implementers' Forum, which includes companies such as Benchmarq, Duracell, and National Semiconductor, says that the standard will make is easier and less expensive to develop the devices and get them to market faster. The smart battery systems (SBS) protocol also covers power monitoring, safety features to avoid overcharging, and the ability to discharge automatically to increase battery life.

3.2.9 Market Overview

Hard Disk Drives

IBM introduced the world's first disk drive in 1956, with 5 MB of storage at a cost of $10,000 per MB. In 1996, worldwide shipments of hard disk drives totaled 105 million units, according to Disk/Trend, with more than 131 million estimated to ship in 1997 and 201 million drives projected for shipment in the year 2000. Figure 3-9 shows worldwide disk drive shipments by vendor in 1996 and 1997.

Figure 3-9: Worldwide Disk Drive Shipments by Vendor

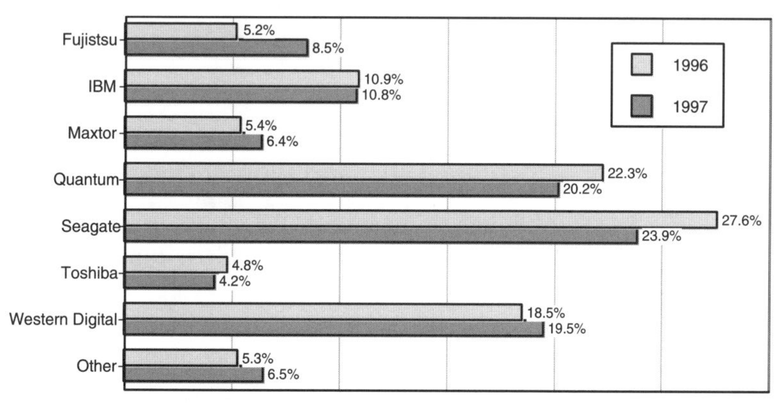

Source: *San Jose Mercury News*, 1997

3.5-inch drives will continue to be the leading drives used in desktop PCs, network file servers, and mainframe storage systems in the near term. Disk/Trend projects that 144.3 million 3.5-inch drives will be shipped in 1999. In addition, the firm reports that sales of 2.5-inch drives, mostly installed in notebook computers, are increasing and projects shipment of 19.4 million units in 1999. However, as demand increases for higher-capacity drives, the 5.25-inch desktop drive market will begin to grow, reflecting a return to a form factor that was used in the 1980s.

Annual increases in disk recording densities, combined with increased demand for data storage, have pushed up disk drive capacities faster than ever. In 1991, average disk drive capacity was 145 MB; in 1997, average capacity was 2.65 GB. Disk/Trend expects the 9-GB and 18-GB 3.5-inch drives, which will debut in 1998, eventually will replace current 4.5-GB and 9-GB models.

Prices are falling drastically as well. In 1993, for example, the least-expensive 500-MB drive cost more than $1,000; in late 1997, in comparison, a 6.4-GB hard disk drive cost about $200.

In terms of actual sales revenues, Disk/Trend projects that the market for RAID subsystems will approach $13 billion in 1997. In late 1997, RAID systems cost about $24,500 for a 54-GB array, compared with $38,000 for a 4-GB array in early 1992. Cost per MB has dropped significantly as well, from $9 per MB in 1992 to $2 per MB in 1994, and $0.45 per MB in 1997.

Figure 3-10 gives IDC's projections for hard disk unit shipments worldwide.

Figure 3-10: Worldwide Hard Disk Units Shipped by Capacity

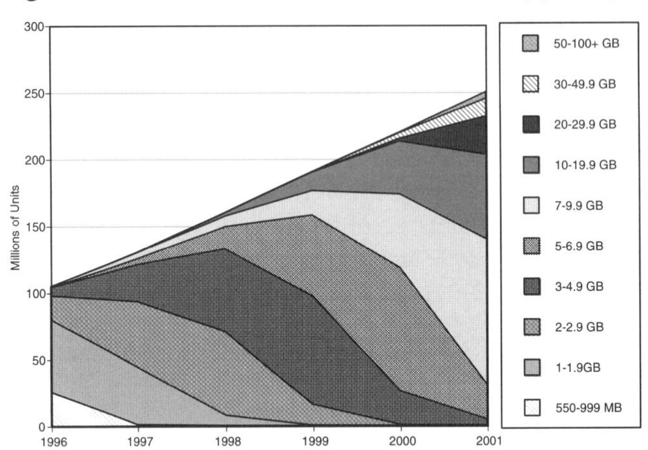

Source: IDC, 1997

CD-ROM, DVD, and Floppy Disk Drives

Although shipments of DVD-ROM drives began in early 1997, they are not expected to surpass CD-ROM drive annual shipments until after 1999, according to Disk/Trend. DVD-ROM prices are falling much faster than initially predicted. Instead of the $325 to $250 per drive many had forecast, some vendors are selling drives for as little as $100, a price point not anticipated until mid-1998. Like CD-ROM drives, DVD-ROM drives were originally introduced at the same speed grades as commercial CD or DVD players. These 1X drives have already begun to give way to higher-margin 2X drives, offering about twice the performance.

The removable disk drive market is expected to grow, fueled in part by the success of Iomega's 3.5-inch 100-MB Zip drive. Sales of high-capacity floppy drives reached 5.1 million units in 1996. By 1999, analysts project that high-capacity removable floppy disk technology will have begun to supplant today's standard, 3.5-inch 1.44-MB drives – depending on whether a single standard has emerged by then and whether the ability to read 1.44-MB floppy disks continues to be needed.

In late 1997, prices for removable disk drives ranged from $200 for SyQuest's 1-GB SparQ drive to $650 for Iomega's new Jaz 2-GB drive. Nomai's new 750.c drive was priced at $250 (internal) or $300 (external) for a 750-MB system that is as fast as the Iomega's original 1-GB Jaz drive. Media costs range from $99 per 3-pack (for SparQ disks) to $150 per 3-pack (Jaz disks) or $70 per cartridge for Nomai's 750.c drive.

Enterprise Storage

Although shipment growth rates in the storage systems market have declined from 1995's 70 percent capacity growth, they are expected to remain robust through the forecast period, producing enormous enterprise user storage portfolios and management headaches. Although the cost of storage will be relatively minor (about $0.07 per MB in 2002 and less than $0.01 per MB by 2007, or 10 percent and 1 percent of current pricing, respectively), automated management across the enterprise will become the major challenge.

IDC reports that in 1990, IBM held 76 percent of the worldwide market for mainframe disk storage. Between 1991 and 1993, EMC's share of the mainframe storage market grew more than four-fold to challenge IBM's position. By the end of 1995, as a result of the success of its Symmetrix products, EMC had surpassed IBM, capturing 41 percent of the market, compared with IBM's 35 percent. In the following years, EMC's market share grew steadily.

EMC and Hitachi will continue to benefit from IBM's decreasing enterprise storage market share, according to the META Group. With the introduction of IBM's new Seascape architecture in 1997 and products that will be released in 1998, however, IBM is expected to re-establish momentum and credibility as a fully integrated enterprise storage vendor. (See Figure 3-11.)

Figure 3-11: Mainframe Storage Market Share

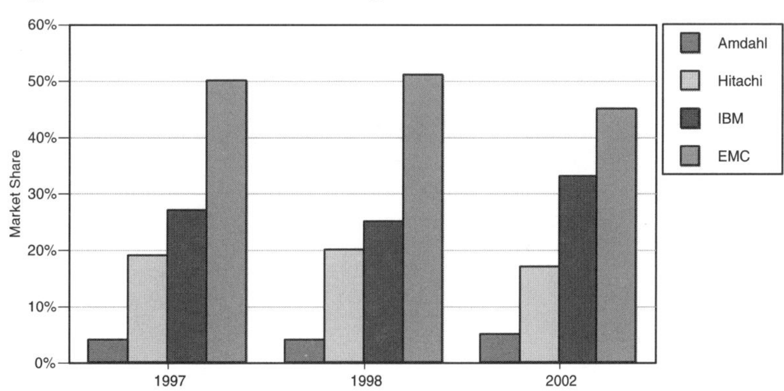

Source: META Group, 1997

The trend toward both physical and logical consolidation of heterogeneous storage carries tremendous opportunity for vendors as well as end users. Sales of storage systems open to multiple server platforms jumped from $5 billion in 1996 to $8 billion in 1997, according to IDC, which calls this the biggest trend in the "open" storage market. Several vendors, including Data General and Sun, strengthened their disk arrays in 1997 to challenge market leader EMC in the open systems storage arena.

Batteries

The worldwide battery market is extremely diverse. Battery uses range from automobiles, to games and power tools, to portable computers. Of the three market sectors – automotive and industrial, disposable consumer, and rechargeable – the high-performance rechargeable sector is growing fastest. Two forces are shaping the rechargeable market: increased sales of portable electronic products and environmental concerns about the chemicals released when primary and NiCd batteries are disposed. Such environmental concerns have led to increased use of NiMH batteries as well as a surge in new battery chemistries.

The increasing numbers of mobile workers who use portable computers and cellular phones translates to dramatic growth in mobile computing devices. Although batteries and charging systems represent less than 5 percent of a notebook computer's total cost, the importance of longer run times makes battery capability a powerful factor in purchase decisions. Battery weight is also a significant issue for users. The laptop/notebook battery market will see some major shifts as manufacturers compete to develop reliable batteries capable of fueling more-powerful notebooks without overheating (or even worse, igniting).

Two potentially significant developments could affect the direction of the battery market during the forecast period. First, airlines are beginning to wire plane cabins for DC power that can be used to plug in a PC. This trend could eliminate much of the pressure for longer battery life in notebook computers because air travel is the main situation in which users are away from AC power for extended periods of time. Second, the role of hand-held computers is becoming more significant, and the consumer expectation is that these devices will run for a long time (at least a day and possibly a month) on a single set of batteries. Unfortunately, once a modem in the form of a PC Card is inserted in a hand-held computer, battery life drops sharply.

Figure 3-12 shows IDC's shipment forecast for portables worldwide.

Figure 3-12: Worldwide Forecast of Portables by Battery Type

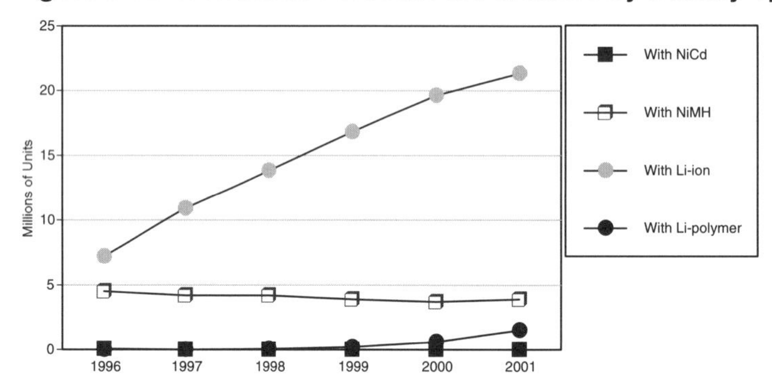

Source: IDC, 1997

Dataquest reported in early 1997 that Li-ion batteries were making inroads in the battery marketplace. In fact, Li-ion has shown tremendous growth, with market share jumping from 1.5 percent in 1994 to a projected 72.1 percent in 1997 and a projected 82 percent in 1998. Dataquest also predicts that NiMH technology will improve in 1998 and will still be incorporated in many lower-priced notebooks.

Lithium-polymer technology is still suffering manufacturing difficulties in the U.S. In the near term, it will command a high price but will offer no cell density advantages over Li-ion.

3.2.10 Forecast

Storage

- Magnetic disks and tapes will continue to be the primary mass storage devices for all processor groups well into the next decade. Magnetic disk drives will be used for active data access; streaming magnetic tape will function chiefly as backup and archival media.

- Areal density of magnetic hard disk drives will increase at an annual rate of about 60 percent.
- Price per megabyte of disk storage will continue to fall by 50 percent every 15 to 18 months. High-end server disk drive capacity will reach 30 GB in 1998.
- Multimedia applications will continue to create growing demand for storage in both the consumer (games and entertainment) and business (training and documentation) markets. Television and motion picture production have emerged as leading applications requiring terabytes of on-line storage.
- Low-cost, removable disk drives from Avatar Systems, Iomega, Nomai, and SyQuest will increase market penetration significantly in both the home and business PC markets. Capacity will grow from approximately 2 GB in early 1998 to approximately 8 GB by the end of the forecast period.
- DVD is expected to make significant inroads into the CD-ROM market beginning in 1998. However, CD-ROM will remain the standard medium for software distribution throughout the forecast period. The adoption of DVD-RAM will be delayed by competition between differing standards.
- The adoption of writable optical media (CD-RW, DVD-RAM) will be forestalled by confusion about multiple media types, the lack of standards, delays in product availability, and the falling prices and increasing capacities of magnetic disk storage (both removable and non-removable).
- Enterprise storage systems will continue to enhance their capabilities, performing functions traditionally associated with mainframe storage management, and offloading them from the host system. These functions include offline backup, off-site storage, hierarchical storage management, data snapshot and remote copy, and are available regardless of the host operating system. These systems increasingly will make it possible to share data among heterogeneous operating systems.
- Enterprise storage systems will evolve into network-attached storage, including capabilities to access data at both the file system and the application (for example, database) level.
- High-end storage systems will shift to using Fibre Channel as their principal means of attaching to hosts.
- Firewire will become the standard for connecting desktop systems to their storage devices by the end of the forecast period.
- The use of RAID technology (particularly RAID 5) will become standard in all systems larger than an individual desktop computer. Operating systems and network and system management products will increase their ability to monitor the status of RAID arrays, alerting operations staff when corrective action is required.
- New PC systems will start to be shipped without conventional (1.44 MB) floppy drives, with higher-density floppy drives (initially in the 100 to 250 MB range) substituted as an alternative.

Batteries

- During the forecast period, Li-ion batteries will remain the leading rechargeable battery type. However, NiMH will continue to be used in low-end notebooks.
- Although lithium is already the technology-of-choice, the more stable lithium-polymer batteries will remain unattractive due to a high price point.

- As laptop use increases, business and consumer demand as well as the need for PC vendors to cut costs will drive the standardization of battery form factors.
- New zinc-air battery technologies will allow users a full day's power using a standard notebook PC on a single charge by 1999.
- As airlines make DC power available in their cabins to notebook computer users, the focus of battery innovation will shift to size and weight.

3.2.11 References

■Articles

Bassock, Gil. 1997. Running on empty? *Mobile Computing & Communications*. June: 76-85.
De Nike, Kristina and Rik Myslewski. 1997. The future of storage. *MacUser*. May: 62-68.
Dillon, Nancy. 1997. CD-rewritable drives take leap forward. *Computerworld*. October 13: 67.
Hitachi Maxell squeezes Li-ion into card size. 1997. *Electronic Engineering Times*. May 19: 54.
Hurwicz, Mike. 1997. Superfast peripherals. *Byte*. March: 89-91.
Lammers, David and Terry Costlow. 1997. MO-drive group scales back next-generation spec. *Electronic Engineering Times*. August 4: 18.
Massiglia, Paul. 1997. RAID revolution: the dawn of enterprise storage. *Database Programming & Design*. June: 42-49.
Poultney, John. 1997. More CD-RW choices coming. *MacWeek*. October 13: 30.
Poultney, John. 1997. The Seagate tapes. *MacWeek*. October 13: 30.
Schenk, Rob and Brian Fikes. 1997. Does CD-RW have the write stuff? *Computer Shopper*. November: 602-605.
Smith, Greg. 1997. Lithium-ion units lead portable charge. *Electronic Engineering Times*. July 14: 114-116.
Stolitzka, Dale. 1997. Smart-battery technologies push design. *Electronic Engineering Times*. January 27: 66-68.
The smart battery is coming of age. 1997. *Electronic Business Buyer*. June: S17.
Wade, Will. 1997. Consortium developing smart battery specifications. *Electronic News*. October 20: 20.
Whillier, Graham. 1997. Breathing new life into SCSI. *Computing Canada*. March 3: 39.

■Periodicals, Industry and Technical Reports

Quantum and TeraStor announce strategic partnership for near field recording disk drives and technology. 1997. EDGE: Work-Group Computing Report. September 22: 10.

■Conferences

Storage Track '96. 1997. Dataquest.

■URLs of Selected Mentioned Companies

3M (now Imation) `http://www.3m.com`
Adaptec `http://www.adaptec.com`
AER Energy Resources `http://www.aern.com`
Amdahl `http://www.amdahl.com`
Ampex `http://www.ampex.com`
Avatar Systems `http://www.avsysltd.com`
Benchmarq `http://www.benchmarq.com`
Brocade `http://www.brocadecomm.com`
Circuit City `http://circuitcity.pic.net/`
Compaq `http://www.compaq.com`
Conner Peripherals (now Seagate) `http://www.seagate.com`
Cygnet Storage Solutions `http://www.cygnet.com`
Data General `http://www.dg.com`
Digital `http://www.digital.com`
Duracell `http://www.duracell.com`
ECMA `http://www.ecma.ch`
EMC `http://www.emc.com`
Encore Computer `http://www.encorecomputer.com`
Epson `http://www.epson.com`
Exabyte `http://www.exabyte.com`
Fuji Photo Film `http://www.fujifilm.com`
Grau `http://www.grau.de`
Hitachi `http://www.hitachi.com`
Hitachi Maxell `http://www.maxell.co.jp`
HP `http://www.hp.com`
IBM `http://www.ibm.com`
Intel `http://www.intel.com`
Iomega `http://www.iomega.com`
JTS `http://www.jtscorp.com`
LG Electronics `http://www.lge.co.kr`

Matsushita `http://www.panasonic.com`
Maxtor `http://www.maxtor.com`
Micro Design `http://www.microdesign.no`
Micropolis `http://www.micropolis.com`
Mitsubishi `http://www.mitsubishi.com`
Mitsumi `http://www.mitsumi.com`
MTI `http://www.mti.com`
National Semiconductor `http://www.national.com`
NCR `http://www.ncr.com`
NEC `http://www.nec.com`
Nomai `http://www.nomai.com`
O.R. Technologies `http://www.ortechnology.com`
Olympus `http://www.olympus.com`
Panasonic `http://www.panasonic.com`
Paramount `http://www.paramount.com`
Philips `http://www.philips.com`
Pinnacle `http://www.pinnaclesys.com`
Quantum `http://www.quantum.com`
RAID Advisory Board `http://www.raid-advisory.com`
Ricoh `http://www.ricoh.co.jp`
Sanyo Electronic `http://www.sanyo.co.jp`
SBS Forum `http://www.sbs-forum.org`
Seagate `http://www.seagate.com`
Sequent `http://www.sequent.com`
Sharp `http://www.sharp.co.jp`
Sony `http://www.sony.com`
Sony Electronics `http://www.sel.sony.com`
Sony Recording Media `http://www.sel.sony.com`
StorageTek `http://www.storagetek.com`
Sun `http://www.sun.com`
Symbios `http://www.symbios.com`
Syquest `http://www.syquest.com`
Tadiran Electronic Industries `http://www.tadiran.com`
Tandberg Data `http://www.tandberg.com`
Tera Stor `http://www.terastor.com`
Toshiba `http://www.toshiba.com`
Ultralife Batteries `http://www.ultralifebatteries.com`
Universal `http://www.universalstudios.com`
Vertex `http://www.vertex.com`
WaltDisney `http://www.disney.com`
Western Digital `http://www.western-digital.com`
Yamaha Systems `http://www.yamaha.com`
Zenith `http://www.zenith.com`

3.3 Smart Cards

3.3.1 Executive Summary

The term smart card collectively refers to several different types of technologies, all of which fulfill a common basic function – the portable, persistent storage of information. Smart cards facilitate secure debit and credit transactions as well as low-value, cashless transactions such as paying for transportation, a telephone call, or a road toll. Some systems do not require physical contact with the card, allowing automated toll collection, for example. Data stored on the card is encrypted, and access requires either a personal identification number (PIN) or an encryption key.

Most current smart cards simply contain a non-volatile memory chip that is used to store information such as a bank account balance securely (so-called stored-value applications). This type of card is known as a memory card and has no intelligence whatsoever; instead, it relies on an application running on a server or terminal. Functionally, the smart card appears as a small disk file system to the application. This setup requires that all transactions occur on-line, which can contribute to significant operational network costs. Increasingly, however, semiconductor smart cards contain a microprocessor and sometimes a specialized cryptographic coprocessor; such cards are known as microprocessor smart cards.

Most older smart cards are almost invariably single-application cards focused on a specific function, such as a debit card. The lack of computing power available also limits the encryption algorithms that can be used, and most single-function cards employ Triple-DES encryption. However, more recent smart cards are multi-application cards that exploit the larger memories and faster processors to support a multi-application operating system (MAOS) as well as concurrent execution of applications.

Several types of smart cards do not use semiconductor storage, but they can be used in the same way as a semiconductor memory card. The familiar magnetic stripe credit and debit cards fall into this group, as do cards storing data as two-dimensional bar codes and optical cards that use a laser-driven storage method similar to a CD-ROM. Some cards are hybrids that combine storage methods; these cards almost always contain both a magnetic stripe and semiconductor memory. Hybrid cards are becoming more common as applications such as credit cards move from magnetic stripe only to one of the semiconductor-based smart card platforms.

A range of different formats are used to package smart cards. The actual chips are embedded in a small plastic carrier module about 1 inch square. Some applications use just the carrier package – in some mobile telephones, for example, where it is referred to as a mini-card. Most familiar, however, are the credit-card-sized smart cards with the obvious gold contacts on the front. This type of package is used for memory and microprocessor cards, which must be inserted into a special reader known as a card acceptance device to function. This type of card draws its power from the card acceptance device through two of the contacts; other contacts are used for communications. A variant of this package is the contactless card, where communication with the reader is established using a radio frequency (RF) link. Some contactless cards are battery powered, and others derive power from the RF carrier wave broadcast by the card reader in the same way as the electronic tags that protect merchandise in retail stores.

Major barriers to the acceptance of smart cards include consumer reluctance and the costs associated with replacing existing magnetic stripe readers and other infrastructure items. The lack of computing power in older smart cards, coupled with the lack of interoperability standards, has meant that in practice, each different application required a different smart card. JavaCard will provide a viable, interoperable multi-application platform, which in turn will allow multiple functions to be combined on a single card. This system lowers deployment costs and will reduce consumer reluctance by replacing several plastic cards or paper-based systems with a single smart card.

3.3.2 Background

From as early as 1961, electronics-based cards were being created and patented in Germany, Japan, and the U.S. However, it was not until the next decade that an institution would support the technology for commercial use. In 1974, Frenchman Roland Moreno, a journalist and inventor, devised a circuit-based memory ring (to be worn on the finger) that would allow secure payments. He presented his invention to the bank consortium, Le Groupement des Cartes Bancaires (GCB), which applauded the idea but convinced him to move the chip off the ring and onto its bank card. That done, Moreno approached the French company CII-Honeywell Bull to manufacture the cards in quantity.

Bull engineer Michel Ugon was attracted to Moreno's basic concept, but thought that for security reasons, a microprocessor solution would be better for bank cards. The banks agreed and decided to back Bull through its development endeavor. Although many deemed Ugon's idea unworkable, in October 1980, Bull produced and patented the smallest microprocessor to be manufactured for a bank card. Shortly thereafter, Philips and Schlumberger released their own standards.

To choose between the three non-interoperable technologies, French banks enlisted three towns, Blois, Caen, and Lyons, to conduct retail trials. The technology winner was Bull for its open-ended architecture and tighter security. Bull went on to develop GCB's standard multiple-application "smart card."

Over the years, the term smart card has been bandied about in a variety of industries, and now it defines a whole range of products that fall into the category of integrated circuit (IC) cards, some of which are "smarter" than others. These include simple memory chip cards with little or no processing power or security, protected memory chip cards with a security overlay, and variations on the original microprocessor cards. (See Figure 3-13.)

Figure 3-13: A Smart Card Family Tree

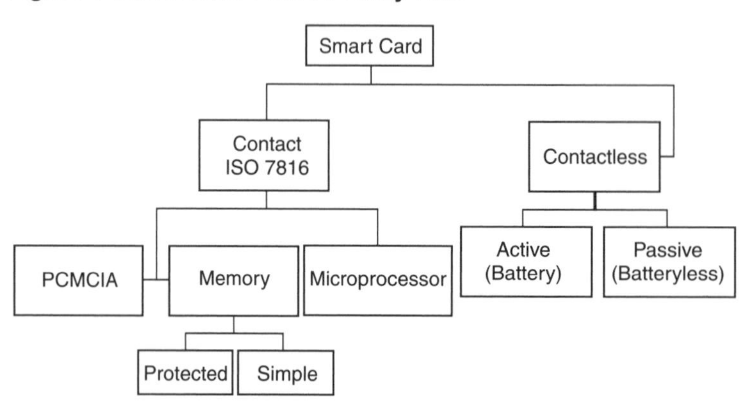

Source: Tower Group, 1997

3.3.3 Contact Cards

Microprocessor "contacts" are simply conduits for information to and from pre-designated applications in the chip itself. Magnetic-stripe terminals use electro-magnetic readers to lift information in sequence off the card stripe, and therefore do not need the microprocessor contacts to act as pointers or pathways.

Smart cards must conform to the physical dimensions of magnetic-stripe terminals even if the chip is not read by that terminal. This situation exists because converted smart cards usually carry magnetic stripes as well – to work internationally or to support existing applications that have not been migrated yet to the chip. For example, a bank card with an electronic purse chip still may use the magnetic stripe debit function.

Standard contact cards are made of polyvinyl chloride (PVC), with a thickness of 0.76 mm. The contact card uses a surface set of eight contacts, six of which carry information about a specific function. The exact layout of the contact plate is defined by the International Organization for Standardization's ISO 7816 standard. Behind the contact plate is the circuit itself on a piece of silicon. The contacts are used to provide power to the card and exchange data between the card and the terminal.

Memory Cards

The memory card is much less secure than the microprocessor-based card because it cannot execute encryption algorithms and other security functions. Therefore, memory cards are used to store information (such as in the German health card program) or for the purchase of very small-value services, such as telephone calls or subway rides. Approximately half of all memory cards are disposable and simply thrown away when the value on the card is used up.

Thanks to their low cost, high demand, and technological simplicity, memory cards have been in production for more than 15 years. Their usefulness encouraged several manufacturers to produce them independently early on, so no standard protocol was put in place for the technology. Several different protocols currently are used on memory cards, each with varying communication patterns, precluding development of a common interface or mass-market production by a single vendor.

About 880 million memory cards are in circulation worldwide today, 160 million of which are simple memory cards and 720 million of which are protected memory cards.

■ Simple Memory Cards

The simple memory card can only store data. It can, however, store ten times more data (2 KB) than the common magnetic stripe card (0.2 KB), and it can reuse memory by writing over existing data. The magnetic stripe card in use today typically uses Write Once, Read Many (WORM) technology; however, data in Electrically Erasable Programmable Read Only Memory (EEPROM) on smart memory cards may be erased and overwritten by a special electronic device. Simple memory cards most commonly are used as prepaid phone cards or in other applications that require minimal security and relatively low-cost (often disposable) media.

■ Protected Memory Cards

The protected memory card is like the simple memory card with a security overlay that allows it to be accessed only when a secret key is presented correctly. The key is a function of the serial number of the card, which impedes penetration by hackers into more than one card. The point of the secret key exchange is to ensure that both the card reader/writer and the card are authorized within the system to make transactions that allow the reader/writer to write new data onto the card. Hackers, by definition, are not part of the authorized system, and therefore, the card will not permit unknown devices to read/write to the card.

The extra security on this card makes it a viable option for a reloadable electronic purse (e-purse) scheme.

Microprocessor Cards

The microprocessor card is more intelligent than the simple memory card. It contains a microcomputer, which has the ability to execute encryption algorithms, process data, and manage data files. The card has an operating system and requires software to be written for it. Microprocessor cards generally are used in locations or applications that require high security, such as government, corporate campuses, or financial services. Many processor cards have a secure filing system capable of storing data within logically separated areas. These cards maintain strict application separation and integrity for the protection of card application providers and for consumer privacy. The architecture of the multi-application German "GeldKarte" smart card is shown in Figure 3-14.

Figure 3-14: German GeldKarte Multi-Application Architecture

(EFT/POS)	**Electronic Purse**		Additional Applications			EEPROM
• offline limit	• purse	keys	....	....	....	
• keys	• keys					
• programs	• programs	data	....	....	....	

Operating system	ROM
•communication with terminal •runtime / access control •data administration functions •security functions (cryptography, sequences)	

Source: Tower Group, 1997

There are three types of memory on the microprocessor smart card: Read Only Memory (ROM), Random Access Memory (RAM), and EEPROM.

- ROM – Used to store the fundamentals of the operating system, such as the program that turns the system on when the power supply is activated and the program that manages the password. ROM is inaccessible and unalterable by the card user.
- RAM – The fastest and most quickly accessed type of memory. RAM is the least stable of the memories, however, and loses information immediately when the power supply is cut. For this reason, RAM is used only as workspace by the processor and for temporary data storage.

- **EEPROM** – Allows data to be written and erased from a special read/write terminal. EEPROM is the foundation for stored-value applications on smart cards. EEPROM is used to store information on the card after the ROM-based operating system is placed in the microprocessor. Increased EEPROM is the basis for multiple applications on the card. As this memory grows and its cost falls, more room is available for additional providers to lease space on the card and implement their services. (See Figure 3-15.)

Figure 3-15: More Memory, Less Cost per Card

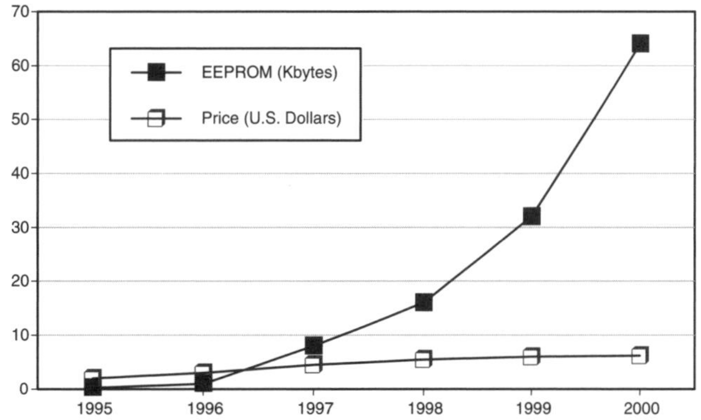

Source: Global Smart Card Advisory Service, 1997

Older microprocessor smart cards are typically based on slow, 8-bit embedded microcontrollers such as the Motorola 6800 and Intel 8051, but the current trend is to produce customized controllers containing a 32-bit RISC processor running at 25 to 30 MHz. If present, the cryptographic processor is a floating-point unit optimized for the bit-shift and exponentiation operations that are the heart of encryption algorithms. Local intelligence in the card coupled with intelligence in the terminal and secure encryption means that transactions can safely occur offline as well as on-line.

Microprocessor smart cards typically contain 1 to 32 KB of memory, partitioned into several areas. The operating system is normally stored in ROM together with any unchanging, hard-wired application code. Current cards also contain a storage area of EEPROM, sometimes referred to as flash memory. Although it is derived from EEPROM technology, which can be erased in place, flash memory is less expensive and more dense. EEPROM is persistent memory that can be wiped and then re-programmed on (privileged) command; therefore, applications held in EEPROM can be updated or replaced. Most smart cards also contain a scratch memory area that consists of a few hundred bytes of non-volatile RAM (NVRAM). NVRAM requires a relatively high power budget, and the scratch memory area is normally kept as small as possible.

EEPROM is the most expensive type of memory, and therefore, the amount of EEPROM on a smart card is kept as small as possible. This situation leads to optimizing cards for particular applications – as many fixed applications as possible are stored in ROM, leaving EEPROM available for dynamic loading of applications. This concept is particularly important to JavaCard, which provide a Java runtime environment that allows Java applications to be executed on the card.

Applications stored in ROM are fixed, and changes to them require that a new card be issued, which is both expensive and time consuming. A mask containing the operating system and applications must be prepared and tested (and in some cases, certified) before card production can begin. Each card must then be personalized and dispatched to the user. Typically, this process requires 3 to 6 months to complete.

Multi-application cards must provide a robust transaction processing environment that guarantees application integrity, and these cards must deal with the same problems of long transactions, atomicity, and rollback of failed transactions that are encountered in client/server transaction processing systems. Some multi-application operating systems segment each individual application in its own isolated partition (rather like running multiple DOS applications under Windows), and others such as JavaCard provide a UNIX-like threaded environment. In both cases, common subsystems may be shared, but application data and access to system resources are normally protected by firewalls that are implemented using a public-key encryption mechanism. An application requiring access to data must have the correct key. More sophisticated operating systems make use of the other facilities offered by public-key encryption, such as authentication and digital signatures, to provide enhanced security.

Multi-application cards control access to stored data by categorizing it into one of three groups: user-only access, third-party-only access, or public access. (Access can be controlled further by granting applications either read-only or read-write access as appropriate.) User-only data encompasses items such as a bank balance; read access requires entry of a PIN or password. Write access normally requires the application to possess the appropriate encryption key. Third-party-only data is used to store information such as medical records. Write access again requires an encryption key; read access for sensitive data such as the medical records will require a PIN or password, but information such as a driver's license may be available openly. Public data is usually unprotected.

■ JavaCard

JavaCard is a specification for a type of microprocessor card that allows execution of Java "applets" (small, self-contained applications) that have been coded in a subset of Sun's standard Java programming language. These applets can be downloaded from the terminal or server – for example, when the card is inserted into an ATM or EPOS terminal, via the Internet, or via telephone systems. Phase 2+ GSM cellular telephone system subscriber identification modules (SIMs) will be based on JavaCard technology. Users will therefore be able to access other applications such as banking facilities from their mobile telephone handsets. The handsets will download the necessary applets using the GSM short message service (SMS) protocols, and these will then execute on the SIM.

The Java language subset definition supported by the JavaCard application programming interface (API) is available freely. Java applets for JavaCard can be developed with standard mainstream Java development tools hosted on a standard PC or other platform. This process will significantly reduce the cost of smart card application development as well as lower the entry barriers by removing the need for special-purpose card programming hardware. Such benefits should generate significant growth in the number of JavaCard applets available, most of which will be third-party applications not written by the card issuer.

Specifications for JavaCard and the supported language subset are managed by the JavaCard Forum, which includes companies from smart card and terminal developers to application groups.

Because Java is a portable language, applets by definition will be interoperable and able to run on any card that uses JavaCard, regardless of manufacturer. Until now, the lack of interoperability standards has been the major barrier to deploying large-scale, multi-application smart card systems.

Several manufacturers are currently producing JavaCard-based smart cards in small quantities and plan to go into volume production in mid-1998. Prices are expected to drop to between $3 and $4 per card, comparable to the current price of memory cards. (See "JavaCard and MultOS" on page 125 for more information.)

■ Card Operating System

An 8-bit microprocessor runs code from ROM, into which the card operating system (COS) has been placed during manufacture. The COS controls runtime (on/off), communications, file management, and security protocols.

- **Runtime** – The function of activating/deactivating the card functions when power is supplied, monitoring voltage to the card and clock, and ensuring overall card integrity before a transaction is allowed to start.
- **Communications** – The communication line to the processor is half-duplex (or able to carry data from card to terminal and back, but in only one direction at a time). Processor communication is at 9600 bits per second, with the protocol handled by the COS.
- **File management** – Managing card files and applications can be handled multiple ways, each tailored to suit a particular environment and dependent on the technology available at the time. The classical smart card architecture may separate each application and its keys in physically distinct regions of the file system. Some systems may share some core data, such as cardholder name and account number, but most allow each application to manage its own data to keep the applications as self-contained and secure as possible.

Object-oriented technology is exploding onto the smart card scene. Java is an object-oriented language that is being modified to fit the smaller environment of smart cards (JavaCard); MultOS is MasterCard's response to JavaCard, but it was created specifically for use in the smart card environment. (See "JavaCard and MultOS" on page 125.) Object orientation encapsulates and protects data such that each object exists as an independent, opaque entity and never interferes with others on the card. Applications are composed by binding objects together with message-passing mechanisms, thereby allowing different applications to share data objects securely.

■ Contact Card Security Issues

Several levels of security are maintained on a microprocessor card. Initial security is built into the COS. At the point of interaction, the card and terminal operating systems first query each other for authenticity using public or private keys. The cardholder is then prompted for a PIN or other digital identification. Finally, the data itself is verified by the receiver through a Message Authentication Code (MAC) and digital signature, providing proof of the sender's identity.

Two types of encryption are commonly used for smart cards, often in conjunction: the private-key Data Encryption Standard (DES), and the public-key RSA algorithm. (See "Encryption" on page 419 for more information.)

Encryption/decryption is used throughout the life cycle of the smart card to guard against the potentially devastating consequences of card theft or fraud. Some ways in which cryptographic technology is used include initial manufacture, verification of data integrity and message non-corruption, and verification of cardholders.

- **Manufacture** – A series of keys is placed into ROM throughout the stages of card fabrication to ensure proper card handling by the manufacturers. When the microprocessor is tested before being placed on the card, the manufacturer will use its own private key to encode its authentication certificate or signature on the chip. The embedding company then decodes the data using the manufacturer's public key to authenticate the manufacturer, and re-encodes it using its own private key. When the issuer receives the card, it decodes the signature with the embedding company's public key to verify that the chip indeed was sent from its purported sender. Once the issuer has authenticated the origin of the card, it issues the PIN and application key(s) encrypted with its own private key, before distribution to the cardholder. The terminals at which the card is subsequently used have the issuer's public keys and can decrypt the PIN and unlock the card applications.
- **Verification of data integrity and message non-corruption** – Smart cards ensure data integrity by providing data checks at each end of a transaction and then comparing one result with the other. When a message is passed from the card to the terminal and received intact, a digital signature built from the sender's private key is created to be used as a receipt for non-repudiation.
- **Cardholder verification** – Verification can be done in a number of ways. Today, a paper-based signature during the sale is accepted as proof of identity, but a digital, transmissible identification is a requirement for the payments process. A PIN, known only to the cardholder, may be used to secure the information stored in the card. To gain access, the cardholder enters the PIN on a PIN pad at the merchant location. The use of PINs is common in today's smart card projects and may offer direct access to a debit or credit account or allow large sums of money to be stored in an e-purse. PINs are sometimes used in special consumer card balance readers to "lock" cards that have stored value, similar to an e-purse. The Mondex scheme is an example of such a program.

  Biometric technology is the next frontier in cardholder ID. Voice, facial structure, retina, and fingerprint recognition systems are becoming increasingly viable for cardholder verification in the payments and security industries. (See "Biometrics" on page 376 for more about their use in security systems.)

Because a microprocessor is included on smart cards, additional security for authentication purposes can be imposed by the card. For example, the Global System for Mobile communications (GSM) uses a scheme that requires entry of a "master" PIN number (of 15 digits or more) by the issuer after 3 incorrect PIN attempts to use the card; after 10 incorrect entries of a master PIN, the card is erased and deemed unusable.

Figure 3-16 illustrates the steps needed to authorize a transaction using a contact smart card.

Figure 3-16: Smart Card Transaction Authorizations

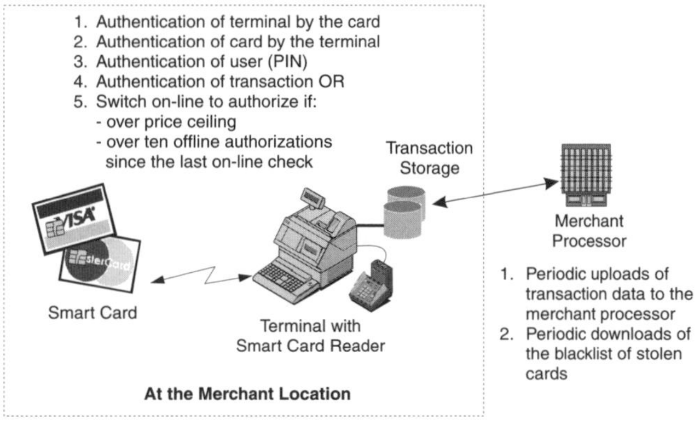

1. Authentication of terminal by the card
2. Authentication of card by the terminal
3. Authentication of user (PIN)
4. Authentication of transaction OR
5. Switch on-line to authorize if:
   - over price ceiling
   - over ten offline authorizations since the last on-line check

Transaction Storage

Merchant Processor

1. Periodic uploads of transaction data to the merchant processor
2. Periodic downloads of the blacklist of stolen cards

Smart Card

Terminal with Smart Card Reader

At the Merchant Location

Source: Tower Group, 1997

3.3.4 Contactless Cards

The contactless card was first used as a security access key to buildings that required entry authorization and logging. When placed in the radio frequency (RF) field of a special card reader, the cards instantly pass back the user identification number. Because they do not require the cardholder to insert the card correctly into a terminal and wait for card and terminal interaction, contactless cards have been recognized as prime candidates for high-traffic transit systems.

The contactless card can work in two ways: as a stored-value card or as an identification card that transmits an ID number linked to cardholder information that resides in a central database.

Today, more than 70 transportation contactless card pilot programs and rollouts are in place around the world, including highway tolls and subway and bus fare collection. The advantages of this system include the convenience of exact change, rapid ticket payment, increased surety of payment and reliability of transaction, reduced cost of extracting cash from vehicle toll booths and ticket terminals, and increased data regarding frequency of transit use by clients. However, the price of contactless cards (nearly three times that of similar contact cards) has not yet fallen to the point where the mass market will adopt the cards for use in all types of applications.

Contactless cards are divided into two categories: active and passive:

- **Active cards** – Feature a long-life battery, typically lithium, that keeps a constant flow of power to the card during transactions. Because active cards cannot be separated from their power source, the chance of card corruption is relatively low. Disadvantages of this type of card includes high card cost, potential battery leakage, and forbidden air travel due to their RF emissions. For these reasons, active contactless cards are not widespread.
- **Passive cards** – Passive cards derive their logic power from a DC voltage produced by converting RF energy from a special card-reading terminal. The antennae coils from the terminal and the card come together to support the flow of modulated RF signals, which are converted to supply power to the card. In a different RF frequency, data is passed back and forth between card and terminal.

  As contactless cards move away from the terminal, the relative power needed to run the processor increases by the distance cubed. Therefore, most contactless cards are memory cards, which have less RAM and need about one-tenth the power of microprocessor cards.

■ Contactless Card Security Issues

As more applications are developed to use contactless smart cards, card security becomes more critical. As with the contact card, manufacture or distribution security has been adopted to ensure that contactless cards are not cloned during the coding or embedding process. A series of encryption key changes, as discussed under manufacturing in "Contact Card Security Issues" on page 113, is used to authenticate the origin of the card.

The second security issue surfaces when reloading value on a card. Because the card is held at a distance from the terminal during the reload cycle, a foreign card placed between the smart card and terminal could intercept the value or keys. The combicard, or contact/contactless card, was developed to guard against such an interception. (See Figure 3-17.) Card reloading is done with a PIN on the contact chip. The value then transfers within the card to the contactless IC, and payment will be made by the contactless portion of the card. Combicards are the next generation of smart cards. A single card will soon be used for payments in a contact telephone or vending machine environment, and automatic fare collection in a contactless transit environment.

Distance-systems such as highway tolls typically use database schemes. The card ID number is simply sent to a centralized (protected) database for prepaid account debit. If suspicious activity is recorded – for example, more than 3 or 4 passages per day, or same-way passage without a return trip in between – the ID can be blocked or changed.

Figure 3-17: Contactless Card and Combicard

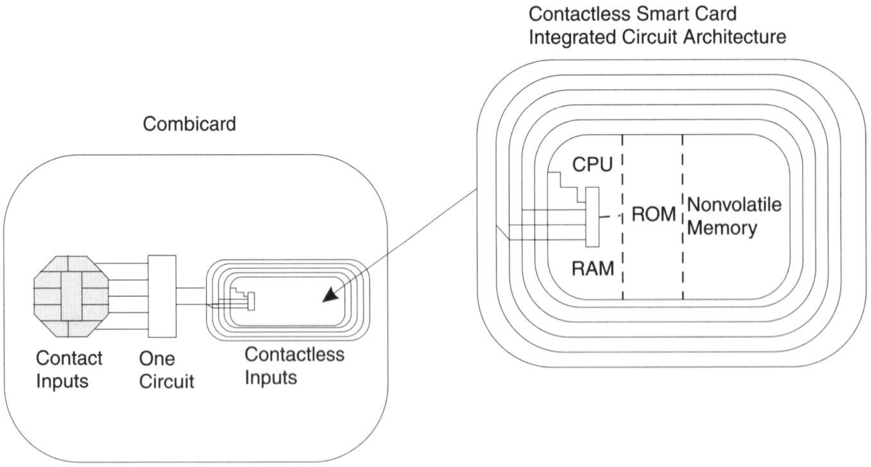

Source: Morotola, 1997

3.3.5 Alternative Identification Technologies

The silicon-based smart card has retained competitive advantage in the data storage and security markets despite an array of alternative products and technologies. The card's ever-growing storage capacity, ability to interact dynamically with the terminal, reasonable cost (as low as $1 for a simple memory card to $13 for a highly secure, multi-application microprocessor card), and ease of manufacture (leveraging core competencies of the established silicon industry) have made the smart card an appealing option for future mass-market applications.

Smart cards compete with a variety of other identification media. They must win market share from even the simplest and least-expensive form of plastic voucher as well as other enabling technologies such as magnetic stripe, bar code, RF identification, and optical laser technology.

Punched Key Cards

The punched key card has been around for more than a century. These plastic cards, punched with a number of strategically placed holes, are used for building or room access, typically for customers in a hotel. Punched key cards are very inexpensive (about $0.10 per card), but can be copied easily.

Magnetic Stripe Cards

Magnetic stripe technology has been standardized by the ISO and has been in use since the 1960s. The necessary infrastructure is well-established in the U.S. and many other countries, and the cards are likewise ubiquitous. Traditional magnetic stripe cards have three tracks, each about 3 mm wide, which carry 200 to 300 bytes of data. Stored data includes cardholder information such as name, account number, and PIN. Magnetic stripe cards are inexpensive (about $0.25 per card), but are easily forged.

Many features of magnetic stripe cards have been improved in recent years. The stripe itself has been made suitable for more data and better security. New watermark tape provides another level of card security. Watermark tape is created through a process of exposing suspended magnetic particles to a series of magnetic fields that are ordered to create a card-specific arrangement of magnetic particles. (A digital watermark serves as a form of authentication for the magnetic stripe itself, independent of the information stored on it.) The watermark tape is very difficult to forge. Finally, imprint magnetic cards have embedded the magnetic stripe inside the card plastic, making tape-lifting and forgery nearly impossible.

Bar Codes

Bar codes have been around since 1929, when a scientist from Westinghouse engineered a way to separate utility bills by region using lines or line combinations to represent characters. A photocell scanner then reads the lines from the reflected light. Bar codes have evolved significantly in recent years. One of the most sophisticated bar codes today is PDF417, a high-density two-dimensional (2-D) pattern developed by Symbol Technologies that can include up to 1,108 bytes of data, store digital images such as fingerprints or photographs, and identify and correct errors of up to 49 percent of the bar damage.

Bar codes are flexible storage media. They can be sized to fit the format of individual industries (most of which use proprietary systems), are low-cost, have no scraping contact with a reader (thereby reducing wear and tear), and can use a wide variety of media (ink/paper, thermal direct/plastic, metal etching, and so on).

Groceries, retail stores, and mail services are the traditional users of bar code technology for inventory tracking and pricing. Today, enhanced capacity has made the bar code popular even with the U.S. Department of Defense (DOD) for secure document authentication. 2-D bar codes are used to identify document tampering by displaying (through a printer or on screen) the document that was written into a symbol. To authenticate a change in a document that has been printed into a symbol, a user would have to re-issue the symbol with its new data.

Some bar code technologies can also encode information using encryption to minimize the risk of fraudulent reproduction of valid symbols and non-authorized access to restricted data.

Other possibilities for the bar code are data storage for patient files and wristbands in hospitals, voter registration documents, loyalty cards, and passports or national identification cards. Leading bar code technologies include Codabar, used by Federal Express; Interface Mechanics, used by the DOD to identify logistics materials; Scandia Imaging Systems for document scanning; proprietary technology from Symbol Technologies, the developer of PDF417; and a proprietary code format developed by the U.S. Postal Service. Recently, there have been continuing attempts to promote a single specification for bar codes for universal interoperability.

Radio Frequency Identity (RF-ID) Tags

RF-IDs have been used for several years by scientists to track farm animals. These cards are passive contactless cards that respond to RFs with a simple ID number. Philips/Mikron and TI are the primary RF-ID suppliers. In recent years, RF-ID technology has become more sophisticated and is now able to include a chip that stores supplementary personal user information. RF-ID cards are similar to contactless cards in their functionality. RF-ID tags use the same micro-packaging and security techniques as contactless smart cards; however, they have a wider range of transmission speeds and distances.

An interesting new trial for RF-ID technology is the SuperTag system currently used in U.K. supermarkets. The tags are stuck to each product, the products are placed in shopping bags, and the bag itself is passed over a sensor. The terminal is able to detect all RF-ID tags, calculate the prices, figure the total, and deactivate the radio frequency electronic article surveillance function associated with each tag.

The advantage of RF-ID tags is that the placement of tag and sensor need not be fixed in the line of sight. This feature makes the tags a good choice for tracking moving animals, and more recently, athletes during a race.

However, RF-ID tags have no widely accepted standards. Once an organization selects an RF-ID tag from a certain vendor, it must remain committed to that vendor. In addition, the tags are more expensive than other options, such as contactless smart cards. Therefore, RF-ID tags are found in environments that have fewer items about which data must be recorded – typically things of some permanence, such as pallets or containers in a warehouse.

Optical Laser Cards

Optical laser cards can hold a great deal more data than any of their competitors. Between 2 MB to 6 MB of data, including digitized graphic images and fingerprints, can be kept on an optical laser card using WORM storage. The data is burned into the card at issuance by tiny laser beams arranged in specific 8-bit patterns. WORM technology allows information to be written permanently to the card and read back numerous times. Any area of the card may be written exactly once. After data has been written to a particular area, that data cannot be erased or altered. The cards have a very long data life – at least 10 years with no fading.

Optical laser cards are the same size and shape as standard credit cards. They are used in environments that require bulk data storage, card durability, and longevity. Medical records storage, electronic banking, and bank card applications are the target markets for optical laser cards. Canon, LaserCard Systems, and Dynacard are among the front-runners in the world of optical cards. Most optical cards cost about $10 each.

3.3.6 Smart Card Applications

The first chip cards were simple prepaid telephone cards (created using memory cards) sold in Europe in the mid-1970s to combat public pay-phone vandalism. The prepaid phone card continues to thrive today among consumers in Europe and has made its way to Asia-Pacific and Latin America. The U.S. uses prepaid phone cards as well, but these generally are not smart cards; instead, most of the cards simply have a dial-in number and PIN written in ink on the plastic instead of a chip. This method is most efficient in the U.S., where the cost of the additional telephone time is negligible and line capacity and reliability are high.

Today, industries all over the world are testing smart card technology and applications. Common applications are found in health care, telecommunications, transit, and financial services, where the cards are used, often offline, as secure payment for goods or services rendered, access keys to buildings or networks, and information storage. However, smart card use has expanded to other, less-obvious fields. For example, retailers could use smart chips embedded in items for sale to control and track inventory.

Pay-per-view is a new smart card application that will be available through a set-top box. Hong Kong Telecom IMS, which has been awarded a video-on-demand (VOD) programming license, will launch an interactive TV service, including VOD, music-on-demand, and home shopping. Other services such as home banking and network games will be added as the market evolves.

A smart-card slot has been incorporated into the company's Digital Smart Box. Households that wish to subscribe to the service can expect to pay a monthly fee of about $150 to $200. VOD and music-on-demand will be charged on a pay-per-view basis, with a movie costing between $8 and $30, and a music recording about $3 to $5. A one-time installation fee may also be levied.

In addition, Internet TV service WebTV is working on using a smart card as a portable subscription timekeeper. WebTV terminals have a built-in smart card port that will eventually let users store time and use it away from home in such places as a hotel room with a WebTV terminal.

Token Cards

The telecommunications carrier France Telecom was the first to attempt a nationwide campaign (on behalf of the French government) to convert coin users to cards. Telephone cards were issued with a varying number of units written into the chip, which allowed the phone itself to read and decrement the chip for the duration of the call. Periodically, the phone would send the units upstream to a central server at France Telecom for settlement.

The smart telephone card application simultaneously reduces pay-phone vandalism and coin collection costs, encourages pay phone use by increasing consumer convenience, and accumulates substantial "float" on the outstanding, unredeemed value on the card, which serves as an interest-free loan to the issuer. Deutsche Telekom, for example, makes 200 million to 300 million DM (US$115 to $170 million) each year on float from prepaid cards. The vending machine, transportation/transit, and laundry industries are all likely candidates for similar prepaid smart card projects. The success of such projects is not assured, however. Other technologies for low-value applications are already available, such as magnetic stripe cards, that work almost as well as smart cards and do not require any investment in new infrastructure.

Offline Debit Cards

Financial services was the second major industry to embrace smart card technology. In 1968, the deferred debit card and network Carte Bleu (CB) was launched in France by five major banks. Deferred debit card purchases were automatically debited from the cardholder's account at the end of each month. The card was originally issued without a magnetic stripe, using a signature for cardholder ID, and a paper hot list for stolen cards.

In 1971, a magnetic stripe was added to allow cardholders to get cash from ATMs, but the cards did not go on-line for authorization due to the high cost (nearly US$0.25 per transaction) and unreliable nature of telecommunications at the time. Through 1987, the lack of card authorizations prompted a rise in fraud. Losses were high enough to warrant the replacement of the entire magnetic stripe infrastructure with a microprocessor-based card system. The French system worked as illustrated in Figure 3-18.

Figure 3-18: French Offline Debit Transactions

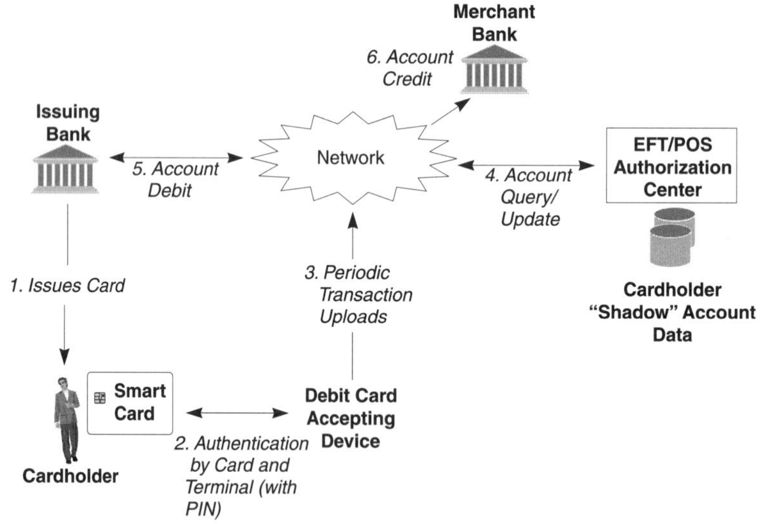

Source: Tower Group, 1997

The microprocessor chip serves as a cardholder ID, into which a PIN is encoded for authentication of the cardholder. When the cardholder inserts the card, the card and terminal perform a "handshake" to verify the other's authenticity, and a PIN is entered by the cardholder for identification. This security process has made counterfeit and ID fraud almost nonexistent.

At its peak in 1987, fraud cost France 546 million FRF (US$100 million), or 0.27 percent of all bank card transactions. Ten years later, in 1997, French fraud had been reduced to 183 million FRF (US$33 million), or 0.018 percent of card transactions. Between 1988 (the introduction of smart cards) and 1997, total fraud savings have been 13 billion FRF (US$2.4 billion). (See Figure 3-19.)

Figure 3-19: French Smart Cards Cut Fraud

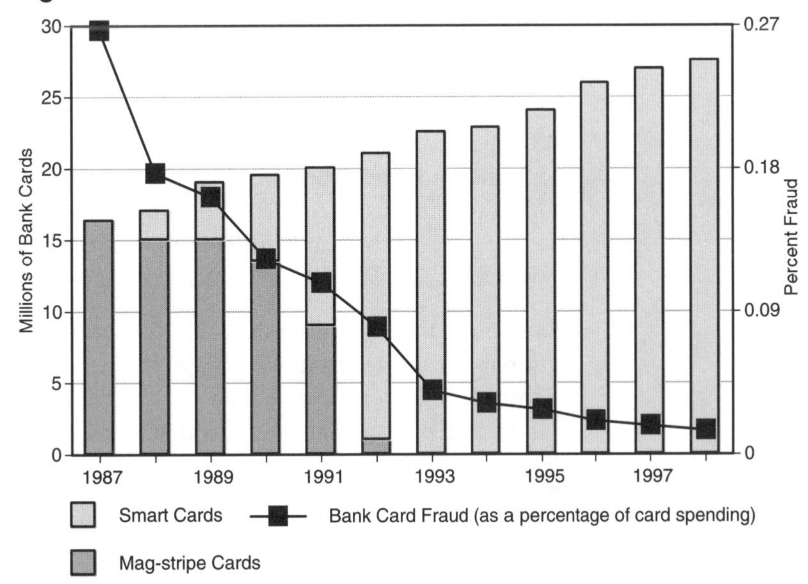

Source: Le Groupement des Cartes Bancaires, 1997

France is a prime example of early smart card adoption. However, France is unique in its motivation for using smart cards and in its method of implementation. In countries such as the U.S., where the market guides commercial applications, similar government-driven standardization and rollout is unlikely. In addition, because U.S. banks, merchants, and credit-card issuers such as MasterCard and Visa have already have spent hundreds of millions of dollars on a nationwide, ubiquitous system of on-line authorizations and transactions, smart cards as payment media at retail locations hold little appeal. Finally, telecommunications costs in the U.S. are low (averaging $0.02 per transaction, versus $0.11 per transaction in most European countries) and telephone lines are generally reliable, making on-line authorization for credit/debit cards inexpensive and easy.

The French rollout ended in December 1992, and no other country has chosen to implement a similar debit replacement system. Telecommunications deregulation has made communications costs fall around the world, and sophisticated fraud tracking systems have enhanced existing magnetic stripe infrastructures to reduce fraud loss. In addition, new applications such as the e-purse have displaced plans to convert old debit systems and prompted a rush to develop new software and processors.

Stored-Value Cards

The e-purse allows the consumer to load value onto the smart card from a checking account, credit card, or cash. The value is then deducted with each purchase at a merchant terminal and the remaining card balance is rewritten on the card.

There are two types of stored-value card implementations: auditable, in which each transaction is balanced back to the issuing bank, and unauditable, in which the transactions are passed from card to card. Almost every stored-value (e-purse) scheme in existence today falls into the former category because of the ability to monitor all transactions for fraud. The only significant e-purse scheme that does not leave an audit trail is Mondex, whose parent company, Mondex

International, is now 51 percent owned by MasterCard. Mondex is the only scheme that supports card-to-card transactions, allowing a cardholder to transfer money from one card to another by way of a calculator-sized reader.

The first major e-purse program was Denmark's Danmønt, a joint venture of PBS, the national payments organization, and TeleDenmark, the Danish telephone company. Danmønt went into pilot phase in 1992, completed the pilot in 1993, and immediately started a country-wide rollout. It hoped to create an effective program by blanketing the country with cards and card acceptance devices for small-value transactions (less than $10 per transaction).

Today, more than 1 million smart cards have been issued in Denmark, and terminals are located in more than 90 cities. Smart card readers appear in ticket machines for Copenhagen mass transit; in stamp, food, and beverage vending machines; in parking meters; in fax machines, copiers, and printers; in public and GSM wireless phones; in laundries and cafés; in PCs for Internet payment at post offices and public libraries; and in retail convenience stores. Smart card readers are even available at recharge stands for electric cars, which tempt customers by offering free parking with smart card payment.

To date, 8 to 10 percent of the Danish population use the e-purse, with an average transaction value of $2. Card-accepting merchants pay $0.03 per transaction, with an annual terminal leasing fee of $200. Figure 3-20 shows the total annual transactions in Denmark of inhabitants using Danmønt cards from 1992 to 1997. The program's success was recognized by Visa in 1995, when it chose Danmønt technology as the basis for its e-purse trials in June 1996 at the Olympic Games in Atlanta, Georgia.

Figure 3-20: Danmønt Cards and Transactions on the Rise

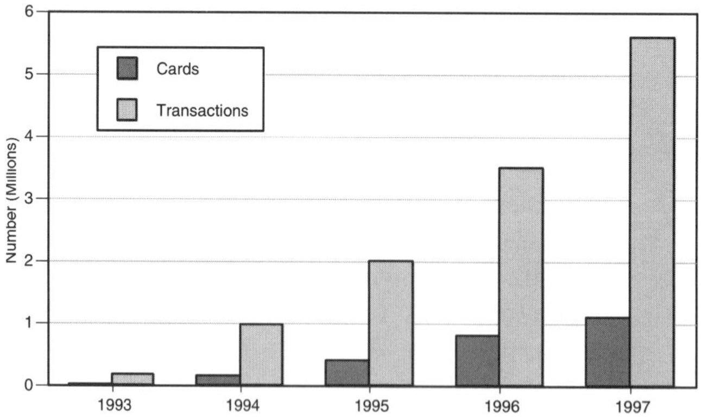

Source: Danmønt, 1997

Electronic purse programs (such as Danmønt) typically rely on the smart cards and terminals themselves to authorize and conduct the transactions. (See Figure 3-21.) In some reloadable e-purse schemes, transactions are checked against a blacklist or negative file stored in the terminal to prevent fraudulent use of lost or stolen cards. Most programs treat the value loaded on the card like cash: If the card is lost, the money remaining on the card is lost along with it.

Figure 3-21: Reloadable Electronic Purse System

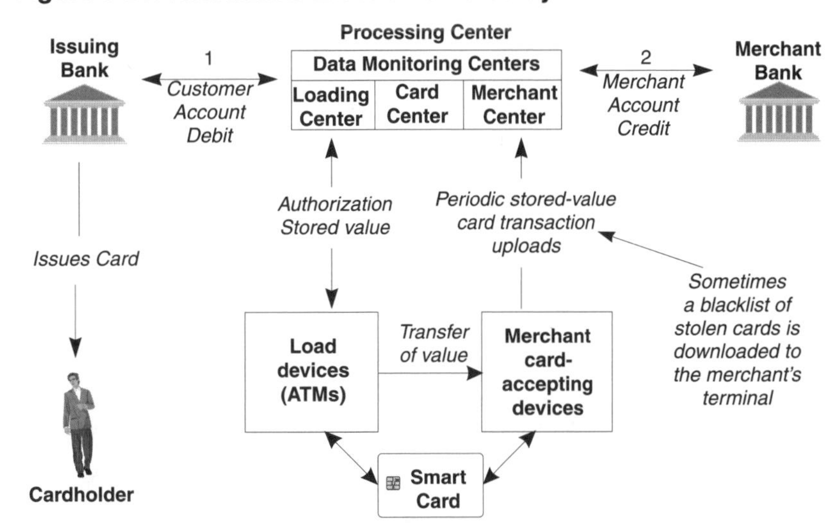

Source: SNI, Tower Group, 1997

3.3.7 Interoperability and Standards

Smart card interoperability is the most widely debated issue in the industry today. Visa and MasterCard, among others, are striving to promote their own operating systems for smart cards. In an open environment that supports a variety of industry players, common standards serve as drivers to the market, allowing basic systems compatibility and enabling each company to build added-value features on top of existing standards. Without interoperability, consumers are frustrated by the lack of merchant outlets that accept their cards, merchants are reluctant to choose a standard for their stores, and the market is restricted by fragmentation and confusion. Such was the world of ATMs and credit/debit cards at the outset. The evolution of each of these markets, however, eventually eliminated competition between technologies by establishing industry consortia or by mergers and acquisitions to create a dominant technology. The smart card, too, will go through these stages.

Industry groups, such as the Global Chipcard Alliance (GCA), are focusing on developing standards for interoperable smart card technology. GCA members (as of September 1997) include American Express, Bell Canada, Deutsche Telekom, Elcotel, Gemplus, GTE, IBM, Landis & Gyr Communications, Microsoft, Nortel, Oracle, PTT Telecom Netherlands/Unisource, SPT Telecom, Telecom Malaysia, Telstra, and US West. The first interoperability agreement was signed by PTT Telecom/Unisource and Deutsche Telekom, followed by a similar agreement between US West and GTE, which was recently expanded to include Bell Canada. According to Zona Research, GCA (founded in 1996) is nearing "critical mass," making it extremely difficult, if not impossible, for the few remaining major vendors – MasterCard, Schlumberger, and Visa – to continue their resistance to its work.

Smart card industry standards groups have produced several significant proposals, including EMV, C-SET, JavaCard, and MultOS.

EMV

In the early 1990s, Europay, MasterCard, and Visa formed the EMV (Europay/MasterCard/Visa) consortium, working together to develop industry-wide chip card specifications to ensure interoperability of smart cards and card-reading terminals. In 1994, the EMV consortium published version 1.0 of the EMV "integrated circuit card (ICC) specifications for payment systems." In 1995, the group published version 2.0 of the EMV specifications; version 3.0 followed in July 1996.

The EMV specification is a series of communications protocols for credit/debit applications on the smart card, with the goal of creating a global payments framework using smart cards. The final version (version 3.0) of the EMV specification consists of three documents: the ICC card specification, the ICC terminal specification, and the ICC application specification. The most recent additions address optional card and terminal processing methods and applications. Fujitsu announced in July 1997 that it is developing three new smart cards and a new card reader/writer that all support the EMV protocol.

The Visa Integrated Circuit Card Specification (VIS) is a customization of the EMV specification. It describes the minimum requirements for Visa chip cards and terminals and defines Visa's requirements for implementing chip card programs for credit/debit products.

Card-Secure Electronic Transactions (C-SET)

C-SET architecture combines MasterCard and Visa's SET standard with chip card technology. The protocol is derived from Secure Electronic Transactions – the MasterCard/Visa standard for electronic transactions that uses direct access to an existing bank account or a virtual bank account. SET simply validates credit cards, but not the holder of the card. C-SET requires users to input a PIN, thereby identifying themselves to the transaction recipient as well.

With C-SET, cryptography on smart cards and card readers attached to PCs form the core security component for sensitive information transmission over the Internet. Security software no longer resides only in the consumer's PC. This new format has a hardware element and a software element. Consumers will be able to carry their cards between card-reading PCs and use them easily for Internet transactions. C-SET can also be used for home banking because it covers all non-face-to-face transactions. (See "Alternatives to SET" on page 577 for more about C-SET.)

C-SET makes it easier for Internet payments to receive the blessing of the authorities. Providing cryptography that is inside a "trusted" hardware environment minimizes potential criminal abuse of the system.

Two C-SET pilots are underway in France. The first involves several Cartes Bancaires banks, along with Europay France, in a pilot to include about 10,000 cardholders. A second project involves other Cartes Bancaires member banks and Kleline, the credit-card subsidiary of Banque Paribas, which sells its own Internet payment system. Across Europe, Europay France is working with Europay International on a C-SET pilot with other European banks. Cartes Bancaires is working with Belgium's Banksys card association to develop the first interoperable C-SET standard, funded by the European Commission.

C-SET is expected to have a significant impact on the future development of Internet-accessing smart cards, although whether it will unify or further fragment the industry is still unclear.

JavaCard and MultOS

As a solution to smart card hardware and software incompatibility, Sun developed and licenses JavaCard, a version of the Java programming language designed for smart card operating systems. The JavaCard specification allows the smart card manufacturer to include the Java Virtual Machine (JVM) in the card operating system, which provides a common interface for any Java applet that might be downloaded onto the card.

The new Java programming language serves several functions. First, it addresses issues of incompatibility because the JVM allows any card platform to support Java. Second, application time-to-market is reduced significantly by allowing anyone who can program in Java to develop applications for the smart card. In contrast, a handful of smart card-specific programmers now strive unsuccessfully to keep up with the demand for new applications. Third, applications will be downloaded easily at the level of the issuer, allowing the issuer to customize particular cards to fit the needs of particular customers.

In October 1997, Gemplus announced that it had created the first smart card to use a 32-bit RISC microprocessor, based on the JavaCard 2.0 specification. The new GemXpresso, scheduled for release in early 1998, is targeted toward developers. Because the most advanced smart card has previously been limited to an 8- or 16-bit microprocessor, GemXpresso's 32-bit platform could ease the integration of the smart card and IT industries.

To take advantage of the new Java technology, Visa formed a JavaCard consortium with Sun and others interested in licensing JavaCard for future versions of their smart cards. In response, MasterCard formed a consortium with key industry players (called MAOSCO) to develop MultOS (Multi-application Operating System). The MultOS platform is being designed as a non-proprietary, open platform and will be EMV-compliant. Card applications can be written in C or C++ and then converted using MultOS Executable Language, an object-oriented programming language similar to C and C++. MultOS, in fact, can support JavaCard 2.0 and a JVM, and will be able to execute Java applets on the card by early 1998.

A pilot project begun in October 1997 in New York by MasterCard and Visa is testing a new hybrid chip/magnetic-stripe card-based system. The system, due in early 1998, is designed to create an infrastructure for more secure, more intelligent debit cards along with a new class of electronic commercial applications to run on them. The system includes MultOS (on a Gemplus card with an 8-KB Hitachi chip), a set of specifications that member banks can use to support the cards, a pair of enterprise applications for bank financial systems (due in 1998), and support and services to help banks deploy card-based applications. The new hybrid cards, which use the established U.S. debit card infrastructure, could make the transition to smart cards easier.

However, Microsoft attempted to garner market share from Sun by announcing its own smart card programming standards. Microsoft's Smart Card Software Development Kit (SCSDK) is based on preliminary standards developed by the ISO and specifications developed by the PC/SC Workgroup, a group of technology companies interested in establishing interoperability standards for smart cards and computers. The PC/SC Workgroup includes Bull PTS (Personal Transaction Systems, a division of Groupe Bull), HP, IBM, Microsoft, Schlumberger, Siemens-Nixdorf, Sun, Toshiba, and VeriFone. Microsoft's SCSDK features device-independent APIs to enable the development of smart card-aware applications. Support for the SCSDK has been announced by Fischer, Gemplus, and Litronic in addition to PC/SC Workgroup members.

3.3.8 Development Issues

Multiple application cards, which are generally media that allow several providers to offer complimentary services to consumers, have demonstrated the most potential for revenue – and for trouble among application operators. A bank, for example, may issue a chip-based electronic purse card and lease space on the card to the local telephone company for use in public phones, or to a retailer with a points-based customer loyalty program. Issuer revenues increase from licensing fees and enhanced customer loyalty. Cost sharing helps decrease the burden to the individual provider, and increase revenue potential for all involved. Selling advertising on a smart card is another possible revenue-generator.

However, designating the sharing of benefits and responsibility for applications can cause problems among partner companies. If the card does not function properly in one of the user environments, who is responsible for problem rectification? Do separate maintenance entities (and therefore call centers, cardholder protection policies, levels of service, and so on) exist for each application? How is the card capacity shared among the participating organizations? Are each of the secondary "lessees" relegated to a small brand icon on the back of the card? Is there confusion among consumers as to the participating retailers? Do cards have to be reissued every time an application is added?

Legal and policy issues are equally difficult to resolve. Control of cardholder information is the issue that has captured the attention of most consumers. Should the primary issuer alone have access to consumer demographic information stored on the card's chip? Are application owners allowed to see how the consumer has been using the card outside their own applications? For example, should Phone Company X be able to see how much coffee the cardholder bought with the card at Café Y? How much information should the issuer be able to gather about the cardholder? Can a bank capture every retail transaction, trace it to its source, and develop sophisticated consumer profiling models based on the information? Could this information then be used to grant or deny credit? The final, and most pressing issue, is the status of electronic cash. Can cards be issued that simply pass value back and forth among themselves, or must each transaction be cleared centrally to guard against fraud? The Mondex system works from card to card, but some issues are still not resolved regarding misuse.

Smart card technology has matured during the last 15 years. However, a continuing lack of interoperability among smart card programs, a lack of standards to ensure interoperability, and a lack of perceived benefits to consumers have become the most significant obstacles to smart card acceptance. Would-be issuers contract with card and terminal manufacturers to use a single set of specifications in the manufacturing process. This agreement ensures card/terminal compatibility for that program, but not interoperability with smart card projects elsewhere. Given the numerous proprietary operating systems promoted on the market today, the choice of smart card interface standards for interoperability is a topic of much speculation and debate.

3.3.9 Market Overview

About $6.8 billion was spent worldwide in 1997 on smart cards and devices. By 2000, that figure will be nearly $12 billion. Today, smart cards are being used in a wide variety of applications, including loyalty programs, identification/access systems, and for information storage.

Loyalty Programs

Loyalty programs, which originated in the airline industry, are becoming more widespread. Until recently, points accumulation was limited to a single retailer, and simple paper or plastic punched or stamped cards were the only tokens to prove frequent participant status. Increasingly, however, inter-industry alliances within a given lifestyle segment are combining loyalty programs to capture market share within that sector. For example, many frequent flyer programs are now linked with frequent stayer (at hotels), frequent auto renter, or frequent diner programs and even telephone loyalty programs.

Paper-based stamp cards are no longer able to separate and track combined loyalty programs with efficiency. Microprocessor smart cards allow individual applications to reside side-by-side on the card, store points from various retailers, and store/transmit enough cardholder information so that retailers can market more effectively.

Many more functions can be combined on the smart card, making it a "lifestyle" card that functions in multiple environments. The card will be marketed to specific consumer profiles such as "business traveler" or "child." American Express, for example, is piloting a new smart card loyalty program that will allow points earned from stays in Hilton hotels to be stored on the card and used interchangeably with frequent flyer miles on American Airlines. At special self-service kiosks in airports and hotel lobbies, cardholders will be able to check their points accumulation, purchase American Airlines tickets on their American Express accounts, reserve their plane seats, store their boarding passes, check into the hotel, request a hotel room key, and pay the hotel bill without ever checking in at the desk. They might even set personal preferences (for example, a non-smoking room on the top floor near the elevator, with a king-size bed) or register complaints. Future industry partners will include other companies that cater to business travelers. A car rental agency, cellular phone company, or Internet service provider might lease application space to expand their markets and help lock American Express customers into their service.

Identification/Access "Keys"

For years, smart cards have served as physical access keys to secure buildings that require entrant identification and traffic flow tracking. Today, in a world that relies increasingly on intelligent network technology and centralized application/information storage, smart cards may serve less as multi-function tools for the end user and more as logical access keys to networks. Although the Internet is the most obvious network environment, corporate intranets and LANs, GSM telecommunications wireless networks, and satellite tracking systems all make use of networks to pass identification information. Smart cards will become increasingly useful in on-the-spot customization of a given computer system for a given user. PC smart cards may store information about the Start Up menu, call up certain applications on the network (such as Microsoft Word or Netscape Navigator), and instantly retrieve documents previously saved by the user.

GSM smart cards are among the fastest-growing market segment. They offer the same kind of customization as PC cards – serving as identification to the GSM network and also personalizing the phone in use at the time. GSM cards act as Subscriber Identification Modules (SIMs) and can support multiple value-added functions such as frequently called numbers and calling-card information. New GSM phones come with a prepaid function on the SIM, allowing anyone to take advantage of the technology. The consumer sends in a check, receives a card, and can make calls until the prepaid value runs out. Reloading (adding value to) the card can be done at special terminals.

Information Storage

Smart cards can store vast quantities of information in a portable and convenient medium. Cardholder health information, for example, is one of the applications most in use today. Several countries in Europe, including France, Germany, and Spain, have rolled out millions of health cards to all citizens. Some of these schemes simply use a smart card ID function to access a central medical database. Some, however, store emergency medical information directly on the card. This setup allows doctors to know instantly (with the help of a hand-held card reader) whether the patient (who may be unable to communicate) has diabetes or is allergic to penicillin, for example. Blood type, allergies, next-of-kin, insurance, primary care physician, and other important data may all be stored on the card.

New Markets

The promise of smart cards has opened up an array of new markets in terms of both industry and geography. Vendors are seeking additional revenues using smart cards as enablers in emerging industries such as Internet-based electronic commerce, partnership loyalty programs, and identification for intelligent networks.

Smart cards are also becoming tools for companies trying to promote their brand across industries. For example, Spain's banks have recently convinced Telefonica, the national telephone company, to accept their cards and pay the banks for the service. As soon as the cardholder loads telephone value onto the card at a bank ATM, the bank transfers the value to Telefonica (so that Telefonica can take advantage of the float). Clearly, though, the banks have their brand name associated with the payment card for subsequent phone calls.

■ The Internet

The most recent development in electronic payments is an Internet payment function, enabling e-purse smart cards to send money over the Internet to pay for small-value items. Using a smart card rather than a PC for Internet payments offers cardholders additional mobility and security. Value loaded onto the card from the cardholder's debit account can be spent incrementally in a secure Web environment. Protected Web pages that cost a few cents for each viewing could be paid for directly from the e-purse, without risk to the vendor and without undue small-value clearing and settling by the bank. Growing Internet payments security should produce more Web sites offering small-value goods and services for sale.

■ Emerging Geographical Markets

In many respects, smart cards are optimal for use in underdeveloped countries, in which telecommunications infrastructure is neither ubiquitous, cheap, nor reliable. The ability to perform secure offline transactions and to transport large cash volumes in a PIN-protected environment without fear of theft are two reasons why smart cards are emerging in Latin American and Africa.

Nigeria, for example, has an e-purse project that allows the cardholder to download up to $120,000 onto the card. The card is then locked and can be opened with the PIN only during transactions. However, using a smart card to store large amounts of value may pose the risk of smuggling and money laundering activities. Card projects launched in these environments must take careful steps to minimize such risks.

Countries such as Thailand are creating national identity cards that serve a variety of functions. Based on magnetic-stripe technology, the new ID cards can be used as ATM cards and also contain information about medical services, taxes, and personal data such as marital status, voter eligibility, and weapons permits. The cards are linked to a central database, and the system is based on Sun servers and a client/server architecture. Eventually, the government plans to allow cardholders to travel throughout Southeast Asia using the ID card instead of a passport.

Industry Players

The smart card industry has ballooned in recent years. Early smart card chip and card manufacturers and systems integrators were primarily French companies, such as Bull, Gemplus, and Schlumberger/Sligos, all of whom have remained at the forefront of the industry. Increasingly, however, Japanese and U.S. companies are gaining market share. Major smart card chip manufacturers include Hitachi, Motorola, Siemens, and a number of small niche market companies such as General Information Systems (GIS) in the U.K. Most chip manufacturers will also supply complete cards and related devices such as terminals and CADs. (See Table 3-1.)

Table 3-1: Smart Card Chip and Card Maker Alliances

Chip Makers	Card Makers
Atmel	Bull
Hitachi	Dai Nippon
Motorola	Giesecke & Devrient
Philips	Mitsubishi
Philips/Mikron	Gemplus
SGS Thomson	OKI
Siemens	Orga
TI	Schlumberger/Sligos SGS Thomson Siemens Sony

Source: Tower Group, 1997

Today, most applications developers are waiting to ensure that standards are in place for interoperability, that costs drop substantially, and that card/terminal ubiquity is achievable before committing to the technology.

There are currently about 1.1 billion smart cards in the world, the vast majority of which are stored-value cards or stored-value substitutes, such as tokens. Current worldwide smart card usage is shown in Figure 3-22. (For lack of an accurate measurement, "smart cards in circulation" includes disposable cards issued in 1997 and reloadable cards issued in 1996 and 1997. Therefore, a 1:1 ratio of cardholders to cards does not exist; instead, there are actually far fewer cardholders than cards.)

Figure 3-22: Worldwide Smart Card Usage: 1997

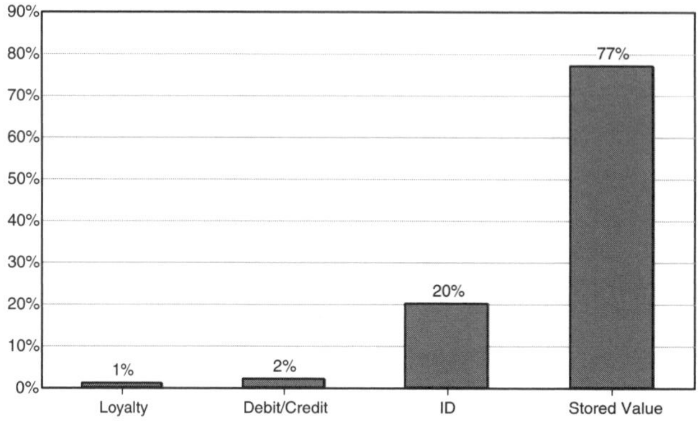

90%
80%
70%
60%
50%
40%
30%
20%
10%
0%
1%
2%
20%
77%
Loyalty
Debit/Credit
ID
Stored Value

Source: Tower Group, 1997

The compound annual growth rate (CAGR) of smart cards worldwide is estimated to be 28 percent over 3 years, resulting in a total of 2.3 billion smart cards by the year 2000. Estimated annual issuer IT expenditure is about $1.9 billion. (See Figure 3-23.) Total card issuer IT spending for new cards in 1997 will reach more than $1.9 billion worldwide, which averages slightly more than $4.40 per new card. According to the Tower Group, this price factors in the costs of the chip, the card, the card operating system, card applications development, and account maintenance.

Figure 3-23: Worldwide Smart Card Issuer IT Spending: 1997

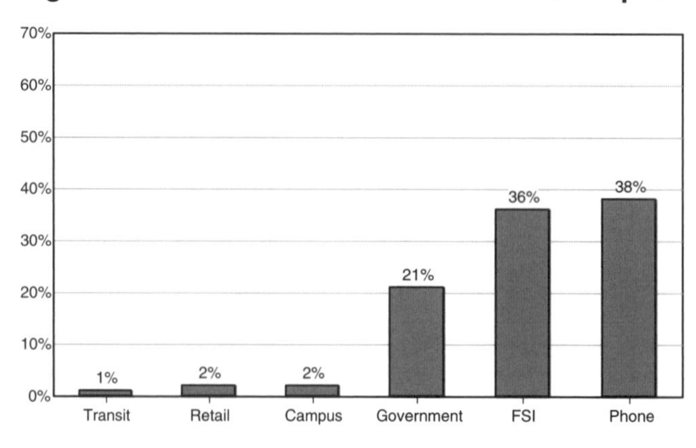

70%
60%
50%
40%
30%
20%
10%
0%
1%
2%
2%
21%
36%
38%
Transit
Retail
Campus
Government
FSI
Phone

Source: Tower Group, 1997

Today, most card schemes are single-application. Worldwide card usage by type of application is shown in Figure 3-24.

Figure 3-24: Cards, Sectors of Activity, and Applications

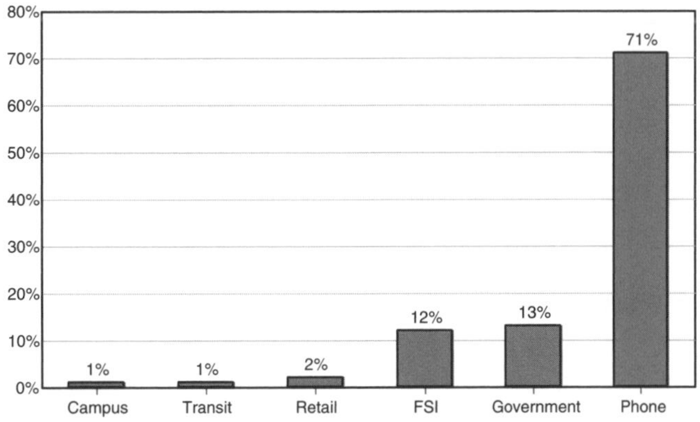

Campus	Transit	Retail	FSI	Government	Phone
1%	1%	2%	12%	13%	71%

Source: Tower Group, 1997

Single-application cards are dominant today, but as issuers struggle with the business case, multiple-application cards are emerging as a cost-sharing, value-added solution. By the year 2000, about 30 percent of all cards will support multiple applications.

By the year 2000, Europe will represent 50 percent of the smart card market, compared with 90 percent in 1997. Asia currently is about 8 percent of the market, and will increase its market share to 35 percent by 2000. Latin America is 0.8 percent of the market today, and the Middle East/Africa is a mere 0.2 percent. North America is lagging far behind other industrialized regions at 1 percent today, but it is expected to grow to 8 percent of the total market by 2000.

3.3.10 Forecast

- The U.S. will continue to lag behind the rest of the world in the adoption of smart card technology, in part due to inexpensive and widely available communications networks that can be used for credit/debit card authorization.
- In the U.S., smart cards will be found mostly in closed systems environments, including university campuses, transit systems, and military bases throughout the forecast period. From the outset, interoperability will not be an issue for most closed systems.
- Several countries (most notably Malaysia) will introduce national multi-purpose smart cards for governmental, medical, and financial functions during the forecast period.
- Smart cards for corporate network security will be the next major market, possibly in connection with the increased adoption of network computing. Smart cards will be used for highly secure IDs and execution of cryptographic functions that permit access to networks, applications, and data.
- Although most debit/credit projects have not been retrofitted to be EMV-compliant, most developers agree that EMV will become the universal standard for future incarnations of the smart debit/credit card.
- Smart card software environments will continue to move into the world of object-oriented technology, taking advantage of the security and flexibility of Java-like programming. In the short-term, operating systems will follow in the footsteps of PCs, becoming more robust, faster, and more secure, with an increasing number of applications available for use on the card. However, the long-term future of smart cards is simplification – to provide an access ID into secure networks, which themselves will store the bulk of applications and data.

- Given the significant advantages of JavaCard, it is likely to become the dominant smart card platform within the forecast period.

- Smart cards will be deployed in connection with the extension of the SET protocol to the realm of in-person credit/debit card transactions, allowing safe and efficient payments by retaining the cryptographic keys and algorithms within the card, thus preventing their compromise by the POS device. This is an application for which smart cards are ideally suited, and will be the driving force behind smart card introduction into the U.S.

3.3.11 References

■Articles

Blythe, Ian. 1997. Smarter, more secure smartcards. *Byte.* June: 63-65.
Microsoft joins smart-card race. 1997. *Computerworld.* August 18: 8.
Moeller, Michael and Scott Berinato. Big boost for smart cards. *PCWeek.* July 14: 1-2.
Pendery, David and Bob Trott. 1997. Microsoft sharpens its smart-card development plan. *InfoWorld.* August 18:21.
Santoni, Andy. 1997. Support for smart cards gets stronger. *InfoWorld.* September 22: 32.
The smart card gets even brainier. 1997. *Business Week.* November 3: 166C.
Watson, Thomas. 1997. Chipcards need standards to get smarter. *Computing Canada.* September 15: 18.
Williams, Martyn. 1997. *Hitachi to launch MultOS products early next year.* Newsbytes News Service. May 16.
Woods, Bob. 1997. *MasterCard formally launches global smart card platform.* Newsbytes News Service. May 15.

■Periodicals, Industry and Technical Reports

The Global Smart Card Advisory Service. 1997. *Smart cards: vital statistics.* The Tower Group and Centum Consultancy.
Weaver, Cynthia. 1996. *Smart cards in banking: the future of money?* The Tower Group. December.
Weaver, Cynthia. 1997. *Smart cards in the U.S., an infrastructure cost analysis.* The Tower Group. February.

■Books

CardTech/SecurTech. 1996. Conference Proceedings, Vol. I & II. Rockville, Md.: CTST Inc.
Banerjee, Ramanuj, Penelope Ody, and Richard Poynder. 1995. *The case for smart cards.* London: IBC Business Publishing.
Allen, Catherine and William J. Barr (editors) 1997. *Smart cards: seizing strategic business opportunities.* Times Mirror Higher Education Group.

■On-Line Sources

Doaré, Hervé. 1997. *International harmonisation of health cards – EU and G7 initiatives.* `http://concord.cscdc.be/conference/abstract/3_1030_5_4.html`
Gemplus announces GemXpresso Rapid Applet Development. 1997. Gemplus. `http://www.gemplus.com/presse/gemxpresso_uk.htm`
JavaCard API 2.0 specification finalized. 1997. Sun Microsystems. October 16. `http://www.sun.com/smi/Press/sunflash/9710/sunflash.971016.2html`
Kirstetter, Jim and Scott Berinato. 1997. *MasterCard puts money on chip card strategy.* PCWeek Online. October 13. `http://www.zdnet.com/pcweek/news/1027/31emcard.html`
Schlumberger selected by Visa to develop the Open Platform card, based on JavaCard API 2.0. 1997. Schlumberger. `http://www.schlumberger.com/ir/news/et-visa1097.html`

■URLs of Selected Mentioned Companies

American Airlines `http://www.amrcorp.com`
American Express `http://www.americanexpress.com`
Atmel `http://www.atmel.com`
Bell Canada `http://www.bell.ca`
Bull `http://www.bull.com`
Canon `http://www.canon.com`
Drexler `http://www.drexlertechnology.com`
LaserCard `http://www.lasercard.com`
Elcotel `http://www.elcotel.com`
Europay `http://www.europay.com`
Federal Express `http://www.fedex.com`
Fujitsu `http://www.fujitsu.com`
Gemplus `http://www.gemplus.com`
Giesecke & Devrient `http://www.gdm.de`

Global Chipcard Alliance http://www.chipcard.org
GTE http://www.gte.com
Hewlett-Packard http://www.hp.com
Hilton http://www.hilton.com
Hitachi http://www.hitachi.com
IBM http://www.ibm.com
JavaCard Forum http://www.javacardforum.org
Kleline http://www.kleline.com
Landis & Gyr Communications http://www.landisgyr.com
MasterCard http://www.mastercard.com
Microsoft http://www.microsoft.com
Mikron http://www.-eu.semiconductors.philips.com
Mitsubishi http://www.mitsubishi.com
Mondex International http://www.mondex.com
Motorola http://www.motorola.com
Nortel http://www.nortel.com
OKI http://www.oki.com
Oracle http://www.oracle.com
Orga http://www.orga.com
PBS http://www.pbs.org
PTT Telecom Netherlands/Unisource http://www.unisource-group.net
Schlumberger http://www.schlumberger.com
SGS Thomson http://www.st.com
Siemens-Nixdorf http://www.siemens-nixdorf.com
Sony http://www.sony.com
SPT Telecom http://www.spt.cz
Sun http://www.sun.com
Symbol Technologies http://www.symbol.com
Telstra http://www.telstra.com
Texas Instruments http://www.ti.com
Toshiba http://www.toshiba.com
U.S. Postal Service http://www.usps.gov
US West http://www.uswest.com
VeriFone http://www.verifone.com
Visa http://www.visa.com
WebTV http://www.webtv.com
Westinghouse http://www.westinghouse.com

4 Communications

4.1 Transmission Media

4.1.1 Executive Summary

Proper choice of transmission media, correct testing and installation of that media, and media-capacity planning significantly reduce network costs and problems. Wiring issues, such as improper use of media for given conditions and distances, loose connections, and the incorrect amount of slack, cause up to half of all network problems.

The two most commonly used types of physical transmission media for voice and data communications are copper wiring (including both twisted-pair wiring and coaxial cable) and fiber optic cable. Copper and fiber in the local area network (LAN) environment compete with and complement each other. Each medium has strengths and weaknesses, and most large networks use a variety of different media types to carry a message from a sender to a receiver.

Newer, faster networking protocols and services are being carried over copper-based transmission media and fiber optic cable. Networking schemes originally designed to run on fiber optic cabling may run on copper and vice versa. For example, Asynchronous Transfer Mode (ATM) runs over copper, as does Gigabit Ethernet.

Fiber optic technology has improved, costs for equipment have decreased, and the installed base of fiber continues to grow. New technologies such as dense wave division multiplexing will increase the number of communications channels within a fiber optic cable, thereby boosting bandwidth without requiring the installation of new cable. However, the costs associated with installing fiber remain an inhibiting factor.

Worldwide, copper has the edge because of the lower cost of enabling electronics; it is also a well understood technology and therefore convenient to use. The large copper installed base continues to influence market offerings and provides attractive economies of scale.

Telecommunications services providers and cable companies have moved almost exclusively to fiber optic cable for long-haul transmission media in the U.S. and increasingly, in Europe. Fiber is being incorporated into the local loop, but running fiber optic cable to the individual desktop or residence will remain the exception. In Asia and parts of the rest of the world, fiber is being adopted as well, but perhaps the largest trend in these areas is the move to wireless systems. Wireless offers an opportunity to cut costs associated with developing a large wireline infrastructure. There has been tremendous growth in new cost-competitive broadcast and wireless services worldwide. High-powered satellites that use small, inexpensive dishes have brought down prices in direct-to-home satellite television services. Other major satellite offerings are in progress and will be available within a few years.

Cable companies in the U.S. are installing hybrid systems – retaining coaxial cable for local distribution and using fiber optic cable for longer distances. Like telephone companies (telcos), cable companies are installing fiber optic cable to reduce maintenance costs by decreasing the number of amplifiers and the number of "hops" necessary for data. Because glass fibers have the ability to transmit light waves long distances without amplification or processing, cable companies can replace the numerous amplifiers required by coaxial cable systems, resulting in improved reliability and better picture quality.

Choice of media affects service offerings and the costs and capabilities of those offerings. Transmission media choices are influenced by installation and materials costs, capacity, room for expansion, life span, and regional regulatory, legal, and geographic factors. For example, costs of materials may come down, but labor costs may remain high, rendering certain choices unfeasible.

4.1.2 Wireline Media Technology

The major wireline transmission media alternatives include twisted-pair copper cable, coaxial cable, and fiber optic cable. (See Table 4-1.) Each alternative has distinct advantages and disadvantages. A combination of twisted-pair cable and fiber is in use for LANs and in the local phone network, coaxial is being used by cable television networks, and fiber media has been the choice for long-distance telephone networks.

Table 4-1: Transmission Media Alternatives

Media	Advantages	Disadvantages	Typical Applications
Twisted-pair cable	Inexpensive Well understood Easy to add nodes	Sensitive to noise Distance limitations Limited bandwidth Security (easily tapped)	Premises wiring for telecommunications and LANs Local loop of the phone system
Coaxial cable	High bandwidth Long distances Noise immunity	Physical dimensions (large diameter) Security (easily tapped)	Cable television systems
Fiber optic cable	Very high bandwidth Noise immunity Very long distances High security Small size	Difficult connections Higher cost of interface electronics	Long-distance telecommunications Network backbones

Source: Allied Telesyn, 1997

Fiber Optics

An optical fiber is a strand of glass as thin as a human hair that is designed to carry information using pulses of light. Fiber cables contain many separate fibers bundled together and encased in an outer sheath. At one end of each fiber, a laser-generated beam of light transmits data into the cable. At the other end, a photodetector picks up the beam and turns it back into electrical impulses. One fiber can handle voice and data traffic at many gigabits per second, whereas its traditional copper-pair counterpart used in the local loop typically transmits just a few megabits per second.

Volume shipments of fiber optic cable, improvements in fiber cable technology, and greater familiarity with fiber installation have contributed to reduced fiber costs. When costs further decrease on photoelectronic equipment such as transceivers that convert electronic signals to light and light to electronic signals, fiber may become a more attractive alternative for network cabling to individual homes or neighborhoods.

The use of light or photons minimizes loss of energy (which can reduce power costs) and loss in speed that can occur during transmission of data. Optical amplifiers can refocus and reinforce the signal without requiring additional power. Today, the most powerful amplifiers are designed to be economical for use in remote locations or where power is unavailable, such as in transatlantic cables or network backbones, but not in residences. Optical systems permit longer intervals of signal transmission than metallic-based systems, allowing distances of 100 kilometers (approximately 62 miles) without amplification and requiring fewer repeaters. (See Figure 4-1.)

Figure 4-1: Repeaters Required for Fiber versus Copper Cabling

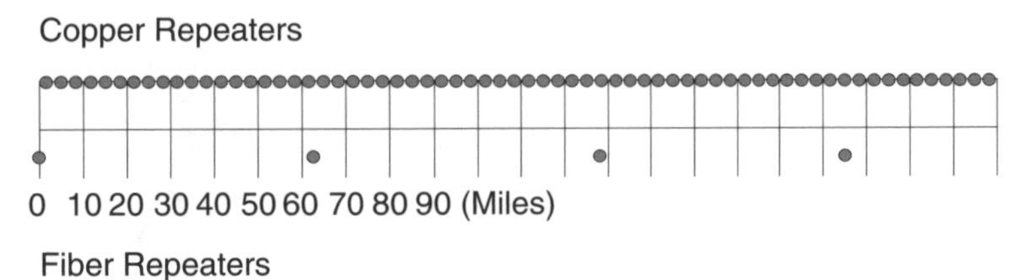

Source: Siecor, 1997

Technologies under development will result in improved fiber optic quality and further increases in bandwidth. Improvements in electronics will result in optical switching and routing at terabit rates.

Optical fiber is actually a relatively complex structure consisting of three layers:

- **Core** – The glass in the center of the fiber; this is the region through which the light pulses travel.
- **Cladding** – The glass that surrounds the core and keeps the light from escaping the core.
- **Coating** – A layer of a soft, plastic-like substance called acrylate (acrylic resin), which surrounds and protects the glass fiber.

Both core and cladding are made from ultra-pure glass. The cladding is purer than the core because when fiber is made, a substance called a dopant is added to the core glass to change its light-guiding properties. The difference in properties between the two glasses keeps the light pulse contained in the core. It creates a glass tunnel of sorts (although the center is not hollow) through which the light travels; when the light within the core interacts with the glass of the cladding, it reflects back into the core. This principle, called "total reflection," keeps the light trapped in the core and allows it to "bend" around curves in the fiber. In this manner, light signals can travel approximately 100 kilometers before they need to be boosted. The pulses of light carry information through fiber using digital signalling – the pattern of "on" and "off" that determines what information the signal contains.

Glass optical fiber is the only medium with proven performance at data rates of 2.5 gigabits per second (Gbps), although the theoretical limit is much higher. A single pair of optical fibers could potentially transmit 10 terabits of information per second. The limiting factor is the capability of the electronics at each end of the fiber optic cable. In most cases, Time Division Multiplexing (TDM) technology is employed by the terminating electronics to aggregate multiple data streams for transmission on a single fiber. Wavelength Division Multiplexing (WDM) and Dense Wavelength Division Multiplexing (DWDM) promise speeds of up to 20 Gbps by carrying multiple frequencies of light on a single fiber.

To understand how fiber can carry so much information, consider that a single laser can be turned on and off up to a billion times per second. An individual pulse of light represents a binary "bit" of information. A single strand of glass fiber can carry multiple wavelengths (different colors) of light simultaneously. These are the qualities that set fiber optics apart from other communications technologies; nothing else comes close to the information-carrying capacity of optical fiber. Figure 4-2 shows examples of optical fiber and copper cable with the same information-carrying capacity.

Figure 4-2: Fiber Cable and Copper Cable

Source: Courtesy of Siecor, 1997

Even with today's techniques, only a fraction of the theoretical capacity of a single strand of fiber is used. Because light is composed of photons, and fiber optics transmit light, photons can represent information. Photons can be represented in watt seconds: 1 photon equals 10^{-19} watt seconds. Today's lasers deliver 10^{-3} watts. The number of photons transmitted in 1 second is 10^{16} photons or 10 million billion photons per second. The receiving-end electronics interpret 10 photons as 1 bit of information. Thus, if the entire theoretical capacity of the fiber could be used, it would be the equivalent of approximately 1 Gbps for each of 1 million users.

Except for plastic fiber, which uses a bright light-emitting diode (LED) light, all current fiber optic systems use light in the infrared region that is invisible to the human eye. Wavelengths are given in nanometers (nm), meaning 1 one-billionth of a meter.

Connectors are used to mate two fibers or fibers to equipment where they are expected to be disconnected occasionally for testing or rerouting. Splices are permanent connections between two fibers made by welding the fibers in an electric arc (fusion splicing) or aligning them in a fixture and gluing them together (mechanical splicing).

■ Loss

Loss is a measure of the amount of light that is lost in a mated pair of connectors, a splice, or in a long length of fiber. It is expressed in decibels (dB), where -10 dB means a reduction in power by 10 times, -20 dB means another 10 times (or 100 times overall), -30 dB means another 10 times (or 1,000 times overall), and so on. Most connectors have a loss of about 0.5 dB. A splice has approximately a 0.2 dB loss.

Fiber optic testing standards have been set by several groups, but most follow the Electronic Industries Association's (EIA's) fiber optic test procedures (FOTPs) for testing. Some of the EIA procedures are called optical fiber system test procedures (OFSTP).

Loss is measured with a source-and-power meter according to two industry standards set by the EIA. FOTP-171 covers testing short jumper cables and OFSTP-14 covers installed cables. All loss tests work the same way – the loss of an unknown cable mated to a known good cable (launch reference cable) is assessed using a source-and-power meter. Optical power is measured in dBm, or decibels referenced to 1 milliwatt of power. Although loss is a relative reading (one power relative to another), optical power is an absolute measurement referenced to standards.

■ Single-Mode and Multi-Mode

Multi-mode and single-mode fiber are the two types of fiber in common use. Multi-mode has a bigger core (usually 62.5 microns – a micron is 1 one-millionth of a meter) and is used with LED sources at wavelengths of 850 nm and 1,300 nm. Single-mode fiber has a much smaller core, only about 9 microns, with laser sources at 1,300 nm and 1,550 nm. By comparison, plastic optical fiber has a large core (about 1 mm, or 1,000 microns) and uses visible light at 650 nm for low-speed applications.

The applications for multi-mode and single-mode are similar, but there are tradeoffs between distance and speed. Overall demand tends to be much larger for multi-mode fiber. Multi-mode is used for horizontal and backbone cabling and is less expensive because it uses LEDs rather than lasers. Multi-mode is also less expensive because it has a significantly larger installed base, and economies of scale favor its development.

Single-mode fiber uses lasers, which are more powerful and permit faster transmission speeds. Single-mode allows repeaters or amplifiers to be placed farther apart from each other with less attenuation. However, the use of lasers makes single-mode fiber more expensive than multi-mode. Single-mode tends to be used in backbone networks or very large campuses, such as military bases, where the organization owns the right-of-way for transmission lines. Because most organizations do not own the rights-of-way across long distances (greater than 1 kilometer), a telecommunications services provider usually is required to provide the long-distance connection. Multi-mode fiber is generally adequate for those segments of the campus that are separated by less than 1 kilometer. Table 4-2 shows some of the differences between the two types of fiber optic cabling.

More than 16 million miles of fiber are installed in the U.S., and this amount is expected to grow rapidly in response to customer demands for faster networks. In addition to installing more fiber, carriers will continue to implement solutions such as Synchronous Optical Network (SONET) and WDM systems, which increase the effective bandwidth of existing fiber.

■ SDH/SONET

Fiber networks originally were deployed using transmission formats optimized for voice telephone traffic carried over copper wires or microwave radio. However, the need to carry mixed types of communications (voice, data, and video) prompted development of higher bit rates. To meet this need, a newer,

Table 4-2: EIA/TIA-568-A-1 Optical Fiber Specifications

Optical Fiber Type	Multi-Mode Fiber	Single-Mode Fiber
Dimensions	62.5 microns – core 125 microns – cladding	8.3 microns – core 125 microns – cladding
Bandwidth – Low-speed (850 nm) – High-speed (1,300 nm)	 160 MHz 500 MHz	 NA >1 GHz
Attenuation – Low-speed – High-speed	 3.5 dB/km 1.0 dB/km	 NA >0.5 dB/km
Backbone cable length	20 km	30 km
Horizontal cable length	100 m	Not typically used in horizontal distribution systems
Applications	Ethernet, Token Ring, FDDI 155-Mbps ATM Baseband video Security systems	FDDI, ATM (1.2 Gbps) Fibre Channel CATV, telephony Broadband video
Connector type: T568-SC	Beige Shroud	Blue Shroud

Source: Anixter, 1997

more robust communications standard – known internationally as Synchronous Digital Hierarchy (SDH) – was established. Three versions of SDH exist: SDH-Europe, SDH-Japan, and SDH-SONET for North America.

SDH systems operate at multiples of 51.85 Mbps (called Optical Carrier-1, or OC-1) to allow for efficient conversion from one data rate to the other. For example, a 2.4-Gbps system (OC-48) comprises 48 OC-1 signals. Lower data rates (for example, a 64-Kbps voice channel, 1.5-Mbps T1, or 2.0-Mbps E1) are carried within the basic OC-1 bitstream on "virtual tributaries" and are multiplexed into the OC-1 bitstream using SDH add/drop multiplexers. (See Figure 4-3.)

Figure 4-3: How SONET Works

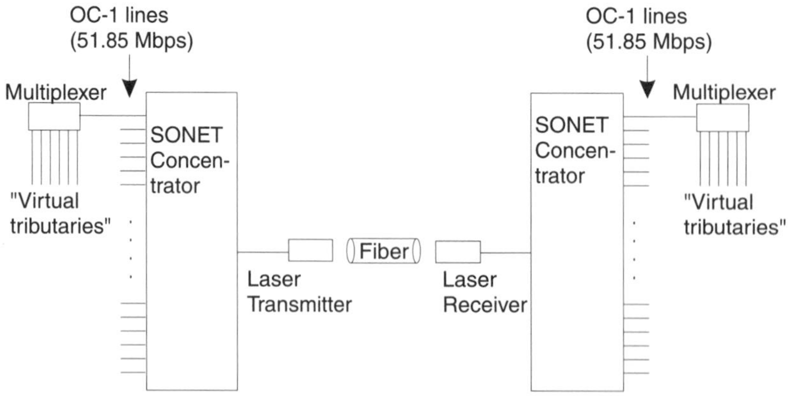

Source: Marc Matheson and Associates, 1996

At present, the OC-192 level of SONET offers the greatest bandwidth (10 Gbps) of any SONET-based system. However, OC-192 does not work satisfactorily with much of the in-place cable. Polarization mode dispersion (PMD) – a shifting of the signal frequency over long distances – limits effectiveness of OC-192 systems. PMD is most prevalent in older cables, cables that have been stressed, and cables that are elliptical in cross section rather than circular.

Although it is still uncertain how much existing cable will have to be replaced to accommodate OC-192, lower levels of SONET are field-proven and likely will continue to be deployed rapidly by the carriers as part of an overall strategy for upgrading network capacity. Advances such as WDM also may be used to achieve or exceed these speeds.

Fiber Optic Technology Advances

Significant advances in fiber optic technology are occurring in WDM and DWDM. Costs are dropping as well, which will make broader applications of these technologies feasible.

■ Wavelength Division Multiplexing (WDM)

In conventional fiber networks, the light signal transmitted on the fiber strand is of a single wavelength. Multiple channels are created by allocating a specific "time-slot" to an individual channel. This technique is called TDM. In contrast, WDM uses simultaneous transmission of light from multiple sources operating at different wavelengths over a single fiber optic line. The result is multiple simultaneous channels over a single fiber. This technique increases the capacity of existing cables and can offer savings in capital expenditures associated with new cables.

Instead of using only one laser, WDM uses multiple lasers operating at different wavelengths on the same fiber. By installing WDM and SONET equipment at each end of a fiber, a local exchange carrier can gain the equivalent of many new fibers without installing new cable. (See Figure 4-4.)

Figure 4-4: How WDM Works

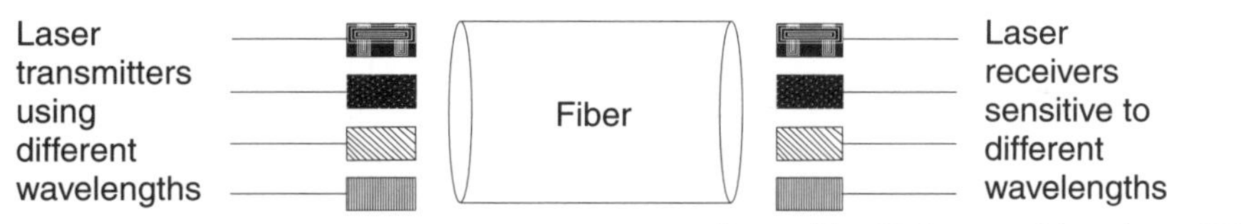

Source: Marc Matheson and Associates, 1996

Advances in TDM bandwidth can come only from faster switching speeds. Because it is limited to transmitting on a single wavelength, however, TDM cannot exploit the full bandwidth capability of fiber optic cable. Current TDM technology's upper limit is OC-192 (10 Gbps). Faster TDM, operating at or over 10 Gbps, could be several years away.

Since its introduction, usage of WDM has grown rapidly. WDM equipment examples include WaveMux 3200, an 8-channel (10 Gbps per channel) system from Pirelli; Optical Line Systems (OLS) from Lucent Technologies, supporting 16 channels; and IBM's model 9729 WDM system, which supports 20 channels at 4 Gbps each. Lucent began selling its 8-channel, 20-Gbps WDM system in 1995. WDM products with 16 channels followed, and products with 32-channel and 40-channel capacities now are becoming available. Looking forward, Lucent has demonstrated an experimental ultra-wideband optical-fiber amplifier that can support 100 or more channels.

Due to ongoing developments in this area, there is no way to predict the long-term cost of WDM technology accurately. Depending on configuration, current systems range from $13,000 to $100,000 per channel. Costs over the next few years should decline, given decreasing manufacturing costs as production levels ramp up. Increased competition as suppliers encounter limited opportunities for differentiation also will be a factor in the decreasing cost over time.

■ Dense Wavelength Division Multiplexing (DWDM)

DWDM also combines multiple optical signals on a single fiber, boosting the capacity of existing fiber cabling. Although no clear standards differentiate WDM and DWDM, channel spacings less than 1.6 nm generally are considered dense.

In 1997, Lucent announced an enhanced laser device for DWDM applications that increases the transmission distance of lightwave systems without the use of signal regenerators. Lucent's earlier electroabsorption-modulated isolated laser module (EMILM) could transmit a signal 600 kilometers without signal regeneration. The newer E2550 EMILM can transmit 320,000 conversations up to 1,000 kilometers without signal regeneration. Most systems in operation today can transmit up to 40,000 conversations on a pair of optical fibers without signal regeneration.

Other vendors in this market include Ciena. Ciena states that its MultiWave Sentry DWDM transport system, aimed at the long-distance telecommunications carrier market, can carry 500,000 simultaneous voice or Internet connections over a single fiber. Ciena also believes that DWDM can be economical at lower speeds and over shorter distances and is taking steps toward delivering DWDM functionality for short-distance applications.

Ciena's MultiWave Firefly product is designed for point-to-point short-haul applications in public networks. MultiWave Firefly is a DWDM transport system that will enable local exchange carriers (LECs) to expand the bandwidth of existing fiber optic cable by up to 24 times its existing capacity. MultiWave Firefly operates point-to-point without amplifiers in routes of up to 65 kilometers. The system multiplexes up to 24 discrete optical channels operating at 2.5 Gbps each over one fiber pair, allowing network operators to provide up to 60 Gbps of traffic on a fiber link.

Ciena's MultiWave Metro, scheduled to be available in late 1998, is for ring-based metropolitan network applications. The MultiWave Metro system will allow service providers to add high-capacity services in the local loop; it is a ring-based metropolitan network system using DWDM technology with up to 16 different wavelengths. The MultiWave Metro system will target interoffice rings and high-bandwidth local loop services. It will aggregate multiple traffic types, including SONET/SDH (at both the SONET levels of OC-12 and OC-48 and the SDH rates of Synchronous Transport Module-4, STM-4, and STM-16), ATM, and fast IP in a ring environment.

■ Improvements in Optical Electronics

Developments in components using photonic energy will continue, allowing for increased performance and precision as well as decreased power requirements. In addition to greater carrying capacity from DWDM developments and optical amplifiers, improvements in laser diode manufacturing and packaging and network chip and board interconnect technology are being made.

Improved manufacturing and packaging techniques create more reliable components with increased wavelength stability; increased precision can mean less degradation over time and smaller variances in channel spacing.

Developments in very large-scale integration (VLSI) photonics are reducing diode sizes. More efficient packaging is being used to integrate these diodes at the chip level. For example, new vertical-cavity surface-emitting lasers (VCSELs) in combination with more efficient plastic fiber made from preflourinated polymers will lower costs at the board level for optical interconnect applications. Conventional lateral-emitting diode lasers use a 5-micron by 1,000-micron geometry for the aperture. The new VCSELs have a round opening and are somewhat easier to package in array configurations.

HP offers 106.2-Mbps and Gigabit Ethernet VCSEL-based transceiver modules, and Vixel offers 1-Gbps VCSEL-based components for fiber optic applications. Sun uses VCSEL technology in its ATM network boards.

Photonic Power Systems uses a photovoltaic device as a power converter that converts optical energy from a laser diode into electrical power. This process enables reliable optical powering of amplifiers and originating or terminating electronics along the cable; low-power circuit designs keep power consumption relatively low, even for high-speed data links.

■ Fibre Channel and Gigabit Ethernet

High-performance network solutions will involve greater use of high-speed fibre channel interconnect technology with gigabit-per-second fiber optic transceivers. Fibre Channel is an ANSI standard (ANSI X3T9.3) developed as a high-speed interface for linking mainframes and peripherals. This high-speed transmission technology offers a scalable data rate beginning at 133 Mbps but ranging to more than 1.06 Gbps. Fibre Channel potentially can support transmission rates of up to 4.268 Gbps. The transmission rate of slightly more than 1 Gbps full duplex yields approximately 100 MB of actual data in both directions. Fibre Channel transmission technology is used in network backbones and theoretically can run on single- or multi-mode fiber, coaxial cable, and twisted-pair wiring. Fibre Channel can be used in point-to-point, arbitrated loop, or switched topologies. (See Chapter 3.2, *Storage and Batteries,* for more information on Fibre Channel.)

Gigabit Ethernet currently uses a version of the ANSI X3T11 Fibre Channel physical layer, instead of the specifications called for in the Fiber Distributed Data Interface (FDDI) standard that Fast Ethernet uses. Users of FDDI can use less-expensive multi-mode fiber as well as single-mode fiber, and FDDI is used extensively in network backbones today. Gigabit Ethernet users installing single-mode fiber will have to invest fairly substantial sums to upgrade network performance to gigabit speed. Therefore, Gigabit Ethernet will be used for backbone networks, unless a need arises for fiber to the desktop.

After the Institute of Electrical and Electronics Engineers (IEEE) 802.3z Task Force has finalized the initial standards for Gigabit Ethernet in 1998, work on a new encoding scheme and standard to run Gigabit Ethernet over Category 5 UTP wire is expected to begin. Copper-based products for Gigabit Ethernet are likely to be announced in 1998. (See Chapter 4.2, *Networking Systems and Protocols,* for more information on networking hubs, network interface cards, and routers.)

Copper Cable

Copper cable consists of insulated copper wires bundled together and surrounded by a sheath. Most cable entering homes consists of two pairs of wires to accommodate two voice circuits. Within the telephone distribution network, cables of 100 pairs or more are used.

Copper is an excellent conductor, and copper cabling used in telecommunications, cable telecommunications, premises, and residential markets is divided into two major types: twisted-pair and coaxial cable.

■ Twisted-Pair

A twisted-pair consists of a pair of copper wires, with diameters of 0.4 to 0.8 mm, twisted together and wrapped with a plastic coating. The twisting increases the electrical noise immunity and reduces the bit error rate (BER) of the data transmission. A twisted-pair cable contains from 2 to 4,200 twisted pairs. Twisted-pair is a very flexible, low-cost media and can be used for either voice or data communications. Its greatest disadvantage is limited bandwidth, which restricts long-distance transmission with low error rates.

The following are types of twisted-pair cables:

- **Shielded twisted-pair** (STP) – STP cables use a thick braided shield. These cables are heavier, thicker, and more difficult to install than their UTP counterparts. Some STP cables use only a relatively thin overall outer foil shield. These cables, called screened twisted-pair (ScTP) cables or foil twisted-pair (FTP) cables, are thinner and less expensive than braided STP cable. However, they are not any easier to install. The minimum bending radius and maximum pulling tension force must be rigidly observed when these cables are installed; otherwise, the shield may tear.
- **Unshielded twisted-pair** (UTP) – UTP cable, on the other hand, does not rely on physical shielding to block interference, but on that any noise induced on one wire will be induced equally but oppositely on the other wire as they are twisted around each other. With properly designed and manufactured UTP cable, this technique is easier to maintain than the shielding continuity and grounding of an STP cable. Because UTP cable is lightweight, thin, flexible, well understood, and inexpensive, millions of LAN nodes have been and continue to be wired with UTP cable, even for higher data-rate applications.

■ Unshielded Twisted-Pair (UTP)

Prior to 1991, telecommunications cabling for premises wiring was controlled by the manufacturers of computer equipment. End users often were confused by manufacturers' conflicting claims concerning transmission performance and were forced to pay high installation and administration costs for proprietary systems.

The telecommunications industry recognized the need to define a cost-effective, efficient cabling system that would support the widest possible range of applications and equipment. The EIA, Telecommunications Industry Association (TIA), and a large consortium of leading telecommunications companies worked cooperatively to create the ANSI/EIA/TIA-568-1991 Commercial Building Telecommunications Cabling Standard. Additional standards documents covering pathways and spaces, administration, cables and connecting hardware were subsequently released in 1995 in the form of ANSI/EIA/TIA 568-1995. The ANSI/EIA/TIA-568-1991 cabling standard was revised in 1995 and again in 1997, and it is now referred to as ANSI/EIA/TIA-568-A-1. The ANSI/EIA/TIA-568-A-1 Commercial Building Telecommunications Cabling Standard defines the expectations and limitations of cable and provides structure and direction for technological advances. With the advent of these standards, virtually all new installations using copper cable have employed twisted-pair technology.

Before the 568 cabling standard was established, cable consisted primarily of unshielded twisted-pair for voice and proprietary coaxial for data. Distinguishing the characteristics of one copper cable from another was difficult, if not impossible, so most people believed that all cable was the same. However, twisted-pair cable construction and electrical performance varied widely among cable manufacturers, and no uniform standard of measurement existed to compare one brand of cable to another. EIA/TIA published the cabling standard that set the baseline for interoperability in structured cabling and provided a consistent platform for networking devices.

The ANSI/EIA/TIA Category Specifications provide for the following cable transmission speeds and distance limitations:

- **Category 1** – No performance criteria
- **Category 2** – Rated to 1 MHz (used for telephone wiring)
- **Category 3** – Rated to 16 MHz (used for Ethernet 10Base-T), up to 100 meters
- **Category 4** – Rated to 20 MHz (used for Token-Ring, Ethernet 10Base-T), up to 100 meters
- **Category 5** – Rated to 100 MHz (used for Fast Ethernet 100Base-T, Ethernet 10Base-T), up to 100 meters
- **Category 6** – Proposed (rated to 350 MHz)
- **Category 7** – Proposed (rated to 400 MHz)

Within the networking industry, many cable manufacturers enhance specifications for high-speed networks. Some examples include the following:

- **Standard Category 5** (UTP)
  - Four pairs wrapped in a thermal plastic insulator, twisted around one another, and encased in a flame-retardant polymer
  - Maximum operating frequency of 100 MHz
  - Suitable for Token Ring, Ethernet, Fast Ethernet, Gigabit Ethernet (after mid-1998), and 155-Mbps ATM
  - Costs about $300 per 1,000 feet
- **Enhanced Category 5** (UTP)
  - Same as Category 5, but the manufacturing process is refined and the cable is of a higher grade
  - Maximum operating frequency of up to 200 MHz
  - Suitable for Token Ring, Ethernet, Fast Ethernet, Gigabit Ethernet (after mid-1998), and 155-Mbps ATM
  - Costs about $360 per 1,000 feet
- **Category 6** (SFTP)
  - Four pairs wrapped in foil insulators, twisted around one another, and encased in an extra insulating shield and a flame-retardant polymer jacket

- Maximum operating frequency of 600 MHz
- Suitable for Token Ring, Ethernet, Fast Ethernet, Gigabit Ethernet (after mid-1998), and 155-Mbps ATM
- Costs about $800 per 1,000 feet

Testing and Certification

Testing and certification are critical in determining the soundness of various implementations. Standardized testing and certification ensure cabling installations will be protected against future problems. For high-speed configurations, as the transmission frequency increases across the cable, the soundness of the cable installation becomes more and more critical.

■ Testing

The ANSI/EIA/TIA-568-A-1 cabling standard specifies testing procedures but excludes any testing of shielding effectiveness. TSB-67 is being widely used and is the only document to address field testing in great detail. It is expected to be the basis for future similar international standards.

Test configurations include channel and basic link. System designers, end users, and installers of complete LAN systems will use the channel test configuration. It will be used to verify the performance of the overall transmission path, including patch cords. If the channel conforms to TSB-67, the user patch cords may be approved for use in that channel only.

A channel consists of a user's equipment cord leading into the host patch panel that is then linked to the user patch panel via a cross-connection followed by a maximum of 90 meters of horizontal cabling. The horizontal cabling is connected to an outlet or transition point followed by a fly lead. The total length of equipment cords, jumpers, and patch cords shall not exceed 10 meters.

The basic link is what an installer might work with, including the wall plate, horizontal wiring, and first cross connection. It can include up to two connections at each end. Thus, the near-end crosstalk (NEXT) and attenuation requirements are different for a channel and a basic link.

TSB-67 specifically requires that the items listed above are measured in the field, and it defines a pass/fail criteria for these items based on the components and cable length in the link. The specifications call for measurement in increments small enough to detect crosstalk in the full frequency-carrying spectrum in the cable. The specified increments are as follows:

- The maximum step size for attenuation is 1 MHz and ranges from 1 MHz to 100 MHz
- The maximum step size for NEXT measurements in the frequency spectrum for 1 MHz to 31.25 MHz is 150 KHz. The maximum step size for 31.26 MHz to 100 MHz is 250 KHz.

By defining the maximum step size, TSB-67 ensures that when a swept frequency measurement is made, a suitable number of points are used to accurately determine worst case conditions.

TSB-67 specifically requires that field tests of NEXT shall be tested at both ends of the basic link and channel and that every pair and pair combination shall be tested for attenuation and crosstalk. This criteria is specified because the longer the basic link and channel, the less the connector at the far end has an effect on the result of the measurement.

■ Certification

Certification of new or existing cable plants provides a safeguard against future problems, which is especially important for companies contemplating the use of high-speed networks. Cabling standards are intentionally application-independent. Cabling standards define the cabling performance requirements common to a broad range of networking applications but do not address all the physical medium requirements for any specific network.

One such network-specific requirement, common to many higher-speed networks (including 100BASE-T4, 100VG-AnyLAN, 622 Mbps ATM, and Gigabit Ethernet), is the need for parallel transmissions – simultaneous transmission on two or more wire pairs in the same four-pair cable. The possibility that the cabling will some day need to support parallel transmissions affects the way that the link should be characterized for internally coupled crosstalk. The commonly used method for characterizing near-end crosstalk (NEXT) of cables carrying multiple signals is "Power Sum." This power summation function sums up the power received, which indicates the decibel or level of loss between multiple pairs of wires. This figure is important for the proposed Gigabit Ethernet over copper standard, which calls for four pairs of wires. If the attenuation is significant enough, high speeds cannot be achieved.

Advances in Transmission Media

Vendors continue to push the limit of copper cabling. Tut Communications, for example, announced 155-Mb capability over existing Category 3 unshielded twisted-pair cable. Gigabit Ethernet vendors are planning to deliver gigabit transmission speeds over Category 5 unshielded twisted-pair cable using a scheme that uses four pairs, each transmitting a 125-MHz signal in both directions. Although the Gigabit Ethernet standard is not yet complete, it probably will include this technique.

■ Standards and Speed in Premises Wiring

New technologies such as ATM and Gigabit Ethernet are pushing the need for greater bandwidth in horizontal distribution. Standards for Category 5 cable and beyond continue to evolve and will determine whether, for example, the Category 5 cable that was installed in 1996 will support planned migration to Fast or Gigabit Ethernet. In some cases, manufacturers design only for what is minimally required, and the need for performance as well as the standard will have changed over time.

Companies such as Anixter have developed additional standards of cable performance. Steps to assess cable performance include looking not only at attenuation and NEXT, but also at the interaction between the two. This interaction, called attenuation-to-crosstalk ratio (ACR), is a measure of the difference in performance between one cable and another. The greater the ACR, the more data capacity in the cable.

Anixter proposed the original standards specifications, which were called levels. Later, TIA changed the designation to category. The original Category 5 (Level 5, by Anixter's definition) specification from 1992 was modified to cover the performance requirements for existing Category 5 cables. The more stringent requirements for what has been called High-End Category 5 or Enhanced Category 5+ cables are referred to as Level 6 in Anixter's proposed standards. Anixter's new Level 7 products are designed to meet at least twice the bandwidth requirements of Category 5 cable, can support multiple applications at different frequencies under one jacket, and will support Gigabit Ethernet for lengths up to 100 meters.

■ Transmission Using Power Lines

In the future, more power companies or utilities may work together with telecommunications service providers to transmit voice and data over power lines. Pilot projects were conducted in 1997 in North America and in the U.K. to test the feasibility of this idea. Previously, small amounts of data were sent using power lines, but there has been no widespread commercial implementation to date. The frequencies between electricity and phone conversations are sufficiently different to prevent interference if they are transmitted over the same line. Other technical and safety issues still need to be resolved, however.

Economic issues include convincing a minimum number of homes in a given area to sign up for this service and the costs associated with conversion. For example, if signal separations or conversions occur at a transformer, this component of the new system would be more expensive in the U.S. because the 110-volt standard requires double the number of transformers than the U.K. standard of 220 volts.

Coaxial Cable

Coaxial cable (coax) comprises two conductors separated by an insulating material and encased in a sheath. The first conductor, a center wire, is surrounded by insulation and the second conductor, in a grounded shield of braided wire. The shield minimizes electrical and radio frequency interference. Coaxial cabling is the primary type of cabling used by the cable television industry and also has been widely used for computer networks. Although more expensive than standard telephone wire, coax is much less susceptible to interference and can carry much more data.

Coax is well suited for applications requiring wide bandwidth transmissions (up to 1 GHz) over short distances, such as cable TV. It can be used for digital as well as analog applications. For example, a TV signal with a bandwidth of 6 MHz can be digitized and reduced to between 1 MHz to 2 MHz to increase effective channel capacity and permit other services (such as video-on-demand) to be offered.

Coax is relatively easy to deploy compared with fiber. However, coax can be expensive and bulky and can exhibit large signal losses over a relatively short distance. Moreover, analog signals are particularly prone to noise and distortion over coax, partly because of the cable's electrical characteristics and partly because of the frequent amplification necessary to maintain signal levels over long runs.

Data transmission over coaxial cable is divided into two basic types:

- **Baseband transmission** – Where each medium (wire) carries only one signal, or channel, at a time. Transmission is through a single analog modulated frequency carrier. The data signal is propagated bidirectionally across the full length of the media. All stations are connected via simple digital to analog transceivers. An example of baseband transmission is 10Base5 (see below).

  The IEEE has issued standards governing baseband transmission in local area networks (LANs).

- **Broadband transmission** – Which enables a single wire to carry multiple signals simultaneously. Transmission is through a digital signal modulated onto a radio frequency carrier (analog). Multiple channels are made possible through frequency division multiplexing. All stations must be connected via radio frequency modems.

10Base5 (Thick Ethernet)

The 10Base5 Ethernet standard is based on a thick (approximately 1 cm or 0.4 inch diameter) and relatively inflexible coaxial cable. It was designed for distances of up to 500 meters. The outer insulation (jacket) of the cable may be plain PVC (yellow color) or Teflon (orange-brown color). Teflon is used for plenum-rated cable. Most commercial buildings have drop ceilings that support lighting systems, air ducts for heating and cooling systems, and wiring for power and communications. The space between the floor above and the drop ceiling is called a plenum. The plenum is continuous across each floor of a building, making it easy for flames to spread during a fire. The National Electrical Code (NEC) requires that plenum wiring be installed inside metal conduit or that more combustible wire insulation be covered with materials that provide adequate fire resistance and low smoke-producing characteristics, as tested and approved by Underwriters Laboratory (UL) or other recognized agencies.

Thick Ethernet coaxial cable must be designed especially for use in Ethernet systems so that the cable meets the IEEE specifications, including a 50 ohms characteristic impedance rating and a solid center conductor. Examples of thick coaxial cables specifically designed for Ethernet are Belden numbers 9880 (PVC) and 89880 (plenum-rated). These cables were historically used for Ethernet backbones.

10Base5 coaxial segments are equipped with male type *N* coaxial connectors at each end. Installing the coaxial connectors onto the cable requires special stripping and crimping tools and must be done carefully to avoid signal problems. For this reason, the correct operation of thick coaxial segments depends on the correct installation of the coaxial connectors.

10Base2 (Thin Ethernet)

The 10Base2 Ethernet standard is based on thinner (approximately 0.5 cm or 3/16 inch) coaxial cable that is more flexible and easier to deal with than the thick Ethernet variety. It was designed for a maximum length of 100 meters. The cable must have a 50 ohm characteristic impedance rating and a stranded center conductor. These specifications may be met by cable types RG 58 A/U or RG 58 C/U, but cable vendors sometimes use these cable numbers for cables with different impedance ratings. The thin coaxial Ethernet standard uses a much more flexible cable that makes it possible to connect the coaxial cable directly to the Ethernet interface in the computer. This capability results in a lower-cost, easier-to-use system that was popular for desktop connections until the twisted-pair media system was developed.

Hybrid Fiber-Coax Systems

Hybrid fiber-coax (HFC) systems use both fiber and coaxial cable. Cable television carriers have been replacing coaxial cable (coax) in their systems with fiber to eliminate many of the repeaters (amplifiers) associated with coax. Carriers want to increase bandwidth so they can offer a greater number of TV, data, or voice channels; decrease operating and repair costs; and improve service to customers.

Initially, fiber replaced coax in trunk lines (the lines between switching systems). Currently, fiber is replacing coax in the part of the distribution system that connects the cable's head end to neighborhood nodes. From there, the signals usually are transferred to existing coax or copper for distribution to the end user. (See Figure 4-5.)

For U.S. cable providers, an HFC system – fiber all the way to a neighborhood and then a single piece of coax routed to several hundred homes – eliminates many maintenance issues and provides greatly expanded bandwidth.

Conversion of signals sent over the coax, among other things, is performed by a network interface unit (NIU) at the customer's home, so power is required there. Most current coax systems were designed to handle traffic in a single direction only, from the carrier to the consumer, and thus use amplifiers that work only in one direction, so there is a problem in handling the return or back channel (used to send signals from the home back to the switching office). Cable companies are in the process of replacing all repeater amplifiers with new units that pass signals both ways.

In the U.S., telephone companies have adopted HFC as one way to hold market share threatened by cable provider infrastructure. HFC allows them to penetrate the cable companies' markets with television programming, video, and pay-per-view services. Companies such as @Home Network in the U.S. chose HFC to deploy their data over cable services. Telephone companies have lagged behind cable companies in HFC deployment. Even the most aggressive telephone companies expect deployment to be only 50 percent complete by the year 2000. In the interim, telephone companies also will offer services based on Digital Subscriber Line (DSL) technologies that permit higher-speed services over copper wiring. (See Chapter 4.3, *Telecommunications Services,* for more information on DSL technologies.)

HFC currently offers potential savings over complete fiber optic networks because the equipment required to convert electric signals to light is shared among many customers.

Figure 4-5: How Hybrid Fiber-Coax Systems Work

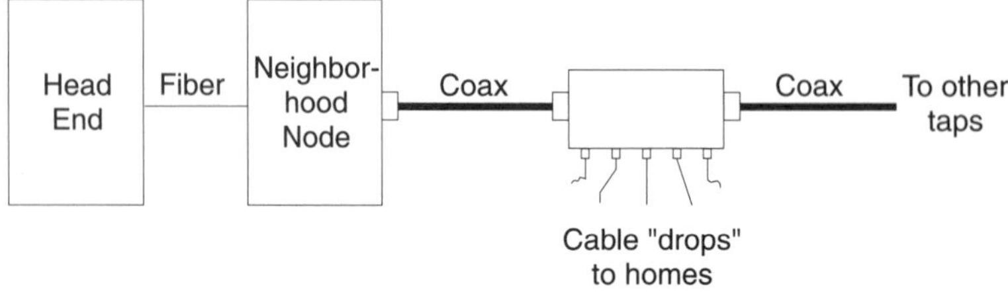

Source: Marc Matheson and Associates, 1996

■ Switched Digital Video

Switched Digital Video (SDV) is a fiber-to-the-curb solution. Fiber is terminated in an optical network unit (ONU) that serves about 30 homes. From the ONU to the home a combination of twisted-pair and coax may be used. This approach is more costly than HFC, but Bell Atlantic and others feel that the economics are good for providing video dial tone. SDV also can be used for digital television. Some are concerned that HFC may not be robust enough for interactive applications. The result is that some companies – such as Bell Atlantic – are pushing ahead with plans to install fiber-based SDV instead of HFC.

Conversion from optical signals to the form required for the user application (digital, audio, video) happens at the ONU, so power is required there. The back channel can be handled over existing copper lines.

Copper versus Fiber

Copper has remained at the forefront of the cabling arena for many years because of price. That situation is beginning to change, however. Vendors and installers claim that these days fiber is "only" 30 percent more expensive than copper.

Media is not necessarily the most expensive component of a cabling solution. Table 4-3 describes a typical network installation for 200 users on four floors and compares the costs of cable, hardware, and installation.

Table 4-3: Network Installation Costs

Component	Fiber Backbone (copper to the desktop)	All Fiber (fiber to the desktop)
Cable and connectors	$14,544	$36,863
Hardware (NICs, hubs, adapters)	$33,043	$79,489
Labor	$51,606	$51,606
Total cost	$99,193	$167,958
Approximate cost per user	$496	$839

Source: *Data Communications*, 1997

The greatest expense in a complete fiber cabling solution is the electronics equipment, followed by the cost of fiber cable. According to Corning, the inventor and leading manufacturer of fiber optic cable, the fiber optic cable (without connectors) in an all-fiber LAN installation should account for less than 5 percent of the total cost of the installation.

Fiber can help improve productivity and save money by reducing downtime caused by cable outages. Installation costs are only one part of the overall investment; for example, operational savings over time can occur as a result of reduced downtime. According to figures provided by AMP and Forrester Research, all-fiber installations result in a savings in operational costs of $174 per user per year. (See Table 4-4.)

Table 4-4: Fiber Savings – Operational Costs per Node per Year

Type of Support	Copper Cabling	Fiber Cabling	Savings
Physical LAN support	$280	$210	$70
Bridge/router support	$110	$83	$27
Outages	$160	$83	$77
Total savings			$174

Source: AMP, Forrester/*Network World*, 1997

■ Advantages of Fiber

Optical fiber provides users with higher reliability, superior performance, and greater flexibility than copper-based systems. The construction of optical fiber makes it immune to many of the factors that adversely affect copper, factors that often become more pronounced at higher data rates, thereby increasing network cost and complexity.

Fiber is immune to ElectroMagnetic Interference (EMI) and Radio Frequency Interference (RFI) signals. Because optical fiber carries light rather than electricity, it cannot be affected by EMI from power (sub-kilohertz), radio (kilohertz to megahertz), or microwave (gigahertz) sources. Optical frequencies are more than 1,000 times higher than microwaves and therefore are completely out of the range of these signals. However, these interfering signals can intrude into copper cables and make it difficult for the receiver to differentiate between the interference and valid data. This type of problem, which tends to happen sporadically, is difficult to troubleshoot and repair. Copper-based systems must be designed and installed carefully to minimize this interference.

Fiber also is immune to crosstalk. Crosstalk occurs when unwanted signals are coupled between conductors. Because the structure of optical fiber nearly completely constrains the light energy to the core of the fiber, signals cannot couple between fibers in a cable, thus eliminating crosstalk. However, crosstalk can occur in copper cabling systems because electromagnetic fields are generated around each signal carrying conductor. Crosstalk can be described as EMI within the cable itself. Crosstalk most often occurs near the source of the transmitted signal (near-end crosstalk, or NEXT), but also can occur at the opposite end of the link (far-end crosstalk, or FEXT) on short-distance links. FEXT is difficult to distinguish from NEXT and can cause confusion in testing. Of critical concern, however, is that the crosstalk performance of a copper cable or connector depends upon operating frequency. Therefore, cabling with sufficient performance at low speeds does not ensure proper performance at higher speeds.

Fiber is considered a more secure method of transmission. For example, fiber cannot be tapped without interrupting service, and changes on a line are easy to monitor and detect. Unlike metallic-based transmission media systems, fiber also makes it difficult to detect a signal being transmitted within the cable remotely.

■ Safety and Reliability Considerations

Fiber provides greater safety and reliability. Unlike copper facilities, all-dielectric fiber cabling systems do not conduct lightning strikes. They also do not conduct currents that can result from differences in ground potential or from improper grounding of the shield used in shielded or screened twisted-pair.

Fiber complies with all U.S. Federal Communications Commission (FCC), European, and Pacific Rim safety, emissions, and susceptibility standards. Copper cables can radiate and receive electromagnetic energy over a broad range of frequencies. The FCC and other regulatory bodies limit the levels of radiated and conducted emissions permitted by electrical devices, primarily to reduce interference with radio and television reception.

In some countries, copper-based systems also must be tested for susceptibility to external EMI. However, passing these tests does not eliminate interference; it merely reduces the probability of interference. If other parameters remain constant, the band of potential interference frequencies broadens as bit rates increase. To combat interference problems, copper-based systems must rely on more complex and costly transmission schemes, including various categories of cabling and increasingly complex coding electronics. By contrast, radiated emissions and susceptibility to external interference and interception are almost entirely eliminated by the inherent design of optical cables.

■ Installation Life Cycle

Fiber networks cost less to maintain than copper networks because they experience fewer network problems and therefore require less time and effort to troubleshoot and correct. For example, troubleshooting problems is easier in optically based premises networks because the only parameter to examine is attenuation – and that is done with only a hand-held power meter and light source. An Optical Time Domain Reflectometer (OTDR) also may be used to help locate the source of attenuation problems.

■ Simpler Cable Configurations

Fiber networks are a testament to the adage that simpler is better. Standard fiber outlets are made from simple, two-fiber appearances: one fiber for transmitting and the other for receiving. In contrast, standard copper outlets use 4-pair wiring. In general, one copper pair is used for transmitting and another for receiving, but the pairs used vary by application. This increased complexity and nonuniformity can result in confusion during troubleshooting. Because of the multi-pair nature of copper wiring, problems can result from incorrect wire mapping with crossed pairs, reversed pairs, and split pairs. These problems can manifest themselves as open or short circuits and lead to erroneous assumptions and wasted effort. If an improper wire map is not the cause of a copper problem, troubleshooting diagnostics must be used to test for excessive length, attenuation, and crosstalk. The list of potential problems and diagnostic tests required to isolate them increases the maintenance costs of copper networks relative to fiber networks.

4.1.3 Wireless Media Technology

A variety of wireless services has developed as a result of an increase in the number and types of satellites and because of microwave and other forms of fixed wireless transmission media. Large networks frequently comprise a combination of wireless and wireline media.

Satellite Communications

Commercial communications satellites have been in operation for more than 35 years. Until recently, however, the mainstay of satellite offerings has been long-haul voice and data retransmission services operated by a relatively small number of companies. Today, that scenario is changing. Many companies – spurred on by deregulation and rapidly changing technology – are vying to reap the rewards from growth in satellite-based networks, providing services such as teleconferencing, mobile communications, and direct broadcast of television signals to homes around the world.

The three major types of satellites are geosynchronous earth-orbit (GEO), medium-earth-orbit (MEO), and low-earth-orbit (LEO). These satellites differ by the distance of their orbits above the surface of the earth. Different satellite systems also may be characterized by the number of satellites required in a constellation or system for global coverage. (See Table 4-5.)

Table 4-5: Three Basic Types of Telecommunications Satellites

Type	Orbit (Miles)	Constellation Size (Typical)	Considerations
GEO	22,300	6	Satellites remain stationary relative to point on earth Handful of satellites needed for global coverage Requires high-powered transmitters Limited frequency reuse/sharing 250 millisecond transmission delay Most expensive to build and launch Longest orbital life (12+ years)
MEO	1,500 - 6,500	10 - 12	Satellites move relative to point on earth Moderate number of satellites needed for global coverage Uses medium-power transmitters Better frequency reuse/sharing than GEO Insignificant transmission delay Less expensive to build and launch Moderate orbital life (6 to 12 years)
LEO	550	48 - 66; 288 (Teledesic)	Satellites move rapidly relative to point on earth Large number of satellites needed for global coverage Needs only low-power transmitters Best potential for frequency reuse/sharing Negligible transmission delay Least expensive to build and launch Shortest orbital life (as low as 5 years)

Source: Marc Matheson and Associates, 1996

Technically, any satellite can be used for any type of service. However, practical considerations – such as power requirements, coverage area, and common frequency availability across multiple countries – dictate that specific types of satellites be used under different circumstances, as described in the following sections.

■ GEOs

A geosynchronous orbit allows a satellite to remain fixed above a given point on the earth's surface and makes it possible for ground station antennas to point continuously at the satellite without the need for a tracking mechanism.

Geosynchronous satellites are a proven technology. Although they are expensive to build and launch, because of their long life span (12 years or more) and their high capacity, the effective cost is relatively low. However, even with signals traveling at the speed of light, their distance translates to transmission delays (250 milliseconds, or 0.25 seconds) and, because of their wide coverage areas, frequency reuse is limited. Furthermore, because radio power is inversely proportional to the square of the distance between transmitter and receiver, GEOs require high transmit power.

Historically, GEO satellites have provided services for television, voice, and data transmission as well as mobile communications to ships at sea, thanks to their relatively seamless global coverage. However, many other uses have developed over the years, expanding the number of possible applications. Some companies plan to provide service to hand-held units from GEO satellites.

One such system – Spaceway by Hughes Communications – is projected to deploy nine GEOs at a cost of $3.2 billion. This project will provide access to a variety of interactive, high-speed, high-quality wideband telecommunications services. The system is scheduled for operation by 2000.

Very small aperture terminals (VSATs) are small, inexpensive satellite dishes (1.8 meters in diameter) with associated electronics and software for providing voice, facsimile, and data transmission/reception capabilities. A typical VSAT communications channel handles 64 Kbps of digital information. (VSATs that support the T1 data rate of 1.544 Mbps are called TSATs.) The channels can be leased from VSAT providers and used for private networks that incorporate telephony, videoconferencing, interactive distance learning, LAN interconnection, real-time inventory management, point-of-sale data gathering, and other services.

■ MEOs

MEOs sometimes are lumped in with LEOs or called Big LEOs in discussions of service offerings because of their similarities. Both MEOs and LEOs orbit closer to the earth than GEO satellites and do not appear at a fixed point over the earth. Both are more simple and less costly, and both have fewer problems with time delays that arise from GEOs being so far out in orbit.

However, MEOs have some advantages over LEOs. MEOs orbit more slowly, can be replaced less often (an estimated 6 to 12 years as opposed to every 5 years), and can cover the earth with a fewer number of satellites and ground stations. MEOs also suffer less from the phenomenon called shadowing, where the lower-orbit LEOs have signals cut off by mountains, buildings, or other topographical features that block transmissions.

Examples of projects that will use MEO satellite systems are Odyssey, a joint venture between TRW and Teleglobe, and ICO Global, a venture developed by Inmarsat, the international satellite communications organization.

ICO Global has more than 55 international strategic investment partners, including Beijing Marine Communications and Navigation Company, DeTeMobil of Germany, Etisalat, ICO Korea Co. Ltd., Inmarsat, OTE of Greece, Satellite Phone Japan, Singapore Telecom, and VSNL of India. ICO was established in January 1995 as a private company to provide satellite-enabled personal mobile global communications services. The ICO system is expected to provide service starting in the year 2000. Once complete, it will be possible to connect a subscriber using a pocket telephone to almost any telephone at any location in the world.

In December 1997, TRW announced plans to abandon its Odyssey MEO satellite telephony project and support ICO Global instead (although it had been suing ICO over patent infringement). TRW now is one of the largest shareholders in ICO, and TRW and ICO grant each other cross-licenses for their respective patents relating to global telecommunications systems. Additionally, TRW receives certain distribution rights in the U.S. for ICO products and services, and the parties dismissed their respective patent litigation suits. The boards of directors of both TRW and ICO have approved the agreement, which is expected to be completed in early 1998.

To make spectrum available that is needed by other global personal communications services, TRW plans to turn back the license it received from the FCC for its Odyssey MEO program.

■ LEOs

LEOs offer some distinct advantages over GEOs. Because their low orbits enable the use of lower-power transmitters, LEOs have lower build and launch costs, and their transmission delay is negligible. On the downside, LEOs are not geostationary, so many more satellites are needed to maintain continuous coverage of a given service area. Lower orbits also result in shorter lives (an estimated 5 years due to greater stresses). A shorter satellite life span results in the need for more launches and redirecting of numerous antennas as satellites enter and leave service. Some of these issues may be addressed by improvements in technology or counter-balanced by lower build and launch costs.

LEO systems classified as mobile satellite services (MSS) will offer voice services to hand-held or cellular phone types of devices. Many large LEO providers intend to offer such services. However, some proposed LEO systems will supply only low-speed data and position-location systems. Because they will not be offering voice services, these "Little LEOs" are excluded from the definition of MSS.

The number of satellites required to provide worldwide MSS services depends on many factors, but a few dozen will be necessary even in relatively simple deployments. More complex systems will require more satellites. Teledesic plans to offer wideband services from 288 LEOs organized into 12 elliptical orbit planes sometime in 2002.

The first generation of LEOs to supply MSS will share a small amount of spectrum made available by the World Administrative Radio Conference of 1992. However, much more spectrum will be necessary to meet projected demand. In July 1995, the FCC announced it would divide a portion of spectrum desired by wireless cable ventures and LEO satellite communications services. This portion (in the 28 GHz range) is split as follows: 850 MHz to wireless cable, 750 MHz for projects such as the Hughes Spaceway Service, and 500 MHz for projects such as Teledesic (dubbed a global Internet system). Frequencies in the U.S. are auctioned by the FCC.

In November 1997, the World Radiocommunication Conference, which meets every 2 years and allocates international radio spectrum, completed the allocation of radio spectrum required for Teledesic to provide two-way telecommunications services through its global, broadband satellite network. A pair of 500-MHz bands of spectrum were allocated internationally for use by non-geostationary fixed satellite services, such as those Teledesic will provide. The spectrum are the same paired bands the FCC licensed to Teledesic in March 1997.

■ Direct Broadcast Satellites

A direct broadcast satellite (DBS) transmits up to 150 channels of video and audio service directly from a geosynchronous satellite to residences equipped with an 18- or 36-inch dish. These satellites transmit a digital signal that typically results in higher-quality reception by the consumer.

Dish receiver costs have come down in price over the last few years to $200 or less. New data services will be available from EchoStar II and DirecTV. AT&T purchased 2.5 percent of GM Hughes Electronics's DirecTV and is aggressively marketing the service in all 50 states in the U.S., especially in bundled packages.

Microwave

Microwave transmission uses electromagnetic waves in the radio frequency spectrum of approximately 1 billion cycles per second (GHz) and above. The definition is sometimes limited to frequencies between 1 GHz and 30 GHz. Microwave is commonly used method of transmitting telephone, facsimile, video, and data for both public carrier and private networks. Microwave signals travel only in straight lines. In terrestrial microwave systems, signals typically are good for 30 miles, at which point another repeater tower is needed. Microwave is the frequency for communicating to and from satellites. Microwave transmission frequencies are used in communications satellites and in line-of-sight systems on earth.

Many users have microwave systems for short-haul campus and metropolitan-area networks. Microwave is available today at up to SONET rates, but obtaining an FCC license to use the T-3 and OC-3 microwave systems may be difficult in some areas.

Microwave can be used with fiber optics transmission media, for example with so-called dark fiber. This is fiber that has been installed but is not being used. To use microwave with dark fiber, a microwave antenna could be placed at the fiber head-end. A second antenna would beam the traffic to the head-end location, where equipment would be located to transfer the microwave traffic onto a router and then onto the fiber. This example would require co-location agreements with the fiber owner or carrier.

4.1.4 Market Overview

Fiber

The fiber market will continue to experience fairly significant growth. According to KMI, worldwide deployment of fiber optic cable will grow from 29 million fiber-kilometers in 1996 to an estimated 65.7 million fiber-kilometers in 2001.

ElectroniCast, a market research firm, estimated in 1997 that the DWDM transmitter market will grow by 34 percent per year until 2001.

■ U.S. Telecommunications

The amount of fiber installed in the U.S. long-distance market is experiencing significant growth due to a combination of factors. First, simple traffic demand is increasing very rapidly. Second, existing operators such as AT&T, MCI, and Sprint, competing on the basis of network reliability, are using up installed capacity by adding "hot standby" circuits (idle circuits that stand ready in case of network failure). Third, new long-distance carriers are building separate networks for resale to third-party wholesalers and retailers of long distance services. And fourth, the Regional Bell Operating Companies (RBOCs) will be allowed to enter the market as a result of the Telecommunications Act of 1996.

These market dynamics are creating opportunities for additional network capacity and players. The long-distance carriers alone may install 40,000 sheath kilometers of optical cable over the next 5 years. (Sheath kilometers refer to a bundle of fibers as opposed to measurement of individual fiber strands, which are known as kilometers of fiber.)

■ Cable Television

Cable television operators have installed fiber systems rapidly over the past 5 years to improve reliability and reduce costs by replacing the old coaxial broadcast network. Competitive forces, industry consolidation, and the desire to increase revenues via new services are driving a number of changes in this industry's infrastructure.

Cable TV systems today exist as islands, independent of one another, receiving programming from a combination of satellite, tape, and live studio feeds and distributing signals to subscribers through a fiber-coaxial tree-and-branch network. These same cable TV systems are beginning to remake themselves through the application of fiber optics. Of particular interest to the utility industry is the developing trend toward head-end consolidation or elimination, also called regional interconnection. This technique involves running fiber between existing head ends in adjacent towns, consolidating expensive signal processing equipment in one location, and transmitting the programming to multiple systems over fiber.

WDM

According to a new study prepared by IGI Consulting, the WDM market is set to expand dramatically in the next 3 years into Local Exchange Carrier (LEC), Competitive Local Exchange Carrier (CLEC), and private networking applications. (CLEC is a term coined for service providers following deregulation, pursuant to the Telecommunications Act of 1996, and could include a variety of telecommunications services providers, CATV providers, power or utility companies, Internet service providers, and cellular carriers.)

WDM technology has been in use by long-distance companies in recent years to expand the capacity of their trunks by allowing a greater number of signals to be carried on a single fiber. AT&T, MCI, Sprint, and WorldCom have made long-term commitments to WDM technology and will be using it to ramp up their trunk speeds from 2.5 Gbps to more than 40 Gbps without adding new fiber.

As bandwidth demands grow in the local exchange, cable TV networks, and private corporate networks, WDM technology will be finding new applications, broadening the WDM market significantly and creating a wider range of opportunities for WDM device and systems vendors. Beside increasing the speeds of trunk lines, WDM technology also can be used to perform optical multiplexing, which can segregate groups of users or particular services into individual transmission channels, making high-bandwidth networks easier to manage, maintain, and provision. Network providers can begin leasing secure wavelengths on their networks or provide optical add-drop multiplexing and routing for more highly reliable and manageable networks.

Today, the WDM market is just beginning to grow, with only a few major suppliers and a number of smaller device manufacturers vying for a limited market. Over the next 5 years, however, the WDM market will expand rapidly, driven by the growth of the Internet and intranets, residential broadband services, and high-speed international data services. As prices for optical amplifiers and WDM devices fall, the market will expand further, particularly into corporate data networks and research networks, where WDM technology offers highly efficient use of existing fiber optic facilities. The WDM market also is growing rapidly in Europe and Asia, and WDM technology already is being deployed into the next generation of international submarine fiber optic systems to provide gigabit or multi-gigabit rates across the world's oceans.

Copper

U.S. shipments of copper wire and cable for telecommunications grew in the 1970s, fell in the early 1980s, and leveled off from 1985 to 1992 at approximately 140,000 metric tons per year. According to the Copper Development Association, the drop in the 1980s was not due to fiber optics (which had not penetrated the subscriber loop at that time), but rather to the use of smaller copper wires and multiplexing. Shipments of copper wire stayed about the same in the early 1990s, with the RBOCs' move back to thicker copper wires helping to balance inroads being made by fiber optic cable in the feeder portion of the subscriber loop. Technologies such as asymmetrical digital subscriber loop (ADSL) and improvements to Category 5 wiring will help create more powerful copper-based systems.

The mix of copper wire and cable varies from country to country. Table 4-6 shows estimated market share for premises wiring types in Europe. Figures are based on estimated cable lengths for 1996 shipments.

Table 4-6: Share of Premises Wiring Types in Europe

Country	UTP	STP	FTP
France	17%	4%	79%
Germany	10%	64%	26%
Italy	80%	13%	7%
Netherlands	65%	10%	25%
Spain	75%	5%	20%
United Kingdom	86%	2%	12%

Source: *LAN Times*, 1996

Satellite

Demand for satellite services has been driven primarily by the fact that satellites can offer services at a lower price, extending reach cost-effectively, especially in rural areas where fiber optic cable or even copper wiring may not exist or may be expensive to install. Satellites can help reduce wait times for phone lines in some of these same areas or can be used for broadband services to address the need for more bandwidth associated with rapidly growing Internet use.

Market growth for satellites offering alternatives to fiber and traditional wireline media will depend on meeting goals and time lines associated with spectrum allocation, launching satellites, and delivering affordable, worldwide access to telecommunications services. These services include linking enterprise networks, broadband Internet access, and forms of interactive multimedia such as videoconferencing. (See Chapter 4.3, *Telecommunications Services,* for more information on satellite-based services and expected growth.)

Microwave

The market for digital microwave radio and services has grown rapidly in many countries. For example, digital microwave radios will be used to connect cellular users in Bombay, India, with GSM base stations in a network operated by Indian cellular operator Hutchison Max Telecom.

In October 1997, NEC and Marubeni received part of a large SDH-based digital microwave system order in China, scheduled to be completed in 1998. The order was awarded by China's Ministry of Post and Telecommunication. The project calls for a 2,200 kilometers communications link between Beijing and Wuhan-Guangzhou and will cross six provinces. It will use a 4-GHz and 6-GHz, 155-Mbps system and will comprise 50 hops between 51 relay stations. The network is the equivalent of 32,000 telephone lines and has a value of approximately $37 million.

4.1.5 Forecast

- Fiber throughput rates will reach the terabit (1 billion bits per second) level by 2005 as a result of considerable growth and innovations in WDM and SONET.
- In the U.S., most telcos do not expect to complete the conversion from copper to fully fiber optic systems before 2010 or 2015 at the earliest. Extension of fiber to the curb, with high-speed, twisted-pair or coax service to the home, will be widely deployed in major residential markets around 2005. Fiber will not reach the home until 5 to 10 years later.
- Although the technological advantages of fiber are compelling, fiber-to-the-curb in residential areas or fiber-to-the-desktop in businesses will not become widespread in the next several years due to high costs and to improvements in interim technologies such as Gigabit Ethernet and ADSL that allow faster transmission over copper lines.
- Copper cabling will continue to be used widely in LANs and in telephone networks due to its cost advantages, large installed base, large number of trained installers, and economies of scale associated with manufacturing.
- Category 5 cabling installations will grow significantly in commercial, high-end, and new residential buildings, offering substantial bandwidth improvements over traditional Category 2 or Category 3 copper wiring.
- The higher cost of Category 6 and 7 cabling for newly installed premises wiring will be more than justified by its ability to run very high-speed data networks such as Gigabit Ethernet, particularly when compared to the cost of installing Category 5 cable initially and then having to replace it. Adoption of standards for Category 6 and 7 cabling will help promote wider deployment.

- The long-haul fiber network operators will continue to push new technology that increases their system capacity and lowers the cost per bit of information. This goal will be accomplished by a transition from TDM to WDM.
- The DWDM transmitter market will grow substantially in the next few years as long-distance carriers carry more voice and Internet traffic over existing fiber optic cable and a growing market demand develops among LECs and private networks. The market for WDM components in general will more than double within the forecast period.
- Premises wiring schemes will continue to use fiber network backbones. Fiber connectivity to desktop devices will increase but will continue to be limited to a small minority of desktops. Copper connectivity to the desktop will continue to dominate.
- Accelerating the speed of local loop connections (using ADSL technology, for example) will increase the demand for bandwidth in their backbone networks for ISPs and carriers.
- The use of wireless transmission technologies, including cellular/PCS, LMDS/MMDS, and satellite, will continue to grow. Improvements in wireless transmission technologies will continue during the forecast period.

4.1.6 References

■Articles

Buerger, Dave. 1997. Fiber is good for you. *Network World.* August 4: 51.
Clark, Elizabeth. 1997. WDM expands fiber's horizons. *LAN Magazine.* March: 1.
Cray, Andrew and Chappell Brown. 1997. Wiring for speed, playing for time. *Data Communications.* April: 75.
DARPA looks to optical interconnects. 1997. *Electronic Engineering Times.* May 26: 35.
Makris, Joanna. 1997. Premises wiring. *Data Communications.* August 21: 199.
Nolle, Tom. 1997. Take a look at the dark side of fiber. *LAN Times.* October 13: 63.

■Periodicals, Industry and Technical Reports

Wireless. (monthly) Cedar Knolls, N.J.: Wireless Publishing Co.
Worldwide optical fiber and fiberoptic cable markets. 1997. Kessler Marketing Intelligence. KMI. January.

■On-Line Sources

Introduction to fiber optics. Siecor. `http://www.siecor.com/introfo.htm`

■URLs of Selected Mentioned Companies

@Home Network `http://www.home.com`
Allied Telesyn `http://www.alliedtelesyn.com`
AMP `http://www.amp.com`
Anixter `http://www.anixter.com`
ANSI `http://web.ansi.org/default.htm`
AT&T `http://www.att.com`
Bell Atlantic `http://www.bellatlantic.com`
Ciena `http://www.ciena.com`
Corning `http://www.corning.com`
DirecTV `http://www.directv.com`
EchoStar II `http://www.echostar.com`
Electronic Industries Association `http://www.eia.org`
Etisalat `http://www.etisalat.com`
GM Hughes Electronics `http://www.hughes.com`
Hewlett-Packard `http://www.hp.com`
IBM `http://www.ibm.com`
IEEE `http://www.ieee.org`
Lucent `http://www.lucent.com`
MCI `http://www.mci.com`
NEC `http://www.nec.com`
Optical Society of America `http://www.osa.org/`
Photonic Power Systems `http://www.photonicpower.com`
Scope Communications `http://www.scope.com`
Siecor `http://www.siecor.com/index.htm`
Sprint `http://www.sprint.com`
Sun Microsystems `http://www.sun.com`

Telecommunications Wire & Cable
`http://www.copper.org/cable/homepage.htm`
Teledesic `http://www.teledesic.com`
TRW `http://www.trw.com`
Vixel `http://www.vixel.com`

4.2 Networking Systems and Protocols

4.2.1 Executive Summary

More than ever before, an enterprise is dependent on how well it executes its overarching local area (LAN) and wide area network (WAN) strategy. It is becoming increasingly difficult to divide LANs and WANs into separate areas of planning, implementation, and management. One of the most pervasive issues information technology (IT) executives face is sustaining network performance in the face of an explosion of network-based applications and traffic volume. The growth of client/server applications, internetworking, intranets, and the Internet have increased the attention paid to overall enterprise network performance. Although security, cost, and legacy systems issues remain crucial, network response time and availability are viewed widely as the key issues for IT managers to solve in the coming year.

Above the physical transmission media layer, networking systems provide services with the intelligence to route and manage data across LANs and WANs. These LANs and WANs can be designed for optimal performance, management, and security for enterprises of all sizes. LANs can be distributed in a building, throughout a campus, across a nation, or between continents and linked together by a private or public WAN. The performance of network systems infrastructure has improved so that the difference between accessing a computer across the hall and one across the country has become less obvious to users. These technology advances have given IT executives more options, adding flexibility – as well as complexity – to enterprise computing solutions.

In many organizations, LANs initially were built so users of personal computers (PCs) could share peripherals such as printers and disk drives. These print and file services dominated network computing applications from the mid-1980s to the early 1990s. As these work-group environments were linked together, demand for more sophisticated devices to expand and connect LANs created new markets for technology from companies such as 3Com, Bay Networks, and Cisco.

Today's distributed applications are not limited to simple data exchange or the sharing of expensive peripherals. Client/server, sales automation, imaging, multimedia, Internet access, and videoconferencing applications now are common in many organizations. With a broader, more diverse application base, more users are being attached to LANs, more LANs are being installed, and more LANs are being linked via WANs. Falling computer prices have resulted in even more users with additional devices in increasingly distributed locations. Firms now can justify having employees telecommute. With dial-up or wireless communications, employees literally carry their network connections with them.

The Internet and the World Wide Web have had a dramatic effect on enterprise networks (intranets), with companies using them for electronic commerce, e-mail, marketing, and other types of information. "Push" software has put more demands on corporate networks as well. Internet multicasting is emerging as an extension that handles broadcasts more efficiently than push technology, but it requires upgrades to routers all along the path from source to destinations to achieve this efficiency. In addition, vendors have contributed to Web traffic by making legacy systems, such as IBM S/390 mainframes on Systems Network Architecture (SNA) networks, accessible via a Web browser.

The rapid growth of the Internet has forced most vendors to focus on improving product performance and diversify their product offerings to attract a wide range of potential users. The Internet also has added to the complexity of enterprise networks by heightening performance, security, electronic commerce, and network management issues. IT executives no longer manage networks in controlled, self-contained environments. The world now wants access to on-line corporate resources, and companies must offer such access to remain competitive.

Managing and maintaining network performance will continue to be a challenge. Response time can degrade if new data- and graphics-intensive applications and an increasing number of network and Web users are not anticipated and managed effectively. Network performance problems are exacerbated by an organization's installed infrastructure – typically a heterogeneous mix of technologies and media that expand on an as-needed basis. Because infrastructure investments are expensive, managers use their initial technology for as long as possible. They also are reluctant to forgo the investments they have made in training. Vendors are working to upgrade network product performance, such as in the area of Layer 3 routing and switching, and to maintain consistent network management capabilities.

Consolidation continues among networking vendors at a steady pace. For large organizations, mergers and acquisitions offer the potential benefit of working with a single vendor for most management needs but may come at the cost of reduced competition.

4.2.2 Types of Network Connections

Because companies vary in their corporate network needs and because vendors have different opinions about the appropriate technology as well as competitive positioning, suppliers have developed a series of networking techniques to connect users that address different business requirements. The alternatives vary in network capacity, bandwidth per user, reliability, the cost of each connection, quality of service, and the ease with which they can be installed and managed.

For companies seeking a low-cost, easy-to-implement network, some variety of Ethernet may be the best selection, with Fast Ethernet and Gigabit Ethernet offering greatly increased throughput. For those whose business involves dealing with complex video images, however, ATM still may be the appropriate choice. However, companies most likely will implement several technologies simultaneously, for example, combining Ethernet or Fast Ethernet to the desktop with a different backbone.

Ethernet

Today, Ethernet technology is nearly ubiquitous. According to IDC, more than 83 percent of all installed network connections worldwide were Ethernet at the end of 1996, and IDC projects that Ethernet dominance will continue beyond 1999. The remaining connections are a combination of Token Ring, Fiber Distributed Data Interface (FDDI), Asynchronous Transfer Mode (ATM), and other protocols. All popular operating systems and applications currently available are Ethernet-compatible.

Some vendors are moving to simplify network technology choices and configuration. For example, in late 1997, 3Com introduced a full enterprise Gigabit Ethernet system. The five products that make up the system also will ease migration by supporting shared and switched Ethernet, Fast Ethernet, FDDI, Gigabit Ethernet, and ATM within a single network and within a single switch.

■ Fast Ethernet

Among the high-speed LAN technologies currently available, Fast Ethernet (100Base-T) has become the leading choice for a smooth, non-disruptive way to achieve 100-Mbps performance. The Fast Ethernet standard, approved in 1995, established Ethernet as a scalable technology. The growing use of 100Base-T connections to servers and desktops, however, is creating a clear need for an even higher-speed network technology at the backbone and server levels. Ideally, this technology also should provide a smooth upgrade path that requires no retraining and is cost-effective.

■ Switched Ethernet and Fast Ethernet

In a switched Ethernet network, the network interface cards (NICs) in each machine in an existing 10Base-T network still can be used. The Ethernet hub is changed to an Ethernet switch, and the medium is no longer shared; instead, dedicated bandwidth on demand runs between any two stations. The switch, which can handle hundreds of megabits per second, allows each user the full Ethernet bandwidth of 10 Mbps to another node.

Switched 100-Mbps Ethernet provides another upgrade path from Fast Ethernet for those users transitioning to higher-speed networking such as Gigabit Ethernet. Just as in the case of the original switched Ethernet, by replacing the hub with a switch, this approach replaces a shared medium that has an aggregate bandwidth of 100 Mbps with a dedicated 100-Mbps connection between any two nodes.

■ Gigabit Ethernet

One developing solution for high-speed networking is called Gigabit Ethernet. Gigabit Ethernet will provide 1 Gbps bandwidth with the simplicity of Ethernet at a lower cost than other technologies of comparable speed. It also offers a natural upgrade path for current Ethernet installations, leveraging existing equipment and expertise. In essence, Gigabit Ethernet extends the scalability of Ethernet even further than Fast Ethernet or switched Ethernet.

Gigabit Ethernet follows the same technology approach as its 10-Mbps and 100-Mbps precursors. All three Ethernet speeds use the same IEEE 802.3 frame format and potentially can operate in full-duplex mode.

Gigabit Ethernet is already well into the standards process. In May 1996, 11 vendors formed the Gigabit Ethernet Alliance, whose goal is to accelerate the development of standards for 1-Gbps Ethernet transmission. Current alliance members include 3Com, Bay Networks, Cisco, Compaq, Intel, LSI Logic, Packet Engines, Sun, UB Networks, and VLSI Technology.

In July 1996, the IEEE 802.3 work group created the Gigabit Ethernet task force (IEEE 802.3z), whose key objective is to develop a Gigabit Ethernet standard, which is expected to be finalized by mid-1998. Some manufacturers will not ship final products until the specification is official. Also, prices of NICs are still high – generally between $1,000 and $2,000 (at the end of 1997); switches are about twice as expensive as for the Fast Ethernet product.

Joining the startups vying for a share of the Gigabit Ethernet market are larger vendors such as 3Com, Bay Networks, Cabletron, Cisco, Digital, and IBM. 3Com was the first company to unveil a dedicated gigabit offering, the CoreBuilder 3500, in September 1997. This multiprotocol backbone switching router, which supports ATM or Ethernet, is capable of handling 56 million packets per second; shipments of CoreBuilder 3500 began in November 1997. 3Com's CoreBuilder 9000 Enterprise Switch, announced in November, has a switching fabric that supports up to hundreds of gigabits of bandwidth; shipments are scheduled for early 1998.

In October 1997, Cabletron announced a series of Gigabit Ethernet products, including NICs. Also in October, Digital announced its fault-tolerant GIGAswitch/Ethernet, a seven-slot chassis with six slots for Gigabit or Fast Ethernet NICs.

Gigabit Ethernet technology has some limitations, however. First, until compatibility issues are resolved, users may have to manage a variety of products that do not interoperate completely. Second, not all computer operating systems work equally well with Gigabit Ethernet, although Solaris and other flavors of UNIX as well as NetWare readily handle performance demands. Third, 1-Gbps Ethernet still will not completely address customer issues such as the need for predictable and sustained performance, fault tolerance, bandwidth management, and quality of service. To date, ATM still provides a better solution to those needs.

Token Ring

Token Ring, which has lagged far behind Ethernet in market share and speed for several years, finally will address the latter issue. In November 1997, the IEEE began to discuss a high-speed Token Ring specification, spearheaded by the High-Speed Token Ring Alliance (HSTRA). The alliance presented a draft standard to the IEEE 802.5 committee for 100 Mbps and gigabit Token Ring (compared with the current top speed of 16 Mbps). The vendors hope to put the standard on a fast track for completion by late 1998, to coincide with product introductions.

Formed in September 1997, HSTRA is an outgrowth of the now-defunct Alliance for Strategic Token Ring Advancement and Leadership. Current members of HSTRA include 3Com, Bay Networks, Cabletron, Cisco, IBM, Madge Networks, Novacom, Olicom, and Xylan.

According to a 1996 survey, the vast majority of Token Ring networks are overloaded, and some network managers may migrate to entirely different technologies such as Fast Ethernet if they cannot afford to deploy ATM backbones. Vendors and analysts say that ATM makes senses for large Token Ring customers, but smaller sites are considering the new high-speed Token Ring technology. One issue is whether to implement 1-Gbps Token Ring natively or encapsulate it in another protocol such as Fast Ethernet. The first alternative appears to be more popular, with quick deployment expected to take priority; standardization efforts on the encapsulation solution will begin toward the end of 1998.

Asynchronous Transfer Mode (ATM)

ATM is a high-bandwidth, low-delay, packet-switching, and multiplexing technology. Usable capacity is segmented into 53-byte fixed-size cells, consisting of header and information fields and allocated to services on demand. ATM is an evolution of earlier methods such as X.25 packet switching and frame relay, which used frames or cells that varied in size. Fixed-length packets (like those used in ATM) could be switched more easily in hardware, however, and thus result in faster transmissions.

The ATM Forum, formed in 1991, currently claims more than 850 members in the aerospace, computer, communications, entertainment, manufacturing, and other industries. One of its key missions is to work with ITU and ANSI standards organizations on ATM specifications. The forum has promoted ATM as a versatile, high-bandwidth technology for a variety of types of traffic and services that can be implemented across network backbones, WANs, LANs, and desktop connections.

Compared with other network technologies, ATM offers several advantages: large increases in maximum supported bandwidth; support for multiple types of traffic such as data, video, and voice transmissions on one communications line; and virtual networking capabilities, which increase bandwidth utilization and ease network administration. ATM also removes the distinction between LANs and WANs. In the LAN environment, ATM can replace or augment LAN technologies such as Ethernet, Token Ring, or FDDI. In the WAN environment, ATM can be used as an alternative to frame relay, X.25, or leased lines.

Each ATM cell contains a 48-byte payload field and a 5-byte header that identifies the cell's virtual circuit. ATM allocates bandwidth on demand, making it suitable for high-speed combinations of voice, data, and video services. Currently, ATM access tops out at 622 Mbps; however, ATM has been doubling its maximum speed every year. Continuing at that rate, ATM could reach 10 Gbps by the year 2000. Although the technologies may switch dominance from time to time, ATM will probably always be faster (but more expensive) than Gigabit Ethernet.

ATM's cell networking topology is beneficial when predictable delay characteristics and reliable throughput are important, such as when carrying video transmissions. Without a constant stream of bandwidth, video transmissions can break up and parts can be lost. Other networks provide bandwidth on an ad hoc basis, so bandwidth cannot be guaranteed. ATM includes a feature, dubbed Quality of Service (QoS), that ensures sufficient bandwidth is available for specific applications, eliminating delayed packets. Currently, Ethernet, Token Ring, and FDDI all lack this capability, although the latter two provide what could be considered limited QoS.

Despite these advantages, companies are only deploying ATM to the desktop in isolated instances to handle high-end applications such as imaging, videoconferencing, or complex graphics. Most applications do not yet require the sophisticated features ATM offers, and the technology has many drawbacks. For example, moving to ATM requires a company to install new adapter cards, wiring hub connections, and network management software. In addition, network technicians have to learn the nuances of a new technology. Equally important, an ATM adapter card, which supports transmission speeds of 150 Mbps, costs from $500 to $750, and a complete connection can cost $1,500 or more. Finally, ATM has a relatively high overhead (5 bytes out of 53 bytes, or about 10 percent), which is too much on slow lines and too expensive and inefficient on fast lines.

ATM is having more success replacing FDDI as a backbone networking technology. Although ATM offers companies a migration path to high-speed transmission rates and eases the convergence of LANs and WANs, it still will have trouble displacing Ethernet at the desktop because the costs of replacing the network connections do not justify the performance improvement for most individual users.

■ LAN Emulation

To add routing features to ATM networks, the ATM Forum developed two specifications. The first technique, LAN Emulation (LANE), carries existing networking technologies such as Ethernet or Token Ring on ATM networks. LANE allows ATM to be deployed on a legacy LAN or with legacy applications. Suppliers began delivering LANE products in 1995, but the first versions were unsophisticated and failed to offer sufficient throughput.

In July 1997, the ATM Forum approved the LANE 2.0 specification, which is expected to accelerate acceptance of the technology for data and multi-service networking. Cisco introduced LANE emulation capabilities for its family of Token Ring products in December 1997, following 3Com, IBM, and Xylan, which began offering Token Ring LANE earlier in the year.

Some analysts, however, believe that LANE will not make ATM a viable transport protocol for LANs. They claim that ATM's point-to-point, connection-oriented framework cannot match the ability of Ethernet and Token Ring to provide connectionless accessibility. They also point out that LANE is an attempt to retrofit ATM to Ethernet, but that no compelling reason exists to put ATM on desktops. They suggest that rather than trying to make ATM act like Ethernet, developers should focus on making applications that take advantage of native ATM.

■ Multi-Protocol Over ATM (MPOA)

The second option for adding routing features to ATM, MPOA (approved in July 1997), routes Transmission Control Protocol/Internet Protocol (TCP/IP) traffic on ATM networks. MPOA essentially does away with the need to maintain a host of routers across large networks. MPOA synthesizes the switching and routing functions of the network into a single process, letting enterprises mix protocols over ATM. For example, some organizations now are using ATM as a scalable backbone for IP protocol transport. MPOA also eases some of the administrative burdens faced by those charged with running large ATM networks.

Vendors already are taking advantage of MPOA's flexibility. For example, 3Com and IBM forged a technology partnership in July 1997 to round out each other's offerings for campus and work-group networks and between ATM, Ethernet, and Token Ring LANs. Under the agreement, 3Com will deploy IBM's Multiprotocol Switched Services Layer 2/Layer 3 switching/routing technology into 3Com adapters, allowing users to extend MPOA functionality to desktop devices.

■ 25-Mbps ATM

ATM backers have recognized its problems at the desktop and have developed a slower, less-expensive version. These 25-Mbps ATM connections run over twisted-pair wiring so a company does not have to run fiber cable to each user's desktop – a requirement for the higher-speed version of ATM.

However, this networking option has several drawbacks that could limit its success. First, although some 25-Mbps ATM adapter cards cost as little as $179 (due to significant price reductions in the last 2 years), these products are still more expensive than Fast Ethernet cards, which are priced at around $100 and run at 100 Mbps. Second, 25-Mbps ATM requires companies to replace their network hubs with ATM switches.

The demise of 25-Mbps ATM has been widely predicted by the media since early 1997. In particular, detractors point to the fact that Whitetree, a desktop 25-Mbps pioneer, was acquired by Ascend in 1997 and now is developing technology for Ascend's existing product line rather than focusing on desktop 25-Mbps ATM.

Frame Relay

Frame relay is a packet-switching protocol used in WANs that has become popular for LAN-to-LAN connections between remote locations. Originally, frame relay access topped out at about 1.5 Mbps; although so-called "high-speed" frame relay now offers around 45 Mbps, this speed is still slow compared with other technologies such as ATM.

Frame relay services are provided by all the major telecommunications carriers in the world. These services employ a form of packet switching analogous to a streamlined version of X.25 networks. The packets are in the form of frames, which are variable in length. The key advantage to this approach is that a frame relay network can accommodate data packets of various sizes associated with virtually any native data protocol. A frame relay network is completely protocol-independent; because it does not undertake a lengthy protocol conversion process, frame relay offers faster and less-expensive switching. Frame relay also is faster than traditional X.25 networks because it was designed for today's reliable circuits and performs less-rigorous error detection.

Because of its variable-length packet architecture, however, frame relay is not the most efficient technology for real-time voice and video. (See "Frame Relay" on page 221 for more information.)

Formed in May 1991, the Frame Relay Forum is an organization of more than 300 frame relay equipment vendors, carriers, end users, and consultants working to speed the development and deployment of frame relay products as well as interfaces with other broadband technologies, such as ATM.

In October 1997, 16 of the industry's leading equipment vendors and service providers announced they would collaborate to promote the use of frame relay's switched virtual circuits (SVCs) in corporate networks. Compared with permanent virtual circuits (PVCs), which establish a permanent, point-to-point connection, SVCs establish a connection across a network on an as-needed basis that lasts only for the duration of the transfer. The specific path of an SVC is determined on a call-by-call basis, based on the end points and the level of congestion in the network. The advantage of this "dynamic" routing is that if some links in the network are overloaded or fail, SVCs can find an alternate route. Although SVCs are provided for in frame relay, they have not been implemented primarily because PVCs handle the routing adequately and are much less complex to provision.

Among the participating vendors in the collaborative SVC frame relay initiative are 3Com, Ascend, Bay Networks, Cabletron, Cisco, Digital, Harris & Jeffries, Hughes Network Systems, LCI, Newbridge Networks, Netrix, Nortel, MCI, Sprint, Trillium Digital Systems, and Unispan. MCI, for example, began providing frame relay SVCs in November 1997, when it introduced its new frame relay service, HyperStream SVC. MCI says that the SVC capability is fully compliant with the Frame Relay Forum's specification for the service.

MCI's service offers speeds between 16 Kbps and 6 Mbps and gives customers the ability to mix and match SVCs and frame relay PVCs on the same ports. This capability could produce large savings for corporate customers with sites that have occasional demands for high-bandwidth data services.

Fibre Channel Interconnect

Fibre Channel Interconnect is an emerging standard for providing high-speed (1 Gbps, bidirectional) connections over point-to-point, loop, or switched networks. Originally a technology exclusive to fiber optic cable, when Fibre Channel became available on copper cable, the International Organization for Standardization (ISO) adopted the French and British spelling of fibre to disassociate the technology from only fiber optic cable.

This standard is being used to create clusters of high-performance servers and workstations and to connect servers and workstations to peripherals, particularly mass storage devices. A Fibre Channel link can deliver between 133 Mbps to 1,062 Mbps up to a distance of 10 kilometers over optical fiber. Over copper cable, Fibre Channel can be run at lower speeds over shorter distances.

As server clustering becomes an increasing option for IT managers, Fibre Channel will compete with other high-speed networking systems, such as ATM, to link the servers into a single, coherent device. This networking technique can be geared to host a variety of high-speed protocols, such as FDDI, High Performance Parallel Interface (HiPPI), Intelligent Peripheral Interface (IPI), and Small Computer Systems Interface (SCSI).

In September 1997, Compaq endorsed Fibre Channel for its next-generation storage solutions. This announcement will build momentum behind the technology, which is expected to take off in the market in 1998. According to HP, which has teamed with Compaq on a Fibre Channel strategy, Fibre Channel is poised to become the leading storage/server interface. Compaq selected HP's Tachyon Fibre Channel controller for its new storage products, and the two companies will work on better performance and interoperability for future generations of Fibre Channel solutions. As part of the agreement, HP will be able to use Compaq's host bus adapter design and suite of software device drivers to create and manufacture its own product, scheduled to be available by the end of 1998. In addition, both companies will work on reinforcing the Tachyon architecture as the commercial standard for Fibre Channel input/output (I/O).

In December 1997, Sequent announce that its NUMA-Q 2000 UNIX servers would come with integrated Fibre Channel network switches instead of a SCSI interconnect. Using a switch instead of an adapter card supplies users with several redundant paths to storage devices, which in turn guarantees high availability (important for large, complicated server clusters). Sequent will use 6-port or 16-port versions from Brocade Communications Systems' Silkworm switch family. The switches add about $1,800 per port to the server cost.

Analysts say this type of agreement between storage makers and server vendors offers users the best way to get Fibre Channel switching. For example, Digital also plans to add Silkworm switches to its StorageWorks disk arrays. Data General, HP, and Sun also have Fibre Channel switch plans in place.

4.2.3 Network Building Blocks

Enterprise networks in medium to large organizations are complex entities. The devices used to construct and manage these networks range from inexpensive, commodity network interface cards (NICs) to high-priced, high-value routers.

A quick snapshot of a typical LAN/WAN is shown in Figure 4-6. Here, a heterogeneous network starts with the backbone of a corporate or public WAN and continues through switches and routers to different kinds of LANs within a campus or building as well as remote user sites.

Virtually all communications protocols used today can be analyzed using a model defined by the ISO called the Open Systems Interconnection (OSI). The OSI model analyzes the communications process in seven different categories and places these categories in a layered sequence based on their relation to the user. Layers 7 through 4 deal with end-to-end communications between the message source and the message destination; Layers 3 through 1 pertain to data transport across the network.

- **Layer 1** – The Physical Layer deals with the physical means of sending data over lines (the electrical, mechanical, and functional control of data circuits).
- **Layer 2** – The Data Link Layer is concerned with procedures and protocols for operating the communications lines. It also has a way of detecting and correcting message errors. On a shared-media LAN, such as Ethernet or Token Ring, it includes a Media Access Control (MAC) sublayer that defines how the network allocates bandwidth to the multiple uses.
- **Layer 3** – The Network Layer determines how data is transferred between computers. It also addresses routing within and between individual networks.
- **Layer 4** – The Transport Layer manages end-to-end delivery of information within and between networks, including error recovery and flow control.
- **Layer 5** – The Session Layer is concerned with dialog management. It controls the use of the basic communications facility provided by the Transport Layer.
- **Layer 6** – The Presentation Layer provides transparent communications services by masking the differences of varying data formats (character codes, for example) between dissimilar systems.
- **Layer 7** – The Applications Layer contains functions for particular applications services, such as file transfer, remote file access, and virtual terminals.

1. Hub: A network device that terminates the connection to each network location, thus providing connectivity between users on the same LAN through the use of a passive or active backplane.

2. Ethernet switch: A network device that uses packet-based technology to switch traffic between workstations and other end nodes, providing dedicated bandwidth to each node. Ethernet switches are by far the most widespread, although Token Ring and FDDI versions are starting to emerge.

3. Router: An internetworking device, primarily software-driven, that routes traffic between two different LAN types or two network segments with different protocol addresses. Operates at the Network Layer.

4. ATM switch: A network device that uses cell-based technology to switch traffic, providing dedicated bandwidth that operates most commonly at 155 Mbps and 622 Mbps.

5. Border switches: An internetworking device that accepts packet-based feeds from traditional LANs and converts them into ATM-based cells for further transmission by an ATM switch or on an ATM-based public network.

6. Network interface card (NIC): The network physical interface between a desktop computer or server and the network.

7. Bridge: An internetworking device that handles traffic exchange between two similar LANs (or two different LAN types). Operates at the Data Link Layer.

Figure 4-6: Networking Devices Overview

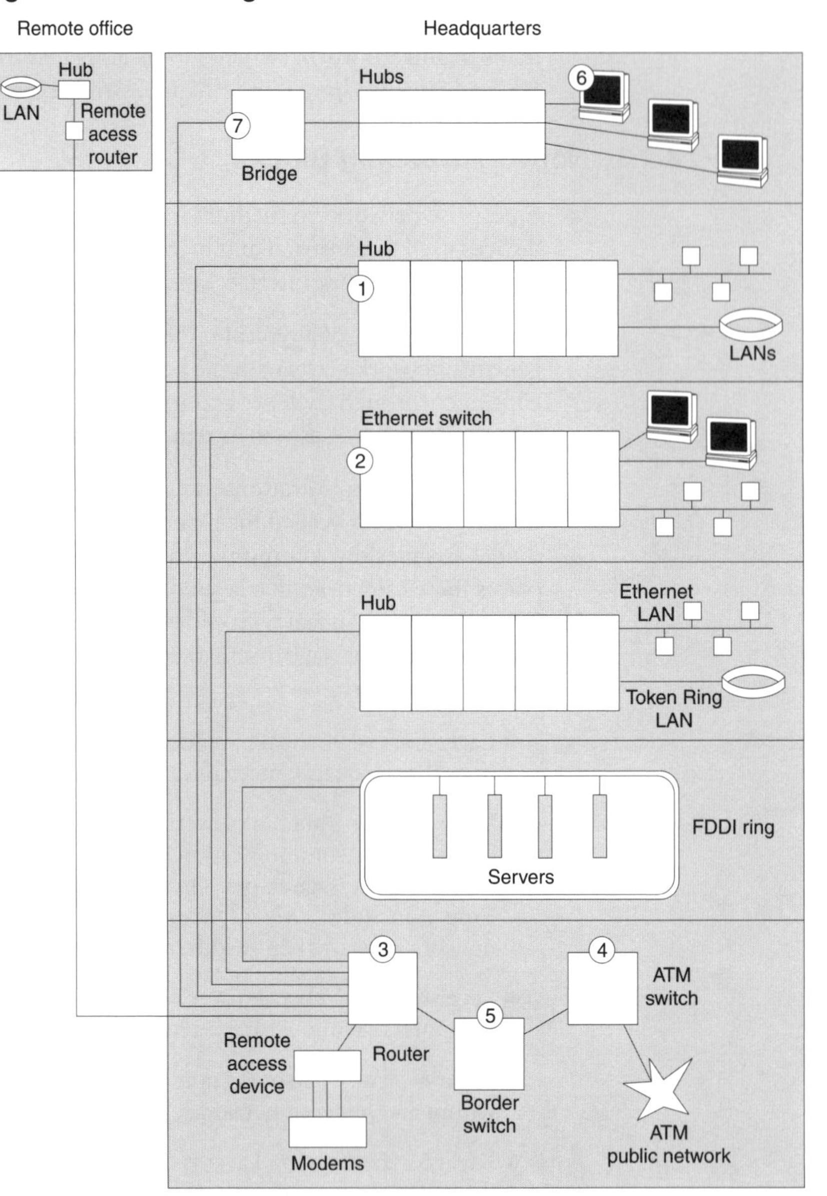

Wiring Hubs and Switches

In the 1980s, most LANs consisted of a series of network devices, such as PCs and printers, connected through an Ethernet network interface card (NIC) or Token Ring hub to a single piece of coaxial cable. By the early 1990s, this arrangement largely had been replaced by the use of wiring hubs. Network growth enabled vendor sales of structured wiring hubs, and vice versa. These hubs let LAN managers replace coaxial cable with the less-expensive and easier to install telephony industry-standard twisted-pair wire. These hubs terminated the twisted-pair connections to each user's desktop in central devices that resembled patch panels. This design made it easier for companies to change their network configurations as needed at the hub instead of by replacing and re-laying coaxial cable.

Hubs also improved network resiliency and have been increasing steadily in sophistication. With an intelligent hub, a network technician could examine the status of each connection and disconnect a faulty system so it would not affect the entire network. Consequently, network uptime and reliability greatly improved.

Although the initial hubs were still logically configured as shared-media network segments, current products such as switches offer each computer its own dedicated bandwidth. Switches have an internal bus that can carry the aggregate bandwidth of multiple network nodes. Another advantage of switches is that users can be moved from one LAN segment to another with a simple software change, whereas previously, a technician had to change the desktop wiring. However, not all switches offer virtual LAN (VLAN) capability. Standards for VLAN technology have not been established to date, and configuration can be fairly complex. (See “Virtual LANs (VLANs)” on page 182.)

Bridges

Bridges are the simplest and easiest way to connect LANs. A bridge is a device that joins physically separate LAN segments (such as two Ethernet cables) into one logical LAN segment. A bridge learns to distinguish traffic that is local on one segment from traffic that should be copied to other attached segments. There are four main categories of bridges: transparent, source routing, encapsulating, and translational. Transparent bridges are used primarily for Ethernet, whereas source routing bridges are used for Token Ring networks. Encapsulating bridges connect two segments of the same media (for example, Ethernet to Ethernet, or Token Ring to Token Ring) over an intermediary such as FDDI. The receiving bridges remove the envelope, check the destination, and send the packet to the destination device. Translational bridges are used to bridge between different types of network media, for example, between Ethernet and FDDI.

Bridges are designed to interconnect multiple LAN segments (such as physical cable or wiring hubs) and make them appear as a single, composite LAN. A composite LAN is limited in its ability to route data selectively and efficiently, to permit robust configuration topologies for the network, and to enforce security. Also, broadcast packets (those destined for all hosts on the network) usually are transmitted to all LAN segments interconnected by a bridge. Thus, a problem on one segment could affect users on all other segments, and all stations on a composite LAN have to process each broadcast, which results in additional overhead on each node. This problem is compounded as the number of interconnected LANs grows. A broadcast storm (the disruption that results when large numbers of broadcasts occur very rapidly, such as when a configuration change to one LAN is sent to all others) can adversely affect performance of all nodes on a composite LAN. This is one reason why companies have avoided relying solely on bridges to construct enterprise networks.

Bridges are inexpensive and simple to install and are more appropriate for networks that are less susceptible to broadcast storms. Token Ring networks, when used in conjunction with source routing, direct traffic better than Ethernet bridges.

Routers

A router interconnects two or more physically and logically separate network segments. In contrast with a bridge, when multiple network segments are joined together by a router, they maintain their separate logical identities but now constitute an internetwork – a group of networks linked by routers.

Although a bridge passes all traffic, regardless of its network protocol type, a router distinguishes data packets according to protocol types, such as TCP/IP, Internetwork Packet eXchange/Sequenced Packet eXchange (IPX/SPX), or SNA. The router then forwards traffic according to network-level addresses (rather than employing the individual device hardware addresses used by a bridge), using information that the routers exchange among themselves to find the best path between network segments. Broadcasts typically are not forwarded by a router, thus providing some isolation from broadcast storms.

Routers also can be used to limit access to a network, in some cases by the type of application (allowing e-mail traffic, but not file transfers, for example), thus providing a certain measure of security. This capability degrades the performance of the router, however, and is not used very often except in creating firewalls, which provide various levels of security when an organization is linked to the Internet. (See "Firewalls" on page 415.)

Although an essential and widely deployed component in the Internet and many enterprises, routers are complex and costly devices that add considerable overhead to the transmission of data. Routers work with packet-switched networks, in which each packet is inspected and then forwarded. Because these packets are implemented in the router's software, performance is further affected. In addition, routers can be difficult to configure and maintain.

Routing technology is moving in two different directions: a switched core surrounded by a layer of edge routing (also known as an overlay) and a new generation of gigabit routing devices. In the former, routing is overlaid on a switched core fabric. In this model, the core typically is composed of ATM switches with external edge routers. Although routers are present at the periphery of the network, ATM protocols rule the network's inner workings, providing QoS and other functions. Leading vendors providing this type of solution include 3Com, Bay Networks, Cisco, Digital, Fore, Hitachi, NEC, Newbridge, and Xylan.

The second path involves a new generation of gigabit routing devices that have been termed switching-routers or routing-switches. These devices have been fitted with customized chips called application-specific instruction circuits (ASICs) that assist in the routing process and have architectures that can handle millions of packets per second (usually only IP). Some of the key vendors in this area include startups Extreme, Foundry, and GigaLabs as well as established vendors such as 3Com, Bay Networks, Cabletron, and Cisco.

■ New High-Speed Routers

The traditional router is giving way to devices that perform routing functions but do not inspect each frame. The first frame is analyzed at the Network Layer (Layer 3), and a route to the destination is determined. Once this route has been calculated, the remaining frames of the message are then forwarded at the Data Link Layer (Layer 2) without the need to recalculate the route each time, thus making this method considerably faster.

For example, Cisco's 20-Gbps to 80-Gbps Gigabit Switch Router (GSR), introduced in September 1997, offers new technology for distributed route-table management, switch fabric design and performance, and line card performance.

The GSR, a product initially oriented toward commercial ISPs, will pioneer technologies that will then be used in the products Cisco sells to end-user organizations: the Cisco 7x00, StrataCom, and LightStream product lines. Innovations in Cisco's GSR include the elimination of address caching by keeping the product's entire forwarding information base on each line card; a 16-port, fully synchronous crossbar switch with 5-Gbps ports and a new scheduling algorithm that can handle 17 virtual output queues per port, raising switch efficiencies; and two generations of GSR line cards. New virtual output queues on the line cards are the basis of GSR's efficient fabric utilization, multicast support, head-of-line blocking avoidance, and any future class-of-service additions. The first generation will include Packet Over SONET (Synchronous Optical Network) at OC-3/12 and ATM OC-12. Other line cards, due in 1998, will support OC-48, Gigabit Ethernet, and Digital Service Level 3 (DS-3).

A half-dozen startups are developing routers capable of handling speeds of 200 million packets per second and eventually, perhaps trillions of packets per second. These new devices may eliminate the Internet's traffic bottlenecks and reduce the traffic loads of large, multi-enterprise extranets. At the heart of the new products are application-specific integrated circuits (ASICs) that can handle up to 1 million packets per second on a single chip. The routers combine the capabilities of routers, ATM, and frame relay switches and can perform advanced, automated mapping of IP addresses. The routers also will support extensive QoS features.

The startups fall into two groups: vendors readying terabit products that combine router, switch, and trunking features (Avici Systems, GigaPacket Networks, and NeoNet) and others developing similar but slower products, targeting Internet service providers' (ISPs') points of presence (POPs) and private networks (Berkeley Networks, Juniper, and Torrent Networking Technologies).

Startups are using a variety of technologies to implement wire-speed routing and to differentiate themselves with features such as QoS. For example, Torrent Networking Technologies is challenging Cisco's 7500 routers with wire-speed Gigabit Ethernet IP routers for enterprises and service providers, introduced in September 1997. Yago Systems (in which Cabletron is a minority investor) is developing a gigabit-speed switching router that uses Layer 4 packet-header information to provide application-level QoS.

Founded by former Cisco, MCI, and Sun engineers, Juniper has garnered the most attention among the startups because in 1997, it received $40 million in funding from major vendors including 3Com, Lucent Technologies, Nortel, and UUNet. Juniper's public-network IP router is intended for applications that combine fiber optic and copper network segments and accommodates both packet routing and optical transmission. The system uses wavelength-division multiplexing equipment from Lucent and other companies and will not rely on fast ASICs to accelerate IP protocols.

Another startup, Pluris, is developing terabit-speed routing by building massively parallel routers. The company plans to use less-expensive, off-the-shelf components rather than the expensive massively parallel computers from which it borrowed the load distribution concept, enabling costs to remain low. Pluris expects to begin beta testing its new router in mid-1998.

Established router vendor Compatible Systems released two- and four-port versions of its new switch/router in June 1997. Compatible runs multiple parallel processors using a crosspoint switch on the backplane. The company also is developing a 16-port version of the switch/router. Each port is 1 GB, making the 16-port version equal to 16 GB of backplane throughput. Compatible plans to price its product at 30 percent less than competitive offerings from Ascend Communications and Cisco.

Analysts continue to warn users against the dangers of buying from any vendor that is not well-known, however, making the future of many startups uncertain at best. More important, despite the intriguing technologies these new companies offer, users will continue to demand good service and support – a harder quantity for small organizations to deliver. In addition, users may restrict the number of vendors used to reduce incompatibility and support problems, a practice that has contributed to the consolidation of the network industry.

■ New Routing Protocols

Routing protocols are at the core of a router's ability to manage dynamic network conditions automatically. A routing protocol determines the best path for routing each packet over a network. Early protocols were designed for smaller networks, making them unsuitable for large-scale enterprise networks, especially those connected to the Internet. Protocols that have been supplanted by versions designed for today's complex networks include Gateway-to-Gateway Protocol (GGP), Interior Gateway Protocol (IGP), Exterior Gateway Protocol (EGP), Inter-Domain Routing Protocol (IDRP), Routing Information Protocol (RIP), and Source Demand Routing Protocol (SDRP).

Currently, all routing protocols fall into one of three classes: distance vector, link state, and path vector. Each class uses its own characteristics and methods to determine the best route to a destination. Distance vector and link state protocols are best suited for interior routing, whereas path vector protocols address the needs of border routes.

- **Distance vector protocols** – In these protocols, each router maintains a table indicating the distance to every other router. It also periodically broadcasts this information to all other routers, allowing them to update their routing tables if any changes have occurred in the network topology. This distance information then is used to determine the best route to any destination. The shortest distance is usually the most preferred; however, it is possible to consider other factors in the decision as well. Distance vector protocols include RIP and Enhanced Interior Gateway Routing Protocol (EIGRP).
  - **RIP** – The pioneer protocol, RIP is still popular for small networks with few routers. However, vendors never have agreed on a standard for the protocol. As a result, AppleTalk, DECnet, TCP/IP, NetWare, and VINES all use incompatible versions of RIP. More important, RIP never was designed to support large networks in which there may be thousands of possible destinations from any given source. RIP works just by counting "hops" and originally was limited to networks containing less than 15 hops. In addition, the volume of traffic generated in a large network by every router broadcasting its distance information every 30 seconds rapidly becomes overwhelming. Although RIP Version 2 (RIP2) added support for subnet zero, classless IP, and some basic authentication, modern networks have outgrown RIP.
  - **EIGRP** – Introduced in 1994 as an enhancement to Cisco's 1988 IGRP, EIGRP is a proprietary protocol available only on Cisco products. (Cisco currently has no plans to release EIGRP to its partners or the industry.) IGRP and EIGRP both calculate their routes based on bandwidth, latency, reliability, and load. This process makes their metric values far more accurate than that of any other protocols. EIGRP also supports load

balancing across multiple paths to allow users to use the full network bandwidth. Also included in EIGRP is the capability to detect loops in network routing.

- **Link state protocols** – Link state is type of routing protocol that includes more analysis in its calculations than the distance vector protocols. Instead of basing routing decisions on a table showing the distance to each other router, all network nodes maintain a complete copy of the network topology map from which to perform a computation of the best routes. The network map is held in a database, where each record represents one link in the network. Link state protocols include the Open Shortest Path First Protocol, Novell's NetWare Link Services Protocol, and the OSI protocol suite's Intermediate System to Intermediate System Protocol.
    - **Open Shortest Path First (OSPF)Protocol** – Introduced by the Internet Engineering Task Force (IETF) in 1989, OSPF was designed to address some of RIP's problems. The Internet Architecture Board currently recommends OSPF as the routing protocol of choice on the Internet. Unlike other protocols, OSPF uses multicasting to deliver route information update packets. However, OSPF has drawbacks, primarily that it only provides for IP and does not accommodate IPX.
    - **NetWare Link Services Protocol (NLSP)** – A link state protocol from Novell to improve performance of IPX traffic in large internetworks, NLSP transmits routing information only when a change occurs somewhere in the network or every 2 hours if there is no change in network topology. Analogous to OSPF in the IP world, NLSP reduces the overhead of frequent RIP broadcasts, which traditionally have consumed scarce WAN bandwidth on large Novell networks.
    - **Intermediate System-to-Intermediate System (IS-IS) Protocol** – The IS-IS protocol is an OSI link-state hierarchical routing protocol based on DECnet Phase V routing, whereby routers exchange routing information based on a single metric to determine network topology. Integrated IS-IS (formerly called Dual IS-IS) includes support for IP or other networks. Integrated IS-IS implementations send only one set of routing updates, regardless of protocol type, making it more efficient than two separate implementations.
- **Path vector protocols** – Although distance vector protocols fit well with the hop-by-hop approach of IP routing, they do not offer sufficient protection against routing loops. Using the link state approach to this problem is unrealistic in extremely large networks with complex topologies because of the amount of routing information that would need to be maintained. Instead, developers invented a new technology called path vector.

  Path vector protocols maintain comprehensive lists of routes that can handle arbitrary topologies and, therefore, can be used for various autonomous systems such as a series of independent networks all connected to the Internet. In these protocols, each router broadcasts the complete list of networks that lie between the source and destination of each route it knows. The most prevalent path vector protocol is the Border Gateway Protocol (BGP).
    - **Border Gateway Protocol Version 4 (BGP4)** – Released in 1994 by the IETF, BGP4 is the only border routing protocol that supports Classless Inter-Domain Routing (CIDR), a modification to the method for assigning IP addresses that removes certain restrictions on address assignment, thus allowing the existing 32-bit Internet address space to be used more efficiently. BGP4 is targeted toward companies that have multiple connections to the Internet or that receive information from multiple ISPs. BGP4 lets a company's interior routing protocols use the most cost-effective route. If a

link to the Internet fails, these interior routes are recalculated to redistribute traffic to a functioning link, and the ISP sees this change. Drawbacks to BGP4 include the lack of support for policy-based routing, which is important for companies that need to route traffic based on where it came from as well as where it is going.

As companies require more stable Internet connections and redundancy becomes more important, BGP4 is expected to play an increasing role in networking systems. Plans to combine portions of the OSI's Inter-Domain Policy Routing (IDPR) into the next version of BGP are underway within the IETF.

Switches

There has been some confusion about the use of the term switching in recent years. LANs are moving from shared media to switched connections, but Ethernet or other LAN-oriented switches function similarly to bridges, sending packets to their destinations based on MAC-level hardware addresses. The current debate between routing and switching generally has more to do with handling traffic over public and private WANs using protocol-specific (Layer 3) addressing.

Switches offer some advantages over bridges and routers. Routers were delivered before ATM was widely available, and most do not take advantage of its key features. Switches are high-capacity, cell-based devices designed to support broadband networking. A switch acts as a faster form of bridge, where operations are performed in hardware using specialized chips, and the users usually do not compete with one another for access to bandwidth. Eventually, switches will be able to integrate access, multiplexing, and switching functions.

Switches are replacing routers in many WANs mainly because of performance advantages. However, Layer 2 switches lack the ability to prevent broadcast storms, provide security, and translate network differences. To address those limitations, suppliers have begun integrating routing functions into their switches.

Layer 3 Switching/Routing

Currently, a half-dozen proposals for new interswitch protocols or modifications of the IP protocol will give users the performance enhancements that come with Layer 2 switching while retaining the services that routers perform, such as security, traffic prioritization, and policy management. A switch that performs optimized IP routing delivers the speed benefits of switching without asking users to rearchitect their IP addressing scheme.

Layer 3 switching, which adds router intelligence to switches, combines high performance and low cost with the ability to determine a packet's ultimate destination. Choosing a Level 3 switching approach is confusing, however, because of the number of diverse strategies used by vendors. Each vendor has its own scheme, and it is difficult to decide as yet which will be the most effective or most likely to be adopted as a standard.

In the meantime, two general approaches are used: adding Layer 3 functions to a Layer 2 switch or adding Layer 2 services to a router. Both will work, but it is generally easier to add Layer 3 functionality to a Layer 2 switching architecture than it is to allow a Layer 3 router also to serve as a Layer 2 device.

Initially, most proposals are targeting ATM networks, including Ipsilon's IP switching, Cisco's tag switching, and the ATM Forum's MPOA. For Ethernet, Bay Networks' acquisition Rapid City Communications has implemented IP routing in silicon, permitting switch-speed routing without introducing any new protocols between switches.

■ IP Switching

The Ipsilon-developed IP switch is a combination of an IP router and ATM switching hardware, which also is being incorporated into several other vendor offerings. The IP switch implements the IP protocol stack on ATM hardware, allowing the device to shift dynamically between store-and-forward and cut-through switching based on the flow requirements of the traffic as defined in the packet header. Data flows of long duration, thereby, can be optimized by cut-through switching, with the balance of the traffic afforded the default ATM treatment, which is hop-by-hop, store-and-forward routing.

IP switches can address high-performance internetworking needs where backbone routers may have run out of bandwidth. One of the advantages of IP switching is the use of IP itself – a mature, well-understood protocol that is widely deployed across a wide range of networks.

However, the IP switch has caused considerable controversy in the industry because it uses a proprietary method for providing high-speed ATM switching of IP, bypassing standards sanctioned by the ATM Forum. In response, Ipsilon has made some of its protocols available, which have been adopted by companies such as Digital, Hitachi, and NEC.

■ Flow-Switching

Flow-switching technologies, represented by Cisco's tag switching (similar to IBM's aggregate routing and 3Com's Fast IP), ultimately may be more efficient and support greater speeds than ATM can – without the overhead. In a flow-switching architecture, a Layer 3 routing switch establishes the fastest, most efficient route of a packet by determining whether the packet is part of a "flow" of data from one source to another. Once a flow has been determined, a virtual connection is established between the two points, and all relevant packets are sent over that path at a much faster rate than if traditional route-calculation was required for each packet.

Cisco's tag switching (also sometimes called switched routing) adds information in the form of a label, or "tag", to each packet as it is loaded onto the network. This information can be decoded for routing purposes without a tag switch having to review routing tables. (A routing table lookup normally is used on each packet, providing the redundancy and fault tolerance for which IP is known.) Although Cisco's tag switching approach adds a small amount of overhead to each packet, the speed of packet processing is improved considerably, especially in complex networks in which multiple routers or switches must act on each packet. Tag switching can be applied to other network protocols besides IP, unlike IP switching. As of late 1997, Cisco's tag switching was being evaluated for standardization by the IETF's Multiprotocol Label Switching (MPLS) working group.

Cisco has provided more than 80 percent of the routers that make up the Internet and plans to leverage those routers along with its own switches in tag switching. If other vendors could agree on a common approach to IP switching, they could prevent Cisco's tag switching technology from dominating the industry.

A number of startups also have appeared to challenge Cisco's position in the market. Among these are FlowWise Networks, which announced a Layer 3 switch family supporting both packet and cell traffic in September 1997. The products let administrators set four static priority levels based on addresses or policies. Another startup, NBase, is using Dynamic Host Configuration Protocol for automatic setup of virtual LANs in a Layer 3 switch that is shipping already.

Layer 4 Switching

In addition to the flood of high-speed routers and Layer 3 switches being introduced in late 1997 is a concept called Layer 4 switching. Layer 4 switches can route and forward TCP/IP traffic based on TCP information in addition to Layer 2 and Layer 3 addresses. For example, users can allocate bandwidth on a Layer 4 switch based on the standard TCP port number of an application.

Layer 4 switching is being promoted just as most customers are beginning to adopt Layer 2 and Layer 3 switching, causing debates about what occurs at Layer 4 (the Transport Layer in the OSI model) and how relevant switching is there. Proponents of Layer 4 switching include Alteon Networks, Torrent Network Technologies, and Yago Systems. However, other vendors, including Cisco, insist that such functionality is available in existing routers operating at Layer 3, the Network Layer.

To make the matter more complicated, some vendors are developing devices they describe as "layerless." In November 1997, a new company called Neo Networks announced its StreamProcessor network forwarding product line. Instead of using UDP and TCP ports, StreamProcessor looks at the data itself to determine the application and makes decisions based on preset policies regarding who has access to which data. Neo Networks' product uses 7 ASICs and more than 1,000 RISC processors to manage networks and to process multiple-gigabit data flows in real time based on information at Layers 2 through 7.

Virtual LANs (VLANs)

LANs have grown larger, more complicated, and more dispersed since they were introduced in the 1980s. Along with this growth has come the requirement for greater ease of administration.

Virtual networks, a LAN hub or switch feature, enable companies to use network bandwidth more efficiently by grouping users who typically exchange information on the same LAN segment. Rather than requiring a technician physically to reconfigure a network, a virtual network enables administrators to perform network adds, changes, and moves electronically.

VLANs allow network managers to create logical LANs that can arbitrarily assign nodes to a given logical network segment, regardless of geographical location. Virtual networking segments the network into logical LANs, decoupling the physical connections from the logical network topology and keeping the two views synchronized with centrally controlled management software. Virtual networking can simplify tasks such as moving or adding workstations, creating work groups, reassigning security levels, and other changes. By using switching technology to segment LAN traffic, organizations can connect more users to the network without reducing performance.

Load balancing (gaining additional bandwidth without making changes to the infrastructure) can be performed with next-generation intelligent hubs, which can support a large number of LAN segments and use switching to let network managers optimize network performance by reducing or increasing the number of users per segment through software controls. Bandwidth can be allocated dynamically as required, without segmenting the logical network, through software that assigns bandwidth to circuits on an as-needed basis.

Several technical obstacles hinder virtual network deployment. For example, deploying VLANs can be difficult. To take full advantage of the technology, corporations need to understand how much network traffic different employees share so they can group employees who exchange a lot of data on the same LAN. Making such a determination requires comprehensive network performance information, but current products offer limited insight into network traffic patterns.

The lack of standards in virtual networking also has been a problem. Vendors have devised proprietary virtual networking techniques that do not interoperate, prohibiting a company from integrating equipment from multiple vendors in a single virtual network. Currently, it is difficult, if not impossible, to use products from different vendors to establish interoperable, multi-vendor VLAN solutions. Recognizing this problem, the IEEE established the 802.1Q committee in 1996 and charged it with developing a VLAN standard. It appears that initial efforts will result in a standard for using explicit tagging to form port-based VLANs. The tagging will enable packets to flow between products from different vendors that support this evolving standard.

The 802.1Q committee has much more work ahead to develop the standard to a point where manufacturers can consider implementing it. Some important issues that still need to be addressed include how network stations should be registered or assigned to VLANs, how frames without a VLAN association should be handled, what effect expanded frames will have on repeaters that are not VLAN-compliant, and what bridges and other network devices may be used in a VLAN environment. The committee's target completion date of May 1998 should allow time to resolve these issues.

The Network Interoperability Alliance (NIA) was formed in 1996 by 3Com, Bay Networks, and IBM to accelerate implementation of these and other standards. Today, the NIA also includes First Virtual, General DataComm, U.S. Robotics (now a division of 3Com), and Xylan in addition to its founders. In October 1997, the NIA formed a customer panel as part of its continuous improvement process. The panelists represent end users across a range of industries and markets, including manufacturing and service industries, government, health care, and universities, and provide input from a broad spectrum of networking requirements.

4.2.4 Virtual Private Networks (VPNs)

A virtual private network (VPN) is a wide area communications network operated by a common carrier that provides what appear to be dedicated lines when used, but that actually consists of backbone trunks shared among all customers as in a public network. Essentially, a VPN allows a private network to be configured within a public network.

For years, companies have worked with major telecommunications carriers to secure guaranteed long-distance bandwidth for their WANs. These VPNs generally used frame relay or Switched Multimegabit Data Service (SMDS) as the protocol of choice because those protocols defined groups of users logically on the network without regard to physical location. In the 1990s, ATM has gained favor as a VPN protocol as companies require higher reliability and greater bandwidth to handle more complex applications. VPNs using ATM offer companies networks with the same virtual security and QoS as WANs designed with dedicated circuits.

The Internet has created an alternative to VPNs – at a much lower cost: the virtual private Internet. The virtual private Internet (VPI) lets companies connect their disparate LANs via the Internet. Users install either a software-only or hardware-software combination that creates a shared, secure intranet with VPN-style network authorizations and encryption capabilities. A VPI normally uses browser-based administration interfaces.

VPN Standards

Standards are expected to be finalized for Internet-based VPNs in 1998. Currently, four standards are being used for VPNs: IP security (IPsec), Point-to-Point Tunneling Protocol (PPTP), Layer 2 Forwarding (L2F), and Layer 2 Tunneling Protocol (L2TP).

■ IP Security (IPsec)

The IETF has proposed a security architecture for the Internet Protocol that can be used for securing Internet-based VPNs. IPsec facilitates secure private sessions across the Internet between organizational firewalls by encrypting traffic as it enters the Internet and decrypting it at the other end, while allowing vendors to use many encryption algorithms, key lengths, and key escrow techniques. The goal of IPsec is to let companies mix-and-match the best firewall, encryption, and TCP/IP stack products.

The IETF's Secure Wide-Area Network (S/WAN) initiative specifies how IPsec is implemented to ensure interoperability among firewall and TCP/IP products. Currently, users and administrators usually are locked into single-vendor solutions network-wide because vendors have been unable to agree upon the details of IPsec implementation. The S/WAN effort tries to remove major obstacles to the widespread deployment of secure VPNs. Specifications such as SSL and S-HTTP could be layered on top of S/WAN implementations, and all these security specifications then could work together. However, S/WAN also could be used instead of SSL or S-HTTP. To guarantee IPsec interoperability, S/WAN defines a common set of algorithms, modes, and options; key sizes range from 40 bits (for exportability) to 128 bits. S/WAN also can be implemented using the Data Encryption Standard (DES).

IPsec allows a partner's network to access a company's network via an encrypted data link across the Internet. As of November 1997, IPsec still was not an official standard, and further work on the proposal by the IETF working group had been delayed until late 1998. Vendors such as Compatible Systems, TimeStep, and VPNet support IPsec because it provides data encryption as well as authentication, and they are planning to release products that implement the protocol. (See Chapter 5.4, *Security*, for more information on encryption and authentication.)

Because IPsec currently does not specify how to exchanging cryptographic keys over the Internet or manage keys, two (incompatible) schemes have arisen:

- **Simple Key Management for IP** (SKIP) – Sun's SKIP technology combines packet encoding and key exchange management. SKIP secures the network at the IP packet level, allowing any networked application to gain the benefits of encryption without requiring modification. SKIP is unique in that an Internet host can send an encrypted packet to another host without necessitating a prior message exchange to set up a secure channel. SKIP is particularly well-suited to IP networks because both are stateless protocols.
- **Internet Secure Association Key Management Protocol** (ISAKMP) – Provides a framework for Internet key management and the specific protocol support for negotiation of security attributes. By itself, ISAKMP does not establish session keys; however, it can be used with various session key establishment protocols to provide a solution to Internet key management.

SKIP and ISAKMP have very different capabilities. SKIP is optimized for a workstation/IP environment only, while ISAKMP can build security associations in a multi-protocol environment (with low overhead). Version 4 of the IP protocol (IPv4) allows either SKIP or ISAKMP to be used, though IPv6 will require ISAKMP. Although the main goal of IPv6 is solving the exhaustion of Internet addresses problem, it also implements an IPsec security layer, securing anything using TCP/IP.

Point-to-Point Tunneling Protocol (PPTP)

The PPTP is an alternate approach to VPN security. Unlike IPsec, which is designed to link two LANs together via an encrypted data stream across the Internet, PPTP allows users to connect to an organization's network via the Internet by a PPTP server or by an ISP that supports PPTP. PPTP was proposed as a standard to the IETF in early 1996. Although it is still unclear whether PPTP will become an approved standard, firewall vendors are expected to support it in the future. Table 4-7 compares the features of IPsec and PPTP.

Table 4-7: Comparison of IPsec and PPTP

Feature	IPsec	PPTP
Support	Network firewall vendors such as Check Point, Raptor, and TIS	Vendors such as 3Com, Ascend, ECI Telematics, Microsoft, and U.S. Robotics
Application	Remote and LAN-to-LAN VPNs	Remote user-to-LAN; LAN-to-LAN in the future
Maturity	As of November 1997, IPsec consisted of 29 IETF drafts	As of November 1997, PPTP consisted of 33 IETF drafts
Available platforms	None	Microsoft Windows NT Server 4.0, Windows NT Workstation 4.0, and an upgrade to Windows 95
Protocols supported	IP	IP, Novell's IPX, NetBEUI

PPTP was developed by Microsoft with 3Com, Ascend, and U.S. Robotics and is currently implemented in Windows NT Server 4.0, Windows NT Workstation 4.0, and Windows 95 (via an upgrade).

The "tunneling" in PPTP refers to encapsulating a message so that the message can be encrypted and then transmitted over the Internet. However, one downside to PPTP is that by creating a tunnel between the server and the client, the protocol ties up processing resources and may create communications bottlenecks.

■ Layer 2 Forwarding (L2F) Protocol

Developed by Cisco, Layer 2 Forwarding Protocol (L2F) resembles PPTP in that it also encapsulates other protocols inside a TCP/IP packet for transport across the Internet or any other TCP/IP network. Unlike PPTP, though, L2F requires a special L2F-compliant router (requiring changes to the LAN or WAN infrastructure), runs at a lower level of the network protocol stack, and does not require TCP/IP routing to function. It also provides additional security for user names and passwords beyond that found in PPTP.

■ Layer 2 Tunneling Protocol (L2TP)

The Layer 2 Tunneling Protocol (L2TP) combines specifications from L2F with PPTP. In November 1997, the IETF approved the L2TP standard, but widespread implementation is not expected until mid-1998. Cisco is putting L2TP into its Internet Operating System software, and Microsoft is incorporating it into Windows NT 5.0.

Its key advantage over IPsec, which covers only TCP/IP communications, is that L2TP can carry multiple protocols. L2TP also offers transmission capability over non-IP networks. However, critics say that L2TP ignores data encryption, an important security feature for network administrators to employ VPNs with confidence. Also, for the protocol to work, ISPs must implement L2TP-enabled hardware at each POP, a potential logistical hurdle.

4.2.5 Remote Access Devices

Remote access devices typically connect off-site users to a corporate network. These users are salespeople, customer support personnel, branch office workers, and other business professionals who often travel or telecommute rather than work in a fixed office location. As the number of employees working at home or in more distributed locations increases, tools that enable users to enjoy fast, secure, reliable access to networks will continue to be in demand.

Today, the carrier-provided telecommunications services typically used for remote access services include analog phone circuits equipped with new high-speed 56-Kbps modems, cable modems, integrated services digital network (ISDN), and digital subscriber line (xDSL) technologies. The growing demand for remote access has not yet resulted in new, high-speed services that are reliable, widely available, or easily deployed, however. Therefore, most analysts believe that analog phone circuits will be the most widely deployed technology throughout the remainder of the decade.

To differentiate themselves from their competitors, vendors are adding features outside those normally associated with remote access devices. For example, Ascend Communications' Secure Access Manager and Secure Access Firewall offer menu-driven installation of firewall security at remote offices for Ascend's Pipeline 75 routers. In October 1997, Ariel Communications introduced a modem rack that can transform a standard PC-based Windows NT server into a remote-access server; the Rascal RS1000 Model 4802 can support as many as 48 simultaneous modem or ISDN calls. Also in October, Sentient Networks announced the addition of a multi-service access switch to its Ultimate family; the Ultimate 1200 can support as many as 30,000 modem ports in one rack.

Remote access vendors also are promising changes in the ease of management of their ISDN router product lines. New features expected by mid-1998 include additional Dynamic Host Configuration Protocol (DHCP) support and improved first-time setup and filtering configuration. In addition, the industry in general is moving toward a more cohesive management console.

■ Modems

Modems operate at different speeds and include various error-checking capabilities. The specific techniques used to encode the digital bit stream into analog signals are called modulation protocols. The International Telecommunications Union-Telecommunications Standards Section (ITU-TSS) has defined the V Series of modem modulation protocols, which include the following:

- **V.32bis** – Supports transmission speeds up to 14.4 Kbps over a standard analog (two-wire) telephone line. (This standard was used in most high-speed modems until V.34 was finalized in 1994.)
- **V.32terbo** – Not a formally approved standard but sponsored by a group of vendors that developed a way to boost the V.32bis transmission speed to 19.2 Kbps. The V.32terbo specification is in the public domain.
- **V.34** – Takes advantage of extended phone line bandwidth to support transmission rates of up to 28.8 Kbps. **V.34bis** is a faster version of the V.34 standard, which supports rates as high as 33.6 Kbps, considered for years to be the maximum modem speed possible over existing analog phone lines.
- **V.42** – A standard for 19.2 Kbps transmission that includes error-checking procedures.
- **V.42bis** – Adds on-the-fly compression at a ratio of 3.5:1 to the V.42 standard; for example, V.42bis can yield speeds of up to 57.6 Kbps with a V.32bis modem. (On-the-fly transmission is useful only in sending data that has not already been compressed.)

V.34 modems are the most widely used today. Whether they actually can achieve throughput of 28.8 Kbps depends on the quality of the phone connection used.

Vendors have begun shipping faster modems that they claim will support 56 Kbps. These 56-Kbps modems employ a radical change in modem architecture: the higher-speed data rates are possible only downstream (from the host to the user); upstream rates (from the user to the host) remain at V.34 or below and require a digital connection at one end. Initial tests of these modems show that although significant speed improvements exist, true 56 Kbps is rarely, if ever, achieved.

Currently, there are two competing technologies – one from Rockwell Semiconductor and Lucent Technologies, known as K56flex, and one from U.S. Robotics (a subsidiary of 3Com), called X2. In December 1997, however, the vendors announced tentative agreement on a compromise standard being referred to as V.pcm (for pulse code modulation). The forthcoming standard will be based on a mix of K56flex and x2 technology. It is expected to receive preliminary ITU approval in February 1998, with adoption of a final standard expected by late 1998. Modems based on both technologies can be upgraded in software when a standard emerges.

56-Kbps modems gradually will replace 28.8 Kbps and 33.6-Kbps modems because all laptop computers (their major market) soon will include either the Rockwell/Lucent or the U.S. Robotics/3Com version. The companies already have shifted most of their modem chip set production to the faster, 56-Kbps products.

■ Cable Modems

A cable modem is a modem designed to connect a computer to a cable television (CATV) broadband cable to provide access to Internet or on-line services. The growth in the usage of cable modems will be dependent on how fast cable television companies can install the necessary infrastructure required to support two-way capability.

Cable modems are really not modems in the conventional sense of the term. They modulate and demodulate signals like a conventional modem, but otherwise, they function more like routers that are designed for installation on CATV networks. These cable modems, as well as the cable operators' plant equipment, even are being managed using the familiar Simple Network Management Protocol (SNMP).

A popular application for cable modems is Internet access to services at speeds 100 to 1,000 times faster than a telephone modem. Like 56-Kbps modems, cable modems are engineered to handle higher-speed downstream data at 10 Mbps, while uploading at as much as 1/10 that rate. Because cable modems share bandwidth with other cable modem users on a given segment of the broadband CATV cable, true data throughput will vary. Other services provided by cable modems may include access to streaming audio, video, and CD-ROM servers or local content such as community information and services.

Cable modem pricing will be influenced in large part by the ongoing standards-setting process that is already underway by several groups, including the Multimedia Cable Network System (MCNS) consortium and the IEEE 802.14 work group. Cable operators want interoperable modems so they can transfer that cost to retail outlets. When standards are established, customers will be able to buy a cable modem in one place, take it anywhere, plug it in, and it will work.

4.2.6 Network Protocols

Once a company installs the wiring, switching, and routing physical infrastructure, network protocols provide the software infrastructure. Initially, network protocols were designed to work with specific types of equipment. For example, IBM's SNA operated primarily with the company's mainframes and other devices, and Digital's DECnet worked with that company's minicomputer systems. These protocols have been in use for many years and include sophisticated features needed to run mission-critical applications. For instance, SNA includes bandwidth prioritization features so an important or delay-sensitive transaction will receive available communications connections before other, less-important applications. Standard on SNA networks is the ability for end-node devices to get bandwidth when they need it for specified applications, for the network to manage this bandwidth for users on the network (taking into account what an acceptable delay might be), and for the network to know how to prevent lost data during times of overload. However, those features are only beginning to appear in WAN and LAN environments that use other protocols such as TCP/IP.

Users have been encouraging vendors to move away from proprietary network protocols and technologies to multi-vendor ones. However, because it is difficult for all necessary vendors to agree on how to add new features and services to a multi-vendor protocol, vendor-specific protocols still can offer a greater level of sophistication. For example, the initial versions of ATM completed by the ATM Forum did not have built-in QoS capabilities. Recent releases of the specification added those features, including parameters for cell-transfer delay and cell-loss ratio. However, interoperability among different vendors' equipment and device performance still need improvement.

The IETF is working on defining certain IP "classes of service" – a rough equivalent to ATM's QoS – as part of its Integrated Services Architecture (ISA). ISA's proposed elements include the Resource Reservation Protocol (RSVP), a defined Packet Scheduler, a Call Admission Control module, an Admission Control manager, and a set of policies for implementing these features (many of the same concepts already outlined in ATM's QoS).

■ TCP/IP

The Internet Protocol (IP) has become the primary networking protocol of the 1990s, in large part because of the success of the Internet, which is based on TCP/IP. TCP/IP is the most common method of connecting PCs, workstations, and servers. IBM offers TCP/IP connectivity on all its products, including S/390 mainframes. It is included as part of many products, including desktop operating systems (such as Microsoft's Windows 95 or Windows NT) and LAN operating systems. To date, however, TCP/IP has lacked some of the desired features needed for mission-critical applications.

The historically most pervasive LAN protocol, IPX/SPX, which underpins Novell's NetWare, is losing ground to TCP/IP. However, Novell has announced that it will incorporate native IP support into the next upgrade to NetWare/IntranetWare (code-named Moab), ending NetWare's need to encapsulate IPX packets when carrying them over TCP/IP connections. Moab is scheduled to ship in mid-1998 (a beta version was released in November 1997). Both UNIX and Windows NT servers use TCP/IP. Banyan's VINES, IBM's OS/2, and other LAN server operating systems also use TCP/IP.

■ IPv4 and IPv6

IPv6 (previously called next-generation IP, or IPng) is a backward-compatible extension of the current version of the Internet Protocol, IPv4. IPv6 is designed to solve problems brought on by the success of the Internet (such as running out of address space in router tables) and adds needed features, including security, auto-configuration, and real-time services similar to QoS.

Increased Internet usage and the allocation of many of the available IP addresses has created an urgent need for increased addressing capacity. IPv4 uses a 32-bit number to form an address, which can offer about 4 billion distinct network addresses. In comparison, IPv6 uses 128 bits per address, which provides for an inconceivably larger number of available addresses. IPv6 networks should start being deployed sometime in 1998. However, it will take several years for the disparate networks that make up the Internet to upgrade to the new IPv6 protocol.

■ Resource Reservation Protocol (RSVP)

Originally developed to enhance IPv4 with QoS features, RSVP lets network managers allocate bandwidth based on an application's bandwidth requirements. Basically, RSVP is an emerging communications protocol that signals a router to reserve bandwidth for real-time transmission of data, video, and audio traffic.

Resource reservation protocols that operate on a per-connection basis can be used in a network to elevate the priority of a given user temporarily. RSVP runs end to end to communicate application requirements for special handling. It identifies a session between a client and a server and asks the routers handling the session to give its communications a priority in accessing resources. When the session is completed, the resources reserved for the session are freed for others to use.

RSVP offers only two levels of priority in its signaling scheme. Packets are identified at each router hop as either low or high priority. However, in crowded networks, two-level classification may not be sufficient. In addition, packets prioritized at one router hop might be rejected at the next.

Accepted as an IETF standard in 1997, RSVP does not attempt to govern who should receive bandwidth, and questions remain about what will happen when several users all demand a large block of bandwidth at the same time. Currently, the technology outlines a first-come, first-served response to this situation; however, the IETF has formed a task force to address the issue. RSVP was developed to give certain data streams, such as IBM's SNA traffic, higher priority than "burst" traffic, creating a certain inherent priority.

Because RSVP provides a special level of service, many people equate QoS with the protocol. For example, Cisco currently uses RSVP in its IPv4-based Internetwork Operating System to deliver IPv6-type QoS features. RSVP is only a small part of the QoS picture, however, because it is effective only as far as it is supported within a given client/server connection. Although RSVP allows an application to request latency and bandwidth, it does not provide for congestion control or network-wide priority with the traffic flow management needed to integrate QoS across an enterprise.

With Gigabit Ethernet standards emerging, some vendors are promising to combine RSVP with Gigabit Ethernet to displace ATM as the leader in network backbones requiring QoS. However, RSVP lacks many of the infrastructure technologies needed to achieve scalability. For example, it cannot achieve end-to-end QoS unless every router in the chain supports RSVP, and currently no routing protocols work with it. Development has begun on a revision of OSPF to include RSVP. Also, large-scale RSVP deployment remains unlikely because routers must maintain state information on every RSVP session they carry, and guaranteed RSVP requires too much processor overhead to implement in high-speed links.

■ Real-Time Transport Protocol (RTP)

RTP is an emerging protocol for the Internet (championed by the IETF's Audio/Video Transport work group) that supports real-time transmission of interactive voice and video over packet-switched networks. RTP is a thin protocol that provides content identification, timing reconstruction, loss detection, and security. With RTP, data can be delivered to one or more destinations, with a limit on delay variation.

RTP and other Internet real-time protocols, such as the Internet Stream Protocol Version 2 (ST2), focus on the efficiency of data transport. They are designed for communications sessions that are persistent and that exchange a lot of data. RTP does not handle resource reservation or QoS control. Instead, it relies on resource reservation protocols such as RSVP, communicating dynamically to allocate appropriate bandwidth.

Companies such as Rapid City Communications, a Gigabit Ethernet startup, are incorporating RTP and RSVP into products such as its Fully Integrated Routing Switch Technology (FIRST) family of 1-Gbps switches. RTP adds a time stamp and a header that distinguishes whether an IP packet is data or voice, allowing prioritization of voice packets, while RSVP allows networking devices to reserve bandwidth for carrying unbroken multimedia data streams. Rapid City claims that FIRST's support for both RSVP and RTP capabilities makes the devices competitive with ATM's QoS capabilities.

Real-Time Control Protocol (RTCP) is a companion protocol to RTP that analyzes network conditions. RTCP operates in a multicast fashion to provide feedback to RTP data sources as well as all session participants. RTCP already has been adopted by several voice-over-IP vendors to circumvent the limits of datagram transport of voice on the Internet and private IP networks. With RTCP, software can adjust to changing network loads by notifying senders and receivers of spikes, or variations, in network transmissions. Using RTCP network feedback, telephony software can switch compression algorithms in response to degraded connections.

■ Mobile IP

Another new protocol for IP is Mobile IP, which provides IP routing for wireless/mobile hosts. The Mobile IP standard involves a two-part solution: mobile IP agents and mobile nodes. These mobile IP agents function as software routers that route all packets destined for registered mobile nodes. A mobile node is any client device such as a laptop or notebook computer that changes its point of attachment from one subnet to another.

Mobile IP already is being implemented in a variety of products. For example, Bay Networks introduced BayStream Dial VPN Service software in February 1997 to help ISPs offer VPN services. BayStream is based on the Mobile IP protocol; later versions will be based on the emerging L2TP protocol. FTP Software's Secure Client software supports IPsec and Mobile IP, allowing mobile users to communicate as they change subnets, without having to change IP addresses.

Although the Mobile IP standard has been completed, revisions are underway, and work on related standards also is continuing. Mobile IP is being tested on existing IPv4 networks, and work is also in progress at the IETF to make Mobile IP work with IPv6; an Internet draft standard ("Mobility Support in IPv6") was completed in December 1997. Until all the foregoing work is finished, users will continue to encounter compatibility problems among products.

Evolving Internet Protocols

Internet protocols are evolving constantly as limitations are uncovered and as new applications are invented. Major changes are on the horizon in TCP/IP networking with the introduction of multicasting protocols, QoS guarantees, and reserved bandwidth. These changes are necessary to support real-time applications such as Internet telephony, Internet fax, videoconferencing, and efficient streaming audio and streaming video. Today's evolving Internet protocols include IP multicasting protocols and networking protocols.

■ IP Multicasting Protocols

Digital voice and video comprise of large quantities of data that, when broken up into packets, must be delivered in a timely fashion and in the right order to preserve the qualities of the original content. Protocol developments have been focused on providing efficient ways to send content to multiple recipients, transmission referred to as multicasting.

Multicasting involves the broadcasting of a message from one host to many hosts in a one-to-many relationship. A network device broadcasts a message to a select group of other devices such as PCs or workstations on a LAN, WAN, or the Internet. For example, a router might send information about a routing table update to other routers in a network.

Several protocols are being implemented for IP multicasting, including upgrades to the Internet Protocol itself. For example, some of the changes in the newest version, IPv6, will support different forms of addressing for unicast (point-to-point communications), anycast (communications with the closest member of a device group), and multicast. Support for IP multicasting comes from several protocols, including the Internet Group Management Protocol (IGMP), Protocol-Independent Multicast (PIM), and Distance Vector Multicast Routing Protocol (DVMRP). Queuing algorithms also are being implemented to ensure that video or other multicast data types arrive when they are supposed to without visible or audible distortion.

■ MBone

MBone stands for the IP Multicast Backbone on the Internet and is an experimental IP multicasting infrastructure composed of subnets interconnected by multicast routers and "tunnels." These subnets form a virtual network, overlaid on the Internet, that is used for videoconferencing, collaborative white board use, scientific data dissemination, and distributed interactive simulation tools.

The first multicast tunnel was established between BBN and Stanford University in the summer of 1988. Tunneling originally was meant to be a temporary measure until routers knew about IP multicasting. MBone relies primarily on software routers and unicast-style "tunnels" to move multicast packets between networks that do not have hardware multicast routers.

IP multicast is an addressing scheme (Class-D) in IP developed at Xerox PARC. MBone-based audio videoconferences have been allocated a certain address range within the broader range of addresses that make up Class-D IP addresses.

Researchers in different countries used this technology to conduct a live IETF meeting in 1992 from remote desktops, and MBone was implemented that same year as an experimental service, with 40 subnets in 4 countries. By 1996, MBone had grown to more than 2,900 subnets and began moving toward a production service. The MBone has been used primarily to conduct multi-party videoconferences over the Internet. Some, but not all, ISPs now support the handling of multicast IP packets.

MBone can deliver real-time multimedia broadcasts in an efficient way from one source to thousands of receivers while saving network bandwidth and CPU cycles. It permits the sender to route information to only those users interested in receiving the information. Today, the public part of the MBone is supported on more than 3,200 public networks and boasts more than 10,000 organizations connected worldwide.

The Internet multicast technology is deployed today in the Internet infrastructure all the way from switches and routers to the operating systems and desktop applications. Some of the commercial vendors of this technology include the following:

- Multicast routers from 3Com, Ascend, Bay Networks, and Cisco
- Multicast switches from 3Com and PACE Switch
- Multicast operating systems such as Windows 95 and Windows NT, MacOS, and all versions of UNIX
- Multicast stack software from FTP Software, ICAST Communications, Microsoft, and NetManage
- Multicast applications from ICAST Communications and Starburst

Commercial versions of multicast delivery have limitations and may not be widely used in the near future. They are not standards-based compared with Internet multicast streaming, and the network delivery mechanism consumes n times more bandwidth and server processor cycles than multicast applications. (N is the number of clients connecting to the server.)

Internet multicasting, on the other hand, is already a standard that has been developed and tested by the Internet community. The IETF standard for Internet Multicasting is described in RFC 1112.

■ Real-Time Control Protocol

The RTP, currently an IETF draft, is designed for end-to-end, real-time delivery of data such as video and voice. It works over the User Datagram Protocol (UDP), providing no guarantee of in-time delivery, QoS, delivery, or order of delivery. RTP works in conjunction with a mixer and translator and supports encryption and security. The RTCP is a part of the RTP definition that analyzes network conditions. RTCP provides mandatory monitoring of services and collects information on participants. RTP communicates with RSVP dynamically to allocate appropriate bandwidth.

■ Resource Reservation Protocol

Internet packets typically move on a first-come, first-served basis. When the network becomes congested, RSVP will enable certain types of traffic, such as videoconferences, to be delivered before less time-sensitive traffic such as e-mail – for a premium price. RSVP will change the Internet's pricing structure by offering different QoS at different prices. Supported by Cisco and Intel, RSVP is expected to become more widespread once standards are completed.

The RSVP protocol is used by a host, on behalf of an application, to request a specific QoS from the network for particular data streams or flows. Routers would use the protocol to deliver QoS control requests to all necessary network nodes to establish and maintain the state necessary to provide the requested service. RSVP requests will generally, although not necessarily, result in resources being reserved in each node along the data path.

RSVP is not itself a routing protocol; RSVP is designed to operate with current and future unicast and multicast routing protocols. An RSVP process consults the local routing database(s) to obtain routes. In the multicast case, for example, a host sends IGMP messages to join a multicast group and then sends RSVP messages to reserve resources along the delivery paths of that group. (The IGMP is used by IP hosts to report their host group memberships to any immediately neighboring multicast routers.) Routing protocols determine where packets are forwarded; RSVP is concerned with only the QoS of those packets as they are forwarded in accordance with that routing.

Efforts are underway to finalize the specification of RSVP. Two Request For Comments documents – RFC 2208 (Version 1 *Applicability Statement, Some Guidelines on Deployment*) and RFC 2209 (Version 1 *Message Processing Rules*) – were published in October 1997.

4.2.7 Factors Affecting Network Performance

As corporate networks become larger and more complex and network traffic increases, bottlenecks can emerge in three areas: the desktop, the backbone, or the server.

Bandwidth to the Desktop

Providing adequate bandwidth to the desktop is a key aspect of ensuring acceptable performance and often has been offered through segmentation. LANs originally were designed to connect groups of workers onto a common system. In the past, for example, a department with 20 employees installed one shared-media LAN, and all users would share the available bandwidth. As the number of users rises on a shared LAN, the volume of network traffic rises, sometimes geometrically. Companies solve this problem by dividing one LAN into multiple LANs, a process called segmentation. One 200-user LAN can be broken into two 100-user LANs or even four 50-user networks.

Upgrading users from a shared LAN to a fully switched LAN provides the ultimate segmentation because these networks provide each user with an individual connection. As noted earlier, switching has proven to be a popular high-speed networking option. Moving from a shared Ethernet LAN to a switched Ethernet LAN, or from a shared Token Ring LAN to a switched Token Ring LAN, requires a company to replace a wiring hub with a LAN switch or change a component in the wiring hub. Such a change can be made inexpensively, and the same local adapter cards, network management software, and wiring usually can be used.

In addition, access to Internet and the graphical content of the Web has increased the traffic per desktop. Although switched, 100-Mbps connections are beyond what most companies currently require, network managers should assess planned applications and prepare for increases in desktop bandwidth accordingly. The price of 10/100 Mbps Ethernet cards is now equivalent to 10 Mbps-only cards, making deployment an easy choice. Users who need higher-speed access can be given individual parts on a 100-Mbps Ethernet switch, with Gigabit Ethernet used to supplement additional requirements as needed.

Backbone Bandwidth Requirements

A backbone is the part of the communications network that carries the heaviest traffic and ties together different departmental LANs. LANs can be attached to the backbone with routers or bridges. Usually, a company runs all its inter-departmental traffic on one backbone network.

Because hundreds or even thousands of users rely upon backbone networks, backbones typically require faster transmission speeds than departmental LANs. The most popular choices for LAN or campus backbones include FDDI, Fast Ethernet, ATM, and switching. Different backbone implementations vary in part according to distance limitations associated with different media. Many small and midsized companies rely on shared and switched Ethernet for their backbones, which relieves them of supporting two or more networking technologies.

Typically only ISPs and telecommunications service providers have actual WAN backbones. Most corporate users have WAN links. For large users, frame relay is the most popular choice. Point-to-point links or circuits and ATM also are used.

Server Bandwidth Requirements

The bandwidth available to servers represents the last area for potential bottle-necks within an enterprise. With the move to distributed computing, companies built applications that divided processing chores among a series of networked systems. As the number of computers accessing a specific server increases, the server's network connection can become overloaded, even if the server itself still has adequate processing power. A company can alleviate this problem by providing a high-speed connection from the server to department LANs or by installing multiple network cards in the server.

No one networking option to increase server bandwidth is preferred. Fast Ethernet is gaining ground, however, because many Ethernet switches were designed to support 10-Mbps switched Ethernet connections to individual desktops but 100-Mbps connections where faster access has been required, such as to alleviate server congestion. For server clusters, Gigabit Ethernet is a likely solution. For mainframes-as-Web-servers, IBM and other vendors such as General Signal Networks have high-speed solutions that feature 20-Gbps throughput.

Another option for addressing server bandwidth limitations is dynamic host assignment, where software monitors how busy each of the application hosts are. When a user requests access to a service, the software gives the address of the least-busy server. This is a technology used on the Internet to help balance the traffic load dynamically across multiple Web servers.

FDDI has not been used widely in this area to date. ATM is a potential solution; however, this approach usually requires that companies mix a couple of networking technologies, such as Ethernet to the desktop and ATM to the server. Products that integrate two networking technologies are new, and some users have been disappointed in their initial performance.

4.2.8 Market Overview

Changing corporate buying patterns have contributed to network market consolidation. Because customers want to purchase as much as possible from a single vendor, network companies are acquiring each other or merging at an incredible rate. More important, the proliferation of new technologies is occurring faster than existing large vendors can handle, and time-to-market pressure means they are forced to acquire new products rather than develop them in-house.

During 1996 and 1997, for example, Ascend Communications acquired Cascade Communications; Bay Networks purchased Rapid City Communications and the digital signal processing modem business of Penril Datability Networks; Cabletron acquired NetLink, Network Express, and three other companies; Cisco picked up Ardent, the xDSL business of Integrated Network, Stratacom, and part of Telebit, among others; Compaq acquired Microcom (a remote access manufacturer); and 3Com bought U.S. Robotics. In November 1997, Cabletron also announced the $430 million purchase of Digital's Network Product Business, which includes switches, hubs, and routers.

Unfortunately, vendor consolidation has not resulted in product rationalization or product integration. Products designed by different companies do not suddenly work together just because two organizations have decided to merge. In fact, some acquisitions may be driven less by complementary technologies than by the race to acquire existing distribution channels and key customers.

Larger companies such as HP and IBM also have moved aggressively into the network market. Both HP and IBM intend to sell end-to-end network systems to companies, by designing and building the products themselves, acquiring other firms, or via OEM agreements.

According to IDC, 1996 worldwide network market revenue was divided as shown in Figure 4-7.

Figure 4-7: Worldwide Network Market Revenue – 1996 (in billions)

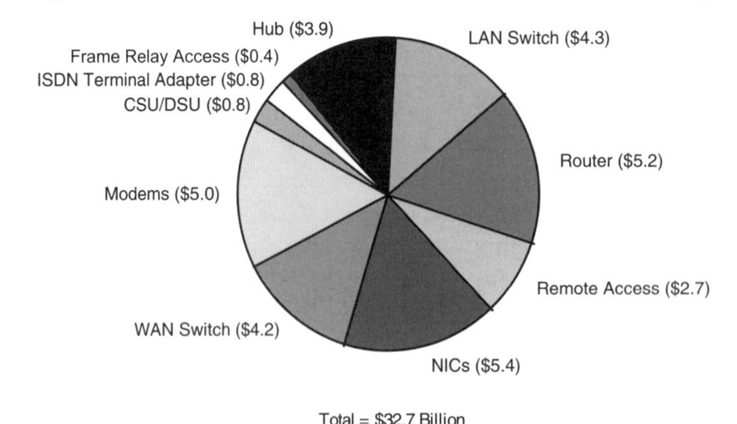

Source: IDC, 1997

ATM

Vendors continue to deliver ATM products that are less expensive and easier to install than previous versions. Even so, 155-Mbps ATM remains more expensive than 100-Mbps Fast Ethernet, and 622-Mbps ATM is costlier than 1-Gbps Gigabit Ethernet. The price differential is shrinking, however, with ATM vendors cutting prices up to 40 percent on switches and NICs in late 1997. The declining price of ATM solutions also will place increasing pressure on Fast Ethernet and Gigabit Ethernet.

Although faster versions of Ethernet are the technology of choice for boosting LAN bandwidth, ATM remains strong among companies seeking the reliability and flexibility Ethernet still does not provide. In early 1998, for example, AT&T will launch an ATM Internet access service that will let users tie into the Internet using existing ATM WAN links. Integrating existing data network connections with carrier-managed Internet access will be a boon to high-bandwidth customers in particular.

ATM's greater potential for use in multimedia, time-critical, and more complex configurations continues to offer significant advantages over Gigabit Ethernet. Key advantages that ATM offers include QoS, designed-in isochronous traffic support, and WAN integration.

Ethernet

In comparison with ATM, Gigabit Ethernet offers raw bandwidth gains (though not necessarily equal to real-world throughput), a natural path for the existing Ethernet LAN installed base, and a straightforward upgrade to networks requiring more data. However, several roadblocks remain before Gigabit Ethernet can be established as the best next step for future LAN environments:

- Formal standardization of Gigabit Ethernet is not anticipated until mid-1998.
- Configuration limitations in technology, product packaging, and network design are significant and probably will remain so for some time.
- Performance testing of large-scale Gigabit Ethernet networks began in early 1997, and results are not yet final.
- Products and related management systems are in the early development stages, making proof of concept and, more important, operations unanswered questions.
- Multi-vendor interoperability is uncertain before standardization, interoperability testing, and installed base expansion. The Gigabit Ethernet Alliance is expected to take a strong position regarding interoperability testing and available product "stamping."

IDC predicts the average sales price of Gigabit Ethernet NICs will drop from almost $2,000 in 1997 to $1,320 in 1998, $932 in 1999, $695 in 2000, and $560 in 2001.

Switching

The LAN market is undergoing some fundamental changes as traditional LAN hub technologies continue to be replaced by more robust and better-performing switching solutions. Table 4-8 shows worldwide hub port shipments by technology, according to IDC.

Growing requirements for more bandwidth from Internet, intranet, and extranet applications are feeding high growth in LAN switching. Today's backbone networks will migrate to Layer 2 switches or switches with Layer 3 functionality. Within the network core, customers will look for switches that provide extremely low latency for both routed and switched traffic, IP multicast, and RSVP prioriti-

Table 4-8: Worldwide Hub Port Shipments by Technology

	First Half 1996		First Half 1997	
	Ports (thousands)	Revenue ($millions)	Ports (thousands)	Revenue ($millions)
Ethernet	23,145	$1,361	25,095	$1,254
Fast Ethernet (100Base-T)	1,391	$225	1,818	$276
Token Ring	2,737	$369	2,525	$316

Source: IDC, 1997

zation support. One of the key differentiators for these next-generation switches will be how they handle Layer 3 forwarding and whether they are able to do so at wire speed.

Table 4-9 compares shipments of ATM and other LAN switches between the first half of 1996 and the first half of 1997.

Table 4-9: Worldwide LAN Switch Market

	First Half 1996		First Half 1997	
	Ports (thousands)	Revenue ($millions)	Ports (thousands)	Revenue ($millions)
ATM LAN switches	270	$396	180	$291
Total LAN switches	8,112	$3,954	8,234	$2,625

Source: IDC, 1997

Remote Access

As demonstrated by the previous list of recent acquisitions, the larger networking companies rapidly are acquiring remote access device companies or related technologies. Because of the growth of the Internet and the increasing use of telecommuting, remote access has become one of the fastest-growing and most lucrative segments of the networking business.

Routers

The router market (projected to total $6.2 billion worldwide in 1997) continues to experience solid growth. Worldwide shipments grew to 625,395 units for the first half of 1997 (compared with 579,676 in the first half of 1996); revenue grew to $2.79 billion in the same period (compared with $2.62 billion), according to IDC. This moderate growth rate is due to the inclusion of routing functionality within Ethernet, ATM, and frame relay switches, which have begun to undercut the router market. Other contributing factors include the general downturn in investment in high-end and midrange routers by enterprise customers as well as ISPs and the increasing acceptance of routers priced at less than $1,500.

Currently, the price for a high-end router that will support from 200 to 2,000 users starts at $20,000. Midrange routers average between $8,000 and $20,000. Low-end devices for remote sales offices with 1 to 9 users can cost less than $8,000, but they offer few management, firewall, or other security features.

The U.S. router market grew more slowly in 1997 than in 1996. As a percentage of the world's router shipments, however, the U.S. held its own against the traditionally higher growth Asia/Pacific region. In Western Europe, the proliferation of

ISDN routers continues strongly. Although there is growing interest in ISDN connectivity in the U.S., adoption of frame relay equals that of ISDN as the preferred "new" alternative to standard leased lines.

The intensifying downward pressure on pricing at the low end and on unit shipments at the midrange and high end will constrain revenue growth, which is projected to grow at a compound annual growth rate (CAGR) of only 5 percent through 2001.

On the revenue side, the U.S. router market will reach its height in 1998. In Western Europe, revenue growth is predicted to be only a modest 8 percent CAGR through 2001 (compared with a CAGR of 27 percent for unit shipments in the same period). Within Asia/Pacific, unit shipments will continue strong (24 percent CAGR), but revenue growth will moderate to 8 percent CAGR through 2001.

The high-end segment of the router market will decline through 2001, showing a -9 percent CAGR, as multi-layer switches take over the router's role as LAN concentrator, intranet backbone manager, and Internet doorway. The midrange segment will increase at a CAGR of 4 percent, although revenues will remain flat until 2001. In 2001, IDC projects that the low-end market will account for 95 percent of all routers shipped. In particular, unit shipments of low-priced routers (averaging less than $1,500) are predicted to grow to more than 1.8 million in 2001, a CAGR of 39 percent.

4.2.9 Forecast

- Some variety of Ethernet will continue to hold the dominant share of desktop LAN connections throughout the forecast period. In 1998, new purchases will be predominantly Fast Ethernet, with some use of Fast Ethernet switching.
- ATM to the desktop will continue to lag behind earlier expectations. Lower-cost, high-speed Ethernet products, such as Fast and Gigabit Ethernet, will limit ATM's desktop adoption to specialized circumstances.
- Gigabit Ethernet will start to be deployed in building and campus backbones early in the forecast period. Usage will grow once a final standard is adopted.
- Use of ATM in the corporate backbone network and in the WAN will increase. ATM will be attractive particularly to companies trying to combine voice, video, and data traffic onto a single WAN infrastructure. However, frame relay will continue to be deployed more widely than ATM in corporate WANs throughout the forecast period.
- Routers are becoming significantly faster, reaching gigabit speeds, but with router intelligence being added to switches, distinctions between the two will blur quickly. Some form of Layer 3 switching will be incorporated into virtually all router products within the forecast period despite the lack of a single standard to date.
- New technologies that allow companies to exploit the Internet as a VPN will start to gain acceptance in 1998 as IT departments become more comfortable with security issues and try to pare down WAN access costs. However, the success of this strategy will be limited by the lack of reliability and guaranteed throughput across the Internet backbone, and the potential replacement of existing Internet connection charges with traffic-sensitive pricing.
- IPv6, which vastly increases the number of possible network addresses and adds quality of service and multicast features, will begin to appear in products in 1998. Migration to IPv6 will be ongoing throughout the forecast period.

- Consolidation among networking vendors will continue, with major mergers and acquisitions occurring in 1998. The major networking vendors will continue to acquire innovative startups to gain access to new technologies. Startups working on very high-speed router products and new Layer 3 and Layer 4 switching technologies are likely to be attractive acquisition candidates.

- One or more of the established telephone equipment manufacturers are likely to make a major acquisition of a networking company to protect their market position and gain access to new technology, as packet-switching makes inroads into traditional circuit-switched telecommunications networks.

- Protocols other than TCP/IP and SNA will continue to decline in market share. Vendor-specific protocols such as AppleTalk, DECnet, and IPX will disappear gradually from use. Encapsulation of SNA traffic inside other protocols (principally TCP/IP) for transport across enterprise networks will gain in importance.

- Adding the ability to assign differing priorities to different types of traffic (known as QoS) will be a major area of development for TCP/IP technologies and products. Products that offer "traffic-shaping" capability, including the ability to assign priorities based on the user's identity and the type of data being transferred (video versus Web pages versus e-mail) will be the focus of startups. These capabilities also will be added to the major vendors' product lines.

4.2.10 References

■Articles

Bort, Julie. 1997. VPN tunnels in. *VARbusiness*. July 15: 63.
Bunal, Ron and Christopher Smith. 1997. SOHO to the enterprise: end-to-end or dead end? *Network Computing*. June 1: 48.
Dutcher, William. 1997. Will IP switching replace ATM? *PCWeek*. September 22.
Girishankar, Saroja. 1997. Internet gets a lift from mega routers. *CommunicationsWeek*. August 3: 1.
Haber, Lynn. 1997. IPv6: is it revolutionary or is it redundant? *Computer Reseller News*. July 7: 73.
Hurwicz, Mike. 1997. Preparing for Gigabit Ethernet. *Byte*. October: 63.
Lawson, Stephen. 1997. Start-ups rush to routing market. *InfoWorld*. September 22: 5.
Lawton, Stephen. 1997. A mirage at Layer 4. *LAN Times*. October 13: 1.
Mandeville, Robert and Deval Shah. 1997. Mix-and-match ATM. *Data Communications*. July.
Moskowitz, Robert. 1997. Secure communications now or later? *Network Computing*. June 15: 35.
Musich, Paula. 1997. High-end routers promise relief from bottlenecks. *PCWeek*. April 14: 1.
Pace, Mark. 1997. Routing protocol design. *InfoWorld*. September 8: 78.
Paone, Joe. 1997. VPNs offer new way to get there. *LAN Times*. June 23: 47.
Swallow, George. 1997. MPOA, VLANS and distributed routers. *The ATM Forum Newsletter*, Vol. 4, No. 3.
Talley, Brooks and Tom Young. 1997. Routing toward the future. *InfoWorld*. September 8: 74.

■Periodicals, Industry and Technical Reports

Business Communications Review. (monthly) Hinsdale, Ill.: BCR Enterprises.
Gigabit Ethernet: accelerating the standard for speed. September 1997. White paper. Gigabit Ethernet Alliance.
Shared hub and LAN switch market. (quarterly) Portola Valley, Calif.: Dell'Oro Group.

■On-Line Sources

Janah, Monua. 1997. LANs: ATM bounces back. InformationWeek Online. November 17. `http://techweb.cmp.com/iw/657/57iuatm.htm`

■URLs of Selected Mentioned Companies

3Com `http://www.3com.com`
Alteon Networks `http://www.alteon.com`
Ariel Communications `http://www.ariel.com`
Ascend Communications `http://www.ascend.com`
AT&T `http://www.att.com`
Avici Systems `http://www.avici.com`
Banyon `http://www.banyon.com`
Bay Networks `http://www.baynetworks.com`

BBN http://www.bbn.com
Brocade Communications http://www.brocadecomm.com
Cabletron http://www.cabletron.com
Cisco http://www.cisco.com
Compaq http://www.compaq.com
Compatible Systems http://www.compatible.com
Data General http://www.dg.com
Digital http://www.digital.com
Extreme http://www.extreme.com
First Virtual http://www.firstvirtual.com
FlowWise Networks http://www.flowwise.com
Fore http://www.fore.com
FTP Software http://www.ftp.com
General DataComm http://www.gdc.com
GigaLabs http://www.gigalabs.com
Harris & Jeffries http://www.hjinc.com
Hitachi http://www.hitachi.com/index95.html
HP http://www.hp.com
Hughes Network Systems http://www.hns.com
IBM http://www.ibm.com
ICAST Communications http://www.icast.com
Intel http://www.intel.com
LCI http://www.lci.com
LightStream http://www.lightstream.com
LSI Logic http://www.lsilogic.com
Lucent Technologies http://www.lucent.com
Madge Networks http://www.madge.com
MCI http://www.mci.com
Microcom http://www.microcom.com
Microsoft http://www.microsoft.com
NEC http://www.nec.com
Neo Networks http://www.neonetworks.com
NeoNet http://www.neonet.com
NetLink http://www.netlink.com
NetManage http://www.netmanage,com
Netrix http://www.netrix.com
Newbridge Networks http://www.newbridge.com
Nortel http://www.nortel.com
Novacom http://www.novacom.com
Novell http://www.novell.com
Olicom http://www.olicom.com
PACE http://www.pace.com
Packet Engines http://www.packetengines.com
Pluris http://www.pluris.com
Rockwell http://www.rockwell.com
Solaris http://www.solaris.com
Sprint http://www.sprint.com/index.html
Sun http://www.sun.com
Telebit http://www.telebit.com
TimeStep http://www.timestep.com
Trillium Digital Systems http://www.trillium.com
U.S. Robotics http://www.usrobotics.com
UB Networks http://www.ubnetwork.com
Unispan http://www.unispan.com
UUNet http://www.uunet.com
VLSI Technology http://www.vlsi.com
VPNet http://www.vpnet.com
Xerox http://www.xerox.com
Xylan http://www.xylan.com

4.3 Telecommunications Services

4.3.1 Executive Summary

The telecommunications industry worldwide continues to undergo changes because of deregulation and the adoption of new technologies. The opening of previously restricted markets to new competition is affecting the pricing and availability of telecommunications services. The U.S. Telecommunications Act of 1996 has led to industry consolidation and some new offerings, but little in the way of rate decreases. Deregulation in the U.S. also has spilled over into the electric utility and power companies, which have announced agreements with telecommunications service providers. In addition, the Act has influenced deregulation efforts in many other countries. For example, January 1, 1998 is the deadline for ending all remaining restrictions on the provision of telecom services throughout most of the European Union (EU).

Improvements in network components and protocols that permit more data traffic at higher speeds have made buying decisions more complex. Frame relay, Asynchronous Transfer Mode (ATM), and Switched Multimegabit Data Service (SMDS) are being incorporated into networking equipment; however, tradeoffs must be weighed with each one. When coupled with technologies enabling cheaper, faster, and smaller devices, these improvements should make high-quality video and other high-bandwidth applications more practical. Faster transmission media will lower the costs of some services and enable the creation of others as well.

The difficulties involved in mixing voice and data are being eliminated gradually. Voice, data, image, and video services now can be distributed on-site and off-site in an increasing number of ways. Now that long-distance, local, and regional carriers and cable, wireless, and satellite vendors can offer a variety of services, it is essential to put some architectural and business ground rules in place before deciding on particular technologies or carriers.

The global market for cellular telephone and personal communications services (PCS) products and services is growing rapidly. Different implementations of standards in wireless networks and products will continue to exist, and multi-function wireless devices and hybrid phones will continue to proliferate.

The growth of the Internet has attracted local and long-distance companies as well as PCS and cable providers. Traditional carriers have entered into alliances with cable and other providers to offer Internet access services and prevent possible lost revenue as interest in phone, video, paging, and other messaging applications over the Internet grows.

4.3.2 Deregulation in the Telecommunications Industry

Deregulation has created opportunities for beneficial change but also pitfalls for the unwary. Following are examples of legislation and deregulation efforts underway in different countries.

The U.S. Telecommunications Act of 1996 and Its Implications

The Telecommunications Act of 1996 deregulated wired and wireless communications in the U.S. The Act allows new players to enter existing markets, encourages new service providers and alliances, and enables new service offerings. Among the possibilities the Act enables are these:

- Local phone companies (known as local exchange carriers, or LECs) can offer cable, wireless, and long-distance services once they face effective competition for local phone service.
- Long-distance companies (known as interexchange carriers, or IXCs) can enter the cable, wireless, and local access telephone business.
- Cable operators can offer wireless and wired telephone and other network or Internet services. Cable operators may find this option attractive to maintain market share and revenue as they face competition from direct broadcast satellite operators, local telephone companies, and even video rental outlets.
- Gas and electric companies and others who own certain infrastructures and rights-of-way may develop alliances with other providers or become service providers themselves.

The Act's objective was to create increased competition, reduce rates in the long term, and generate new services. However, long-distance carriers have not entered the local phone market, and the Regional Bell Operating Companies (RBOCs) therefore have been prevented from entering the long-distance market, as originally expected. Instead, most IXCs have focused on the more lucrative business and long-distance markets rather than the less-profitable residential market.

Nevertheless, there are signs of change. Cable companies now offer telephone services, telephone companies now offer Internet access, and computer companies are entering the cable market. For example, US West has taken advantage of deregulation to enter the cable TV market aggressively and is now the third-largest cable operator in the U.S. In addition, agreements between telephone services providers and electric utility companies have occurred.

Congress left the Telecommunications Act's language vague in many areas, preferring to let the Federal Communications Commission (FCC) make some of the more detailed implementation decisions. The FCC has been trying to implement the provisions of the Act, but progress has been slowed as companies battle in the courts over details in the regulations.

The industry also has consolidated as major players merge and smaller companies are acquired. Some examples of mergers include the following:

- On April 1, 1996, SBC and Pacific Telesis announced a merger agreement. The merger was the first between two former Bell System companies and became effective on April 1, 1997. SBC and Pacific Telesis Group reported combined 1996 revenues of $23.5 billion. SBC offers, through its subsidiaries, Southwestern Bell, Pacific Bell, Nevada Bell, and Cellular One brands; local and long-distance telephone service; wireless communications; paging; Internet access; and cable TV and messaging. SBC has more than 31 million access lines and 4.7 million wireless customers in the U.S. and has investments in telecommunications businesses in nine countries.

- In November 1996, the merger of Bell Atlantic and NYNEX was approved by the companies' shareholders. In mid-1997, the FCC entered into an agreement with the two companies that permits them to merge, provided they meet certain conditions. The FCC agreement also makes it easier for competing local service providers to begin offering service in the Bell Atlantic/NYNEX region and will help later when the companies seek permission to offer in-region, long-distance service. In August 1997, the FCC announced its official approval of the merger.
- British Telecommunications PLC and MCI Communications announced they had signed an agreement to become one company in November 1996; final approval was expected in late 1997. However, on October 1, 1997, WorldCom also made a bid for MCI, which was followed on October 15, 1997 by a bid from GTE. (MCI previously had complained to the FCC that it would lose $800 million in 1997 because the Bell companies have blocked its efforts to begin providing local phone services.) On November 10, 1997, MCI announced it had accepted WorldCom's bid. With the addition of MCI, WorldCom boosts its share of the long-distance market from approximately 5 percent to 25 percent.
- In October 1996, Cable & Wireless PLC announced plans to merge with its subsidiary, Mercury Communications PLC, and three cable companies, NYNEX CableComms PLC, Bell Cablemedia PLC, and Videotron Holdings PLC.

Legislative Update

At least a few major changes in the FCC's interconnection rules have occurred already. For example, a U.S. federal appeals court removed the FCC's right to regulate the prices local phone companies can charge for interconnections to their networks. Some LECs, led by GTE, also filed petitions to block implementation of the FCC regulations enabling competition in the local phone markets. The U.S. District Court of Appeals in St. Louis invalidated key national pricing guidelines developed by the FCC for interconnection of potential competitors with the local networks operated by the RBOCs and GTE. This ruling means potential rivals to the local telephone monopolies must revert to using state-mandated pricing rules or privately negotiated interconnection agreements. Concerned that lack of national rules for opening local telephone markets will further slow local competition, the FCC appealed this decision to the U.S. Supreme Court in November 1997. Sprint and WorldCom joined in the FCC's petition.

The Supreme Court agreed to take up the FCC's appeal at its January 23, 1998 conference. If the Court decides to hear the case, arguments will be held in the ensuing months and a decision could be released by early summer. This timetable represents an accelerated process for the Court; otherwise, the case would have stood little chance of being reviewed until fall 1998.

In one of the most interesting developments of 1997, SBC Communications filed a lawsuit in mid-year challenging the constitutionality of the Telecommunications Act. In its suit, SBC alleged that the law violates the U.S. Constitution because it singles out the five RBOCs by name and curbs their entry into long-distance and other markets. (Prior to filing suit, SBC's bid to offer long-distance service in Oklahoma was rejected by the FCC.) US West joined in the suit later in the year.

In response to the suit filed by SBC and US West, a federal court in Texas struck down portions of the Telecommunications Act on January 1, 1998 that prevent local phone companies from offering long-distance service to customers in their regions. The ruling (which is subject to appeal) thus pits the courts against the FCC regarding the law that governs local phone markets. As recently as late December 1997, the FCC rejected a request by BellSouth to get into the long-distance business from its local calling region of South Carolina. The FCC claimed the company had not met the requirements of the Act by opening its local markets to competition.

Convergence and Competition

As individual markets are opened to competition under the provisions of the Act, LECs – which include the RBOCs – will have the opportunity to offer services previously restricted to IXCs and cable companies. Similarly, IXCs can offer local exchange or cable services. Cable companies also have the chance to become part of the local exchange infrastructure or to provide long-distance services if they so desire.

Initially, many service providers will continue to build on their current strengths while they buy and resell the excess capacity of other providers. For example, IXCs will resell LEC access services, and vice versa. This approach allows providers to establish their presence in new service areas while building (or acquiring other companies for) their own infrastructures. A key issue in the regulations implementing local phone service competition is the prices charged to new LEC entrants for use of existing RBOC facilities.

Those who have benefited from protected markets – the LECs, IXCs, and cable television (CATV) operators – now are scrambling to hang onto market share by combining many services (local, interexchange, Internet, and wireless) into a consolidated, "one-stop-shopping" offering.

From an IXC perspective, IXCs have been forced to pay up to 40 percent of long-distance revenue to the LECs who control the local loop. A key objective of the IXCs will be to circumvent these access charges by using low-cost wireless, CATV, or competitive access providers (CAPs), now often known as competitive LECs (CLECs). Another way to avoid such charges would be for an IXC to acquire or merge with an LEC. In January 1998, for example, AT&T announced it had signed a merger agreement with Teleport Communications Group (TCG) as part of its effort to provide end-to-end communications services to businesses. TCG will become the foundation on which AT&T will build a new unit with accountability for the company's local services.

In addition to mergers and acquisitions, more open competition will result in alliances and situations where former competitors may buy each others' services and resell them.

Deregulation Outside the U.S.

As deregulation spreads across the globe, the privatization of government-owned telcos (known as Post, Telegraph, and Telephone administrations, or PTTs) is picking up speed. Also of note is the increasingly complex ownership of these former monopolies, with investments occurring across national and international boundaries. Joint ventures are becoming increasingly attractive during the transition to full privatization. BT has an investment in France's Minitel and is part of several alliances with other network and service providers around the world. For example, in October 1997, BT announced a joint venture with Bharti Enterprises in India to offer satellite voice and data services.

■ Europe

Europe is undergoing similar changes to the U.S. as part of the movement toward a unified European economy. Deregulation mandated by the Commission of the European Communities (CEC) is supposed to occur as of January 1, 1998. However, countries such as Finland, Sweden, and the U.K. are moving more quickly and either have, or soon will have, competition at all levels as a result of government support for deregulation.

Perhaps the biggest uncertainties are the decisions of regulators. The European Commission is pushing individual governments to make large-scale changes that would encourage new rivals. However, many governments remain torn by conflicts of interest. For example, the French government still owns a majority stake in its national phone company, France Telecom, and is selling shares in public offerings; the first IPO occurred in May 1997. The European Commission began legal actions against seven countries in 1997, including Germany and Italy, claiming that they were not moving to open their markets to competition.

Denmark, France, Germany, Italy, and the Netherlands have stated they will begin to introduce some level of competition by 1998. Germany is furthest along, with privatization of Deutsche Telekom occurring in 1996, when it began trading on the German stock exchange. However, some observers claim the company is being favored by the German regulators and that the German government is protecting Deutsche Telekom instead of protecting the market. As an example, they cite the government's approval of Deutsche Telekom's plan to reduce prices charged to businesses, without passing any savings to private customers. In addition, three large telecommunications companies that have been unable to reach interconnection agreements with Deutsche Telekom have complained to government regulators about its high access charges. In August 1997, an administrative court in Germany ruled in favor of the three negotiating companies; however, Deutsche Telekom is expected to appeal the decision.

In September 1997, Tele Danmark, the former state telco of Denmark, announced it would acquire part of the government's shareholding. The complex business transaction is a first in European telecom privatization history and will involve the telco buying around 30 percent of its shares from the government for a fixed price in 1998.

Privatization of France Telecom, originally scheduled for April 1997, was delayed at the last minute because of President Jacque Chirac's decision to launch a snap general election in France during May. The long road toward privatization of France Telecom may be coming to an end, however, because in October 1997, the government sold 10 percent of the telco to the public and 10 percent to Deutsche Telekom.

Scottish Telecom, a product of worldwide deregulation and privatization, has spent years laying fiber cable and now is selling trunk lines and switched services to corporate clients – competing head-on with the former monopoly, BT.

In October 1997, BT and MCI announced they had made several strategic investments in Portugal Telecom as part of the third phase of that country's privatization process. BT will provide management assistance, and MCI will consult on telecommunications opportunities. The alliance also calls for Portugal Telecom to distribute Concert global services.

To complicate matters further, joint ventures are being forged specifically to attack established carriers. BT has the most consistent and aggressive strategy: It teams up with at least one powerful local company in each country, usually a utility or railroad, to offer a full range of residential, business, and wireless services. For example, in Germany, BT has joined with Viag AG, the Bavarian power company, and Telenor, the dominant telecom carrier in Norway to offer services that will compete with Deutsche Telekom. Similarly, in France, BT has joined with Generale Des Eaux, the water utility, and SBC to form Cegetel, which will offer services beginning on February 1, 1998 competing with France Telecom.

U.S. companies other than SBC also are joining in. WorldCom, together with newly acquired MCI, is building high-speed, fiber optic networks across Europe. AT&T has a stake in Germany's Mannesmann Arcor, and Ameritech has bought into the Belgian and Danish phone companies.

In response, established vendors are drastically reducing prices, usually in areas competitors will target first: domestic long-distance and international calls. For example, in mid-December 1997, Deutsche Telekom announced rate cuts totaling nearly $1.3 billion (approximately 4.5 percent).

■ Asia/Pacific

Many companies already have active alliances in Asia. For example, Nokia Mobile Phones Inc. is working with Singapore Telecom to provide wireless services. Ericsson has signed a major Global System for Mobile Communications (GSM) system order from Guangdong Mobile Communications to provide service in China. (See "GSM" on page 215.) Germany's Deutsche Telekom recently acquired a 10 percent stake in Technology Resources Industries Bhd, a deal that eventually will give Deutsche Telekom 21 percent of the Malaysian cellular telephone operator.

Wireless markets in Hong Kong and Japan have experienced tremendous growth. Taiwan opened its telecommunications markets to competition in 1996, resulting in a great deal of activity as service providers, manufacturers, and investors scrambled for opportunities. In the latest step in Taiwan's deregulation of its telecom market, which is scheduled to be liberalized by 2001, eight paging licenses were issued by the Ministry of Transportation and Communications. Three Hong Kong companies were among those being granted licenses: HK Telecom, which partnered with First International, one of Taiwan's biggest computer companies; Hutchison, which teamed up with Southern Telecom; and ABC Paging, part of the Express Link group. Both HK Telecom and Hutchison were unsuccessful in winning a license in Taiwan's cellular mobile market (estimated at $900 million) in early 1997.

In 1997, the Ministry of Posts and Telecommunications (MPT) in Japan abolished requirements that forced users to make contracts with alternate long-distance carriers before they could use these carriers' services. Telephone users were restricted to using the long-distance service of Nippon Telegraph and Telephone (NTT). NTT agreed to split into three separate domestic carriers in exchange for permission to enter the international telecom market. NTT supplied the telephone line and all local calls in Japan, unless a contract was signed with one of three competing long-distance carriers. In the international market, restrictions were lifted so that callers can choose a carrier by merely dialing a different international access code. Some restrictions remain for cellular telephone users, who must complete a contract with an international carrier to make international calls.

Deregulation also removed the distinction between Japanese domestic and international operators and is allowing companies to compete in both markets. For example, Japan Telecom, a major long-distance carrier, and International Telecom Japan (ITJ) are merging operations and plan to compete in both markets; Kokusai Denshin Denwa (KDD) has announced marketing agreements with DDI and Teleway Japan, both long-distance carriers.

Under its agreement with the World Trade Organization, Thailand must open its telecom industry to overseas competition by 2006. The agreement means that the country will have to end monopolies in telex, facsimile, mobile, and fixed-line telephone service. Thailand's mobile phone market has long been dominated by companies such as Motorola, Nokia, and Siemens; to boost local telecommunication equipment and expertise, the government has issued a new telecom master plan that includes support for local manufacturers as well as research and development activities.

The Australian communications market was deregulated as of July 1, 1997. Telstra, the second-largest telco in the Asia region (after NTT) and Optus Communications, its major competitor, both are working in partnership with Indonesia's Indosat. The three companies recently introduced a new cable link that will cater to Internet, high-speed voice, data, and multimedia traffic between Australia and Indonesia.

In New Zealand, where deregulation has been in place for several years, some of the anticipated benefits are beginning to emerge: International call charges have dropped 55 percent over 5 years. Companies such as Clear Communications and BellSouth have entered the local and mobile phone markets in New Zealand, and Telstra has gained part of the international call market.

■ The Americas

In the Americas (Central, Latin, and South), national infrastructures are improving, aided by the privatization of government-owned telcos everywhere but Brazil and by investments in local telcos by U.S. carriers in countries such as Mexico.

Mexico opened its telecommunications market to competition on January 1, 1997 – a move that was promptly embraced by U.S. telephone and cable television companies, including AT&T, MCI, SBC, and Sprint. For example, Telefonos de Mexico (Telmex) has formed a joint venture with Sprint to market long-distance service to the business and residential Hispanic market in the U.S. Avantel, the joint venture by MCI and Grupo Financiero Banamex-Accival, was the first Mexican company to apply for and receive a license to compete in Mexico's long-distance market. Avantel is competing with Telmex for the Mexican business community's high-volume long-distance market.

In September 1997, the government of Guatemala announced the privatization of Empresa Guatemalteca de Telecomunicaciones (Guatel) and invited potential investors to buy between 51 and 95 percent of a newly formed Guatemalan corporation. As the successor to the majority of Guatel's operations, the company will be the principal provider of telecommunications services in Guatemala and will own and operate all the public telecommunications network facilities.

■ The Caribbean

Privatization of the Puerto Rico Telephone Company (PRTC) was authorized in August 1997 by the Commonwealth's legislature. PRTC had been acquired from ITT in 1974. The PRTC is the principal subsidiary of the Puerto Rico Telephone Authority (PRTA) and is the island's largest provider of local exchange services to the business and consumer markets – including wireless, ISDN, and Internet access services – as well as related products such as pagers and cellular phones. The privatization of the PRTC is part of a larger agenda that includes large-scale economic reforms calling for the privatization of operations and the sale of assets currently owned by the government of Puerto Rico.

Internet Services

The number of Internet users doubled every year between 1985 and 1996. Telecommunications service providers, eager to capitalize on a rapidly growing number of Internet users, or at least eager not to be left behind, are offering more Internet services to customers. LECs, IXCs, cable operators, wireless providers, and others have offered new on-ramps to the Internet. Telcos are emphasizing that they offer a range of business services on the Internet, including Web site hosting. For example, BT relaunched some services in September 1997 and introduced BT Internet ISDN LAN (BTiL), a local area network (LAN) Internet facility designed to operate over BT's integrated service digital network (ISDN) service. At least in the U.S., much of the Internet backbone is run by telcos such as GTE (BBN Planet), MCI, UUNet, and WorldCom.

Even though the market in low-cost Internet voice services currently is limited because of quality issues and the extra procedures or equipment required to achieve full capability, the service remains a competitive threat to the IXCs. However, growth also is expected in Internet-based telephony and videoconferencing, particularly if the FCC continues to refrain from regulation in these areas. In addition, new technologies and restructuring of the Internet eventually will help ease traffic congestion. (See Chapter 4.4, *The Internet and Intranets.*)

4.3.3 Transmission Technologies

Deregulation and demand for new services have created a need for more network carrying capacity, or bandwidth, particularly in the "last mile" to the customer. In response, current technologies are being improved and new technologies are being introduced to satisfy this need.

Wireline Media

Copper remains the predominant transmission medium at the local loop level (between the customer and the carrier's central office). Except for large customers, typically located in business districts, copper still serves the vast majority of customers worldwide. Fiber optic systems have replaced copper in network backbones, but the immensity of the task and the unattractive economics of running fiber all the way to the home or desktop make the elimination of copper a long-term proposition. Hybrid fiber-coax systems that use both fiber optic and copper cabling have been implemented by cable television companies to offer new services, and switched digital video systems also are being implemented. (See Chapter 4.1, *Transmission Media,* for a description of wireline media, including copper and fiber optic cable.)

The local loop or local access part of the network traditionally has used analog technology. In recent years, digital transmission over copper wires has become more cost-effective for a broader range of customers.

■ Integrated Services Digital Network (ISDN)

ISDN is a circuit-switched, internationally standardized narrowband service that provides a continuous digital transmission path for the duration of the call. ISDN is ideal for applications such as voice, videoconferencing, remote access to office networks, Internet access, and fax and image transmissions. It is relatively inexpensive and requires only moderate upgrades to existing infrastructures.

ISDN uses two or more "B-channels" that operate at 64 Kbps each. B-channels can be combined for higher data rates, if needed. ISDN comes in two standard offerings:

- **Basic-rate interface (BRI)** – Two B-channels and one 16-Kbps "D-channel" for exchanging call setup information. BRI was designed for single users or small offices.
- **Primary-rate interface (PRI)** – 23 64-Kbps B-channels and 1 64-Kbps D-channel (30 64-Kbps B-channels plus 1 64-Kbps D-channel in Europe), for a total of 1.544 Mbps (2.048 Mbps in Europe), the same bandwidth as a T1 (or E1). PRI was designed to increase the efficiency of corporate telecommunications systems, such as the private branch exchange.

Although most modems today have a top downstream transmission speed of 56 Kbps, ISDN transmission rates begin at 64 Kbps and reach as high as 1.5 Mbps in North America and 2 Mbps in Europe. An ISDN connection features two types of channels: The B-channel carries the digital information, and the D-channel carries information for each call (and also can carry data at up to 9.6 Kbps). The D-channel reduces call setup time and improves management of telephone connections. In addition, it is possible to allow a B-channel to drop while the network connection persists at a lower speed through the D-channel.

Although ISDN services have been widely available in Australia, Europe, Japan, New Zealand, and Singapore – both for local and international connections – U.S. telephone companies have been slower to deploy ISDN. For example, ISDN service was widely available in Europe in 1987, with a continent-wide telecommunications standard in place by 1994.

In the U.S., the rising interest in business usage of the Internet for commercial purposes has created increased demand for higher-speed services, and the telcos have responded with ISDN. A number of data communications equipment suppliers have developed BRI ISDN remote access devices to connect remote users to corporate networks or Internet service providers. These devices can cost as little as a few hundred dollars and support 128 Kbps of bandwidth. Growth of BRI-based ISDN offerings also results in increased use of PRI-based equipment by service providers and large corporations to aggregate the growing number of BRI and analog lines. However, ISDN is seeing increasing competition from

56-Kbps modems in the customer premises (which are less expensive and easier to install) and, in some communities, from cable modems (which can transmit data at much faster rates than PR1-type ISDN devices).

■ Digital Subscriber Line (DSL)

Digital Subscriber Line (DSL) technologies permit wideband transmissions over existing twisted-pair copper wiring already in place in the local loop. These technologies allow carriers to offer high-rate services (typically at 1.544 Mbps or faster) more economically and with shorter installation cycles than T1 or E1 services. A DSL connection uses a DSL interface device on the customer premises and a DSL interface device at the central office, which often is integrated with a concentrator and called a DSL Access Multiplexer, or DSLAM. Distance limitations of 12,000 to 18,000 feet between a customer and a central office have limited these services to metropolitan areas.

High-speed DSL (HDSL), the most mature of these technologies, permits two-way transmission over two or three pairs of copper wire to form a T1 or E1 connection. HDSL is being used for wideband applications formerly serviced by T1 lines. Because it uses two or three pairs of wire, it has been used less for end-user data applications. Single-pair HDSL, also known as symmetric DSL or SDSL, uses a single pair of copper wire to transmit data at roughly half the speed of a T1 line (768 Kbps) and is used where similar amounts of bandwidth may be needed for incoming and outgoing traffic. Table 4-10 shows data rates for various digital subscriber line technologies.

Table 4-10: DSL Technologies

Acronym	Digital Subscriber (DSL) full name	Maximum Data Rate	
		Downstream	Upstream
IDSL	ISDN DSL	128 Kbps	128 Kbps
CDSL	Consumer DSL	1 Mbps	128 Kbps
HDSL	High-data-rate DSL	1.544 Mbps	1.544 Mbps
SDSL	Single-line DSL	768 Kbps	768 Kbps
ADSL	Asymmetric DSL	9 Mbps	640 Kbps
RADSL	Rate-adaptive ADSL	7 Mbps	1 Mbps
VDSL	Very high-data-rate DSL	52 Mbps	2.3 Mbps

Source: *InfoWorld*, 1997

Asymmetric DSL (ADSL) uses a single copper pair and allows a telephone line to handle high-speed data such as VCR-quality video transmissions. ADSL uses asymmetrical transmission rates, typically transmitting from the customer to the telephone company at rates up to 640 Kbps and from the telephone company to the customer at 1.544 Mbps (T1) and higher (up to 8 Mbps). Unlike HDSL, ADSL carries voice at a lower frequency so that it does not affect data throughput, allowing for simultaneous voice and data transmission. Rate-adaptive ADSL (RADSL) is another new version of ADSL that will provide more control over different speeds and degrees of symmetry required for different applications. RADSL adjusts the speed based on signal quality, providing downstream rates from 600 Kbps to 7 Mbps and from 128 Kbps to 1 Mbps upstream.

Very High Speed Asymmetrical DSL (VH-ADSL) is a variation of ADSL developed by GTE Laboratories. VH-ADSL transports digital signals at speeds ranging between 25 Mbps and 52 Mbps over distances of a few thousand feet. Cost estimates (per subscriber basis) for deploying this technology indicate it might be comparable to hybrid fiber-coax.

In November 1997, Nortel and Rockwell jointly announced plans to promote Consumer DSL (CDSL), which permits simultaneous voice calls and Internet access on one phone line. The line operates with a 128-Kbps upstream link and up to 1 Mbps downstream to the customer.

Wireless Cable TV

Multi-channel, multi-point distribution service (MMDS) uses wireless technology to distribute video services over microwave frequencies in the 2,500 MHz to 2,700 MHz range. Some U.S. operators use two additional channels in the range of 2,150 MHz to 2,162 MHz. Transmitters send the line-of-sight signals to antennas at subscribers' homes where set-top boxes decode and decompress the signals for viewing on a normal television. Because MMDS microwave transmissions bounce off objects in their path, service may be affected by cars, tree leaves, and rain and other weather-related factors.

The total number of channels available depends on the bandwidth of each channel (determined by the compression algorithm and the country's broadcast standard). Eventually, MMDS technology is expected to mature to the point where operators can offer up to 300 high-quality channels to customers within a 40-mile radius.

Early on in the U.S., MMDS was treated as a stepchild of the cable television industry. Prior to the Cable Deregulation Act of 1993, MMDS operations were often undercapitalized and unable to arrange for quality programming, a fact that made it difficult for operators to obtain capital financing.

Today, MMDS deployment is beginning an upward trend in the U.S., and various companies are investing heavily in it. In emerging countries, MMDS has been deployed quite successfully. Because its wireless nature eliminates the high cost of installing fiber or coax, MMDS also is enjoying success in Latin America and Asia.

A variant of MMDS, called local multi-point distribution service (LMDS) is being promoted in the U.S. by a consortium of large telecommunications companies and manufacturers. LMDS provides greater upstream bandwidth than MMDS and other wireless services and also is competitive with conventional CATV. LMDS will share the 28-GHz band with the Teledesic satellite system. A pilot program by CellularVision in Brooklyn, New York, is using LMDS to deliver high-quality digital video to 6-inch dish antennas mounted in customers' windows. (Because LMDS uses line of sight, it requires a transmitter every couple of miles.) The system also is designed to provide local voice and high-speed data services.

Wireless Telephone Technologies

Generally, wireless telephones are serviced by radio equipment that covers either broad areas or small cells. Satellites are good examples of equipment that provides wide-area coverage. However, more wireless telephones are served by cellular, land-based equipment.

The term cellular refers to a type of service where the service areas are divided into cells. Each cell is serviced by a low-power transmitter that is only powerful enough to communicate with the cellular devices within its cell. Although each cell can operate on all available frequencies, interference is avoided by not reusing the same frequencies in adjacent cells.

Cell transmitters are controlled by a mobile switching office that performs the following functions:

- Identifies which cell the cellular phone is in
- Routes calls from cell to cell (known as handoff)
- Interconnects to the landline telephone network
- Monitors calls for billing

Traditionally, cellular telephones used analog signaling. However, the trend is toward digital signaling throughout the world to provide for increased call capacity within the same spectrum. Digital signaling is currently much more widespread in Europe than in the U.S.

In a more narrow sense, cellular refers to wireless telephone networks operating in the 800 MHz or 900 MHz portions of the radio spectrum. These systems use cells with diameters of 2 miles to 25 miles. Such systems are commonplace throughout the U.S., Europe, Asia, and other parts of the world.

Cells are also integral to certain types of services offered by providers of personal communications services (PCS) – called personal communications network (PCN) in Europe. The cells are smaller in diameter (typically 100 feet to 1/2 mile) and often are called "microcells." Not all PCS services are cell-based; some services, such as wireless private branch exchange, have service areas defined by a building rather than a cell.

Europe has well-established standards for both analog and digital cellular telephone networks. However, the U.S. market is standardized only for analog cellular systems because the FCC has not dictated standards for digital-based systems, preferring to let each carrier choose its own technology. As a result, U.S. service providers are attempting to sort out and decide between three different digital standards as they transition from analog to digital cellular technology.

■ AMPS

The majority of North American cellular systems use an analog technology known as the Advanced Mobile Phone System (AMPS). AMPS also is used elsewhere in the world, including parts of Asia, Australia, Scandinavia, and South America. With AMPS, 30 MHz of the 800-MHz spectrum is divided into channel pairs, and each call totally occupies one pair. Although this technology is well established, the number of calls that can be made through a cell site at one time is essentially limited to the number of available channels. The dramatic growth of the cellular market has the industry looking for ways to multiply capacity cost-effectively.

As analog (AMPS) cellular systems near capacity, cellular operators plan to migrate to digital systems to avoid losing market share to other providers. Customers will be enticed by the digital network's increased flexibility and better performance.

■ D-AMPS and TDMA

Digital-AMPS (D-AMPS) uses the same radio channels as analog systems but employs a technology known as Time Division Multiple Access (TDMA). TDMA digitally encodes speech so it can be transmitted in short bursts. Between bursts, transmissions to or from other users can occur. In this way, a single channel pair can accommodate more than one user. The D-AMPS capacity is three times to four times greater than its analog counterpart, with greater densities likely to be achieved as the technology matures.

Handsets incorporating D-AMPS technology as well as analog technology can detect whether a network is digital or analog and switch modes accordingly. For AMPS-based systems operators, D-AMPS has distinct advantages. D-AMPS offers three times to four times the revenue from a given amount of spectrum. D-AMPS also offers the ability to migrate channels gradually from analog to digital.

Dual-band handsets now available permit AMPS and D-AMPS operation in both the 800-MHz band and the new 1900-MHz PCS/PCN band. Such units create new opportunities for international roaming.

■ GSM

The Council of European PTTs (CEPT) began development of a digital standard in 1982. The goal was to permit development of a European network that would permit international roaming. Dubbed the Global System for Mobile communications, the first GSM networks were put into service in 1992 in the 900-MHz band. Today, more than 80 countries have implemented, or are planning to implement, GSM networks. Globally, 166 GSM or digital European cordless telecommunications (DECT) networks are planned or in operation. (DECT is the pan-European standard based on advanced TDMA technology and used primarily for wireless private branch exchange systems, telepoint, and residential cordless telephony today; potential uses for DECT include paging and cordless LANs.) A number of network service providers in the U.S. market also are implementing GSM. (Digital cellular systems using GSM also are called PCS and are in use worldwide. GSM cellular phones using the 1.9-GHz frequency are called PCS phones in the U.S.)

Another European standard, called DCS 1800, is a variant of GSM and operates in the 1800-MHz band. DCS systems have been operational since 1993. The new European PCS-1900 standard uses the proven GSM-based technology, this time in the 1900-MHz band, to offer data and fax communications and the ability to deliver short, text-based messages to a user's phone.

■ CDMA

An alternative standard, Code Division Multiple Access (CDMA), first proposed in the late 1980s, uses a different approach than TDMA to convert analog signals into digital for transmission. CDMA (also called spread spectrum) works by splitting a voice signal into tagged fragments that are sent across a broad part of the spectrum shared by fragments of other calls. To complete the call, the fragments are reassembled into the original pattern. CDMA must be implemented across a large set of channels simultaneously, whereas TDMA can be implemented on a channel-by-channel basis.

CDMA systems can offer up to 20 times more call-handling capacity than conventional cellular systems by assigning a special electronic code to each call signal, allowing more calls to occupy the same space and be spread over an entire frequency band. Although increased capacity is a major benefit of CDMA, the technology also provides three features that improve system quality:

- **"Soft" handoffs** – Ensures a call is connected before the handoff is completed, reducing the possibility of a dropped call
- **Variable-rate vocoding (voice coding)** – Allows speech bits to be transmitted at only the rates necessary for high quality, thereby conserving battery power in the subscriber unit
- **Multi-path signal processing** – Combines power for increased signal integrity

Qualcomm, which holds several patents on products using its CDMA implementation, is beginning to see its early commitment to the technology vindicated by the marketplace. Approximately 9 years after Qualcomm first demonstrated the viability of CDMA, some 57 percent of new digital wireless systems will be based on it. CDMA already is being used in inside-building wireless private branch exchange conversations by companies including SpectraLink.

In September 1997, 3Com, Qualcomm, and Unwired Planet formed an alliance to promote CDMA as an alternate way to connect to the Internet. The companies are using quick net connect technology (developed by 3Com and Qualcomm) to extend CDMA and predict it will offer much faster access than landlines. Quick net connect will be available first to users of Unwired Planet's UP.Browser software on Qualcomm CDMA telephones. For example, Qualcomm's 5-ounce, palm-sized Q Phone "computer" uses CDMA for wireless access and also boasts a Web browser and an e-mail client.

3Com and Qualcomm have been working with other manufacturers to develop CDMA standards, in part through the CDMA Development Group (CDG). The CDG is a nonprofit trade association formed to foster the worldwide development, implementation, and use of CDMA, representing more than 80 companies to date, including many of the world's largest wireless operators and equipment manufacturers. According to CDG, more than 80 percent of CDMA growth is occurring outside North American in areas such as China, India, Japan, Korea, and other high-growth wireless markets.

In Korea, for example, CDMA has technically prevailed over TDMA in wireless telephony. Korean companies announced the successful deployment of 700,000 CDMA phones based on Qualcomm licenses. Korea Mobile Telecom reported that the CDMA system outperformed both analog and GSM phones in terms of capacity, uptime, voice quality, and power usage.

■ Cellular Digital Packet Data

Cellular Digital Packet Data (CDPD) is a digital transmission technology that sends data (including fax and e-mail) over the existing cellular infrastructure using a wireless modem. The modem finds an unused channel, sends a packet of data, then clears the channel. Because it makes use of idle time on several channels, CDPD does not need a dedicated channel (though some carriers may choose to use one).

Modems for CDPD generally do not work through existing analog cellular phones; instead, they include their own transceivers. However, some newer phones will be CDPD-ready.

CDPD is best for short, bursty transmissions such as those used for e-mail. It offers digital encryption and uses a standards-based protocol that can support wireless formats such as AMPS and CDMA. Current modems operate at a maximum of 19.2 Kbps, but their effective data throughput can be significantly lower.

■ PCS/PCN

Personal communications services (PCS) – referred to as personal communications network (PCN) in Europe – is a set of wireless technologies that provides various services, such as the following:

- **Low-mobility telephony service** – Characterized by pedestrian-speed handoff, implemented through the use of microcells, primarily in major population centers.
- **Fixed wireless local loop service** – Competes with LECs by providing telephone service to the home and office. For example, the 32-Kbps fixed wireless local loop (offering mostly voice applications) is available from PCS vendors as an alternative to CAPs, particularly for residential service.
- **Wireless private branch exchange** – Permits intrabuilding cordless telephone service.
- **Wireless LANs** – Allows wireless connections for data transmission.

PCS networks can use CDMA, GSM, TDMA, D-AMPS, and other technologies.

PCS digital networks will make more digital services possible on cellular networks. Digital phone services, over both cellular and PCS networks, provide better sound quality and permit a mix of voice and data as well as features such as caller ID. Short text or numeric messages can be displayed on the phone, and because many new phones support the IP protocol, they can be used to access e-mail, the Web, and other applications on data networks.

At the end of 1997, suppliers were only beginning to roll out PCS networks and initially were concentrating on voice services; vendors still must address application-deployment issues and end user hardware costs. In addition, acceptance of PCS may come slowly for a number of reasons. Currently, PCS carriers are concentrating on voice services (rather than data services) because they have been so widely accepted in the analog cellular market.

■ Antennas

Cellular operators need to install tall towers to hold cellular antennas to provide new services such as PCS. When antennas are required in residential areas, equipment manufacturers also conceal antennas in structures such as bell or clock towers perched atop churches or other buildings. Whereas cellular typically operates in the 800 to 900 MHz range, PCS would operate in the 1.5 to 1.8 MHz range. Because PCS/PCN systems use very high frequencies, high-gain antennas come in small packages that blend into their surroundings. The many antennas needed to establish a PCS/PCN system are less noticeable when they are part of light poles, traffic signals, telephone and power poles, and street signs. Combining several antennas into one package can help reduce visual antenna clutter and deployment costs.

Drawing on research funded by the U.S. Department of Defense, some manufacturers are working on adaptive array antennas that can lock onto a user's signal and follow it electronically (rather than by mechanical means or by switching to a different antenna). These "smart antennas" may play a significant role in PCS applications because they offer the following potential benefits:

- Increased cell capacity through greater frequency reuse
- Reduced number of cells (and associated infrastructure and deployment costs)
- Improved service quality, including fewer dropped calls
- Use of smaller, less-costly handsets

Wireless Terminals

The line between telephones, pagers, and computers is blurring rapidly. In the never-ending quest for a bigger slice of the telecommunications pie, manufacturers seem to be turning out new wireless terminal designs every day. Integration of digital paging, Internet access, and e-mail with handsets already has taken place, and many variations are available from manufacturers such as Ericsson, Motorola, Nokia, and Sony.

Personal organizers, personal digital assistants (PDAs), palmtop computers, and hand-held Web browsers also are becoming wireless using PCS. A wide variety of products already have been announced, many of which integrate two or more technologies. Table 4-11 contains a sampling of products that became available in 1997 and illustrates the diversity of wireless products.

Table 4-11: Examples of Wireless Products

Manufacturer(s)	Function of Unit
HP	Traditional hand-held HP organizer with built-in Web browser
IBM	Handset with special high-resolution screen used with a mirror to give the impression of viewing a 10-inch video monitor
Motorola, Sony	Pen-based "personal information communicators" for exchanging wireless e-mail; accessing databases; sending faxes and alphanumeric pages; receiving Internet news, sports, and weather information; and connecting to on-line shopping services
Nokia	Cellular-like handset flips open to reveal keyboard and screen for browsing the Web, accessing databases, and sending and receiving e-mail and faxes

Paging and EPS

Paging systems have been around since the 1960s. Early U.S. systems offered only a tone page; later, multiple tones became available to indicate various callers. Eventually, voice messages were added and then alphanumeric displays. For many years, systems typically operated in the VHF (30 MHz to 50 MHz and 150 MHz to 170 MHz) and UHF (around 450 MHz to 470 MHz) bands, where other mobile radio services exist.

Until recently, paging was a one-way proposition; there was no assurance that a transmitted signal had been received by the intended pager. However, the allocation of narrowband PCS frequencies in the 930-MHz spectrum (auctioned in 1994) opened the door to two-way paging using enhanced paging services (EPS).

One service that EPS enabled allows pagers to transmit back a response, allowing the customer to indicate a "yes," "no," or any other preprogrammed response. Another service permits the generation and transmission of messages directly from the pager.

Mixing EPS with other PCS services presents more possibilities. For example, voice telephony can be combined with paging, voice mail, faxing, and Web browsing. Many products currently on the market provide various combinations of these capabilities.

■ Expanded Specialized Mobile Radio (ESMR)

In 1991, the FCC adopted rules – proposed by Nextel – for Enhanced Specialized Mobile Radio (ESMR). ESMR completely remodels the old Specialized Mobile Radio (SMR) concept epitomized by the dispatch systems in use since the 1960s and 1970s by adding interconnection to the landline telephone network, packet-switched data, and paging services to the dispatch function. Eventually, ESMR also will provide fax, vehicle location, and emergency service.

Although ESMR can be analog or digital, it uses new equipment that borrows techniques such as time division multiplexing from digital cellular and PCS. Hand-held units or mobile units resembling traditional SMR radios can be used for ESMR. However, ESMR's potential to merge many services has given rise to the development of cellular-like handsets, with push-to-talk switches like walkie-talkies, that can combine some or all the services noted previously for those who switch to digital.

4.3.4 Telecommunications Services

Voice service over copper wires using analog signaling has been the traditional business of the world's major carriers for many years. However, the last few years have seen the rapid development of new services to keep up with rapidly changing user requirements as well as new competitive conditions.

In many telecommunications service areas, key trends will be determined more by marketing and availability than by the merits of the technology. Services that may be more desirable technically may lose out to products that are quickest to deploy, lowest in cost, or de facto standards or to companies that have products well-positioned and marketed.

For example, GSM is well on its way to being a dominant world cellular technology and de facto industry standard. Some U.S. providers already are implementing the GSM standard currently used in many other parts of the world (although the U.S. providers are implementing the standard in a different frequency range, so compatibility will continue to be an issue). Most analysts predict that GSM will dominate the digital cellular telephone market well into the next decade, although they concede the newer CDMA technology is superior to GSM in many respects. With a much smaller installed base to date, however, CDMA is not expected to catch up with GSM before both technologies are replaced by newer, third-generation cellular systems.

New technology is enabling the development of products that adapt readily to the needs of the marketplace. For example, dual-mode cellular handsets can detect whether a network is analog or digital and adapt accordingly. This flexibility is important to service providers that want to change their networks to digital technologies to take advantage of lower costs (and higher profits) from higher channel densities. It is also important to consumers who worry about buying equipment that will become obsolete or unusable when roaming outside their home area.

Wireline Digital Communications Services

Demand for high-speed digital services is growing rapidly. Private and semi-public networks are migrating quickly toward services that can support their LAN interconnection, videoconferencing, and multimedia requirements.

Networks can be either circuit-switched or packet-switched. There are two types of packet-switching: connectionless and connection-oriented (virtual circuit). Table 4-12 compares the characteristics of circuit switching and packet switching.

Table 4-12: Circuit Switching versus Packet Switching

Circuit Switching	Packet Switching	
	Connectionless	Connection-Oriented (Virtual Circuit)
Dedicated transmission path	No dedicated path	Dedicated bandwidth along a predetermined path
Continuous transmission	Transmission of packets	Transmission of packets
Fixed bandwidth transmission	Dynamic bandwidth	Dynamic bandwidth at call setup

Source: *Broadband Communications*, 1995

■ Private Line Services

A private line is a non-switched line; that is, one end is connected permanently to the other end. Organizations with sufficient traffic to justify the cost often use these commonly available lines in their own private data networks.

Standard data rates for private-line digital data services (DDS) are between 2.4 Kbps to 56 Kbps (capable of supporting data); a minimum rate of 64 Kbps is required for voice and video applications. However, T3 lines (which offer 45-Mbps speeds) are coming into greater use (for Internet connections, for example).

One of the most common private line offerings is a T1 circuit (E1 in Europe). Another offering, fractional T1, permits users to buy a portion of a T1 in multiples of 1/24 (64 Kbps) of a T1 line. Two other types of lines, DS1 and DS3, are services based on digital standards; however, they provide the same data rates as T1 and T3. A fractional DS1 service (called DS0) is also available. Strictly speaking, T1 refers to the twisted-pair line, regenerators, and terminating equipment that carry a 1.544-Mbps (2.048-Mbps for E1) digital signal between two points. A T3 line (E3 in Europe) operates at 45 Mbps (34 Mbps for E3).

The major IXCs also offer switched services from one point-of-presence (POP) to another at data rates between 56 Kbps and 45 Mbps. Generally, for users who need connectivity only a few hours per day, switched services are more economical than non-switched services. (See Figure 4-8.)

Figure 4-8: Switched Lines Can Be Cost-Effective

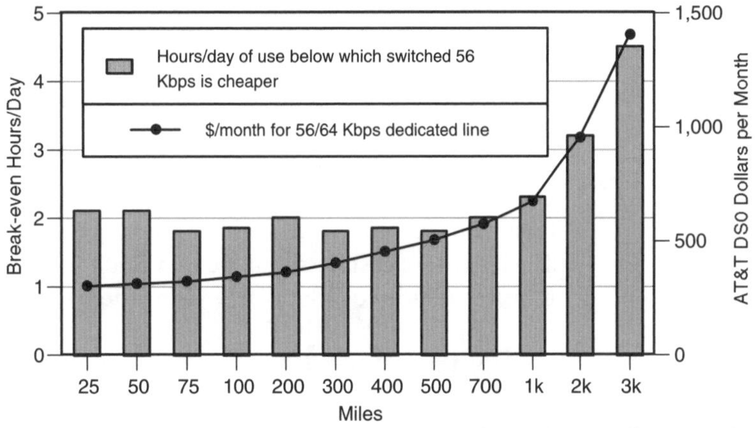

Source: *Business Communications Review*, 1995

■ ISDN

At present, ISDN is available to a majority of the U.S. population, but usage has been relatively low, probably because residential implementation has been somewhat complicated, and pricing structures have been confusing.

Euro-ISDN is the preferred signaling standard in Europe; it differs from North American National ISDN in that Euro-ISDN is very limited in the options it offers. In the U.S., ISDN comes with many options, including multiple call appearances, conference calling, call forwarding-variable, call forwarding-busy, call forwarding-no answer, voice mail with indicator, two secondary directory numbers, and so on. However, users can call between the U.S. and Europe and complete ISDN calls.

■ Broadband ISDN

Broadband ISDN (BISDN) can be described as the set of high-performance, end-to-end digital services that will be offered on the public network of the future. It is a network concept that requires multiple technologies (including ATM and SONET) and network-wide fiber optic media for its implementation. BISDN standards provide speeds of 155 Mbps and higher. Given that major network infrastructure and methodology changes are also necessary for implementation, it is unlikely that BISDN will be a practical reality before the year 2000.

■ Frame Relay

Frame relay is a packet-switching technology that packages data into frames (a type of packet) so they can be relayed from node to node. Processing at a node is limited to switching coordination, translations of addresses, and error detection. Designed to run over digital lines, frame relay is much simpler than previous packet-switching protocols (such as X.25) designed for slower, noisy, error-prone analog lines.

Frame relay can support both permanent and switched virtual circuits. A virtual circuit appears to be dedicated (from the customer's perspective), but it is really only a defined path through a shared facility. In a permanent virtual circuit, a connection is always up between two nodes. In a switched virtual circuit, a connection can be established on request.

Typically operating at 56 Kbps to 1.5 Mbps, frame relay is a good choice for applications such as LAN-to-LAN connections where the amount of traffic may vary significantly over time, with short bursts of very high bandwidth required. Some vendors offer 45 Mbps frame relay. Pricing varies from carrier to carrier and depends on specific parameters such as port access speed, number of virtual circuits, distance, traffic volume, and committed information rate (the guaranteed amount of bandwidth available at all times to a given customer). Customers can contract for either guaranteed or non-guaranteed bandwidth. The latter case costs less, but it could result in delays or suspensions of data transfers if the network becomes congested.

■ Asynchronous Transfer Mode (ATM)

Asynchronous Transfer Mode (ATM) uses small, fixed-length (53-byte) cells that can be switched at tremendous speeds. Because of the low latency and predictable throughput ATM provides, it is ideally suited to mixing different data types (such as voice and video) on a network. Applications include traditional data communications, imaging, full-motion video, and multimedia.

ATM was conceived originally as a service that would run over the broadband public telephone network. Although ATM was designed primarily for the transfer of high-quality video and multimedia content, the technology so far has been used mainly for data transmission. ATM has been adopted by the computer industry as a technology for use in private corporate backbones and wide-area networks (WANs). (See Chapter 4.2, *Networking Systems and Protocols.*) ATM has been especially popular among telcos whose customers need additional bandwidth beyond that available with frame relay to meet their data transmission requirements.

More than half of the major U.S. telecommunications carriers currently offer ATM services. ATM services are available at access speeds up to 622 Mbps (although equipment is available to support lower speeds up to 45 Mbps). The backbone carrier networks currently operate at speeds as high as 2.5 Gbps. The ATM edge and core backbone switches operate at very high speeds and typically contain multiple buses providing aggregate bandwidth of as much as 200+ Gbps. Long-distance carriers AT&T, MCI, and Sprint already have implemented high-speed ATM backbones. RBOCs such as Ameritech, NYNEX, and Pacific Bell also offer ATM-based services. European telcos have been moving toward adopting ATM as a long-term strategy because they believe the higher bandwidth will be necessary to handle evolving multimedia applications.

ATM addresses the communications industry's increasing need for specific Quality of Service (QOS) levels. This is especially important for delay-sensitive applications such as audio and video transmission or perhaps Internet telephony. ATM is flexible in that it can be used to transmit data based on IP, IPX, SNA, and other upper layer protocols. ATM's scalability is also an attractive feature for growing organizations.

■ Switched Multimegabit Data Service

Frame relay and ATM are connection-oriented services. Switched Multimegabit Data Service (SMDS) is a connectionless packet service that offers speeds up to 45 Mbps. It originally was intended to fill the gap for broadband services until BISDN could be developed. Because the infrastructure for BISDN is not yet in place, some users have chosen SMDS.

In 1993, BT upgraded an existing academic network in the U.K., JANET, to SMDS. The new network that links the major academic institutions is called SuperJANET. The 10-Mbps system is used for medical image transfer, distance learning, and other bandwidth-intensive applications. Other countries where SMDS services are offered include Belgium, France, Germany, Italy, Sweden, Switzerland, and some parts of the U.S. (at least within an LEC's territory). Providers in Austria, Ireland, and Portugal also have plans to offer SMDS in the near future.

Competitive LECs (CLECs)

CLECs are companies that offer alternative access to the public switched network, bypassing the LEC's network. Typically, in the U.S., CLECs were used by businesses to connect directly to their IXC, thus eliminating the cost of local network access charges for long-distance phone calls. Metropolitan Fiber Services (MFS Communications), now owned by WorldCom, was one of the first companies to provide such a service through its network of fiber installed in many major metropolitan service areas.

Lower cost and better service are CLEC's main attractions to business users. In addition, an IXC can partner with a CLEC to provide easy entry into the local telephone market.

Cable Television Communications Services

Over the past few years, there has been a lot of talk about video-on-demand, digital set-top boxes, and high-definition TV, but those new service offerings have been slow to materialize.

Cable operators in the U.S. have been raising rates significantly in an attempt to recover from more stringent rate regulations imposed in 1992. The effects of this regulation were somewhat lessened in November 1994 by a decision by the FCC. This decision allowed cable companies to raise rates when adding channels or when covering cost increases and gave operators a way to increase revenues. In 1996, Congress agreed to deregulate cable rates in 3 years. By the end of this period, however, the maturity of other services (such as direct-broadcast satellite TV) may force rates to go down in response to competition. With the advent of deregulation, many U.S. cable operators want to broaden services by offering local telephone and Internet access services. Adding to an already complicated situation, many of the LECs are gearing up to offer video-on-demand. Bell Atlantic, for example, has been particularly active in this area.

■ Cable Telephony

The introduction of two-way switched voice traffic (telephone) over cables that already carry a wide range of radio frequency signals (mostly TV) is no easy feat, as some cable operators have discovered. In most cases, their infrastructures were not originally designed to provide priority for voice and data, so major reengineering efforts are necessary to offer those services.

Traditionally, cable systems in the U.S. have been one-way services, combining video signals at the head end and sending them down miles of cable and through numerous amplifiers and splitters to subscribers. Modifying this infrastructure to allow for sending voice and data in the reverse direction is extremely challenging, both technically and economically. Even if existing fiber and coaxial cable can be used, almost everything else in the distribution part of the system must be replaced, modified, or heavily augmented to permit two-way traffic.

Most new designs have focused on using frequencies previously not active on the cable, especially those in the lower part of the spectrum. However, major technical challenges lie ahead. One of these issues is the need to overcome interference from a wide variety of sources, including static generators such as appliances and motors. Interference from commercial, amateur, and CB radio equipment is also an issue.

■ Internet Connection

Cable-based data services, such as Internet access, are another way cable operators hope to retain market share. However, the CATV infrastructure must be modified for two-way traffic on the cable, and cable-specific modems are required for end users. Even so, some networks have been reworked, and about 20 major manufacturers are shipping modems. (Some of these devices use ordinary telephone lines for their back channel, avoiding the complex reengineering of their infrastructures required to support full two-way transmission.)

The market for Internet connections is seen as a priority for many cable operators, causing them to give video-on-demand and voice a much lower priority.

Telephone Switching Facilities

Traditional telephone private branch exchanges (PBXs) are providing more services as computer equipment and telephone equipment become increasingly integrated.

■ Private Branch Exchanges

Although PBXs have continued to add features, they actually have become physically smaller in size. Most of these changes are attributable to improvements in computing technologies. What once required the footprint of a large mainframe to support up to 600 telephones now can be compressed into two, 6-foot by 2-foot cabinets. Prices, however, have remained in the $600 to $1,000 per line range over the past 10 years. PBX technology continues to be proprietary, expensive, and relatively slow to innovate, compared with the computer industry.

■ Computer Telephony Integration

Using computer telephony integration (CTI) to improve customer service has become less costly and much simpler to implement.

Automatic call distribution (ACD), a function of either a PBX or the phone company's central office switch, allows sequencing and uniform distribution of calls among multiple customer service agents. In addition, call setup information can be passed through a PBX, and CTI facilitates the interface between a customer information database and the PBX's ACD system. Incoming calls that include automatic number identification first can be routed to the database for identification and sorting by category. Then, those calls can be allocated to the next available agent in a given category along with the customer's account information (displayed before the agent answers the call). All of this process can occur instantaneously or be held in priority queues, subject to the volume of calls versus the number of available customer service agents.

■ Wireless Telephones

Office and manufacturing buildings, hospitals, apartment buildings, and other densely occupied structures can make good use of wireless telephone technology. Workplaces can be equipped with unobtrusive radio equipment that enables low-power wireless telephones to be used in place of wired telephones. This approach provides mobility within a building. The wireless phone access points deployed through the structure then are linked to a wireless PBX that provides switching within the building and to external lines.

Some companies have created a wireless phone system that works both on- and off-site. Using a dual-mode handset, the device attempts to contact the wireless PBX system first. If a connection cannot be established, the call is then sent via the cellular network. Companies have agreements with local wireless companies to support their employees when they exit the local company site. When off-site, users pay the typical cellular rates, but while on-site, they are not charged.

4.3.5 Market Overview

Cable Data Transmission

The market for cable modems has continue to grow, albeit slowly. The leading proponent of cable data transmission services, @Home Network, offers services in 15 states from a half dozen cable companies, including Cox Cable, Comcast, and TCI. Without standards, it has been difficult to sell cable modems to customers who cannot be guaranteed that what they buy from a cable company in Connecticut will work on a cable operator's network in California. Therefore, cable operators prefer to rent modems to customers for a modest fee and then charge a premium for Internet access at higher speeds.

Although fewer than 250,000 cable modems were expected to ship in the U.S. in 1997, increasing to only 750,000 units in 1998, the future looks more promising. A recent market study indicates that cable modems will win the majority of the North American residential access market, growing to an installed base of more than 7 million units by the year 2002, with prices possibly falling as low as $100. Declining prices are already in progress: Bay Networks' Generation 4 cable modem, expected to ship in 1998, will cost less than $200. The availability and range of related services also will dictate cable modem sales and could affect pricing significantly.

Wireless Telephone Technologies

In the U.S., the wireless telephony market is projected to reach $100 billion within 5 years, and these high stakes are causing the proponents of standards such as TDMA and CDMA to try to establish dominance.

In 1996, Europe had 16 million digital wireless phones versus 1.5 million in the U.S.: Lack of agreement in the U.S. about which technology should reign has resulted in a fragmented implementation of digital wireless phone systems, so that handsets that work in one market may not be able to access digital services in a different geographical area. (This is a particular problem with PCS, which typically provides no AMPS capability.)

PCS/PCN

PCS and digital cellular systems together will drive down mobile communications costs. IDC predicts that the installed base of cellular and PCS subscribers worldwide will grow from 132 million in 1996 (up 59 percent from 83 million in 1995) to 364 million in 2001. Equipment prices are expected to continue to fall, resulting in a worldwide decrease in equipment revenue over the same period.

Paging

Demand for pagers continues to grow at a fast pace, partly because new services are being added continually. For example, messages can be sent via e-mail, over the Internet, to a paging customer. Interactive e-mail pagers allow users to generate as well as receive e-mail messages. Pagers can receive stock quotes and sports scores (for an additional fee) and voice mail.

Expanded Specialized Mobile Radio (ESMR)

The market for ESMR is expected to grow steadily, though prices are high for hand-held digital handsets ($800 to $1,000) and air time ($25 to $30 per month for 75 minutes of service, compared with $16 to $23 for analog). Nationwide ESMR services are possible through roaming agreements with companies such as OneComm.

Frame Relay

All major U.S. IXCs offer frame relay services, as do many LECs. Bell Atlantic and US West are the largest LECs offering the service. Reduced administrative costs and aggressive pricing are contributing to the service's increasing popularity.

SMDS

Although SMDS is not offered universally by IXCs and LECs in the U.S., usage continues to rise. (See Figure 4-9.) IDC expects SMDS service revenue to increase from $42 million in 1996 to $148 million in 2001. MCI remains the major U.S. IXC offering SMDS. The other IXCs have passed over SMDS in favor of frame relay and ATM. Similarly, only a handful of LECs offer SMDS. In Europe, however, BT has made SMDS a cornerstone of its strategy for a high-bandwidth network, and several other European telephone operators have implemented SMDS as a trial run for ATM.

ATM

Overly optimistic growth projections hurt ATM for a few years due to slower growth than anticipated. Demand for ATM is expected to grow substantially over the next 10 years. According to a 1997 report by Frost & Sullivan, the U.S. ATM WAN equipment and services market will have an estimated value of $1.87 billion by 1998. The report (which includes both ATM switches and carrier services) predicts the market will grow to more than $4 billion by 2000. The Yankee Group predicts that the worldwide revenues of U.S. ATM service providers will increase in value from $32 million in 1995 to $672 million in 1999.

Figure 4-9: Packet/Cell-Based Service Revenue by Segment

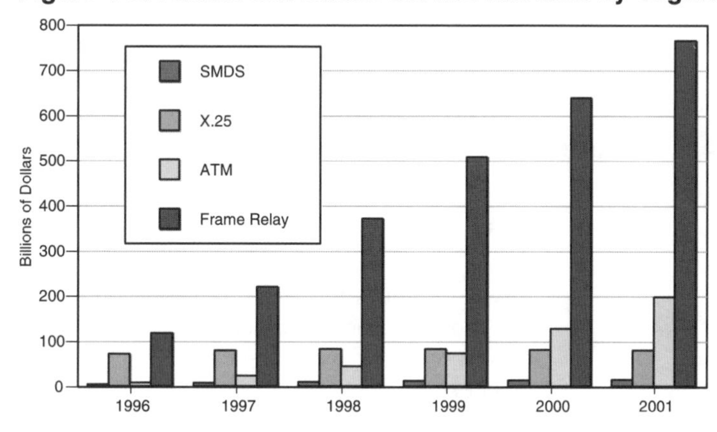

Source: IDC, 1997

ISDN

Pricing structures vary from provider to provider, but in the U.S., an ISDN line typically costs around $50 per month, plus usage charges; however, those costs are expected to decrease to $30 per month by 2001. IDC estimates that total ISDN lines in the U.S. will grow at approximately 41 percent per year through 2001.

ISDN is being deployed rapidly in Europe, with most European carriers offering the service. Usage in Europe is expected to grow 40 percent per year until the year 2000. Figure 4-10 shows anticipated growth in BRI lines.

Figure 4-10: ISDN BRI Usage Growth

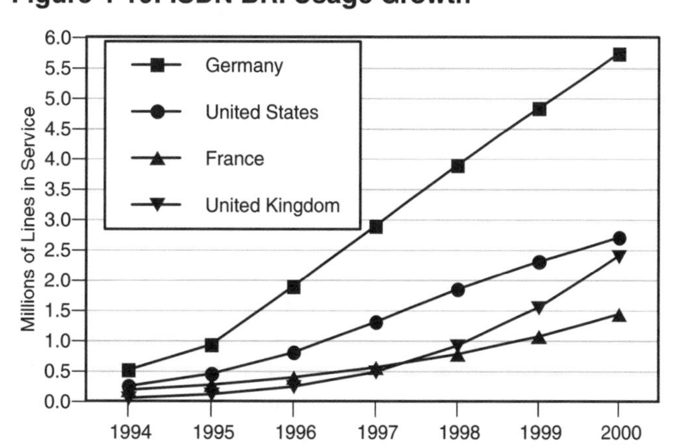

Source: Dataquest, 1996

Table 4-13 shows the line and projected growth rates in the U.S. for BRI and PRI revenues.

Table 4-13: U.S. BRI and PRI ISDN Revenue Model and Forecast

ISDN Lines and Revenue	1996	2001	CAGR (percent)
BRI lines (000)	730	3,903	39.82%
Average revenue per line	$600	$360	-9.71%
Total BRI revenue	$438 million	$1.4 billion	26.24%
PRI lines	31,000	370,000	63.81%
Average revenue per line	$6,000	$3,600	-9.71%
Total PRI revenue	$188 million	$1.33 billion	47.90%
Total local ISDN revenue	$626 million	$2.73 billion	34.29%

Source: IDC, 1997

4.3.6 Forecast

Deregulation

- In the U.S., further consolidation will occur among telecommunications service providers as a result of the Telecommunications Act of 1996. This trend will allow business customers to purchase an increasing range of telecommunications services from a single provider. The benefits to residential consumers will be slower to appear.
- In much of the rest of the world, the primary effect of deregulation will be the opposite: an increase in the number of service providers as new competitors enter markets previously served by a single (often state-owned) carrier. However, alliances and mergers between carriers in different countries will allow for global "one-stop shopping."
- In the U.S., the entry of IXCs into the markets of LECs and vice versa will continue to depend on the outcomes of legislative, judicial, and regulatory actions.

Transmission

- Growth in Internet use among consumers and in telecommuting will continue to increase the demand for higher-speed remote access throughout the forecast period. This trend will lead to the wider adoption of existing services such as ISDN, and foster the development of new services such as DSL and cable data transmission.
- Availability of DSL and cable data transmission in the U.S. will be stimulated by efforts on the part of major computer industry vendors, who see the lack of high-bandwidth services in the residential market as a key inhibitor to the growth of new Internet services and thus of the demand for their future products.
- Use of fixed wireless transmission will expand to provide an alternative to wireline media, especially in areas not served by a modern, high-capacity wireline infrastructure. Fixed wireless will also be attractive to CLECs, who would find the cost of duplicating the existing wireline infrastructure of the incumbent carrier prohibitive.

Service

- Aggressive pricing in an increasingly competitive wireless environment in the U.S., especially as PCS/PCN products enter the market, will drive down prices to the consumer. New services, aggressive marketing, and reductions in operating costs will be key for wireless service providers. Some PCS providers are offering nationwide service at no additional charge, eliminating the high roaming charges that historically have been assessed.
- Different digital and PCS network standards will continue to cause confusion in North America as some providers implement GSM and some choose TDMA, CDMA, or other technologies. This lack of standardization will confront subscribers with a situation where their PCS telephones may not offer nationwide roaming coverage because PCS service compatible with their handset may not be available in every market.
- Internet access services will be offered by LECs, IXCs, and a range of other providers. Because the Internet is seen both as a threat and an opportunity, nearly all telecommunications providers, including CATV operators, will position themselves as ISPs to maintain market share (sometimes by purchasing backbone ISPs). In response to increased competition, the independent ISPs who traditionally have provided the access will continue to add services, look for new market niches, or form alliances with other providers.
- Wireless data services will become a major PCS/PCN service offering, but its effects will not be felt profoundly before 2000.
- Data traffic will become a larger market than voice by the end of the forecast period.
- Global roaming will be enabled, in part, through LEO systems scheduled to begin narrowband service in 1998. LEO-based broadband service is scheduled to become available shortly after the end of the forecast period.

4.3.7 References

■Articles

Andrews, Edmund L. 1997. Phone market is opening up for Europeans. *The New York Times.* December 29.
Brown, Jim. 1997. DSL: Stuck in the slow lane. *Network World.* November 3: 55.
Clark, Elizabeth. 1997. *Real-world ATM.* Network. July: 42.
Grover, Ronald and Peter Burrows. 1997. US West can't take its eyes off TV. *Business Week.* March 24: 61.
Horwitt, Elisabeth. 1997. Latin America: it might be only 3% online today, but fasten your seat belt. *Computerworld.* September 29: SA17.
Lawson, Stephen and Kujubu, Laura. 1997. Nortel, Rockwell add CDSL voice support. *InfoWorld.* November 24: 45.
Liebmann, Lenny. 1996. ISDN unplugged. *InformationWeek.* April 22: 65-70.
Matthews, John. 1997. Wide priorities. *Computerworld.* September 29: SA5.
McClelland, Stephen et al. 1997. Asia: the global telecom dynamo. *Telecommunications.* September: 42.
Sandoval, Ricardo and Howard Bryand. 1997. Telecom deregulation taking shape in Mexico. *San Jose Mercury News.* May 11: 1E.
Weinberg, Neil. 1997. This gorilla wants to dance. *Forbes.* September 22: 97.

■Periodicals, Industry and Technical Reports

America's Network. (monthly) Santa Ana, Calif.: Advanstar Communications.
Business Communications Review. (monthly) Hinsdale, Ill.: BCR Enterprises Inc.
Telecommunications. (monthly) Norwood, Mass.: BPA International.
Gillott, Iain. April 1997. *Worldwide cellular and PCS infrastructure market assessment, 1996-2001.* IDC.
Shuchat, Rona and Caroline Robertson. April 1997. *ISDN forges ahead.* IDC.

■On-Line Sources

Court ruling opens long-distance market to Baby Bells. 1998. `http://www.techweb.com/wire/story/TWB19980101S0005`

■URLs of Selected Mentioned Companies

3Com http://www.3com.com
ABC Paging http://www.abcpaging.com
Ameritech http://www.ameritech.com
AT&T http://www.att.com
Bay Networks http://www.baynetworks.com
Bell Atlantic http://www.bel-atl.com
Bell Cablemedia http://www.bellcablemedia.co.uk
BellSouth http://www.bellsouth.com/index.shtml
BT http://www.bt.com
Cable & Wireless http://www.cwplc.com
CDMA Development Group http://www.cdg.org
Cegetel http://www.entreprises.cegetel.fr
CellularVision http://www.cellularvision.com
Clear Communications http://www.clear.co.nz
Comcast http://www.comcast.com
Cox Cable http://www.cox.com
DDI http://www.ddi.co.uk
Deutsche Telekom http://www.dtag.de
Ericsson http://www.ericsson.com
First International http://www.fic.com.tw
France Telecom http://www.francetelecom.fr/indexus.html
Grupo Financiero Banamex-Accival http://www.banamex.com/eng
GTE (BBN Planet) http://www.bbnplanet.com/annual/planet/index.htm
GTE http://www.gte.com
Guandong Mobile Communications http://www.gmc.guangzhou.gd.cn
HP http://www.hp.com
HK Telecom http://www.hkt.net
Hutchison http://www.hutchnet.com.hk
ITU Americas Regional http://www.itu.org.br
Lucent Technology http://www.lucent.com
Mannesmann Arcor http://www.cni.net
MCI http://www.mci.com
Mercury Communications http://www.mercury.co.uk
Metropolitan Fiber Services http://www.uu.net
Minitel http://www.minitel.fr
Motorola http://www.mot.com
Nextel http://www.nextel.com
Nippon Telegraph and Telephone http://www.ntt.co.jp
Nokia http://www.nokia.com
Nortel http://www.nortel.com
NTT http://www.ntt.co.jp
NYNEX http://www.nynex.com
Optus Communications http://spinnaker.optus.net.au/
Pacific Telesis http://www.pactel.com
Personal Communications Industry Association http://www.pcia.com
Puerto Rico Telephone Company http://www.prtc.net
Qualcomm http://www.qualcomm.com
Rockwell http://www.rockwell.com
SBC http://www.sbc.com
Siemens http://www.siemens.com
Singapore Telecom http://www.singtel.com
SMDS Interest Group http://198.93.24.25/
Sony http://www.sony.com
Sprint http://www.sprint.com/index.html
T1 Telecommunications http://www.t1.org/
TCI http://www.tci.com
Tele Danmark http://www.teledanmark.dk
Telefonos de Mexico (Telmex) http://www.telmex.com.mx
Telenor http://www.nextel.no
Teleport Communications Group http://www.tcg.com
Telstra http://www.telstra.com.au
Texas Instruments http://www.ti.com
Unwired Planet http://www.uplanet.com
US West http://www.uswest.com
UUNet http://www.uu.net
WorldCom http://www.wcom.com

4.4 The Internet and Intranets

4.4.1 Executive Summary

The Internet and the World Wide Web (also known as the Web) form a massive electronic communications pipeline between businesses, consumers, government agencies, schools, and other organizations worldwide. Perhaps equally important, the Internet has opened up exciting new possibilities that challenge traditional ways of interacting, communicating, and doing business.

On-line services have incorporated the Web, and the distinction between on-line service providers and Internet service providers (ISPs) has become blurred. Internet applications and on-line services involve interactive databases, three-dimensional graphics, virtual reality, animation, audio and video, Java applets, and other emerging technologies. Developers are putting these technologies to work by building innovative content and services for business, commerce, entertainment, education, information exchange, and numerous other uses.

The pace of new products, software and content development, and service innovation on the Web has been incredibly fast. However, the explosive growth of the Web has made finding specific information in its millions of pages a daunting task. As a result, a variety of solutions and tools for navigating the Web have emerged, including directories, search engines, agents, Web robots, and other approaches. The popularity of search and directory Web sites has made them focal points for advertisers as well as for communications about what is happening on the Web.

Deployment of Web applications is still in its infancy. Internet and Web technologies are being used for business transactions and already have become de facto standards for global communications and commerce. Companies are using their Web sites to market and sell products and services, establish brand names and corporate identities, gather market and competitive research, provide help-desk functions, and test products. Barriers to business use of the Web still include security issues (not all sites have or make effective use of security technologies); a lack of public trust in the safety of on-line transactions and the reliability of the Internet in general; Internet performance and access and download time on the Web; and a technology lag between the availability of new technologies and the ability of the installed base of Web browsers to support them.

Many corporations have included Internet technologies in their internal networks or intranets and as the basis for establishing links with their trading partners in extranets. Although not all applications are robust enough for production environments, Web browser-based intranets can offer easy-to-use gateways to organizational information.

The Internet continues to evolve. Work is underway on new protocols that will enable the use of a new class of applications over the Internet. The use of the Internet to carry real-time data, such as voice and video, in addition to traditional forms of data is on the horizon. Although Internet telephony is still a relatively small market, as streaming audio and video applications become more commonplace, traditional voice carriers will face increased competition from ISPs in the transmission of voice and facsimile.

Given the rapid pace of development, Internet and Web product life cycles tend to be short. What is considered a "hot" and unique technology today may well be either passé or widespread in 3 to 4 months. The rapid pace of innovation on the Web is likely to continue during the forecast period.

4.4.2 History of the Internet

The Internet grew out of a U.S. Defense Advanced Research Projects Agency (DARPA) project and originally served a largely technical audience composed of the military, government agencies, and academic researchers and scientists. The original goals of the project were to allow researchers to share computing resources and exchange information, regardless of their locations, and to create a resilient, fault-tolerant wide area network (WAN) for military communications.

During the mid-1990s, commercial enterprises and individuals discovered the benefits of being connected to the Internet, eventually creating a mass-market phenomenon. Today, although access is limited in some areas, most countries have ties to the Internet.

Taking advantage of the Internet is relatively inexpensive for many people with access to a personal computer (PC). Today, almost anyone with a PC or alternative device, a phone line and modem, and the proper software can gain entry to cyberspace, with access typically costing $20 per month or less. Small business content developers can rent space on someone else's Web server, design their own Web sites, and have them up and running for less than $10,000 per site. In some cases, space may be obtained for free from an ISP. However, more sophisticated and complex Web sites can cost more than $1 million when all the development and operations costs are included.

Convergence of the Internet, Web, and On-Line Services

One of the first on-line services, CompuServe, was launched in 1979. Prodigy and AOL followed with offerings targeted toward home computer users. This focus contrasted with the Internet's original orientation, which was geared to the academic and research community, and that of on-line research services such as Lexis-Nexis (a division of Reed Elsevier), which target large corporations.

ISPs originally were established to provide connectivity, not content. Many ISPs now offering dial-in Internet access to consumers initially were set up to provide dedicated Internet connections to educational and commercial organizations. Others (such as NetCom) began by supplying dial-in access to Internet-connected time-sharing systems (before the Web). UUNet started by providing dial-up connections for routing of e-mail and Usenet news (discussion groups devoted to specialized areas of interest) between non-Internet-connected sites. The number of ISPs grew rapidly following widespread use of the Web and now includes major telecommunications service providers as well.

Differences between ISPs and on-line services (OLSs) have diminished over time. Offering access to the Web became an effective way for OLSs to attract and retain customers, particularly non-technical customers who found dealing with a consumer-oriented OLS easier than dealing with an ISP. As OLSs have evolved, they have repositioned themselves as ISPs that also offer access to value-added services or content not available elsewhere. At the same time, OLSs have made much of their content available to anyone using the Web, not just to their subscribers. OLS content has become more integrated with the rest of the Web via hypertext links to Web sites outside the OLSs' own content.

Today, consumers can connect to the Internet using either an OLS or a traditional ISP to access the Internet for e-mail and Web sites. For example, AOL calls itself an "Internet On-line Service," and both Microsoft Network (MSN) and Prodigy have been relaunched as Internet services. MSN is now an ISP with some proprietary content. Prodigy was sold in May 1996 by previous owners IBM and Sears to International Wireless and now calls itself Prodigy Internet; new users access it through a standard Internet ISP connection rather than through its proprietary access network.

Some remaining differences exist between OLSs and ISPs. OLSs can act as content aggregators, while ISPs typically offer little or no content of their own. In addition, OLS membership structure and billing systems allow subscribers to be charged for particular content via a debit to the subscriber's monthly bill. In contrast, ISPs are set up to bill only for Internet connectivity, so subscribers must make separate arrangements with Web site operators to pay for access to any fee-based content.

In September 1997, AOL announced it would acquire CompuServe's OLS with approximately 3 million consumer subscribers after WorldCom (the parent company of ISP UUNet) bought CompuServe from parent company H&R Block for $1.2 billion in stock and then spun off CompuServe's interactive services division to AOL. WorldCom also is paying $175 million to acquire ANS Communications, AOL's networking arm, and will assume from AOL the responsibility for providing dial-up access to AOL customers. AOL, which wants to be seen as a media company, intends to run CompuServe as a separate entity focusing on small businesses and professionals. The transaction sharpens the lines between companies that provide telecommunications infrastructure and those that deliver on-line content.

According to Computer Intelligence, although the number of people in the U.S. connecting to OLSs grew to more than 25 million by mid-1997, the number of people in the U.S. directly accessing the Internet via an ISP (17.6 million) exceeded the number of subscribers to OLSs using the Internet (16 million). Some people are using both services.

The number of ISPs will continue to grow around the world, with eventual consolidation occurring among ISPs, OLSs, and telecommunications service providers. Gartner estimates that the number of European ISPs grew to approximately 2,000 by the end of 1996 and that the European ISP market will continue to grow at an average annual rate of approximately 20 percent per year through 1999.

In 1997, the Indian Government decided to privatize Internet access in the country. Videsh Sanchar Nigam has been the sole ISP, with 40,000 subscribers. The National Association of Software and Service Companies sees that number growing to 125 service providers, although the Department of Telecommunications projects that there will be around 1.5 million subscribers in India by 2000.

The World Wide Web

The technology underlying the World Wide Web was created by Timothy Berners-Lee, who in 1989 proposed a global web of hypertext documents that would allow physics researchers to work together. This work was done at the European Laboratory for Particle Physics (known by its French acronym CERN) in Geneva. In December 1990, the first Web server and the first client hypertext browser/editor were released within CERN and then to a wider audience on the Internet in mid-1991.

The Web is a multimedia system that links computer resources around the world. It provides hypertext access to Web-enabled information for computers on the Internet. The combination of a simple addressing scheme called Uniform Resource Locators (URLs), a communications protocol called HyperText Transfer Protocol (HTTP), and a standard software authoring language called HyperText Markup Language (HTML) provided a standardized way to publish and access information using the existing infrastructure of the Internet. Berners-Lee originally envisioned a group working an a single document; however, Web documents are read-only documents, and collaborative software is required for shared work.

Many people believe that the Web is virtually synonymous with the Internet, which is not true. The Internet functions as the transport mechanism, and the Web is an application that uses those transport functions. Other applications also run on the Internet, the most widely used of which is e-mail. Other services include file transfer using the file transfer protocol (FTP) and a variety of information access and location services such as the Wide Area Information Service (WAIS).

At first, the Web was text only. Then, in 1992, researchers at the National Center for Supercomputing Applications (NCSA) at the University of Illinois at Champaign-Urbana developed Mosaic, the first graphical Web browser. (Lynx, another early Web client, is a line-by-line, text-only browser.) The Mosaic browser provided a graphical front end that enabled users to point-and-click their way across the Web. Graphical content on the Web has expanded enormously in the 6 years since the introduction of Mosaic.

Web browsers became a means of universal access because they deliver the same interface on any operating system under which they run – Windows, Windows NT, OS/2, MacOS, or UNIX. In addition to providing platform-independent access to Web content, browsers also provide platform-independent interfaces for applications.

The emergence of commercial, graphical browsers that access documents written in HTML created a mass medium that allowed large numbers of people without sophisticated computer skills not only to access information, but also to publish their own content on the Internet. Newcomers could enjoy direct, interactive access to the Internet's contents without having to use system commands or contend with terminal emulators. HTML was easy to learn, and it was easy to develop authoring tools for the language. Aided by desktop publishing tools that produce HTML pages from standard documents and Internet connection kits, HTML "programmers" set up their Web sites hoping to attract traffic from and communicate with millions of potential Web users worldwide.

From its beginning, the Web has been the focus of intense and rapid innovation. The Web now attracts the attention, resources, and technological innovation of the entire information technology (IT) industry. Although OLSs and the Internet took several decades to achieve prominence, the Web has accomplished as much or more in only 6 years.

The Internet Today

The number of computers connected to the network continues to grow rapidly. These computers have been set up voluntarily to conform to the Internet's set of nonproprietary standard protocols. The power of the Internet rests in this uniform, open architecture. (See Chapter 4.2, *Networking Systems and Protocols,* for more information on underlying network protocols such as TCP/IP.)

Currently, the U.S. government pays only a small percentage of Internet costs. The physical network backbone of the Internet is now largely provided by commercial communications companies. The U.S. government continues to contribute some funds to essential administrative processes, such as standards development and the domain name system (DNS), through contracts. The National Science Foundation (NSF) also pays for certain high-performance portions of the network backbone.

The rest of the Internet infrastructure is supplied by backbone providers, such as ANS (part of WorldCom), BBN Planet (part of GTE), MCI (soon to be part of WorldCom), Sprint, UUNet/WorldCom, and others. Businesses and individual subscribers connect to the Internet through access providers. In some cases, the access provider also may be a large backbone provider, and in other cases, it will be a smaller (often local) company connecting to a backbone provider. Both backbone and access providers are referred to as ISPs. Service providers charge customers for various combinations of bandwidth, traffic, and access time.

Backbone networks must be connected to one another and to access providers. The various backbone networks that compose the Internet transmit information to one another on a reciprocal basis, meaning that each carrier agrees to transport traffic originating on another carrier's network, historically without charge. Data transmission on the backbone networks is not currently metered according to usage volume, time, or distance. However, this practice is beginning to change because of growing application demands for bandwidth and the need to finance expansions in network capacity to accommodate increasing Internet traffic.

In the U.S., ISPs use large public hubs, called Metropolitan Area Exchanges (MAEs) or Network Access Points (NAPs), as points of access into the Internet. These generally are located in urban areas with heavy traffic. Hubs also are known as points of presence (PoPs). Increased congestion at these hubs may mean more hubs need to be added, and it also has resulted in an increasing number of private peering agreements between larger ISPs.

Private peering agreements establish direct connections between ISPs and allow them to exchange traffic without going through crowded public network access points. These direct connections help improve performance and offer higher quality of service (QoS). Service providers then can charge different rates based on different degrees of QoS. These changes have raised concerns about smaller service providers having to pay higher fees for network access to the larger providers in the future. Others believe that smaller ISPs have not been paying their full share of infrastructure costs in the past.

4.4.3 Internet Challenges

Internet challenges include making the Internet more suitable for electronic commerce transactions, rapid turnover of new technologies, evolving standards and regulatory frameworks, and a growing need for additional bandwidth.

Barriers to developing electronic commerce on the Internet include the security of transactions and the slow propagation of strong encryption engines outside the U.S. due to government restrictions. (See Chapter 5.4, *Security,* for more information on government regulations and encryption.) Other factors include the lack of consumer trust in the reliability and integrity of the Internet, Internet performance and wait times Web users experience when downloading information, and the technology lag between when technology is introduced and when it becomes widely available.

Various initiatives are underway to enable secure commerce. Both Microsoft's Internet Explorer and Netscape's Navigator include the Secure Sockets Layer (SSL) – an industry standard for secure electronic commerce on the Web. (See Chapter 7.1, *Electronic Commerce.)*

In addition, vendors are adopting new technologies more rapidly than many users and customers can implement them. For example, one of the most popular Web browsers is Netscape's Navigator. According to Zona Research, Navigator was installed on more than 60 million computers around the world in early 1997. Many of the most innovative sites on the Web use Java applets, interactive three-dimensional (3-D) graphics, and video and audio clips; commercial sites often rely on SSL. However, a large percentage of Navigator users are still on Version 1.x, which does not support any of these technologies. To access these Web sites and take advantage of their innovative content, users need to have Navigator Version 3 or better. (The same is true for Microsoft's Internet Explorer, which requires users to have Version 2 or better to run Java applets.) Although a massive upgrade of Web browsers clearly is needed for users to access the most innovative Web sites, that is not occurring as rapidly as Microsoft, Netscape, and others are pushing the technology into the marketplace.

Internet Regulation

Technical organizations such as the Internet Engineering Task Force (IETF), World Wide Web Consortium (W3C), and others have played an important role in the evolution of the Internet and the Web. These organizations are not formally charged in any legal or operational sense with responsibility for the Internet; however, they define the standards that govern the Internet's functionality. Hardware and software vendors also have been instrumental in submitting specifications for consideration to standards bodies and in creating de facto standards of their own.

Recent government attempts to regulate the content of Internet-connected computers have generated concerns about privacy, security, and the legal liability of service providers. Some content providers have addressed these issues with filters, ratings, and restricted access; however, it is difficult to regulate content across international borders. Regulation of services such as gambling also has been debated.

In addition to issues surrounding who should dictate standards for infrastructure and content, many issues surround authority for Internet domain naming.

■ Domain Names

Controversy has arisen over who should be in charge of registration and administration of domain names on the Internet, the relationship between trademarks and domain names, the assignment of popular names, and the creation of new top-level domain names. The number of organizations involved in registration and the decision-making process has increased. Participants include businesses, service providers, governments, nonprofit organizations, and standards bodies.

Domain names are widely used instead of the equivalent but more cumbersome numeric IP addresses. Top-level domain names are the suffixes such as `.com`, `.org`, or `.net` in an Internet host name or URL. Most of the more than 1 million domain names in 1997 ended in `.com`; in 1992, in contrast, most of the 7,500 registered domain names ended with `.edu` or `.gov`.

Domain names, like easily recognizable or heavily promoted phone numbers, have commercial value in themselves. Domain name registration in the countries where a company is doing business also helps prevent others from using a trademark in their domain names. Most customers recognize second-level domain names, which are frequently company or organization names, such as "ibm" in `ibm.com`.

Country-specific top-level domain suffixes such as "de" for Germany or "us" for the United States are determined by International Organization for Standardization's specification ISO-3166 TLD. Administration of these domain names is performed independently by the country in question, and two-letter top-level domain codes are reserved for future ISO 3166 country-code allocations.

Country suffixes have not been used consistently. For example, because of the Internet's start in the U.S., ".us" is not usually added as a country suffix; instead, the absence of a specific country suffix implies U.S. origin. As use of the Internet has grown outside the U.S., some believe names such as `.com` should become `.com.us`, just as `.com.ca` and `.com.it` tell people that they are doing business with a company in Canada or Italy, to eliminate this built-in assumption. Alternatively, since existing top-level domains already are used by multinational organizations headquartered anywhere in the world, others believe these should be available for global assignment.

The current system for allocating domain names was established in 1983. As of late 1997, assignments of second-level domains within `.com`, `.net`, and `.org` are administered by a company called Network Solutions (NSI) under a contract with the U.S. National Science Foundation. NSI's contract to register second-level domain names expires in March 1998. Funding also is scheduled to run out for the Internet Assigned Numbers Authority at the University of Southern California, which handles assignment of top-level domains such as `.com` and `.gov`.

NSI had proposed to continue to administer `.com` and allow new registrars to administer new top-level domain names. However, there was concern that this setup would grant NSI a monopoly. Some participants in the discussion were opposed to having only a single registrar located in the U.S. and suggested several different registrars or the involvement of an international governing board that could handle disputes.

The International Ad Hoc Committee (IAHC) was established to study the issue and proposed a plan to increase the number of top-level domains. The plan was ratified in Geneva in May 1997 by 80 organizations, but not the U.S. government. The plan called for an unlimited number of organizations to assign domain-name registries and for seven new generic top-level domain names (gTLDs). (See Table 4-14.)

Table 4-14: New Generic Top-Level Domain Names

`.firm`	Businesses
`.store`	Businesses offering goods for sale
`.web`	Web-related sites
`.arts`	Cultural and entertainment
`.rec`	Recreational activities
`.info`	Information services
`.nom`	Personal sites

The IAHC subsequently was replaced by the Interim Policy Oversight Committee (IPOC). Formed to administer the original IAHC proposal, the IPOC governs worldwide registration for the new gTLDs and runs a nonprofit association in Switzerland. Under that plan, the World Intellectual Property Organization would moderate trademark disputes over domain names.

The Internet Society and some Internet founders have been supportive of IPOC's efforts. However, IPOC has been controversial with those who do not favor governmental involvement in the management of the Internet. Such a model conflicts with the fundamental structure of the Internet as a network of private networks, some argue. Another objection is that an unlimited number of registries may be confusing if different organizations can register multiple domain names (`.com`, `.store`, and `.firm`, for example).

Regardless of concerns, the list of top-level domains is expanding to accommodate growth of the Internet. In early October 1997, a group of 73 registrars called the Council of Registrars (CORE) was formed to develop policies and procedures and maintain a repository for the domain name registration system. On November 3, 1997, the CORE announced it had signed a letter of intent with Emergent Corp. to develop and operate the Shared Registration System. Emergent will maintain the list of second-level domains in CORE's seven new gTLDs. The creation of these new Internet domains is governed by the Generic Top-Level Domain Memorandum of Understanding (gTLD-MoU). (For more information, see *`http://www.gtld-mou.org`*.) Unlike the system in place for the existing generic top-level domains, Internet users will be able to register new Internet domain names with their choice of more than 80 registrars worldwide.

Internet Expansion

The Internet was not designed to provide a mass-market interchange of high-density information. As a result, the massive growth of Internet traffic has strained some elements of the network. This strain manifests itself as slowdowns in retrieval time, unreliable transmission of streamed data, and denial of service by overloaded servers.

The Internet's mesh design, with many potential transmission paths, is in theory highly resistant to outages caused by failed links. In practice, the Internet often is affected by software problems.

A wide range of factors contribute to congestion or slowdowns. Problems include improperly configured networks, overloaded servers, very dynamic Internet usage patterns, congested hubs, problems with routing tables, and too much traffic for available bandwidth. Approaches to solve these problems include installing high-speed transmission media to accommodate large amounts of data; bigger, faster routers and more sophisticated load balancing and management software to handle peak traffic periods; local caching of frequently requested Web pages to improve response times; and more reliable tiers of service to those willing to pay for them. (See Chapter 4.2, *Networking Systems and Protocols.)*

Browsers use caching to store data on disk so that it does not need to be downloaded every time, which improves performance. Network caching takes this one step further by storing multiple copies of Web sites in various places on the Internet. In April 1997, Network Appliance announced NetCache server software, a caching technology that filters data requests and stores that information closer to the user. In September 1997, Cisco announced plans for a Web Cache, a large system that could store 25 million pages in one location on a network. In October 1997, Inktomi announced Traffic Server, network management software for caching that can store up to a trillion bytes of data on a network and uses parallel computing technology in its search engines. In 1998, similar announcements from other networking and system vendors attempting to improve performance and response times are likely. Some ISPs, such as Demon Internet in the U.K., provide priority caches for their customers.

Low-speed modems can cause large numbers of customers to experience slow access to Web sites even if there is no congestion on the Internet itself. Many users still have 14.4-Kbps modems, and the number of individuals who have upgraded to modems beyond 28.8 Kbps or 33.3 Kbps is still very small.

■ Internet2

The academic research community, which the Internet was originally intended to serve, now finds that system too slow for data-intensive applications such as transmitting supercomputer model data or telemedicine. In 1996, a consortium of universities began establishing a faster network with limited access devoted exclusively to research purposes. Closed networks offer a model for large organizations with intensive data transfer needs to bypass the Internet's congestion.

Announced in October 1996, Internet2 grew from 34 to more than 110 U.S. research universities in one year. Goals include developing advanced networking capabilities and new network PoPs capable of transmitting gigabits (billions of bits) of information per second. Member universities, federal agencies, nonprofit organizations, and corporate partners are developing tools for scientific research and higher education in the future. Universities are contributing dollars and resources and corporations such as Cisco, IBM, and Newbridge Networks are collectively contributing millions of dollars in network equipment and technology.

Internet2 grew out of the very high-speed Backbone Network System (vBNS), a fiber optic network provided by MCI that links five supercomputing research centers. By October 1997, the first of two large PoPs at MCNC and Georgia Tech were linked. Georgia Tech is completing an OC-3c (155 Mbps) link to the vBNS PoP in Austell, Georgia, and is sharing the facility with MCNC. The physical connections involve Caronet, Time-Warner Cable, and WorldCom. A DS3 circuit between MCNC and Georgia Tech directly connects Cisco asynchronous transfer mode (ATM) switches at each location, with each switch connected to Cisco routers. Internet2 is a project of the University Corporation for Advanced Internet Development.

Network Computers and Internet Access Devices

More consumers are buying PCs and other devices that are equipped with Internet access capabilities, helping to increase Internet and Web traffic. Network computers (NCs) are similar to PCs but offer low-cost Internet access capability, provide minimal (or no) local storage and rely on servers for applications and data storage. Network computing devices may be less expensive, easier to install, and less expensive to service over time. (See "Network Computers (NCs) and Network PCs (NetPCs)" on page 359.)

Consumer-oriented Web access devices are becoming more affordable, with prices dropping below $300 (at the end of 1997). Lower prices are likely to increase mass-market acceptance, especially among customers who do not want or need a PC, although PC prices are dropping as well. Television-based Internet devices (such as WebTV) allow users to access the Web through their televisions and could further broaden Internet use. New types of devices with an IP address and phones – such as the Ideo screen phone and the Nokia 9000, with a small keyboard and screen designed for Internet access – will continue to proliferate.

International Developments

The Internet is becoming more global over time. Generally speaking, countries with a higher gross domestic product per person will have higher rates of PC ownership and higher Internet adoption rates. Political, cultural, and regulatory barriers slowed the rate of Internet adoption in some countries.

Internet growth is expected to continue to be high in what Gartner calls "Internet leaders," countries such as Finland, Germany, the rest of Scandinavia, the U.S., and the U.K. Countries such as Brazil, China, and Malaysia have the potential to experience rapid Internet growth as their economies develop and access limitations ease.

IDC has categorized Internet usage by country and made predictions for changes by 2000. U.S. Web users, currently 72 percent of the world's Web users, are expected to make up just less than 50 percent by 2000.

Internet Multi-Culturalism

The Web is still largely U.S.-centric, although this orientation may change over time. Currently, about 60 percent of all Internet host computers reside in the U.S. Overall, less than 2 percent of the world's adults have access to the Internet. Although this situation represents tremendous opportunities for growth, cultural pitfalls need to be overcome. Censorship still limits content in some parts of Asia, such as Singapore and Vietnam. Germany, for instance, caused CompuServe to suspend its service there temporarily because of regulations governing adult content. Japanese consumers apparently are slower to embrace the Web because of a preference for domestic technology. In France, expansion has been hampered by a preference for the local Minitel system. Although not comparable to the Web, Minitel's content is entirely in French.

Internet Language Problems

Today, 90 percent of the content on the Web is in English. Most current software allows transmission only of characters from Western European languages. An existing standard – Unicode – allows a character set of 65,000 characters, which is sufficient to represent most human languages. Unicode, however, is far from being widely implemented. Indeed, many older systems do not even recognize the Extended Simple Mail Transport Protocol (ESMTP), which allows for transmission in an e-mail message of the diacritical marks that are used in languages such as French, German, and Spanish. Alis Technologies and Accent Software offer browsers that recognize the Unicode character set. Extensible Markup Language (XML) also supports Unicode.

4.4.4 Internet Standards and Protocols

Technically, the Internet is defined as an interconnected set of networks, all running the TCP/IP protocol. Internet applications such as the Web call on services provided by TCP/IP to transport data across the Internet. The most important protocols that make up the Internet and the Web – the Internet Protocol, the HyperText Transport Protocol, HTML, and XML – are being updated frequently and will continue to evolve.

The Internet Protocol (IP)

The underlying data transmission protocol for the Internet is the Transmission Control Protocol/Internet Protocol, or TCP/IP. The Internet today is based primarily on Version 4 of the Internet Protocol (IPv4), which is limited in its support for advanced functionality such as QoS, multicast, and mobile computing. Also, computer systems connected to the Internet must be assigned a specific IP address. Although Dynamic Host Configuration Protocol (DHCP) frequently is used to allocate shared IP addresses dynamically, a limited number of addresses are available for a growing base of users.

A DHCP server assigns client IP addresses on an as-needed basis. When a new client starts up, it obtains its IP address from the DHCP server. This process eliminates the need to assign each client PC its own IP address.

IPv6 is the successor to IPv4. IPv6 is designed to help ease these address limitations by using an expanded addressing scheme of 128 bits compared with today's 32 bits. IPv6 also includes provisions for additional security and real-time communications and multicasting. Many vendors are including both IPv4 and IPv6 compatibility on their routers and switches and will eventually discontinue IPv4 support. (See "IPv4 and IPv6" on page 189.)

HyperText Transfer Protocol (HTTP)

HTTP is the communications protocol used to connect to servers on the Web. It establishes a connection between a client browser and a Web server and sends HTML pages to the client browser.

Heavy use of HTTP in recent years has revealed the need for improvements in performance, scalability, and extensibility of the protocol. The W3C has been working closely with the IETF to improve HTTP. HTTP Version 1.1 (HTTP 1.1), which became an IETF proposed standard in late August 1996, is the most recent result of this collaboration.

Most HTTP 1.0 implementations use a new TCP connection for each request/response exchange. In many cases, there are multiple HTTP exchanges per Web page. In HTTP 1.1, a connection may be used for multiple request/response exchanges. This process results in a reduction of network traffic and faster response time to the end user because it allows TCP to adjust to the characteristics of the network link.

The W3C now wants to gain experience on how HTTP 1.1-compliant software and applications perform on the Internet. The W3C is actively promoting the deployment of HTTP 1.1 and providing an HTTP 1.1 Implementer's Forum.

HTML

HTML was derived from the more complex Standard Generalized Markup Language (SGML), a text-based language for describing the content and structure of digital documents. This meta-language (a language for specifying markup languages) was first developed by IBM, then adopted by the ISO in 1986. Some organizations use SGML for production of complex, highly structured documents. Web developers bypassed SGML in favor of HTML, a simpler subset of SGML.

HTML is widely used and continues to become more sophisticated. In January 1997, the W3C endorsed the latest HTML specification, HTML 3.2. Developed during 1996 with vendors including IBM, Microsoft, Netscape, Novell, SoftQuad, Spyglass, and Sun, HTML 3.2 incorporates into the standard features such as tables, applets, text flow around images, superscripts, and subscripts. Additional HTML 3.2 features support meta-information describing link relationships and document properties such as authorship, content rating, and copyright statements. HTML 3.2 also provides backward compatibility with the existing HTML 2.0 standard.

Version 4.0 of HTML, approved as a specification in December 1997, incorporates extensions to HTML for multimedia objects, scripting, style sheets, layout, forms, higher-quality printing, and display of mathematical notation. A new element (an object tag) in Version 4.0 opens HTML to outside formats and languages.

■ Cascading Style Sheets

Cascading Style Sheets (CSSs) are an enhancement to HTML that add page layout features to Web documents. A CSS acts as a template that defines the appearance or style (such as size, color, and font) of an element of a Web page, such as a box. Once it has been defined, the CSS is invoked with specific content (such as the text to be displayed in the box), creating a graphic element that can be placed on the page.

The use of CSSs has several benefits. Because basic HTML mark-up commands can be used only to format text, graphic elements typically are created as separate graphics files (such as .GIF files) and then downloaded by the Web browser as needed for display. A CSS can replace the GIF file and display the same image, but the CSS file typically will be much smaller than the GIF file, allowing the contents to be downloaded more quickly. Because the same CSS can be reused to create multiple elements on the page, this process eliminates the need to download multiple GIF files and thus improves performance.

CSSs also can be used by Web page designers to enforce a consistent appearance and format across a series of different Web pages. In addition, the CSS defined by the page designer can be overridden by a special CSS defined by the user. For example, this capability would allow a visually impaired user to specify a larger font size than the one originally specified.

In December 1996, a Cascading Style Sheets proposal (CSS1) became an official W3C standard. Both Microsoft and Netscape have adopted versions of this standard in their browser software. Proposed style sheet extensions include CSS Positioning, which would allow authors to position HTML elements on a page in three dimensions.

■ Dynamic HTML (DHTML)

DHTML is a set of extensions to HTML that allows Web pages to be updated in response to actions taken by the user, without the need to re-connect to the Web server to download a new version of the page. For example, data in a table could be re-sorted at the user's request, or new visual elements (such as a pop-up text box) could appear when the mouse is positioned at a particular point on the page.

DHTML also gives the Web page designer greater control over how an HTML page appears in the user's browser window; for example, the designer can specify the exact position where a graphical object should appear. Web pages created with DHTML also can have contents that change automatically without action by the user, for example, by having a graphic element move across the page or perform special effects such as animation or morphing. Because all these actions can be carried out by a DHTML-enabled browser, they utilize the processing power of the client computer. This process eliminates the load on the Web server that occurs when it is required to send the browser a new version of the page each time the appearance is updated. It also gives the user a faster response because the latency imposed by a round-trip network exchange between the browser and server is eliminated.

Microsoft and Netscape included their own incompatible versions of DHTML features in Internet Explorer 4.0 and Netscape Communicator, respectively, in 1997, even though the W3C had not yet issued a specification on DHTML. DHTML also will be incorporated into Microsoft's ActiveDesktop.

■ Extensible Markup Language (XML)

XML is another method for describing the markup of different types of documents and a data format for structured document interchange on the Web. Like HTML, XML is a subset of SGML. Unlike HTML, however, XML allows user-defined tags and attributes to provide greater control over presentation characteristics and application functionality. XML supports Unicode, thus providing support for most of the world's written languages. In addition, XML is much better than HTML at drawing data from heterogeneous data sources and displaying that data in a consistent format.

Even if XML is used widely in large-scale Web sites, HTML will continue to be used for small-scale Web sites. However, XML will meet the requirements of large-scale Web content providers in industries that already use SGML and for those developing "push" content. XML will assist in the generation of Web pages for large structured databases such as retail catalogs.

In December 1997, the W3C issued a proposed XML 1.0 standard that is compatible with both HTML 4.0 and DHTML. The XML standard is designed to bring the key benefits of SGML to the Web in a way that is easier to implement and understand while remaining fully compliant with the SGML ISO standard.

Microsoft and Netscape are working on including XML support in their browser software. Microsoft's Channel Definition Format (CDF), included in Internet Explorer 4.0, is based on XML standards and will be used for scheduling content delivery. Netscape's Resource Description Format (RDF), which also channels content, is similar to CDF.

■ Document Object Model (DOM)

The W3C has proposed a DOM that defines an object-oriented application programming interface (API) for accessing and changing Web pages. A draft DOM specification was released in October 1997. DOM provides a hierarchical tree-like structure for working with HTML and XML documents and would allow document management and publishing functions to exist more independently. Microsoft and Netscape currently each have different object models they are using. (See Chapter 5.2, *Software Component Architectures,* for more information on object models.)

4.4.5 Browsers

The Web primarily is accessed through software applications called browsers. At a minimum, a browser is capable of communicating via HTTP, parsing and rendering HTML, and displaying certain data types, such as GIF and JPEG for graphics and Microsoft Windows WAV for sound. Browser capabilities have been expanding at a rapid rate, with new extensions to HTML standards to enable tables, frames, and other features. The race between Microsoft and Netscape to provide the richest possible user experience by introducing proprietary extensions to standards, and then trying to get those extensions adopted as standards, has been the source of many new browser capabilities. This situation has been extremely beneficial for users, providing them with highly capable applications at almost no cost, although the race to add value has resulted in some diverging standards.

For example, Netscape added support for the <LAYER> tag to Netscape 4.0. This tag allows users to place elements in specific screen coordinates and provides overlapping. With layers, HTML and JavaScript can be used to create dynamic animations on a Web page and simulate a desktop publishing environment.

Another result of the foregoing competition between Microsoft and Netscape has been a divergence in strategy for the two companies.

Microsoft's strategy is to minimize the importance of the browser as a distinct application by building browser functionality into Office 97 and other applications as well as into Windows itself. Microsoft is breaking its browser up into components that can be integrated into applications and even into its new Windows operating system, Windows 98, due in mid-1998.

Netscape's strategy is to make the browser the core of a compelling suite of applications for corporate users. Netscape has pursued the approach of developing a Web-based suite of communications services, including groupware offerings for intranets.

Although Microsoft and Netscape will continue to dominate the browser marketplace, there are, in fact, other browsers. A browser technology also is available to developers that can be embedded in other applications. For example, Spyglass develops browser components that it resells to companies such as IBM, Microsoft, Oracle, and WebTV Networks (acquired by Microsoft in 1997).

Browser technology is useful for providing simple interfaces to new and existing applications where full-blown browser features are not necessary.

Browser vendors have been competing for market share by adding an expanding array of features to their products. As browsers have evolved rapidly, there have been concomitant decreases in stability and security as well as increases in complexity, disk and memory requirements, and the need for sophisticated configuration. Additional plug-in products add to browser complexity.

Microsoft Internet Explorer

Faced with the tremendous lead in the browser marketplace that Netscape established, Microsoft embarked on a strategy to gain market share and penetrate the installed base: It gave Internet Explorer away for free and bundled it with its Windows operating systems. This approach has had some success but also resulted in scrutiny by the U.S. government.

In October 1997, the Justice Department accused Microsoft of violating a 1995 antitrust consent decree and filed a petition in a federal court to prevent the company from requiring PC manufacturers to bundle Microsoft's Internet browser software with Microsoft's Windows 95 operating system. At issue was whether Microsoft tried to monopolize the Internet browser software business by refusing to let PC makers license the Windows operating system unless they also ship their PCs with Internet Explorer. Microsoft maintained that Internet Explorer was an enhancement of Windows, not a separate product, and that the company therefore was not violating its antitrust settlement. According to Microsoft, Internet Explorer's tight integration with Windows offers users the advantage of "one-stop computing."

Internet Explorer Version 4.0 presents a new desktop and file management model, allowing documents in the local file system and a remote Internet site to appear and behave the same. The file system is beginning to converge with the Web browser. In November 1997, Microsoft also released beta versions of Internet Explorer 4.0 for the Windows 3.1 and Solaris operating system platforms.

Netscape Communicator

Netscape added significant functionality and embedded its Navigator browser in the Communicator suite to emphasize the latter's new role in corporate environments. Communicator is not just a browser like its Navigator predecessor. It is a multipurpose suite that handles news, e-mail, audio and videoconferencing, and more. The Communicator suite is comprised of Navigator 4.0, Composer (HTML authoring tool), Messenger (an e-mail client), Collabra (a group discussion client), and Conference (a collaboration client); a professional edition also includes Calendar (group calendaring).

Communicator uses Internet standards such as the Lightweight Directory Access Protocol (LDAP) for directory services, Internet Messaging Access Protocol (IMAP) for e-mail replication, HTML for rich text authoring, and Network News Transfer Protocol (NNTP) for news groups and discussion threads.

Netscape has different versions of Communicator available for those who may want certain features and not others:

- **Netscape Communicator Internet Access Edition** – Aimed at new users who want an Internet-access solution. The product offers a choice of ISPs for new accounts and comes with 30 days of free Internet access.
- **Netscape Communicator Deluxe Edition** – Bundles Web tools and software including Adobe Acrobat Reader, Asymetrix Neuron, Excite Personal Access List (PAL), Inso Word Viewer, Norton Antivirus Internet Scanner, and Sybase FormulaOne/NET.
- **Netscape Communicator Professional Edition** – Includes host connectivity, such as an IBM 3270 terminal emulator, and calendaring.
- **Netscape Publishing Suite** – Includes Netscape Communicator and NetObject's Fusion software for individuals and businesses that want to create their own Web pages and content.

Under some pressure by users in 1997, Netscape also decided to continue to offer its Navigator browser as a stand-alone application.

Browser/Desktop Integration

Increasingly, vendors are integrating browser and desktop features. For example, Microsoft is trying to make Internet access a seamless part of a desktop operating system. DHTML is incorporated into Microsoft's ActiveDesktop with Internet Explorer 4.0, allowing users to receive "push" broadcasts from content providers such as MSNBC, PointCast, and others. The ActiveDesktop is the client side of Microsoft's Active Platform initiative, which incorporates clients and servers for Web-based applications.

Version 8 of the Macintosh operating system (MacOS 8) makes Internet access more transparent to the user. For example, URLs can be accessed directly from the desktop. A personal Web sharing feature in MacOS 8 allows easy publishing of Web pages on local networks, providing, in effect, a rudimentary intranet capability.

Aside from the regulatory issues discussed previously, there is likely to be continued integration between desktops, operating systems, and browsers.

Future Developments

Web browsers are easy to operate; however, as currently designed, they do not overcome the inconvenience of slow transmission, which may stem from a variety of factors. Browsers cannot speed up the transmission of data over the network but they can optimize the order in which contents of a page are downloaded, and features such as pre-fetching and caching of data are helpful. All browsers already perform some level of caching. For example, selecting the back arrow button to return to a previous Web page causes a browser to fetch that page from a cache on a disk. Products that build on the idea of page caching include PeakJet from PeakSoft. PeakJet determines which pages a person is likely to visit and fetches them when a user's modem is idle. It fetches all the pages linked to the current page, and can fetch pages that are commonly visited. Like a browser, PeakJet maintains its page cache on disk; using the product accelerates the process of viewing pages.

Browsers also do not help users understand the structure of a collection of information or filter interesting content from the mass of available data. These deficits are giving rise to the development of "intelligent" browser features such as enhanced bookmarks, which enable users to create catalogs, active Web monitors, and personal agents.

4.4.6 Information Retrieval and Delivery

As the amount of information available on the Internet grows, new mechanisms for delivering it to consumers are being developed. Since its inception, the Web has been based on a "pull" model of information access. In this model, the Web user must actively seek out information by specifying a page to be "pulled down" to the desktop by typing in a URL, following a hot link, or using the search results from a Web search site. However, an alternative "push" model of information delivery has emerged. In this model, information is "pushed" to the user's desktop automatically by a process running on either the user's desktop or a network server.

Push Technology

Passively placing content on a Web site and waiting for people to come browse is not well-suited to establishing and fostering strong relationships with customers or prospects. The concept of content push addresses the fact that with millions of Web sites available for browsing, the only way to guarantee that users receive certain content is to send or "push" it to them. (See Figure 4-11.) Push client packages typically are given away free, and the companies that publish them rely on advertising for revenue.

Figure 4-11: Push Technology

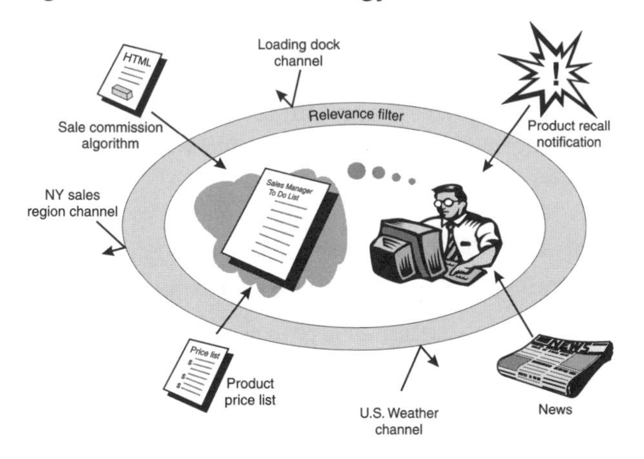

Source: Forrester, 1997

■ PointCast

One of the earliest products embodying the push model was the PointCast Network from PointCast. PointCast developed software that uses the Web browser as a platform and displays news and other information on the user's screen. The user can customize the information to be displayed by creating a personal interest profile. PointCast's SmartScreen technology allows the PointCast client software to run as a screen saver, creating an electronic billboard that displays information on the screen when the computer is not in use.

PointCast makes its client software available as a free download from the Internet and has licensed news and other informational content (such as stock price quotations, sports scores, and weather forecasts) from a variety of sources. This information is distributed for free from PointCast's Web site. PointCast sells advertising, which appears on the screen alongside the informational content. PointCast 2.0 allows a Web developer to create a small file that allows PointCast to push any Web page to the user's desktop as a subscribe channel. This feature means that any page can be brought down to a PointCast user's desktop, not just the licensed content provider's.

PointCast also offers the PointCast I-Server, which can be installed locally on a corporate network. Using the server reduces the traffic caused by multiple desktop clients all connecting through the company's Internet gateway for updates.

■ Marimba

A slightly different approach has been taken by Marimba. Marimba's Castanet product is designed for third parties that want to distribute their own content. Castanet uses a metaphor based on the world of television, with the Castanet "tuner" (client software) able to access content broadcast by "transmitters" (servers) on a variety of "channels." These can be public channels available to anyone with access to the Web or private channels available only to specific individuals (for example, to corporate employees accessing content over their intranet).

The Castanet tuner can download Web-based content – consisting of stand-alone Java applications, Java applets, or pages from a Web site – to the user's hard disk. Once downloaded, the content is "self-updating" because the Castanet tuner will download any changes from the original Web site automatically as they become available. In addition, the Castanet software is self-updating because any changes to the tuner also are downloaded automatically. Marimba believes that in addition to providing Web users with automatic updates to the information to which they have tuned, Castanet technology also will make users more productive by eliminating the time they spend waiting for pages to be downloaded to their browsers.

■ BackWeb

BackWeb provides a method of delivering messages and content directly to users so that notification of arrival actually pops up on their screens. Users choose their method of notification: alerts that flash on-screen, a full screen saver, or wallpaper. Using the BackWeb server, a content provider creates a BackWeb channel. Users download either a generic BackWeb client or a client that has been customized by the channel provider. The generic client allows the user to subscribe to and manage multiple BackWeb channels. Tools are provided for gathering profiles from users and allowing them to customize how each BackWeb channel functions on their machines. BackWeb has a consumer-oriented list of channels, including music, PBS, and business news.

■ Push Technology Integration

In November 1996, Netscape announced it would include technology from both PointCast and Marimba as part of its Constellation initiative. Constellation, planned for a future version of Netscape's Communicator suite, will allow messages, applications, and Web pages to be sent automatically to the user's desktop. As of August 1997, both Standard and Professional Editions of Netscape's Communicator came with Netscape Netcaster push client software.

Microsoft's Active Platform initiative is geared toward developing new Internet and intranet applications. ActiveDesktop, the client side of Active Platform, uses Microsoft's ActiveX and Internet Explorer browser technology. In December 1996, Microsoft announced a partnership with PointCast to distribute information from the PointCast Network via ActiveDesktop. Microsoft's Internet Explorer 4.0 includes support for PointCast and other push technologies.

■ Push Technology in the Workplace

Push technology can provide timely, prioritized distribution of information in a corporate network environment. Channels can be oriented to different levels and departments to focus attention on important communications. In 1997, IDC suggested that short-term success in the push market will depend on selling to business (rather than consumer) end markets. For example, Wayfarer Communications plans to ship a new feature for its Incisa broadcaster that will call an employee's pager if a priority bulletin is received.

■ Push Technology in the Consumer Market

Push-enabled software promises to invigorate traditional Web advertising. Users no longer need to find advertisements; instead, the user's attention can be directed to them. In addition, the quality of presentations can be improved by tuning channels specifically to the user's platform and connection speed.

Innovative services such as personalized news or information feeds become increasingly possible with push technology. For example, Berkeley Systems' After Dark 4.0 screen saver product now delivers news, sports scores, and stock quotes to idle computer screens monitors. Users can determine the type of information delivered and the frequency of updates.

■ Push Technology and Software Distribution

Push channels can be used for software delivery and updates. A 1997 Forrester Research survey of 50 IT professionals found that less than 10 percent of software currently is being delivered on-line. However, many of those surveyed felt that this figure would be considerably higher in 2 years.

Microsoft has twice included features in its operating system software that enabled automatic updates on-line. In both cases, controversy ensued. Some end users were resistant to software that "calls Redmond (Microsoft's headquarters) at night." Forrester's survey concurs, finding that 68 percent of those polled feel that software updates will never be delivered by push technology.

■ Challenges in Push Technology

As with any innovation in network applications, agreement on common standards has surfaced as an issue in push technology. Some initial vendor-defined standards already have been eliminated by the entry of Microsoft and Netscape into this market, and further consolidation of standards is inevitable.

The Channels feature in Microsoft's Internet Explorer 4.0 and Netscape Communicator's Netcaster both can be used to receive push content. The two products differ slightly in terms of setup options, scheduling formats, e-mail notification, preview and channel bar options, and other features, but most content viewed using standard HTML looks similar in both. Microsoft is working with Marimba on a proposed standard called Open Software Description for the transmission of software. Netcaster is included in Netscape Navigator 4.0 (or higher) and in Netscape Communicator 4.02 (or higher) and can be used to display Marimba Castanet channels.

Users can subscribe to any Web site as a channel using Internet Explorer and Communicator, but in both cases, they work as somewhat limited offline browsers. In addition, systems generally slow after launching push components.

Both products provide information about the channel through the use of files downloaded from the channel provider's site. Microsoft's Internet Explorer uses the Channel Definition Format, based on XML, which acts as a guide for scheduled downloads of pages. Netcaster relies on JavaScript to provide the same information.

Search Engines

Search engines are programs that return a list of Web sites or pages (designated by URLs) that match some user-selected criteria such as "contains the words cotton and blouse." To use one of the publicly available search sites, the user navigates to the search engine's site and types in the subject of the search. Table 4-15 gives the addresses for some popular Internet search sites.

Table 4-15: URLs for Internet Search Sites

Alta Vista	`http://www.altavista.com`
eXcite	`http://www.excite.com`
Infoseek	`http://www.infoseek.com`
Inktomi	`http://www.hotbot.com`
Lycos	`http://www.lycos.com`
Open Text	`http://www.opentext.com`
WWW Worm	`http://wwww.cs.colorado.edu/wwww`
WebCrawler	`http://www.webcrawler.com`
Yahoo!	`http://www.yahoo.com`

■ Search Engine Databases

Search engines for large numbers of Web pages, such as those that attempt to cover the entire Internet, do so by maintaining databases that model the Web's structures. Through a combination of information-trolling robots that collect information automatically about Web pages and developer registration, search engines select a large number of Web sites to be indexed. Their databases are then populated with information about the contents of each page deemed useful. The Internet search site Yahoo!, the largest U.S. search site by volume (with more than 25 million unique U.S. users per month in September 1997, according to Mediamark Research), served an average of 50 million page views per day in the same month. Many search engines also are used within a given Web site.

Some engines search not only Web pages, but also Gopher sites, FTP resources, or Usenet news groups.

Search sites select pages for inclusion in their databases in two primary ways:

- **Web crawlers** – These traverse the Web automatically, collecting index data as they do so. Depending on its programming, the crawler will search either depth-first, following only the links that are deemed relevant to a topic, or breadth-first, collecting the entire network of links from a given starting point regardless of the page contents. (Web crawlers also are called spiders, ants, robots, bots, and agents.)
- **Registration** – Most search sites allow Web developers to register their sites or pages by submitting a form. This process enables developers to ensure that their sites eventually will be included in the search index.

Most sites use a combination of crawlers and registration to guide their coverage of the Web. No search engine covers every single page on the Web. Also, many Web sites that require user registration are not indexed.

Search engine databases index Web pages (and other Internet resources) by keywords, by the most significant words, or by all words. Web developers can designate the keywords that will call up their pages either by including HTML meta tags or through the site registration process. However, some search engines ignore developer-assigned keywords or do not permit registration. As an alternative to programmed keywords, some search engines extract the significant words on each indexed page. Methods for selecting significant words vary and are used in different combinations on different search engine sites. Some methods include counting word frequency, either on the whole page or within a standard-sized portion of the page, and focusing on emphasized text, such as in headings or heavily punctuated lines. Open Text's Open Text search engine provides full-text search: It not only examines the entire page, but also indexes all words and phrases on the page.

■ Query and Results

When a user enters a search query, the engine searches its database for relevant Web pages. It assembles a list of pages sorted by "relevance" or other, user-specified weighting factors. Some sites also remove redundant references to pages from the list. Search results are returned as a list of relevant pages that then can be retrieved via hyperlinks. Different tools can produce results that vary widely, ranging from not finding critical pages (poor recall) to finding tens of thousands of documents with few that are relevant (poor precision ratio). Users are becoming aware of this problem and must choose the search tool that is right for them.

All search engines offer a simple query facility on single words that is easy for novice users to operate; however, it typically returns references to too many pages. Some search engines also offer complex queries or weighting factors that influence the results. More complex queries can include the Boolean operators AND, OR, and NOT as well as proximity operators such as NEAR. They may narrow the search to Web pages posted within certain dates or to Web pages that definitely do or do not contain certain literal strings or that are in certain languages.

Search results and relevance ratings can be counter-intuitive and confusing because some methods of rating relevance have little to do with the topic conceptually and depend instead on text characteristics without sufficient context. Search results can include seemingly irrelevant references because the engines use "intelligent" algorithms such as Soundex to compensate for spelling errors and employ thesauri and fuzzy logic to draw associations between words.

How each search engine measures a page's relevance varies with the vendor. Below are some sample heuristics (guidelines) used by vendors to determine page relevance:

- Developers who have registered their pages have the most relevant contents.
- Pages most often linked to by other pages are considered more relevant.
- The more often the word is used in a page, the greater the relevance.
- Pages that were updated most recently have greater relevance.
- For queries involving more than one word, how close the keywords are to each other within the text increases relevance.

Search engine results typically are displayed in the order of relevance, with pages having the highest relevance scores listed first. Some sites, such as Excite, allow users to select a different sort criteria, such as the domain name of the site. More advanced results sorting is beginning to emerge. The search engine offered by Inference, for example, automatically clusters its listings by URL or type of site, removes redundant references, and discards results with low relevance ratings.

■ Information Filters

Filters – automated methods of sorting relevant from irrelevant information – are being developed to help people access information with more precision. As the information available over the Internet mushrooms, users increasingly need to narrow the content through which they wish to search. For example, the majority of postings to Usenet groups are now low-quality, unsolicited advertising, which has led to the inclusion of increasingly sophisticated filtering options in browsers and news readers. In addition, server administrators and newsgroup moderators are seeking automated ways to eliminate such traffic.

Programs to screen out adult content from Web browsers, intended for home markets, began to appear commercially in 1997. Examples of Internet screening software include CyberPatrol, Net Nanny, Solid Oak Software's CyberSitter, and Spyglass' SurfWatch. These programs prevent access to a list of sites deemed unacceptable by the company providing the software. Customization and automated on-line updating of the list are possible. Services such as AOL allow parents to block adult chat areas or real-time requests for chat.

In response to concerns and to preempt possible federal regulation, AOL, Disney, Microsoft, Netscape, and other companies are supporting the Platform for Internet Content Selection (PICS), a specification for labeling Web content. PICS embeds labels in HTML page headers to rate different Web sites. Microsoft's Internet Explorer software allows parents to block categories and set a password. A similar feature is planned for Netscape's Navigator browser.

A more active method of filtering information uses agents. Agent technology is a concept that has been around for some time. The goal of agents is to create applications that automatically carry out tasks for users without their intervention – other than initial configuration and updating with new requests. Much of the functionality originally subsumed under agent technology, such as filtering and user profiling, has migrated to the push and pull technologies previously discussed, opening the way for the creation of new capabilities.

■ Clipping Services

The number of publications, traditional and electronic, available on-line continues to grow. In digital format, publications are easily amenable to efficient or automated clipping. Excite now offers NewsTracker, a free clipping service. Users can track up to 20 news topics and retrieve articles from a database of more than 300 publications. Revenue is generated by advertising.

■ Metasearch Engines

An inevitable extension of the Web search technologies discussed previously is a Web page that automatically enters search queries into a number of other search programs and returns the results. Such sites are known as metasearch engines. Examples of metasearch sites include All4one, Metacrawler, and Starting Point.

Personalized Web Services

Personalization refers to the use of information about the user, information about activity during the current or previous visits to a Web site, the type of browser, or browser preference settings to generate Web content that is personalized. The ability to let a site visitor define a custom home page is a type of personalization because this home page is generated dynamically, based on the user's previous setup. This feature has become a requirement of any on-line news service, appearing in the Web sites of publications ranging from *The Wall Street Journal* to *Wired*.

The rapid growth of various digital broadcast channels has been accompanied by several announcements about new personalized services and delivery options. For example, Microsoft offers a Personal Delivery Service to members of the MSN that enables them to receive a personalized daily version of MSNBC news topics via e-mail as well as MSNBC News Offline and News Alert news delivery.

■ Offline Browsers

Off-line browsers (or pull products) enable a user to retrieve pages automatically from Web sites at predetermined times, often during the night. The latest pull products work with a user's existing browser.

The ForeFront Group's WebWhacker is an offline browser that allows users to define a group of sites by their URLs and then download text and images from those sites to their local storage. WebWhacker lets the user determine how much of a Web site to retrieve – title pages only, any linked pages, or all pages. The product's search engine lets users perform Boolean searches and bookmark items. ForeFront has announced a server version of WebWhacker, targeted at corporate users, that will enable businesses to store information from frequently visited Internet sites on their intranet server. This approach uses a form of caching, designed to speed up access.

Traveling Software's WebEx, like WebWhacker, lets users gather content from specified Web sites and store it on their hard disks for future reference. WebEx acts as a proxy server, fooling a browser into thinking the user is still on-line. To speed delivery and conserve disk space, WebEx captures the newest content at the selected sites and automatically overwrites the previous version. Users can customize downloads by specifying how many layers to drill down in a Web page, limiting download size, skipping links outside of the site's domain, or requesting text-only delivery.

■ Collaborative Filtering

Collaborative filtering is an innovative new category of personalization services exemplified by Firefly Network (formerly Agents), an advertising-supported Web site. Upon registering in Firefly, the user's ratings of films and music of various kinds are combined with other users' preferences to predict the new user's tastes and offer other music and movies he or she might like. The longer an individual uses Firefly, the more accurate the predictions become, and consequently, the more personalized the information on the site.

The Interactive Internet

All Web pages are to some degree interactive in the sense that if a user clicks on a hyperlink, some action will result. Interactive pages are defined as pages that actually change in response to a user's action. For example, various mouse movements may rotate an object or change the viewer's relative position in a three-dimensional (3-D) space. Some interactive pages are primarily two-dimensional (2-D), performing scrolling and zooming actions, whereas 3-D pages actually portray graphical representations of 3-D objects or spaces.

Java has been instrumental in allowing animated content on the Web. Some additional products include Macromedia's Shockwave and Director and a range of virtual reality tools such as Caligari's Fountain.

■ Shockwave and Director

Shockwave is a family of multimedia players from Macromedia designed to provide a range of multimedia experiences on the Web. Director is Macromedia's content creation tool. Shockwave players display and play back multimedia content from interactive games, to multimedia user interfaces, to on-demand streaming audio, including live concerts and radio. Shockwave was introduced to the Web in 1995. Since then, more than 17 million Shockwave players have been downloaded, and thousands of sites have been designed using Shockwave multimedia content, including those of Apple, GM, Intel, Kodak, Microsoft, and Nissan.

Shockwave Director creates dynamic, fully interactive Shockwave multimedia. Director creates multimedia projects, ranging from presentations and interactive advertising to kiosks, titles, and Web pages containing motion and interactivity. Director is based on a theater metaphor, with a Stage where users view the Director File. Behind the scenes is the Cast Window that stores the media objects, which can be sounds, 2-D and 3-D graphics, animation, digital video, text, and database objects. The Score window synchronizes media elements and provides layers of elements on screen.

■ Virtual Reality Modeling Language (VRML)

VRML is used to create 3-D worlds through which users can navigate within their Web browsers using a mouse or other device. These worlds are platform-independent and viewable in any VRML-capable browser. Elements of a VRML world are described in the VRML language, which is interpreted by a VRML-capable browser. The VRML language continues to evolve and enable an increasing variety of user-browser interactions. VRML content creation and management is complex; worlds need to respond to movements the user makes in a 3-D space. Hyperlinks can take the person to other sites or another virtual world.

Although VRML is not widely used today on commercial Web sites, future applications include virtual business or shopping districts, special interest communities, and entertainment arcades. VRML worlds need to be visually compelling and well designed so they are easy to navigate through without getting lost. Worlds also need to be fast-loading and fast-rendering.

Tools for creating VRML worlds include Caligari's Fountain, ParaGraph International's Home Space Builder, Radiance Software's Ez3d, Silicon Graphics' Cosmo Worlds and WebSpace Author, and Virtus' 3-D Website Builder and Walk-Through Pro.

■ Java

Because Java applets have the ability to respond to user input, Java can be used to design interactive pages. Its abilities currently are limited but should improve with future releases of the language and with tools specifically designed to produce interactive Java pages. (See Chapter 6.1, *Application Development Environments,* for more information on Java.)

"Cookies" and Internet Privacy

Netscape, Mosaic, and Internet Explorer browsers support "cookie" technology. A cookie is a MIME-type (Multipurpose Internet Mail Extensions) header that can be used to exchange information automatically between a server and a browser without a user seeing what is being transmitted. The cookie is a small quantity (a text string) of information with an expiration date. Transmission of the cookie is initiated by the server, which can request that the cookie data be stored in or retrieved from the browser. Cookie data is discarded after a given expiration date.

The cookie is useful in tracking users' actions and preferences. For example, cookie data can be used to save the values of data entered into a form. The Web developer can use the cookie and associated data to collect information on a particular browser or user over multiple sessions, without requiring the user to log in. This background information can then be used to customize the Web content that is given to the user. For example, a user who has expressed interest in certain car models can be given pages with advertisements about those models.

A drawback of cookie-only personalization is that the cookie is associated with the specific browser installation. When the user installs a new version of the browser or accesses the Web from another PC, the original cookie is not available. Also, if more than one person is using the same computer and browser, cookie information is shared. To some users, the automatic creation and retrieval of cookies raises privacy concerns. Some browsers allow users to disable the cookie function. (See "Cookies" on page 408.)

4.4.7 Web Authoring

Web sites have become important creative media, as visually appealing as any other visual media with the added benefits of multimedia and dynamic database-driven content. Tools for page and site design range from ASCII text editors to full-blown integrated development environments. Some new tools can help turn Web sites into push channels.

The limiting factor that underlies all layout and design tools is what the most commonly used Web browsers can display. Standard HTML, which is constantly evolving and being extended with proprietary enhancements by browser vendors, is the common denominator. Graphics files in the CompuServe Graphics Interchange Format (GIF) format are common, as are graphics in the Joint Photographic Experts Group (JPEG) format. Browsers can be extended with additional capabilities through plug-in applications and software components that are able to display other types of content, ranging from text formatted more richly than HTML allows, to animated graphics, audio tracks, video clips, and more.

Web Authorizing Tools

Hundreds of layout and design tools will continue to be available on the market for the next few years as new refinements and features are added. Interactive and animated elements have increased, and many packages have made Web page creation faster and simpler by offering a choice of customizable templates or forms. The following sampling of tools illustrates the range of capabilities available for Web page design and electronic publication.

■ Adobe's PageMill

PageMill 2.0 is a popular graphical Web page development product. It allows visual layout and editing of a Web page as well as editing of the underlying HTML directly. Tags for positioning, buttons, graphics, and other elements are inserted automatically, and links can be added by dragging and dropping them from other pages. A spell check tool is built in.

■ FutureTense's Texture

One of the early applications of Java technology is Texture, a layout program that produces pages that are delivered to clients as Java applets. Texture Web-authoring software is a complete Web presentation and document creation system. Texture has two advantages. First, the designer can have tighter control over the look and feel of the page because attributes such as font faces or font or screen colors are not subject to the defaults users set in their browsers. Second, the pages that Texture produces can be very dynamic, including far greater intelligence, richer content, and more interactivity than would be possible with conventional HTML.

■ Kobixx Systems' EZine Publisher

EZine Publisher, written in Java, is intended for publishers creating electronic magazines. It provides the content management features needed to design and publish an electronic magazine, including professionally designed templates for different types of articles, automatic table of contents generation, keyword search capability, and bad-link detection and correction. An add-on package handles subscriber management, demographic surveys, and the automatic generation of a "new issue" e-mail notice to subscribers that can include the table of contents or excerpts from articles. Publisher Pro also includes other production management capabilities, such as generating customized content for different groups of subscribers, tracking what subscribers actually read to help fine-tune the publication, and notifying contributors and advertisers by e-mail of schedules and deadlines.

- **Microsoft's FrontPage**

Another graphical Web page authoring tool, this product offers many powerful features. These include a command to find and track broken links and support for Common Gateway Interface (CGI) scripts used to link Web pages to databases and other programs. FrontPage supports frames and tables as well as the ability to add new HTML elements as they are developed. FrontPage 98 adds support for CSS and DHTML and allows publishing of content with Microsoft's CDF.

- **Netscape's Navigator Gold and Composer**

The Navigator Gold edition of Netscape's popular browser includes a basic Web publishing interface aimed chiefly at intranets. Composer is the Web publishing module included with the Netscape Communicator suite.

- **RandomNoise's Coda**

Coda is a Java-based development tool designed to produce entire Web sites in the form of Java applets, without the designer having to write any HTML. Coda allows intranet developers to construct interactive Web pages containing options that offer much more than animations or catchy graphics. Coda attempts to overcome a lot of HTML's limitations in areas such as positioning, layering, and fonts. It addresses visual page design as well as requirements for adding logic to a Web site. Coda also provides a variety of dynamic objects that can be dragged onto a page, including database access objects and animation controls.

4.4.8 Internet Voice, Video, and Multimedia

Not too long ago, it was assumed that given the typical speed of Internet connections at 28.8 Kbps, the ability to stream audio, let alone video, would be out of reach for most consumers. However, by using sophisticated compression algorithms, buffering, and network bandwidth utilization monitoring, products are delivering usable audio and video today, and quality is expected to improve steadily.

Streaming Audio and Video

Streaming allows Internet users to see and hear data as it is transmitted from the host server instead of waiting until the entire file is downloaded. Streaming audio and video is transmitted in packets that are played as a computer system receives them, meaning customers do not have to download files to their hard drives. For example, RealNetworks' RealAudio allows a Web site to deliver live and on-demand audio over the Internet and can work over connections as slow as 14.4 Kbps. Traditional video formats such as QuickTime and AVI must be downloaded to the computer before the user can view the file.

To date, streaming video technologies are not yet ready for real-time interactive applications such as videoconferencing, and it may be several years before vendors overcome technical limitations. Despite these limitations, streaming video has other applications, including training, entertainment, communications, advertising, and marketing. Streaming audio and video are being used to deliver market sensitive news and other "live" status reports to stock traders, to brief sales people on new products and to deliver corporate news to employees, and to view TV commercials for approval. More efficient codecs and better flow control mechanisms will continue to improve video quality.

Streaming audio already has enabled the broadcast of radio programs, music, press conferences, speeches, and news programs over the Internet. In the future, streaming audio and Internet telephony use will overlap and complement one another.

Increasingly, streaming audio and video are merging as vendors offer products with enhanced Internet broadcast capabilities. Streaming audio and video vendors include Apple, Microsoft, RealNetworks, VDOnet, Vivo, Vosaic, and Worldwide Broadcasting Network.

■ Apple

Although not primarily a streaming protocol, QuickTime's conferencing feature can send and receive streamed audio and video. In addition, Version 3.0 supports the DVCam digital video format, which lends itself to streaming. QuickTime has evolved into a comprehensive architecture for the authoring and playback of synchronized video, audio, graphics, and text. More than 20,000 sites on the Web now offer QuickTime content.

Popular with CD-ROM developers, the QuickTime format is used in half of all video on the Web, according to a survey in *New Media* in September 1997. The second most popular video format is MPEG, which also can be played by QuickTime applications. Apple has expanded QuickTime to the Windows platform and estimates that Windows users have downloaded more than 1 million copies of QuickTime for Windows since July 1997. Developers soon will be able to author QuickTime movies under Windows, not just play them.

■ Microsoft

Microsoft's ActiveMovie, which replaces Video for Windows as Microsoft's alternative to QuickTime, offers QuickTime-like features.

In response to the popularity of products such as RealAudio, Microsoft has developed NetShow 1.0, a platform for streaming audio and video delivery and live IP multicasting. The components include On-Demand Player, an editor, and a software developer's kit. Although the On-Demand Player currently works only with Internet Explorer, it is free, as is the NetShow server. Microsoft's Active Streaming Format (ASF) editor tool creates files in Microsoft's streaming format and allows users to import WAV and graphics files. Microsoft would like ASF to become the VHS of video streaming.

Today, it is necessary to download different clients to view Web-based video. In 1997, Microsoft announced an agreement with (and acquisition of 10 percent of) Progressive Networks (now RealNetworks); and in August 1997, Microsoft acquired VXtreme and stated that its product line would be folded into NetShow 2.0. If future versions of Microsoft's NetShow support RealAudio/RealVideo content, and RealNetworks supports Microsoft's Active ASF in future versions of its software, the need to download different clients would decrease.

■ RealNetworks

One of the early leaders in streaming audio, RealNetworks (then Progressive Networks) introduced RealVideo in mid-1997. RealVideo delivers newscast-quality video and playback that is near TV broadcast quality at higher bandwidth. The product includes several interactive features (for example, clickable regions of video can be defined to link to other video files or Web sites). The Real Player plug-in, which plays RealAudio and RealVideo, is available as a free download from RealNetworks' Web site.

■ VDOnet

An early leader in providing streaming video capabilities to the Web, VDOnet claims to have the largest market share, with more than 500 Web sites, and includes among its content providers CBS, Fox, News Corp., PBS, Preview Travel, and Sportsline USA. VDOLive uses scalable technology that allows the VDOLive server to measure the bandwidth available to the user at any given point and optimize picture quality accordingly.

VDOLive supports both on-demand and live broadcasting. The VDOLive Broadcast Station/Server is available for live transmission on the Internet. Viewers need only one player to see both on-demand video and live video broadcasts. The VDOLive Player comes in various forms – as a plug-in for Netscape Navigator, an ActiveX control for Internet Explorer, and a "helper app" player for Windows 3.x, Windows 95, and PowerMac.

VDOLive tools can be integrated into other video editing tools such as Adobe Premiere, run on both Macintosh and Windows platforms, and allow for individualized content creation. VDOnet has already implemented industry standards such as H.323 and H.324 (the ITU-TSS interoperability standard for videoconferencing over analog telephone).

■ Vivo

Originally a developer of codec technology, which provided an early advantage in the streaming video market, Vivo has focused on content creation and on adding value to a video stream. For example, Vivo Events can drive actions elsewhere on a page, bringing forward various page elements in synchronization with a video. Vivo will be working with various partners to help them enable their content creation tools to use video content created with Vivo's tools.

■ Vosaic

Java-based streamers, such as Vosaic's Java TV Station server, have emerged as cross-platform alternatives where clients are not required. Vosaic, an MPEG streaming solution, uses special flow control mechanisms to adjust video bit rates and maintain audio quality. Using sophisticated techniques for analyzing content and network traffic, Vosaic can deliver video at the high frame rates necessary for life-like movement. Future releases of Vosaic will further improve its capabilities, particularly in low-bandwidth environments. Moving from video to audio, Vosaic offers Vosaic Audio for Java. The technology enables any Web server to broadcast an unlimited number of audio streams across the Internet, intranets, and other TCP/IP networks. According to Vosaic, more than 1 million people tuned in to see NASA's coverage of the Mars landing using the company's Java solution.

■ Worldwide Broadcasting Network (WBN)

Streaming video content provider/developer WBN is working in partnership with BBN and using the Multicast Backbone on the Internet (MBone) and RSVP protocols to broadcast video content. WBN is limiting its use to sites with access to MBone or to corporate LANs routing IP multicast protocols. Early adopters of WBN's content have been companies that need to provide training for employees at multiple locations.

Real-Time Audio and Video

In real-time audio and video, the source is live or only slightly delayed. Although in streaming audio and video the data can be compressed before transmission, real-time audio and video has to be compressed on the fly. This requirement is a significant technical challenge, though far greater for video than for audio.

Real-time audio and video applications include point-to-point conversations between two people; conferences among more than two people; collaborative "white boarding" and shared hypermedia documents; live broadcasts of news, talk shows, or sporting events; and broadcasts of music and concerts.

Computer networks offer less predictable bandwidth for a real-time conversations than telephone networks. The H.323 standard, developed by the ITU, helps address this problem by guaranteeing communications over packet-switched networks such as the Internet when the quality of the network cannot be determined or if it may fluctuate. An H.323 gateway connects computer telephony devices with telephone and other networks, making computer-to-telephone calls possible. The standard also helps guarantee interoperability between different products. Most current products also limit the number of concurrent users to avoid response time degradation.

The ITU's T.120 series of standards support the negotiation and establishment of multi-user conferencing sessions. The T.120 suite handles error correction and security and contains specifications for collaborative data applications such as white boards and application sharing and viewing. It also ensures different vendors' products work together over different network topologies and protocols.

Real-time audio and video vendors include Intel, Microsoft, and White Pine Software.

■ Intel

Intel's Internet Video Phone is designed to support standards such as H.323 and to interoperate with Microsoft's NetMeeting. The device allows users to see and talk to each other while using other Internet applications. It takes advantage of the latest PC technology, gaining performance through faster processors. Intel's objective is to accelerate market demand for more powerful hardware, stimulating a replacement market for older PCs. Intel's Video Phone software comes bundled with PCs from Compaq, NEC, Toshiba, and other vendors.

■ Microsoft

NetMeeting 2.0 has a variety of capabilities, including real-time audio and videoconferencing. The product is designed to support various types of collaboration and interaction, ranging from white board, to chat, to Internet telephony.

Microsoft has been working closely with Intel to produce interoperable, multimedia conferencing tools based on H.323 and the T.120 series. In addition, Microsoft and Intel have cross-licensed each other's conferencing technologies. Videoconferencing capabilities are limited in Version 2.0, however, which delivers approximately 5 frames per second – a speed not yet suitable for commercial videoconferencing.

■ White Pine Software

CU-SeeMe originally was developed as a video tool for the Macintosh, and it has been enhanced as a commercial product by White Pine Software. Enhanced CU-SeeMe, coupled with its Reflector technology, offers group collaboration and videoconferencing. With Enhanced CU-SeeMe, a group of people can share documents and information interactively, with integrated collaborative solutions such as white board, chat window, and Web browser support. CU-SeeMe uses the Internet Protocol to establish a connection.

Enhanced CU-SeeMe provides up to eight participant windows for videoconferencing and an unlimited number of participants for audio and chat. It provides a caller ID feature, which is a message alert box for incoming connections. Enhanced CU-SeeMe also has a white board for collaboration during conferences. A phone book allows the user to save, add, and edit participant addresses and Reflector sites. Security features include password, caller ID, and other conference and inbound call security features. Plans call for CU-SeeMe to work in conjunction with MBone in the future.

The Reflector is a server-based application that allows CU-SeeMe clients to have group conferences. It accepts multiple CU-SeeMe connections and reflects the video, audio, and additional data to all participants concurrently. Depending on the hardware and network configuration, as many as 100 active participants can be supported on one UNIX workstation. Multiple Reflector sites can be linked to create a network for larger group conferences or video broadcasts.

Internet Telephony

Internet vendors now are providing products that emulate traditional Public Switched Telephone Network (PSTN) applications. Internet telephony lets users talk across the Internet to any PC equipped to receive the call – even around the world – for the price of only the Internet connection. Internet telephony and higher bandwidth videoconferencing applications are still experimental, but they are gaining user acceptance as quality and reliability increase.

Many experts believe that the Internet currently lacks sufficient bandwidth to handle increased telephone access. Predictable service quality and latency issues also exist. Standards may help address that problem. Proposals include the H.323 standard for low-bandwidth audio and videoconferencing, and the International Multimedia Teleconferencing Consortium's Voice Over Internet Protocol (VOIP) to deal with server-to-server communications. Although most users can connect only if they have the same Internet phone software, the development of H.323-compliant products brings the prospect of interoperability between different telephony products much closer.

Phone-to-Phone Internet telephony carriers began operating commercially for the first time in the second half of 1997. This is an application that does not rely on PCs but instead uses local dial-up numbers to route long-distance calls over a packet-switched backbone. (See Figure 4-12.) Because these services do not require customers to purchase or operate any equipment beyond ordinary telephones, they circumvent several of the barriers to effective Internet telephony.

Figure 4-12: Internet Telephony

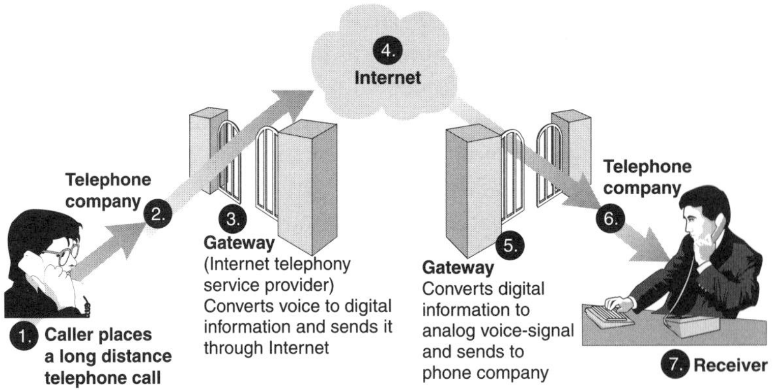

Source: *San Francisco Chronicle*, 1997

■ Internet Telephony Products

Internet telephone software is available from a variety of vendors. Most products work with standard PC sound hardware and Internet connections (either via network or modem), and all support both full-duplex (simultaneous two-way) and half-duplex (alternating send/receive) transmission. Many of the phone "packages" themselves are free and can be downloaded from the Internet.

Many Internet phone packages now also include video and conferencing tools, causing a convergence with the products discussed previously. For example, Microsoft's NetMeeting 2.0 Internet phone includes collaboration tools, features for multi-user conferencing, and one-on-one H.323 video calls and audio. Some of these products may be suitable for intranets. For example, IBM's IC Phone also works over stand-alone networks, letting users connect via a local intranet. IBM is preloading IC Phone on all its multimedia PCs.

Further adoption of and compliance with the H.323 standard will allow for increased interoperability between products. Intel's Internet Video Phone, Vienna Systems' Vienna.way, and several other products already support the H.323 standard, facilitating interoperability with other H.323-based software products. Microsoft licensed Internet Video Phone to integrate its capabilities into Microsoft's Internet Explorer browser and Windows operating system. The product uses commercial "white pages" directory servers. (See "Directory Access" on page 263.)

Business-oriented telephony features include NetSpeak's WebPhone's call management options, which let users handle up to four point-to-point calls at a time, send voice messaging via e-mail, block or restrict voice mail, and entertain callers on hold with MIDI-based music. Premiere Technologies' Orchestrate allows customers to retrieve all their messages from their Web browsers and respond to or forward messages from their Web browsers as well.

Vienna's Vienna.way server-based IP telephony product brings voice, data, and video together on a corporate LAN. The server software establishes and manages calls and provides telephony features such as call hold and call forwarding. It switches traffic between different IP networks and, through its gateway module, between an IP network and the public telephone network.

Multifunction products such as Vienna's Vienna.way could serve as an enhanced teleconferencing facility on corporate IP networks or a Web site and as a server for other telephony applications. Customers can speak to other Vienna.way users, to users on the public telephone network, or to users of other H.323-compliant IP telephony products. To the outside world, the user's telephone number looks like any other telephone number. The user gains features such as call forwarding, conferencing, transferring, redirection, integrated voice and e-mail.

■ Licensing Agreements and Alliances

Several important licensing agreements have occurred between smaller Internet telephony startups and larger vendors. For example, Voxware's MetaVoice technology has been licensed by multi-player gaming vendors and already is incorporated into other Internet telephone vendor's products, such as Netscape's CoolTalk. Intel is using Voxware technology.

In March 1997, VocalTec and the Internet Software Products Division of Motorola announced a memorandum of understanding for a licensing and distribution agreement. The Motorola division formed a network equipment development and systems integration business unit to deliver Voice-over-Internet Protocol technologies for commercial use. Motorola worked with VocalTec to optimize IP telephony application performance over enterprise networks. Deutsche Telekom has purchased VocalTec technology and a share in the company.

Other companies have developed alliances such as Vienna Systems, which is an affiliate of Newbridge Networks. Bell Atlantic, Microsoft, and US West have invested in VDONet.

■ Internet Telephony Service Providers

Delta Three, Global Exchange Carrier, and USA Global Link were among the first to offer consumers Internet phone-to-phone calling. These providers offer rates for international calls that are typically one-fifth the rate charged by traditional long-distance services. Quality is good but variable, and reliability remains a concern.

Improvements in underlying technology should boost quality over time. Refinements in Internet telephony software user interfaces and the adoption of consistent standards and directories should increase ease-of-use for consumers in the future. AT&T and MCI were running trials for IP phone service in 1997. Additional announcements from service providers are expected throughout 1998. (See Chapter 4.3, *Telecommunications Services,* for more information on traditional carriers and service offerings.)

■ Directory Access

Just as phone directories are available in certain standard formats to help consumers locate telephone numbers for businesses and individuals, directory standards in Internet telephony and conferencing play an important role as well. LDAP can be used as a standard format to determine who is on-line, available, and willing to accept calls. LDAP is a simpler version of the X.500 open directory services protocol, an ISO and ITU standard. LDAP has received fairly widespread support from organizations such as the IETF, Netscape, and Microsoft. LDAP enables the creation of standardized phone-book-like user directories that can span multiple intranets and the Internet and allows browsers and client applications to access schemas within these directories. However, incompatibilities still exist in LDAP implementations that need to be addressed. (See "Lightweight Directory Access Protocol (LDAP)" on page 391.)

A metadirectory or universal standard to access information across operating systems in the same way that the Internet Domain Name System (DNS) allows users to find specific servers in a huge group of IP addresses, or in the way that search engine sites help users find what they need, is desirable but not likely in the forecast period.

Internet Fax

The use of the Internet for real-time fax transmission is emerging as an application that may serve as a precursor to an increasing shift of traditional analog traffic from the PSTN to the packet-switched Internet. Essentially, Internet fax works by sending a stream of fax data from a network fax server or fax machine through a device that places an envelope around the data and sends it to an Internet gateway. The envelope carries addressing information that routes the packets of fax data to a remote gateway where the envelope is opened, the telephone number of the destination fax is decoded, and the fax is sent to its final destination. This application is useful because faxes can be sent long distances at local telephone rates and delivery can be guaranteed through store-and-forward mechanisms.

Internet fax also is a precursor to Internet telephony and a time when an increasing volume of voice calls will be routed over the Internet. Although some Internet fax products are designed for low-volume use, the significant products are those designed for ISPs to provide high-volume service for corporate customers.

The following are high-end Internet fax products and services: Brooktrout Technology's IP/FaxRouter, Castelle's FaxPress, Ibex Technologies' Fax-From-Web, JFax's Fax/Voice to Internet forwarding, NetCentric's FaxStorm, and Open Port Technology's Open Port Harmony.

4.4.9 Intranets and Extranets

An intranet is a private network that uses Internet software and TCP/IP protocols. In essence, an intranet is a private Internet, or group of private segments of the public Internet network, reserved for use by people who have been given the authority to use that network. Companies increasingly are using intranets – powered by internal Web servers – to give their employees easy access to corporate information. Intranets also are being viewed as an effective medium for application delivery. Although communications traffic is restricted to corporate LANs or WANs, key partners and suppliers often may be part of an extended intranet as well. This outward-facing extension of an intranet is often called an extranet.

The growth of TCP/IP-based intranets has followed the move to distributed computing, with new protocols, additions to graphical user interfaces (GUIs) in the form of client browsers, and increased demands for bandwidth. Web browsers increasingly are being used to access many corporate applications. Intranet software often is a add-on to groupware, enabling electronic communication, collaboration, and access to information within an organization to facilitate productivity.

Web browsers enable intranet deployment because they provide a ready-made GUI client and thus offer an inexpensive means to develop new systems. Network novices can set up basic Web servers. The cost of tools for authoring documentation, prototyping applications, and deploying small-scale databases on the Web is relatively low. These tools allow people to develop on-line documentation easily that is available instantly to everyone on the intranet. HTML documents also can contain forms, and numerous database interface packages have been introduced that make it easy to pass data to small-scale databases such as those implemented in Microsoft Access or FileMaker Pro. Although some of these tools may be available at no charge or bundled with other products, "free" browsers and Web servers do not include support. More substantial costs may be associated with content management over time.

Larger-scale applications are being implemented for intranets that involve more complex requirements and are generally more expensive. Building and interfacing with larger, industrial-strength databases such as those based on relational database management software (RDBMS) from IBM, Informix, Oracle, and Sybase requires the purchase of software, license fees, and middleware to connect the database to the Web; labor for professional database design and programming; and ongoing administration and maintenance.

Intranets present an alternative to existing groupware products such as Lotus Notes. However, early intranet offerings have lacked some important security features, document organization features, structured e-mail, workflow capabilities, and other capabilities offered by Notes. Lotus has responded to the rapid growth in the Internet and intranets by introducing its Domino server software and its InterNotes Web Navigator browser. These products allow Notes to act as a Web server and browser.

More sophisticated Web site development tools are providing additional functionally organized, preformatted Web pages, or site templates from which users may choose. These offerings may extend beyond color and style choices to include suggested links between organizational groups, such as marketing, accounting, and manufacturing departments. A variety of tools provide ways to monitor traffic and manage the site on an ongoing basis, which is especially important for sites that will have a large number of pages and site content that will be updated often.

Extranets

Extranets are private WANs that are accessible to business partners and customers as well. They also may have links to public sites. Opening up corporate intranets to key customers, contractors, and suppliers is a way to improve customer service, foster close partnerships by sharing more information, reduce the costs of doing business, and increase the visibility of a company's products and services.

Companies have extended access to corporate intranets to distributors and retailers. Partners can access information on products, order status, delivery schedules, service, selected accounts, and other information, usually by employing a user ID and password to gain entry.

The tools required to establish an extranet basically are similar to those for administering intranets, but they offer increased access to outside users. Opportunities exist for vendors to position their products as being amenable to extranet construction. Netscape is pursuing this strategy, and work-group software vendors are marketing their products for use as extranet servers.

4.4.10 Web Server Software

A Web server program implements the HTTP protocol to communicate and exchange messages between the server and the requesting client program, usually a Web browser. In the simplest case, the software receives a request from the browser in the form a of URL, locates the corresponding HTML file, and send its contents back to the browser for presentation.

The file that is located this way need not be an HTML page. The page could be any file type because the HTTP protocol allows for the returning stream of data to be prefixed with a header that tells the receiver what type of data is contained. Hence, a simple ASCII text file or a more complex video clip encoded in MPEG format can be sent to the browser, which then interprets the type of data represented and presents it to the user.

The term Web server can refer to a physical host machine or platform that includes the hardware, operating system, network software, and Web server software. Web server also can be defined more narrowly as the Web server software itself. In this discussion, the term Web server host will be used to designate the physical system or platform to avoid confusion with the server software.

Web browser and server software programs were first developed by the NCSA and the W3C. NCSA developed the first freeware HTTP or Web server, known as httpd, which ran only on UNIX. Web servers now run on a broad range of computing platforms and operating systems, ranging from the largest IBM mainframes to desktop PCs. Although many low-cost shareware or even freeware Web server software solutions are available, corporate customers have opted for commercial products that now offer additional features, support, and integration with other applications.

The Role of HTTP

HTTP is considered a stateless protocol in that once the server has fulfilled the request, it closes the connection initiated by the browser and does not retain any information about its interaction or "state" with that browser. Although this model is well-suited to simple Internet information access requests, it is less appropriate for Web browsers performing more complex information access and transaction processing applications, where transaction integrity is difficult to guarantee. For example, statelessness means that a Web search engine cannot automatically send the "next page" of search results when asked to do so. Instead, the search engine creates a complex URL which is used to request that the query be run again, this time returning a different subset of the search results.

Web site developers currently use several mechanisms to compensate for the stateless quality of HTTP and to ascribe a state to transmitted data. These mechanisms include having the browser maintain state information and return it to the server via the URL with each new HTTP connection, the use of certain attributes in HTML (the *type=hidden* attribute of HTML forms), the use of cookies containing state data, and the use of a database to store current state information.

Several Web middleware development efforts are underway either to strengthen HTTP or to replace it. Distributed Computing Environment (DCE) vendors are working on middleware that will run HTTP over DCE's secure remote procedure call (RPC) infrastructure rather than directly over TCP. Some vendors also are suggesting that HTTP might be replaced by something such as the Internet Inter-ORB Protocol (IIOP).

In addition to the HTTP protocol, several other standards and APIs are important, including the Common Gateway Interface (CGI) standard. The most common approach to gain access to legacy data has been to run CGI scripts on Web servers in response to a query to invoke a back-end database or transaction processing system.

■ Common Gateway Interface (CGI)

CGI was one of the first extensions to the Web server. CGI is used when a browser's request for information or a transaction needs to access a corporate database. CGI defines the way programs communicate with the Web server but also may refer to the programs or scripts themselves. When it receives a URL containing "cgi-bin," the Web server runs a CGI script, specified by the URL, to execute a script or program and pass the results to the server for transmission back to the requesting Web browser. A CGI script can pass a query obtained from a Web browser to a full text indexed database, format the results into HTML, and send the formatted results back to the requesting Web browser. It can access a relational database to display and update the contents of a record or create HTML on the fly to customize the Web pages being presented by the Web browser.

A CGI script can be written in any programming language that can be compiled for or interpreted by the Web server's operating system. Examples of commonly used languages are C++, JavaScript, or Perl.

The biggest request by corporate information systems managers of Web server vendors is that access to existing data be made easy to implement and transparent to users. Improvements are being made to CGI to make it more robust for high volumes of transactions or traffic requiring fast and reliable access to multiple databases.

Web Server Architectures

Web server architectures consist of a foundation layer that defines how server program and data modules will interact, along with a rich set of software components that provides services needed across many applications. Typically, these architectures also provide a development environment.

The server architectures that have garnered most attention and market share to date are from major software vendors such as Lotus, Microsoft, Netscape, and Oracle. Their application architectures are being implemented by businesses building corporate intranets and extranets and by a variety of organizations constructing public Web sites.

■ Lotus' Domino

Domino is a Web server that builds on Lotus Notes, the industry's leading groupware product, with its strong integrated authentication, encryption, document-level access control, user directory services, collaborative content authoring, replication, data content storage, and application development environment. Domino also provides typical Internet server facilities such as full text searching and on-the-fly conversion of Notes documents into HTML. Unlike other Web servers, Domino does not provide a CGI interface, but it does include its own Visual Basic-like scripting language, which integrates closely with the server's Notes API for functions including access to relational databases, IBM's MQSeries message-oriented middleware, and other systems. (See Figure 4-13.)

Figure 4-13: Lotus Domino Architecture

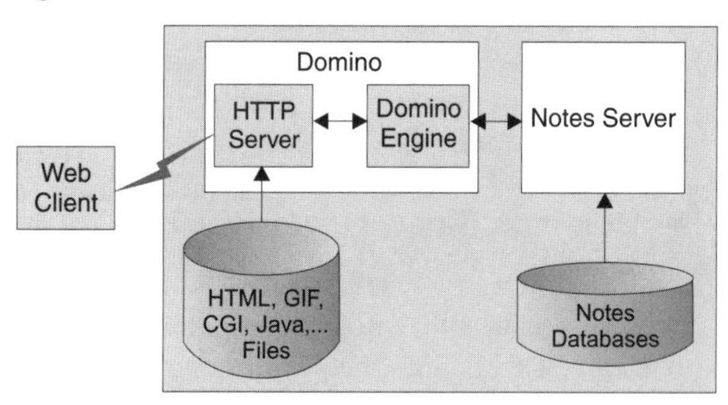

Source: Lotus, 1996

Lotus Domino 4.6 Web Server software, released in late 1997, supports LDAP, Java and ActiveX applets, and JavaBeans. It also provides a set of integrated services, including a workflow engine, server agents to automate frequently performed processes, and enterprise integration to link with real-time and batch-level IT systems and packaged applications. An advanced messaging process acts as a server for SMTP/MIME, X.400, and cc:Mail and can be accessed from a variety of mail clients such as POP3, IMAP4, and MAPI.

Domino is a mature environment with a large installed base and comes packaged with a variety of application frameworks (Notes templates) for threaded discussions, user registration, electronic publishing, marketing, a retail storefront with payment processing, a customer service application, and an internal employee information application.

■ Microsoft's Internet Services Platform

ActiveX is the foundation layer for Microsoft's architecture and is based in turn on the Distributed Component Object Model (DCOM) architecture. Microsoft's Internet Information Server (IIS) is an HTTP server that is tightly integrated with Windows NT. IIS, like Netscape's HTTP server, includes an API (Internet Server API, or ISAPI) for invoking programs without the overhead of CGI. IIS also includes a search engine, streaming multimedia capabilities, and log file analysis tools. Release 4.0 of IIS provides support for multiple Web sites, improved administration and management features, security enhancements, and transaction server integration, making it attractive for ISPs and application developers. It also adds support for HTTP 1.1 so clients can improve performance by pipelining or persistent connections.

The Commercial Internet System is the set of servers that extend IIS to implement directory, mail, news, and other services and to integrate with other Microsoft BackOffice components such as Exchange and SQL Server. Architecturally, this suite corresponds almost component for component to Netscape's Commerce Platform, although it is limited to running on Windows NT. Microsoft also complements its platform with development tools such as FrontPage for HTML authoring; Visual Basic, C++, and J++ for programming; and Studio for interactive page layout. Application frameworks for retail (Merchant, based on technology acquired from eShop) and publishing are built to take advantage of Commercial Internet System services. (See Figure 4-14.)

Figure 4-14: Microsoft's Commercial Internet System

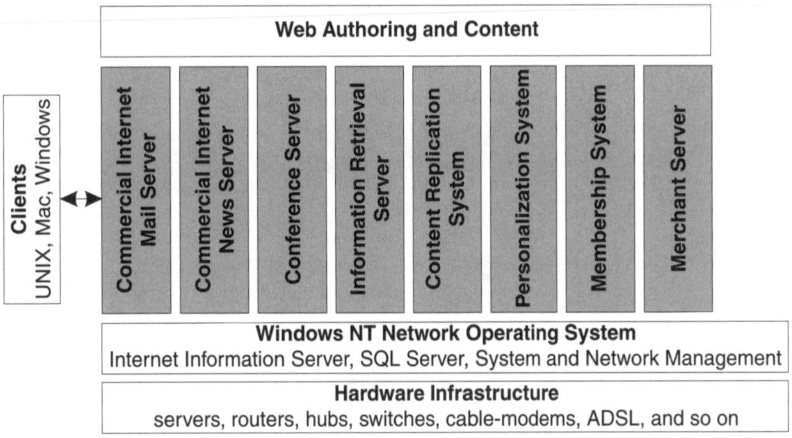

Web Authoring and Content

Clients
UNIX, Mac, Windows

Commercial Internet Mail Server
Commercial Internet News Server
Conference Server
Information Retrieval Server
Content Replication System
Personalization System
Membership System
Merchant Server

Windows NT Network Operating System
Internet Information Server, SQL Server, System and Network Management

Hardware Infrastructure
servers, routers, hubs, switches, cable-modems, ADSL, and so on

Source: Microsoft, 1996

■ Netscape's Open Network Environment (ONE)

Netscape's ONE is based on a set of APIs and class libraries being used to build transaction-oriented Internet and intranet applications. ONE is a foundation layer that specifies how HTML pages, Java, JavaScript programs, plug-ins, and other components and objects will interact with one another in a distributed computing environment. It also specifies interactions with Common Object Request Broker Architecture (CORBA) objects (via IIOP, a protocol that allows objects to communicate over the Internet) and with ActiveX components.

On the server side, ONE specifies the Netscape API (NSAPI), which allows the HTTP server to invoke program components with less overhead than CGI. Several services are specified in the ONE architecture, including the basic HTTP server, directory, mail, news, search, security, and payment. The Internet Foundation Classes (IFC) are Java programs that provide interfaces to the browser GUI as well as to the server components that provide the ONE services. (See Figure 4-15.)

The ONE architecture also includes development tools. Netscape's Enterprise Server comprises a development environment that supports JavaBeans and Java. Netscape's SuiteSpot server family includes Navigator Pro for HTML development and LiveWire and LiveWire Pro for Java for database access.

The ONE architecture and Enterprise Server are extended into a Commerce platform, which in turn supports application frameworks for electronic publishing (the Publishing System), connected communities (the Community System), and retail applications (the Merchant System).

Figure 4-15: Netscape's Open Network Environment Architecture

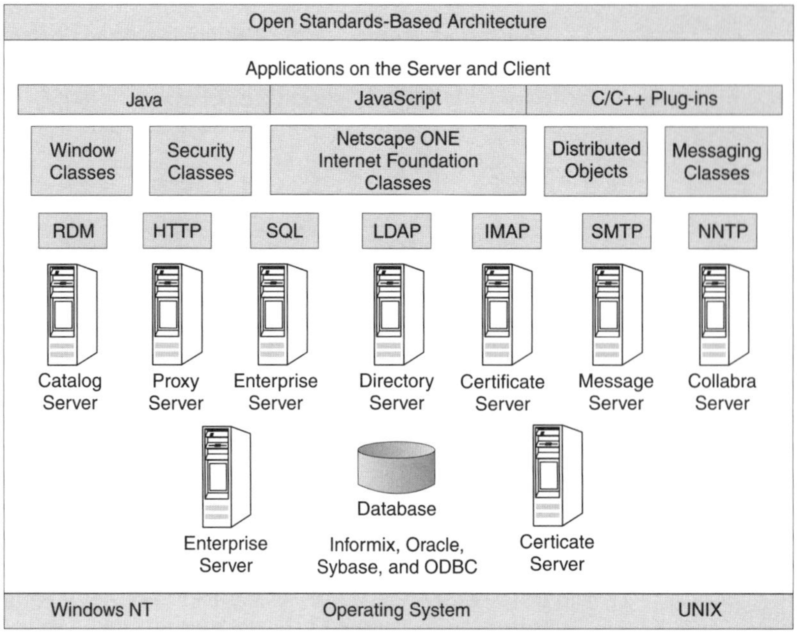

Source: Netscape, 1997

The Netscape SuiteSpot family is a suite of integrated server software for communications, access, and information sharing on an intranet or the Internet. SuiteSpot provides user services such as information sharing and management, communications and collaboration, navigation, and application access. SuiteSpot also provides network services. Network services include directory services that offer centralized "white pages" for an intranet and implement the LDAP standard, replication services that help make data and resources available transparently across the Internet, security features, and management capabilities.

SuiteSpot Standard Edition includes the Calendar Server, Collabra Server, Directory Server, Enterprise Server, LiveWire Pro, and Messaging Server. SuiteSpot Professional Edition contains all the components in the Standard Edition, plus the Certificate Server (to create and manage digital certificates for security), Compass Server (to provide search and indexing capabilities), and Proxy Server (to replicate content throughout an enterprise). SuiteSpot Hosting Edition contains additional functionality for ISPs.

Netscape's FastTrack Server software is a low-end Web server aimed at individuals or small work groups that supports HTTP 1.1. FastTrack Server software is bundled with Communicator for content creation and management, utilizes LDAP for centralized user or group management, and provides a platform for building applications with server-side Java or JavaScript.

■ Oracle's Application Server

Oracle's Internet architecture is oriented toward Web applications built around an RDBMS. The HTTP server itself is robust but otherwise relatively generic; the critical component of Oracle's architecture is the Web Request Broker (WRB). (See Figure 4-16.) The WRB is compliant with the CORBA specification and manages system activities such as load balancing, allocating service resources, and transaction management. Oracle's server is tightly integrated with its database product. Data retrieved from an Oracle database is formatted by the server into HTML.

Figure 4-16: Oracle's Web Request Broker Architecture

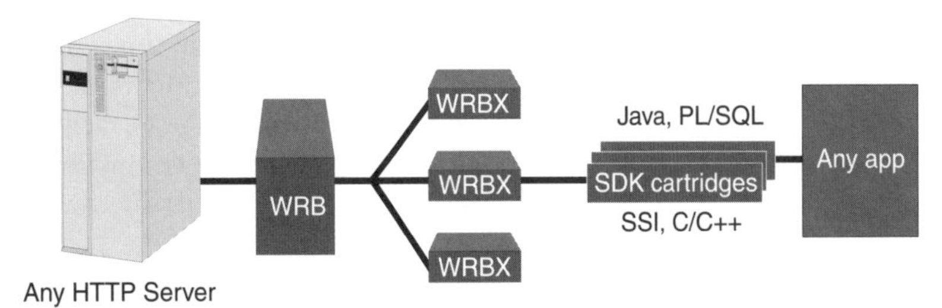

Source: *LAN Times*, 1996

Oracle's Network Computing Architecture supports multi-tier configurations and enables application logic to reside on Web servers, database servers, and application servers. (See Figure 4-17.) To Web-enable an application, developers writing code to the WRB API must build a cartridge, a container for holding a module of code that communicates with the application. Oracle has defined three tiers of objects and related containers within its component framework: the client tier, the middle (or application server) tier, and the data logic (or universal server) tier. The Inter-Cartridge Exchange facility is a CORBA-based object bus that acts as an inter-container protocol to provide services and object brokering among different cartridge types. It provides bridging services such as Java mapping for both CORBA and COM, thereby allowing Oracle's architecture to co-exist within a JavaBeans, Microsoft COM, or other CORBA-based models. Cartridges that support Java, Perl, VRML, and database access are provided with the product. Others are available from third-party vendors.

Figure 4-17: Oracle's Network Computing Architecture

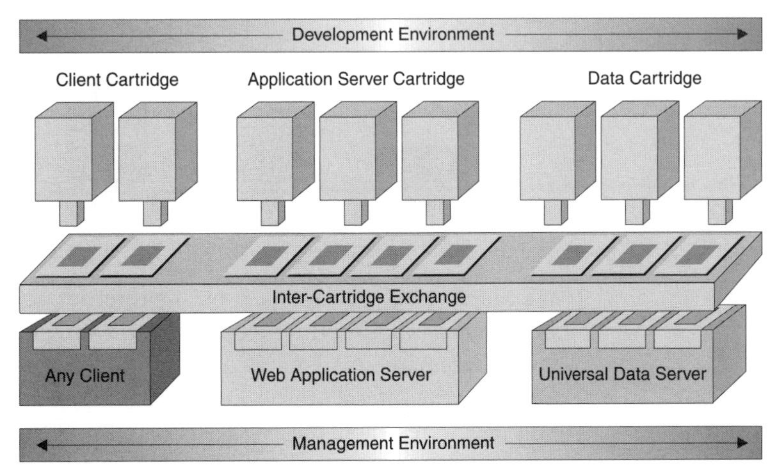

Source: Oracle, 1997

The latest release of Oracle's Application Server is aimed at delivering mission-critical enterprise applications to thin-client or browser-based desktops. Version 4.0's key features include simplified transaction management and WRB administration, support for both Oracle and non-Oracle data sources, and compatibility with other HTTP servers. It is compliant with industry standards, including CORBA 2.0, IIOP, DCOM, and Enterprise JavaBeans.

4.4.11 Market Overview

The Internet market is characterized by growth, investment, creativity, hype, skepticism, and most important, opportunity. Internet and Web market growth will continue as long as businesses can use technology to conduct international marketing at reduced cost, gather more in-depth information about customers' needs and preferences, cut production time and costs associated with development and distribution of paper documents, develop low-cost distribution channels for software and other products, facilitate business communications and transactions, conduct research, demonstrate products, recruit employees, and for other uses that cut costs or increase revenues.

The expansion of the Internet can be viewed from many perspectives. In January 1997, the number of Internet hosts reached 16.1 million, which represents a 25 percent growth rate in 6 months, according to figures from CyberAtlas and Network Wizards. For comparison, the growth rate over the preceding 6 month period was 36 percent, suggesting that the rate at which computers are being linked to the Internet has slowed somewhat. (See Figure 4-18.) The increase in the number of Web pages continued unabated, however: the number tripled (to about 80 million) in the 6-month period ending January 1997, having only doubled in the 6 months prior to that.

Figure 4-18: Growth in Web Hosts

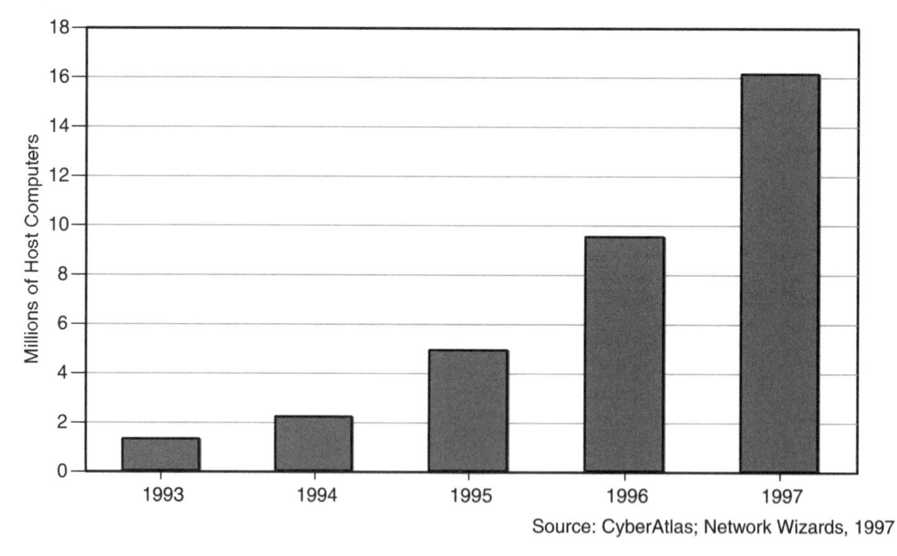

Source: CyberAtlas; Network Wizards, 1997

In a March 1997 special report on the Internet, *Scientific American* estimated there were 54 million Internet users worldwide, with an additional 14 million having access to e-mail only. In the U.S., 16 percent of the population older than are 15 use the Internet or on-line services. The average Internet user in the U.S. has come to resemble the average person more closely, though significant differences remain. For example, a May 1997 survey by the Graphics, Visualization, and Usability Center (GVU) at the Georgia Institute of Technology found that 30 percent of those using the Internet worked in computer-related fields.

Estimates of this population vary considerably, but researchers agree that Web use still is growing and well below its potential size. Estimates in 1997 for Internet users just in the U.S. ranged from 35 to 45 million. Exact numbers for current and projected users still are difficult to determine. (See Figure 4-19.)

Figure 4-19: Internet Users

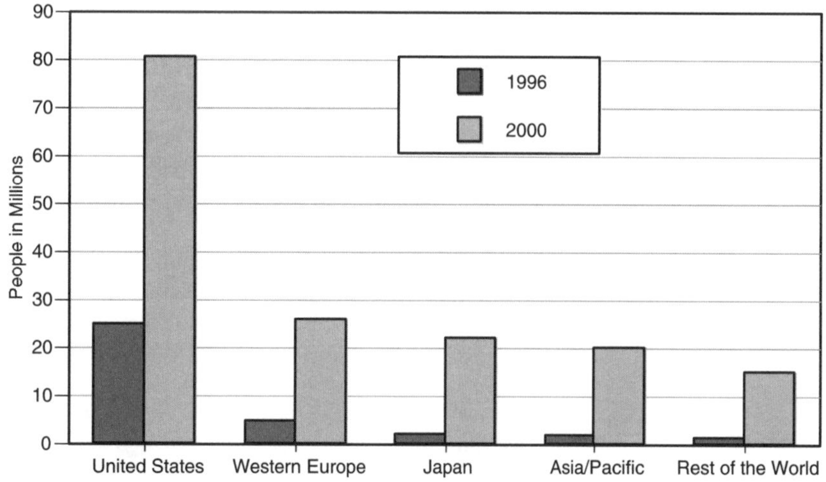

Source: IDC, 1996

Internet services constituted a $35 billion market at the end of 1996, according to Zona Research. The major share, 58 percent, went to providers of the communications infrastructure. Internet computer software and hardware accounted for an additional 22 percent of that market.

Market researchers are predicting an enormous market for Web commerce. Predictions about the size of the market for on-line shopping for goods and services vary widely, in part because of what is included in this definition. In a May 1996 report, Forrester Research predicted that Internet services revenue would grow to $30 billion – including transactions, hosting, and access – in the U.S. by the year 2000. Of that amount, revenue from transaction services will account for the smallest segment, just $1.5 billion.

In June 1996, Forrester predicted that in the year 2000, 20 percent of all U.S. adults will be on-line and that revenues will begin to exceed costs for content providers. At the typical Web content site, advertising will continue to dominate revenue streams.

Not only is mass-market penetration of the Internet continuing, but the adoption of diverse services within the Internet also is growing, which will continue to provide opportunities for innovative marketing of new information services. According to an IDC survey, searching for pull-type information and exchanging e-mail remain by far the largest uses of Web browsers. Although these activities will continue to predominate, there is room for expansion of alternative models such as electronic publishing and push distribution. Challenges to overcome include the necessity of finding revenue streams other than advertising. The GVU survey reports widespread reluctance among Internet users to pay for access to Web-based information.

Zona predicts a fourfold increase in revenue for content providers (including on-line services) between 1996 and 2000. The market for Web authoring and application development tools is expected to double over the same period.

Market Segments

The Internet market can be divided into several different tiers, or value chains, each of which has multiple players. Technology providers form one tier that includes companies providing hardware and software as well as the related services necessary to build the Internet and the applications that run on it. The next tier of the market comprises the distribution channel and the integration services associated with technology deployment. This segment includes dealers, resellers, systems integrators, and consultants. The third tier comprises the service providers, including ISPs, and providers of related services such as Web hosting, credit card transaction processing, digital cash, and digital certificates. Participants in this tier purchase products and services from the lower two tiers, add value, and provide higher-level services to the top tier.

At the top tier are the companies that use the Internet for business-to-business transactions as well as companies selling goods and services to consumers. Also included are companies that provide aggregation points, such as information broadcasters, clearing-houses, and shopping malls.

The growth of the Internet has served as a driver for server, software, and networking purchases over the last few years. The early beneficiaries of the Internet's rapid growth have been the suppliers of infrastructure technology, tools, and server technology, including software and hardware. This will remain a viable market as demand continues to increase for bandwidth, driven by new audio, video, and telephony applications and with new forays into electronic commerce.

Evolving International Markets

Early Web usage numbers have been highest in the U.S., which is why an estimated 80 to 90 percent of the content on the Web is in English. The Internet and many supporting telecommunications technologies were developed in the U.S., graphical Web browsers were first launched and widely adopted there, and more homes in that country have PCs. Web use is expected to grow in the rest of the world as browsers adopt support for extended character sets such as Unicode, as telecommunications services become more accessible and inexpensive, and as more content in other languages becomes available. (See "International Developments" on page 240.)

Browsers

Revenue for this segment of the Internet market is insignificant compared with other areas such as server software, connectivity, content development tools, and content.

The network browser market has grown rapidly, but revenue numbers have not always been an accurate reflection of market share. The browser market has matured and changed since the first graphical browser, NCSA's Mosaic, was introduced in 1992. Netscape's strategy of distributing and frequently enhancing products for little or no money enabled the company to capture 70 to 90 percent of the browser market by the end of 1995. Netscape used its Navigator browser to establish market presence quickly and hoped to collect more revenue for server software later.

In 1995 and 1996, however, users had additional ways to obtain low-cost or no-cost browser software. For example, Microsoft announced it would bundle Internet Explorer into its Windows 95 operating system. Before Microsoft released Internet Explorer 3.0 in August 1996, many analysts believed that Netscape had an enduring stranglehold on the Internet market. Netscape appeared to lead other browser vendors not only in market share but also in feature development, and content developers seemed well advised to focus on the Netscape browser first, if not exclusively.

Instead, the release of Internet Explorer 3.0 heightened competition over browser features and began a downward trend in Netscape's market share. Analysts began to predict that Microsoft eventually would dominate the browser market.

In a series of monthly *Computer Reseller News* surveys during the first half of 1997, market share numbers for Netscape fluctuated between 53 and 67 percent, often as new features or other news was announced. Numbers for Microsoft's Internet Explorer fluctuated between 28 and 36 percent. In these surveys, the overall gap in market share between Microsoft and Netscape diminished, and the use of smaller vendors' browsers decreased. The survey findings also pointed out that some users were tending to standardize on one browser and de-install any others.

In September 1997, Zona published the results of a survey on corporate browser usage. Of those respondents saying that their companies encouraged the use of a specific browser (about 52 percent), 55 percent were using Netscape Navigator and 32 percent were using Internet Explorer. (Approximately 13 percent of respondents selected "unknown.")

In November 1997, Dataquest estimated browser user market share between just Netscape and Microsoft at 57.6 percent and 39.4 percent, respectively.

One significant trend influencing the development of browsers and extensions is the increasing corporate use of intranets. The overall browser market will grow due to the intranet boom over the next few years, and vendors who can capitalize on this trend will gain market share. IDC predicts the number of installed intranet browsers will grow from 20 million in 1996 to almost 100 million by the end of the year 2000. By that time, revenue from the use of these products in an intranet environment will outstrip Internet usage by a factor of six to one.

For 1997 and beyond, however, licensing revenue from browsers is expected to grow more slowly than usage. Zona Research forecasts that revenues from browser shipments will increase to $292 million in 1997, $351 million in 1998, and $387 million in 1999. (It may become more difficult to break out some revenue, for example from Microsoft's Internet Explorer, if it becomes part of future operating systems.) Even so, this level of revenue is insufficient to fund the level of support users will demand as Web applications become more complex and more critical to their daily lives. Therefore, Web browser support may come from third parties rather than from browser vendors. This trend is similar to what occurred in the PC market, where corporations needed to provide their own integration and support as PCs became commodity products.

Analysts expect revenues from other sectors of the Internet market to dwarf browser income. Ultimately, the winners of browser market share (as quantified by usage, not revenue) will be those vendors who can attract support from the most popular content providers.

Search Tools

The market for network search tools is divided into two parts: providers of search-and-retrieval sites on the Web and providers who sell search technology for inclusion in other products and services.

Popular search Web sites – Digital's Alta Vista, Excite, Infoseek, Lycos, and Yahoo! – use a business model that allows users to search for free. This model is based primarily on advertising revenue and its success has been tracked carefully by industry observers. These companies have two main sources of revenue: advertising from traditional advertisers and from Web sites that wish a preferential position in the search site's table of contents. When companies such as Yahoo! went public and became profitable in 1997, this model became much more accepted.

Beginning in 1996, several of these companies stepped up their marketing of search engine technologies for use in other Internet and intranet sites. For example, Excite made its search engine software available for download from the Internet, hoping to build market share that would lead to technology licensing fees and added popularity for the Excite site. Some sites, such as Alta Vista, rely on pure search capabilities, while others, such as Yahoo!, offer a variety of entries under subject category or directory headings.

Most of the directory-oriented companies expanded their tables of contents for the Web, adding portions that organize the content by topics of interest, such as news, sports, and hobbies. Consumers visiting these sites can use either the search engine or the table of contents. In essence, these sites positioned the companies as Web content aggregators: central locations from which consumers could find whatever they were looking for.

Search engines for corporate intranets are less useful when indexed content is limited to HTML pages and information from other corporate databases is unavailable. Vendors are enhancing their products to accommodate the need to search legacy data using a Web browser. Advances include support for different types of data, ODBC compliance, and improved administrative control. Public Web-search companies have improved their intranet offerings. A new group of startup companies, such as Autonomy, WiseWire, and Perspecta, are attempting to merge search engines with collaboration and push technology.

Providers of Web search engine technologies also expanded their reach by forming alliances with major system and service providers. Fulcrum Technologies formed alliances with CompuServe (now part of AOL), Fujitsu, Microsoft, and Novell; Verity allied with AT&T, Attachmate, Individual, Lotus, Netscape, Network News, Questal Online, and Tandem. Yahoo! initially adopted Open Text as its search technology provider, then switched to Digital's Alta Vista technology software in 1996.

Forrester Research predicts that search sites increasingly will become the navigation hubs for the Web. As content aggregators, they already attract partnerships with popular content providers in addition to registration from Web sites wishing to draw traffic from the search engines. With the resulting traffic from millions of Web surfers who want to find the most useful and entertaining content, these sites stand to become Web centers with considerable sway with advertisers and content providers.

In 1997, push receivers were included as standard components in the two most popular browsers: Microsoft bundled the PointCast client with Internet Explorer 4.0, and Netscape included a Castanet Tuner with its Communicator product. Microsoft plans to integrate its browser technology with Windows 98, thereby providing a desktop window for push reception. In 1997, Forrester Research predicted corporate push revenues of more than $300 million by 2001, with an installed base of 140 million push-enabled browsers by then.

Internet Voice, Video, and Multimedia

The embryonic state of these technologies currently precludes accurate estimates of future markets. Too much hinges on critical, as yet unanswered, questions of bandwidth limits, standards and interoperability, reliability, consumer acceptance, regulatory responses, and competitor reactions. As a result, forecasts of growth in these markets remain variable and should be regarded as preliminary. For instance, although Forrester predicts no foreseeable near-future market gains for Internet telephony from telcos, Killen & Associates predicts $63 billion in revenues within 5 years.

Intranets and Extranets

In a study of corporate intranets, IDC has found returns on investment in excess of 1,000 percent. Many key players therefore predict massive growth rates for intranet and extranet adoption. Forrester, for instance, predicts nearly half a million intranet sites by 2000. IDC and most market analyst firms believe the number of intranet servers installed by 2000 will far exceed the number of Internet servers in use then. Zona figures show a high rate of overall growth especially for Windows NT-based servers. (See Figure 4-20.)

Figure 4-20: Internet and Intranet Server Shipments

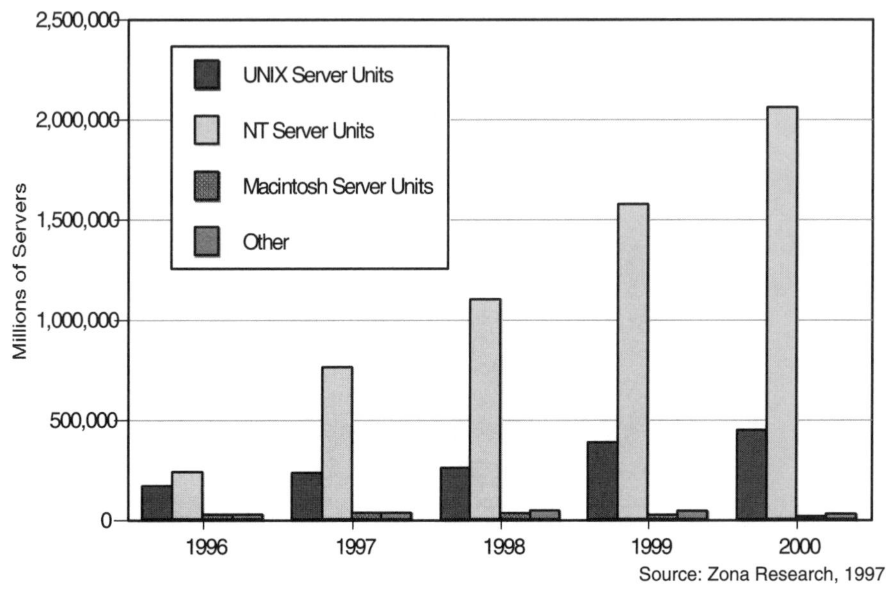

Source: Zona Research, 1997

4.4.12 Forecast

General

- New applications of Web technology will appear continually. Due to the low cost of entry and the ease with which new applications can be developed and modified, new products and services will appear daily. Experimentation will continue throughout the forecast period.
- Web use has yet to reach the saturation point. During the forecast period, Web commerce will reach the majority of high-income, technically savvy users in the U.S.
- Traditional network vendors and service providers will play an increasingly important role as suppliers of new Internet services.

- Growth of the number of computers connected to the Internet and the number of people accessing the Internet remains substantial. As Internet access, particularly access at higher speeds, becomes available to more users worldwide, the value of the Internet as a media for entertainment, commerce, and communications will continue to grow.

- The demand for higher reliability and security in Web-based applications will grow as more critical applications move to corporate intranets and the Internet. This demand will be met through improvements in network infrastructure as well as new protocols and standards.

- Consolidation among ISPs and OLSs will continue throughout the forecast period.

Internet Standards and Technologies

- Rapid evolution in standards and protocols will continue, driven by the need to standardize innovative technologies created and released by vendors and by the demand for improved services. Given the rapid rate of new product development, inconsistent approaches by different vendors trying to address the same need will continue to cause problems in areas ranging from development environments and browser plug-ins to directory services.

- XML will begin to be used in 1998, with one of the initial applications being Web sites offering consumer-oriented financial services due to XML's ability to allow the user to manipulate data within the browser. XML also will provide collaboration facilities for use primarily in the scientific or technical communities. However, adoption of XML will not eliminate the need for downloadable applets.

Server Software

- Servers will continue to add support for personalization and customization of the Web.

- Web server software increasingly is becoming bundled with other server software, including e-mail, news, proxy, directory, security, and database server functionality. These products will be marketed as a "server in a box" for specific storefront or e-commerce applications, or for intranet use.

- The concept of Web servers as unique entities will disappear as the HTTP protocol becomes a standard functional component embedded in most servers. Tools for site creation, administration, and management will be incorporated.

- High-end server software vendors will focus on improved scalability and reliability to be able to handle enterprise-level transaction volumes.

- Wider deployment of HTTP 1.1-compliant servers will allow developers to make effective use of proxy caching and persistent connections. For corporate intranets, this capability will reduce overhead and decrease utilization of network resources, as caches of retrieved or "pushed" content from the Internet are collected and stored inside the firewall.

Browsers

- Browsers will become standard offerings on all client computing platforms and increasingly will be integrated with system software and popular applications. The degree of bundling or integration will be influenced by the outcome of the U.S. Justice Department's case against Microsoft.

- More applications will evolve that use the Web browser as their delivery platform. This trend has the potential to change the computer platforms and operating systems customers purchase. Differences between client operating systems could become less important in this environment.
- For the next few years, the development of new browser features increasingly will be influenced by the demands of corporate intranets, with the emphasis on office document sharing, workflow, and work-group collaboration. An increasing number of corporate applications will include interfaces to the Web that can be accessed by these customized browsers.
- Web pages increasingly will become dynamic and animated with the incorporation of technologies such as DHTML, Java, and ActiveX.
- Netscape and Microsoft will remain the leading vendors of Web browsers. Microsoft could ultimately capture a majority of the market in the next few years due to the bundling of Internet Explorer with Windows and the planned integration of Internet Explorer into Windows 98. Much will depend on the outcome of anti-trust actions taken by the U.S. government.
- Innovation in browsers will encourage the rapid adoption of new HTML standards. However, development of new nonstandard extensions to HTML will continue, threatening the credibility of HTML standards as they lag behind. Browsers will incorporate support for XML, but XML primarily will be used by highly sophisticated Web sites.
- In 1998, new mark-up language technologies for content creation and distribution (HTML 4.0, CSS, DHTML, and XML) will be used in various combinations. As they become better understood and their usage more widespread and less experimental, their relative strengths and weaknesses will emerge and appropriate deployment decisions will be made.
- Browser features and extensions that help users navigate the Web, beyond simple bookmarks, will develop and flourish. These features are likely to include Web site-map visualization and outlining aids, intelligent agents, and client-side search engines. Functionality incorporated into browsers to support push technology will continue to improve.
- The distinctions between Web content and applications and between content that resides on the desktop, intranets, and the Internet will continue to blur from the end user's perspective.
- Browsers will reach out to a new audience via Internet appliances, increasing the difficulties of authoring for the widest possible audience. New gaps between the features of different browsers will appear as some extensions reach toward the living room audience and others toward the desktop.
- The increasing complexity of browser configuration (where plug-ins and helper applications are required) will prove an obstacle to consumer use of the Web and a challenge to ISP and corporate technical support organizations.

Search Engines

- Query features will improve to encompass natural language capabilities, more incremental searches, and the ability to delimit the scope of a search. Search queries for particular Web sites also will be enhanced by expert system-like dialogs that help users hone queries, especially in areas such as vendor technical support, where a high degree of precision is required.

- Presentation of search results will become more intelligent, with refined heuristics for rating the relevance of material and with consolidation of content from the same or similar Web sites.
- Distributed models for indexing the Web will emerge, enabling Web sites to create and control the indexes of their contents that are used by search engines. This trend will create additional opportunities for differentiating between search engine brands.
- Searching visual, multimedia, and interactive content will remain largely text-based, with search engines relying on the textual descriptions attached to these media, within the forecast period.

Internet Multimedia

- Internet telephony will remain primarily a consumer technology. Its growth initially will be hindered by product incompatibilities and usability problems (quality, reliability, performance, and ease of use).
- Transmission of faxes over the Internet (rather than the telephone network) to avoid long-distance charges will experience significant growth. For business users, this will be a more significant use of the Internet than voice.
- Demand for multimedia and video content will remain high, but bandwidth limitations will restrict their use to internal or specialized networks. Wider adoption of these technologies will remain tied to the spread of 56-Kbps modems, ISDN, and new higher-speed access technologies such as cable modems and xDSL.

Intranets and Extranets

- The use of intranets will continue to expand significantly for the same reasons that spurred the growth of the Internet itself: open protocols, widely available products, and ease of use. Over the next few years, vendor revenues from intranet use will surpass those from Internet use.
- Driven by the competitive advantages of extranets, more organizations will open their networks to contractors, suppliers, and customers. An increasing understanding of network security issues will assuage doubts about increasing outside access.

4.4.13 References

■Articles

DeVoe, Deborah. 1997. When push comes to shove. *InfoWorld.* February 17: 1.
The Economist telecommunications survey: a connected world. 1997. *The Economist.* September 13.
Kalish, David E. 1997. Spinning Webs in the living room. *San Francisco Examiner.* September 16: C2.
Levitt, Jason. 1997. Rift over HTML's dynamics. *InformationWeek.* April 7: 94.
Maddox, Kate. 1997. Online data push. *InformationWeek.* February 24: 61.
Marshall, Jonathan. 1997. Internet phone calls getting easier. *San Francisco Chronicle.* August 19: C1.
Schwerin, Rich. 1997. XML in center stage. *PC/Computing.* November: 435.
Tools: what you should know about HTML 4.0. 1997. *Wired.* August 11.
Web gold! 1997. *Computer Currents.* September 16: 51.

■Periodicals, Industry and Technical Reports

Cove-Brigham, Joan. March 1997. *Web browser market share, March 1997.* IDC.
Directions 1997: follow the consumer. April 1997. IDC.
GVU Internet survey. 1997. Graphics, Visualization, and Usability Center, Georgia Institute of Technology.
Internet and intranet: 1997 markets, opportunities, and trends. 1997. Zona Research.
McKenzie, Matt. 1997. *Cascading style sheets: coming soon to a Web page near you.* Patricia Seybold Group. June: 3.

Roussel, A. 1997. *Internet adoption worldwide, Part 1.* Gartner. March 25.

Sullivan-Trainor, Michael and Ted Julian. 1997. *Reversing the Web: maximizing the value of Webcasting.* IDC. April.

■Books

Hahn, Harley. 1996. *The Internet yellow pages.* Berkeley, Calif.: Osborne/McGraw Hill.

Minoli, Daniel. 1997. *Internet and intranet engineering.* New York: McGraw Hill.

■On-Line Sources

Excite launches free Web clipping service. 1997. Reuters. February 4.

Patel, Jeetil J. September 1997. *iNumbers, Vol. 2, Issue 4.* The Hambrecht & Quist Internet Research Group. `http://www.iword.com/iword24/inumbers24.html`

■URLs of Selected Mentioned Companies

Accent Software `http://www.accentsoftware.com`
Adobe `http://www.adobe.com`
Alis Technologies `http://www.alis.com`
All4one `http://all4one.com`
ANS Communications `http://www.ans.com`
Apple `http://www.apple.com`
Asymetrix `http://www.asymetrix.com`
AT&T `http://www.att.com`
Attachmate `http://www.attachmate.com`
BBN Planet `http://www.bbnplanet.com`
Bell Atlantic `http://www.bellatlantic.com`
Berkeley Systems `http://www.berkeleysystems.com`
Brooktrout Technology `http://www.brooktrout.com`
Caligari `http://www.caligari.com`
Castelle `http://www.castelle.com`
CBS `http://www.cbs.com`
CERN `http://www.cern.ch`
Cisco `http://www.cisco.com`
Compaq `http://www.compaq.com`
CyberAtlas `http://www.cyberatlas.com`
CyberPatrol `http://www.cyberpatrol.com`
Delta Three `http://www.deltathree.com`
Demon Internet `http://www.demon.net`
Disney `http://www.disney.com`
Dow Jones `http://www.dowjones.com`
Emergent `http://www.emergent.com`
eShop `http://plaza.msn.com/msnlink/index.asp` (The Microsoft Plaza)
Firefly Network `http://www.firefly.com`
Fox `http://www.fox.com`
FTP `http://www.ftp.com`
Fujitsu `http://www.fujitsu.com`
Fulcrum Technologies `http://www.fulcrum.com`
FutureTense `http://www.futuretense.com`
GM `http://www.gm.com`
H&R Block `http://www.hrblock.com`
HTTP 1.1 Implementer's Forum
`http://www.w3.org/pub/WWW/Protocols/HTTP/Forum`
Ibex Technologies `http://www.ibex.com`
IBM `http://www.ibm.com`
Individual `http://www.individual.com`
Inference `http://www.inference.com`
Informix `http://www.informix.com`
Inso `http://www.inso.com`
Intel `http://www.intel.com`
International Multimedia Teleconferencing Consortium `http://www.imtc.org`
Internet Engineering Task Force `http://www.ietf.org`
Internet2 `http://www.internet2.edu.`
Kobixx Systems `http://www.kobixx.com`
Kodak `http://www.kodak.com`
Lotus `http://www.lotus.com`
Macromedia `http://www.macromedia.com`
Marimba `http://www.marimba.com`
MCI `http://www.mci.com`
MCNC `http://www.mcnc.org`
Mediamark Research `http://www.mediamark.com`
Metacrawler `http://www.metacrawler.com`
Microsoft `http://www.microsoft.com/microsoft/htm`
Motorola `http://www.mot.com`
MSNBC `http://www.msnbc.com`
NASA `http://www.nasa.gov`
National Science Foundation `http://www.nsf.org`

NEC *http://www.nec.com*
NetCentric *http://www.netcentric.com*
NetCom *http://www.netcom.com*
NetNanny *http://www.netnanny.com*
Netscape *http://www.netscape.com*
NetSpeak *http://www.netspeak.com*
Network Appliance *http://www.networkappliance.com*
Network Wizards *http://www.netwizards.com*
Newbridge Networks *http://www.newbridge.com*
News Corp. *http://www.newscorp.com*
Nissan *http://www.nissan.com*
Norton *http://www.norton.com*
Novell *http://www.novell.com*
Open Port Technology *http://www.openport.com*
Open Text *http://www.opentext.com*
Oracle *http://www.oracle.com*
ParaGraph International *http://www.paragraph.com*
PBS *http://www.pbs.org*
PeakSoft *http://www.peaksoft.com*
Perspecta *http://www.perspecta.com*
PointCast *http://www.pointcast.com*
Premiere WorldLink *http://www.premiereworldlink.com*
Preview Travel *http://www.previewtravel.com*
Radiance Software *http://www.radiance.com*
RandomNoise *http://www.randomnoise.com*
RealNetworks *http://www.realnetworks.com*
Scientific American *http://www.scientificamerican.com*
Silicon Graphics *http://www.sgi.com*
SoftQuad *http://www.softquad.com*
Solid Oak Software *http://www.solidoak.com*
Sportsline USA *http://www.sportsline.com*
Sprint *http://www.sprint.com*
Spyglass *http://www.spyglass.com*
Sun *http://www.sun.com*
Sybase *http://www.sybase.com*
Tandem *http://www.tandem.com*
The Wall Street Journal *http://www.wsj.com*
Time-Warner Cable *http://www.time-warner.com*
Toshiba *http://www.toshiba.com*
Traveling Software *http://www.gowebex.com*
US West *http://www.uswest.com*
USA Global Link *http://www.usagloballink.com*
VDOnet *http://www.vdo.net*
Verity *http://www.verity.com*
Virtus *http://www.virtus.com*
Vivo *http://www.vivo.com*
VocalTec *http://www.vocaltec.com*
Vosaic *http://www.vosaic.com*
WAIS *http://www.wais.com*
Wayfarer Communications *http://www.wayfarer.com*
WebTV Networks *http://www.webtv.com*
WiseWire *http://www.wisewire.com*
World Wide Web Consortium *http://www.w3.org*
Worldwide Broadcasting Network *http://www.wbnet.com*

5 Systems and Architectures

5.1 Operating Systems Environments

5.1.1 Executive Summary

An operating system is a layer of software that resides between computer hardware and applications such as e-mail systems, word processors, databases, and so on. An operating system manages processes (such as scheduling print jobs) and the flow of data to connected devices (such as a disk drive). It also offers various services that applications can access via application programming interfaces (APIs). On desktop computers and departmental servers, an operating system typically has a graphical user interface (GUI).

For the sake of simplicity, analysts and users generally divide the operating system market into three categories: desktop, departmental, and enterprise. Several factors influence an operating system's acceptance. On the desktop, these factors typically include applications support and network integration. In addition to these criteria, departmental and enterprise operating systems usually are chosen for scalability, reliability, and performance.

On the desktop, migration from 16-bit operating systems to modern 32-bit operating systems is underway within most (but certainly not all) businesses. Most 32-bit operating systems support networking, multitasking of applications, basic security, and Internet connectivity via bundled Web browsers. In some cases, operating system user interfaces are moving from the established "desktop/file/folder" metaphor to a browser metaphor. The latter metaphor lets a user retrieve data whether it resides on a local hard drive or on a remote server, such as a Web server. Most operating systems also are gaining a Java Virtual Machine (JVM), which lets users run programs written in Sun's Java language.

5.1.2 Departmental and Enterprise Server Operating Systems

Departmental server operating systems are consolidating rapidly around Microsoft's Windows NT Server, Novell NetWare (formerly called IntranetWare), and several versions of UNIX. Additional offerings, such as IBM OS/2 Warp Server, maintain a strong presence in traditional IBM markets such as banking and insurance.

The Internet continues to shape the departmental server operating system landscape. Most such operating systems now come bundled with Web server software and support for TCP/IP (the networking standard of the Internet). Networking, file, and print services are standard, although departmental server operating systems also support application services for running mail software, databases, and so on.

In the coming years, departmental server operating systems will attempt to scale to higher workloads by gaining improved support for clustering, message queuing, transaction processing, symmetric multiprocessing, secure electronic commerce, and 64-bit processors.

UNIX is widely considered the most scalable departmental server operating system. Digital, HP, IBM, the Santa Cruz Operation (SCO), and Sun's SunSoft subsidiary (among others) each offer slightly different UNIX products. Most vendors selling UNIX-based systems (with the notable exception of Sun) are developing Intel systems that support Windows NT Server. However, UNIX sales continue to grow. SunSoft's Solaris, HP-UX, and IBM's AIX are particularly popular as a database and Internet/intranet server, and SCO's OpenServer and UnixWare are the volume UNIX offerings for the Intel marketplace.

Windows NT Server continues to gain popularity. More than 1 million units were sold in 1997, making Windows NT Server the highest-selling departmental server operating system during that period. Windows NT Server is a general-purpose operating system that supports file, print, application, and Web services. To date, it lacks the scalability, reliability, and performance of top-tier UNIX offerings, but planned enhancements from Microsoft and its partners (Compaq, Digital, and HP, among others) could help Windows NT Server approach UNIX scalability by the year 2000 or shortly thereafter.

Novell NetWare remains the most widely deployed departmental server operating system, with more than 55 million desktops linked to Novell servers. NetWare continues to offer unsurpassed file, print, and directory capabilities, but so far it lacks adequate applications and will not gain native TCP/IP support until mid-1998. Novell also hopes to offer Java application services and advanced clustering capabilities within the same time frame.

IBM continues to enhance OS/2 Warp Server (like Windows NT Server, it offers file, print, and application services). In some ways, OS/2 Warp Server goes beyond Windows NT Server because it offers extensive, bundled systems management tools and mainframe connectivity. (Windows NT Server requires add-on products to provide those features.)

Apple is expected to roll out Rhapsody sometime in mid-1998. Rhapsody is based on UNIX-like technology acquired from NeXT. The operating system is expected to target desktop as well as modest departmental servers based on the PowerPC microprocessor.

IBM continues to dominate the enterprise operating system market with OS/390 and OS/400. Additional offerings include IBM VM and VSE as well as Digital OpenVMS. IBM has transformed OS/390 and OS/400 successfully from traditional operating systems to true client/server operating systems, which let personal computers (PCs) and other desktop systems access distributed graphical applications. Although some server operating systems, such as Solaris, have scaled into the enterprise operating system market, MVS remains unmatched in terms of its ability to manage large volumes of data and transactions. IBM has repositioned OS/390 and OS/400 as open platforms that support traditional UNIX applications. Digital OpenVMS remains a viable offering for established Digital customers as is Digital UNIX, but Digital's stated long-term direction is enhancing Microsoft's Windows NT Server.

5.1.3 Operating System Basics

All programmable and extensible computing devices, from set-top boxes to mainframes, use an operating system. Hundreds of niche operating systems exist, including embedded operating systems used in factory machinery. For simplicity, this chapter focuses on high-volume offerings, including the following:

- **Desktop operating systems** – DOS, Windows 3.x, Windows 9x, Windows NT Workstation, OS/2 Warp, MacOS, Rhapsody, and UNIX
- **Departmental server operating systems** – NetWare, Windows NT Server, OS/2 Warp Server, and UNIX

- **Enterprise operating systems** – MVS, OS/400, and OpenVMS

An operating system offers a layer of abstraction that frees independent software vendors (ISVs) from worrying about the details of particular hardware. This goal is achieved through a set of operating system APIs, which an application can use to request services from the operating system, rather than make direct calls to hardware and handle low-level details, such as which type of disk or disk controller is installed on a computer. For Windows 3.x, Windows 9x, Windows NT, OS/2, and MacOS, the API also enables applications to manage windows, menus, icons, and other elements of a GUI. A single operating system can support multiple APIs, and the same API set can be supported by multiple operating systems.

Operating System Facilities

Desktop, departmental server, and enterprise operating systems provide several services to the computer. Generally, the range of services and their robustness increase when moving from the desktop, to a departmental server, and finally to an enterprise server. (See Table 5-1.) In effect, a desktop operating system generally offers a subset of the functionality provided by a server operating system and an even smaller subset of the features found in a server/host operating system. (The notable exception is the user interface, which is most sophisticated on clients and least sophisticated on host operating systems.) Enterprise server operating systems are generally the most mature operating systems and were designed from the beginning to support thousands of users and transactions simultaneously. By contrast, traditional departmental server operating systems typically support from a few dozen to a few hundred users at best. Desktop operating systems are generally less mature and therefore less stable and were designed to support a single user or a small work group of users.

Table 5-1: Characteristics of Major Operating Systems

	Preemptive Multitasking	Multi-threading	SMP	Virtual Memory	Memory Protection	Advanced Security	Basic Networking	Window System	Portability
MS-DOS	No	No	No	No	No	No	No	No	No
Windows 3.x	No	No	No	No	No	No	No	Yes	No
Windows 9x	Yes (Win32 only)	Yes	No	Yes	Yes (Win32 only)	No	Yes	Yes	No
Windows NT	Yes	Yes	Yes	Yes	Yes	Yes	Yes	Yes	Limited
OS/2 Warp	Yes	Yes	Yes	Yes	Yes	Yes	Yes	Yes	No
MacOS	No	No	No	Yes	No	No	Yes	Yes	No
Rhapsody	Yes	Yes	Yes	Yes	Yes	Yes	Yes	Yes	Yes
UNIX	Yes	Yes	Yes	Yes	Yes	Yes	Yes	Yes	Yes
NetWare/ IntranetWare	Limited	Limited	Yes	mid-1998	mid-1998	Yes	Yes	mid-1998	No
OS/390	Yes	Yes	Yes	Yes	Yes	Yes	Yes	No	Limited
OS/400	Yes	Yes	Yes	Yes	Yes	Yes	Yes	Yes	No

The basic services provided by an operating system include process management, memory management, cache consistency, I/O, file system management, and clustering support. More details of other services are provided in Table 5-1.

Process Management

Process management means managing the program or programs running on the processor at a given time. In the simplest case, such as MS-DOS, the operating system provides a mechanism for a program to load its executable image from a storage device into main memory and then begin executing, with the program continuing to occupy resources and execute until it relinquishes control of those resources. However, more-robust operating systems offer several more sophisticated forms of process management:

- **Multitasking** – The management of two or more processes or tasks running on the computer system at the same time. On a single-processor system, multitasking is made possible by sharing processor time among a series of programs. Each program is run in sequence for a very short time. Hence, the system gives the appearance that it is managing several processes or tasks at the same time. There are two types of multitasking: preemptive and cooperating.
  - **Preemptive multitasking** – Allows a program that is currently running to be involuntarily suspended and another program run instead, either because the first program's processor time has expired or because a second program with a higher priority has interrupted it. Most modern server operating systems support preemptive multitasking. However, the systems vary widely in their ability to exploit additional processors.
  - **Cooperating multitasking** – Allows multiple programs to run simultaneously but relies on the program to voluntarily relinquish control periodically. Because the system has no way to preempt the running program, there is no guarantee that other programs actually can run on the processor simultaneously. Windows 3.x, Windows 9x (for 16-bit applications), Novell NetWare, and MacOS support cooperating multitasking.
- **Multithreading** – A form of multitasking that focuses on running multiple tasks within a single application simultaneously. To support multithreading, a program itself must be designed to support multiple threads of execution. For example, a word processor application must support multithreading in order to edit one document while another document is being spell checked. Most operating systems, excluding MS-DOS and Windows 3.x, support multithreading; the MacOS 8 supports only a very limited form of multithreading.
- **Multiprocessing** – A computer system with two or more processors can run more than one program or thread at a given time by assigning them to different processor units. To create a multiprocessing system, the operating system must be extended to manage the assignment of processes of threads to each of the processor units. There are two major forms of multiprocessing: symmetric and asymmetric. (See Chapter 5.3, *Computing Platforms,* for more information on multiprocessing systems.)
  - **Symmetric multiprocessing** (SMP) – Allows any processor in a system to execute any instruction. Most operating systems discussed in this chapter, except desktop-only systems such as MS-DOS, Windows 3.x, Windows 9x, and MacOS, support SMP.
  - **Asymmetric multiprocessing** – Allows only a single "master" processor unit to handle privileged operating system code, leaving other processors to run ordinary user code. Because the privileged code typically includes the handling of all I/O instructions, the master processor unit easily can become a bottleneck in an asymmetric multiprocessing system. This kind of multiprocessing is fairly rare today.

■ Memory Management

Storing data and programs in memory, keeping track of them, and reclaiming the memory space after use requires memory management services. Today's operating systems go a step further and offer several sophisticated memory-management extensions:

- **Virtual memory** – Simulates more memory than actually exists in the computer system. Virtual memory lets a program behave as if it had access to the full "address space" provided by the computer's design architecture (typically 2 GB or more on a 32-bit processor), rather than just to the amount of physical memory installed on the specific computer system in use.
- **Memory protection** – In some operating systems, such as MS-DOS, any program can read or write to any area of main memory. This feature can allow programs or data to be corrupted by ill-designed or misbehaving applications. More-sophisticated operating systems offer protected memory, which lets the operating system itself control what areas a program can read from or write to.

■ Cache Consistency

In a multiprocessor system, each processor may have its own local memory cache to store copies of instructions and data it is currently using. As each processor executes, the data held in its local memory cache will become different from the original version of that data held in main memory. If another processor were to access the original version of the same data from main memory, the data retrieved would differ from the contents of the first processor's cache. Therefore, an operating system designed to run such multiprocessor systems needs to have special "cache consistency" instructions that synchronize the contents of each processor's cache with main memory.

■ Input/Output

In simple operating systems, such as MS-DOS, any application can directly access any peripheral device such as a disk drive. In contrast, in a more sophisticated operating system, only the operating system itself can access such I/O devices. This feature prevents multiple programs from trying to access the same I/O device at the same time, which could corrupt incoming data or data on the output device.

■ File System

The operating system is responsible for managing the arrangement of and access to data held in storage devices, including disks and tapes. The operating system creates and manages a directory structure that allows files to be created and retrieved by name, and it also may control access to those files based on permissions and access controls. Sophisticated operating systems often permit multiple types of file systems to exist on the same computer system. For example, a PC may use one type of file system on its internal hard disk and another type to read data from a CD-ROM.

■ Clustering Support

To enhance scalability and system reliability, departmental server and enterprise operating systems increasingly support clustering. The most basic form of clustering is "failover" between two servers. If one server in the cluster fails, users and applications transparently are redirected to the secondary server in the cluster. A more advanced type of clustering is known as application load balancing. There are two types of application load balancing: dynamic and passive.

- **Dynamic application load balancing** – Automatically switches users and application transactions to the least-busy server within a cluster.
- **Passive application load balancing** – Requires a server administrator automatically to set the workloads of each server within a cluster.

Although some forms of UNIX have long supported failover and application load balancing, Microsoft introduced failover support in Windows NT Server Enterprise Edition only in 1997. Load balancing will not be introduced by Microsoft until 1999.

Additional Features of Operating Systems

In addition to the basic functions mentioned previously (process management, memory management, cache consistency, I/O, file system, and clustering support), several other key features often are discussed when comparing operating systems.

■ Security

Security provides the ability to control access to computer resources and data. Security involves measures to minimize risks to availability, confidentiality, and integrity of electronic systems. For an operating system to be secure, it typically needs to provide protected memory, manage all I/O devices, and maintain adequate access control on files in the file system. It also needs to manage privileged code, keep track of users and their authority level, and audit changes to security permissions. There are several types of security, including authentication, integrity, confidentiality, access control, and availability/fault tolerance. (See Chapter 5.4, *Security,* for more information about authentication, integrity, confidentiality, and access control.)

■ Availability and Fault Tolerance

Fault tolerance is the ability of a system to produce correct results and continue to operate even in the presence of faults or errors. Fault tolerance can involve error-correcting memory, redundant components, and hot-swappable system elements and related software that protect the system from hardware, operating system, or user errors.

■ Networking

Networking is the connection of a computer system to other devices by a communications medium. Beyond managing the physical networking interface card and the communications protocols that run over it, an operating system must provide APIs that allow programs to use network resources, such as also finding a device or resource on the network or translating Internet host names into network addresses. (See Chapter 4.2, *Networking Systems and Protocols.)*

■ Windowing Systems

Older character mode or block mode interfaces have been replaced largely by windowing systems. Most UNIX offerings support the Motif graphical interface. In the PC world, support for GUIs and window management became available via add-on programs (such as Microsoft's Windows, which runs on MS-DOS) or were integrated into the operating system (such as IBM's OS/2 Presentation Manager and Windows 95). In the Macintosh world, a windowing system has been part of the basic operating system since Apple's first Macintosh.

■ Character Sets

The extended ASCII character set used in most PCs included 256 characters, with each character stored in an 8-bit byte. Because computers need to support languages with different or larger character sets, new character sets have been invented that store each character in 16 bits (2 bytes), thus allowing up to 65,536 characters. One such character set, Unicode, provides support for extended alphabets and contains more than 20,000 Han characters, which are used to represent whole words or concepts in languages such as Chinese, Japanese, and Korean. The first 128 codes of Unicode are identical to ASCII.

■ Portability

Historically, most computer operating systems were written in low-level assembly languages and tightly tied to a particular hardware (processor) platform. UNIX was the first operating system written to be portable across a large variety of hardware types, such as Intel's x86 and RISC processors. Microsoft's Windows NT (both the Workstation and Server editions) also was designed to be portable and is available for 80x86 and Digital Alpha processors. To the extent that the JVM can be seen as an operating system, it represents the latest effort in widely portable operating systems.

■ Interapplication Communications

Operating systems increasingly support new standards and higher-level architectures that permit applications to communicate among themselves on a single machine or across a network. These interapplication communication technologies include Microsoft's Component Object Model (COM), Distributed COM+, and ActiveX; IBM's System Object Model (SOM) and Distributed SOM (DSOM); and CORBA, an object standard supported by several hundred software companies.

Microsoft announced COM+ in September 1997 as an evolutionary extension to COM to make it easier for developers to create software components in any computing language using any programming tool. COM+ supports various services, including transactions, security, message queuing, and database access. Beta versions of COM+ for developers are scheduled to be available by the end of 1997. (See Chapter 5.2, *Software Component Architectures.)*

Departmental and Enterprise Operating System Facilities

Departmental and enterprise operating systems generally have to support all the services offered by desktop operating systems, with the notable exception of GUIs. They also must provide services to help maintain and manage computer networks and to run large applications that can be accessed by hundreds or even thousands of users. These additional services include the following:

- **Job management and scheduling** – The operating system carries out job control instructions, which describe the programs to be run and the resources they require. This feature is used primarily when transactions are processed in batches. Scheduling of jobs is the method used to plan jobs for execution. Some examples of application criteria include priority, time-in-job queue, prerequisite events, complete with/by constraints, and resource availability.

- **Task management** – The operating system is responsible for the concurrent operation of multiple programs. The hardware systems are designed to overlap operations, and data can move simultaneously in and out of the computer through separate channels while the operating system governs the overall operations. Task management can occur on a single processor or across multiple processors. It identifies which task the processor should work on next, as opposed to multiprocessing, where the tasks are divided between processors and addressed simultaneously.
- **Storage management** – Storage management capabilities, such as compression, often are built into the file system. Windows NT's New Technology File System (NTFS), for instance, provides file-level compression. File locking is another example of storage management: It allows one user to access a file while keeping all other users locked out.
- **Clustering** – Described previously. (See "Clustering Support" on page 289.)

Network Services

Support for network hardware and basic network protocols either is built into an operating system or layered on top of an operating system. These network services provide the service interface for applications seeking to communicate across a computer network. Commonly supported network architectures include Transmission Control Protocol/Internet Protocol (TCP/IP) and Systems Network Architecture (SNA). Commonly supported network services include the following:

- **Directory services** – Allow a user, program, or administrator to locate resources on the network. Directory services are like an electronic phone book to help network clients find objects and services. There are several designs, including the X.500 standard, the Domain Name System (DNS), Novell Directory Services (NDS), and Microsoft Active Directory (scheduled for delivery with Windows NT Server 5.0 in 1998). NDS, for example, is a type of database that stores user rights and tracks every object on a network, from a user's telephone number to the location of a networked printer.
- **Network file access** – Provides remote access to files stored on another system. NFS, developed by Sun and shipped as part of most commercial UNIX systems, is among the most popular standards for remote file access. In the case of a true network file system, the distinction between local and remote files is transparent, and all operations supported on local files (creation, deletion, modification, renaming, and so forth) can be applied to a remote file as well. SunSoft's Solstice Network Client offers this kind of transparency by permitting Windows desktops to view and access files that reside on a UNIX server. Similar services are provided by NetWare and Windows NT. A host of third-party products also offer these capabilities.
- **Remote log-in** – Allows users to access multiple systems via the network rather than requiring them to have a dedicated terminal or PC for each system they use. Remote log-in capability is available on most server and enterprise operating systems that allow multiple, concurrent, interactive users.
- **Remote job transfer and manipulation** – For transferring, scheduling, executing, and controlling jobs on remote systems.

5.1.4 Major Desktop Operating Systems

Major operating systems that run on desktop PCs include MS-DOS and the Windows family of products from Microsoft, the OS/2 family from IBM, and the Macintosh operating system from Apple. UNIX operating systems are covered here as well as in "Major Departmental Server Operating Systems" on page 301.

MS-DOS

MS-DOS was one of the original operating systems for the IBM PC – a non-multitasking, 16-bit operating system tightly coupled to the Intel 80x86 processor. The operating system provides no built-in networking, no memory protection, no windowing system, and no security. MS-DOS has a major limitation in that programs must execute in the lowest 640 KB of the processor's address space, an area shared with the operating system itself, device drivers, and network drivers. MS-DOS is still widely available, although Microsoft may discontinue its sale as a separate product within the next few years.

IBM also sells a version of DOS called PC-DOS, although future upgrades are unlikely. Caldera has relaunched DR-DOS (formerly owned by Novell, which acquired it from Digital Research) as OpenDOS and made it freely available on the Internet.

The Microsoft Windows Desktop Family

Microsoft's Windows desktop product family includes Windows 3.1, Windows for Workgroups 3.11, Windows 95, Windows NT Workstation 4.0, and Windows CE (for hand-held computers, set-top boxes, and Windows terminals). Forthcoming upgrades include Windows 98, Windows NT 5.0, and the so-called Windows 2000 initiative, which calls for all desktop Windows releases (except Windows CE) to be based on the Windows NT kernel.

■ Windows 3.1 and Windows for Workgroups 3.11

Prior to Windows 95's arrival in 1995, Microsoft's mainstream windowing systems were Windows 3.1 and Windows for Workgroups (WFW) 3.11. Windows 3.1 and WFW are not operating systems but GUIs that operate on top of MS-DOS. Windows 3.1 and WFW extend the capabilities of MS-DOS. For instance, Windows 3.1 supports cooperative multitasking, and WFW supports peer-to-peer networking and includes some 32-bit code to speed disk access. Windows and WFW inherit many limitations found in the underlying MS-DOS operating system. These limitations include a lack of memory protection. For instance, one misbehaved application can crash another or even the entire operating system. Windows and WFW were designed for 16-bit processors; they cannot take full advantage of modern 32-bit microprocessors nor of emerging 64-bit microprocessors.

■ Windows 95

Microsoft's Windows 95 was designed as the upgrade path for business and consumer PCs running Windows 3.1, WFW, or MS-DOS. Windows 95 now is installed on virtually all new consumer PCs and most new business PCs. As of late 1997, its installed base topped 100 million units worldwide.

Still, many businesses continued to run Windows 3.x because they lacked the hardware required for Windows 95 or Windows NT Workstation or because of the perceived cost of this transition, especially because many Windows 3.x applications did not run correctly on Windows 95. Complicating matters, many customers today still are not sure whether Windows 95 or Windows NT Workstation is the better operating system for their environment.

Microsoft is aware of this confusion and has sought to address it with the release of Windows 98. (See "Windows 98" on page 295.)

The initial release of Windows 95 supported several features not found in Windows 3.1 or WFW:

- Preemptive multitasking for 32-bit applications written for Windows 95 and cooperative multitasking for 16-bit Windows applications.
- A new desktop GUI metaphor that allows users to launch applications directly from their desktops. It also includes a task bar that more easily allows users to switch back and forth between applications.
- A new file system, based on the File Allocation Table (FAT) file system, which is backward-compatible with the MS-DOS file system but supports file names up to 254 characters in length, thereby eliminating the 8-character file name/3-character file extension limitation found in MS-DOS, Windows 3.1, and WFW.
- Plug-and-Play, a specification that can ease or automate the installation of new hardware by enabling the operating system to recognize new hardware and install the necessary software (called device drivers) automatically.
- Advanced Power Management (APM), a standard that allows laptops and mobile computers running Windows 95 to maximize the length of time they can operate between battery recharges.
- Built-in networking, including integrated support for TCP/IP, SPX/IPX, and dial-up networking.
- Remote access awareness built into the file system – critical for remote users.

Most new PCs now ship with a Windows 95 update, commonly known as Windows 95 OSR2. (However, Microsoft decided not to sell OSR2 as a product to update existing Windows 95 systems.) OSR2 includes bug fixes, patches, and minor enhancements that Microsoft has introduced since Windows 95's initial release:

- Support for wireless, line-of-sight infrared (IR) connections. This feature allows a mobile computer equipped with an IR transmitter/receiver and running Windows 95 to share documents with or send them to another device equipped with an IR transmitter/receiver, such as a printer.
- Support for NDS in a NetWare 4.x network.
- Support for TCP/IP multihoming, which allows more than one TCP/IP stack for connections to more than one network. Under Windows 95, each TCP/IP address requires a separate network interface card (NIC); under Windows NT Workstation, each NIC can support up to five TCP/IP addresses.
- Support for automatic data recovery, which will attempt repair at reboot any damaged files caused when a user turns off a Windows 95 system incorrectly.
- Support for FAT32, which lets a user format a hard drive larger than 2 GB as a volume with a single file system.

Windows 95 was intended to offer new features while maintaining backward compatibility with the vast majority of 16-bit programs written for MS-DOS and Windows 3.1 – a goal that imposed certain constraints on Windows 95's design. For instance, all 16-bit applications run in the same application space under Windows 95, meaning that one misbehaved 16-bit application can crash another 16-bit application. By contrast, 32-bit applications written for Windows 95 run in separate memory spaces, meaning that one 32-bit Windows application cannot crash another 32-bit Windows application in Windows 95.

In mid-1997, Microsoft enhanced Windows 95 with Internet Explorer (IE) 4.0, an Internet browser that competes with browsers such as Netscape's Communicator and Navigator. Formerly code-named Nashville, IE 4.0 integrates Windows 95 with the Internet and offers so-called Active Desktop features. Active Desktop is a customizable dashboard that allows icons and HTML elements to be placed on the Windows 95 desktop. Active Desktop also supports "channels" that permit content providers (such as a news service) to "push" content to a user. (See "Push Technology" on page 247.) It is unclear whether businesses actually want tight integration between Internet browsers and operating systems. Businesses that do not want such capabilities can easily deactivate Active Desktop and simply use IE 4.0 like a traditional browser.

■ Windows 98

The next Windows 95 release, known as Windows 98 (code-named Memphis), is slated for delivery in mid-1998. Rather than a major upgrade, Windows 98 offers minor refinements, bug fixes, and enhancements previously available in Windows 95 OSR2 and via the Internet. Microsoft has indicated that during the next 3 years, Windows 98 increasingly will be positioned for consumer PCs. Future releases of Microsoft's highest-end desktop operating system, Windows NT Workstation, will be positioned for corporate desktops, workstations, and even laptops.

Highlights of Windows 98 include the following:

- Supports Active Desktop and IE 4.0, which marries the Internet and HTML links to Windows 98's user interface. For example, Windows Explorer (the Windows 9x utility for file system navigation) includes Forward and Back buttons (which borrow from the Internet browser metaphor). The buttons allow users to navigate between folders, documents, and Internet sites by single-clicking on icons.
- Support for new devices and standards, including DVD (also referred to as Digital Versatile Discs or Digital Video Discs). DVDs, nearly identical to existing CD-ROMs, provide 4.7 GB of storage. Future versions will support up to 17 GB of storage. (See "DVD" on page 90.)
- Support for Multimedia Extensions (MMX) in 80x86 processors. This capability should speed the performance of Windows-98-compatible multimedia applications and games that use MMX technology.
- Support for Universal Serial Bus (USB). USB is a new bus designed to replace the keyboard and mouse connectors and serial and parallel ports. USB transmits data at up to 12 Mbps, and it can daisy chain up to 127 devices off a single port. Support for FireWire also has been announced. FireWire (officially, the IEEE 1394 specification) is a faster alternative to USB with the ability to handle speeds of up to 400 Mbps. Greater speeds of 1,600 Mbps are planned.

- Support for up to eight monitors on a single PC. Users drag and drop, resize, and move desktop items such as windows, folders, icons, and applications from one monitor to the next. This capability requires one video card per monitor.
- An upgrade path from Windows 3.x and Windows 95.
- Additional enhancements include new backup utilities and disk defragmenter capabilities.
- The Win32 Driver Model (WDM) will permit developers to write a single device driver for both Windows 98 and Windows NT 5.0, which should improve the latter's device driver support. Also promised is better management of application components and of dynamic link libraries (DLLs).

■ Windows NT Workstation 4.0

Initially introduced in 1993, Windows NT was designed as an operating system for high-end desktops and workstations. Three versions of Windows NT are available: Windows NT Workstation, Windows NT Server, and Windows NT Server Enterprise Edition. This section discusses Windows NT Workstation.

Unlike Windows 95 and Windows 98, Windows NT was written from the ground up as a 32-bit operating system. It includes integrated support for symmetric multiprocessing; preemptive multitasking of 32-bit and 16-bit Windows applications; portability across Intel, Digital Alpha, PowerPC, and Mips microprocessors (Microsoft has announced future releases will not support Mips or PowerPC); networking; and file-system security.

Microsoft has shipped two major upgrades and one minor upgrade to Windows NT Workstation since the operating system's initial release. The latest upgrade, Windows NT Workstation 4.0, arrived in 1996. It provides Windows NT with the Explorer interface from Windows 95. It also includes support for DCOM, the distributed version of Microsoft's interapplication communication standard as well as e-mail and Internet connectivity. (See "DCOM" on page 332.)

Windows NT uses a microkernel design. The microkernel takes requests issued by the applications program, validates the request according to a security object model, and then executes the request (if validated) on behalf of the application. The microkernel is so small that one copy is executed on each processor in a multiprocessor system, thereby providing Windows NT with true symmetric multiprocessing capabilities. In contrast, most operating systems use only a single kernel even when running on a multiprocessor system.

■ Windows NT Workstation 5.0

Microsoft has indicated it will release a major Windows NT upgrade, Version 5.0, in late 1998 or early 1999.

Microsoft considers Windows NT 5.0 its most critical software upgrade to date. Windows NT 5.0 is intended to be a robust, full-featured, desktop operating system with many features of Windows 98. Microsoft has said that Windows NT Workstation 5.0 will include the following features:

- Power management via the Advanced Configuration and Power Interface (ACPI), which requires ACPI-aware hardware. ACPI supports advanced network-oriented power management capabilities not offered in the original APM specification for Windows 95. Windows 98 will support both ACPI and

APM; however, Windows NT 5.0 will support only ACPI. Therefore, Windows NT 5.0 requires new ACPI-compliant hardware to activate the operating system's power management features.

- Client support for Active Directory, Microsoft's new directory service in Windows NT Server 5.0.
- Software to take advantage of the Accelerated Graphics Port (AGP), a high-performance, component-level interconnect for three-dimensional graphical applications as defined in the AGP Interface Specification by Intel. Microsoft will support AGP hardware via the Microsoft DirectDraw API.
- DirectX 5.0, an API for writing multimedia applications and computer games.
- Distributed File System (DFS), which permits multiple disk drives to be grouped into a single, virtual volume.
- The FAT32 file system, which improves compatibility with Windows 95 and Windows 98. (Windows NT 4.0 supports only FAT and NTFS, making Windows NT 5.0 the first NT release with FAT32 support.)
- Enhanced security via file-level data encryption and Kerberos, an industry standard for user authentication.
- The Microsoft Management Console (MMC), a GUI that supports third-party management utilities.
- Plug-and-Play (discussed previously in connection with Windows 95).
- Enhanced backup to tape and optical drives.
- The Win32 Driver Model (WDM), which allows software developers to write a single driver for both Windows 98 and Windows NT Workstation 5.0.
- An automated upgrade path from Windows 95, Windows 98, Windows NT Workstation 3.51, and Windows NT Workstation 4.0.

Windows NT 5.0's advances, particularly its support for Plug-and-Play and power management, could prompt many businesses to use Windows NT Workstation 5.0 rather than Windows 9x on corporate desktops and even laptops. Until now, Windows NT on laptops has been difficult because of lack of Plug-and-Play (and lack of power management), which is needed when PC Cards (formerly PCMCIA cards) are inserted and removed.

Although Windows 98 and Windows NT 5.0 have numerous similarities in their features, their architectures are substantially different. Windows 95 and Windows 98 still are based internally on an MS-DOS-like kernel, rather than on the fully multitasking, protected-memory kernel found in Windows NT. Windows 95 and Windows 98 generally offer better backward compatibility for 16-bit Windows applications and demand less RAM and hard drive space than Windows NT Workstation 4.0 or test releases of Windows NT Workstation 5.0.

In contrast, Windows NT Workstation 4.0 and 5.0 offer superior scalability, portability, security, and stability because they support SMP, Digital Alpha and Intel microprocessors, NTFS, and a hardware abstraction layer (HAL). The HAL prevents applications from making direct calls to hardware, which can crash an operating system, and it provides portability across different processors.

The basic features and functionality of Windows 95, Windows 98, Windows NT Workstation 4.0, and Windows NT Workstation 5.0 are outlined in Table 5-2

Table 5-2: Windows Operating-System Features

Feature	Windows 95	Windows 98	Windows NT Workstation 4.0	Windows NT Workstation 5.0
Minimum realistic RAM	12 MB+	24 MB+	32 MB+	48 MB+
Required disk space	80 MB	100 MB (estimate)	120 MB	200 MB (estimate)
Minimum realistic processor	486	fast 486	Pentium	Pentium Pro
Processor support	x86 only	x86 only	x86, Digital Alpha, Mips, PowerPC	x86, Digital Alpha
Symmetric multiprocessing	No	No	Yes	Yes
Win32 preemptive multitasking	Yes	Yes	Yes	Yes
Win16 preemptive multitasking	No	No	Yes	Yes
MS-DOS/Win16 application support	Excellent	Excellent	Good	Good
File systems	FAT, FAT32	FAT, FAT32	FAT, NTFS	FAT, FAT32, NTFS
Plug-and-Play	Yes	Yes	No	Yes
Power management	Yes	Yes	No	Yes
Secure file system	No	No	Yes	Yes
Win32 driver model	No	Yes	No	Yes

After Windows NT 5.0 ships, Microsoft has indicated it will develop a 64-bit version of Windows NT for Digital Alpha processors and Intel's Merced chip, a 64-bit microprocessor designed by HP and Intel that is expected to debut in 1999. At the same time, Microsoft has indicated it will not merge Windows NT Workstation 5.0 and Windows 98 into a single code base. Instead, the company's Windows 2000 initiative calls for all future Windows releases (excluding Windows CE) to be based on Windows NT's kernel.

■ Windows CE

Windows CE is the latest addition to Microsoft's operating system family. Introduced in 1996, Windows CE is a 32-bit operating system designed for so-called hand-held PCs (HPCs), subnotebook computers, Windows terminals, and set-top boxes. Windows CE includes extensive connectivity with Windows 9x and Windows NT. Windows CE uses a subset of the Win32 API and has a multithreaded kernel that supports preemptive multitasking. It supports processors commonly found on desktop computers (Intel 486 and Pentium, Motorola/IBM PowerPC processors, and other processors used in HPCs, such as the Mips 80 and the Hitachi SH3). Casio, Compaq, and HP, among others, offer HPCs based on Windows CE. Microsoft has indicated that it also intends to position Windows CE-based systems against network computers, JavaStations from Sun, and other thin clients. Analysts estimate that fewer than 200,000 Windows CE devices have been sold to date worldwide.

In late 1997, Microsoft announced a major Windows CE upgrade, Version 2.0, which should be widely available by early 1998. Windows CE 2.0 will add the following features:

- Built-in device support for keyboards, touch panels, and other interface devices

- Serial communications (often used to synchronize data between HPCs and desktops)
- PC Card and Socket Services for removable or built-in storage cards
- An improved object store that manages file-system and registry information
- Gray-scale LCD and color VGA displays
- Improved communications via TCP/IP, Windows Sockets, Secure Sockets Layer (SSL), and remote connectivity
- Pocket IE, which supports HTML 3.2, frames, tables, forms, ActiveX controls, and Netscape plug-ins

IBM's OS/2 Family

IBM has offered a 32-bit, multitasking, protected-memory operating system for Intel-compatible PCs since OS/2 Version 2.0 was released in 1992. The current OS/2 release, OS/2 Warp Version 4.0, builds on this 32-bit foundation with additional features such as voice recognition software (IBM VoiceType, also available for Windows 95 and Windows NT), support for a built-in JVM, and integrated connectivity to mainframes and minicomputers.

In general, OS/2 is more comparable architecturally to Windows NT Workstation than to other products in the Windows family. OS/2 is more robust that Windows 3.1 and arguably more stable than Windows 95 when running 16-bit applications.

OS/2 Warp 4.0 can run MS-DOS, Windows 3.x, and some Win32 applications as well as 16- and 32-bit OS/2 applications. IBM provides support for 32-bit Windows applications via Citrix Systems' WinFrame, a multi-user technology that lets OS/2 Warp desktops access 32-bit Windows NT applications. In the first half of 1998, Citrix will ship multi-user technology (code-named Picasso) that will permit OS/2 desktops to access 32-bit Windows applications running on Windows NT Server. Windows NT Server also must be running Microsoft's Windows-based Terminal Server software (code-named Hydra). Using Picasso and Hydra together, OS/2 desktops give the appearance of running 32-bit Windows applications (such as Office 97) locally, when in fact, the applications actually reside on a Windows NT Server.

Rather than deliver major upgrades, IBM plans to enhance OS/2 Warp 4.0 with regular service packs. Ultimately, the company hopes to make OS/2's Java support second to none. Although OS/2 Warp is not a high-volume consumer operating system, it enjoys continued popularity within large IBM accounts, many of which have a significant installed base of custom-designed software written for OS/2.

Macintosh Operating System

Apple's Macintosh operating system (MacOS) Version 8 was released in mid-1997. Like previous MacOS releases, Version 8 is a 32-bit operating system. It supports extensive Internet integration, 32-bit applications, virtual memory, cooperative multitasking, and integrated AppleTalk networking. It also supports various multimedia standards such as QuickTime, QuickTime VR, QuickDraw 3D, and MPEG. It runs on hardware manufactured by Apple and by companies Apple had licensed to make Macintosh-compatible systems for a brief period ending in 1997. (As of early 1998, only Umax was actively selling Macintosh clones.) MacOS 8 runs on PowerPC and Motorola 68040 processors; however, it does not support Motorola's 68030 or earlier processors.

MacOS 8 lacks many of the higher-end features provided by OS/2 and Windows NT, including protected memory, preemptive multitasking, and symmetric multiprocessing. Apple has promised to advanced capabilities in a new operating system code-named Rhapsody, which will be based on code from NeXT, a company Apple acquired in 1996.

Rhapsody is expected to ship in 1998. Apple has announced that Rhapsody will offer a fully preempted and memory-protected multitasking environment, with support for symmetric multiprocessing and Java. Apple's existing multimedia technologies (the QuickTime Media Layer family) will be optimized for Rhapsody. Apple has indicated that Rhapsody will have the look and feel of MacOS and will run Intel and PowerPC processors.

Apple has suggested that Rhapsody initially will be positioned for departmental servers and high-end desktops used for publishing, video editing, application development, and Internet content creation capabilities. Apple will gradually make Rhapsody available for mainstream desktop systems; the slow rollout is intended to ensure that Rhapsody can run the vast majority of existing MacOS applications. Over the short term, Rhapsody is likely to have some problems running selected MacOS applications (similar to Windows NT's initial limited support for Win95 applications). Apple hopes that users will embrace Rhapsody-compliant applications over the next few years, which will encourage ISVs gradually to broaden Rhapsody's appeal beyond workstations and servers and onto mainstream desktops. This is the same formula that Microsoft used to transform Windows NT from a niche to a mainstream desktop operating system.

UNIX

UNIX provides many sophisticated desktop features, including symmetric multiprocessing and preemptive multitasking. The very low end of the UNIX workstation market faces growing competition from low-cost, Intel-based Windows NT workstations, and Microsoft has been encouraging ISVs to port their desktop UNIX applications to Window NT. However, several versions of UNIX – particularly HP's HP-UX, IBM's AIX, and SunSoft's Solaris – remain popular on mid- to high-end engineering workstations and specialized desktops. These UNIX offerings run on RISC processors, which offer the superior floating-point performance required by CAD/CAM and graphical applications.

Several additional versions of UNIX, such as Linux and SCO OpenDesktop, also compete in the desktop operating system market. A University of Helsinki (Finland) student named Linus Torvalds developed Linux. The operating system's source code is available freely on the Internet and has gained popularity within other universities and some corporate networks. Several companies, including Caldera and Red Hat Software, sell versions of Linux with extensive networking enhancements. Linux runs numerous Web sites on the Internet and gradually is building applications support from third parties.

SCO continues to sell and support OpenDesktop but has indicated that its long-term strategy is to focus primarily on the UNIX server market. Toward that end, SCO is preparing UnixWare 7 (formerly code-named Gemini). UnixWare 7, scheduled for release in early 1998, is a merger of SCO's OpenServer and UnixWare operating systems. (See "Major Departmental Server Operating

Systems" on page 301.) SCO has not indicated whether it will offer a desktop version of UnixWare 7, but the operating system will support OpenDesktop applications.

JavaOS

Sun's JavaOS operating system executes programs written in the Java language without the need for a traditional operating system. It runs on SPARC, Intel x86, and Digital StrongARM processors, among others. It is designed for intranet and Internet applications and embedded devices. It runs on a microkernel and includes a memory manager, device drivers, multithreading, the JVM, windowing systems, networking, and support for Java APIs. Sun says applications written for JavaOS also can run in Java-enabled browsers and operating systems. JavaOS is positioned for hand-held products and network computers. JavaOS requires about 2.5 MB of RAM and 4 MB of disk space or ROM.

5.1.5 Major Departmental Server Operating Systems

The major operating systems currently available for departmental servers include several versions of UNIX, Microsoft Windows NT Server, Novell NetWare, IBM OS/2 Warp Server and OS/400, and Banyan VINES. Several UNIX versions, including Digital UNIX, are full, 64-bit technologies, but most departmental server operating systems are 32-bit. In general, most operating systems in this category also include features such as symmetric multiprocessing and integrated networking.

UNIX

UNIX is a multi-user operating system written in the late 1960s at AT&T Bell Laboratories (now part of Lucent Technologies). Several different versions of UNIX are available for Intel, Digital Alpha, HP-PA RISC, Silicon Graphics Mips, Motorola/IBM PowerPC, and Sun SPARC microprocessors, among others.

Many companies, including Digital, HP, IBM, and Sun, have licensed the rights to UNIX and added vendor-specific extensions to it. As a result, the UNIX market is fragmented; applications written for one vendor's version of UNIX need testing and perhaps modification before they can run on another vendor's version of UNIX.

Generally, all versions of UNIX include multitasking, TCP/IP networking, network management, and directory services.

■ Versions of UNIX

UNIX is a popular operating system for running enterprise databases, large-scale corporate applications, Web servers, and other Internet applications. It also may emerge as a popular operating system for hosting network computers. The following sample UNIX versions (or "flavors") are widely installed within corporations.

- **Digital UNIX** – Formerly OSF/1, Digital UNIX runs on Digital's RISC-based Alpha workstations and servers. Digital UNIX supports 64-bit computing and a wide variety of standards including POSIX and the Single UNIX Specification (formerly known as Spec 1170). Digital has announced plans to port Digital UNIX to a next-generation 64-bit processor under development by HP and Intel, code-named Merced.

- **HP-UX** – HP/UX runs on HP 9000 workstations and HP servers based on the PA-RISC processor. HP-UX is derived from UNIX System V Release 3.2 (SVR3.2), which is an older version of the base UNIX code released by UNIX Systems Labs in the early 1990s. HP is enhancing HP-UX to run on the forthcoming Merced processor.
- **IBM AIX** – AIX (Advanced Interactive eXecutive) runs on IBM's RS/6000 workstations and servers, ES/9000 mainframes, and PCs. It is based on SVR3.2.
- **SCO OpenServer/SCO UnixWare** – Formerly called SCO UNIX, OpenServer is the market leader in terms of sales and installed base for UNIX for Intel-based hardware. OpenServer is based on UNIX SVR3.2. Open Server historically has been used by smaller businesses to turn an Intel-based PC into a multi-user system that could be accessed through dumb terminals. It has strong support among ISVs targeting this market segment. SCO acquired UnixWare from Novell in 1995. UnixWare is based on UNIX V Release 4.2. SCO plans to deliver UnixWare 7, a merger of OpenServer and UnixWare, in early 1998. The company also intends to deliver targeted UnixWare 7 builds throughout 1998 that support key server segments, including departmental databases, application servers, e-mail, intranets, and enterprise applications.
- **Silicon Graphics IRIX** – IRIX is a 64-bit implementation based on UNIX System V Release 4 (SVR4) that runs only on SGI Mips processors. Its popularity generally is limited to imaging and intensive graphics uses, including special effects in commercial movies.
- **Sun Solaris** – Solaris 2.6 is available for Sun's SPARC microprocessors and Intel x86 workstations. Solaris is based on a merger of SVR4 and SunOS, Sun's original version of UNIX based on work done at the University of California, Berkeley. In late 1997, Sun announced plans to port its Intel version of Solaris to the 64-bit Merced chip and to pursue the Intel server market aggressively with this version of Solaris.

Windows NT Server

Microsoft's Windows NT Server offers almost all the features in Windows NT Workstation plus some additional enhancements that specifically target the departmental server market. For example, Windows NT Server is a 32-bit operating system that supports symmetric multiprocessing (up to four processors). The product (formerly known as Windows NT Advanced Server) was originally released in 1993.

Like UNIX, Windows NT was designed to be portable. As a result, Windows NT Server is available for Digital Alpha and Intel microprocessors. However, unlike UNIX, the source code for Windows NT Server is controlled by one company, Microsoft; who can deploy it on what platforms is still a closed, proprietary decision.

Windows NT Server did not generate significant sales until 1994, with the release of the second major version, Windows NT Server 3.5. Microsoft delivered another major release, Windows NT Server 4.0, in 1996. It included support for the Windows 95 interface, cryptography APIs, improved scalability on four-processor servers, and integrated Web server software, called Internet Information Server (IIS). Windows NT Server often is used as a file-, print-, application-, and Web server.

Windows NT Server 4.0 Enterprise Edition

Microsoft released a new product bundle, Windows NT Server 4.0 Enterprise Edition, in late 1997 (the base Windows NT Server 4.0 product still is available separately). The Enterprise Edition offers several new features that bolster scalability and reliability, including the following:

- **Microsoft Cluster Server** (MCS) – Formerly code-named Wolfpack, MCS allows two servers to be clustered together so that applications failover automatically to the secondary if the primary server in the cluster fails. Microsoft has indicated that MCS will gain support for four or more servers and dynamic application load balancing in 1998. The latter capability will permit application loads to be moved from server to server automatically.
- **Microsoft Transaction Server** (MTS) – Formerly code-named Viper, MTS permits software developers to write applications that offer reliable transactions across a Windows NT network.
- **Microsoft Message Queue Server** (MSMQ) – Formerly code-named Falcon, MSMQ ensures messages are delivered even when a server crashes or portions of a network fail.
- **4 GB RAM tuning** – Allows developers to allocate 3 GB of memory to applications (leaving 1 GB for Windows NT Server). The traditional Windows NT Server 4.0 has a 2-GB memory limit for applications (with 2 GB reserved for the base Windows NT). This feature is important because it can allow very large applications (up to 3 GB) to run on a Windows NT system.

Microsoft released an additional product bundle, BackOffice Small Business Server, in late 1997. It includes a specialized version of Windows NT Server 4.0 for 2 to 25 users. It also includes Microsoft Proxy Server (which caches Web content locally), IIS, Microsoft SQL Server, and Microsoft Exchange Server.

Microsoft also intends to offer multi-user capability, code-named Hydra, in Windows NT Server during the first half of 1998.

Windows NT Server 5.0

Microsoft began testing its next major departmental server upgrade, Windows NT Server 5.0, in September 1997. Windows NT Server 5.0 attempts to eliminate some of Windows NT Server 4.0's limitations, including the lack of advanced directory services. Windows NT Server 5.0 is expected to ship in mid-1998, although analysts say delivery could be delayed until late 1998. Windows NT Server 5.0 includes all components in Windows NT Workstation 5.0 as well as several major enhancements:

- **Active Directory** – A new hierarchical directory service, Active Directory supports three key capabilities: single sign-on, a single point of administration for all network objects, and the ability to query any attribute of any object, such as whether a printer can print two-sided output. Like NDS, Active Directory can be replicated across multiple servers, partitioned, and extended to support new objects, such as multimedia data. A Directory Service Migration Tool (codeveloped by Computer Associates' Cheyenne Division) discovers NetWare resources and migrates them to Active Directory.
- **Intellimirror** – Intellimirror is a new Microsoft technology that automates software management between Windows NT Server 5.0 and Windows NT Workstation 5.0. Intellimirror is part of Microsoft's Zero Administration for Windows (ZAW) initiative, which attempts to reduce a network's total cost of ownership (TCO). ZAW is targeted toward Windows NT Workstations.

Intellimirror copies Windows NT Workstation 5.0 data, applications, system files, and administrative settings to a Windows NT Server 5.0. If a Windows NT 5.0 desktop crashes, all its applications, preferences, and administrative policies still can be retrieved from Windows NT Server 5.0. Microsoft is considering whether to offer Intellimirror on Windows 98.

IBM's OS/2 Warp Server

In 1996, IBM shipped OS/2 Warp Server 4.0, which merged IBM's LAN Server file and print services with the 32-bit OS/2 Warp operating system. OS/2 Warp Server runs on Intel x86 processors and supports IBM's System Object Model (SOM), which is compatible with the CORBA object standard.

Like Windows NT Server, OS/2 Warp Server supports file, print, and application services. It includes integrated systems management and backup software. In late 1996, IBM introduced symmetric multiprocessing support for OS/2 Warp Server. IBM also offers add-ons for OS/2 Warp Server, including Directory and Security Services (DSS) for OS/2. In August 1997, IBM announced WorkSpace On-Demand (formerly code-named Bluebird), which lets network computers, terminals, and PCs access applications running on OS/2 Warp Server. IBM's long-term strategy is to make OS/2 Warp Server a premier departmental server operating system for Java desktops.

Novell NetWare

Novell NetWare has the largest installed base of all departmental server operating systems for sharing files and printers among computer users. Novell says more than 55 million users are linked to Novell IntranetWare and NetWare networks (as of September 1997).

Like most other server operating systems, NetWare is 32-bit. In September 1996, Novell released NetWare 4.11 (officially called IntranetWare 4.11 and formerly code-named Green River). It supports symmetric multiprocessing, integrated Web software, and advanced directory services.

In September 1997, Novell demonstrated its next major NetWare release 5.0, code-named Moab. NetWare 5.0 is slated for delivery in mid-1998 and will support the following services, according to Novell announcements:

- TCP/IP as a core protocol, replacing IPX/SPX
- Expanded application server capabilities, including memory protection, virtual memory, and support for applications based on Java, CORBA/IIOP, and ActiveX
- Dynamic Host Configuration Protocol (DHCP) and Domain Name Service (DNS) integrated with NDS
- A new GUI
- An optional clustering service, code-named Orion

Banyan VINES

VINES is an enterprise network operating system based on UNIX that offers among the most robust file, print, and directory services for large networks. Banyan shipped VINES Version 7.1 in March 1997. It supports TCP/IP and graphical administration tools for Banyan's StreetTalk directory service. Aware of the fierce competition facing VINES, Banyan has branched out by making StreetTalk for Windows NT its flagship product.

IBM's OS/400

OS/400 is IBM's operating system for the AS/400 line of computers. IBM delivered the current OS/400 release (Version 4, Release 1) in late 1997. IBM says OS/400 4.1 offers a 30 percent performance boost over previous OS/400 releases.

OS/400 4.1 includes a fully integrated database and network and systems administration functions. It also supports new, Internet-oriented AS/400 applications such as the IBM Firewall for AS/400 and the IBM Internet Connection Secure Server (ICSS) for AS/400. The latter provides security for sending confidential information over the Internet and corporate intranets. SSL also is supported. IBM is shipping Net.commerce, its electronic commerce applications tool set, with its latest AS/400s. IBM markets versions of the AS/400e with pre-loaded applications from J.D. Edwards, SAP, and SSA. IBM says OS/400 will support Lotus Domino, Windows NT, and Java in early 1998. Windows NT support will require dedicated Intel hardware.

OS/400 4.1 offers backward compatibility with previous OS/400 releases, which include OSF's DCE services such as authentication, directory services, and remote procedure calls (RPCs). The OS/400 family also supports TCP/IP and an Integrated Language Environment (ILE) C/400 programming language, which lets applications perform two to three times faster compared with previous C applications written for the OS/400.

5.1.6 Enterprise Operating Systems

Enterprise operating systems – UNIX; IBM's MVS, VM, VSE; and Digital's OpenVMS – generally run on mainframes and minicomputers. The continued popularity of enterprise operating systems (particularly in the age of corporate intranets and client/server systems) and the hardware they run on can be attributed to several factors, including the superior manageability and stability of enterprise operating systems.

IBM's MVS and OS/390

Multiple Virtual Storage (MVS) and Virtual Machine (VM) from IBM dominate the mainframe operating system world. MVS was introduced in 1974. It has since been enhanced to support on-line applications (such as Customer Information Control System, or CICS), UNIX APIs, and electronic commerce. MVS also can run on plug-compatible mainframes from Amdahl, Hitachi, and others.

MVS is designed to handle very large systems with thousands of users and many gigabytes of data. It is far more scalable than departmental server operating systems. IBM's flagship version of MVS now is known as OS/390 Release 2. In September 1997, IBM delivered OS/390 Version 2 Release 4, which supports these services:

- **Digital certificates** – Digital certificates are signed electronic documents that contain information uniquely identifying the user. They authenticate a Web browser's access to a Web server. OS/390 Security Server accepts the authenticated digital certificate from the Domino Go Webserver (the user does not need to enter a user ID or password). The user then is able to access OS/390 data and transact business on the Internet. (See "Certificate Authority" on page 393 for more information on digital certificates.)

- **OS/390 Firewall Technologies** – These new security technologies build upon existing S/390 security capabilities and offer customers additional protection for network applications by controlling user access to specific servers inside and outside a network. A separate firewall component also screens every piece of data to determine whether the data is permitted to enter or leave a protected network. In addition, OS/390 firewall technologies support Virtual Private Networks (VPNs), which allow data to flow securely between branch offices and central offices. (See "Firewalls" on page 415.) VPNs also work with the S/390 Cryptographic Coprocessor, a hardware-embedded security function of IBM S/390 Parallel Enterprise servers. (See "Virtual LANs (VLANs)" on page 182.)
- **Domino Go Webserver Version 4.6 for OS/390** – A Web server with full-text search and dynamic workload capabilities, Domino Go Webserver 4.6 exploits S/390 Parallel Sysplex clustering technology (see below) and incorporates the functionality of Internet Connection Server (ICS) for OS/390 and other security technologies.
- **Java for OS/390** – Offers Version 1.1.1 of the Java Development Kit (JDK), which includes a JVM that lets OS/390 operate as a full Java platform. IBM intends to make all its operating systems support Java. Java applications can scale up and down IBM's entire hardware and software line.
- **Net.Commerce Version 3.0 for OS/390** – Offers electronic commerce capabilities for customers who do high volumes of financial and business transactions. Net.Commerce 3.0 is an MVS enhancement slated for delivery in the first half of 1998. It can be used with Domino Go Webserver to set up an electronic Internet presence that displays products and services.

For additional reliability and scalability, OS/390 can run Parallel Sysplex, a technology that can harness multiple S/390 servers into a single, logical enterprise server. Using Parallel Sysplex, workloads can be distributed dynamically to any available processor.

IBM's VSE

Virtual Storage Extended (VSE) from IBM is a multitasking operating system for smaller mainframes built on the System 370 architecture. VSE is the operating system installed on one-third of existing 4381s and many 9370s. It supports transaction- and batch-processing services. Once known as DOS or DOS/VSE, VSE has dropped the DOS portion of the name because of confusion with the multiple versions of MS-DOS available for PCs.

The current Visual Storage Extended/Enterprise Systems Architecture (VSE/ESA) 2.2 release is Year 2000 compliant. IBM began shipping a new release of VSE/ESA, Version 2 Release 3, for small to midsized S/390 customers in December 1997. The company also will offer VisualAge 2000, which helps customers ensure that their own applications are Year 2000-compliant.

VSE/ESA 2.3 includes TCP/IP for VSE/ESA, which offers the basic TCP/IP protocol stack and the Sockets API, plus File Transfer Protocol (FTP), Telnet (TN3270) as well as support for IBM Network Stations and client/server applications. It also includes new network computing capabilities such as Web server software, a JVM, a GUI, and access to VSE data from Web browsers. This version also supports guest operating systems (such as OS/390 Parallel Sysplex testing) and UNIX as well as connections to IBM's network computers.

Moreover, an enhanced Language Environment for VSE (LE for VSE) adds an X/Open-compliant implementation of the Sockets API for C. LE for VSE, formerly an optional product for VSE/ESA, is now a component of the basic VSE/ESA 2.3 package.

IBM has stated that VSE/ESA 2.3 supports VSAM files greater than 4 GB (the theoretical limit is nearly 500 GB), multiple catalog backup and restore, and automated daylight-saving time updates. A new S/390 Service Update Facility (SUF) provides an Internet-based GUI utility to help S/390 customers (VSE, VM, and OS/390) obtain corrective and preventive service.

IBM has indicated it will ship a new CICS transaction server based on CICS/ESA from OS/390 in the next VSE/ESA release – Version 2 Release 4 – scheduled for late 1998.

IBM VM

IBM's VM/ESA supports an interactive, time-sharing environment called Conversational Monitor Systems (CMS). Applications for CMS include office productivity tools, software development, database management, decision-support tools, and a large library of third-party software.

The current VM/ESA release, known as VM/ESA Version 2 Release 2.0, provides Year 2000 support and offers enterprise-class performance, better connectivity to Microsoft's Windows family, and various Internet capabilities.

Digital OpenVMS

OpenVMS, formerly known as VMS, is a general-purpose, multi-user operating system for Digital's VAX and Alpha computers. It supports networking and multiprocessing and includes a management console (called OpenVMS Management Station) that allows administrators to manage printers and print queues across multiple OpenVMS Cluster systems. OpenVMS also supports OSF Motif, POSIX interfaces, and DCE services.

Digital continues to enhance and support OpenVMS as one of three strategic operating systems (Digital UNIX and Windows NT Server are the other two). Digital's strategy is to improve the integration of OpenVMS and Windows NT while maintaining its technology lead in the UNIX market. Longer term, Digital expects Windows NT to be installed on the majority of servers it sells, with Digital UNIX holding a solid second position and OpenVMS declining (though OpenVMS will continued to be supported, and much of its technology will be available for Windows NT in the future). For instance, in the fourth quarter of Digital's fiscal year 1997, Digital's Windows NT-based product sales grew 50 percent, UNIX server sales increased 7 percent, while OpenVMS sales declined. IDC expects OpenVMS revenue to decline by an average rate of 2.6 percent through the year 2000.

The current OpenVMS includes a multithreaded kernel, 64-bit addressing, TCP/IP, and some integration with Windows NT. In 1998, Digital expects to deliver Galaxy, a combination of symmetric multiprocessing and clustering capabilities for OpenVMS (and for Windows NT). Galaxy will let a single Digital system perform like a partitioned cluster, in which each processor has its own memory and database.

5.1.7 Market Overview

Desktop Operating Systems

The transition to Microsoft's 32-bit Windows operating systems (Windows 95 and Windows NT Workstation 4.0) is well underway. Microsoft has discontinued active sales of Windows 3.x.

OS/2 Warp maintains its role within IBM accounts, although IBM has begun marketing Windows NT hardware and software aggressively as well. An overview of the desktop operating systems market is shown in Table 5-3.

Table 5-3: Worldwide Desktop Operating System Sales (millions of units)

Operating System	1996	1997	1998	1999	2000
Windows 3.x	39.6	20.8	6.1	1.0	0.4
MS-DOS	4.2	2.0	1.0	0.5	0.2
MacOS/Rhapsody	4.8	5.3	6.0	6.7	7.5
UNIX	1.4	1.5	1.6	1.6	1.6
OS/2	1.7	1.9	2.0	1.8	1.6
Windows 9x	45.6	64.3	67.9	64.9	70.0
Windows NT	4.1	12.5	26.4	41.8	52.5

Source: IDC, 1997

■ Microsoft Windows Family

Microsoft has shipped more than 100 million copies worldwide of Windows 95 as of late 1997. The Windows 95 operating system is preinstalled on virtually all consumer PCs sold to consumers in North America and most new corporate PCs. This situation represents a change from 1996, when most new corporate PCs still came with Windows 3.x preinstalled. However, many existing corporate PCs have not been upgraded from Windows 3.x to Windows 95 because of migration issues, support issues, or cost factors.

Businesses running Windows 3.x have several upgrade options: They can deploy either Windows 95 or Windows NT Workstation 4.0 or wait for the mid-1998 release of Windows 98 or Windows NT Workstation 5.0.

Microsoft has indicated it will position Windows 98 for consumers and Windows NT Workstation 5.0 for new corporate computers, including laptops. Microsoft, therefore, is urging Windows 3.x business customers to bypass Windows 95 and Windows 98 and instead adopt Windows NT Workstation 4.0 (or Windows NT Workstation 5.0 when it arrives), as long as their applications and hardware support Windows NT technology. Acceptance of Windows NT Workstation is growing rapidly. Microsoft estimates the installed base of Windows NT Workstation at more than 5 million units as of late 1997, with shipments accelerating monthly. As of late 1997, Microsoft claimed Windows NT Server had sold more than 1 million units worldwide during the preceding 12 months; this is more units than any other departmental server operating system during the same time period.

Several factors may continue to hinder Microsoft's attempt to make Windows NT Workstation a standard for corporate desktops, however. First, analysts recommend that Windows NT Workstation be used on PCs equipped with at least a Pentium processor and 32 MB of RAM. Second, for security and stability purposes, Windows NT does not run Win16 and MS-DOS applications that use "real mode" device drivers to make direct calls to hardware. Also, some Windows 95 applications do not run on Windows NT Workstation. Third, Windows NT Workstation's list price is more than twice that of Windows 95. Fourth, upgrading a PC from Windows 95 to Windows NT Workstation 4.0 requires reinstalling the applications currently installed on the Windows 95 PC. This requirement is the result of Windows 95 and Windows NT Workstation having incompatible registries (a system database containing software and application settings). Microsoft says Windows NT Workstation 5.0 will correct some of these problems and will provide an upgrade path from Windows 95, Windows 98, and Windows NT Workstation 4.0.

Windows CE, meanwhile, has found a niche as an HPC operating system. In early 1998, Microsoft is expected to announce a new hardware specification (so-called Windows Terminals) that runs Windows CE and competes with network computers from IBM, Oracle, and Sun.

■ IBM's OS/2 Family

IBM continues to introduce service packs and minor enhancements for OS/2 Warp 4.0. OS/2 Warp 4.0 had sold roughly 1 million units worldwide as of late 1997, bringing OS/2's total installed base to 15 million units, according to IBM.

OS/2 was among the first operating systems to support Sun's Java, and IBM plans to continue enhancing OS/2's Java support. Ideally, Java applications will be able to run on any IBM operating system, from OS/2 to OS/390. However, further development of native OS/2 applications (other than Java applications) is no longer a top priority, and IBM will not enhance OS/2 to support Microsoft's Win32 APIs natively.

■ Apple Macintosh Operating System

MacOS 8, released in mid-1997, has achieved popularity among traditional Macintosh customers. However, it does not include certain advanced features found in Windows NT, as discussed previously.

Apple also has stopped licensing MacOS to third-party hardware vendors because of its belief that Macintosh clones were taking sales from Apple rather than expanding the overall Macintosh market.

■ UNIX

UNIX is a niche solution for business desktops that require advanced capabilities. The UNIX workstation market grew 10 percent during the first quarter of 1997, according to Dataquest. However, sales of high-end Intel workstations running Windows NT grew 242 percent during the same period. Most UNIX vendors have chosen to focus on the UNIX departmental server market rather than the desktop market. The notable exception is Linux, the freeware version of UNIX, which has remained primarily a desktop and workstation product. Linux now is available on Intel, SPARC, and Digital Alpha platforms.

Departmental Server Operating Systems

The leading UNIX offerings as well as Windows NT Server, Windows NT Server Enterprise Edition, Novell NetWare, IBM OS/2 Warp Server, and Banyan VINES are included in the definition of the departmental server operating system market, as is IBM's OS/400. The relative worldwide market share of the major operating systems on servers is shown in Table 5-4.

Table 5-4: Worldwide Customer Revenue by Server Operating System

Server Operating System	1996		2000		CAGR
	$millions	Market Share	$millions	Market Share	
UNIX	$22,639.1	37.6%	$32,778.5	38.0%	9.3%
OS/390	$8,154.5	13.5%	$9,315.4	10.8%	3.5%
OS/400	$5,181.5	8.6%	$5,618.5	6.5%	1.7%
NetWare	$4,902.7	8.1%	$6,065.9	7.0%	4.5%
Windows NT	$3,691.0	6.1%	$15,074.6	17.5%	38.2%
OpenVMS	$393.9	0.7%	$350.5	0.4%	-2.6%
Other	$15,226.8	25.3%	$17,050.9	19.8%	2.5%
Total	$60,189.4		$86,254.1		9.1%

Source: IDC, 1997

■ UNIX

UNIX remains popular as a platform for deploying large relational databases and enterprise-wide applications. It continues to gain momentum as a platform for Internet servers. In particular, Sun's Solaris has emerged as the preferred UNIX version for hosting Web servers and database applications. Leading ISVs such as Netscape, Oracle, and Sybase typically develop their applications first for Solaris and Windows NT before porting them to additional UNIX versions.

Although thousands of applications are available for UNIX, it remains difficult for ISVs to write a single application that can run on all versions of UNIX because no two UNIX versions are exactly alike.

The Single UNIX Specification was supposed to give software developers a single set of APIs that supported every major version of UNIX. Initially, UNIX vendors were slow to offer support for the Single UNIX Specification, and its original goal of creating a market for shrink-wrapped UNIX applications remains largely unfulfilled.

Many UNIX vendors now are looking to Java to offer cross-platform application capabilities. Java's "write once, run anywhere" promise may someday allow Java applications to be written once to run on any operating system, including all versions of UNIX, Microsoft's Windows family, and enterprise operating systems.

■ Microsoft's Windows NT Server and Windows NT Server Enterprise Edition

Windows NT Server 4.0 release supports file, print, application, and Web services but lacks robust directory, transaction processing, and clustering capabilities. Microsoft introduced Windows NT Server 4.0 Enterprise Edition in late 1997 to correct some of these limitations. A more ambitious Windows NT Server release, Version 5.0, is expected to ship in mid- to late 1998 with support for advanced directory services.

Like Windows NT Workstation, Windows NT Server is available for Digital Alpha and Intel x86 hardware (Microsoft has discontinued future development for PowerPC and Mips processors). More than 95 percent of Windows NT systems run on Intel x86 hardware. Windows NT software vendors must recompile their Intel x86 applications to run on Digital Alpha, but few vendors have performed the recompilation because Digital Alpha is used primarily on very high-end servers rather than volume departmental servers. Digital's fx!32 product provides automatic translation of Intel programs that lets them run natively on Alpha computers, albeit more slowly.

Many UNIX application developers, including IBM, Oracle, Sybase, and Informix, now offer applications for Windows NT Server. IBM, in fact, offers more Windows NT software than any other company and has indicated it wants to become the number-one provider of Windows NT software and hardware solutions. IBM develops software, such as Lotus Domino (formerly Notes), to run on Windows NT and IBM AIX first before adding incremental support for additional operating systems. This approach represents a change from IBM's traditional practices. Previously, Lotus Domino and other IBM applications were written first for OS/2 and later modified for Windows NT.

Many enterprise hardware and software providers, including Computer Associates, Digital, HP, IBM, and Tandem (now a division of Compaq) are seeking to diversify their revenue streams by adding value to Windows NT. Third-party enhancements include clustering, directory services, and transaction processing systems.

■ IBM's OS/2 Warp Server

IBM OS/2 Warp Server remains widely installed within IBM accounts. OS/2 Warp Server now supports multi-user operation, network computers, and Java technology. IBM remains committed to enhancing OS/2 Warp Server with incremental maintenance releases.

■ Novell's NetWare

Novell's NetWare remains a leading choice for file and print servers, and NetWare's directory service is popular for managing networks. NetWare lacks proven application and Web services, although Novell plans to address these shortcomings via Moab, a NetWare upgrade slated for release in 1998. Novell also has formed a new company with Netscape, called Novonyx, that will bring Netscape's Web software to NetWare. At the same time, Novell is attempting to diversify its product line by enhancing NetWare and porting its directory service to UNIX and Windows NT Server.

■ Enterprise Server Operating Systems

Enterprise server operating systems remain a critical component within corporate data centers. IBM has successfully positioned OS/390 and OS/400 as platforms for hosting Internet, intranet, and graphical database applications as well as electronic commerce. IBM also has delivered Java support across its entire operating system line. In theory, application developers can develop a single Java application that will run on any of IBM's operating systems.

5.1.8 Forecast

Desktop Operating Systems

- Most new PCs purchased by corporations now include 32-bit operating systems. Although millions of older corporate PCs continue to run 16-bit operating systems, this number will decline throughout the forecast period.
- Specialized devices such as cellular phones and set-top boxes increasingly will use Sun's JavaOS or Microsoft's Windows CE, although other alternatives still will be available.
- User interfaces will move beyond the desktop/file/folder metaphor and will gain tighter Web integration. In the long term, voice recognition technology will become integrated into the operating system. IBM already has bundled voice recognition technology with OS/2 and has since ported the technology to Windows 95 and Windows NT. Likewise, Microsoft is developing voice recognition technology for all its Windows platforms and has a small group of developers exploring social interface technology that may debut shortly after the year 2000.
- Sun's Java could emerge as a platform for writing applications that run on any operating system. However, the promise of "write once, run anywhere" remains widely unfulfilled because JVMs so far differ from operating system to operating system. Also, Sun and Microsoft disagree over Microsoft's plans for implementing Java in Windows NT and Windows 9x.
- Cost of ownership will continue to be an issue during the forecast period. Microsoft will make some progress with its ZAW effort, while Intel focuses on its Wired for Management initiative.

Microsoft's Windows Family

- Microsoft will deliver Windows 98 (a maintenance release for Windows 95) and Windows NT Workstation 5.0 (a major upgrade to Version 4.0) in late 1998. Microsoft will position Windows 98 for consumers and position Windows NT Workstation 5.0 for all new business desktops, including notebooks, NetPCs, PCs, and workstations. Windows 98, however, will be used by many businesses not needing the additional features provided by Windows NT, not wanting to provide the additional hardware resources it requires, or requiring backward compatibility for Windows 3.x and DOS applications.
- Windows NT Workstation will pressure UNIX in the low-end workstation market. Most UNIX workstation vendors, including HP, IBM, and Silicon Graphics either have announced or delivered low-end Windows NT workstations that complement their respective high-end UNIX desktop solutions.
- A 64-bit Windows NT release will debut after the Merced processor is released in late 1999. The release will be designed to run on Merced (a forthcoming 64-bit chip from HP and Intel) and also is expected to run on Digital Alpha processors.

- Within the forecast period, Microsoft plans to release additional Windows CE versions for wallet PCs, subnotebooks, Windows Terminals, and automobiles. Future Windows CE versions also will support voice recognition features.
- In 1998, Windows CE is expected to become the next operating system for WebTV, a hardware/software platform that allows televisions to access Internet content.
- By the close of this century, Microsoft hopes to phase out the base technology in Windows 95 and Windows 98 and will develop all future Windows releases (except Windows CE) using Windows NT's kernel.

IBM's OS/2

- No major OS/2 upgrades are expected during the forecast period. IBM will deliver maintenance releases on an ongoing basis that further improve OS/2's Java support, among other things.

Apple's MacOS and Rhapsody

- Apple has embarked on a dual-operating system strategy. The current MacOS 8 will remain Apple's volume operating system for the foreseeable future. A second operating system, code-named Rhapsody, is expected to ship in late 1998. It will run on PowerPC and 80x86 processors.
- MacOS will continue to face mounting pressure from Windows 9x and Windows NT Workstation, even within established Macintosh markets.

Desktop UNIX

- Desktop versions of UNIX will remain popular for niche uses on very high-end workstations that run CAD/CAM, engineering applications, and high-end graphical programs, such as those used for video special effects.
- Windows NT Workstation will replace UNIX on many low-end workstations as applications become available for Windows NT.
- Industries with significant existing commitments to desktop UNIX (such as securities trading, animation, and high-end engineering) will not consider migrating to Windows NT until after a 64-bit version for Merced is available.

Departmental Server Operating Systems

- By 1998, all major departmental server operating systems will support Java as well as bundled or third-party directory services.
- By 1998, most major departmental server operating systems will come with optional electronic commerce software that goes beyond basic Web server software.
- By late 1998, optional support for enterprise security standards will be widely available for all major departmental server operating systems.

Server UNIX

- UNIX will remain the primary choice for enterprise-wide packaged corporate applications and for applications off-loaded from mainframes.
- UNIX will remain popular for deploying high-end Web servers and mission-critical databases.

- By 1999, the UNIX market will begin to consolidate. The remaining UNIX versions will continue to gain high-end features historically associated with mainframes.
- Most major UNIX vendors will port their version of UNIX to the 64-bit Merced processor from Intel.

Windows NT Server

- Microsoft will continue to expand its Windows NT Server product line with product bundles targeted at specific market segments, such as small businesses, department networks, and large enterprises.
- Microsoft intends to ship a major upgrade, Windows NT Server 5.0, by late 1998. It is expected to support an advanced directory service, 64-bit very-large memory applications on Digital Alpha hardware, the server component of Zero Administration for Windows, and other high-end enhancements.
- By late 1998, Windows NT's clustering technology will move beyond basic two-node failover and will support application load balancing and multi-node connections.

Novell NetWare

- NetWare will remain popular for file and print services and will gain increasing interest from network administrators seeking directory services.
- Novell will deliver a major NetWare upgrade, NetWare 5.0, in 1998. It will use TCP/IP as its native protocol and offer improved application services and optional clustering capabilities.
- Novonyx (a startup jointly owned by Netscape and Novell) will port portions of Netscape's SuiteSpot bundle to NetWare.
- As demand for traditional file and print services declines, Novel increasingly will position itself as a networking services provider by selling NDS for additional platforms. NDS for Windows NT, for instance, shipped in November 1997.

IBM's OS/400

- OS/400 will gain more electronic commerce and Internet features as part of IBM's eCommerce strategy.

Enterprise Server Operating Systems

- Enterprise server operating systems, particularly MVS, will continue to anchor large corporate networks because these systems offer unmatched scalability, reliability, and security.
- Enterprise server operating systems will run more Internet, intranet, and client/server applications.

IBM's OS/390, VM, and VSE

- OS/390, with its superior manageability, security, and scaleability characteristics, will remain an attractive platform for deployment of additional applications within existing S/390 sites.
- OS/390 will add features for electronic commerce.

- CMOS hardware prices will continue to decrease, thereby improving the price/performance position of OS/390.

VM and VSE

- The number of VM licenses will continue to decline, but customers likely will upgrade to new VM releases to gain Year 2000 compliance, TCP/IP capabilities, POSIX interfaces, and advanced support for Internet applications such as Lotus Domino.
- VSE increasingly will be run under VM, and VSE will not gain POSIX compatibility.

Digital OpenVMS

- Digital continues to enhance and support OpenVMS, but Windows NT Server is now Digital's primary platform for revenue growth.
- Many features first developed for OpenVMS will be moved by Digital to Windows NT Server. For instance, Digital Galaxy, a new clustering technology, will support both Windows NT and OpenVMS.

5.1.9 References

■Articles

Baltazar, Henry. 1997. Keeping the faith: Novell's "Moab" beta inspires NOS loyalty. *PCWeek*. December 1: 1.
Greenemeier, Larry. 1997. V4R1 upgrade offers greater RISC, lowers risk. *Midrange Systems*. September 26: 3.
Greiner, Lynn. 1997. Windows 98 takes the Internet to heart. *Computing Canada*. November 10: 62.
Hayes, Mary. 1997. UNIX rules the high end. InformationWeek. March 24: 44.
Johnston, Stuart J. and John Foley. 1997. Betting on the new NT. *InformationWeek*. October 13: 18.
Matzkin, Jonathan. 1997. Windows 98 brings home the Web. *Computer Shopper*. December: 614.
Mohta, Pushpendra. 1997. Windows shopping. *HP Professional*. September: 24.
Moran, Joseph. 1997. Get ready for Windows NT 5.0. *Windows Sources*. October: 156.
Ouellette, Tim. 1997. IBM smooths OS/400 upgrade. *Computerworld*. November 24: 67.
Panettieri, Joseph C. 1998. Delivery day near for Windown 98, NT 5.0 Beta 2. *Windows Magazine*. February: 47.
Symoens, Jeff. 1997. Moab looks up to the challenge. *InfoWorld*. December 8: 153.

■URLs of Selected Mentioned Companies

Amdahl `http://www.amdahl.com`
Apple `http://www.apple.com`
AT&T Bell Labs `http://www.belllabs.com`
Caldera `http://www.caldera.com`
Casio `http://www.casio.com`
Cheyenne Software `http://www.cheyenne.com`
Citrix Systems `http://www.citrix.com`
Compaq `http://www.compaq.com`
Computer Associates `http://www.cai.com`
Digital `http://www.digital.com`
Digital `http://www.digitalresearch.com`
HP `http://www.hp.com`
Hitachi `http://www.hitachi.com`
IBM `http://www.ibm.com`
Informix `http://www.informix.com/infmx-cgi/webdriver`
Intel `http://www.intel.com`
J.D. Edwards `http://www.jdedwards.com`
Lucent Technologies `http://www.lucent.com`
Microsoft `http://www.microsoft.com`
Motorola `http://www.mot.com`
Netscape `http://www.netscape.com`
NeXT `http://www.next.com`
Novell `http://www.novell.com`
Novonyx `http://www.novonyx.com`
Oracle `http://www.oracle.com`

Red Hat Software `http://www.redhat.com`
SAP `http://www.sap.com`
SCO `http://www.sco.com`
Silicon Graphics `http://www.sgi.com`
Sun `http://www.sun.com`
SunSoft `http://www.sun.com/software`
Sybase `http://www.sybase.com`
Tandem `http://www.tandem.com`
TCI `http://www.tci.com`
Umax `http://www.umax.com`
WebTV `http://www.webtv.com`

5.2 Software Component Architectures

5.2.1 Executive Summary

A few years ago, distributed computing was a cutting-edge technology used only by early adopters or information technology (IT) groups with very specific technological needs. Today, due to the increasing demand for Internet computing and the enormous complexity of enterprise-wide computing infrastructures, distributed computing has become very much an accepted solution. Distributed computing, however, is still in its infancy when it comes to the technology that supports it. This status is evident in the lack of mature standards and uniform approaches to common issues.

In terms of distributed software components, the competing standards are the Object Management Group's Common Object Request Broker Architecture (CORBA) 2.1, Microsoft's ActiveX, and Sun's JavaBeans. No other standards for distributed components are likely to be successful in today's market. Previous technology proposals such as OpenDoc, Taligent Frameworks and NeXT Objects have come and gone.

Although each technology favors a different aspect of enterprise computing, CORBA 2.1 is the most capable of supporting enterprise computing today. Meanwhile, Microsoft desktops long have used ActiveX technology (formerly OCX) to perform inter-application communication and, recently, computing of a simple, quasi-distributed nature. The JavaBeans specification was released only in 1997, so it is difficult to assess its long-term future. Because JavaBeans relies on the Java Virtual Machine (JVM), its success is wholly dependent on the acceptance of Java as a programming language and execution environment. Both ActiveX and JavaBeans technologies are likely to develop extensions that allow them to interoperate with CORBA object request brokers (ORBs). If Microsoft and Sun, respectively, do not provide these extensions, they will undoubtedly be provided by third parties. Likewise, some kind of interface mechanism between ActiveX and JavaBeans components should be expected.

Meanwhile, enterprises that need enterprise-wide and wide-area communications of a different sort have continued favoring middleware and transaction-processing (TP) monitors. Specifically, companies needing reliable, real-time data distribution, such as financial services firms, have made extensive use of message-oriented middleware (MOM). Firms with large volumes of transactions have continued to rely on TP monitors to secure the integrity of transactions as they span multiple platforms and databases.

Increasingly, however, users of middleware are looking for transaction capabilities, and users of TP monitors are looking for faster pure communications capabilities. As a result, middleware and TP monitors are converging in terms of feature sets and capabilities. Both categories of products lack universally accepted standards for application interfaces. Therefore, each package has its own proprietary feel. If standards were universally adopted and the convergence of features were to continue, TP monitors and messaging middleware would look a lot alike by the end of the forecast period.

As the Internet way of computing pushes deeper into the enterprise, new software developed around the principles of distributed computing will emerge. ORBs that comply with the CORBA specification should become an accepted and pervasive part of most computing environments within the forecast period. The extent to which CORBA is accepted will be the key factor that determines how far into the mainstream distributed computing will go.

5.2.2 Technology Background

Until the 1990s, multi-user computing consisted primarily of a single large processor (generally a mainframe or a midrange system) supporting multiple users through a time-sharing scheme. Users interacted with the processor through a terminal and a keyboard. Terminals were called dumb terminals then because they performed no computing. Instead, they simply displayed character output generated by activities performed on the main processor.

This style of computing was known as host-based computing, but it frequently is referred to as single-tier computing because computation occurs in only one location: the main processor. Single-tier computing has the important advantage of being inexpensive. Adding users to a system can be done at minimal cost and adds very little resource management to the computing infrastructure. Today, single-tier computing commonly is found in mainframe-based applications and in small offices (where it typically runs on a high-end personal computer operating some version of UNIX) where common, simple transactions such as data entry predominate.

The advent of the personal computer (PC) and its arrival on business desktops began changing this computing model because PCs were intelligent devices capable of performing many computational tasks (such as spreadsheets) locally. Soon, businesses discovered that PCs not only performed these tasks, but also could be used as stores for local, small databases. To avoid the duplication of data on multiple PCs, and in particular the problem of having multiple local databases that were out of synch with each other, business sites turned to networking PCs so that common data could be shared.

Initially, networking between PCs followed a peer-to-peer model: PCs could access data and documents on other PCs, but there was as yet no centralized repository for all a company's data. Although this process avoided some of the problems of data duplication, it did not solve them, and it created unwanted complexity in managing data that was distributed between large numbers of PCs at an IT site.

Eventually, the model that emerged was to place all data on a central server and have the PC clients access the data as needed. Processing still would be performed locally on the PCs, but all documents and data would rely on a central file server. Eventually, this file-sharing model evolved to a system in which the file servers (now known as database servers) also performed computational work. This model came to be known as client/server computing and constituted the advent of two-tier architectures. The client tier consisted of PCs, and the server tier included servers of all types. In strictly PC environments, the server tier often was a souped-up PC running Novell NetWare; in larger environments, the server could be a traditional mainframe or midrange system.

The client/server model characterized most PC networks until the late 1980s. However, the advent of graphical systems and of sophisticated database needs began to make this model impractical. Moving hundreds or thousands of database records to an individual PC so it could select those needed to prepare a single report began to consume significant resources. Specifically, the shipment of these records consumed significant network bandwidth. In addition, once the records arrived at the PC, the PC often was overwhelmed with the computing tasks, especially if they also involved complex, graphical representations of the subject data. Therefore, in the early two-tier computing models, network bandwidth often was used wastefully, and PCs often were tied up for long periods performing complex processing.

Eventually, client/server computing evolved to perform many data management tasks on the server itself. This transition conserved network bandwidth and, for certain applications, freed client PCs to do other work while waiting for the results of a query or a report.

Predictably, this approach encountered the problem of overloading servers, which now were saddled with the task of providing database access and performing much of the enterprise computation. (Interestingly, the now-freed resources of the PC were quickly put to use by new versions of the operating system and by business applications such as PowerBuilder that required significant processing power from the client.) Eventually, a third tier appeared in enterprise computing: the application server. In this scheme, the processing logic for business applications was performed on an application server that interacted with a separate database server; meanwhile, the client handled data presentation and some data validation. There were now three distinct tiers: client PCs, application servers, and database servers. This model proved successful and frequently is found today at businesses of all sizes.

Three-tier computing quickly lost its conceptual simplicity, however. The two driving causes of this change were the increasing heterogeneity of computing in IT departments and the advent of large, multi-site enterprise computing architectures. The latter driver was given further impetus by the integration of the Internet into IT needs. Soon, multiple application servers were accessing multiple database servers (often running databases from different vendors) while being accessed by a variety of clients such as dumb terminals, PCs and Macintoshs, UNIX workstations, and so on. Computing now was distributed throughout the enterprise, and rather than the clean division of labor of the three-tier model, most enterprises were better described as having *n*-tier models, where *n* represented any number larger than three. In many cases, *n* was a fairly large number.

To address the difficulties of developing applications for such a widely disparate architecture and to handle the management of so many tiers, two different technologies evolved. The older technology is messaging middleware, which arose from the success of TP monitors. The newer technology, which is seen as the next wave of enterprise computing, is distributed objects.

Middleware relies on a generally proprietary communications mechanism that rides over standard networking protocols, and it has been successful in sites where this sort of structure can be implemented. Distributed objects, in contrast, rely on standardized interfaces and universal transport protocols. As such, distributed objects are ideally suited to Internet/intranet computing in addition to computing within the enterprise. Figure 5-1 illustrates how these approaches to client/server computing have built upon each other.

Figure 5-1: The Three Waves of Client/Server Computing

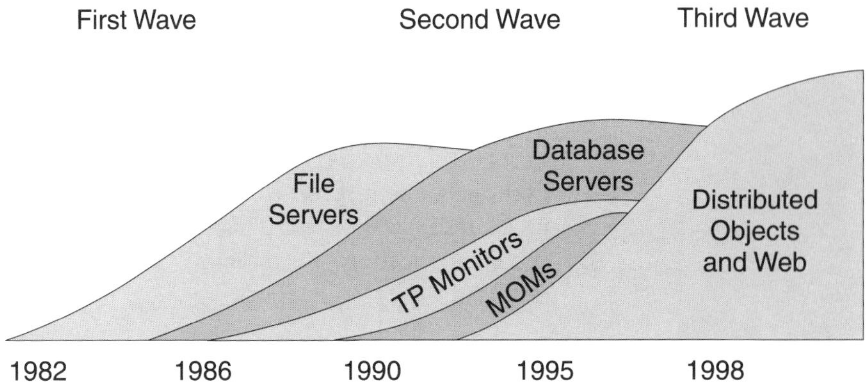

Source: *Instant CORBA*, 1997

Middleware

Middleware originally was designed as a solution to the problem of developing software for heterogeneous and *n*-tier computing infrastructures. Specifically, the question was how to write software that could access databases such as Microsoft SQL Server, Oracle, and Sybase while supporting clients running Windows, MacOS, and UNIX and while interfacing with application servers that ran on UNIX boxes. Beyond the problem of writing the application for such a heterogeneous environment was the question of maintaining the application if a new database server was added or a new client operating system was introduced.

The solution middleware offers is to provide developers with a uniform interface through which their programs can access enterprise resources. The developers interact with only the middleware layer, and the middleware then performs the necessary translations for the respective databases, operating systems, and clients. In this sense, the middleware presents the developer with a single, consistent interface that masks the complex computing infrastructure. The middleware package – generally a large, intricate, and expensive piece of software – is installed in the enterprise between the application servers and the clients on one side and the database servers on the other.

Of the many categories of middleware, modern TP monitors come closest to realizing the original goal of the solution. TP monitors initially were devised to ensure the integrity of transactions in high-volume, on-line transaction processing (OLTP) environments. Since then, they have been extended to be effective tools for masking client and database heterogeneity. However, TP monitors are primarily transaction-oriented and rarely are used as a pure communications medium across different processes in the enterprise. This latter goal – the ability to communicate data across disparate platforms, multiple sites, and through or within firewalls – has become the main driver for generic middleware. Pure middleware products now relegate themselves to the primary task of providing the communications infrastructure for data processing within the enterprise.

Middleware originally was built on a technology called remote procedure calls (RPCs), but it has evolved to a message-oriented approach, which is the most common implementation today.

■ Remote Procedure Calls

RPCs allow a program running on one computer to call a procedure (similar to a function or subroutine) that executes on a remote machine and performs a single, discrete task. Generally, the RPC sends a request in the form of some data to the remote procedure and waits until it obtains a reply. RPCs long have been a part of the UNIX operating system, and today they function in many high-end client/server operating systems such as Windows NT.

This approach is conceptually simple and can be learned by any experienced programmer. However, it suffers from several drawbacks. First, the program containing the RPC call must wait for a response to the call before continuing. This approach (known as request-reply) depends on synchronous operation: The request must be responded to before processing can continue. This process requires guaranteed communication and suffers from difficulties if the remote machine is busy, offline, or if it cannot return a response quickly.

Second, RPCs are very tedious to write. Due to their conceptual simplicity, numerous RPCs must be generated for even straightforward work to be done between computers. Each of these RPCs must be written individually, requiring considerable effort on the part of developers. Finally, RPCs allow point-to-point communication between machines but do very little to mask the details of implementation from the developers. Therefore, although RPCs are effective communication devices, they are not suited to serve as the principal form of middleware.

RPCs are generally not sold as stand-alone packages. UNIX provides one model of RPC, Windows NT another (although similar) model. The primary third-party RPC implementation comes from The Open Group as part of the Distributed Computing Environment (DCE). DCE is a massive, enterprise-wide package that is supposed to provide a comprehensive set of services for distributed computing specifically optimized for business processing. It includes modules for security, clock synchronization, distributed file systems, and many other purposes. However, the lack of commercial success of DCE has thwarted attempts to assess its capabilities to deliver on these claims. A notable exception to the absence of third parties is Noblenet, whose product RPC 3.0 provides an RPC infrastructure and numerous software development tools that ease programming for RPCs.

■ Message-Oriented Middleware

MOM uses a different approach from RPCs. Messages are sent asynchronously to a destination. That is, they are sent there, but the sending program does not require a reply before continuing operations. This approach generally is implemented using one of two mechanisms: publish-and-subscribe or message queuing.

In publish-and-subscribe, the message is sent (or published) and the target destination awaits receipt of messages specifically addressed to it or to a destination in which it has expressed interest (subscribed). In this model, many different targets could subscribe to the same data. For example, several programs could be waiting for real-time data to come in from the factory floor. Rather than having the real-time data collection process figure out who is interested in its data, the program simply publishes the data to a predetermined target, whereupon all subscribers would receive the data. Another example would be sending the same HTML page to numerous intranet users.

Publish-and-subscribe does not work well in "store-and-forward" situations, where data must pass over several network segments and be stored safely at each segment before being sent on.

Under the message queuing approach, the message is sent to the target's queue. Then the target, on its own schedule, can check the queue for messages it needs. This approach is suited to communication with devices that might be offline, such as computers of mobile users. The messages pile up in the queue and await connection with the target to begin delivery.

In both cases, the middleware package manages the messages, making sure they are delivered to the correct queue or target in real time and that they are sent only once.

How does the original program obtain an answer if its computation is predicated on a return message? It checks its own queue for messages from the original destination system. If it cannot proceed without the specific reply, it will continue checking the queue until the reply is received (or until some preset interval has been established, whereupon it views the delay as an error condition). In this manner, if the program needs the reliability of a request-respond model as is provided by RPCs, it can implement it with MOM, but the program has the additional asynchronous capability of sending messages that do not require a response.

The absence of an immediate response does not create a reliability issue for asynchronous operations based on MOM. If for some reason the MOM cannot deliver the message, the MOM will perform some predetermined action. The two most popular options involve use of a callback: Either an error function is called by the MOM to inform the sending program of the error, or the connection uses multiple threads for communication, one of which is used to signal the error. Therefore, a built-in mechanism ensures notification if a message cannot be delivered or if some other error occurs.

In practice, MOM has not been able to fulfill the original mission of providing a uniform interface across heterogeneous clients, operating systems, and databases. The main reason for this failure has been the near-exponential growth in complexity of most enterprise computing infrastructures. Instead, MOM has emerged as a messaging infrastructure: It provides enterprise-wide communication of data, but third-party products then perform the translation to the individual databases or applications as needed. An example of this arrangement is TSI International's Mercator translator package that plugs into IBM's MQSeries MOM. Many sites choose to write their own custom interfaces to the MOM.

This problem highlights one key drawback of messaging middleware: No standards for the message formats exist. Every vendor has its own format and design for messages. As a result, a key criterion when assessing a specific middleware package is the extent of available third-party software that can plug directly into the MOM.

Transaction Processing Monitors

TP monitors have their roots in early mainframe processing. The first widely implemented TP monitor was IBM's Customer Information Control System (CICS), which was used to ensure the integrity of on-line transactions. At the time, most database engines were hierarchical and used numerous indexed files to hold data. The primary technologies were the Indexed Sequential Access Method

(ISAM) and the Virtual Storage Access Method (VSAM). ISAM- and VSAM-based transactions often updated several files simultaneously and therefore needed some process to ensure the transactions were recorded correctly across all files. CICS provided this mechanism to developers who could make calls to CICS from within COBOL programs whenever they were ready to commit a transaction to permanent storage. CICS is used today primarily in mainframe sites and by COBOL programmers. However, its capabilities have been expanded significantly, and it is capable of handling complex transactions that can involve databases rather than just ISAM files.

Modern TP monitors provide the same transactional integrity across numerous client platforms and servers and interface with most leading databases. In addition to verifying the transaction, they enforce numerous other data-processing policies, such as security, transaction and error logging, data replication, and so on.

Traditionally, TP monitors have used synchronous communication: For every action they perform, they require verification that the action has occurred correctly. This requirement is inherent in the need to provide data and transactional integrity. Although this function is suitable for OLTP, the response delay built into this model means that TP monitors have not functioned well as pure message-passing tools (such as for real-time data feeds) in the same manner that MOM currently is used. Recently, however, TP monitors have added asynchronous communications to their feature list and now provide message queuing similar to most MOM. This functionality is provided separately from the TP monitoring activity. However, this development underscores the increasing convergence of MOM and TP monitors.

TP monitors have retained a singular commitment to making transaction integrity in a heterogeneous environment as easy as possible for programmers to develop. Most TP monitors, despite supporting a wide variety of clients and servers, use only 30 or 40 application programming interfaces (APIs), so most developers can avail themselves easily of TP monitors' capabilities. In addition, The Open Group has published a set of standard APIs for TP monitors, making their appeal to developers even stronger.

5.2.3 Distributed Computing

Although RPCs, messaging middleware, and TP monitors provide a communications mechanism by which business processes can be performed across a wide variety of platforms, they share a common drawback in that the middleware layer typically is proprietary. This layer is defined and maintained by one vendor, and all tools and applications must integrate with the layer to be able to derive its benefits.

This model may work well within a single enterprise, where a uniform environment can be created; however, it begins to be difficult if Internet computing and substantial remote access are necessary. Dial-in customers, extranets, and Web site visitors may need to perform business transactions with a company, but they cannot be expected to have chosen the necessary software to plug into the messaging middleware used by the vendor site. Therefore, companies with these needs increasingly are looking to distributed objects to allow global access to resources while hiding platform and database heterogeneity.

A method gaining increasing popularity in this context is based on ORBs. These software products allow software components to interact across the enterprise. To understand how ORBs work requires examining the nature of objects and components.

How Components Work

In their simplest form, components are parts of a program that can be called by the main program to perform certain specific tasks. Some of the earliest component-like technologies were dynamic link libraries (DLLs). These libraries, common in UNIX for more than 10 years and in the Windows family of operating systems since the early 1990s, work in the following manner. A DLL will consist of a series of related functions, for example, functions that will render three-dimensional (3-D) graphics on the screen. This DLL is known to exist on the system at some predefined location (on UNIX, this location is almost always */etc/lib*; on Windows systems, the location can change, but it must be somewhere on the execution path as specified in the user's environment). When a function in the DLL is called, the DLL is read into memory and linked into the main program as if the DLL always had been part of the original program. Once linked this way, any and all functions in the DLL can be called and executed without further steps.

Components used in distributed computing need to possess several key characteristics to work correctly. The two main traits are borrowed from the world of object-oriented technology: encapsulation and data hiding. (See Chapter 6.1, *Application Development Environments,* for a discussion of other aspects of objects.) To see how encapsulation and data hiding work requires an examination of how programming was done prior to object orientation (OO) and how OO has affected it.

For example, in writing a program that computes the physical size of drawn figures, a useful calculation would be obtaining the circumference of a circle. In pre-OO days, programs often had distinct data and instruction sections, as shown in the left panel of Figure 5-2. In the data section, π would be defined as having the value 3.14159…. Sprinkled throughout the instruction sections would be lines of code that would compute circumferences. Some might use the formula, *circum* = π *x diameter*; others might prefer *circum* = π *x 2 x radius*. Either way, the correct result was generated.

The problem with this approach was that if the formula had to be changed (always to use the radius, for example), the entire program had to be reviewed to look for calculations of the circumference. However, users never could be sure they had found all occurrences because a programmer might have used variable names that did not easily suggest this particular calculation was being performed.

OO encapsulates the routines that perform discrete functions. In an OO program (see the right panel of Figure 5-2), an object has been defined; pass it the diameter, and it returns the circumference. All calculations of circumference are required to use this particular object. As a result, the complete function is encapsulated in this object, which has a distinct interface to the world. Now, if the interface has to be changed to accept a radius rather than a diameter, only the code in one place of the program must be changed, and all instances are immediately updated. This benefit is known as encapsulation.

Data hiding addresses a different problem. Suppose a programmer performs some maintenance on the old-style, non-object program and is unaware that π already has been defined in the data section. The programmer might define a new variable and use a different value (perhaps rounding off π after three digits). Now, the program might give slightly different results for the same calculation. Data hiding places data needed by an object's functions within the object (see the right panel of Figure 5-2) where it can be accessed only by specially designated functions in the object itself. Data hiding is a critical trait of distributed objects: they must carry with them all the data they will need when called. Note that the hidden data is by design limited to context-insensitive information. Variables that depend on context (such as the radius needed for calculating a circumference) generally are sent to the object when it is invoked.

Figure 5-2: Traditional Program Functions versus Objects

Traditional Programming:

```
Data Section
        π = 3.14159
        . . . .
Instructions
  . . .
CIRCUM = π x diameter
  . . .
  . . .
CIRCUM = π x 2 x radius
  . . .
```

Object-Oriented Programming

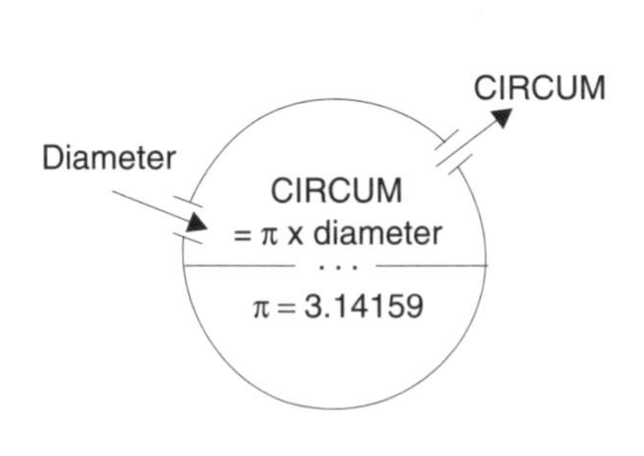

Source: *Instant CORBA*, 1997

In this context, components can be defined as being a group of related functions encapsulated with data into an object that can be called by a running program to perform a specific task. The calling program can be on the same machine, on a different machine within the enterprise, or on a completely different continent connecting to the calling program across the Internet. Examples of objects typically found on the same machine are ActiveX components and JavaBeans; objects found across the enterprise would be CORBA objects plus the previous two; objects found across the Internet would be CORBA objects and Java applets. These are just the typical settings, however. Any one of these component types can be found anywhere in the enterprise.

CORBA and ORBs

For components to work successfully across the enterprise, many vendors felt it was necessary to band together and form standards by which components could be defined to a central authority and invoked with the knowledge that any standard-compliant component could be found and activated successfully. A group of vendors came together in the early 1990s as the Object Management Group (OMG) to fashion a set of guidelines for distributed computing with components.

Unfortunately, the group chose to use the term "object" rather than "component," forever causing confusion. OMG-defined objects are really components. Components tend to be larger than simple objects – often, they are built from objects – and they tend to lack some characteristics peculiar to objects. In addition, components need not be written with object-oriented programming languages. Nonetheless, because the term "object" is used by the OMG to refer to components, discussion of distributed computing often uses these two terms interchangeably.

The first document produced by the OMG was CORBA Version 1.0. Since its first release, the document has been updated several times. CORBA 2.1 was released in late 1997.

The CORBA model uses ORBs to handle requests from programs to components and from components to each other. ORBs are programs that serve as traffic cops: They handle the requests, find the requested components, call them, and return the results. To enable these operations, CORBA defines a standard way for components to specify their interfaces: the kind of data they expect to be handed and the kind of data they expect to return. The standard way of defining interfaces is known as the CORBA Interface Definition Language (IDL). Using IDL, a programmer can identify that a component will accept an integer (for example, the diameter in centimeters) and return an integer (the circumference, rounded to the closest centimeter). These interfaces are in IDL (the syntax of which looks similar to C++). A translator program then generates code for the component: client stub code and server skeleton code. This code is the basis on which objects can speak to each other. Once the IDL definition is written, the interface definition can be placed in the Interface Repository, a database where all object interfaces that a given ORB can recognize are stored. For an ORB to interact with a component, the component's interface information (technically, its metadata) must first exist in the Interface Repository.

When a program needs to invoke a given component, it can use one of two approaches. The most common approach is to use the client stubs generated by the IDL translator to specify to the ORB which object the program wants. The ORB then will locate the object, once again relying on its interface definition. Once the component is located, the ORB begins routing information back and forth between the client and the component until the processing is finished. (Note that the component may rely on other components it will need to access through the ORB. The component then becomes a client for the extent of those communications and goes back to its work at the behest of its original client.) This approach is known as static invocation. The client program knows *a priori* of a specific object with which it wants to interact and gives this information to the ORB to handle.

A second, less-common, and slower approach is called dynamic invocation. Using this approach, a client looks in a yellow pages-type database to see which objects currently are available through the ORB. It then will contact one on the basis of the stored interface definitions and begin interacting with it through the ORB. Because the client program will not know *a priori* what objects are available, it cannot have client stub code for the object available. In this context, the presence of the Interface Repository and the generally self-describing nature of CORBA objects becomes extremely useful. (Figure 5-3 illustrates the two ways of invocation.) Dynamic invocation might be useful with Internet- or Web-based computing, for example, where a program may not know ahead of time what components will be available.

Figure 5-3: Two Ways of Invoking Objects Using a CORBA ORB

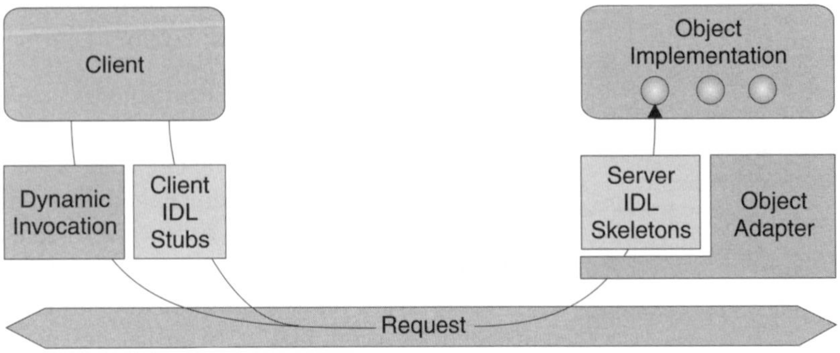

Source: *Instant CORBA*, 1997

The object adapter (shown on the right side of Figure 5-3) is a piece of software that controls how a component is invoked. For example, it may be necessary to invoke the component by spawning another thread on the remote system. This task would be identified and executed by the object adapter.

IIOP

Introduced with the CORBA 2.0 specification in 1996, the Internet Inter-ORB Protocol (IIOP) is a key extension to the CORBA environment. It solves the problem of how ORBs communicate with each other. Prior to CORBA 2.0, the OMG had begun designing a protocol for inter-ORB communication. This protocol came to be known as the General Inter-ORB Protocol (GIOP). It specified the format of seven key data messages ORBs would need to share data, and it detailed how data should be sent between ORBs. Specifically, the data needed to follow the formatting rules of the Common Data Representation (CDR). CDR rules compensate for the fact that different processors use numbers in different formats (for example, on Intel machines, a two-byte number is stored with the low-value byte first; on RISC machines, the high-value byte generally is stored first).

The first implementation of the GIOP specification became IIOP, which is a slightly modified version of TCP/IP. In fact, IIOP is TCP/IP with some extra message formats added.

Interestingly, the CORBA 2.0 document allows for special inter-ORB protocols for environments that have specific needs not handled optimally by IIOP- or GIOP-style protocols. These Environment-Specific Inter-ORB Protocols (ESIOPs) are found in ORBs from Digital (Object Broker) and HP (ORB Plus).

5.2.4 Java-Based Component Architecture

JavaBeans is a specification for components that can be deployed on any Java platform. Like Java applets, JavaBeans components (Beans) can be used to give Web pages or other applications interactive capabilities such as computing interest rates or varying page content based on user or browser characteristics. JavaBeans provides a component development environment to extend Java's cross-platform "write once, run anywhere" capability to the development of enterprise-class applications with the features of component reusability. JavaBeans has emerged as a component definition to create distributable, reusable objects within the Java environment. The Java-based component model has followed a slightly different evolutionary path than Microsoft's Distributed COM (DCOM), although both had their foundations in client-side graphical user interface (GUI) component development.

The Java model did not emerge from an RPC-based component definition that grew to include distribution and object interoperation. Instead, the JVM concept and associated Java programming language created an environment where applets could be written and then fetched across the Internet and run anywhere. As the applet concept grew and Java began to take hold as a cross-platform development language, JavaSoft (a division of Sun) sought to offer a component model by providing JavaBeans as a component definition. The actual component architecture model is a compilation of facilities from both SunSoft (a subsidiary of Sun) and JavaSoft, the key to which is Java. Fundamental to a working Java-based component architecture model are the Java-based APIs that provide the operating services on top of the Java language environment. The collective component model includes Java, Java-based APIs, and JavaBeans.

In essence, JavaBeans is a component specification for the Java language. Initially intended for the Sun-originated Java and aimed at NEO (Sun's first commercial ORB, which Sun abandoned in January 1998), it has developed greater compliance to CORBA standards and broadened its interface capability to become a true, stand-alone component framework.

JavaBeans

A Bean is a piece of Java-based software that can range from a small visual control such as a progress indicator to full-fledged applications such as spreadsheets. A Bean is a specialized, primarily visual component Java class that can be added to any application development project and manipulated by the Java builder tool. Beans can be combined and interrelated to build Java applets or applications and as such must be executed within a larger application (the hosting application is called a container). Beans also can be used to create new, more comprehensive, or specialized Beans.

Beans can be flexibly manipulated, but not executed, without the requirement for a dedicated object container. The properties of a Bean are attributes, which describe the state of the Bean such as color, position, and title, and methods, which are the actual functions to be performed by the Bean. Bean methods include functions that can be performed on the Bean itself to manipulate its internal state or to interact externally with other objects. Methods also can respond to events, which are actions performed when special internal or external signals occur such as a mouse click. Beans also can implement conceptual and nonvisual objects such as business rules. In these regards, JavaBeans do not differ much from other components, except for their requirement of a Java implementation.

Bean features are exposed to development tools through a process known as introspection, which analyzes the Bean at design time or runtime to determine the properties, methods, or events the Bean supports. BeanInfo classes provide advanced control over a Bean's features. A separate introspector class is used by the builder tools to analyze the design patterns of a Bean and its associated BeanInfo.

To build a component with JavaBeans, Java programs are written to include JavaBeans statements that describe component properties such as user interface characteristics and events that trigger a Bean to communicate with other Beans in the same container or elsewhere on the network.

When the components, or Beans, are in use, the properties of a Bean (for example, the background color of a window) are visible to other Beans. In this way, Beans that have not "met" before can learn each other's properties dynamically and interact accordingly.

Beans also have persistence, which is a mechanism for storing the state of a component to disk. This capability would allow, for example, a component (Bean) to "remember" data that a particular user had already entered in an earlier user session.

Enterprise JavaBeans (EJB)

EJBs are specialized, nonvisual JavaBeans that run on a server. The EJB 1.0 specification defines a component model to support multi-tier, distributed object applications. EJBs extend JavaBeans from its origins as a client-side GUI component set to a server-side component model, thus simplifying the process of moving application logic to the server by automating services to manage components.

Server components execute within a component execution system, which provides a set of runtime services for the server components, such as threading, transactions, state management, and resource sharing. This requirement means that EJBs do not need the standard JavaBeans container context to execute.

EJB provide hooks for enterprise-class application developers to build reusable software components that can be combined together to create full-function applications. The hooks include interfaces to common development environments and richer class libraries geared toward the enterprise.

The EJB model supports implicit transactions. To participate in distributed transactions, individual EJBs do not need special code to distinguish individual transactions and their operations. The EJB execution environment automatically manages the start, commit, and rollback of transactions on behalf of the EJBs. Transaction policies can be defined during the deployment process using declarative statements. Optionally, transactions can be controlled by the client application.

The EJB model defines the interrelationship between an Enterprise JavaBean component and an Enterprise JavaBean container system. EJBs do not require the use of any specific container system. Any Enterprise JavaBean can run in any application execution system that supports the EJB specification. At minimum, the EJB Server must provide a container for Enterprise JavaBeans, called an Enterprise JavaBeans Container (EJB Container). The EJB Container implements the management and control services for a class of EJB objects. (See Figure 5-4.)

Figure 5-4: The Architecture of Enterprise JavaBeans

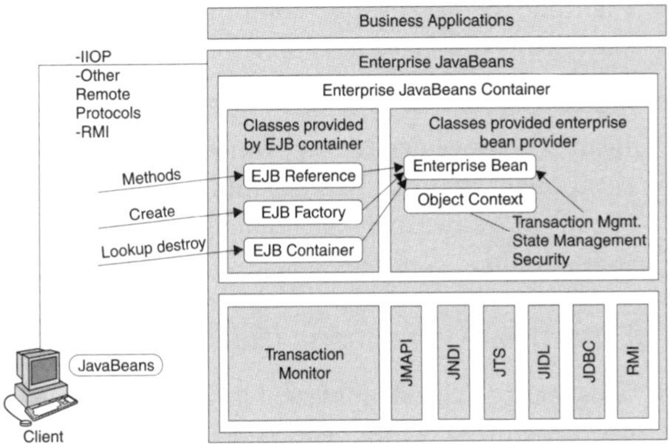

Source: JavaSoft, 1998

JavaBeans and its Enterprise implementation are all part of JavaSoft's Java Platform for the Enterprise, involving the migration of Java beyond the desktop to network and application servers. Enabling Java on the server allows for development of Java applications and JavaBeans that can be deployed on either a desktop or server level, extending the "write-once, run-anywhere" concept of the Java platform.

The EJB model plays a pivotal role in the Enterprise Java Platform. The EJB environment provides life cycle services for enterprise application components. The Enterprise Java Platform itself is not a component container implementation. It is based on an extended API set geared to support enterprise-level transactions, and it provides a common interface to a variety of system services such as naming, directory, messaging, management, inter-process communication, database, and transaction. The enterprise API leverages existing Java APIs (explained later) such as Java Database Connectivity (JDBC), the Java Remote Method Invocation (JRMI), and the Java Interface Definition Language (JIDL) plus the Java Naming and Directory Interface (JNDI). The extended features include the Core Reflection API, a low-level collection of classes that supports the introspection mechanism; the Object Serialization API for maintaining object persistence as well as improvements in the Java Event Model; and the Java Abstract Windowing Toolkit.

EJBs are based on Version 1.1 of the Java Development Kit (JDK 1.1), which includes the enterprise-level APIs. The JDK comprises an event model, a mechanism that objects use to interact dynamically in response to external or internal conditions, providing different classes for varying events and event sources and listeners. The EJB model utilizes the Java security services supported in JDK. A major change in JDK 1.1 is the addition of database access classes in the form of JDBC APIs.

The primary APIs used in the Enterprise Platform include the following:

- **Java Transaction Service (JTS)** – JavaSoft has adopted the OMG's Object Transaction Services (OTS) as the basis for JTS. JTS is used to ensure interoperability with various transaction management services and is geared toward use by TP monitor vendors to create a higher-level API for their specific TP monitors.

- **JDBC** – JDBC is a set of Java classes containing a new SQL call level interface that contains a set of methods for loading database drivers. JDBC also connects to a database, prepares and executes SQL statements, and terminates the session.
- **JRMI** – This API provides inter-object communication within a pure Java environment. JRMI is an extension to the Java language that enables method invocation on remote objects using native Java features such as object serialization. JRMI is primarily a Java-to-Java ORB equivalent that uses native facilities of Java for inter-object exposure, discovery, and method invocation without using an IDL. JRMI stubs provide client proxies for server objects, prepare the data that will be passed in the remote calls, and then make the calls to the server. The JRMI server skeleton receives the remote call, separates the parameters, and then calls the appropriate implementation class. JRMI invocation passes local objects in parameters by value, rather than by reference. JRMI also can pass an object's state and behavior – effectively a code-passing procedure not found in more conventional ORBs.

  JRMI runs over both the native Java Remote Method Protocol (JRMP) and IIOP. Adding IIOP support to JRMI creates a single object transport protocol for use by both JRMI and CORBA-based objects.
- **JIDL** – JIDL used to define inter-object communication between Java objects and CORBA objects. A CORBA/Java ORB is a CORBA IIOP ORB entirely written in Java for portability. The ORB must be able to generate Java language bindings from CORBA IDL, and any code generated by the IDL compiler must be in pure Java. Using a Java ORB, ordinary Java applets can invoke methods on CORBA objects directly using the IIOP protocol over the Internet.
- **JNDI** – Provides a common interface to a variety of naming and directory services such as NDS and LDAP.

JDBC, JRMI, JTS, and JNDI are all APIs that, coupled with JavaBeans, form a middleware framework of which Beans are the component and Java the defining language.

5.2.5 Microsoft's Component Object Model (COM)

Microsoft's Component Architecture is based on Microsoft's COM, which was originally designed to provide interoperability between Windows applications (Microsoft Office in particular). Microsoft's COM supports development of components in the form of ActiveX controls, compound document services and frameworks in the form of DocObjects, and service components in the form of COM objects (OLE) and their automation.

The COM model is extended by Distributed COM (DCOM), allowing clients to access components across a network and supporting client-to-server and server-to-server connections between Windows 95 and Windows NT systems. The COM model is extended further by Component Object Model Services (COM+), which supports platform-independent development of componentized applications over standard protocols such as TCP/IP.

Microsoft bundles COM/DCOM along with a series of other framework facilities and development tools into Windows NT Enterprise, taking the view that component models and frameworks are integral elements of the operating system. Like many component architectures, COM/DCOM is dependent on object communications brokering. However, unlike other component architecture implementations, Microsoft's approach must communicate across three networking protocols (NetBIOS, Novell IPX/SPX, and TCP/IP), which reflects of the mixed heritage of PC component technologies and the lack of industry standards for PC desktops. Microsoft is using Microsoft Transaction Service (MTS) as an abstract layer behind which the three protocols can be bridged. Although not explicitly part of DCOM, MTS is an important element in the Microsoft view of distributed applications and comes bundled in Windows NT Server, Enterprise Edition.

COM and ActiveX

COM principally provides the architectural abstraction for OLE objects. COM supports an implementation typing mechanism centered around the concept of a COM class. A COM class has a well-defined identity, and a repository (the system registry) that maps implementations (identified by class IDs) to specific executable code in the form of ActiveX controls. (ActiveX controls were formerly known as OLE controls, or OCXs.) COM Automation allows applications such as Microsoft Internet Explorer, Microsoft Office, and development tools such as Visual Basic to access COM objects.

From a COM point of view, an object typically is a subcomponent of an application that provides a public interface to other parts of an application or to other applications. OLE objects usually are document-oriented and are tied to some visual presentation metaphor. The typical domain of a COM object is a single-user, multitasking, GUI-based desktop running Microsoft Windows. The primary goal of COM and OLE is to expedite collaboration and information sharing among applications using the same desktop, as in the case of the Microsoft Office suite. This view of COM is different from the CORBA perspective in which an object is an independent component providing services to client applications.

COM components can run within a stand-alone process as an executable or within a container process as a DLL. Most COM interfaces are defined in Microsoft Interface Definition Language (MIDL).

COM supports an extension referred to as OLE Automation, in which compatible interfaces can be described in an Object Definition Library (ODL) and, optionally, registered in a binary TYPE library. Essentially, automation interfaces have both properties and methods, whereas COM interfaces have only methods. OLE automation provides more classic object-like capabilities to the COM environment, such as the ability of one application to start up another (a primary benefit).

DCOM

To extend COM's reach into distributed environments, Microsoft extended the COM model with DCOM, allowing clients to access components across a network. To access a COM object on some other machine, the client relies on DCOM. DCOM transparently transfers a local object request to the remote object running on a different system.

To pass COM requests between systems, Microsoft uses an implementation of DCE RPC technology to enable DCOM. Although providing higher performance than DCE RPCs, Microsoft's RPCs still are bound by the synchronous nature of the RPC programming. Middleware originally was built on a technology called RPCs, but it has evolved to a message-oriented approach, which is the most common implementation today. (See "Remote Procedure Calls" on page 321.) DCOM's core network protocol is Object Remote Procedure

Call (ORPC), which comprises a series of extensions to DCE RPCs that supports DCOM operations, such as providing a primitive data type to support object references. (See Figure 5-5.)

Figure 5-5: Microsoft's DCOM Architecture

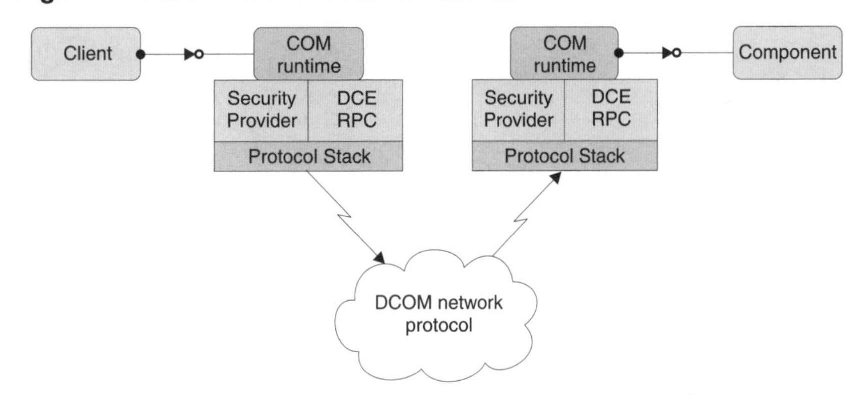

Source: Microsoft, 1997

DCOM provides functions similar to the naming, security, persistence, and versioning services found in CORBA. For directory services, DCOM relies on Active Directory, essentially a combination of Domain Name System (DNS) and Lightweight Directory Access Protocol (LDAP).

Because DCOM uses a proprietary implementation of the DCE RPC for its component architecture, COM/DCOM is limited to use within a homogeneous Microsoft environment. In fact, a major stumbling block in distributed computing is the absence of real interoperability between COM/DCOM objects and CORBA components. Even getting these two component models to communicate with each other has proven difficult. Limited communication between these component types is possible through some third-party products. The notable exception to this rule is Object Bridge from Visual Edge Software, which provides true interoperability between CORBA and COM/DCOM (most other packages simply allow CORBA-to-OLE Automation interoperability). The difference is that the latter approach requires use of CORBA's dynamic invocation function, which is significantly slower than the more common static invocation used among CORBA components.

Realizing the inherent limitations of the DCE-based MS-RPC in DCOM, Microsoft has developed two additional wire protocols rather than create a more extensible wire protocol to service all levels of brokering required within a framework. These protocols are Microsoft Transaction Server (MTS) and Microsoft Message Queue (MSMQ). MTS architecture is transport independent, and DCOM can exist behind a transport abstraction layer of MTS, thus allowing Microsoft to replace DCOM with MTS as its ORB protocol. Microsoft also is promoting the asynchronous MSMQ protocol and an enriched MTS protocol for high-volume transaction environments.

Microsoft Transaction Server

MTS is a component-based TP system for developing and deploying transaction-oriented processing on Windows NT Server. MTS also defines an application programming model for developing distributed, component-based applications, and it provides a runtime infrastructure for deploying and managing these applications.

MTS was written using COM and uses DCOM to communicate with resource managers and other MTS servers on a network. When COM components are installed into MTS, any client and server components that are distributed across more than one node and that must use DCOM to communicate are managed automatically.

MTS is bundled into Windows NT Enterprise Edition and acts as a container for COM service components. It defines classic services such as multithreading, concurrency control, and shared resource utilization, and it relies on native Windows NT services to provide security, naming, and systems management services. Other MTS services include pooling, queuing, connection, and security, all architected within the COM/DCOM model.

Figure 5-6: Architecture of the Microsoft Transaction Server

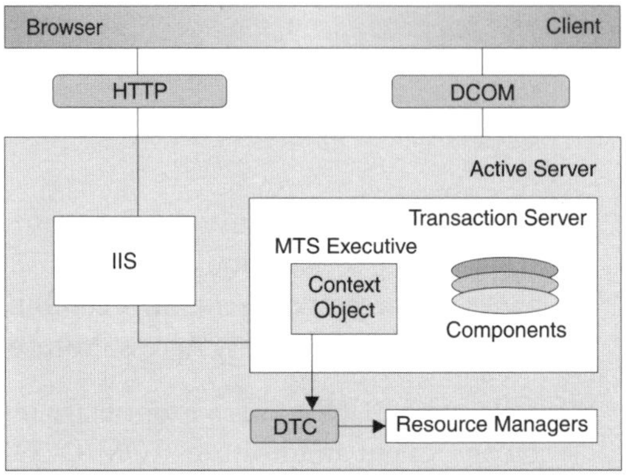

Source: Hurwitz Group; Patricia Seybold Group, 1997

The MTS protocol is designed to accept automatically COM/DCOM-based applications with some additional configuration and DLL packaging. Therefore, it acts as the primary bridging protocol for applications written directly to WinSock, Remote Automation, direct MS-RPC, or COM/DCOM specifications. The MTS protocol specifically allows asynchronous application operation, using MSMQ technology as an enabling protocol.

MTS is enabled by MSMQ – an operating system service providing store-and-forward connectivity between two application programs. MSMQ enables two programs to communicate with one another using queues and includes a queue manager responsible for routing messages to their destination. It also supports transactions, namely transaction delivery using the Distributed Transaction Coordinator (DTC) as the transaction manager.

DTC is a general-purpose transaction manager architected originally to supply two-phase commit to Microsoft's SQL Server relational database. DTC has been extended to provide general-purpose interfaces capable of providing transaction management for program-to-program interactions under MTS. Operating as a runtime middleware environment within DCOM, DTC hosts application program modules and endows them with transaction semantics.

COM+ and Active Server

COM+ supports the development of platform-independent, advanced applications over standard protocols such as TCP/IP. The primary goal of COM+ is to provide additional tools to the COM/DCOM infrastructure to ease management of object technology development, further advancing the COM/DCOM model into a model more generically aligned to other software component models. COM+ provides security and memory management as well as a data-binding feature that allows binding between object fields and specific database fields. COM+ enables components to redirect their service requests dynamically by calling various services at runtime rather than being bound to a single implementation of service. In addition to receiving and processing events related to instance creation, calls and returns, and errors and instance deletion, interceptors also provide the mechanism to implement transactions and system monitoring, further intertwining the COM/DCOM and MTS architecture.

COM+, in conjunction with MTS and the underlying COM/DCOM architecture, forms part of Microsoft's Active Server, an assembly of server-class capabilities geared toward delivering transaction and courier services to any COM-interfaced component, including Java applets. Active Server supports all three wire protocols in Microsoft's component architecture plus HTML and ActivePages.

Given the popularity of the Java language, particularly in developing Internet/Intranet applications and services, Microsoft has embraced Java in its overall development environment. However, the Java language is the only aspect of Java that Microsoft is embracing because the remaining elements of the Java environment (such as Sun's implementation of the JVM, JavaBeans, and the Java system APIs) are viewed as direct competition to Microsoft's operating systems.

Microsoft does not include Java Foundation Classes (JFC) in its implementation of Java, replacing them instead with its own Application Foundation Class (AFC) libraries, which require developers to write to the Windows APIs. An additional facility to tie Java into the Windows environment is J/Direct, a feature within Microsoft's implementation of the JVM that supports direct calls to the entire set of Win32 APIs. With J/Direct, developers can write Java applications that depend on Win32 DLLs without using native methods, thereby subverting the "run-anywhere" aspect of the Java promise.

5.2.6 Market Overview

Message-Oriented Middleware

Wintergreen Research expects that the MOM market will reach $1 billion by the year 2001. The firm also expects sales of add-on tools that interface to MOMs and provide integration with individual applications linked by the MOM will be as high as $8 billion by then. As of late 1997, the total MOM market was estimated to be $173 million.

IBM's MQSeries owns approximately a 45 percent share of the MOM market (based on revenue); next is TIBCO, with 25 percent market share. The rest of the market is divided among numerous vendors. Microsoft expects to enter this market in early to mid-1998, when it releases its MSMQ messaging product.

■ IBM's MQSeries

MQSeries is the market leader in MOMs. It implements the message queuing model of middleware. The package can be run across nearly 20 different platforms.

Numerous vendors provide interfaces between their products and the MQSeries, either offering translation tools from MQSeries to specific applications or providing enterprise management tools that use MQSeries as their communications infrastructure. In the latter group, the leaders are BMC Software's Patrol Knowledge Module for MQSeries, Boole & Babbage's Command MQ, Candle's Candle Command Center for MQSeries, and NasTel's MQControl and Visual MQ products.

■ TIBCO

TIBCO is an independently operated subsidiary of Reuters that has been selling its Information Bus middleware for nearly 10 years. Information Bus is based on the publish-and-subscribe model, which TIBCO claims to have invented. TIBCO currently controls somewhat more than 25 percent of the middleware market. Historically, it has specialized in selling to the financial community; however, it recently announced plans to create a new division to sell to other industries.

■ Other Vendors

Microsoft's Message Queue Server (originally code-named Falcon) is a product that was in late stages of beta testing at the end of 1997. The major concern voiced by users is how extensively MQS will support platforms other than Microsoft's own Windows NT.

XIPC from Momentum Software is a message-queuing MOM that supports most PC operating systems and most versions of UNIX.

Pipes 3.8 from PeerLogic is a message-queuing MOM that specializes in being able to adapt to changes in network configuration at runtime. It can reroute messages transparently, providing added ability to handle network failures. It is also one of the few MOMs to support Novell IPX/SPX.

SmartSockets from Talarian is a high-speed middleware package that implements the publish-and-subscribe model. It runs on most commercial operating systems. Talarian was one of the first independent companies to resell a middleware package and one of the few other than TIBCO to implement the publish-and-subscribe model.

Vcom from Veri-Q is a message-queuing MOM developed by Volvo for its internal use. Vcom then was productized and is now resold by Verimation (the parent company of Veri-Q), a Volvo spinoff. The product is one of few that support MacOS and Apple Open Transport, among other platforms.

TP Monitors

TP monitors had their roots in CICS, IBM's early entry into ensuring transactional integrity in OLTP environments. Today, the leading TP monitoring products are CICS, BEA Systems' Tuxedo, and NCR's Top End. This category also includes TP monitors such as Tandem's Pathway that are closely associated with specific hardware vendors.

IDC estimates total revenue for TP monitors will grow to $577 million by the year 2000, up from $208 million in 1996. (See Figure 5-7.)

Figure 5-7: Worldwide Distributed Transaction Middleware Revenue

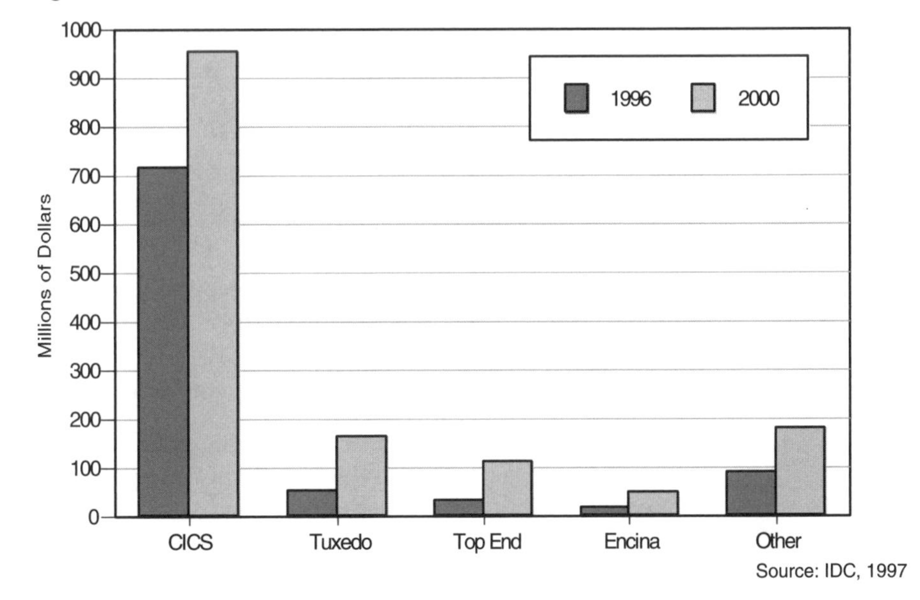

Source: IDC, 1997

■ BEA Systems' Tuxedo

Tuxedo first was developed by UNIX Systems Labs (the unit of AT&T's Bell Labs that also developed UNIX). It was purchased by BEA Systems from Novell, when Novell purchased UNIX Systems Labs. Tuxedo Version 6.3 supports all major XA-compliant databases (CA-Ingres, DB2, Informix, Microsoft SQL Server, Oracle, and Sybase) and platforms. (XA is a standard promulgated by The Open Group for interaction between TP monitors and resources in the enterprise.) It is the only product to implement XATMI, The Open Group's API standard for TP monitors. In fact, the XATMI standard is based on BEA's Application-to-Transaction Manager Interface (ATMI). Tuxedo also provides numerous hooks to fourth-generation languages, third-generation languages, and high-end programming environments.

Tuxedo is the core of BEA's Distributed Application Framework, which supplements the TP monitor function with pure middleware functionality such as asynchronous message queuing.

BEA's success with Tuxedo is evidenced by the company's strong financial performance. Revenue for the 12-month period ending in October 1997 was $131 million, up from $41 million the previous year. Sales soon should reflect the integration of Tuxedo into Release 7 of PeopleSoft's corporate applications product line.

■ NCR's Top End

Top End provides most of the functionality found in Tuxedo. However, it provides additional capabilities targeted at sites with a high volume of transactions. For example, Top End can provide workload balancing and can interface with high-performance database servers such as NCR's Teradata and Oracle's Parallel Server. In addition, Top End provides management of mainframe communication interfaces such as IBM's LU6.2 and the UNIX Transport Layer Interface (TLI). Although Top End does not use the X/Open API for development, it has APIs for C, C++, and COBOL. It soon will support Java clients as well. Top End typically is found at large sites in the retail marketplace, reflecting NCR's long-standing commitment to this industry.

■ IBM's CICS

CICS is the original TP monitor. It still thrives today in many installations where it is used primarily for managing transaction processing on IBM mainframes. Although it can support the major database packages by the use of embedded SQL calls and some custom drivers, CICS primarily is associated with interfaces to VSAM and IBM's DB2 database.

Unlike other TP monitors that are adding functionality traditionally provided by middleware, CICS has remained a purely TP product. For example, CICS does not provide asynchronous message queuing in the style of MOM, nor does it provide the site management aspects found in Top End such as load balancing. However, it is likely that with IBM's purchase of Transarc and its Encina product, these functions could be added to or bundled with CICS in the future.

CICS supports a wider range of programming languages than most of the other products, including C, C++, COBOL, Java, PL/1, S/390 assembly language, and REXX. For the last three languages, CICS is the only TP monitor available.

ORBs

ORBs are sold today by systems vendors (HP, IBM, and Sun, among others) and by third-party vendors. Third-party products tend to have no particular vendor orientation, and they have seen healthy growth during 1997. IDC reports that the total worldwide market for ORBs was $52 million for 1996, representing 23 percent growth from the previous year. The key third-party vendors were Iona ($15.3 million), TCSI ($10.8 million), Visigenic Software ($5.7 million), and Expersoft ($4.5 million).

IDC predicts the market for ORBs will not attain any significant level of maturity until 1998. This view rests on the premise that ORBs are not mainstream tools. The release of CORBA 2.1 should accelerate acceptance of ORBs, however.

■ Iona's Orbix

One of the original CORBA implementations, Iona's Orbix has led the company to several straight years of impressive growth. It is the most widely deployed ORB, with installations at 3,000 sites. For the 9-month period ending September 30, 1997, Iona reported earnings of $32 million compared with $14 million for the previous year. Most of this revenue was generated from sales of Orbix.

■ Expersoft's CORBAPlus

Expersoft also was one of the first companies to ship a CORBA ORB. Because the company is privately owned, sales and installation figures are not available. Expersoft also sells a CORBAPlus ActiveX Bridge, which will enable bridging between CORBA ORBs and Microsoft ActiveX components. This capability will be important for many sites migrating their infrastructures from Microsoft platforms to distributed computing.

■ TCSI

TCSI sells its ORB product as part of a software bundle called SolutionCore, which is sold primarily to the telecom industry. It sports numerous features that cater to the specific needs of this clientele, such as compliance with the Telecommunications Management Network (TMN) specification.

■ Visigenic's VisiBroker

Visigenic's ORB is sold directly to customers (more than 700 sites) and also to OEM vendors. For the fiscal year ending March 31, 1997, Visigenic's revenue was $17 million compared with $5.7 million the year before. On November 18, 1997, the company announced it had agreed to be acquired by Borland, which expects to bundle VisiBroker with its high-end software development tools.

■ Visual Edge's Object Bridge

Visual Edge provides the only true interoperability between CORBA components and Microsoft's COM/DCOM model of ActiveX components. Through Object Bridge, ActiveX components can be made to appear as genuine CORBA components and can be invoked statically, rather than through use of the slower dynamic invocation method. The privately held company generated $12 million in revenue in 1997 from its Object Bridge technology. Prior to its entry into distributed computing, Visual Edge was the market leader in UNIX interface design tools.

5.2.7 Forecast

MOM and TP Monitors

- Vendors of MOM and TP monitors will enjoy the benefits of an entrenched customer base. Most customers are heavily dependent on their middleware for the operation of their sites. Having made the commitment, they will stay with what they have and continue investing in their chosen infrastructure. However, most customers will start pushing for greater Internet/Web and intranet support from their middleware packages.
- MOM and TP monitors will see increasing convergence of their feature sets: TP monitors already are providing the pure communications capabilities associated with MOM; meanwhile, MOM is enhancing its offerings to provide more transaction-oriented features.
- In early 1998, vendors will begin to ship a new technology called object transaction monitor (OTM), which will provide integrity to object interactions and transactions. The appearance of OTMs leads to a convergence of ORB-based computing and TP-monitor-oriented approaches.
- The availability of MTS and MSMQ will provide the tools necessary to build robust transaction processing enterprise applications on Windows NT.

Distributed Components

- Enterprises, especially those with heterogeneous environments, increasingly will adopt CORBA-oriented programming for their distributed computing infrastructure. This adoption will be accelerated by increased dependence on the Internet for enterprise computing.
- Microsoft's DCOM will remain popular in all-Microsoft sites and in desktop applications.
- Inter-object communications among JavaBeans will move from using the Java-specific JRMI to using industry-standard ORB technology.
- ActiveX and JavaBeans will become important drivers for distributed computing. The growth of the market for software components built with these technologies will parallel the adoption of distributed computing.

- The current incompatibilities between JavaBeans and Microsoft's ActiveX components will subside as bridging technologies are designed and then implemented.

5.2.8 References

■Articles

Adhikari, Richard. 1997. Lost in the translation. *InformationWeek*. December 15: 111.
McHugh, Josh. 1997. Old software, new money. *Forbes*. December 15: 248.
Spinner, Karen. 1997. The middleware explosion. *Wall Street & Technology*. July: 94.

■Books

Orfali, Robert, Dan Harkey, and Jeri Edwards. 1997. *Instant CORBA*. New York: John Wiley & Sons.

■URLs of Selected Mentioned Companies

BEA Systems `http://www.beasys.com/products`
BMC Software `http://www.bmc.com`
Boole & Babbage `http://www.boole.com`
Borland `http://www.borland.com`
Candle `http://www.candle.com`
Expersoft `http://www.expersoft.com`
IBM `http://www.ibm.com/products`
Message-Oriented Middleware Association `http://www.moma-inc.org`
Microsoft `http://www.microsoft.com`
Momentum Software `http://www.momsoft.com`
NasTel Technologies `http://www.nastel.com`
NCR `http://www.ncr.com`
Noblenet `http://www.noblenet.com`
PeerLogic `http://www.peerlogic.com`
Talarian `http://www.talarian.com`
TIBCO `http://www.tibco.com`
Transarc `http://www.transarc.com`
TCSI `http://www.tcsi.com/ProdServ/`
Veri-Q `http://www.veriq.com`
Visigenic `http://www.visigenic.com`
Visual Edge Software `http://www.visualedge.com`

5.3 Computing Platforms

5.3.1 Executive Summary

Smaller, cheaper, faster, and more capable – these words summarize the general direction of computer hardware systems. Major technology trends are driving down the price of computer hardware to the point where hardware is estimated to make up only 10 to 25 percent of total information technology (IT) expenditures. Worldwide, 300 million computers were estimated to be in use in 1996 with more than 100 million in the United States alone. In terms of units, personal computers (PCs) account for more than 95 percent of the U.S. computers used for both business and home markets.

Two significant advances in semiconductor technology were announced during 1997. IBM announced breakthroughs that permit the use of copper to replace aluminum in the wires that connect individual circuits on a chip. Copper lowers the electrical resistance of the wires by 40 to 45 percent compared with aluminum, allowing the wires to be thinner and reducing the chip's operating voltage. IBM's CMOS 7S process technology will allow the company to move to a 0.13-micron feature size. Motorola subsequently announced similar technology that uses a dual-inlaid metallization technique, which like IBM's comprises six layers of copper, and is expected to lead to 0.15-micron channel lengths or lower. Intel and LSI Logic are expected to announce more details about their plans for the use of copper wiring in 1998.

The other major advance was Intel's announcement of its StrataFlash non-volatile memory chip, which allows 2 bits of data to be stored in a single memory cell. Intel's announcement followed an earlier announcement by SanDisk. By allowing the level of charge in a memory cell to take on four possible states, rather than only the binary states of on and off, twice the amount of information can be stored in each memory chip. This development initially will affect devices that use flash memory instead of disks to store persistent data; these include digital cameras and the PC Card memory modules used with some small, hand-held computers.

For computer platforms, the rate of core technology advances is the easiest aspect to predict. Currently, storage capacity, processor speed, and data transmission speed are improving at rates of 60 percent per year, doubling in speed or capacity every 18 months, or 110 times per decade.

Lower component pricing and increased competition among vendors are reducing PC pricing and vendor profit margins significantly. The lower pricing and component standardization mean that any dealer can enter the PC assembly business – albeit at low profit margins. This situation contrasts with 1980 when computer designers were scarce, and only a few companies worldwide had the capability to engineer and manufacture the complex systems of the day. To cope with the downward price pressure, vendors traditionally have maintained system price levels (at around $2,000 for a desktop unit) by enhancing hardware capacity, speed, and capabilities each year. Thus, in any given year, a fixed amount of money buys more capability than the year before – larger storage capacity, faster performance, and new capabilities. In 1997, the price point has been challenged, especially by Compaq and IBM, who have introduced systems around the $1,000 price point that are targeted exclusively for home use.

The microprocessor is continuing its climb up the ladder of computer systems, with Intel the largest vendor. Desktop computers always have been microprocessor-based. Mainframe and midrange systems also are becoming microprocessor based, clearly indicating future trends. Although Intel products do not necessarily deliver the highest performance, they are close enough to make software compatibility the dominant decision factor for many purchasers. Although Intel is intent on delivering its next-generation microprocessors as quickly as possible, the key factor is often how quickly Microsoft and other software companies can design software to utilize hardware advances.

The year 1997 saw several important developments in the client device arena. Network computers (NCs) and network PCs (NetPCs) were in the industry spotlight due to their new computing paradigms popularized by the advent of the Internet and the notion of thin-client computing. The industry became interested in easier-to-use, easier-to-manage devices for network and Internet access. The NC, proposed by vendors including IBM, Oracle, and Sun, is a diskless, low-priced machine that downloads programs written in Java from the server and then executes them locally, making upgrades less expensive than for regular mobile and desktop PCs. The NetPC, proposed by Intel and Microsoft, is a slimmed-down, sub-$1,000 client computer aimed at the commercial market. A NetPC costs more than an NC to purchase and manage; however, it includes a hard disk and can have programs installed and run locally. In both cases, reducing the total cost of ownership, not just the hardware cost, was the motivating factor.

Midrange computers are facing greater competition from PCs powerful enough to perform as servers. As a result, midrange computers are gaining features such as multiple processors, larger storage subsystems, and high availability to make them increasingly competitive with traditional mainframes.

Mainframes have been rejuvenated through the use of complementary metal oxide semiconductor (CMOS) integrated circuits. This technology allows mainframes to have lower operating costs and to be manufactured at lower prices that make them more competitive with non-mainframe systems. Year 2000 issues are a pressing reason to extend the life of mainframe systems in large corporations, except for sites that have the luxury of rewriting and moving their applications to other hosts. In addition to these factors, the greater manageability, security, and stability of the mainframe have helped retain customer loyalty and justify renewed investment. However, it is unclear whether these factors actually bring new buyers to the mainframe market.

5.3.2 Memory Chips

Memory chips store data. A variety of types of high-speed semiconductor memory is used in today's computers. This discussion will concentrate on dynamic and static random access memory (DRAM and SRAM, respectively), which are used to hold instructions and data that the computer's processor needs to access while a program is running. Other types of memory, such as read-only memory (ROM), various forms of programmable read-only memory (PROM), and flash memory, are used to hold persistent data even when the computer is not turned on.

Dynamic random access memories (DRAMs) are the most widely used memory chips and are the largest single semiconductor product category. Data stored in DRAM is volatile; it needs to be recharged or refreshed hundreds of times per second in order not to be lost. DRAM is used almost universally as the main working memory in every type of computer system.

Because DRAM chips have a simple design that has not changed fundamentally over time, progress has largely taken the form of increased density, with greater and greater amounts of data being stored in a single chip as improvements in semiconductor process technology allow more and more transistors to be etched onto each chip. Over the last 10 years, the size of the most commonly used memory chip has increased from 256 Kb to 1 Mb to 4 Mb and now stands at 16 Mb, with a further increases to 64 Mb and then 256 Mb within the next 3 years.

The other main form of memory is SRAM, which is static in the sense that the data does not need to be refreshed constantly while in use; however, data held in SRAM still is volatile and will be lost if the power is shut off. SRAM is more expensive than DRAM, but it is significantly faster, operating with cycle times approximately five to ten times as fast as DRAM.

Although DRAM chips have increased significantly in capacity during the last 10 years, they have not become substantially faster. Meanwhile, processor clock speed has increased substantially. As a result, there has been a growing performance gap between memory subsystems and processors, with the memory unable to supply data as fast as the processor needs to use it. Due to this imbalance, system designers supplemented the main DRAM memory with smaller amounts of faster cache memory, while processor designers developed a number of innovations in how a microprocessor works to ensure it is not waiting idly for the data and instructions to be fetched from memory. (See "Microprocessor Architectures" on page 344.)

Most systems today have a hierarchy of different memory types. The fastest but smallest is a Level 1 (L1) cache that consists of memory cells built right into the microprocessor itself. The L1 cache, typically on the order of 64 KB in size, holds instructions and data that the processor is expected to need in the immediate future. Access to the L1 cache is effectively immediate because the information already is stored right on the chip. The Level 2 (L2) cache, which in PCs can range in size from zero to 512 KB or so, consists of SRAM. The largest but slowest step in the memory hierarchy is the main memory, consisting of DRAM chips. Of course, even DRAM, whose operating speed is measured in nanoseconds (billionths of a second), is almost infinitely fast compared with magnetic disk storage, with access times measured in milliseconds (thousandths of a second).

Meanwhile, after years of relative stability in memory design, DRAM manufacturers have been developing new memory technologies to address this imbalance. Most memory used in PCs during the last few years has been Fast Page Mode or Extended Data Out (EDO) DRAM. This memory now is being replaced by two new types of DRAM, known as synchronous DRAM (SDRAM) and burst EDO (BEDO). Both types allow sequential reading of an entire page of memory, once an initial memory address has been specified, instead of waiting for the external memory controller to supply the next address within the page. SDRAM is expected to be more popular than BEDO and will be used increasingly in PCs in 1998.

Starting in 1999, two other memory technologies, SyncLink (SLDRAM) and Rambus DRAM (RDRAM), will come into use in PCs to meet the system design requirements as microprocessors and system buses continue to get faster. SLDRAM was designed by a consortium of major DRAM manufacturers and submitted to the IEEE as a proposed standard. RDRAM was developed by a start-up called Rambus and uses a high-speed interface that transfers data over a bus called the Rambus Channel. Of the two, RDRAM is likely to be more widely adopted – in part because RDRAM product already is shipping and is in use in both Silicon Graphics workstations and Nintendo's 64-bit video games – and SLDRAM is a new technology with no product yet. Perhaps more important, Intel has announced plans to make RDRAM the standard memory type that will be supported for use with Intel processors and chip sets, starting in 1999.

5.3.3 Microprocessor Architectures

An instruction set is the collection of machine instructions that a processor recognizes and can execute. Two processor architecture design styles are currently in use: Complex Instruction Set Computing (CISC) and Reduced Instruction Set Computing (RISC). Intel has announced a new architecture type, Explicitly Parallel Instruction Computing (EPIC), which will debut with the release of the IA-64 processor in 1999.

The basic difference between CISC and RISC is whether the set of instructions that the processor understands is large, with each instruction performing a very specific action, or small, with each instruction being used in combination with other instructions to obtain the desired outcome. CISC architectures have been compared to a desk calculator that can perform 300 arithmetic functions and has a separate button for each one. In comparison, a RISC processor might be a calculator with only 50 buttons that could be used in various combinations to invoke the same 300 functions.

■ Complex Instruction Set Computing

CISC architecture was used in almost all computers built before the late 1980s and is still used today in both mainframes and Intel microprocessors. It implements a large (sometimes as many as 300), complex set of instructions. On a CISC processor, different instructions may require anywhere from one to many clock cycles to execute. Because of the variable number of cycles needed to execute different instructions, taking advantage of advances in computer architecture, such as pipelining, initially proved to be difficult.

■ Reduced Instruction Set Computing

RISC, initially developed at IBM in the early 1970s and first commercialized by MIPS and Sun in the late 1980s, was an alternative design philosophy to CICS.

RISC involves a set of interrelated techniques for designing instruction sets and creating microprocessors that execute those instructions efficiently. RISC architecture is based on the concept that a processor with a smaller number of instructions would be easier to design and could better take advantage of architectural advances. Although many of these techniques since have been adopted in Intel processors, for historical reasons "RISC processor" means architectures such as the IBM POWER and PowerPC, Digital Alpha, HP PA-RISC series, Sun SPARC, and SGI MIPS. Table 5-5 shows the comparative features of these RISC processors in comparison with Intel's CISC processor, the Pentium II.

Table 5-5: Representative Workstation Processors

	Digital 21164	HP PA-8200	IBM P2SC	Intel Pentium II	MIPS R10000	PowerPC 604e	Sun Ultra-2
Clock rate	600 MHz	236 MHz	160 MHz	300 MHz	250 MHz	350 MHz	300 MHz
Issue rate	4 issue	4 issue	6 issue	3x86 instr	4 issue	4 issue	4 issue
Pipe stages	7 stages	7-9 stages	5 stages	12-14	5-7 stages	6 stages	6/9 stages
Memory bandwidth	~400 MBps	768 MBps	2.2 GBps	528 MBps	539 MBps	~180 MBps	1.3 GBps
IC process/layers	0.35μ/4	0.5μ/4	0.25μ/5	0.28μ/4	0.25μ/4	0.25μ/5	0.29μ/4
Die size	209 mm^2	345 mm^2	255 mm^2	203 mm^2	197 mm^2	47 mm^2	149 mm^2
SPEC95b (integer/floating point)	17.0/27.0	16.3/23.0	7.1/23.6	11.9/7.9	13.7/22	12/7	10.4/17.2

Source: *Microprocessor Report*, 1997

■ Explicitly Parallel Instruction Set Computing

An EPIC architecture can be viewed as a hybrid that allows programs to execute on different versions of a processor but that still allows the compiler to provide information about parallel execution by bundling instructions.

EPIC instructions are grouped explicitly into 128-bit bundles for parallel execution. The compiler can specify, for example, a bundle of three instructions that will not use or modify the same data and that the processor can safely assign to three separate arithmetic units. This capability is important because the extra hardware that would otherwise have to detect those dependencies has been one of the things limiting clock speed in previous RISC and Intel processors. In EPIC, not only does the processor do more work on one clock cycle, but the clock cycle itself is shorter. (See "The IA-64 Architecture" on page 347 for more information on Intel's implementation of EPIC.)

Intel x86 Architecture

Virtually all PCs (with the exception of the Apple Macintosh), PC servers, and a growing number of midrange systems are built around microprocessors from the Intel 80x86 family. The initial chips in this family, the 8086 and 8088, were used in the original IBM PC in 1981. These chips were followed by the 80286 (in 1982); the 80386, the first 32-bit processor in the Intel family (in 1985); and the 80486 (in 1989). After the 486, Intel switched to using names rather than numbers to identify its chips. The fifth-generation Pentium processor was released in 1993, followed by the sixth-generation Pentium Pro in late 1995.

Although the original processors in the 80x86 family were designed along classic CISC lines, over time Intel has introduced features historically associated with RISC processors into newer processors in the family. The Pentium processor introduced superscalar capabilities, and the Pentium Pro added branch prediction and speculative execution.

Currently, Intel has two mainstream processor families: the Pentium processor, which is used in notebook computers and low-end desktops, and the Pentium II, which is used in high-end desktops and servers. The Pentium II, released in May 1997, does not represent a new processor generation; instead, it uses the same basic design as the Pentium Pro but was implemented using 0.28-micron process technology as opposed to the 0.6-micron process used with the Pentium Pro. In fact, the Pentium Pro still is being used in certain multiprocessor systems because it has a higher-speed interface to the L2 cache than the Pentium II does. In addition, the accompanying chip set provides support for four-way multiprocessor systems as opposed to only two-way for the Pentium II. However, during 1998, Intel is expected to phase out both the Pentium and Pentium Pro families and replace them with the Pentium II, even in entry-level PCs.

Intel generally follows the initial implementation of a new processor generation with subsequent versions that run at higher clock speeds. For example, the initial Pentium chips ran at 60 MHz and 66 MHz; Pentium chips used in today's desktop and notebook computers run at speeds between 166 MHz and 266 MHz. One of the ways Intel does this is by moving to a new generation of semiconductor process technology. For example, the initial Pentium processors were produced using 0.8-micron technology but were migrated to newer process technologies over time, with 0.35-micron technology in use during 1997. The final versions of the Pentium processor, the 233 MHz and 266 MHz "Tillamook" chips (released in September 1997 and January 1998, respectively), are produced using 0.25μ technology. In addition, even within the same process technology, the processor clock speed can be increased somewhat over time as the manufacturing process is fine-tuned.

Although Intel officially does not announce new products in advance of their actual ship dates, the company does provide general information about forthcoming processors. The following are the anticipated future versions of the Pentium II processor (with release dates as predicted by Intel):

- **Deschutes** – The first 0.25-micron version of the Pentium II, expected to operate at clock speeds starting at 333 MHz and eventually reaching 450 MHz. The initial version of Deschutes, released in late January 1998, is designed for desktop computers and will run at 333 MHz; it will be followed by 350 and 400 MHz versions in the second quarter of 1998. A subsequent version, known as the Mobile Pentium II, will be the first sixth-generation Intel processor designed for notebook computers; release is expected in the second quarter of 1998, with initial clock speeds of 233 MHz and 266 MHz.
- **Covington and Mendocino** – Lower-cost versions of the Pentium II processor, expected during the second half of 1998. Covington will have no L2 cache, which should make it considerably slower than a standard Pentium II; Mendocino will have 256 K of L2 cache built into the processor rather than on a separate piece of silicon. Both are designed for PCs selling for less than $1,000.
- **Katmai** – The first processor to incorporate an additional set of multimedia (MMX) instructions; expected to ship in the first half of 1999. (See "Multimedia Extensions" on page 347 for more information on MMX.)
- **Willamette** – A new implementation of the Pentium II architecture, with performance-enhancing features such as additional arithmetic logic units (ALUs) and possibly using 0.18-microprocess technology; expected to ship in late 1999.

At any point in time during the next several years, Intel typically will be selling several versions of the Pentium II chip, which will be characterized by different clock speeds, amount of L1 and L2 cache memory, power consumption (a significant issue in designing notebook computers), bus interfaces, and other features. The highest-performance (and highest-priced) processors typically will be designed to use a larger amount of L2 cache and have a dedicated high-speed bus interface to the L2 cache than the lower-performance products. Chips used in notebook computers typically will run at slower speeds than those used in desktops.

■ Multimedia Extensions

Intel's multimedia extensions (MMX) to the 80x86 instruction set are a set of 57 new instructions that enable a new family of Pentium, Pentium II, and compatible processors to run multimedia applications more efficiently. Intel designed these instructions to accelerate the execution of operations frequently encountered in multimedia applications, such as multiplying an array of numbers. The MMX instructions use Single Instruction, Multiple Data (SIMD) processing, which allows the processor to execute one command on many data elements at a time. For example, data representing color graphics often is stored in groups of three numbers representing the levels of red, green, and blue. MMX instructions perform arithmetic on the three separate numbers simultaneously, even though they may be packed into a single word. Similarly, certain three-dimensional (3-D) graphics transformations actually operate on 4x4 arrays of numbers, and MMX instructions can do more of those individual arithmetic operations in parallel.

Intel benchmarks indicate that MMX should accelerate calculations common in audio, two-dimensional (2-D) and 3-D graphics, video compression/decompression, and speech synthesis by as much as eight times. However, software must be rewritten to take advantage of the new instructions for there to be any benefit at all. MMX-enhanced chips from Intel and others will encourage the development of a new generation of multimedia applications. In addition to consumer multimedia, MMX technology also has potential in corporate markets, enhancing videoconferencing capabilities.

The initial Intel processors implementing the new MMX instructions were the Pentium MMX, released in January 1997, and the Pentium II, released in May 1997.

Intel also plans to add an additional set of 70 multimedia instructions, designed for 3-D graphics, floating-point calculations, and speech recognition. These have been referred to in the press as "MMX2," but Intel so far has referred to them only as the "Katmai new instructions" because they first will be implemented in Katmai, an upcoming version of the Pentium II processor.

■ The IA-64 Architecture

In 1994, Intel and HP announced plans to work jointly on a next-generation 64-bit processor that would be able to run object code written for both Intel x86 and HP PA-RISC processors. This processor, code-named Merced, will be the first implementation of Intel's new 64-bit architecture, IA-64. Because this will be the seventh major generation of Intel processors, it has sometimes also been referred to as P7.

Relatively little information was available about Merced's design until October 1997, when HP and Intel made the first public disclosures about the chip's architecture. Merced will be based on an EPIC architecture and will bundle three IA-64 instructions into a single 128-bit instruction word. IA-64 will include new features such as the following:

- **Predicated execution** – Multiple instructions can be loaded in parallel, but the processor can be told to execute them only if certain conditions are true, which can eliminate code branches and the resulting delays if the outcome of the branch is mid-predicted.

- **Speculative loads** – Allow memory accesses to be started early without waiting for the outcome of previous instructions that may affect the contents of memory. The Merced chip also will have an increased number of general and floating-point registers, compared with previous Intel chips.

Merced is expected to be available in late 1999. It will be manufactured using 0.18μ microprocess technology. Merced will be followed by other processors in the IA-64 family, such as Flagstaff, expected to be available in 2000 or 2001, which could have a clock speed as high as 1,000 MHz (1 GHz).

Intel intends the Merced chip to be used only in servers and very high-end workstations for at least several years after its release. Virtually all the major UNIX vendors plan to port their versions of UNIX to the Merced chip, and Microsoft plans to release a 64-bit version of Windows NT for Merced. (See Chapter 5.1, *Operating Systems Environments.*) Intel will continue to develop new 32-bit chips in the Pentium II family for use in desktop and notebook computers even after the first IA-64 chips are released.

5.3.4 Computing Platforms and Topologies

A variety of platforms must work together in corporate environments. Mainframes, enterprise and local servers, and client devices such as PCs, NCs, and terminals all play a role in corporate networks of systems. A useful categorization of computing platforms is to rank and classify systems according to performance and cost. (See Figure 5-8.) The overlap of the different computing platforms shows encroachment from the bottom as low-end machines gain added capability. Distinctions between categories no longer are clear-cut; as in the past, this trend is a result of the accelerating development of computer hardware. However, by the year 2002, computing models will be aligned differently: Two parallel tracks – clients and servers – will follow each other up the graph. (See "Realignment of Computing Platforms as Projected for 2002" on page 367.)

Figure 5-8: 1997 Computing Product Segments

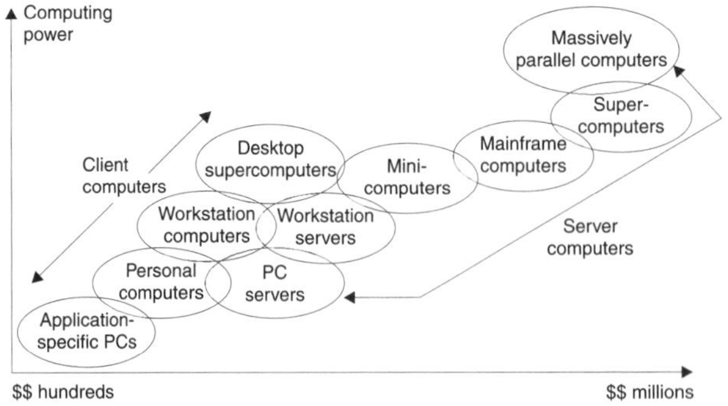

Source: Egil Juliussen, "Computers" in *IEEE Spectrum*, © 1997 IEEE

Mainframes

Ten years ago, people predicted the replacement of mainframes with the less-expensive client/server systems. Instead, mainframes evolved to become cheaper, more powerful, and smaller; they are still the platform of choice to run many large-scale corporate applications and large volumes of complex transactions. Contrary to early expectations about the lower cost of client/server computing, mainframe computing is often less expensive per user than distributed client/server architectures. Applications such as electronic commerce are driving mainframe-based Internet and intranet applications – they demand high security and reliability, and they frequently must access data residing on mainframes. Mainframe vendors also provide cohesive service and support and a single point-of-contact.

Mainframes are not just faster because of faster processors. Bandwidth of the data bus to the processor, cache architecture, and overall system bandwidth also have significant effects on total system throughput.

Traditional mainframes – defined as an IBM or plug-compatible computer running MVS, VSE, VM, or OS/390 Open Edition – are still the dominant industry workhorse. The instruction-set architectures have not changed for years, although parallel processing has been added. However, mainframes have been transformed through the use of CMOS integrated circuits. This technology allows mainframes to be manufactured at lower prices and compete against microprocessor-based designs. By fundamentally transforming the price/performance ratio for mainframe computing, CMOS is enabling new mainframe applications and justifying IT directors' continued investment in upgrading older systems.

■ Complementary Metal Oxide Semiconductor (CMOS) Mainframes

CMOS is a semiconductor technology widely used for processors and RAM. Over the past few years, manufacturers of mainframes, led by IBM, started to offer powerful CMOS processors and multiprocessing configurations. The performance of the new CMOS processors has approached that of the older and more expensive emitter-coupled logic (ECL) processors.

CMOS microprocessor-based systems not only are easily expandable, but they also cost less than the big machines and consume less energy. Customers can add the computing power they need in far smaller increments than previously – making upgrades more cost-effective.

According to the Gartner Group, CMOS mainframes are viewed as an effective cost-reduction strategy by most large users. Integrated-logic CMOS computers are changing the economics of computing at all performance levels, and mainframe costs are plummeting, according to IDC. Due largely to CMOS-based systems, IBM's S/390 hardware prices are dropping at a rate of 35 percent per year.

The use of CMOS and expected improvements in speed and size over the next few years will extend the lives of existing mainframe product lines. Progress in semiconductor fabrication technology has allowed the entire instruction set of an IBM S/390 to be placed on a single chip. This development further reduces the costs of mainframe manufacture.

Other economies of scale derive from the chip manufacturing and sourcing among mainframe vendors. According to Gartner, IBM manufactures CMOS chipsets for Hitachi and Unisys mainframes. Fujitsu provides chip sets to Amdahl (Fujitsu merged with Amdahl in 1997), ICL, and Siemens Nixdorf. NEC supplies the technology for Groupe Bull's high-end DPS9000/900 series.

Reinventing the Mainframe

The critical role of mainframes is attributed to the large installed base, not of units (which is minuscule compared with desktops), but of the number and size of the applications and the associated investment. Each organization has a suite of applications, many developed internally, for running the corporation, often providing essential support for core business processes. These applications are seldom portable. A tremendous investment in data-center system software allows the mainframe vendor to offer a unique, highly integrated system with guaranteed compatibility, performance, and robustness – characteristics particularly important to major corporations and government organizations.

The widely discussed movement of applications off the mainframe, once described as rightsizing, is not occurring at the predicted rate. Corporations have found implementing client/server computing strategies to be a longer, costlier, and more daunting task than expected. In other cases, businesses are upsizing their installations and consolidating data processing sites into super data centers to achieve better economies of scale. This movement is encouraged by the improved availability and reduced prices of wide-area networking, which has spurred renewed interest in mainframes among corporate users.

Mainframe suppliers also have repositioned their systems for new roles in distributed client/server environments within computer networks. The traditional, centralized mainframe that performed only batch data processing and on-line transaction processing (OLTP) will all but disappear over time. Mainframes will act as giant servers to networks of PCs, workstations, and midrange systems.

IBM's CMOS Mainframe Technology

In 1994, IBM introduced Parallel Enterprise Servers (PES), Parallel Transaction Servers (PTS), and Parallel Query Servers (PQS), its first mainframes based on the CMOS technology. In June 1995, IBM introduced its second-generation CMOS processor, utilized by the newer PES mainframes. These air-cooled systems increased the performance range of the first-generation CMOS-based configurations by a factor of 2.7. Up to 32 of these machines can be clustered using IBM's Parallel Sysplex technology, which makes all 32 machines appear as one logical enterprise server. In late 1996, the third generation of S/390 processors (known as G3) further enhanced the line.

In June 1997, IBM announced the fourth generation (G4) of its CMOS mainframe family, called the S/390 Parallel Enterprise Server – Generation 4. At the same time, the company announced the end of bipolar processor technology usage in its mainframe products. Generation 4 offers 14 models that surpass performance levels of Generation 3 by up to 33 percent. They are powered by up to 10 CMOS microprocessors with performance rated at 63 Mips and offer 16 GB of memory in a single enclosure.

In late 1998, IBM is expected to ship the fifth generation (G5) of this architecture. The primary enhancements that are promised are a 50 percent increase in processor performance, doubling the performance of Parallel Sysplex clustering technology, and an end to the limit of 256 input/output (I/O) channels. Analysts predict G5 will accept up to 1,000 I/O channels.

■ Hitachi's Mainframe Technology

Hitachi Data Systems has been developing fast mainframe uniprocessors for its Voyager Family since 1995. The faster performance is enabled by a semiconductor technology called Advanced CMOS-ECL (ACE), which combines ECL and CMOS on the same chip. ECL is used where speed is critical, and CMOS is used in other areas as a cost-saving device.

The Skyline Series of the Voyager Family was enhanced in June 1997 with nine new models. The company claims Voyager has the fastest, single-engine speed in the world. Six of the new models featured faster engines, increasing throughput capacity by 25 percent. A total of 19 models are offered in the Skyline Series, with four engine speeds. The models range from 1- to 8-multiprocessor configurations and up to a 1,000 MIPS performance level. In addition, the models are fully Parallel Sysplex-capable for up to 16 Inter-System Coupling Channels.

Midrange Servers

Mainstream IT in large corporations has been adopting midrange systems aggressively to support midsized server applications, such as transaction processing and database-oriented applications, along with LAN server functions. Previously, these systems appealed to midsized business customers and to large businesses with dedicated applications. Midrange systems fall into two categories: Those that run a vendor-specific operating system and those that run a version of UNIX or Windows NT. The proprietary systems, including IBM's AS/400 and Digital's VAX and Alpha, support operating systems that offer specialized management tools and broad application software availability.

■ IBM's AS/400

The AS/400 has been a very popular midrange machine since its initial offering, and it still commands a major portion of the corporate server market. It is easy to install and manage, and it has a rich set of applications available – all traits that endear it to small business sites and branch or plant locations in larger companies. In 1995, IBM introduced the PowerPC-based AS/400 Advanced Series. These systems support symmetric multiprocessing (SMP) and are capable of running software written for AIX, IBM's version of UNIX, as well as for the standard OS/400. At the low end, the AS/400 Advanced Entry model provides small businesses with low-cost, high-performance computing as well as proven effectiveness in accessing applications, managing software and data, and connecting to a variety of systems. At the high end of the AS/400 family, the AS/400 Advanced System and Server offers large user populations access to client/server, data warehousing and Internet-server application environments. In August 1997, IBM announced the AS/400e series of servers, tailored to fill the role of a Web server, with security features, expanded memory, storage, and processing capabilities. The AS/400e servers employ 8- and 12-way microprocessors working in parallel and providing up to 4.6 times the processing capabilities of previous AS/400 models. Storage capacity was doubled and memory was five times larger at the high end of the series.

Increasingly, the AS/400 is found in many local-area network (LAN) computing environments, supporting NetWare and Lotus Notes. IBM's continued commitment to the AS/400 line, more LAN connectivity options, and a growing body of application software signals an extended life for the AS/400. Moves such as enhancing the AS/400 with Windows NT interoperability strengthen IBM's positioning of the AS/400 as a server instead of a proprietary midrange system.

A key development in the AS/400 sector is IBM's transition of the architecture from 32 to 64 bits. The capability of the new microprocessor (PowerPC AS) to handle 64-bit addresses and words provides two advantages: First, the larger instruction-length word permits more than 4 GB of memory to be addressed directly, allowing large databases to be entirely memory-resident (this capability is called very large memories, or VLM); and second, more data can be processed in clock cycle.

Neither of these advantages accrues to 32-bit software running on a 64-bit machine, so vendors have provided numerous ways to help their customers transition their software to 64 bits: a compatibility box, source-code recompilation, instruction translation, and so on. The AS/400 division of IBM provides a unique approach: Technology-Independent Machine Interface (TIMI). When AS/400 applications are compiled, they do not produce executable binary code directly. Rather, they generate a program template, from which binary code is generated. This template is saved along with the program and the binary code. The new AS/400s can read the program templates and generate 64-bit binaries directly, so existing applications can be converted automatically to 64 bits by OS/400, the AS/400 operating system. Only if the template is missing will other steps be necessary. This investment protection in the AS/400 has made the platform especially attractive to midsized companies.

■ UNIX- and Windows NT-Based Servers

Vendors in the open systems arena are straddling both UNIX and Windows NT Server (NTS) as operating system offerings for their midrange hardware platforms. UNIX-based midrange systems have been using RISC-based central processing units (CPUs) for the past decade; symmetric multiprocessor (SMP) configurations and clustering frequently are offered for added performance and scalability. Vendors that offer large SMP UNIX-based machines include Digital, HP, IBM, NCR, Pyramid, Sequent, Silicon Graphics, Sun, and Unisys. Among these, NCR, Sequent, and Unisys offer Intel-based servers in this market segment; the rest rely on RISC microprocessors.

For the past several years, most UNIX system vendors (with the exception of Sun) have been announcing support for and developing server products to transition to Microsoft's NTS as a strategic platform for the future. Microsoft needs the assistance and cooperation of the midrange server hardware vendors to extend its penetration into the enterprise server market. These hardware partners are anticipating eventually having a sizable market share for Windows NTS in the midrange and enterprise server markets. Two key partners in Microsoft's enterprise push are Digital and Tandem (a division of Compaq).

■ Intel-Based Servers: Wolfpack Clustering Architecture

Compaq, Intel, and Microsoft are leading the effort to establish a standard, called Virtual Interface (VI) architecture, for clustering Intel-based servers and workstations. Clustering allows multiple servers to be linked together to share the processing load and so the workload of a failed server can be transferred automatically to another member of the cluster. The VI standard, if adopted, should accelerate the availability of clustering solutions for Intel-based servers.

Clustering has been a strong point for RISC-based UNIX computers for some time, and adding clustering functionality to Intel-based servers is a prerequisite for them to take on increasingly large workloads.

In 1995, Compaq, Dell, Microsoft, Mitsubishi Electric PC Division, Tandem, and others started investigating the feasibility of taking Intel-based servers and tying them together in a cluster to improve reliability and performance. Microsoft has seven early adopters of Wolfpack, the product arising from this effort: Compaq, Digital, HP, IBM, Intel, NCR, and Tandem. As of late 1997, Wolfpack allowed two servers to be linked using a Fibre Channel interface running at 100 Mbps. Wolfpack can perform what is known as a failover operation. Failover is the process whereby a server's transactions are assumed automatically by a second machine in the event of a system failure. As it stands today, Wolfpack can cluster only two servers, but multi-node is scheduled to become available in Phase II of Wolfpack in 1998 or 1999, as is load balancing.

The clustering architecture supports two models: shared-nothing and shared-disk. The main differences between these models are in the handling of resources, such as data storage devices and memory. Wolfpack has been designed as a shared-nothing model.

In the shared-nothing model, each system of a cluster is allocated its own subset of storage resources. Only one system may own and access any particular storage device, although in the event of that system's failure, a predesignated alternate system can take ownership of the system's users and data and resume processing.

In the shared-disk model, any system in a cluster can access the resources connected to any other system in the cluster. A distributed locking mechanism is implemented to guarantee that data integrity is maintained despite potential access by multiple systems in the cluster. (See "Shared Disk" on page 483 for further discussion as to how these models of clustering affect databases.)

Windows NT clusters will provide three key advantages to the user: Improved availability by continuing to provide a service during hardware or software failure; increased scalability by allowing new processors to be added as system load increases; and simplifying the management of groups of systems and their applications by allowing the administrator to manage the entire group as a single system.

■ Multiprocessor Architectures

Like mainframes, midrange systems long have used multiprocessor architectures for performance, scalability, and availability benefits. As the competition between Intel-based servers running Windows NT Server and RISC-based servers running UNIX intensifies, vendors will be differentiating midrange systems based on the varied multiprocessor architectures used: Symmetric Multiprocessing (SMP), Clustering, Non-Uniform Memory Access (NUMA), and Massively Parallel Processing (MPP).

■ SMP

SMP midrange computers, such as Compaq's ProLiant, Digital's AlphaServer, and HP's HP9000, group multiple processors in a single system. Currently, most SMP systems can accommodate up to 32 processors, which are optimized for running multiple tasks simultaneously. Users can begin with a small system and scale up quickly by simply adding more processors to their existing system. Users also can allocate more processors to a task without having to change the application or the database to account for the added processors. Beyond 32 processors, SMP's performance gain becomes less dramatic because adding additional processors contributes very little to the net gain in processing power. The biggest bottleneck is the memory bus bandwidth. As the processor count increases, so does the amount of traffic on the bus. This situation eventually causes system performance to flatten. SGI's recent Gigaswitch technology, obtained in its acquisition of Cray Computers, is one of the first technologies to address this bottleneck. Other vendors are sure to follow with their own implementations.

Bandwidth also can be increased by clustering systems. In theory, clustering can produce significant performance gains with each additional node. However, because each node has its own copy of the operating system, memory pool, and system bus, programming has to be done to coordinate running applications and sharing of data across the nodes in the cluster. Also, performance improvements when adding a system to a cluster are not linear, which means that users do not realize 100 percent of the computing resource when adding a node (1 plus 1 does not equal 2; it is less than 2). The bandwidth of the interconnect between the nodes themselves is often insufficient to sustain high-performance gains. Some vendors such as Digital with its Memory Channel interconnect and Tandem with its ServerNet technology offer specialized interconnect technology to overcome this bottleneck.

■ Non-Uniform Memory Architecture

NUMA is a type of hybrid between SMP and clustering. NUMA arranges small groups of multiple processors in a server. For example, a 16-processor server can be arranged into four nodes of four processors each. This technique allows users to tie together more processors in one enclosure than SMP does. Sequent's NUMA servers, for instance, can support up to 252 Intel processors in one system enclosure. In some respects, NUMA makes it easier to build large SMP computers. As with SMP, users do not have to alter application or databases each time additional processors are added. NUMA avoids SMP's memory bus bottleneck issue by essentially breaking the bus into several smaller buses. The processors within each node in a NUMA server communicate with one another using those smaller SMP buses, and each of the nodes communicates with the others using fast interconnect technologies. However, NUMA is not as mature as SMP, and much of its success depends on how quickly and easily each node in the system can communicate and share data. NUMA's name derives from how it assigns memory. Each group of processors has its own private, local memory but each group also can access memory from the other groups using a defined access method that avoids memory corruption.

■ Massively Parallel Processing

MPP systems are constructed by interconnecting a large number of processors (generally, more than 32) with a specialized processor connect network. In theory, hundreds, if not thousands, of processors can be tied together. Each has its own memory and bus and is capable of handling its own tasks and communicating with the others via highly specialized interconnects and switches. The system is designed to optimize the execution of a single application across all, or a large subset, of processors. The application, of course, must be designed to run in a parallel environment. MPP systems solve problems by breaking them into pieces and computing them simultaneously rather than solving the same problem one step at a time. Users have virtually unlimited node expansion capability and a single system view to the application. MPP gets past the interconnect and bus bandwidth bottlenecks of some other architectures through brute force, using very fast switching and scalable interconnect technology. Drawbacks are that MPP systems can be significantly more expensive than other multiprocessor architectures, and both applications and databases have to be altered extensively to run optimally. For example, the applications themselves often have to be "parallelized" so specific tasks are allotted to specific processors or sets of processors based on computing requirements. Thus, MPP systems suffer from the difficulty and higher expense of developing applications and the resulting shortage of application software. There is also a shortage of middleware because most standards-based software is architected for uniprocessor or SMP environments and therefore must be rearchitected (and rewritten) for MPP. As a result, most MPP products have had limited commercial success, and several vendors have exited the market.

Supercomputers

Supercomputers, such as the Cray T90 series and Sun UltraHPC, are the fastest calculating engines that computer technology, at any given moment, can produce. Traditionally, government agencies such as the U.S. Department of Defense, research scientists, universities, and other research foundations use supercomputers. With defense cutbacks and the resulting decline in the fortunes of defense and aerospace companies, supercomputer makers have expanded into the commercial marketplace. Customers use these systems to run a wide range of applications, including database mining, image processing, and meteorology. Other computer vendors continually monitor the multiprocessor architectures that supercomputer designers develop since this product segment is frequently a breeding ground for architectures that move into widespread use in the commercial computer sector. Cray and Convex, two well-known supercomputer companies, were acquired by Silicon Graphics and Sun, respectively, during the past few years. Fujitsu and NEC (with its SX-4 family of models) are also players in the supercomputer market.

Client Devices

Computers that access the information housed on servers have been dubbed "client devices." By their nature, client devices are able to access servers by connecting to networks, including both LANs or wide area networks (WANs) such as the Internet.

A wide variety of client devices are in use: workstations, desktop PCs, notebooks, and personal digital assistants (PDAs). New classes of client devices were the subject of a great deal of debate during 1997. These were the network computer (NC) and the network PC (NetPC). Both these newly proposed platforms have less locally resident capability than mobile or desktop PCs because they employ a different style of client/server computing that relies on the network more heavily than the familiar Intel PC or Macintosh do: Applications and data often reside on the server rather than on the client device and are downloaded as needed by the client device.

Vendors interested in providing these new classes of client devices have been cooperating to establish standards before introducing new platforms to the market. Product introductions have been slow to appear in 1997, and one – the NetPC – already has been withdrawn.

Introduced years ago, PDAs have shown low adoption rates until 1997, when adoption rates by business professionals began to show promise. After many attempts and many mistakes, the vendors appear to have discovered and now offer viable products for commercial usage.

■ Workstations

Historically, engineering workstations have been distinguished from PCs by both hardware and software features. Workstations included built-in hardware support for networking and ran operating systems that provided advanced features (such as UNIX). PCs ran operating systems that lacked built-in networking, virtual memory, and multitasking (such as MS-DOS).

Events of the last several years have blurred this distinction. PC systems are now available with built-in network interfaces, and current PC operating systems include virtual memory, multitasking, and network support. Workstations are used mainly for engineering and software development. They continue to offer higher levels of performance than PC systems.

At the high end of graphics workstations, unique features are still available, mainly in the ability to handle 3-D graphics, animation, and video editing. However, for mainstream applications in science and engineering – the main market for workstations – high-end PC systems often can do the job at a lower cost. For this reason, UNIX/RISC workstations are under assault from Intel machines running Windows NT. Compaq unveiled its PC workstation lineup and highlighted plans to use graphics technologies from partners such as ELSE and Intergraph. Digital and HP announced second-generation Windows NT/Intel workstations, actively endorsing Windows NT as the mainstream solution for technical users. They were joined later by Silicon Graphics, which announced in late 1997 that a Pentium II-based desktop computer was under development that was designed to run Windows NT 5.0. Sun, the only major UNIX vendor not to have an Intel hardware offering, announced in December 1997 that it would be porting its Solaris x86 version of UNIX to Intel's Merced chip. Many industry watchers saw this move as Sun's recognition that it could not endure in the long term using an exclusively RISC approach.

Momentum is building for Windows NT in the technical workstation market because high-speed processors and applications are available. Pentium Pro and Pentium II performance rivals that of RISC chips (except in floating-point arithmetic, which is used heavily in scientific and engineering applications). Also, Windows NT's multiprocessor capability allows hardware vendors to apply up to four processors to handle demanding applications, much like UNIX workstations have done since the early 1990s. Most important, many engineering applications have been ported to Windows NT by vendors such as Autodesk and Lightscape. Companies are expecting lower hardware, software, and support costs by maintaining one environment (Windows NT) rather than the multiple UNIX platforms they have supported hitherto.

■ Desktop PCs

Up through the 1970s, the desktop machine was a terminal connected to a mainframe computer, but in the 1980s, PCs became ubiquitous on corporate desktops. In the 1990s, the focus shifted from stand-alone PCs to PCs networked with a local or enterprise server (often a mainframe). The PC is best suited for the knowledge worker who needs significant local storage and access to a large suite of applications.

More-powerful desktops and communications devices are in demand as users connect to the Internet and on-line services. Multimedia capability is becoming increasingly important for the business desktop. Multimedia previously was used for entertainment and games, but businesses are looking at corporate training and videoconferencing as key uses for this technology.

For years, Apple was the multimedia platform of choice for image, sound, and video applications. The Apple Macintosh continues to dominate high-end graphics applications for desktop machines. However, 1997 saw an accelerated shift to the PC platform by businesses due to financial and technical problems at Apple and to the addition of multimedia features to many PCs.

Along with the growth in corporate network infrastructures, network access is a requirement for desktop PCs. Businesses are beginning to order PCs with preinstalled network interface technologies and preinstalled, pretested networking driver software so the PCs are more easily and quickly added to the corporate network.

■ PC 98

In 1997, Microsoft and Intel codeveloped the PC 98 Hardware Design Guidelines for next-generation PCs. Previously, the two companies did not coordinate their plans, which resulted in out-of-sync PC hardware and software developments. PC 98 features include digital I/O ports and modular device bays that will allow components to be plugged and unplugged quickly and easily.

The PC 98 guidelines lay out a 2-year roadmap for several models of PCs ranging from low-end home systems to high-end corporate workstations. The relationships between these models is shown in Figure 5-9.

Figure 5-9: The PC Models Defined in the PC 98 Specification

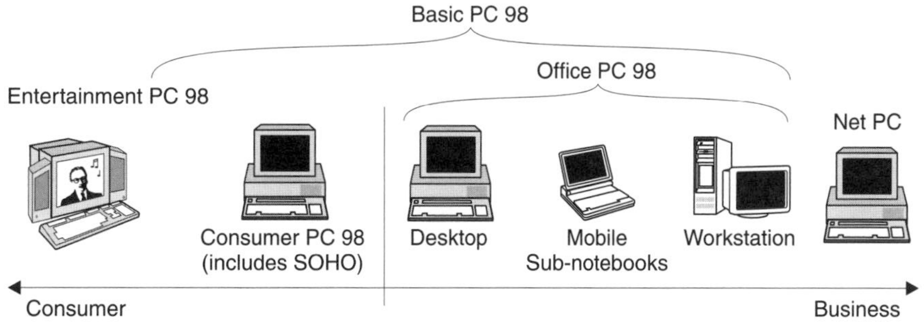

The guidelines specify the basic core configuration of the PC as well as many of the I/O technologies needed for future PC applications. The changes from the current PC model involve the introduction of several new technologies, mostly geared toward better handling of multimedia data types. Table 5-6 shows representative technologies for home and office PCs, including the following:

- **OnNow** – A design that allows PCs to start much more quickly by placing them in a dormant power-saving state rather than turning them off completely.
- **Peripheral Component Interconnect (PCI)** – The primary bus standard in use today for adding expansion cards and peripherals to PCs.

- **Universal Serial Bus** (USB) – A replacement for today's serial and parallel ports. It can move data at 12 Mbps and will be used to connect low-speed peripherals such as printers and scanners.
- **IEEE 1394 (informally called FireWire)** – A faster bus designed to connect high-speed storage devices such as disk drives. IEEE 1394 supports speeds of 400 Mbps, with 1,600 Mbps planned.
- **Accelerated Graphics Port** (AGP) – A high-speed video port increasingly common on PCs today., 199

Table 5-6: Representative Hardware Defined in PC 98 Specification

Hardware	Consumer PCs		Office PCs	
	Required	Recommended	Required	Recommended
System	OnNow		OnNow	
System buses	USB and PCI	IEEE 1394	USB and PCI	IEEE 1394
I/O devices		USB game pad		
Graphics acceleration	2-D in hardware, 3-D in software	3-D in hardware	3-D in software	2-D and 3-D in hardware
Video		AGP video port		AGP video port; TV, DVD video and MPEG2 playback
Audio		Support music synthesis		
Communication	33.6-Kbps modem	Network card	Network card	Internal 33.6-Kbps modem
Storage	Bus mastering 8X CD-ROM	DVD-ROM with DVD video playback	Bus Mastering 8X CD-ROM or DVD-ROM	
Minimum features of Basic PC System are a 200 MHz Pentium with MMX extensions, 32 MB RAM (64 MB recommended), 256 KB L2 Cache, ACPI power management and Plug-and-Play capability.				

Source: *PC 98 Hardware Standard*, 1997

■ Notebooks

Mobile computing, especially with notebook machines, is pervasive among knowledge workers. The weight, screen clarity, and keyboard quality have improved significantly during the past several years. These portable computers often are "parked" in docking stations when used in the office, then pulled out to use away from the desk and outside the office environment. With subnotebooks and notebooks, the goal is to create a desktop-compatible computer that can go anywhere and still provide users with the same access to corporate information.

Mobile computers appeal to professionals whose jobs mean frequent travel or to salespeople who give off-site presentations and have to carry their equipment with them. Also, professionals with limited desk space and those who want to work both in the office and after-hours at home find a notebook especially attractive.

Notebooks offer most of the same technology found in desktop PCs: the most current processor technologies, large amounts of RAM and hard disk capacity, CD-ROM and floppy disk drives, color monitors, high-speed modems, and the common office-productivity software. Users still must compromise on screen sizes and clarity, keyboard size and touch, pointing technologies such as the mouse or track ball, and battery usage (typically limited to 2 hours).

■ Network Computers (NCs) and Network PCs (NetPCs)

The complexity of PCs and the resulting need for high levels of technical support have created an extremely high cost of ownership for business users. Total cost of ownership (TCO) of PCs in businesses is estimated by analysts such as Gartner to be between $10,000 and $14,000 per year. To address this cost, a variety of vendors have proposed stripped-down client machines, specifically the NC and the NetPC.

The NC specification evolved from work done by IBM, Oracle, and Sun. NCs are diskless machines that download programs from the server and then execute them locally. This setup makes NCs less expensive to support than regular mobile and desktop PCs because there is no locally resident software to configure or upgrade. NCs are processor- and operating-system-neutral; rather than being defined by a specific processor and operating system, an NC is defined by its features, which include a built-in TCP/IP protocol stack, a Web browser, and a Java Virtual Machine (JVM). Even though it is dependent on a server for software, the NC is still a true computer, capable of executing applications locally.

NCs are suited for single-purpose or fixed-function environments such as data entry, a bank teller workstation, or a hotel desk clerk. The NC is geared toward transactional workers who run a limited number of applications and do not require large local storage. Many industry analysts concluded that NCs will replace terminals, rather than PCs, in the near term.

The NetPC was proposed by Intel and Microsoft in response to the NC proposal. The NetPC is basically a product-line extension of the existing PC and includes a Pentium-class processor and the Windows operating system. Intel and Microsoft plan to add functionality to reduce support cost through a zero-administration workstation feature that is supported by the operating system and the hardware. The NetPC will have no floppy or CD-ROM drives and no long-term persistent storage; however, it will use a hard drive for temporary storage while running applications from a server.

NCs and NetPCs have disadvantages that will need to be addressed for them to gain the wider acceptance they hope for:

- The devices offer no mobile support or portability because they are dependent on servers for software.
- A critical, if not the critical, factor for successful deployment is the network infrastructure, rather than the client device itself.
- Network and server dependency are increased, so when a server goes down the effects ripple through the organization much more dramatically.
- Current PC users may resist the new devices, perceiving them as less robust and detracting from the user's autonomy and independence from the corporate IT department.

Table 5-7 compares the features and functions of NCs, NetPCs, and Windows terminals.

■ Hand-Held PCs and PDAs

Palmtop computers, or PDAs, have been available since 1991. The most well-known product entry was Apple's Newton. Historically, PDAs have not been successful due to several factors. The devices were too expensive, and they lacked

Table 5-7: Features of Network Computers versus NetPCs

Feature	NCs	NetPCs
TCP/IP Protocol Stack	Yes	Yes
Web browser	Yes	Yes
Java Virtual Machine (JVM)	Yes	Yes
Terminal-emulation software	Yes	No
Remote systems configuration	No	Yes
Pentium-class processor	Optional	Yes
Quantity of local RAM	Medium	Maximum
Network Interface Card (NIC)	Yes	Yes
Smart card slot	Yes	Yes
Floppy drive	No	No
CD-ROM	No	No
Most appropriate user	Transactional worker	Knowledge worker

Source: Adapted from IDC, 1997

useful, nontrivial software applications. Early PDAs also suffered from hardware shortcomings: Hookups to desktop computers and the Internet were very slow, input devices – particularly handwriting recognition – were crude, and battery life was very short. These factors combined to limit adoption rates of PDAs until late 1996, when the Palm Pilot from 3Com finally began to gain market acceptance by addressing these problems.

Market entrants such as 3Com (formerly U.S. Robotics), Motorola, Psion, and Sony, have enhanced functions, lowered prices, and added applications. Vendors have worked to refine the design of PDAs by improving connectivity, performance, search capabilities and lowering power consumption needs. Vendors such as Casio, HP, and Philips Electronics introduced new hand-held PCs that use Windows CE (a slimmed-down version of Windows 95 that Microsoft released in 1996 for this class of computers). Tiny keyboards still are an inhibiting factor as is handwriting-recognition software that must either learn a user's character strokes (such as on the Apple Newton) or impose on the user a specific series of strokes for letter formation (such as Palm Pilot's Graffiti Software).

Uses of PDAs center around date book, address book, to-do list management, and memo pad features. Popular products run on two or four AA or AAA batteries for 8 to 24 hours of continuous use and for approximately 1 month on the more typical intermittent use. Modem cables or infrared ports are used to back up and synchronize PDAs with PCs. Fax modems or network ports (generally provided on PCMCIA cards) let users connect to a LAN or the Internet. Word processor, spreadsheet, and database applications are resident on some devices. As of late 1997, prices had been lowered to the $300 level for entry-level products, although top-end products, such as the Newton 2100, still commanded $1,000.

5.3.5 Market Overview

The overall worldwide factory revenue for computer systems is expected to grow about 18 percent per year from 1996 through the year 2000. (See Figure 5-10.) The forecast from Dataquest most likely reflects the robust world economy experienced the past few years and its positive effect on spending for computer platforms.

Figure 5-10: Worldwide Computer Factory Revenue: 1996-2000

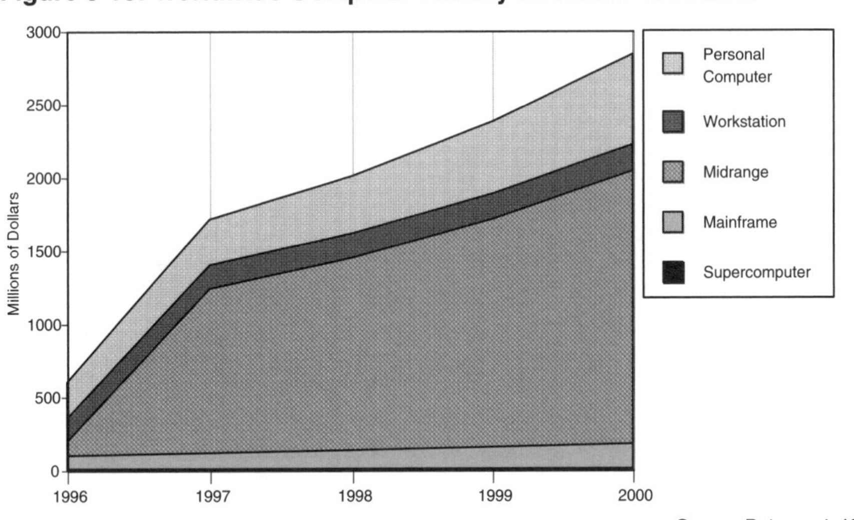

Source: Dataquest, 1996

Mainframes

Mainframes have regained popularity as enterprise servers. This popularity is facilitated by the use of CMOS technology in mainframes. The overall market, however, is slowly shifting away from traditional mainframes to smaller systems. Almost all mainframe vendors have experienced slowed growth rates in their mainframe businesses during the past decade. This trend is beginning to plateau, with mainframe factory revenue forecast to represent a stable market share of about 7.8 percent of non-PC revenues worldwide.

Although the overall large-scale systems market has experienced slow growth, the rate of the decline in revenue growth has slowed due to several factors. Vendors shifted to CMOS in their mainframe products to meet the demand for increased price-performance in the face of stiff competition from multiprocessor midrange systems. Mainframe price declines are continuing, removing the incentive for customers to migrate to alternative platforms just to save on hardware costs. IBM's software-licensing practices also have helped make mainframes more cost-effective.

Mainframe MIPS shipped in 1997 showed growth of approximately 20 percent. This growth does not represent only new system sales because most increases were attributed to upgrades and purchases of used equipment.

IBM is the dominant mainframe vendor in the worldwide market. Plug-compatible manufacturers (PCMs) that offer IBM-compatible systems include two Hitachi resellers – Comparex and Hitachi Data Systems (HDS) – and Amdahl (now merged with Fujitsu). According to IDC, Hitachi gained market share in 1996, growing from a 15 percent market share to 20 percent in terms of revenue, primarily because of strong sales of its high-end Skyline processor. Other mainframe vendors include Fujitsu, Groupe Bull, ICL, NEC, NCR, Siemens/ Nixdorf, and Unisys.

Midrange Servers

According to IDC, the worldwide midrange market grew 27 percent in 1996. The top five worldwide midrange vendors by revenue in 1996, according to IDC, were IBM, HP, Sun, Silicon Graphics (SGI), and NEC. Their respective market shares are shown in Figure 5-11.

Figure 5-11: Revenue Market Share for Midrange Vendors – 1996

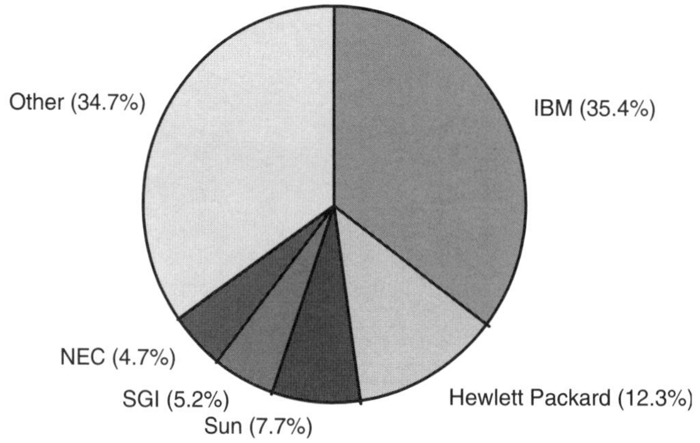

Other (34.7%)
IBM (35.4%)
NEC (4.7%)
SGI (5.2%)
Sun (7.7%)
Hewlett Packard (12.3%)

Source: IDC, 1997

Factors producing growth in the midrange market include the following:

- Widespread implementation of the Internet and corporate intranets
- Prospect of thin clients (such as NCs) driving greater server sales
- Near universal endorsement of client/server

For LAN servers, the Asia/Pacific region showed impressive growth during 1996, fueled by Japan's continued rapid adoption of PC servers. In the U.S., the small- to midsized business market and departmental purchases continued to drive sales, according to IDC.

More price competition will be seen in the midrange server market. The server is becoming a commodity, as high-end features such as high availability, multiple processors, large storage subsystems, and redundant, high-performance storage in the form of RAID enable PC vendors to compete with RISC-based midrange computer systems. (See "RAID" on page 81.) Midrange vendors increasingly have turned to offering Intel-based Windows NT server hardware to remain competitive price-wise.

Client Devices: Workstations

The rapid move by independent software vendors to port technical applications to attractively priced and increasingly powerful Intel-based Windows NT platforms has left UNIX vendors under pressure to retain market share and maintain high-profit, volume shipment of technical workstations. As shown in Figure 5-12, IDC projects substantial workstation growth, but mainly from the Intel-based platforms.

Within the traditional, UNIX-based RISC workstation business, Sun grew at the expense of all competitors except IBM. Although Sun's market share in units rose from 38 percent to 41 percent, Digital, HP, and Silicon Graphics all saw their market shares decrease.

Because Intel-based workstations are expected to increase their market share, all the major RISC UNIX vendors also are shifting their focus toward higher-profit product lines. During 1996 to 1997, all five major workstation vendors announced strategic server product plans. This shift away from the workstation market is important especially for Silicon Graphics and Sun because these two vendors have relied on workstations as their primary offerings.

Figure 5-12: Worldwide Workstation Sales by Type

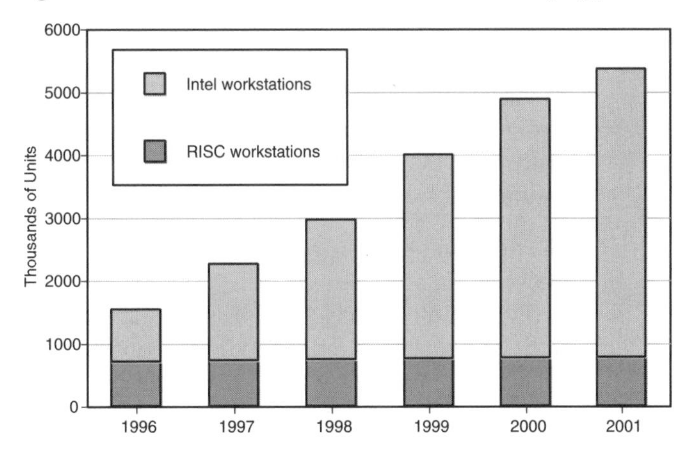

Source: IDC, 1997

Client Devices: Desktop PCs

The greatest changes are occurring on the desktop. The current worldwide PC market is being led by the rapid transition to Pentium and Pentium II processors coupled with big boosts in system features and performance, increasing use of the Internet, the small office and home office (SO/HO) market, and the mass adoption of multimedia technologies. The market continues to benefit from a healthy upgrade cycle of commercial desktops, driven by large gains in price/performance and the migration of businesses to Windows 95 and Windows NT. (See Figure 5-13.)

Another significant trend is the advent of home computers priced in the sub-$1,000 range. These machines, a comparatively new phenomenon attributable to interest in the Internet by consumers who previously had not owned PCs, are being marketed actively by all major PC vendors. IBM was the original pioneer in this market, but by mid-1997, all major vendors were competing for market share.

Figure 5-13: Worldwide Revenue from PC Sales

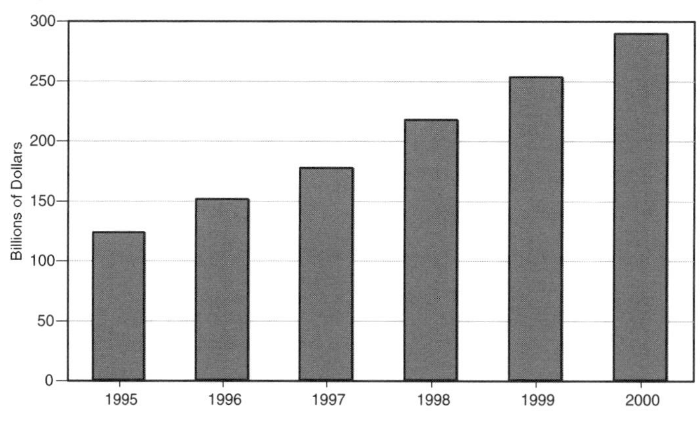

Source: IDC, 1997

By the end of the decade, Dataquest expects annual worldwide PC shipments to grow to more than 131 million units. The desktop PC still will represent the majority of the market in terms of total unit volume: 62 million desktops, 42 million desksides, and the remainder being portables of all kinds.

The division of the PC market between IBM-compatibles and the Macintosh showed dramatic changes during 1997 as Apple lost significant market share to the PC. Most estimates of Apple's share stood at less than 5 percent of the market at the end of 1997. Apple's sudden discontinuation of its clone licensing program in late 1997 probably will shrink market share even further.

Mergers and acquisitions remain an important factor driving the PC industry. Beginning with the Packard Bell/NEC merger, the trend accelerated with announcements from TI and Acer, Samsung and AST, and Olivetti. In addition, ALR was purchased by Gateway 2000. NEC merged its PC business with Packard Bell to give it a range of DOS-based systems in addition to its PC-98 system. PC 98 is a PC model sold by NEC, not a reference to the PC 98 standards discussed previously. (See "PC 98" on page 357.) NEC dominates the Japanese market, where its PC-98 series of PCs (unrelated to the PC 98 standard) passed 15 million unit sales at the end of 1996.

Client Devices: Notebook PCs

Notebook PCs represent the most actively competitive product category in the U.S. market. Market performance was hampered by shortages of components for the products. The segment is marked by strong price competition and the entry of several new, well-capitalized vendors such as Fujitsu and Hitachi. Furthermore, there is increased attention from large vendors, especially Compaq and IBM. (See Figure 5-14.)

Figure 5-14: Worldwide Shipments of Mobile Computers

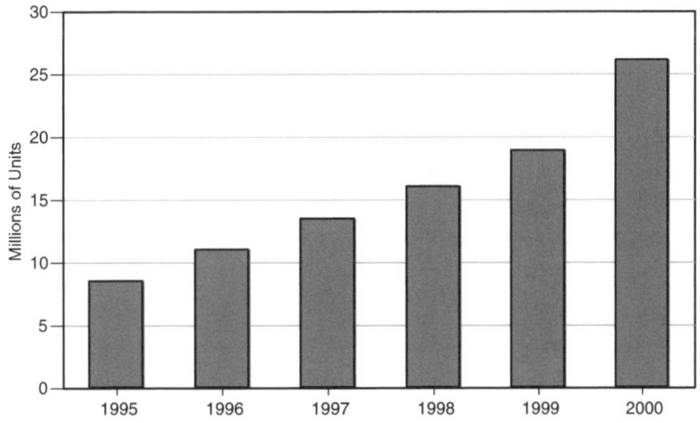

Source: IDC, 1997

Client Devices: NCs and NetPCs

Although many industry market research organizations attempted to forecast the size of market for NCs and NetPCs, the estimates showed extreme variations. This suggests that the market size for these devices is not able to be forecasted accurately at this time.

The entry and then rapid withdrawal (6 months later) of the NetPC product in 1997 (due to a lack of demand) did not bode well for NetPCs.

The prospects for widespread adoption of both NCs and NetPCs is still uncertain and depends on significant technologies to still come to market. These technologies include faster JVMs and redesigned applications suites suited for this kind of thin-client computing.

Client Devices: PDAs and HPCs

Products in this category have overcome many of their earlier deficiencies, and modest growth is expected for PDAs and HPCs. (See Figure 5-15.)

Figure 5-15: Worldwide Unit Sales of PDAs and Hand-Held Computers

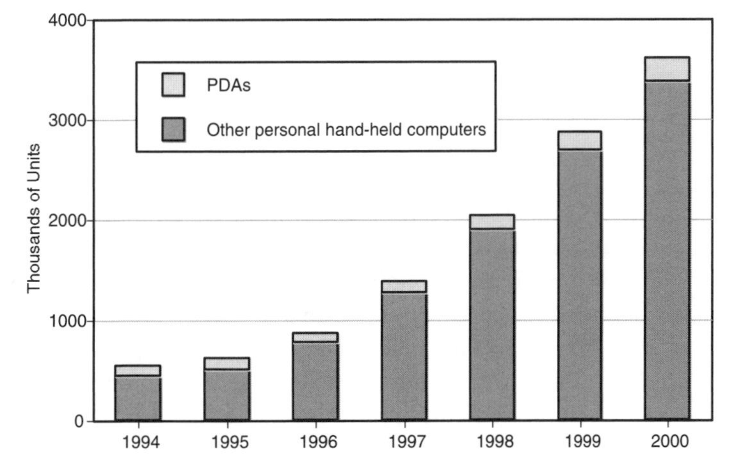

Source: IDC, 1997

Memory and Microprocessors

In terms of memory chip sizes, the current highest-volume chip is 16 Mb, and that will remain true until 1999. Newer 64-Mb chips will ramp up during 1998 and surpass 16 Mb in total number of bits shipped in 1999 and in total number of chips shipped in 2000 and will remain the highest-volume product throughout the forecast period. 128-Mb chips will start to be used in servers and workstations during 1998 and 1999 and will start to be used in PCs by the year 2000. By 2000, 256-Mb chips will be shipping. Looking farther ahead into the future, experimental 4-Gb DRAM chips were demonstrated at the International Solid-State Circuits Conference (ISSCC) in January 1997.

Unit shipments are projected to grow approximately 60 percent per year throughout the forecast period, with average price per bit falling by 20 percent per year.

National Semiconductor's 1997 acquisition of Cyrix enabled National to enter the microprocessor market, extending its computer-on-a-chip business, and to accelerate development of its 0.35- and 0.25-micron processes. To Cyrix's benefit, National's fabrication facilities change Cyrix's position as a fabless operation. It has obtained the production capability to double or triple its market share.

Another acquisition announced in 1997 was the Intel plan to purchase Chips and Technologies (C&T) for $420 million. Intel previously had licensed some of C&T's graphics technology and launched a program to develop a 3-D graphics chip. C&T is known especially for its notebook graphics, a market Intel has announced it wants to further develop. Intel also could benefit from C&T's sales force, which can provide inroads to the graphics-card vendors Intel has not dealt with as part of its desktop business. This acquisition is expected to be finalized in late January 1998.

IBM continues to be a large producer of the PowerPC chip, which has been a Macintosh microprocessor. IBM has used the chip in its own line of RISC systems. After Apple announced it would acquire Power Computing – a user of PowerPC chips and the largest Apple clone producer – and Apple's announcement that it no longer would provide future versions of the Macintosh operating system to Macintosh cloners, Motorola withdrew from its agreement with Apple. Apple announced in 1997 that some of its systems will be based on Intel processors in 1998.

On October 27, 1997, Digital and Intel announced a multi-year agreement that included the sale of Digital's semiconductor manufacturing operations to Intel for approximately $700 million, cross-licensing of patents, a supply of both Intel and Alpha microprocessors, and development of future systems based on Intel's 64-bit (IA-64) microprocessors. Intel will purchase Digital's semiconductor operations including state-of-the-art facilities in Hudson, Massachusetts, as well as development operations in Jerusalem, Israel, and Austin, Texas, for approximately $700 million. The two companies also indicated that Digital will retain its Alpha and Alpha-related semiconductor design teams to continue to develop future generations of Alpha, and Intel will serve as a foundry for Digital for multiple generations of Alpha microprocessors.

5.3.6 Forecast

During the next 5 years, computer systems increasingly will be divided into two categories: clients and servers. Multiuser systems will be classified increasingly as servers, and palmtops, PCs, and workstations will be viewed as part of the client track. Once platforms divide into these two tracks, distinctions within them will begin to have less meaning. A useful way of viewing this rearrangement is shown in Figure 5-16, a projection of the way platforms will evolve by the year 2002. This chart should be compared with Figure 5-8, which shows how these systems are categorized in 1997.

Figure 5-16: Realignment of Computing Platforms as Projected for 2002

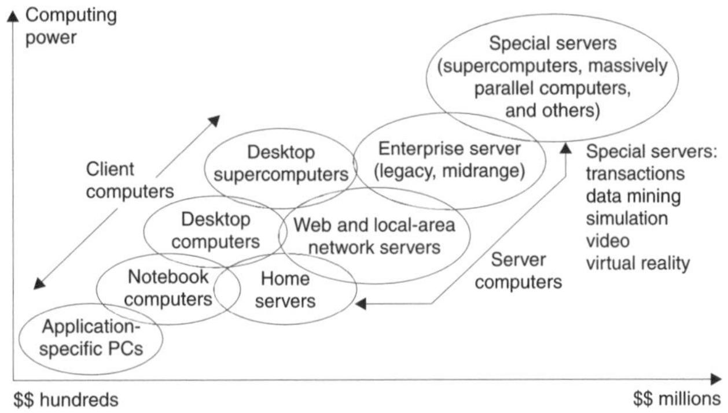

Source: Egil Juliussen, "Computers" in *IEEE Spectrum*, © 1997 IEEE

Mainframes

- Their superior I/O capabilities and bandwidth will continue to provide a role for mainframes, which will be used for transaction processing, data warehousing, Internet/intranet application servers, and network hubs. This market will grow modestly.
- For customers with large investments in customized applications, the business case for conversion from the mainframe to alternative hardware platforms will become less attractive. New CMOS technology and competitive forces will continue to lower the purchase price per MIPS for mainframes as well as the ongoing operating costs, making a conversion to another platform difficult to justify on economic grounds.
- Mainframe software features will continue to provide advantages in robustness, security, and manageability as compared with other platforms. These features will support a premium price.

Midrange Servers

- After the release of Intel's 64-bit Merced processor, expected in late 1999, several vendors whose servers are currently based on their own RISC processors will shift their product lines to Merced-based systems.
- Virtually every major version of UNIX, as well as Windows NT, will be available for Merced-based systems.
- SMP will remain popular for two- and four-way multiprocessor systems based on Intel processors. Beyond four-way, NUMA will be the leading approach to multiprocessing using Intel processors by the end of the forecast period.

- Midrange server vendors will focus on increasing the system bus speeds, memory bandwidth, and I/O throughput of their products to support high-volume transaction processing applications.
- The outlook for midrange systems calls for healthy growth in unit shipments; prices will continue to fall dramatically.
- IBM's AS/400 will continue to be a popular business platform due to its wide selection of packaged applications.
- Intel-based servers using SMP and Wolfpack clustering increasingly will be competitive with RISC systems as midrange servers running both various versions of UNIX and Windows NT through the end of the forecast period.

Client Devices: Workstations

- After years of industry predictions that PCs would offer strong competition to workstation products, this finally will come to pass. Pentium Pro and Pentium II systems running Windows NT will capture the low end of the UNIX workstation market for engineering applications. The top UNIX workstation vendors – Digital, HP, IBM, Silicon Graphics, and Sun – will introduce low-cost systems to compete.
- UNIX workstations will continue to dominate the most-demanding technical applications, such as complex 3-D graphics and large simulations, throughout the forecast period.

Client Devices: Desktop PCs

- Sales of sub-$1,000 PCs will be brisk in 1998. As more major players push into this market space, discount PC manufacturers will see their market share erode. Compaq, Dell, HP, and IBM already have drawn clients away from the low-price clone vendors.
- By late 1998, the typical business desktop configuration will become a 300 MHz Pentium II-based machine with 48 MB of RAM, 2 MB of VRAM, and 6 GB of hard disk storage. Also included is a DVD-ROM drive, a 56-Kbps modem, and two USB ports.

Client Devices: Notebook PCs

- Notebook PCs with a 13.3-inch or 14.1-inch thin film transistor (TFT) display, full-size keyboard, and multimedia capabilities will become a popular choice in 1998 for mobile professionals who need access to the full range of PC capabilities while traveling. For such individuals who have a permanent office as well, the notebook will be used with a docking station to replace their office PC.
- The low end of the notebook market will face increasing competition from hand-held PCs and other Windows CE-based devices, beginning in late 1998. They will be used primarily where the application requirements are limited to basic e-mail, connectivity, and simple productivity applications.

Client Devices: NCs and NetPCs

- NCs and NetPCs both will be slow to gain large-scale deployment in commercial environments.
- The major market for NCs will be to replace the remaining installed base of terminals (and PCs used primarily for terminal emulation) and in fixed-function applications such as bank teller workstations, hotel desks, and airline ticket agents.

- NetPCs will lack strong demand, and few models will be introduced in 1998 because they lack manageability as compared with the NC and lack functionality as compared with the PC. However, all network-attached PCs will incorporate certain aspects of the NetPC design, particularly improved manageability features designed to lower the total cost of ownership.

Client Devices: PDAs and HPCs

- PDAs and HPCs will become more widely adopted during the forecast period as the products develop more applications and adequate connectivity.
- PDAs and HPCs that provide e-mail, paging, Web browsing, and phone service via wireless networks will sell well in the commercial market towards the end of the forecast period.
- PDAs will be successful as extensions to, rather than as replacements for, the desktop or notebook PC. The focus will be on personal information management applications; the ability to synchronize data easily with the user's PC will be critical to these products' acceptance.
- HPCs running the Windows CE operating system will become an attractive platform for software developers because they can use the Windows CE APIs to develop an application that then can be delivered for HPC products from multiple vendors. The availability of third-party applications in turn will make these devices more popular with business users.

Microprocessors

- Although Intel and other x86 chip vendors will continue to have the overwhelming share of the microprocessor market, RISC architecture chips will remain important in midrange servers and high-end workstations through the end of the forecast period. The major RISC vendors (Digital, HP, IBM, Silicon Graphics, and Sun) will continue to deliver enhanced versions of their processors throughout the forecast period. However, some of these products will be used primarily as embedded controllers rather than in general-purpose computer systems.
- Intel will ship a series of new versions of the IA-32 Pentium II processor during 1998, including the first Pentium II designed for notebook use and the first Pentium II designed for low-cost desktop PCs. In 1999, Intel will ship Katmai, the first Intel processor incorporating an extended set of multimedia instructions, and Willamette, an enhanced implementation of the Pentium II architecture. Intel's leading competitors in the IA-32 processor market – AMD, Cyrix, and IDT – also will ship enhanced versions of their processors throughout this period.
- At any point in time, Intel and its competitors will offer a variety of different processors implementing the IA-32 instruction set. These will differ in their clock speeds but also in other factors determining performance and cost, such as the amount of L1 and L2 cache incorporated into the processor or supported off-chip, the existence of a dedicated high-speed bus for cache access, their packaging, and whether they are designed for desktop or notebook computers.
- Intel will add 70 new instructions extending the MMX instruction set to accelerate 3-D graphics and other multimedia operations to the IA-32 architecture in 1999. However, AMD, Cyrix, and IDT will ship chips with their own implementations of similar instructions as much as 6 months earlier than Intel. The existence of two slightly different sets of new multimedia instructions will require vendors selling graphics cards and similar components to develop multiple versions of their software drivers.

- Intel will ship Merced, the first chip implementing the IA-64 architecture, in late 1999, followed by subsequent implementations toward the end of the forecast period. However, use of the IA-64 processors will be confined to servers and high-performance workstations through the end of the forecast period. Despite the debut of IA-64 processors, the IA-32 processor will still be used in virtually all desktop and notebook PCs through the end of the forecast period, and Intel will continue to release enhanced IA-32 processors during that time.

5.3.7 References

■Articles

Brambert, Dave. 1997. Network clustering. *Network VAR.* April: 16.
Clustering: a Windows NT service of related parts. 1997. *Computer Reseller News.* June 9: 90.
Comerford, Richard and Egil Juliussen. 1997. Computers. *IEEE Spectrum.* January: 49.
Fawcett, Neil. 1997. Rhubarb and clusters. *Computer Weekly.* May 22: 48.
Francis, Bob. 1997. Sub-$1,000 PCs now have more extras. *InformationWeek.* December 8: 30.
Garvey, Martin. 1997. Speed in clusters. *InformationWeek.* January 27: 112.
Jacobs, April. 1997. PC 98 Design Guidelines get mixed reviews. *Computerworld.* April 21: 43.
MMX power for desktop PCs. 1997. *Byte.* July: 106.
Manes, Stephen. 1997. What your next PC will look like. *PC World.* July: 352.
Mossberg, Walter. 1997. All the basics you'll need to know when buying a PC. *The Wall Street Journal.* April 10: B1.
Raduchel, Bill. 1997. Beyond the desktop. *InformationWeek.* June 16: 70.
Sager, Ira. 1997. NetPCs are have a hard time booting up. *Business Week.* September 22.
Santoni, Andy. 1997. PC 98 spec pushes CPU speeds, moves away from ISA bus. *InfoWorld.* April 21: 31.
Simpson, David. 1997. Are mainframes cool again? *Datamation.* April: 46.
Vijayan, Jaikumar. 1997. Scaling up your warehouse. *Computerworld.* April 21: 71.
Workman, Will. 1997. PC 98 standard calls for quieter computers. *Computer Shopper.* August: 91.
Yamaha, Ken. 1997. Consortium eyes clustering standard. *Computer Reseller News.* April 28: 84.

■Periodicals, Industry and Technical Reports

Apfel, A. 1997. *Network computers: the early market.* Gartner Group. (Report PTP-072597-02) July 25.
Cappuccio, D. 1997. *Network computers: panacea and Pandora's box.* Gartner Group. (Report PTP-042597-01) April 25.
Card, David, John Gantz, Diana Hwang, Bruce Stephen, and William Zinsmeister. 1997. *The enterprise network computer market.* IDC. (Bulletin #13015R) February.
Donovan, Bryan, Craig Friedson, and Dana Thorat. 1997. *IT vendors announce strong fourth-quarter results despite slower growth.* IDC. March.
Eckerson, Wayne. 1996. *"Network computer" – technology in search of a solution.* Patricia Seybold Group. July.
Hanney, Laura K. 1996. *The future of personal digital assistants in distribution.* IDC. (Bulletin #11947) August.
Henkel, T. 1997. *Servers or workstations? UNIX vendors face tough choice.* Gartner Group. (Report M-142-224) March 18.
Lee, Gene. 1997. *1996 midrange server year in review.* IDC. (G22D) January.
Norris, D. 1997. *How long will mainframes remain profitable?* Gartner Group. (Report KA-071-239) April 18.
O'Brien, Eileen G. 1997. *Network computers: an expanded choice for users.* IDC. (Bulletin #13858) July.
Oltsik, Jon, H. Waverly Deutsch, and Beth Edwards. 1996. *The mainframe's last hurrah.* Forrester Research. May.
Richmond, Howard. 1996. *Will your mainframe be your all-purpose server?* Gartner Group, (Report KA-163-141) August 21.
Segervall, Laura. 1996. *Worldwide Workstation Forecast, 1995-2000.* IDC. March.
Sullivan-Trainor, Michael. 1997. *Internet and network computers: what are the software requirements?* IDC. (Bulletin #13010) February.
Why nobody understands NCs or NetPCs. 1997. Forrester Research. March.

■URLs of Selected Mentioned Companies

3Com `http://www.3com.com`
Acer `http://www.acer.com`
ALR `http://www.alr.com`
Amdahl `http://www.amdahl.com`
Apple `http://www.apple.com`

AST `http://www.ast.com`
Autodesk `http://www.autodesk.com`
Casio `http://www.casio.com`
Compaq `http://www.compaq.com`
Comparex `http://www.comparex.com`
Convex `http://www.hp.com/wsg/products/servers/servhome.html`
Cray `http://www.cray.com`
Dell `http://www.dell.com`
Digital `http://www.digital.com`
Fujitsu `http://www.fujitsu.com`
Gateway 2000 `http://www.gateway2000.com`
HP `http://www.hp.com`
Hitachi `http://www.hitachi.com`
IBM `http://www.ibm.com`
ICL `http://www.icl.com`
IEEE `http://www.ieee.org`
Intel `http://www.intel.com`
Lightscape `http://www.lightscape.com`
Microsoft `http://www.microsoft.com`
Mitsubishi Electric `http://www.mitsubishielectric.com`
Motorola `http://www.mot.com`
NCR `http://www.ncr.com`
NEC `http://www.nec.com`
Novell `http://www.novell.com`
Olivetti `http://www.olivetti.com`
Oracle `http://www.oracle.com`
Packard Bell `http://www.packardbell.com`
Palm Pilot `http://palmpilot.3com.com`
Psion `http://www.psion.com`
Pyramid `http://www.siemens-pyramid.com`
Samsung `http://www.samsung.com`
Sequent `http://www.sequent.com`
Siemens Nixdorf `http://www.siemens-nixdorf.com`
Silicon Graphics `http://www.sgi.com`
Sony `http://www.sony.com`
Sun `http://www.sun.com`
Tandem `http://www.tandem.com`
Texas Instruments `http://www.ti.com`
Unisys `http:// www.unisys.com`
U.S. Robotics `http://www.usrobotics.com`

5.4 Security

5.4.1 Executive Summary

Many tools and procedures can add an important measure of security to a corporate network, but few if any can provide a complete guarantee against break-ins or breakdowns. The steady proliferation of products that focus on some aspect of security – encryption, firewalls, digital signatures, and security tokens or smart cards – illustrates that there are still basic problems with installed security systems and the way they are deployed. However, corporate users are becoming aware that security cannot be added after the fact. Instead, security must be built into operating systems, into all applications that use these systems, and into the design of the network itself – especially for distributed environments.

During the last year, several trends have emerged that highlight the movement toward a comprehensive security approach. Dramatic improvements have occurred in security component functionality in areas such as firewalls, for example. Distinctions between traditional market segments are blurring or dissolving entirely as firewall vendors integrate their products with anti-virus vendors' products, resulting in the addition of important new features and capabilities to existing product lines.

The increasing use of the Internet – as an inexpensive virtual private network (VPN) for electronic commerce, for ordinary World Wide Web sites, and for e-mail – has raised additional concerns about network security in a world of virus propagation, hostile Java and ActiveX applets, "spam" e-mails, IP address spoofing, denial-of-service attacks, mobile user remote access, and Web site or intranet "hacking."

The continued effort by the United States government to regulate the export (and even some domestic use) of strong encryption has driven some U.S. manufacturers to partner with developers outside the U.S., agree to the government's proposed key recovery scheme, or devise new ways to circumvent these restrictions.

The lack of standards and the controversy over which security options will dominate – and when final versions of protocols will be available – have become obstacles to services such as the use of secure Internet e-mail. One response has been the formation of new vendor alliances and industry associations – such as the Financial Services Technology Consortium and the Information Technology Industry Council – that are focusing on the development and promotion of interoperable security products and secure network applications.

Several unsolved areas remain the focus of development efforts: securing remote access, securing the Internet as a virtual private network (VPN), developing alternatives to reusable passwords, devising better authentication methods that incorporate biometric technologies and security tokens, and enhancing security for the Internet itself.

5.4.2 Changing Business Security Needs

Today's business environment has different security requirements than traditional commerce. Enterprise networks are no longer defined by the physical boundaries of a single company location but often encompass remote sites and include mobile users and telecommuters around the world. Also, virtual organizations often use many contractors who are not employees, thus introducing numerous additional security concerns. Security requirements for corporations are therefore complex.

Companies also have begun to rely on shared public networks such as the Internet rather than private leased lines for e-mail and electronic commerce. In addition, various distributed security components to support client/server computing, networks, and network computers must interoperate seamlessly and securely and must support a unified, integrated management policy.

New Security Holes

Corporate information systems are more vulnerable than ever because data is no longer centralized and because the Internet and company intranets provide new avenues for intrusion. According to a March 1997 report by the Computer Security Institute and the Federal Bureau of Investigation (FBI), 47 percent of the U.S. organizations surveyed had been attacked through the Internet, up from 37 percent in the 1996 survey. More important, the report contradicted the belief that the majority of attacks come from within an organization: although 43 percent of respondents reported attacks from within, 47 percent experienced external attacks. The report concluded that the threat from outside had risen dramatically due to increased Internet connectivity. Reported losses (by the 59 percent that could quantify them) totaled about $100 million.

A July 1997 *InformationWeek* survey echoed these responses. Its key findings were that security breaches are on the rise, the Internet increases corporate vulnerability, viruses remain a threat, and industrial espionage is a reality, especially because the Internet makes it easier to steal and exchange information.

New Security Alternatives

Security is a continuous process, rather than a short-term project or a half-dozen disconnected products, and companies are centralizing security administration. Today, security requires ongoing investment and monitoring. In response, companies are expanding security staffs, allocating more resources, centralizing their information technology (IT) security, and developing business plans that include incident management strategies in case of break-ins.

One alternative is to outsource an Internet connection and its security (as part of a complete security plan) to a managed Internet security service (MISS) provider. Such vendors typically develop a standard model for Internet security that is administered with minimal customization for multiple enterprises, thereby providing economies of scale. Many established Internet service providers (ISPs) already are offering or plan to offer specialized MISSs, and some startups are entering the market as well. (See Table 5-8.) Services usually include firewall implementation and management plus real-time vulnerability monitoring; other services being tested or offered by only a few vendors include virus scanning, secure Web site and extranet hosting, and disaster recovery.

An Internet security outsourcing solution is suitable for midsized companies with simple access control needs but little in-house security expertise.

As with other security approaches, however, the effectiveness of MISS outsourcing may be hampered by a market composed of evolving standards, technologies, and products that solve one piece of the puzzle but do not interoperate very well.

Table 5-8: Managed Internet Security Service Providers

Vendor – Product	North America	Europe
ANS – ANS SecureConnection	Yes	No
BBN Planet – SitePatrol	Yes	Yes
IBM – SecureWay	Yes	No
Information Resource Engineering – SafeNet	Pending	Pending
MCI – Concert Internet Plus	Yes	Pending
Pilot Network Services –Secure Internet	Yes	No
Quza – QuzaSafe (Quza is owned by Racal and Integralis)	No	Yes
Sprint – IP Security Services	Yes	No

Source: Gartner, 1997

5.4.3 Information Security Components

Security is generally considered to have three core components:

- **Confidentiality**, or the prevention of unauthorized disclosure of information
- **Integrity**, or the prevention of unauthorized modification of information
- **Availability**, or the prevention of unauthorized withholding of information or resources.

No matter what technology or business application is deployed, the basics of IT security remain authentication, authorization, administration, auditing and accountability, and data integrity.

Authentication

Authentication is the verification of a user's claimed identity (typically when logging into a system) by passwords, personal challenge-response calculators, random password generators, or biometric identification devices. Increasingly, however, the term authentication is being used in relation to authenticating messages to ensure that they are genuine. The former definition involves the use of passwords, smart cards, and biometrics to ensure security, while the latter definition involves digital signatures and certificates.

Authentication methods are usually based on something users know (such as a password), something users have (such as security tokens or smart cards), or something users are (biometrics). Two-factor authentication – using two of the foregoing methods – provides a higher level of security than simple authentication and is rapidly gaining ground within the corporate world. In general, the sophistication of the authentication process should be related directly to the level of business risk associated with an unauthorized person gaining access to the system.

For example, reusable passwords are not an appropriate authentication method for users on insecure networks. Remote users logging in from an Internet kiosk in an airport or from the network provided by a hotel have no way of knowing who may be monitoring their transmissions and must assume that their passwords will be compromised. To ensure security in such environments, users need to employ a different (and more trustworthy) authentication method – a one-session password, which automatically changes with each use via a security token, or an authentication technique such as Kerberos, which is not subject to "capture/replay" (where a hacker intercepts a password and uses it to enter a secure system by impersonating the legitimate user).

Challenges in establishing authentication systems include the lack of uniform standards, the need for an "umbrella" application programming interface (API), certificate management issues (certificates are public keys used to verify digital signatures), including the need for a legal and financial framework for certificate authorities, and the lack of widely deployed smart card readers. (See Chapter 3.3, *Smart Cards.)*

■ Security Tokens and Smart Cards

Security tokens are becoming a popular authentication solution because they can eliminate the drawbacks of reusable passwords. For example, Security Dynamics' SecurID two-factor hardware authentication system combines something the user knows – a personal identification number (PIN) – and something the user possesses – a security token resembling a credit card that generates a new, random password every 60 seconds. The user enters a PIN, and if correct, SecurID generates the password, which the user manually enters via the PC keyboard. Two variations of the SecurID system address other authentication issues: SoftID, an all-software version that dispenses with the hardware token; and WebID, a browser add-on that gives remote users Internet access to the company network.

Many vendors are devising solutions based on existing computer hardware. For example, Fischer makes a module called Smarty that converts the 3.5-inch diskette drive on a PC into a smart card reader. Smarty includes a SmartDisk that provides a security token used for authentication of the smart card and data encryption to prevent theft. Smarty enables any PC to become a secure home banking system.

Other offerings include HP's Net Vectra PC series, announced in June 1997, which offers an optional smart-card-ready keyboard that can be used to provide password-free access control to any similarly equipped PC and allows enterprise-wide deployment of HP's ImagineCard security system; Key Tronics' next generation of PC keyboards, which will incorporate smart card readers; and SCM's SwapSmart reader, a PC Card (PCMCIA) smart card reader/writer that provides a portable, universal, and secure bridge between smart cards and mobile PCs.

In an effort to facilitate compatibility and security, a consortium of vendors (including IBM, Netscape, Oracle, and Sun) announced a new standard in April 1997 that sets interoperability requirements for smart cards, readers, and applications. The OpenCard framework incorporates a public-key industry standard (PKCS-11) for cryptographic tokens for authentication. Users need the smart card and a PIN to authenticate transactions.

■ Biometrics

Biometric technology actually encompasses many technologies, such as hand geometry, fingerprints, iris and retina scans, voice recognition, and face recognition. The goal of biometric identification is to assist in providing a log-in system with security surpassing password and token systems. This goal is achieved because access is provided only to the specific individual, rather than any holder of the access card.

Biometric techniques usually involve an automated process to verify the identity of an individual based on physical or behavioral characteristics. First, biometric templates of the particular characteristic, such as a voiceprint, are collected in a database. The template data then is recalled during a verification process when the biometric template of the voiceprint is compared with the offered characteristic. Depending on the computer and network technologies used, verification takes only seconds.

Biometric techniques fall into two categories: physiological and behavioral.

- **Physiological biometrics** – face, eye, fingerprint, palm print, hand geometry, or thermal images
- **Behavioral biometrics** – voiceprints, handwritten signatures, and keystroke/signature dynamics

Biometrics are used in three general situations: to provide absolute assurance of identity that cannot be compromised by the authorized person "sharing" his or her password or token; to avoid the inconvenience of needing a token; and to avoid the inconvenience of users forgetting their passwords. Historically, biometrics are used most frequently to provide uncompromising security, as in the first situation.

Although many biometric technologies are becoming affordable enough for general-purpose use, acceptance often is based on other less-tangible factors, such as convenience. For example, iris or retina scans can be uncomfortable and could meet resistance if implemented; however, they also may be among the most foolproof processes. An individual iris structure has 400 discriminations available, making it 6 times more unique than a fingerprint, with 60 different forms of variations.

In contrast, face recognition – one of the fastest-growing niches – meets most of the criteria for the ideal solution: It is fast, easy to perform, moderately convenient, affordable, and nonintrusive. Hand geometry already is being used at several major airports and by the U.S. Immigration and Naturalization Service.

Fingerprint biometrics probably will be the first widespread biometric because the technique builds on the fingerprint identity systems already employed by law enforcement agencies. Authentication methods based on ear or lip shape, knuckle creases, sweat pores, "vascular tattoos" (heat spots created by veins and arteries in the face), blood vessel patterns in the hand (the U.K. firm BTG holds a patent on this technology), and body odors (Mastiff Electronic Systems' Scentinel) are also in development. Representative biometric vendors and products are shown in Table 5-9.

Table 5-9: Representative Biometric Technologies and Vendors

Biometric	Pros	Cons	Vendor	Product(s)
Face recognition	Easy; fast; inexpensive	Subject to spoofing; awkward lighting in the image can affect authentication	Identification Technologies	One-on-One
			Keyware Technologies	Face Guardian
			Miros	TrueFace CyberWatch
			Viisage	Viisage Gallery
			Visionics	FaceIt

Table 5-9: Representative Biometric Technologies and Vendors (Continued)

Biometric	Pros	Cons	Vendor	Product(s)
Fingerprints or finger scanning	Inexpensive; very secure	Latent prints possible; cuts and dirt can mar image and prevent the system from authenticating the user	Mytec Technologies	Touchstone
			Oracle/Identix	TouchNet
			Thomson-CSF	FingerChip
			Veridicom	Veridicom
Iris/retina scanning	Extremely difficult to fool	Intrusive and inconvenient	EyeDentify	EyeDentify Retina Biometric Reader
			IriScan	IriScan/IrisCode
Palm scanning or hand geometry	Tiny storage requirements; intuitive operation	Slow; less accurate than finger scanning	Recognition Systems	ID-3D Handkey
Signature recognition or signature dynamics	Inexpensive	Can be affected by physical condition or emotional state	Advanced Recognition Technologies	smARTwriter
			PenOp	PenOp
Thermal imaging	Extremely difficult to fool	Requires expensive infrared cameras	Unisys	"thermogram" technology
Voiceprint	Inexpensive; good for remote access	Slow; can be affected by physical condition or emotional state	Keyware Technologies	Voice Guardian
			Veritel	VoiceCrypt

A key problem with the adoption of biometrics is the lack of industry standards regarding how these systems connect to the networks they are meant to protect. However, market analysts predict that pressure for standardization will inevitably develop as the technologies become more widely implemented.

Industry consortia may again play a major role in standards development. A group called the Biometric Consortium, established in the early 1990s and housed at the National Computer Security Association (NCSA), focuses on "promoting the science and performance of biometrics" for the U.S. government. The consortium includes state welfare agencies, driver's license bureaus, the Immigration and Naturalization Service, the Social Security Administration, the Internal Revenue Service, and other government departments. The consortium plans to police inflated claims of accuracy through product testing at government evaluation centers.

NCSA also manages the Commercial Biometrics Developer Consortium (CBDC), composed of biometric vendors, integrators, and resellers. In July 1997, NCSA introduced a certification program for commercial biometric products on behalf of the consortium. All biometric devices – fingerprint, hand geometry, speech recognition, iris scanning, and facial recognition – are certified using the same criterion to ensure operability and test claims. As the test criteria evolve over the next few years, NCSA expects that biometric testing will become more stringent, perhaps including environmental variables, tighter error tolerances, expanded test populations, and one-to-many performance testing using large, live-scan databases.

One major concern is whether the use of biometrics violates personal privacy, given the possible development of centralized biometric databases. To avoid such dangers, biometrics vendors are pushing so-called "one-to-one" matching systems where a smart card, not a database, holds the biometric information, which is then matched to the appropriate human characteristic.

Authorization

Authorization is the process of determining how an authenticated user is permitted to use specific resources. An authorization mechanism automatically enforces a management policy regarding resource object use. Resource objects vary based on the nature of what is being protected. For example, in a computer system, resources typically include data files, operator commands, transactions, I/O devices, and program processes. The specific rules for authorizing access to data objects usually enforce confidentiality and integrity by either granting or denying access to read, modify, or create data records, and by controlling the creation or deletion of data objects. For a communications network, resource objects usually include specific operator commands and routing destinations. The authorization rules for networks generally limit each network operator's available commands for status inquiry or starting/stopping devices in local and remote network domains. Authorization rules typically are expressed in the form of access control lists (ACLs).

Examples of authentication/authorization products include HP's Praesidium Authorization Server; IBM's RACF, DCE Security Server, and Global Sign-On; and Sun's Solstice Security Manager, SKIP, and Sunscreen SPF 100.

Administration

Administration is the process of defining, maintaining, and deleting users, resource objects, or the authorized privilege relationships between users and objects. Administration translates business policy decisions into a format that an IT system can use. The resulting internal definitions can be enforced, as appropriate, at the point of entry, throughout the network, and in client/server or host computing systems. Security administration is an ongoing effort because business organizations, their systems, and their users are constantly changing.

Examples of products with administration features include CA's CA-Unicenter TNG (The Next Generation), IBM's Tivoli TME 10 Security Management and RACF, Mergent's Domain/DACS and PC/DACS, and Vanguard Integrity's RACF Administrator.

Auditing and Accountability

Auditing is the process of data collection and analysis that allows administrators and others, such as IT auditors, to verify that the users and authorization rules are producing the intended results as defined in a company's business and security policy. Individual accountability for attempts to violate the intended policy depends on monitoring relevant security events, which initiates the auditing feedback reporting loop. The monitoring process can be implemented as a continuous automatic function, as a periodic check, or as an occasional verification that proper procedures are being followed. The auditing information may be used by security administrators, internal audit personnel, external auditors, government regulatory officials, and in legal proceedings. The auditing and individual accountability monitoring functions also apply to documenting the activities of security administrators and auditors to ensure that they are not abusing their authorized capabilities. Most access control or administration products – CA's CA-Unicenter TNG, IBM's RACF, Internet Security Systems' Internet Scanner, Mergent's Domain/DACS and PC/DACS, OpenVision's AXXion, Tandem's Safeguard, and Vanguard Integrity's RACF Administrator – also provide auditing capabilities.

In addition, freeware may provide solutions for companies with minimal security requirements. Freeware such as Computer Oracle and Password System (COPS), Tiger, NFSBUF, and Security Administrator Tool for Analyzing Networks (SATAN) also provide audit and logging capabilities that can assist organizations in examining and correcting UNIX systems security.

5.4.4 Security in Distributed Environments

With computing resources typically dispersed throughout today's multinational corporate environments, managing a variety of users, systems, and procedures has become a significant administrative and security challenge. Security in a distributed environment includes many components, such as single sign-on services, Kerberos protocols, the Secure European System for Applications in a Multi-Vendor Environment (SESAME), and the Distributed Computing Environment (DCE) architecture.

Single Sign-On (SSO)

SSO is the ability to access multiple computer systems or networks after logging in once with a single user ID and password. This setup resolves the common situation where separate passwords and user IDs are required for each application. SSO has three major advantages for organizations: user convenience, administrative convenience, and improved security.

The benefits of SSO have been widely demonstrated. Having only one sign-on per user instead of ten makes administration easier. It also eliminates the possibility that users will write down their ten passwords in order to remember them all, thereby compromising security. Finally, SSO enhances productivity by reducing the amount of time users spend gaining system access. For example, a survey by the Securities Industries Association found that users spent an average of 44.4 hours per year logging on to (an average of) four applications per day. Using SSO, they would spend only 11.1 hours per year logging on – a savings of 33.3 hours. If the average employee's per-hour wage is $25 and a company has 1,000 users, it will save $832,500 per year with SSO.

SSO can be implemented in various ways, either by traditional mainframe-based software in a host environment or by Kerberos in a distributed environment. The trend is toward enterprise-wide, integrated SSO products such as these:

- **CKS's MyNet** – Currently available for IBM MVS mainframe systems and scheduled for release in early 1998 for Windows NT and UNIX servers, MyNet offers SSO, central administration, authentication, encryption (including support for a variety of security tokens), and audit across multiple platforms.
- **Fischer's Watchdog** – Combines passwords at the local area network level; Fischer also has teamed up with CKS to create a global sign-on product for release in late 1997 or early 1998.
- **IBM's Global Sign-On (GSO)** – Released in July 1997, GSO supports OS/2 and Windows clients as well as mainframes, LAN Server, OS/2 Warp, NetWare, Windows NT, IBM AIX, and Sun Solaris servers, various databases, and smart cards.
- **Millennium's FirstStep SSO** – Allows SSO to applications and network gateways; available for Windows, Windows NT, and Macintosh clients to connect with HP-UX, Sun Solaris, IBM AIX, and other IBM, Unisys, and Digital servers and mainframe systems.

Kerberos

Another mechanism to solve the SSO problem, Kerberos is an authentication system designed to let multiple systems send information about a user's identity and access privileges between systems in such a way that no information that could be used to impersonate a user (such as his or her password) is ever sent across the network in "the open" (that is, as plain text).

Kerberos uses secret-key ciphers for authentication and encryption and is becoming a de facto standard for remote authentication in client/server environments. Kerberos' client/server design centralizes the authentication process for multiple systems in a secure way, an important consideration in networked environments. The Kerberos protocol provides authentication in The Open Group's Distributed Computing Environment, a de facto industry standard. (See "Distributed Computing Environment (DCE)" on page 382.)

Kerberos 5 release 1.0, introduced in December 1996, adds features such as inter-realm authentication and credential-forwarding. Inter-realm authentication lets users authenticate with other realms throughout the Internet. Credential-forwarding allows credentials (tickets) obtained at a user's workstation to pass to a remote host automatically when the user logs in (a form of SSO). Commercial implementations of Kerberos 5, such as Cygnus Solutions' KerbNet (released in April 1997), which integrates SSO into the Windows NT log-in procedure, are already available.

Figure 5-17 illustrates how Kerberos works. For a user to gain permission to access a server, the user creates an authentication request, which is encrypted with his or her password to request a ticket from the security server. The security server validates the request by comparing it to the same request encrypted with a stored version of the user's password; if the two match, the user is authenticated, and the server sends back a "ticket-granting ticket." The client then requests a server ticket to access server X. The security server then returns a server key, encrypted using the client's private key. Using this ticket, the client can then request access to server X.

Figure 5-17: Kerberos Security

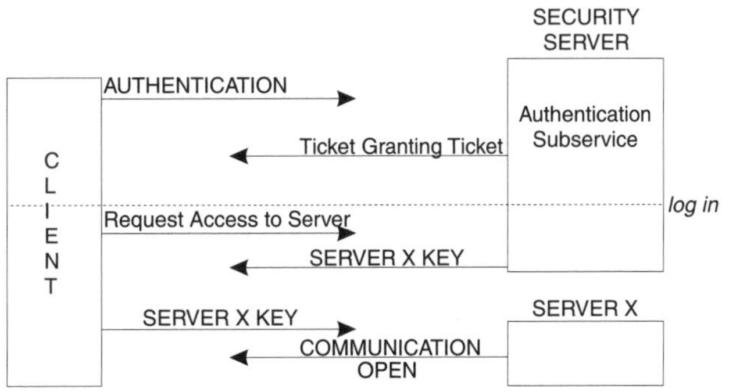

Source: Gartner, 1995

Comparing Kerberos or public-key encryption systems to determine which offers greater functionality and which is better suited for Internet usage is not an accurate "apples-to-apples" comparison. Kerberos is a security application designed to handle user authentication and access control in distributed systems; a public-key encryption system is designed to provide message confidentiality, integrity, and authentication. Many vendors (such as Gradient Technologies,

IBM, and Microsoft) are supporting both technologies and are working on integrating them, enabling a user to be authenticated to a Kerberos server with a public-key certificate or creating a common API that provides access to both a Kerberos and a public-key infrastructure.

Products that feature multiple authentication methods are beginning to appear. For example, Gradient Technologies' WebCrusader V3, released in October 1997, uses two-way, two-level call authentication based on Kerberos to prevent Web spoofing (where an attacker creates a fake Web site to lure users in the hope of stealing their credit card numbers or other information). Because the typical enterprise has many existing applications, each with its own method of verifying the identity of users and trusted servers, WebCrusader V3 integrates multiple authentication mechanisms: well-established user name/password schemes (such as Kerberos or Secure Sockets Layer for transmitting passwords securely), emerging schemes including public-key certificates (such as Entrust's Public-Key Infrastructure), and two-factor authentication (such as security tokens or smart cards).

Another, and more serious, concern, however, is maintaining the interoperability of Kerberos implementations. Different implementations of Kerberos send different authorization information – a key component of interoperability. One development threatening full interoperability among all other Kerberos implementations is Microsoft's introduction of Windows NT 5.0, which features a version of Kerberos that will send only NT-specific security identifiers.

SESAME

SESAME is a European superset of Kerberos. SESAME leverages the Kerberos work by the Massachusetts Institute of Technology (MIT), uses the Kerberos 5 protocol and data structure in addition to SESAME-specific structures, and provides developers with a framework on which to build SSO functionality. SESAME has a trusted third-party security services architecture (like Kerberos) but uses public-key cryptography instead of private-key cryptography.

Basically, SESAME adds heterogeneity, sophisticated access control features, scalability of public-key systems, and better manageability, audit, and delegation to Kerberos. SESAME implementations include Groupe Bull's Integrated System Management AccessMaster (ISM AccessMaster) and ICL's Access Manager. Although based on Kerberos authentication, Siemens Nixdorf's TrustedWeb Internet security software (a beta version was released in August 1997) adds SESAME extensions to provide "role-based" access control (groups of users are assigned access rights because they share similar tasks).

European work on the specification stopped in 1996, however, and there is little familiarity with SESAME elsewhere.

Distributed Computing Environment (DCE)

The Open Group's DCE provides an alternative technology to Kerberos and SESAME for user authentication. The security component of DCE is based on Kerberos 5.0. DCE security comprises authentication, authorization, data integrity, privacy, and audit capabilities. The biggest difference between Kerberos and DCE is that DCE has other integrated services beyond that of security, such as a distributed file system and remote procedure calls (RPCs), thus allowing all these services to invoke the same security server.

Figure 5-18 illustrates how DCE security works.

Figure 5-18: DCE Security

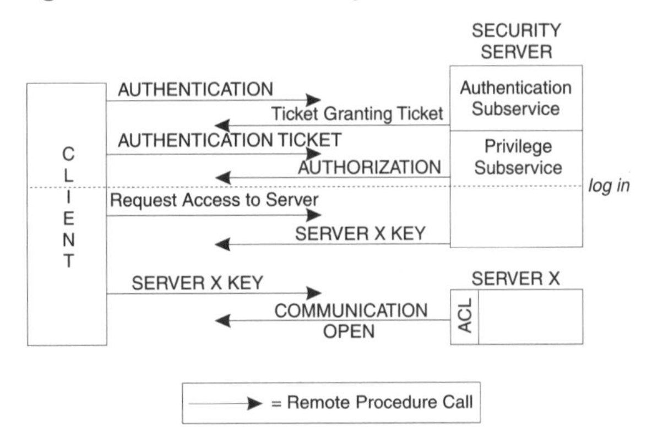

Source: Gartner, 1995

In April 1997, The Open Group announced the release of source code for servers that bring DCE-based security and services to Web sites and applications. Secure Enterprise Web extends DCE's authentication, authorization, integrity, and naming mechanisms to Web documents and Web-based services. The offering consists of a Web/DCE server, a security gateway, and a proxy server; Secure Enterprise Web is available for versions of UNIX from Digital, HP, and IBM and for Windows NT.

5.4.5 Security Evaluations and Standards

In the past, purchasers of computer and security products have had concerns regarding the level of trust placed on vendors' claimed security features. As a response to these concerns, countries such as Canada, Germany, the U.S., and the U.K. defined evaluation criteria so that products could be certified as having attained particular levels of security. Security definitions, such as those provided by the Trusted Computer System Evaluation Criteria (TCSEC, originally issued by the U.S. Department of Defense and often referred to because of the color of its cover as the Orange Book), were issued in each country by the government agency responsible (in the U.S., by the National Computer Security Center).

The European Community's Information Technology Security Evaluation and Certification (ITSEC) scheme, established in 1990, is similar to TCSEC but rates functionality (F) and effectiveness (E) separately. TCSEC and ITSEC continue to be the methods of choice for verifying that vendors' products fulfill specific security objectives and that vendors' claims about product security features are valid. (See Table 5-10.)

In February 1997, the U.S. National Security Agency (NSA) debuted a security technology evaluation program called the Trust Technology Assessment Program (TTAP) to rate low-end commercial software for C2-level certification. The program was scheduled to remain in pilot mode for 2 years before a permanent test process was developed based on the Orange Book and "emerging international common security criteria."

In October 1997, however, the NSA and the National Institute of Standards and Technology (NIST) announced new security classifications called Common Criteria (CC) that will replace existing classifications such as the Orange Book. CC is an international standard for enterprise network security developed to align criteria from Canada (CTCPEC), Europe (ITSEC), and the U.S. (TCSEC). CC compliance testing will be performed by licensed commercial labs, and the

Table 5-10: TCSEC and ITSEC Product Evaluation Criteria

TCSEC	ITSEC	Definition
Level A1	F-B3, E6	System is characterized by a mathematical model that can be proven; highest security – used in computers holding classified information
Level B		Level B provides mandatory access control and multilevel security
B1	F-B1, E3	Security labeling of all users, resources, and objects
B2	F-B2, E4	Guarantees path between user and the security system; provides assurances that system can be tested and clearances cannot be downgraded
B3	F-B3, E5	System is characterized by a mathematical model that must be viable
Level C		Level C provides discretionary access control
C1	F-C1, E1	Requires user log-in, but allows group to share an ID
C2	F-C2, E2	Requires individual user log-in with password and an audit mechanism
Level D	E0	A nonsecure system

test criteria include standards adherence for interoperability. According to the Common Criteria Project, a family of "protection profiles" is being developed primarily for commercial operating systems. (The traditional criteria were for a single system only, excluding a network, so the new criteria are an attempt to extend to something more accurately resembling today's computing environment.) Draft profiles are already available that cover firewalls and databases; profiles still being drafted as of November 1997 cover computers, smart cards, crypto components, network components, and additional areas (Year 2000 conformance, payment terminals, accounting systems, and payment guarantors).

Web Site Evaluation and Monitoring

The critical need for a public Web site to safeguard personal or financial information entered in it by the public has been widely discussed in the media. Various services are available that claim to provide the due diligence required to protect Web sites from claims of negligence in the protection of an individual's confidential information; however, such certifications probably would not provide sufficient legal protection because the entire industry does not yet support a single, recognized standard.

For example, the NCSA, an international membership organization founded in 1989 as an educational resource and clearing-house, is promoting an industry standard to certify commercial Web site security and assure users that the information they provide to a site will be secure. (NCSA already offers certification of anti-virus products and firewalls.) The plan is to certify sites that meet published standards, as determined by a combination of physical site visits, audited site documentation, and remote testing. Similar Web site evaluation services are available from the Big Six accounting firms and vendors such as Trusted Information Systems (TIS). Typically, sites certified by a particular service such as NCSA carry its logo on the home page.

Other available Web services that relate to security include these:

- **CPA Web Trust** – A program to certify the security and business practices of commercial Web sites from a joint task force of the American Institute of Certified Public Accountants and the Canadian Institute of Chartered Accountants. After submitting to an audit of areas including customer information protection and security, certified sites are marked with a seal from VeriSign that links to information on the certification, including the accountants' reports.

- **Internet Emergency Response Service** – A commercial service for corporate networks, intranets, and Web sites from IBM that is similar to a home alarm monitoring system. The service includes real-time intrusion detection, vulnerability evaluation by "attempted intrusion," audit reports, and incident control and recovery. The service debuted in the U.S. in late 1997, with worldwide availability scheduled for 1998.

- **True Site** – A site-verifying service from Application Programming and Development. To address the problem of faked Web sites (copies that are made and posted at a different address for malicious or political purposes), this service maintains a searchable list of valid sites.

Standards Initiatives

Several initiatives focus on developing standards for interoperability:

- **Open Platform for Secure Enterprise Connectivity** – Composed of more than 80 networking and security companies, OPSEC is an industry-wide initiative designed to integrate and manage all aspects of network security through an extensible management framework using a combination of published APIs, industry-standard protocols, and a high-level scripting language.

- **Enterprise Security Initiative** (ESI) – Led by Cisco, this alliance of vendors is intended to streamline network management by implementing a single security policy for in-house, dial-up, and firewall access. The proposed ESI standard adds individualized security clearance levels to OPSEC's basic framework and will help create easy-to-manage VPNs.

5.4.6 Encryption Techniques

Cryptography enables information to be sent across potentially insecure communication networks without losing confidentiality or integrity. Using a digital signature, cryptography also can improve user authentication. For example, Lotus Notes uses cryptography both for message confidentiality and to verify the sender's identity to the recipient. When the computer system or network cannot be trusted fully, cryptography is the only technology that computers can use to provide such assurances electronically.

An encryption algorithm is used to transform plain text into a coded equivalent for transmission or storage. The coded text is subsequently decoded (decrypted) at the receiving end and reverted to plain text. The encryption algorithm uses a key, which is a binary number that is typically from 40 to 128 bits in length (for single-key systems) or 512 to 2,048 bits or more (for public-key systems). The data is "locked" for sending by combining the bits in the key mathematically with the data bits. At the receiving end, the key is used to "unlock" the code, restoring it to its original binary form. Theoretically, the longer the key, the harder it is to crack the algorithm and decode the encrypted message. The effort required by an unauthorized person to decode the unusable scrambled bits into meaningful data is typically a function of the basic algorithm, the length of the cryptographic keys, and in some algorithms, the length of the data.

Two types of algorithms are in use today: shared single-key (also known as symmetric-key) and public two-key.

Single-Key Algorithms

In single-key algorithms, the same binary number is required to encrypt and decrypt the data. This single key must be kept secret for the information to remain secure; thus, a different shared key is required for each pair of users. The system is considered to be symmetric in that the same key and the same algorithm are used for both encryption and decryption. The total number of keys needed for n individuals or organizations to communicate is $[n(n\text{-}1)]/2$. The number of keys increases rapidly as n grows. A population of 10,000 people would require almost 50 million different keys using any secret-key system. Fortunately, most business-to-business relationships involve a relatively small number of participants, so the number of keys remains manageable.

The Data Encryption Standard (DES), which officially became a standard in 1977, is the leading single-key algorithm, with the standard specifying a 56-bit key. In 1996, the Business Software Alliance sponsored research into what it would take to crack the DES algorithm using then-current processor technology. As shown in Table 5-11, its report supports the argument by cryptography experts that the current exportable 56-bit key length for DES-based products is inadequate. Many experts consider longer key lengths of at least 90 bits necessary for the future. (See "Government Regulation of Cryptography" on page 396 for more information about exportable key lengths.)

Table 5-11: Estimated Time Needed to Crack DES Encryption Key

Type of Attacker	Budget	Key Size	
		40 Bits	56 Bits
Casual hacker	$400	5 hours	38 years
Small business	$10,000	12 minutes	556 days
Corporate department	$300,000	24 seconds	19 days
Large company	$10,000,000	7 seconds	13 hours
Intelligence agency	$300,000,000	.0002 seconds	12 seconds

Source: Business Software Alliance, 1996

Businesses already are beginning to use Triple DES and the International Data Encryption Algorithm (IDEA) encryption rather than products based on only 56-bit keys:

- **Triple DES** – Encrypts information three times using two different keys (the "left" key, which encrypts the data, is used twice), thus increasing the effective key sizes of DES such that they are computationally more secure, and therefore, more difficult to break.
- **IDEA** – Encrypts information using a 128-bit key and 8 rounds. IDEA is considered secure, with no algebraic weaknesses that might make it susceptible to being broken. IDEA can be implemented in software or hardware and has similar performance characteristics to DES.

Public-Key Algorithms

The other major type of algorithm in popular use is public-key encryption, which is based on two keys: one to encrypt the data and another to decrypt the data. The algorithm is not symmetric, so knowing the public encryption key is no help in being able to decrypt a message. Users wanting to receive confidential information can freely announce their public key, which is then used by the sender to encrypt data to be sent to them. Typically, public keys are located in some standardized directory. The data can be decrypted only by the holder of the corresponding private key. This type of algorithm eliminates the complexity of handling the large number of secret keys needed for single-key algorithms, but it requires a process to ensure that the public keys are authentic and really belong to their announced owner. In addition, the problem now is managing a large number of public keys and making them widely available. However, interest in and use of public-key cryptography continues to grow rapidly because of its potential to facilitate electronic commerce using the Internet.

The most commonly used public-key mathematical algorithm, which is based on the difficulty of factoring large numbers (typically, 129 or more bits), was invented by Ron Rivest, Adi Shamir, and Leonard Adelman at MIT and published in 1978. This algorithm, known as RSA (for its inventors), is covered by a patent in the U.S. that expires in the year 2000. RSA Data Security (RSADSI, the company formed by the inventors of the RSA public-key cryptosystem) licenses the use of the RSA algorithm to vendors wishing to use it in their products.

A specific assessment of the security of 512-bit RSA keys shows that in 1997, one key may be factored for less than $1 million in cost and 8 months of effort. With the advent of new factoring algorithms and distributed computing, 512-bit keys no longer provide sufficient security and should not be used after 1998. RSADSI's recommended key sizes are now 768 bits for personal use; 1,024 bits for corporate use; and 2,048 bits for extremely valuable keys such as the key pair of a certificate authority. RSADSI expects a 768-bit key to be secure until at least the year 2004. Recommended key length schedules are published by RSADSI on a regular basis (`http://www.rsa.com/rsalabs`).

Public-key algorithms are considerably slower than symmetric-key algorithms. Some software products, such as Pretty Good Privacy (PGP), developed by Philip Zimmermann, use a combination of the RSA public-key and IDEA algorithms to generate a unique-message symmetric key and then use that key for that one message. The message-specific DES or IDEA key is then encrypted with the addressee's public key and sent along with the encrypted message. This method combines the ease of use of public-key encryption with the performance of symmetric-key encryption and works at a significantly faster rate than an all-public-key system. In addition, companies such as Atalla offer hardware encryption devices that support both public- and symmetric-key encryption. Hardware-based encryption products are viewed as having higher performance capabilities than software-based products such as PGP.

PGP 5.5, which became available in December 1997, was developed in cooperation with MIT and is available for Macintosh, Windows 95, Windows NT 4.0, and UNIX operating systems. It also is available for use with Internet e-mail packages such as Claris Emailer for Macintosh, Microsoft Exchange, and Qualcomm's Eudora.

Other public-key techniques include the Diffie-Hellman key exchange and the Digital Signature Standard (DSS). Diffie-Hellman is a pioneering public-key cryptographic technique, and RSA is a system based on it. However, RSA is able to support the use of digital signatures, which Diffie-Hellman does not. (See "Digital Signatures" on page 394.) Although the two use different mathematical algorithms, experts say little difference exists between them when it comes to speed of operation and overall security.

Cylink, a competitor of RSADSI, is the commercial licensing agent for Stanford University, where the public-key Diffie-Hellman algorithm (and related Hellman-Merkle algorithm) were researched. In January 1997, Cylink and RSADSI settled a long legal dispute over patents with a cross-licensing agreement that gives RSADSI full rights to the "Stanford" patents controlled by Cylink; however, the Diffie-Hellman patent expired in September 1997, and the Hellman-Merkle patent expires soon as well.

The adoption of non-RSA public-key encryption algorithms is increasing; in September 1997, Microsoft announced that it will include Cylink's security technology (based on Diffie-Hellman and DSS) in Internet Explorer 4.0 and Windows NT 5.0. The Cylink technology also will be implemented in Microsoft's CryptoAPI architecture, enabling developers easily to write security software that uses Diffie-Hellman and DSS-based encryption.

Key Length

One of the most hotly debated topics in the security arena today is key length, which refers to the number of bits in a particular encryption key. Key length is an issue with both DES and RSA encryption algorithms.

Because the difficulty of breaking a key decreases as computers get faster, there has been increased demand for legislative overhaul. Many global companies have said they will purchase stronger foreign encryption products if U.S. products do not meet their security needs. Countries with growing encryption software and hardware development industries include Australia, Canada, China, Israel, Japan, New Zealand, Switzerland, and Russia. The implication is that stronger encryption produced in those nations could pose a competitive threat to manufacturers from countries (such as the U.S.) that prohibit the export of similar products.

Key length continues to be a controversial issue. In June 1997, the DES cipher, which uses a fixed-size, 56-bit encryption key, was cracked by a team using a "brute force" attack. (A brute force attack involves trying every possible key for a particular system until the right one is located.) The 40-bit and 48-bit RSA keys had been cracked previously.

Encryption Trends

Several recent trends in encryption may alter the balance in the industry:

- **Increasing use of Triple DES** – Although the majority of problems with DES relate to its key length, Triple DES may provide an effective solution. Benefits include no known attacks that have succeeded in breaking its two 56-bit keys, easy incorporation into existing systems, and that it is a standards-based algorithm. Drawbacks include the computing power required (three times that of normal DES) and the difficulty of managing and distributing keys.

- **Development of an Advanced Encryption Standard (AES)** – In September 1997, the NIST issued a request for nominations for a new algorithm, AES, that will eventually supplant 56-bit DES for U.S. federal government use. Nominations are due by June 15, 1998. Proposed requirements are that the algorithm be publicly defined and use a symmetric block cipher equal to or stronger than Triple DES, with "significantly improved efficiency."

- **The rise of Elliptic Curve Cryptography (ECC)** – First theorized by Victor Miller and Neal Koblitz about 15 years ago, ECC is based on the same algorithm problem as Diffie-Hellman but uses a different method for calculations. To break RSA encryption, a hacker must find the factors of a large number. To break ECC, a hacker must find a theoretical point on a curve that has been defined mathematically. ECC promises stronger security for fewer bits than both RSA and Diffie-Hellman. For example, supporters claim that a 160-bit ECC key provides the same security as a 1,024-bit RSA key. Besides shorter keys, the mathematical operations required by ECC are easier to perform, enabling ECC systems to function much faster than equivalent RSA systems.

  In January 1997, RSADSI announced it would include ECC algorithms in version 4.0 of its BSafe security toolkit (due in early 1998). RSADSI has since posted notices on its Web site about the discovery by Nigel Smart, a researcher at Hewlett-Packard Laboratories in the U.K., of a fatal security flaw in a group of elliptic curves previously accepted for cryptographic implementation. RSADSI says this flaw calls into question the industry's current state of knowledge and research around ECC and "warrants continued caution and further research toward a more complete understanding of these complex systems."

  Although the basic ECC algorithm is not patented, several firms (including Certicom and Cylink) own patents for specific implementations. In May 1997, Certicom announced its CE2 Security Builder encryption software, the first offering in its Elliptic Curve Engine (CE2) product line, designed to provide high-end cryptography in low-end devices such as smart cards.

  As of October 1997, ECC standards were nearing completion with both the IEEE and the ANSI. The IEEE draft standard for public-key cryptography (P1363, with sections on elliptic curve digital signature methods and key agreement schemes) is in the final stages and could be published as an official IEEE standard in early 1998. Two ANSI drafts (X9.F1, Security Tools for the Financial Industry, and X9.63, EC Key Establishment) are nearing ballot stage. These standards, in turn, have formed the basis for the development of an International Organization for Standardization (ISO) draft standard on digital signatures (ISO 14888-3), which includes elliptic curve-based digital signatures. (See "Security Evaluations and Standards" on page 383 for more on standards initiatives.)

- **The government's use of commercial IT for electronic commerce** – In June 1997, NIST proposed a revision to the DSS that would allow federal agencies to use commercial information technology (such as RSA's digital signature technology or ECC) to secure electronic commerce transactions. The inclusion of ECC may indicate the potential advantage it holds over RSA and other encryption technologies because of its ability to operate at much faster levels. DSS never has been used heavily outside government circles because it is a signature-only system. Software such as DSS that provides only authentication is exportable; however, software that uses RSA for general encryption is heavily restricted. (See "Government Regulation of Cryptography" on page 396.)

- **The introduction of RPK** – Developed in New Zealand in 1995 by William Raike, the RPK public-key cryptosystem is now being offered as a free download via the Internet. RPK can be used in full public-key form to transmit messages. Its developer claims that RPK is significantly faster than other public-key systems such as RSA, that the RPK system can be implemented easily in hardware and software, that the system is flexible enough to be customized, and that it can be implemented inexpensively. Most important, RPK is not subject to U.S. cryptography restrictions because it was developed outside the U.S. RPK already has been adopted by a Swiss consortium called Adesa for use in its Hawk smart card.

- **Development of other public-key encryption schemes** – For example, IBM is working on a new encryption scheme, announced in May 1997, that uses the mathematical "unique shortest vector" approach to generate keys randomly. This approach mathematically proves that all chosen keys are equally strong, whereas certain keys in RSA can be "weak." IBM says the system will require a good deal more development before commercial applications and products can be based on it.
- **The branching of RSADSI into other areas** – Its private-key encryption technology is slated to be used in a chip (called the 7711), due in October 1998 from Hi/fn (a Stac Electronics spinoff). The chip will be an evolution of Hi/fn's current coprocessors and will include compression and authentication as well as encryption.

Public-Key Infrastructure (PKI)

A public-key infrastructure (PKI) is a dynamic system – the underlying framework that allows security technologies to work together. In most cases, a PKI will include digital certificates and public-key algorithms, integrating them with access-control policies and directories and allowing someone within a company or at a third-party service to manage, control, and modify security policies on an as-needed basis.

Such services could be provided by government agencies (postal authorities), by a third party such as GTE or VeriSign (a spinoff of RSADSI), or by a business organization that uses products such as Nortel's Entrust. Supported applications could include secure e-mail, payment protocols, electronic checks, electronic data interchange (EDI), IPsec network security, electronic forms, and digitally signed documents. (See Chapter 7.1, *Electronic Commerce,* for more information about payment protocols and EDI.) Products are currently available from a variety of vendors, including GTE, Motorola, Nortel, and VeriSign.

Integral to a PKI are components such as smart cards for authentication and encryption, X.500 and X.509 for secure directory services, the Lightweight Directory Access Protocol (LDAP) for secure interoperation of directory servers, and the Simple Distributed Security Infrastructure (SDSI).

■ Smart Cards

Smart cards can be used as security devices for both authentication and cryptographic processing. Most activity in smart card applications is occurring outside the U.S. in countries without established credit card authorization systems (such as Visa and MasterCard). For example, HP and Informix have formed a joint venture with Gemplus to develop a national "flag card," or smart card specific to an individual country's security policies, followed by a cryptographic smart card reader that would be built into PCs, notebooks, or other devices for electronic transactions such as cash registers or point-of-sale terminals. The ImagineCard user authentication system (which began shipping in April 1997) is based on HP's ICF architecture and uses RSA encryption. It enables companies to issue customized smart cards to preregistered customers, allowing them to authenticate themselves for applications such as on-line purchasing, electronic banking, and home training via the Internet. ImagineCard also provides access to the company's intranet for employees working at remote sites.

The Finnish post office is currently testing a cryptographic smart card that can be used for encrypting and digitally signing e-mail. The card holds the private key and performs the cryptographic processing; it interfaces with e-mail software written by ISOCOR and supplied by the post office. The post office also acts as a certificate authority and maintains an X.500-based directory that can be used to store and look up public keys. The card can be used for other applications as well, and the post office is planning to sign up local businesses to develop applications that use the card. The Swedish post office is testing similar technology. (See Chapter 3.3, *Smart Cards.*)

■ X.500 and X.509

X.500 is the ITU-TSS international standard for a directory service that permits applications such as e-mail to access information about individuals and network resources. X.500 provides a way to distribute e-mail directory information through servers situated at strategic points throughout the network. These X.500 servers then exchange directory information so that each can keep its local mail directory information current.

Although X.500 can be used to support X.400 and other messaging systems, it is not restricted to e-mail usage. X.500 provides a hierarchical structure that mirrors the hierarchy of organizational units (for example, countries, states, cities, individuals) to create a directory that can be used globally.

From a security perspective, X.500 has two areas of particular relevance. First, X.500 directory services can be used in the storage and delivery of security-related information for other applications, such as public keys and their associated certificates. For example, Nortel's Entrust certificate product uses an external X.500 directory for storage of keys, certificates, and certificate revocation lists (CRLs). Second, because of the critical nature of X.500 data to the operation of a network, the directory information itself must be secured.

Within the standards for directory services, X.509 defines the directory authentication framework and describes public-key authentication, digital signature techniques, certificates, CRLs, and management procedures. For example, the Netscape Certificate Server in SuiteSpot lets companies create, sign, and manage public-key certificates for internal use based on the X.509 standard. The certificates attest to users' identities and, when used in conjunction with the directory server, give users access to the specific resources for which they are authorized.

■ Lightweight Directory Access Protocol (LDAP)

The IETF's LDAP is emerging as a way to connect X.500-based directory servers to the Internet and have them interoperate. LDAP originated in 1989 at the University of Michigan as a way to access and update directory information in a client/server environment. LDAP supports authentication (both plain text and Kerberos), allowing users authenticated access to sensitive information in the directories.

Because LDAP is designed for IP networks and is easier to implement than other protocols, vendors such as Lotus and Microsoft already give users an LDAP interface to their proprietary directories, making it unlikely that the companies will implement X.500. In July 1997, Netscape announced it would use Java and LDAP to incorporate new server management capabilities into the 1998 release of its next-generation SuiteSpot servers. Code-named Lava, the technology combines the management and security features of LDAP with a Java-based GUI. Lava supports SSO as well as encryption-authentication capabilities in an extranet environment. Meanwhile, IBM has been shipping an LDAP server for AIX and other platforms.

There is some concern that LDAP is being loaded down with too many features, rather than remaining "lightweight," as originally intended. Limitations of LDAP include that it does not assist with access to legacy directories and that it may not interoperate with SET. Critics also say LDAP version 3.0 (due to be finalized by the end of 1997 or early 1998) lacks the replication and synchronization features that would let users do more than browse with data contained in operating-system- and applications-specific directories. Strong (cryptographic) user authentication also is not present in the current version of LDAP but may be included in the final draft of version 3.0.

■ Simple Distributed Security Infrastructure (SDSI)

An SDSI that uses public-key cryptography combined with mechanisms for defining groups and group membership certificates was announced by Ronald Rivest of MIT in July 1996. Although still in the process of definition and review, SDSI can be considered an alternative to the use of X.509, which requires global certificate hierarchies. SDSI emphasizes linked local namespaces rather than hierarchical global namespaces. SDSI combines a public-key infrastructure design with a means of defining groups and issuing group-membership certificates. SDSI's groups provide simple, clear terminology for defining access control lists and security policies. The goal of SDSI is to provide a simple framework for ACLs and security policies in the distributed environment.

U.S. Government Public-Key Proposals

In late 1996, NIST announced an initiative to develop the elements of a public key infrastructure in partnership with companies such as AT&T, BBN, Cylink, Motorola, and VeriSign. The goal of the alliance is to develop a minimum interoperability specification for the technical components of a PKI. NIST then plans to issue an interoperability suite for vendor testing of PKI products.

NIST announced a critical component of the government's PKI framework for digital signatures in March 1997. The new Federal Information Processing Standard (FIPS) 196 for authentication technology allows agencies to extend their use of digital signatures as an alternative to the use of passwords and personal ID numbers for verification.

In April 1997, the General Services Administration announced it had set up six pilot tests with various government agencies to develop secure PKI applications as a way of jump-starting its stalled PKI initiative. (A 1996 plan to launch PKI pilots for the Social Security Administration and the IRS fell through.)

Certificate Authority

One important difference between secret-key and public-key cryptographic systems is the way the keys are managed, and the critical issue is how to store and validate public keys. One solution to this problem has been to have a trusted third party vouch for the authenticity of the public key, either by storing it in a centralized, on-line database or by distributing it with a certificate. The certificate binds the identity of the key holder with the public-key value. The organization or body that performs this binding is known as a certificate authority.

A certificate is similar to an identity card with a notary seal on it. It is valid for a stated period of time and is subject to cancellation by being included on a CRL. CRLs are basically "hot lists" that identify certificates that have been withdrawn, canceled, compromised, or should not be trusted for other specified reasons. If a certificate has been revoked, for example, the key to which it is attached should be treated as suspect. Because a certificate authority cannot force the destruction of all copies of a certificate, anyone who plans to rely on a certificate should check it against a current CRL to ensure its validity.

The issue of CRLs and CR management is becoming an increasing focus of attention. For one thing, checking the validity of a certificate is not straightforward: The user must open a network connection to the issuing authority, find the CRL, and submit the certificate for checking. Companies such as ValiCert offer CR management tools and services. The ValiCert Toolkit allows developers to embed certificate-validation capabilities into applications; the ValiCert Server builds a certificate revocation tree from a CRL; and ValiCert Services (available by mid-1998) will act as a clearinghouse for checking the validity of certificates. For the ValiCert approach to work effectively, however, it must be incorporated in various electronic commerce applications, all major certificate issuing authorities must adopt it, and electronic commerce vendors must choose to turn to ValiCert for checking on certificates.

Individual companies can act as certificate authorities; however, in a public arena, multiple levels and entities may be involved in the certificate process. The need to define certificate authority structures and levels of trust is an increasing concern as the use of certificates proliferates.

Commercial certificate authorities such as GTE and VeriSign are endeavoring to fill the need for trusted third-party services in electronic commerce by issuing digital certificates. VeriSign, for example, issues and manages a number of levels of digital IDs, differentiated by the level of assurance or trust associated with the ID. The assurance level typically is dependent on the level of diligence the certificate authority applies to establishing the relationship between an individual or entity and its public key.

VeriSign defines three classes of certificates that offer different levels of assurance, moving from Class 1 (establishes a consistent presence but does not guarantee that someone is a real person), to Class 2 (issued after a check of credit databases), and then to Class 3 (requires the applicant to have identification checked and notarized before a certificate is issued). A South African certificate authority, Thawte Consulting, plans to offer a higher grade of certificate that will require the certificate holder to meet personally with a representative of the company; these "cybernotaries" will be local entities who can vouch for individuals.

Many large companies also are creating certificate authority solutions. IBM, for instance, is developing certification software (the IBM Registry) that enables the use of digital certificates to authenticate parties involved in a transaction and the use of public-key cryptography to protect the confidentiality and integrity of the transaction based on an electronic framework. The IBM Registry is based on Nortel's Entrust. Outside the U.S., some postal authorities are assuming the role of a certificate authority.

Digital Signatures

Although symmetric cryptography generally is limited to maintaining the privacy of information, public-key cryptography has wider applications, including that of digital signatures. Digital signatures allow the receiver of a digitally signed electronic message to authenticate the sender of the message and verify the integrity of the signed message. Most important, digital signatures are difficult to counterfeit and easy to verify, making them superior even to handwritten signatures.

One obstacle to the acceptance and use of digital signatures is their uncertain legal standing. According to a September 1997 digital signature law survey, parties in most countries are free to agree on concluding contracts by means of digitally signed electronic documents, and these generally are accepted as legal evidence. The survey concluded that the status of digital signature legislation worldwide varies from country to country:

- **Belgium** – A draft law, begun in May 1997, is expected to be finalized by the end of the year.
- **Denmark** – The objective of a proposed Danish bill is to equate digital signatures with handwritten ones.
- **Finland** – Finnish law places digital signatures and electronic documents on a par with paper equivalents.
- **Germany** – The Digital Signature Law, approved in June 1997, provides a technical framework for the use of digital signatures. However, the Law on Revocation of Contracts Concluded Door-to-Door (which allows consumers to cancel agreements concluded door-to-door within a prescribed time) poses a possible threat to the use of digital signatures; currently under debate is whether this right also applies to on-line transactions concluded by consumers within their own homes.
- **Ireland** – Electronic documents are considered hearsay evidence and are not admissible in Ireland.
- **Italy** – A digital signature law provides that electronic documents and digital signatures are legally valid; further criteria will be defined in forthcoming regulations.
- **Japan** – The draft Certification Authority Guidelines, released in April 1997, covers specific services such as certificate issuance but does not appear to include digital signatures.
- **Malaysia** – In June 1997, the Malaysian Parliament passed the Digital Signature Act, making a digital signature legally equivalent to a handwritten signature.
- **Netherlands** – Dutch law contains a provision that will probably allow digital signatures on certain documents.

- **Singapore** – In July 1997, Singapore announced the creation of its own certificate authority for Internet commerce as well as plans for supporting legislation that would give legal status to electronic signatures and other forms of digital transaction.
- **Spain** – Like Finland, Spanish law places digital signatures and electronic documents on a par with paper equivalents.
- **South Korea** – A draft Bill on Promotion of Trade Business Automation contains a provision stating that electronic documents with digital signatures shall be regarded as properly signed.
- **Sweden** – A 1996 report on Electronic Document Handling, presented by a special government committee, suggests considering digital signatures equal to traditional signatures as long as they are verifiable.
- **U.K.** – The Civil Evidence Act of 1995 resolved the admissibility of electronic documents as evidence. In a March 1997 paper, the Department of Trade and Industry included a section on digital signatures that asks whether contract law could suffice for parties to agree on accepting digital signatures or whether additional legislation is needed.
- **U.S.** – Until recently, the acceptance or denial of digital signatures for legally binding purposes has been left up to individual states, meaning that electronic contracts signed in one state may not be enforceable in another. As of September 1997, 29 states had enacted, proposed, or drafted digital signature legislation. In late 1997, however, two bills were introduced in Congress relating to specific uses of digital signatures in the U.S.

  In November 1997, the Electronic Commerce Enhancement Act of 1997 (H.B. 2991) was introduced, which would make federal forms available to U.S. citizens over the Internet and enable them to submit such forms electronically to federal agencies using digital signatures. Also in November, the Electronic Financial Services Efficiency Act of 1997 (H.B. 2937) was introduced, which would recognize digital signatures if they meet certain criteria. This bill also would establish a National Association of Certification Authorities and require all certificate authorities wanting to issue certificates in the U.S. to register with the Association.

Other international developments related to digital signatures include these:

- The United Nations Commission on International Trade Law (UNCITRAL) is working on a model digital signature law; a working group met in February 1997 to discuss a draft report.
- The Organization for Economic Cooperation and Development (OECD) adopted guidelines concerning cryptography, including digital signatures, in May 1997.
- The European Commission (EC) has issued a call for a study on the legal aspects of digital signatures, which will give a summary of national and European Union (EU) policies, existing and proposed rules and regulations, and practices concerning digital signatures in member states and EU trading partners.

  In October 1997, the EC issued a communication urging EU member countries not to pass digital signature legislation that would hamper electronic commerce in the EU; the EC says it will issue guidelines on digital signatures in 1998.

- European Research in Marketing (EURIM), a U.K. parliamentary/industry group, has formed a working party to explore removing obstacles to electronic commerce in the U.K. and across Europe. The group will focus on network security (including encryption policy) and electronic data legal issues (including the status of electronic signatures and contracts).

For sites that are interested in centrally managed security without use of public keys, a new technology of interest comes from TriStrata Security Systems. A central security server (implemented on a pair of hardened Windows NT servers) authenticates users and provides them with one-time keys for encryption. The encryption method relies on a very fast implementation of the Vernam cipher, a technique originally developed for the telegraph that generates near-unbreakable one-time pads. In addition, the TriStrata system uses encrypted access signature keys that are 256 KB in length. The company claims its product, which was first deployed in late 1997, is capable of encrypting 36 MBps or the equivalent of 1,000 transactions per second.

5.4.7 Government Regulation of Cryptography

The main focus of the industry is on the prospects for using strong encryption to promote secure electronic commerce and on the need to be able to export cryptographically strong products to be competitive. The motivation for export controls on strong cryptography, however, is the fear that if security is too strong, national defense could be compromised (when such encryption is used by criminals or terrorists) and law enforcement could become more difficult or impossible.

In 1997, for example, the FBI moved beyond supporting export controls to proposing to outlaw the use of strong encryption (even within the U.S.) that did not contain key recovery provisions. Currently, the situation appears to be a standoff: although Congress is unlikely to lift current export regulations on encryption entirely, the FBI is unlikely to succeed in outlawing the domestic use of strong cryptography. (See “U.S. Legislative Update” on page 401 for more information.)

U.S. Export Issues

U.S. vendors long have claimed that the export key limitation on cryptography puts them at a competitive disadvantage in the worldwide marketplace. For example, in January 1997, BioData of Germany teamed up with U.K. security company Portcullis to create a low-cost firewall system using 112-bit DES encryption developed wholly outside the U.S.

The current status of government regulation of cryptography and related technologies in various countries is shown in Figure 5-19.

Figure 5-19: Government Regulation of Cryptography and Related Technologies

	Encryption Import/Use	Encryption Export	Digital Signatures	Data Privacy
Australia	◔	◔	◑	◕
Canada	◔	○	○	●
China	●	◕	○	○
France	◕	◕	◑	◑
Russia	●	◕	○	○
Singapore	●	●	◔	○
United Kingdom	◕	◑	◔	◑
United States	○	◕	◕	◔

Degree of Government regulation: ○ ◄—► ● Light Heavy

Source: Gartner, 1997

Efforts to circumvent current U.S. restrictions already are strong and varied. Some vendors are forming partnerships with non-U.S. manufacturers. For example, in May 1997, PGPI signed an agreement with French smart card supplier Schlumberger to develop cards using PGP 128-bit encryption that will be manufactured in Europe. PGPI admitted that export regulations prompted it to seek international partners.

Other companies may decide to license and sell encryption software developed by non-U.S. companies. Entegrity Solutions announced in April 1997 that it will offer a customizable array of plug-in modules for user authentication, cryptography, SSO, digital certification, and secure Internet/intranet Web transactions. Entegrity's cryptography software was wholly developed by a Swedish subsidiary, COST, so its strong encryption is not subject to U.S. export restrictions.

Similarly, Sun announced in May 1997 that it plans to import Enduser SKIP (Simple Key Management for IP), which uses 128-bit and Triple DES encryption, from Russian software developer Elvis+ (Electronic Computing Information Systems), repackage it as part of its SunScreen series, and offer it worldwide via third-party distributors. The situation is complicated because Elvis+ technology is built from Sun's SKIP encryption and key management protocol, and Sun owns a 10 percent interest in Elvis+.

Exemptions to export restrictions are becoming more common. For example, restrictions often are waived for U.S. companies that have overseas branches. PGPI was granted a license in May 1997 to export its 128-bit software without key recovery to the overseas branches of more than 100 large, U.S.-headquartered companies, provided that the offices were not located in embargoed countries (Cuba, Iran, Iraq, Libya, North Korea, Sudan, and Syria). PGPI acted as a "broker" by shepherding the applications of these companies through the export process.

Vendors are still being granted export approval without key recovery plans if their products are for use by financial institutions, continuing a long-standing exception to regulation. In June 1997, for example, Netscape received approval to export Communicator client software and SuiteSpot server software, both with 128-bit encryption, to certified banks worldwide.

Other developments focus on isolating the encryption portion of a product. For example, in early 1996, a message created with the 40-bit RSA encryption algorithm used by Secure Sockets Layer was cracked. In response, Microsoft developed an enhancement to SSL called Private Communications Technology (PCT), which separates authentication from encryption. Applications using PCT for authentication, such as credit card validation programs, can take advantage of a much stronger security algorithm that is not restricted. Microsoft plans to include PCT in an upcoming release of its Internet Information Server.

Privacy, Key Escrow, and Key Recovery

Court-authorized wiretapping in the U.S. is routinely used by law enforcement to conduct criminal investigations. The increased use of cryptography over networks could make it more difficult for law enforcement to conduct such wiretaps, however, because the intercepted communication would be unintelligible without the keys needed to decrypt it. Key escrow can be thought of as an encryption system (with a backup decryption capability) that allows authorized people, such as company officers or government officials, to decrypt the encrypted text with the help of information supplied by one or more trusted parties who hold special data recovery keys. These data recovery keys are not the same as keys used to encrypt and decrypt the data but rather provide a means of determining the data encryption/decryption keys. The term key escrow refers to the safeguarding of these data recovery keys.

The history of key recovery proposals in the U.S. includes the following initiatives: Capstone/Clipper, Key Recovery, and Key Encapsulation.

■ Capstone/Clipper

In the early 1990s, concerned about the inability of law enforcement to intercept messages on public or private networks that pertained to terrorism or criminal activity, the U.S. government sought to find a way to ensure access to encrypted data. It focused on enforcing a key escrow policy by pursuing a chip-based technology for all devices using encryption, providing a "back door" decryption capability for legitimate government use via a hardware chip known as Capstone for data applications or as Clipper for voice or telephone implementations (announced in 1993). In 1994, a Bell Labs researcher found a serious flaw in the Clipper chip's integrity, and the Clipper initiative fell apart.

■ Key Recovery

Seeking an alternative, the government announced a new initiative in late 1995 based on key recovery management (KRM) rather than key escrow. KRM was quickly (and incorrectly) termed Clipper II, but the methodologies differed.

Unlike key escrow, key recovery allows a corporation to hold its own keys or store them with a third party, but not necessarily one mandated by the government. Some critics believe the new proposal actually weakens personal privacy even more than the Clipper proposals because it requires trusted third parties to release keys under far more circumstances. To date, only two companies have been approved by the Bureau of Export Administration of the U.S. Department of Commerce to serve as trusted third parties under KRM regulations: SourceFile and TIS.

Many corporations are in favor of key recovery (as opposed to key escrow) because it would make data accessible if an employee holding the keys is unavailable or the keys are lost. KRM is a mechanism that puts the owning organization in control of the key. Another alternative might be to share ownership of the recovery. For example, RSADSI offers a data encryption product called RSA SecurPC that allows recovery keys to be held jointly by several people, with a certain subset of keys being sufficient to decode the message.

■ Key Encapsulation

Although key escrow techniques are based on the government or a trusted third party holding a set of keys, key encapsulation makes a cryptographically encapsulated form of the key (not the key itself, but a way to recover the key) available to a service provider. Key encapsulation is the basis for the most popular KRM techniques employed today. It is used to encode a temporary or "session" key, not the message, which then is used to encrypt the message and transmitted along with it. This technique requires modification to existing communications protocols.

■ Recent Developments

The U.S. government's key recovery scheme proposals have met widespread opposition from the IT industry as well as privacy advocates. In May 1997, a report authored by Whitfield Diffie (one of the inventors of public-key encryption) and Ronald Rivest (the "R" in the RSA cryptosystem) claimed that the deployment of a global key recovery-based encryption infrastructure to meet law enforcement's objectives would result in greatly increased costs to end users and substantial sacrifices in security. These objectives are defined as the ability of the government, with a court order, to access encryption keys without the notice or consent of the user; around-the-clock access; and access to encrypted communications traffic. The report concluded the government's requirements differ from, and are in fact diametrically opposed to, any system business would find useful.

Some vendors who promise to incorporate key recovery features are currently being granted export exemptions. (See "U.S. Export Issues" on page 396.) In February 1997, Cylink, Digital, and TIS all received permission to export products with 56-bit DES encryption to U.S.-friendly countries; the vendors will add key recovery capabilities to their products within 6 months. TIS's RecoverKey Cryptographic Services software has an embedded key recovery feature that protects against the "locked-and-lost" phenomenon. Similarly, in June 1997, VPNet Technologies received approval to export VPN products, such as VSU-1000X, which will incorporate key recovery. In December 1997, Cylink received approval from the U.S. Department of Commerce to export its hardware-based Triple-DES encryption technology to banks in central Europe. Cylink's technology provides 168-bit encryption, considerably tougher than the standard 56-bit encryption rate. In addition, Cylink will develop key recovery products for the Department of Commerce.

Other vendors are finding unique ways to package their offerings to meet export restrictions. For example, NetDox received approval in June 1997 to export its ePackage global digital delivery service with 128-bit encryption (and GTE CyberTrust digital certificates). NetDox encapsulates a user's 40-bit encrypted message in a 128-bit wrapper; the wrapper can be removed without the user's encryption key, if necessary, making the unwrapped message vulnerable to attack by brute force.

In addition, some vendors are developing products to help embed key recovery into applications. In May 1997, IBM announced its SecureWay Key Management Framework; the first products are due by the end of 1997. Secure Way uses IBM's KeyWorks technology, which performs recovery breaking the key into chunks for storage in separate locations or with separate people. A message can be decrypted only if all the pieces are combined physically. The KeyWorks toolkit gives vendors a set of APIs based on Intel's Common Data Security Architecture (CDSA).

Companies are even developing new architectures that support government key recovery requirements and are therefore not regulated. For example, in November 1996, HP announced an International Cryptography Framework (ICF) that calls for embedding encryption technology onto a chip set that can be exported freely; the architecture includes key recovery. ICF partners include Gemplus, Informix, Intel, Microsoft, Netscape, RSADSI, TIS, and VeriFone; the architecture has received approval from France, the U.K., and the U.S. The ICF will interoperate with the open standard for cryptography API technology being developed by the Platform-Independent Cryptography API (PICA) alliance.

In addition, Intel announced in April 1997 that it will develop a hardware encryption mechanism based on the ICF that will consist of a separate coprocessor, mounted on the PC motherboard, that would be dedicated to encryption functions.

Some variations on key recovery schemes are being allowed in products for export. For example, in May 1997, Sybase announced that the Department of Commerce had waived the third-party key recovery agent requirement for some of the company's software equipped with 56-bit DES. Sybase's key recovery mechanism does not require a third-party agent, is voluntary, and is for stored data only, not encrypted network communications.

Joint vendor initiatives to develop a standard, interoperable key recovery scheme include the Security Alliance, which claims more than 40 members, ranging from AOL to IBM, Novell, and Price Waterhouse. For example, IBM's SecureWay Framework will interoperate with the Security Alliance's initiative. The Alliance also has proposed a key recovery plan that will enable an encrypted message to be sealed with strong algorithms and opened only if trusted third parties agree to combine pieces of the key to decrypt the message.

The Key Recovery Alliance is a group of more than 60 international companies dedicated to strong encryption around the world that is helping define a policy framework for businesses and institutions. The Alliance's committees and missions are as follows:

- **Technology Requirements Committee** – Works to achieve interoperability of key recovery technologies while supporting a wide range of existing industry solutions
- **Policy Committee** – Reports on cryptographic regulations worldwide
- **Deployment Committee** – Identifies requirements for worldwide deployment of key recovery and identifies means to expedite that deployment
- **Business Scenarios Committee** – Identifies global business requirements for key recovery
- **Outreach Committee** – Disseminates clear, concise, understandable information about key recovery

U.S. Legislative Update

In March 1997, the Clinton Administration extended its data encryption plans in a legislative proposal called the Electronic Data Security Act of 1997. It wanted a key recovery policy that applied not only to encryption export, but also to data encrypted inside U.S. borders. Law enforcement officials would be able to access data based on a simple request (rather than with a court order or warrant).

In June 1997, however, the House International Relations Subcommittee on International Economic Policy and Trade proposed legislation lifting most encryption export restrictions. The Security and Freedom through Encryption Act of 1997 (SAFE, H.R. 695) would allow U.S. manufacturers to export encryption products freely.

This position is in direct opposition to a compromise encryption bill passed in June 1997 by the U.S. Senate. The Secure Public Networks Act (S. 909) aimed for a middle-of-the-road approach: It would relax encryption controls by allowing a 56-bit key length for companies that promised to install key recovery features within 2 years (for those refusing, the export limit would remain at 40 bits). The bill also would require encryption software used by the government or purchased with federal funds to include key recovery software.

Although initially drafted as a way to loosen export control on encryption, the SAFE Act was amended in early September 1997 in response to requests from the FBI and other law enforcement agencies that want "wiretap" access to encrypted e-mail and other digital files. In late September, however, the House Energy and Commerce committee approved the SAFE Act without the proposed Oxley-Manton (pro-law enforcement) amendment. Instead, the committee voted in favor of another amendment that prohibits domestic encryption controls, strengthens penalties for criminal use of encryption, and requires a 6-month study of key recovery technology. This topic remains the subject of ongoing debate and legislation.

Perhaps the most important event in the export debate occurred in August 1997, however, when U.S. District Court judge Marilyn Patel of the Northern District of California ruled (in *Bernstein v. the U.S. Department of State*) that the U.S. government's regulations on the export of encryption software were unconstitutional. The judge subsequently issued an injunction barring enforcement of the export regulations. That ruling has been suspended, pending appeal by the government.

European Legislative Update

Support for the control of data encryption does not extend much beyond the borders of the U.S. Many countries believe that import/export restrictions on cryptography act as a barrier to free trade, and this issue will be addressed by the World Trade Organization and other international bodies in 1998.

In May 1997, a U.K. Department of Trade and Industry's green paper on using government-developed encryption technology was criticized by the Institute for the Management of Information Systems. IMIS said that encryption policy should balance government control with the requirements of legitimate businesses. These conclusions mirror the findings of the May 1997 Diffie/Rivest report on the U.S. government's key recovery proposal.

The 29-nation OECD group has refused to adopt the U.S. position of greater government access to electronic communications via restrictions on encryption and key recovery systems. Instead, the OECD issued its Guidelines on Cryptography (finalized in May 1997), which are intended to promote international cooperation in cryptography policy and balance the needs of governments and users of cryptography. It recommends that national policies be based on open markets, voluntary choice, and privacy safeguards. The Guidelines include principles on trust, choice, market development, and standards in cryptographic methods; protection of privacy and personal data; lawful access; liability; and international cooperation.

Some OECD members, including France and the U.K., have either outlawed or tightly regulated the use of encryption products and support the U.S. encryption/key recovery proposal. Japan is said to be leaning toward the U.S. policy as well; however, it already has liberalized its laws to allow the export of 56-bit encryption without key recovery. Germany is divided on encryption: the Interior Ministry would prefer curbs be established, but the Economics and Justice Ministries oppose encryption. In the U.K., a proposal requiring key escrow of any service offering public verification of keys is opposed by the ruling Labor Party.

In October 1997, the European Commission (EC) rejected the U.S. government's encryption proposal, questioning the effectiveness and methodology of the proposed key escrow system. In response, the U.S. said that work was underway with individual European countries to "work out agreements on export controls and encryption techniques." However, France has proposed new laws, which it submitted to the EC in late October, that would allow businesses operating in that country to encrypt data but would require that decryption keys be given to a French government-approved entity.

5.4.8 Security Risks in Networked Environments

Security weaknesses in the Internet itself as well as in Web browsers and servers have opened up a variety of new risks and created the need for new solutions. Today, users face several new types of risk to the security of network systems (including, in some cases, the Internet itself): system-modifying attacks (viruses or "hostile" applets), those that consume a machine's resources or make them unavailable (known as "denial of service"), those that invade a user's privacy (such as "cookies"), and those that antagonize a user (such as "spam e-mail").

Viruses

A virus is a program designed to perform some malicious action that is unknowingly triggered by an innocuous event (such as a user action, a certain date being researched, and so on). The universal behavior of all viruses is that they replicate themselves.

The number of viruses is escalating at an alarming rate. *Virus Bulletin* reports that from 1987 to 1997, the number of viruses doubled approximately every 10 months; if proliferation continues at the same rate, the total number of viruses will reach 24,000 in 1998. According to *PC Magazine*, new viruses appear at the rate of more than 200 per month.

Although traditional viruses were operating-system-specific and infected only program files, viruses based on macro languages can infect multiple systems and hide in data files. The Internet has enabled users to exchange e-mail attachments that contain documents with macros, templates, and other potential infection vectors. The macro virus, which usually is transferred via an e-mail attachment, has nearly replaced the boot sector virus as the most damaging source of system infection. Macro viruses are easy to create and spread because documents are exchanged more often than programs. However, new anti-virus offerings include features to stop macro viruses.

New services are available to help protect e-mail recipients from macro viruses. In August 1997, Canadian-based Electric Mail Company introduced E-mmunity, a service that purges companies' incoming e-mail of infected files (including macro viruses). Using Trend Micro's Interscan Viruswall, the service checks for infected attachments; if found, the recipient is sent a "warning of infection" notice with the original e-mail (minus the attachments).

One problem with virus-prevention programs is that they need to be updated continually as new viruses are detected. Most of these programs work by looking for specific data (called a "virus signature") in a suspect file. Each virus may have a different signature, so as new viruses are detected, their signatures must be added to virus detection programs.

Anti-virus software typically extends beyond the desktop, with software designed for firewalls, e-mail gateways, and groupware packages as well as Web browser snap-ins to scan materials reviewed on-line. For example, Trend Micro's MacroTrap security software is able to generate new solutions to each macro virus and variant on-the-fly. Whenever a document is opened, MacroTrap examines the embedded macro commands and if any appear capable of acting maliciously, they are "trapped" (quarantined in a secure area) and examined.

To keep users up-to-date on the latest viruses, the NCSA's ftp site (`ftp.ncsa.com/hub/wildlist`) publishes "The Wild List," a cooperative listing of verified virus sightings reported by 46 virus professionals worldwide; the "Zoo" contains samples of several thousand computer viruses. Products that include a wide range of management features (such as remote-control applications) as well as virus detection are most suitable for enterprise networks. Cheyenne Software's InocuLAN/AntiVirus is one example. On the desktop, Network Associates' (formerly McAfee's) WebShield and GroupShield software and Symantec's Norton's Antivirus for Internet E-mail Gateways are two leading products.

IBM researchers have developed a way to use the Internet to hasten cures for computer viruses – a new technology that can discover and diagnose viruses and immediately issue a prescription to destroy the virus through the Internet. Modeled on the immune system of the human body, the technology is known as the Immune System for Cyberspace. The process starts with the detection of a virus on a desktop PC. The virus then is sent through a company's network to an Internet connection with IBM's automated virus diagnostic lab. The lab receives the sample of the virus, and compares its characteristics to thousands of virus profiles stored in a central database. Once a cure is devised, it is sent back to the company and routed around the network, giving desktop users a vaccine before the virus can reach them. The new technology is currently in pilot tests, and IBM says a commercial product for enterprise networks is scheduled for 1998. The company already has been granted a patent on the software, which uses an artificial intelligence technology called neural networks that mimics the way people learn, interpret information, and apply lessons from experience.

Hostile Applets

Many products that have been introduced to protect against Internet-borne viruses have been extended to protect against hostile applets as well. For example, CyberMedia's Guard Dog, introduced in October 1997, is advertised as a "comprehensive personal Internet security and privacy software product." The utility protects against "rogue" Java applets, hostile ActiveX controls, and "cookies;" it includes complete anti-virus protection using technology licensed from Trend Micro. Network Associates' (McAfee's) WebScanX, introduced in July 1997, protects against viruses downloaded or attached to Internet e-mail as well as destructive applets built with Java and ActiveX controls. Alternatively, eSafe (the U.S. division of Israel-based EliaShim) has introduced "anti-vandal" software. eSafe's Protect does not block unknown code, Java applets, or ActiveX controls, but instead isolates the downloaded files from the rest of the hard drive and then monitors them when opened. (See Chapter 5.2, *Software Component Architectures,* for more information on Java and ActiveX.)

■ Java

Java's platform independence and security awareness have been strong inducements for its adoption. Java programs enforce security in several ways, two of which are built into the language. The first (described later) checks the Java program for internal integrity and to ensure it has been compiled correctly. The second aspect is termed the "sandbox." When a Java program is running, it can be forced to execute in a "constrained execution environment" (the sandbox). In this environment, the program can be barred from a range of activities such as writing data to the local disk or deleting files that could damage the local system. By enforcing the sandbox restrictions, a user can execute a Java program with the confidence of knowing that it cannot damage the local system. Early on, several security holes were found in the sandbox model; however, these have been corrected by Sun. Currently, the biggest threat to the sandbox security model is that developers feel it is too restrictive.

In April 1997, Netscape introduced a Java object-signing model for Communicator that lets users expand Java's sandbox by setting access privileges for individual applets. The Netscape Java security model allows the creation of "libraries" with set security policies for Java applets based on where they come from and the resources they request. This approach may have flaws, however. For example, a "trusted" applet can evade the sandbox if it uses a class library that may be from an "untrusted" source and, therefore, destructive.

Java security is based on the byte-code verifier, the applet class loader, and the security manager, each of which must perform properly for the security model to function correctly:

- **Byte-Code Verifier** – When a Java program is compiled, it is converted into byte codes. These codes are a machine-independent representation of the instructions the programmer has written. (With regular programming languages, the compiler's output is a series of instructions that are specific to a particular microprocessor, which is why Macintosh programs cannot run on a PC without special translation products.) To execute the instructions encoded in the byte codes, a Java Virtual Machine (JVM) is called. It reads the byte codes and sequentially executes the individual instructions. Just before acting on the first Java byte code, the JVM calls the byte-code verifier. The verifier reads through the program byte codes and performs a security check. This check primarily ensures that the byte codes have the proper structure; that is, that they have not been modified by a hacker or by a virus. Byte-code verification also is performed as individual classes are loaded for execution. These, too, are oriented toward ensuring the integrity of the classes before they are executed.

- **Class Loader** – Typically supplied by a browser vendor, the class loader loads all applets and the classes they reference. The class loader determines when and how an applet can add classes to a running Java environment and ensures the applet does not install code that replaces important components of the Java runtime environment.
- **Security Manager** – Consulted whenever a dangerous operation is attempted, the security manager has the ability to veto the operation. Decisions are based upon which class loader loaded the requesting class; built-in classes are given greater privileges than classes loaded over the network.

In response to user criticism of Java, JavaSoft has begun to redo the Java security model; the first enhancement to the Java Development Kit (JDK 1.1) provides for signed applets to contain the creator's encrypted signature to help determine whether they can go beyond the sandbox. Trusted applets could read or write to local storage. Signing, however, does not stop hostile applets; it just identifies whom to blame (similar to ActiveX). JDK 1.2 (in beta release in late 1997) includes a new security model that eliminates the concept of the "sandbox" entirely; instead, a flexible permissions scheme will allow users to assign specific capabilities to each applet.

■ ActiveX

ActiveX is a component architecture that lets a program or component interact with other programs over a network such as the Internet. ActiveX uses Microsoft's Common Object Model (COM) and Distributed COM (DCOM) communications standards.

The ActiveX security model relies on the ability to verify the identify of the source of the ActiveX control; it cannot protect the client from malicious or "buggy" controls. What ActiveX does is give the user the ability to screen out controls based on their authorship. Because ActiveX controls function like any other executable code, they can cause serious damage, including deleting an entire hard disk.

ActiveX security is provided by Microsoft's digital-signing program called Authenticode. Authenticode allows the user who built the components to stamp them with a signature, ensuring that the code has not been corrupted. Authenticode's identification process is enabled by digital signatures and digital certificates certified by a certificate authority; currently only VeriSign is implementing Authenticode. However, a certificate does not mean the code does not perform destructive actions; it only identifies the person or business that signed the application. In June 1997, Microsoft enhanced Authenticode by adding time-stamping to digital signatures to authenticate when they were made. In addition, the ability to revoke a certificate if an author writes malicious code is incorporated in Internet Explorer (IE) 4.0. Authenticode can be used by Java applets and plug-ins, ActiveX controls, and other executables downloaded over the Internet.

The other component of ActiveX's security model is the "bouncer," or IE's own security screens, which let users set different security levels that allow unsigned controls to be accepted. The highest security setting (the default), blocks the download of any unsigned ActiveX controls; the medium setting warns of an unsigned control and lets the user decide how to proceed; the lowest setting lets the browser automatically download any control, signed or not, without notifying the user. This system leaves it up to the user to decide how much of a threat each ActiveX control poses. However, the model provides no protection against any signed control.

Addressing these issues will require changes to ActiveX, particularly a mechanism to distinguish between applications loaded locally over an intranet (which can be assumed to be trusted) and from the Internet. The easiest way might be to attach different digital signatures to different applications automatically, depending on their origin: internal server, local hard disk, or outside the company across the firewall. Microsoft also needs to change ActiveX so that users can run applications with different permissions; for example, applications loaded from the Internet would be given only limited access to files and directories on a user's system.

Trend Micro was the first vendor to ship a product with applet protection in its WebProtect for Microsoft Proxy Server and InterScan VirusWall for Microsoft Windows NT. Others will follow.

Resource Grabbers

At the heart of a corporate security policy is the need to implement measures that enable an enterprise network to remain up and running and to recover quickly from a break-in or breakdown. Resource "grabbers" such as denial-of-service attacks are among the biggest threats to reliable computing.

■ Denial-of-Service Attacks

This term describes several different methods of making system resources unavailable and shutting down service:

- **E-mail "bombs"** – Consist of hundreds of duplicate messages and large files that consume server space, thus potentially filling up systems and making them unavailable for valid use.
- **"SYN flooding"** – Inundates a network with requests to open new connections that carry invalid Internet Protocol (IP) addresses, tying up the server (specifically, the memory buffers in the network protocol stack) as it tries to acknowledge unknown or nonexistent addresses; SYN refers to the TCP message used to initiate a new connection.

  Although it is almost impossible to prevent a SYN attack, newer firewalls are designed to protect against them by inspecting all incoming communications before they enter the host to ensure that the source IP addresses are valid (Check Point's SYNDefender); by shortening the time the host waits for return acknowledgment from the sender (Livermore Software Laboratories' N.O.A.H. add-on to its Portus firewall); or by looking at each packet of a message to spot deviations (LanOptics' Guardian firewall).

  Software packages such as Haystack Lab's Web-Stalker and Internet Security Systems' RealSecure monitor a network for SYN flooding attempts and break persistent connection requests.
- **"Ping of Death" attacks** – Crash network servers or firmware by overloading them with illegally large ping packets. ("Ping," for Packet INternet Groper, is an Internet utility used to determine whether a particular IP address is on-line. It is used to test and debug a network by sending out a packet and waiting for a

response.) The newest version of this attack modifies the IP header to indicate that more data is in the packet than there really is, causing the server to "hang."

- **Specialized attacks** – Trigger automatic firewall alarms designed to close down connections when attacked or cause other system shutdowns.

Invasion of Privacy

Privacy typically is protected worldwide through data protection legislation. The EU Data Protection Directive of October 1995 specified that data processing systems must safeguard people's fundamental rights and freedoms, "notably the right to privacy." The Directive defines personal data as any information relating directly to individuals or to their physical, mental, economic, physiological, cultural, or social identity. It requires EU member states to ensure that such data is collected for specified, explicit, and legitimate purposes; processed fairly and lawfully; and safeguarded during the collection and holding process. In addition, it specifies that personal data may be processed only if the subject has "unambiguously" given consent.

Some individual countries have developed legislation to support the EU directive:

- **Germany** – The Teleservices Data Protection Act is part of the newly adopted regulation of information and communications services; collection, processing, and use of data is permitted only if approved by law or with the user's consent.
- **Greece** – The Data Protection Act, passed in April 1997, establishes a data protection authority plus guidelines, principles, and rules relating to the use, processing, storage, and export of personal data in electronic and manual files.
- **Italy** – Law 975/96 on protection of personal data, passed in May 1997, establishes an independent regulatory authority (the Authority for Information Technology in the Civil Service) with inspection powers and concurrent judicial protection for the individual on civil and criminal grounds.

New concerns about privacy are helping promote the formation of standards:

- **Platform for Privacy Preferences (P3)** – An Internet commerce standard that includes principles being developed by the World Wide Web Consortium (W3C). The goal of P3 is to develop an interoperable way of expressing privacy practices and preferences by Web sites and users, respectively, backed by the requirement of a digital signature. P3 can be incorporated into browsers or into "proxy" products that sit between the client and the server. With P3, users can develop a profile to inform a merchant exactly what information will be made available and how it can be used. A final draft is expected by mid-1998.

  Several vendors have announced product support for P3. In July 1997, Microsoft announced it would implement P3 in a variety of products, including IE 4.0.

- **Open Profiling Standard** – The OPS initiative (begun by Netscape and Firefly and now supported by Microsoft and others) is based on the vCard specification for digital signature and electronic business cards exchanges. The vCard standard was developed by the Versit consortium (established by Apple, AT&T, IBM, and Siemens) and continued by the Internet Mail Consortium (an international group of hardware and software vendors). vCard attempts to standardize electronic business cards and facilitate the exchange of cards via Web pages and e-mail. vCard supports multiple languages and multimedia files.

  With OPS, the user completes a personal profile that controls what information is transmitted and to whom. OPS uses a digital certificate to establish the credentials of vendors who will receive profile information. OPS was reviewed by the W3C in mid-1997 and is an existing specification. Microsoft and Netscape have both announced support for OPS.

P3 is influenced, in part, by OPS. P3 initially was focused on enabling the expression of privacy practices and preferences, while OPS's focus was on the secure storage, transport, and control of user data. When completed, P3 will probably become part of OPS.

- **TRUSTe** – Introduced in June 1997, the TRUSTe initiative (originally called eTRUST), a spinoff of the Electronic Frontier Foundation, is backed by sponsors including AT&T, CommerceNet, IBM, Netscape, Oracle, and Tandem. TRUSTe implements a labeling system of "trustmarks," by which Web sites inform consumers how personal information collected by the site will be used.

  TRUSTe defines three types of privacy, each with its own name and logo:

  - **No Exchange** – Indicates that the site uses no personally identifiable information
  - **One-to-One Exchange** – Indicates that the site collects data, but only for use by the site owner
  - **Third-Party Exchange** – Indicates that the site may share, transfer, or sell personal information to other parties

New technologies called "anonymizing proxies" also are being developed to protect Internet user privacy. In August 1997, Lucent demonstrated its Lucent Personalized Web Assistant (LPWA), which allows users to use one or more LPWA-created identities that are generated with cryptography. Web sites cannot infer the user's true identity from the LPWA-created identity, thus providing anonymity to the user. The LPWA acts as a proxy server and also can be used to prevent "spammers" (junk e-mail senders) from building mailing lists of site visitors. (See ""Spam" E-Mail" on page 409.)

Cookies

A "cookie" is a feature of the HTTP (HyperText Transport Protocol) used on the Internet that lets Web site developers place information on the user's computer for later use. A cookie allows the server to both store and retrieve information from the client. Cookies were designed to facilitate repeat visits to a given Web site by creating a user profile that would shorten or eliminate the user identification part of the log-in process or eliminate the need to store user profile information on the server.

However, cookies are placed on a PC without the user's knowledge, creating concerns about privacy through electronic trespass. In addition, cookies place information on the local hard disk, thus creating a security concern because the Web server that reads the cookies can read other files of the hard disk as well.

Three types of third-party cookie managers are available to help users monitor cookies and specify the degree of control they want to enforce:

- **Blocks all cookies without asking the user** – Cookie managers such as SoftDD's Complete Cleanup get eliminates all cookies, cache files, and history files; the drawback is that users cannot choose which items to eliminate; they either keep them all or get rid of them all. (Users also can block all cookies manually by making the cookie file "read only.")

- **Allows cookies to arrive and deletes them all at one time** – Tools such as Scott McDaniel's Crumbler 97 periodically check a user's system for cookies and then automatically delete any they find; however, these managers do not allow users to delete cookies selectively.
- **Blocks cookies from specific sites and deletes them individually** – More advanced utilities, such as PGPI's PGPcookie.cutter, allow users to decide who can access their private information by establishing a set of "trusted" sites. It also distinguishes between "blocking" cookies and "refusing" cookies because it collects cookies so that users can examine them and decide whether to delete or accept them. The newest version of PGPcookie.cutter employs multiple levels of trust to determine which cookies the user will allow or block even from trusted Web sites.

Browsers such as Microsoft's Internet Explorer or Netscape's Navigator have options for blocking cookies. Both can be set to show a warning whenever a site wants to leave a cookie, letting the user choose whether or not to accept it. However, the large number of cookies used by many sites makes this feature difficult to use.

Another type of option is offered by privacy advocate Junkbusters Corp. – a free tool that lets users get rid of "cookies," advertising, and "spam" e-mail. The Internet Junkbuster is a proxy that sits between the user's browser and the Internet and blocks requests for URLs (typically banner ads) that match its "blockfile," thus preventing those files from being downloaded and displayed on the user's browser. Users create their own blockfiles, specifying what items or Web sites to block. Internet Junkbuster also deletes cookies and other unwanted identifying header information that is exchanged between Web servers and browsers. The source code (for UNIX, Windows 95, and Windows NT) is available as a free download for users to copy, use, modify, and distribute (*http://www.junkbusters.com/ht/en/ijb20.zip*). Junkbusters says an executable version of the tool may be available in early 1998 for Windows 95 and Windows NT.

Antagonistic Attacks

Unlike denial-of-service attacks, antagonistic attacks are meant merely to annoy a user. They can range from playing unwanted sound files to displaying obscene pictures on a monitor or Web site. Because of the ease of forging Internet e-mail return addresses and the ability to relay e-mail traffic through third-party sites to disguise their origin, antagonistic attacks are increasing. The most frequent type of attack is "spamming" (sending junk e-mail).

■ "Spam" E-Mail

The flood of junk e-mail making its way from the Internet to corporate systems has prompted the proposal of new legislation in the U.S. Three bills were introduced in Congress in 1997 to address this problem:

- **The Commercial Electronic Mail Choice Act** – This Senate proposal (S-771) lets citizens "opt out" of receiving unsolicited junk e-mail and would require senders to include the word "advertisement" in the subject line of the mail.
- **The Electronic Mailbox Protection Act** – This Senate proposal (S-875) would make it illegal to send unsolicited bulk e-mail from unregistered domains or use fictitious e-mail addresses to disguise the origin and prevent replies.

- **The Netizens Protection Act** – This House bill (H.R. 1748) attempts to extend the U.S. Telephone Consumer Protection Act of 1991, which made it illegal to send unsolicited commercial faxes, to cover unsolicited e-mail as well.

In a July 1997 survey about junk e-mail conducted via AOL, nearly 75 percent of respondents favored some type of government regulation of junk e-mail (up from 64 percent in a September 1996 survey). Of those responding to the survey, 34 percent favored the Commercial Electronic Mail Choice Act; 28 percent favored the Electronic Mailbox Protection Act; and 21 percent favored the Netizens Protection Act.

Big "spammers," such as Cyber Promotions (which sends up to 20 million junk e-mails per day), are feeling the effects of consumer backlash. In September 1997, its Web site suffered a ping attack. A month later, AGIS (Cyber Promotions' ISP) temporarily discontinued the spammer's service when AGIS came under attack by its own subscribers for carrying spam over its network. Cyber Promotions has since found a new ISP and is creating its own network backbone.

Service providers that tolerate or fail to regulate spamming also are hearing from disgruntled users. In August 1997, UUNet received a so-called Usenet Death Penalty (UDP) attack from a group of system administrators who claimed newsgroup users receive too much spam. The UDP threatened to cancel all traffic from UUNet until the ISP reduced the amount of spam coming through its service to Usenet newsgroups. In an ironic turn of events, a cleverly forged (and intentionally malicious) message claiming to be from UUNet itself was subsequently posted to Usenet. It said the company was aware of the problem but had no time to deal with it and that complaining was a waste of time – the last things users wanted to hear. Such incidents are a reminder that users should not assume the authenticity of messages routed via the Internet.

Literally dozens of anti-spam utilities and services are available to discourage e-mail abuse. For example, Software.com's Post.Office Internet messaging server software includes a filtering mechanism that blocks e-mail from IP addresses or domains of known spam offenders. The software is designed for small and midsized ISPs. Many mail packages (such as E-Corp's eMail '97, which includes tools allowing managers to block out spam e-mail) now offer similar protection. More important, large ISPs such as AOL have been winning court injunctions barring specific e-mail spammers from mailing to their subscribers.

Web Browser Security Issues

Web browser security is a serious issue for any company giving staff access to the Internet or for a user implementing an intranet or extranet. Security "bugs" or "holes" in Web browsers are particularly dangerous because they open a user's entire system (or even a company's network) to possible misuse. Recent problems discovered in the most popular browsers, Microsoft's IE and Netscape's Navigator, include these:

- The ability of a Web page to download and run potentially malicious programs on a PC, bypassing existing security controls. Such a page could, for example, delete an important file from the computer's Windows folder; alternatively, a hacker could create a Web page that reformats a hard drive.

- The ability to use a specific application (such as Norton Utilities 2.0's System Genie) to install a scriptable control (OCX) on a system's hard disk that could be manipulated using a Microsoft scripting tool such as Visual Basic.

Both of the foregoing problems were quickly addressed with software patches issued by the browser vendor. Browsers also are being strengthened with additional security enhancements:

- **IE 4.0** – New features include Security Zones (which let users set classes of Web sites and assign security restrictions for all sites in those classes) and a capabilities-based Java security model with enhanced Authenticode support.
- **Navigator 4.0 (part of the Communicator suite)** – The New Security Advisory dialog lets users set all security-related features; manage personal certificates, SSL, and other encryption-related settings; and shut off Java, scripts, and cookies. In the future, a tool called Mission Control will allow administrators to create custom browser versions with company-wide security settings.

5.4.9 New Security Solutions

Internet security issues continue to be a source of widespread concern, although more security tools are available than ever before. Firewalls are more robust, administration and audit tools are more plentiful, and developing standard security procedures will still prevent most outside attacks while creating an audit trail for any that do get through.

Explosive use of the Internet and the drive toward electronic commerce have generated significant interest in the development of standards for securing these environments. These environments share the common characteristic that they must operate between organizations, not just within a single organization, and often involve the use of networks that are themselves insecure. For example, in early 1996, Visa and MasterCard drafted the Secure Electronic Transaction (SET) standard for securing credit-card payments across the Internet. Other security-related standards – X.500 and X.509 for secure directory services; PEM, S/MIME, PEM-MIME/MOSS, and PGP for securing e-mail messages – have started to result in the availability of products. Although these standards are critical to secure transactions and the deployment of EDI and other applications that leverage the Internet, few are widely accepted yet. Difficult choices remain as to which standard is more effective in solving security issues and which will become an accepted or official standard in the future. (See Chapter 7.1, *Electronic Commerce,* for more about SET and other payment initiatives.)

E-Mail Encryption

Unlike e-mail on a private system, which goes directly to the server and waits there until retrieved, Internet e-mail moves from server to server on its way to the recipient. This process makes the transmission channel impossible to secure and provides numerous opportunities for interference. The only way to ensure e-mail privacy is to use encryption.

■ Privacy-Enhanced Mail (PEM)

The PEM Internet standard provides a specification for secure e-mail over the Internet. PEM was designed to work with the standard Internet e-mail format and includes functions such as encryption, authentication, and key management using both public-key and single-key (secret-key) schemes. Numerous cryptographic functions such as encryption, digital signatures, or hash functions can be used by specifying them in the e-mail header. PEM uses DES cryptography for message text encryption with session-specific keys; RSA encryption is used to transmit the DES key securely along with the message. X.509 certificates form the basis for PEM's public-key management. PEM provides security features such as message origin authentication, content confidentiality, content integrity, and non-repudiation. Although PEM has been available for several years, it has not been implemented widely, and the standard may be superseded by other more popular standards such as S/MIME, MOSS, or PEM-MIME.

■ Secure/Multipurpose Internet Mail Extensions (S/MIME)

MIME is an extension to the original Internet e-mail standards that define how the body of an e-mail message is structured. The major advantage of MIME is that it allows the message text to include binary data (such as images or file attachments) rather than being limited to ASCII text. However, MIME as originally defined does not provide any security services. S/MIME adds digital signatures and encryption to MIME messages. S/MIME is viewed as taking a middle ground between the rigidity of PEM and the looser requirements of PEM-MIME. S/MIME currently has heavy industry support and has been endorsed by vendors such as Lotus, Microsoft, Netscape, Novell, Qualcomm, and VeriSign.

A draft of the new S/MIME 3.0 specification was completed in October 1997 by the Internet Mail Consortium and by government contractors familiar with the security requirements of government networks. Proposal of the draft as an Internet standard is scheduled for early 1998. S/MIME 3.0 features new security enhancements such as signed receipts, labels indicating a message's sensitivity, and the ability to send encrypted messages to a mailing list. Version 3.0 also defines how to encrypt a message's content and how to verify the sender's identity with a digital signature.

Security vendors are promising authentication and encryption software standards to provide better compatibility between their products so that different e-mail systems can exchange encrypted or digitally signed e-mail messages, something generally not possible today. For example, vendor-specific e-mail – such as cc:Mail, Exchange, Lotus Notes, and MS-Mail – either do not support encryption at all or employ encryption algorithms that do not interoperate with each other or with Internet mail.

In August 1997, RSADSI unveiled a new certification program designed to strengthen and standardize testing of S/MIME. Participants' products are put through a variety of tests, including digital certificate handling, digital signature creation and verification, and encryption. Products receiving certification will earn an "S/MIME-enabled" seal, which designates product interoperability. All testing is performed at RSADSI's S/MIME Interoperability Center Web site (*http://www.rsa.com/rsa/S-MIME/html/interop_center.html*) to provide a public record of results and a master matrix of products tested by version number. Vendors that have earned the S/MIME-enabled seal to date include Baltimore Technologies, Deming Internet Security, Entrust, Microsoft, Netscape, NTT Electronics Technology Corp. (NEL), OpenSoft, and Software and Systems Engineering Ltd. (SSE), a Siemens company.

In an effort to create a common protocol for encrypting e-mail, RSADSI published an outline of its proprietary RC2 encryption algorithm for review by the IETF (considered a step in the adoption process). RC2, a key part of S/MIME, provides for message privacy and interoperability between domestic and export versions of products using S/MIME. In August 1997, however, the IETF dismissed RSADSI's proposed specification in favor of considering technology from Pretty Good

Privacy Inc. (PGPI). S/MIME is based on patented technology and trade secrets owned by RSADSI, and the company's claims to ownership of both the technology and the S/MIME trademark may have soured the IETF on its proposal.

Instead, PGPI has submitted a specification to the IETF based on the Diffie-Hellman key management technology. Called OpenPGP (also known as PGP/MIME), the proposal is favored because it would be a nonproprietary standard. Because PGPI will claim no proprietary rights to its technology (unlike RSADSI), it seems likely that OpenPGP will become an official standard, perhaps even replacing S/MIME eventually. The IETF OpenPGP working group was scheduled to submit a draft for comment in October 1997, with a May 1998 target date for submission of the proposed standard.

■ PEM-MIME, or MOSS

PEM-MIME, or MIME Object Security Standard (MOSS), is another Internet draft standard that may be the successor to PEM. MOSS supports X.509. PEM-based security services are added to MIME messages in much the same manner as in S/MIME. However, different security services can be added to each part of the body message. If the MIME message body comprised two parts, one digitally signed and the other plain text, a recipient could read the plain text message without a MIME-compliant mail reader. Using a PEM-MIME-compliant mail reader, the recipient also would be able to verify the digital signature. Additionally, different parts of the message body could be encrypted using different keys and algorithms; some parts could be signed, while others were not. Such flexible requirements have resulted in criticism of PEM-MIME as being too complex to implement, which may be why S/MIME is gaining support instead.

Web Security

Server-to-server authentication and server-to-user authentication can be achieved with SSL and Microsoft's PCT. In addition, the SET specification is designed to provide increased security for financial transactions across the Internet, and other products are in development that leverage certificates and public-key cryptography to provide higher levels of assurance than previously available.

■ HyperText Transport Protocol

The primary protocol used between Web clients and servers is HTTP. Current HTTP implementations have only modest support for the cryptographic mechanisms appropriate for Web-based transactions.

■ Secure Sockets Layer (SSL)

Netscape designed the most widely used security technology on the Web, the Secure Sockets Layer (SSL). SSL is used by developers to add advanced security within TCP/IP applications and has become a de facto standard for encryption between browsers and servers. SSL provides end-to-end security between browsers and servers, always authenticating servers and optionally authenticating clients. In providing communications channel security, SSL ensures that the channel is private and reliable and that encryption is used for all messages after a simple "handshake" defines a secret key. The message transport includes message integrity using Media Access Control (MAC), the protocol that controls access to the physical transmission medium on a LAN. Secure hash functions are used for MAC computations.

Secure hash functions produce a digital fingerprint from a set of input data. It is computationally infeasible to take a hash value and work backward to find other input data that would generate the same value. Thus, if someone sends a file and a fingerprint generated from the file, and if the recipient runs the same hash algorithm on the file received and gets the same result, both parties can be virtually certain that the file was received intact.

SSL secures connections at the point where the application communicates with the IP protocol stack, rather than the application level, so it can encrypt, authenticate, and validate all protocols supported by SSL-enabled browsers, such as FTP, telnet, e-mail, and so on.

When an SSL-compliant client (usually in the form of a Web browser) wants to communicate with an SSL-compliant server, the client initiates a request to the server, which then sends an X.509 certificate back to the client. The certificate includes the server's public key and the server's preferred cryptographic algorithms or ciphers. The client then creates a unique session key, encrypts the key with the public key sent by the server, and sends the newly created session key to the server. After it receives this key, the server authenticates itself by sending a message encrypted with the key back to the client, proving that the message is coming from the proper server. Once this "handshake" process is completed, which results in the client and server agreeing on the security level, all data transferred between that client and that server for a particular session (HTTP, telnet, ftp, and so on) is encrypted using the session key.

Both Microsoft's Internet Explorer and Netscape's Navigator browsers indicate whether a particular Web site is secure (that is, they prevent unauthorized access to personal data) by displaying an icon on the user's screen. When using Navigator, the first part of the URL at the top of the browser also will change from `http://` to `https://` when a secure link has been created.

In addition to server authentication via SSL, a second phase, client authentication, may take place for extra security measures. SSL supports several cryptographic algorithms to handle the authentication and encryption routines. One algorithm, which was developed by RSADSI, is RC4, a variable key-size cipher. Other cryptographic algorithms supported by SSL include DES and Triple DES.

SSL version 3.0, completed in 1996, adds a thorough definition of client certificate requirements to the original protocol. The SSL protocol is being built into a range of hardware- and software-based security products. For example, to meet the demand for server-embedded security, Atalla announced new PCI-based Internet security processor products in April 1997 for SET transactions and general-purpose cryptography. The WebSafe 2/PCI Internet security processor provides strong physical protection for private encryption keys and manages a full range of security functions such as RSA, DES, and SSL 3.0. on Attala's secure PCI-card.

A new product that allows developers to add SSL communications security to client devices with limited resources (such as palm-top computers, PDAs, and Internet "appliances") was announced in June 1997. Terisa Systems' Thin SSL Client option for its SecureWeb Toolkit will be used by WebTV as an upgrade to its set-top box for conducting secure electronic commerce transactions over the Internet.

Specific implementations of SSL may not be secure enough, however. In 1996, for example, two independent groups found a 40-bit RC4 encryption key (RSADSI's proprietary algorithm) within Netscape's SSL-enabled Web browser and managed to crack the code.

■ Secure HyperText Transport Protocol

Secure HTTP (S-HTTP) extends the basic HTTP protocol to allow both client-side and server-side encryption. S-HTTP provides secure communications between a browser and an HTTP server to enable commercial transactions for Web-based applications.

S-HTTP is created by SSL running under HTTP. The protocol was developed in 1994 by Enterprise Integration Technologies as an implementation of the RSA encryption algorithm. Although SSL operates at the Transport layer, S-HTTP supports secure end-to-end transactions by adding cryptography to messages at the Application layer. Therefore, although SSL is application independent, S-HTTP is tied to the HTTP protocol. S-HTTP ensures security of only HTTP-based messages, whereas SSL secures a whole Internet connection. In addition, although SSL encrypts the entire communications link between a client and a server, S-HTTP encrypts each message on an individual basis. Terisa Systems' SecureWeb Client 2.0 and Server Toolkit 2.0 are security tools that support both SSL and S-HTTP and are designed to enable companies to build secure Web servers.

Several cryptographic message format standards may be incorporated into S-HTTP clients and servers, including PEM and PGP. S-HTTP supports interoperation among a variety of implementations and is compatible with HTTP. Message protection on S-HTTP may be provided on three dimensions: signature, authentication, and encryption. Any message may use any combination of these (as well as no protection).

Multiple-key management mechanisms are provided in S-HTTP, including password-style manually distributed shared secret keys, public-key key exchange, and Kerberos ticket distribution. In particular, provision has been made for prearranged (in an earlier transaction) symmetric session keys in order to send confidential messages to those who have no established key pair.

■ Transport Layer Security

A major trend in security is the migration beyond simple secure transactions to secure electronic messaging through techniques such as Transport Layer Security (TLS, also sometimes called Transaction Layer Security). TLS based on SSL 3.0 is in development. This standard involves enhancements to session authentication, such as the inclusion of digital signature blocks, to transcend the traditional site certification currently in use. TLS also incorporates some PCT technologies. The IETF released the TLS Protocol Version 1.0 in May 1997. Under a draft proposal, co-authored by CyberSafe and submitted to the IETF in July 1997, TLS would enable SSL to support Kerberos encryption for the first time.

Firewalls

Network firewalls enforce a site's security policy by meditating traffic between two or more networks (often between a corporate network and an external network such as the Internet). The firewall system provides both a perimeter defense and a control point for monitoring access to and from specific networks.

Firewalls can control access on the basis of protocol attributes (for example, IP source address, IP destination address, TCP or UDP port number). Access control can be both inbound (a traditional function, where the firewall is used to protect a corporate network from being accessed from the Internet), and outbound (a more recent innovation, where the firewall is used to block access to Web sites thought to be objectionable).

Firewalls can defend against attacks ranging from unauthorized access, IP address "spoofing" (a technique where hackers substitute friendly addresses to gain access to an intranet), session hijacking, viruses and rogue applets, and rerouting of traffic. Firewalls can protect a site against some Denial of Service (DOS) attacks; however, there are inherent weaknesses in certain services and protocols that a firewall cannot remedy. Firewall protection mechanisms vary, but traditionally include strong authentication (one-time passwords) and encryption.

Firewall topology depends on the services needed and on the enterprise's Internet connection type. In addition to physical topology, firewall software supports implementation strategies that fall into three categories:

- Application-level proxies
- Circuit-level gateways
- Packet-filtering gateways

Application-Level Proxies

Application-level proxies are special-purpose code that relays traffic for each specified application and service (for example, Telnet, FTP, SMTP, HTTP, and the Network News Transport Protocol, NNTP). These proxies disassemble and reassemble packet data, making it impossible to have a direct connection between the user and the requested service on the Internet. This model guards against data-driven attacks; however, it cannot deal with session stealing, and requires strong UNIX skills to install properly. The Firewall Tool Kit (FWTK) was developed in 1993 by Marcus Ranum (for an ARPA project) and created the firewall standard for application gateways.

TIS's Gauntlet is the commercial version of the FWTK. Gauntlet is an application-based firewall with several advanced features, including ActiveX and Java signature verification to protect against viruses in executable files and applets. Other major application-level proxy firewalls include those from Raptor and Milkyway Networks.

Circuit-Level Gateways

Circuit-level gateways act as a connection service by copying bytes back and forth between the internal network and the Internet, thus creating a TCP virtual circuit across the gateway. Circuit gateways are more general implementations of application-level proxies. Typically, circuit gateways are used when a specific proxy application does not exist.

Packet-Filtering Gateways

First-generation packet-filtering gateways examine the source and destination addresses of IP packets and IP options. Packets that are received by the router are permitted or denied based on an access control list (ACL). The ACL is the "rule base" used to implement the site's security policy (for example, "deny all inbound

Telnet sessions by blocking TCP port 23"). A firewall based on packet-filtering technology permits at least some level of direct packet traffic between the Internet and the hosts on the protected networks.

Stateful inspection, developed in 1995, represents the latest generation of packet-filtering firewalls. Rather than examining the contents of each packet using a proxy, stateful inspection firewalls take a "snapshot" of the entire packet and quickly compare it to a "state table." This table keeps track of inbound and outbound connections and discards packets that are not part of a valid connection. This technique is more efficient than proxying specific applications. Advantages of stateful inspection are that it works well with complex protocols, supports new services easily, and works best where security is a concern, but throughput is more important. Stateful inspection became the model for firewalls such as Check Point's FireWall-1 and Cisco's Pix Firewall.

■ Firewall Trends

Important trends in firewalls include the addition of new functionality and the overlap of firewalls with other security measures such as encrypted virtual private networks (VPNs). Sophisticated packet-level security features are also making their way into network infrastructure components such as routers and switches. These companies are promoting their new internetworking products with built-in firewall protection capabilities such as stateful inspection and encrypted tunneling for VPNs. (See "Virtual Private Networks (VPNs)" on page 183.)

- **Network management** – Many firewall vendors are trying to turn their management consoles into enterprise security systems. For example, Milkyway Networks has integrated support for X.500 directories into its product, so that if a company issues digital certificates to users, they can use the directory both internally and over the Internet.

  Other vendors have introduced network auditing tools, reflecting the growing importance of firewall-management tools in intranets. For example, Check Point's OPSEC platform integrates and manages enterprise security through published APIs, support for standard protocols, and a scripting language.

- **Applet monitoring** – In May 1997, Finjan Software announced the release of SurfinGate to members of the Java Security Alliance (JSA), enabling the integration of Java applet-inspection capabilities into existing security frameworks such as firewalls. The technology provides Java applet scanning at the gateway, permits or denies applets based on the corporation's security policy, and allows end users and network managers to monitor applets as they run; it also includes a database of known applets. JSA alliance members include Check Point, Cisco, Digital, Milkyway, Network-1, Raptor, Software and Technologies, TI, and TIS. Each vendor will implement Finjan code differently, in keeping with its own product development strategy.

- **Virus scanning** – Many vendors have already added or plan to combine virus-scanning software with firewalls so that Web downloads will be checked automatically for viruses.

- **IP address spoofing** – Because IP address spoofing requires the ability to guess the sequence numbers of TCP packets, some firewall vendors (such as Cisco) make this process difficult, or impossible, by using a randomizing algorithm for their generation. Routers also can prevent a host "outside" an enterprise from claiming a source address that belongs to the internal network.

- **Encryption** – Firewalls may provide encryption services to create an encrypted data stream across the Internet. Some vendors now supply modules that provide DES, manual IPsec, and SKIP support. For example, Cisco's PIX Firewall features DES encryption that helps turn the Internet into a VPN, providing secure communications among LAN-based clients.
- **Integration** – Several efforts have begun to unify the security landscape:
  - **Content Vectoring Protocol** – An initiative, led by Check Point, is underway to make the CVP a standard method for integrating firewalls with anti-virus software and content monitoring systems. CVP allows third-party vendors to create firewall plug-ins to provide additional functionality, such as virus checking. Anti-virus vendors that have announced support for the CVP include Cheyenne Software, Integralis, Network Associates (McAfee), Symantec, and Trend Micro.
  - **Java-based management interfaces** – Such interfaces are being added to firewall systems by Raptor and TIS, and will enable developers to plug in security components (such as management, real-time monitoring, content scanning, and authentication services) in the form of JavaBeans.

5.4.10 Market Overview

In addition to those already discussed, key market trends include an increase in security-related product releases by major system vendors (such as HP, IBM, Microsoft, and Sun) and the vertical integration of market segments (such as firewalls and anti-virus products), reducing the total number of vendors and products. Some major acquisitions and mergers in the security sector are shown in Table 5-12.

Table 5-12: Recent Mergers and Acquisitions in the Security Sector

Date	Buyer	Seller	Seller's Focus
April 15, 1996	Security Dynamics	RSA Data Security	Encryption tool kits
May 28, 1996	Secure Computing	Border Technologies	Firewalls
June 14, 1996	Vasco Data Security	Lintel	Tokens
June 26, 1996	Secure Computing	Enigma Logic	Tokens
June 28, 1996	Vasco Data Security	Digipass	Tokens
July 1, 1996	Pretty Good Privacy	Lemcon Systems	Message encryption
November 15, 1996	Pretty Good Privacy	Privnet	Privacy software
January 17, 1997	Axent Technologies	Digital Pathways	Tokens
March 24, 1997	Pretty Good Privacy	Zoomit	Directory services
April 10, 1997	CyberGuard	TradeWave	Certificate Authority tools
May 19, 1997	Spyrus	Terisa Systems	SET encryption tool kits
December 1, 1997	Network Associates (formerly McAfee)	Pretty Good Privacy	Cryptography

Source: Piper Jaffray, 1997

Gartner identifies 15 distinct market segments within the security marketplace, with varying product cycle times, ranging from 1 to 3 months for anti-virus products, 12 to 28 months for firewalls, 2 to 5 years for single sign-on products, and 3 to 15 years for cryptography or certification offerings. However, Gartner cautions that market segmentation within the security arena is misleading because some products are consumed only by other vendors (such as cryptography); some products are developed for end users, but acquired and maintained centrally (such as single sign-on); and still others are used by the same group that acquires and maintains them (such as auditing and assessment tools).

Projected growth in the security market is shown in Figure 5-20.

Figure 5-20: Projected Growth in the Security Market

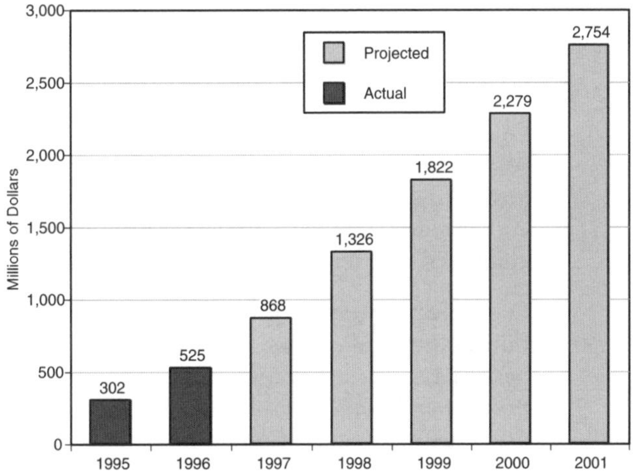

Source: Piper Jaffray, 1997

Cryptography

This area will see strongest growth in e-mail and secure financial transaction applications. The need to enforce confidentiality will continue, but it will be supplemented by the need for non-repudiation and validation of authenticity provided by digital signatures and certified public keys. RSA will continue to dominate the public-key market, but newer technologies such as ECC and RPK will begin to gain ground.

■ Certificates

The most successful vendors will be those providing the toolkits and infrastructure to handle certificate issuance, revocation, and comprehensive key management. Gartner predicts that early leaders will be Entrust and GTI Government Systems in the U.S., Dynasoft in Sweden, and Utimaco in Germany. VeriSign will lead the service-based certificate vendors.

■ Encryption

A variety of new hardware and software solutions has been introduced for different market segments:

- **New Java encryption kit** – In June 1997, RSADSI announced it had licensed J/CRYPTO, a Java encryption library developed by Baltimore Technologies of Ireland, for incorporation in its J/SAFE encryption toolkit. The kit will enable developers to build more secure applets that implement public-key cryptography, enabling both privacy and authentication.

- **New encryption hardware** – In July 1997, IBM unveiled a new PCI cryptographic coprocessor/adapter card to provide hardware-based encryption for bankers, merchants, and users conducting business on the Internet; the card supports SET, DES, RSA, and digital signature processing.
- **New file encryption software** – In August 1997, PGPI introduced PGPdisk file encryption software, which uses 128-bit encryption to securely store data and files on portables, desktops, and networks. PGPdisk functions at the operating level to ensure that no unencrypted data is left on the disk.

Authentication

Several solutions have been proposed to address the issue of reusable passwords:

- **One-time password generators** – Security Dynamics has unveiled SoftID, a software-only version of the company's security system that provides one-time password authentication based on a two-factor system. SoftID is based on an encryption system that generates a new access code every 60 seconds that, when combined with a user's secret PIN, authenticates the user.
- **Dynamic password technology** – Secure Computing's SafeWord is a verification, authentication, and audit system. SafeWord combines something a user has (a password-generating token), with something the user knows (a PIN number and user ID) and a way to prevent someone else from assuming the user's identity (encryption).
- **Biometric techniques** – Many of the newest security systems that authenticate remote users are based on biometric techniques such as fingerprint scans and voice verification.
  - **Fingerprints** – National Registry's NRIdentity Pass for Portables, released in April 1997, is a biometrics-based finger-image authentication system made up of a mini-scanner (about the size of a computer mouse), an image digitizer, and proprietary software. The user scans a fingerprint on the portable scanner and sends the image over wired or wireless connections to be compared with one stored on a database at the corporate LAN. The system costs less than $500.
  - **Voice verification** – Intelitrak's Citadel GateKeeper, announced in June 1997, is a voice authentication gateway for remote network access applications, including voice-protected Internet/intranet servers. The system employs a two-key approach in which a single-use, time-sensitive, session key is generated upon successful voice verification using an ordinary telephone or other input device. A potential user then has a set window of time to gain network entry before the passcode expires.

Biometric revenues are expected to more than triple, from nearly $16 million in 1996 to $50 million by 1999. The number of biometric devices in use is expected to jump from 8,550 in 1996 to more than 50,000 by the year 2000.

Other devices that require authentication, such as ATMs, are making use of new chip technologies based on biometrics. For example, SGS-Thomson announced in May 1997 that it had developed the first microchip that records fingerprints electronically on contact, generating higher-quality images than other personal ID methods. The chip works by detecting variations in the electrical currents running along the ridges and valleys of the skin; through direct physical contact, the sensor array "grabs" a fingerprint pattern without using an optical or mechanical adapter such as a scanner or camera. Potential security applications for the sensor include PC access, electronic security for Internet transactions, physical identification for access to automobiles and buildings, and PIN number replacement. Veridicom has unveiled a similar chip based on technology developed at Bell Labs; the reader, which is the size of a postage stamp, will cost several hundred dollars less than current camera-scanning systems.

Firewalls

The price of firewalls has dropped significantly in the last year. For example, TIS slashed prices by 56 percent on its Gauntlet line, and Raptor introduced a plug-and-play firewall for less than $1,000 for up to 25 users. Some small vendors are marketing firewall software for as little as $100.

Vendors such as I-Planet offer robust multi-function Internet connectivity devices that include a gateway application server and built-in firewall for $5,000 (additional security is provided by seven built-in proxy servers); its new IPS-SMG device designed for the small office/home office, introduced in July 1997, sells for only $2,500. Whistle Communications' InterJet, at $2,000, is similar in design; it includes a built-in firewall with SSL security and packet filtering.

Anti-Virus

According to IDC, the worldwide anti-virus market is a $437 million software segment that includes vendors such as Cheyenne, Dr. Solomon, IBM, Network Associates (McAfee), Symantec, and Trend Micro. (See Figure 5-21.)

The anti-virus market has not reached saturation levels, and although overall revenue growth is beginning to slow, it is still expected to be strong through the forecast period. IDC says that increased pricing competition is the biggest challenge, which will keep growth in revenue much lower than overall shipment growth. The desktop anti-virus market will remain viable, although the anti-virus server market is growing at a slightly faster rate.

Figure 5-21: Anti-Virus Vendor Worldwide Market Share – 1996

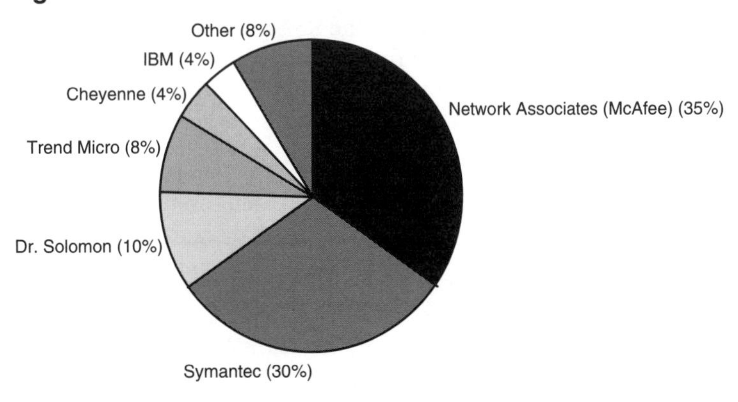

Source: IDC, 1997

Enterprise Security

New technologies are available for network administrators that help find holes in information protection, alert users to vulnerabilities, and advise how to fix them. For example, Netect's Netection system simulates attacks on three levels – the operating system binary, security management, and human – and generates a report. The system is database-driven, with updates delivered by push technology or via the company's Web site. Netection is available for Solaris and Windows NT.

Comprehensive network security "environments" designed by alliances of vendors promise to protect whole enterprises. Network Associates (McAfee), RSADSI, Security Dynamics, and VeriSign announced plans in July 1997 to offer the SecureOne enterprise security framework. The framework will combine Network Associates' (McAfee's) anti-virus protection, VeriSign's digital IDs, Security Dynamics' authentication/access control, and RSADSI's encryption. No release date has been announced.

Multi-Platform Security

Multi-platform security will continue to be an issue for system and network managers as the move to distributed applications continues and the demand for Internet connectivity increases. Organizations continue to struggle to find mainframe RACF or ACF2 equivalent products for client/server systems. Companies such as CA, HP, IBM, Novell, Terisa Systems, and Tivoli offer products that assist in managing security in the distributed environment, but there is still no single complete answer. Single sign-on products continue to evolve as well as new products that are integrated with X.500 directory services.

5.4.11 Forecast

Authentication

- Biometrics are becoming more cost-effective and user-friendly and will be integrated within two-factor authentication packages within the forecast period. Face recognition, hand geometry, fingerprint and voiceprint technologies are the most likely to receive rapid consumer acceptance and implementation. Even so, efforts will be needed to address lingering public resistance about using biometric techniques in consumer contexts such as an ATM.

Cryptography

- Alternatives to the RSA public-key encryption algorithm – such as ECC and RPK – will find early adoption in applications requiring high-speed encryption. However, the installed base of RSA technology and related software support will limit wide-scale acceptance of alternatives during the forecast period.
- Expiration of the U.S. patent on the RSA algorithm in 2000 will reduce resistance to specifying that algorithm as a part of Internet standards.
- Major government and corporate organizations will establish themselves as certificate authorities. Outside the U.S., this movement will be led by postal administrations; inside the U.S., by banks and other financial institutions.
- The lack of product interoperability and of established processes to authenticate certificate authorities and relieve users of liability will limit acceptance of certificates and certificate authorities during the forecast period.
- Digital signatures will receive widespread legal standing and be freely accepted in most countries within the forecast period. However, within the U.S., the lack of consistent policies at the state level will hinder the implementation of digital signatures as legally binding.
- Attempts by governments to regulate the import and export of strong encryption will become less effective as vendors successfully circumvent restrictions in a variety of ways.

Security Standards

- Too many competing standards will continue to hamper product interoperability as well as rapid development and deployment.
- Schemes such as ITSEC that certify product claims will not substitute for initiatives that focus on creating enterprise security solutions (such as OPSEC).
- RSADSI's S/MIME and PGPI's OpenPGP (PGP/MIME) will vie for selection as the IETF-sanctioned protocol for encrypted and authenticated e-mail, with the latter gaining approval because of its lack of dependence on vendor-specific technology.
- Vendors will be forced to take greater responsibility for ensuring security as part of the product design process, rather than trying to add security features after release in response to problems (or to issue updates to fix security problems that escaped notice during software development and testing).

Firewalls

- Firewall vendors will move toward unbundling services, allowing users to pick which features they want. Firewall features also will be embedded into operating systems, thereby solving any interoperability problems, albeit perhaps with a less-robust solution.
- Firewall "appliances" (integrated hardware and software packages) may replace Windows NT-based units as the "easiest-to-use" solution (compared to UNIX implementations) within the forecast period. These will be low-cost, minimal-configuration firewall appliances that will sacrifice flexibility in favor of easy configurability but that will retain the same functionality as firewalls running on general-purpose platforms.
- Firewalls of the future will allow users to pick and choose between those services that can or should be proxied (such as mail, the Internet, and financial transactions) and those that should not (data-intensive services such as video).
- Real-time intrusion detection will grow in importance as a means of validating the firewall, allowing network administrators to detect break-ins as they happen.
- Due to concerns about the weakness of ActiveX security, Java will become more widely used than ActiveX for applications that must operate over the Internet. ActiveX use will be relegated to secure intranets that have adopted Microsoft-only Web solutions (both Web browsers and servers).
- The threat of destructive actions by network-downloaded software components (Java and ActiveX) will lead to the development of products that scan this software for possible threats. This capability will be integrated into firewall products.

Anti-Virus

- Anti-virus protection will become embedded in a wide variety of products, from firewalls to groupware, during the forecast period.
- Consolidation in the anti-virus sector will become more pervasive as the smaller providers become unable to keep pace with market leaders such as Network Associates (formerly McAfee Associates) and Symantec.
- The anti-virus market will be re-energized during the forecast period because of macro viruses, connectivity to the Internet, and the increasing number of applications supported via the Web.

- The Internet and e-mail will continue to provide easy conduits for the transmission of viruses, prompting the adoption of stronger anti-virus measures by corporations and boosting the number of vendors that provide such services on an outsourcing basis.

5.4.12 References

■Articles

Bell, Stephen. 1997. Kiwi encryption guru unfazed by U.S. moves. *Computerworld*. June 2.
Berinato, Scott. 1997. Biometrics tools guard networks. *PCWeek*. April 14: 6.
Davis, Ann. 1997. The body as password. *Wired*. July.
Golden, Ed. 1997. IBM encryption scheme holds significant promise. *InfoWorld*. May 12: 21.
Hudgins-Bonafield, Christy. 1997. Bridging the business-to-business authentication gap. *Network Computing*. July 15: 62.
Kay, Russell. 1996. Security gets a new face. *Byte*. October: 39.
Karve, Anita. 1997. Certifying your Internet identity. *LAN Magazine*. February: 73.
Kosiur, David. 1997. Role of digital certificates looks secure... *PCWeek*. April 28: 115-117.
Paone, Joe. 1997. Legal peril on the Web. *LAN Times*. May 26: 1.
Phillips, Ken. 1997. Biometric identification looms on landscape of network log-ins. *PCWeek*. March 24: 99.
Wayner, Peter. 1997. Who goes there? *Byte*. June: 70-78.
Willis, David. 1997. Villains in the vault. *Network Computing*. January 15: 52-65.
Yager, Tom. 1997. All the Web's a stage. *Byte*. August: 135.

■Periodicals, Industry and Technical Reports

Hopkins, W. 1997. *Making cyberspace safe for marketing*. Gartner. (ECS: KA-197-1190) May 29.
O'Reilley, J. 1997. *Managed Internet security: the state of the market*. Gartner. (ISS: M-INT-139) May 29.
O'Reilley, J. and M. Zboray. 1997. *Managed Internet security: deciding to make the leap*. Gartner. (ISS: KA-INT-140) May 29.
Paperless signature. 1997. Edge: Work-Group Computing Report. January 27: 22.
Wheatman, V. 1997. *Certificate authority liability in a litigious world*. Gartner. (SPA-EC-131) April 23.

■Books

Rutstein, Charles B. 1997. *Windows NT security: a practical guide to securing Windows NT servers and workstations*. New York: Mc Graw-Hill.
Thomas, Scott L. and Brad J. Hoyt. 1997. *Lotus Notes and Domino 4.5 architecture, administration, and security*. New York: McGraw-Hill.

■On-Line Sources

Open Platform for Secure Enterprise Connectivity (OPSEC) architecture. White paper. `http://www.checkpoint.com/opsec/architect.htm`
Summary of Electronic Commerce and Digital Signature Legislation. `http://www.mbc.com/ds_sum.html`
van der Hof, Simone. May 1997. *Summary of the report on the legal status of digital signatures in Europe*. `http://cwis.kub.nl/~frw/people/hof/digsig2.htm`
van der Hof, Simone and Bert-Jaap Koops. November 1997. *Digital signature law survey, version 2.0*. `http://cwis.kub.nl/~frw/people/hof/DS-lawsu.htm`
Welcome to SESAME. 1997. `http://www.esat.kuleuven.ac.be/cosic/sesame/`

■URLs of Selected Mentioned Companies

Advanced Recognition Technologies `http://www.artcomp.com`
AGIS `http://www.agis.net`
American Institute of Certified Public Accountants `http://www.aicpa.org`
ANS `http://www.ans.com/home.html`
ANSI `http://web.ansi.org/default.htm`
AOL `http://www.aol.com`
AT&T `http://www.att.com`
Atalla `http://www.atalla.com`
Axent Technologies `http://www.axent.com`
BBN `http://www.bbn.com`
BBN Planet `http://www.bbnplanet.com`
Bell Labs `http://www.belllabs.com`
Biometric Consortium `http://biometrics.org:8080`
Border Technologies `http://www.border.com`
BTG `http://www.btg.com`
Business Software Alliance `http://www.bsa.org`
Certicom `http://www.certicom.com`
Check Point `http://www.checkpoint.com`

Cheyenne Software http://www.cheyenne.com
CICA http://www.cica.ca/
Cisco Systems http://www.cisco.com
Claris http://www.claris.com/index.html
CommerceNet http://www.commerce.net
Common Criteria Project http://csrc.ncsl.nist.gov/nistpubs/cc/
Computer Associates http://www.cai.com
Computer Security Institute http://www.gocsi.com/homepage.htm
CyberGuard http://www.cyberguardcorp.com
CyberMedia http://www.cybermedia.com
CyberSafe http://www.cybersafe.com
Cygnus Solutions http://www.cygnus.com
Cylink http://www.cylink.com
Deming Internet Security http://www.deming.com
Digital http://www.digital.com
Digital Pathways http://www.digitalpathways.com
Dr. Solomon http://www.drsolomon.com
Dynasoft http://www.dynas.se
E-Corp http://www.e-corp.com/home.asp
Electronic Frontier Foundation http://www.eff.org
Entegrity Solutions http://www.entegrity.com
Enterprise Integration Technologies http://www.eit.com
Entrust http://www.entrust.com
eSafe http://www.esafe.com
Financial Services Technology Consortium http://www.fstc.org
Finjan Software http://www.finjan.com
Firefly http://www.firefly.net
Fischer International Systems http://www.fisc.com/default.asp
Gemplus http://www.gemplus.com
Gradient Technologies http://www.gradient.com
Groupe Bull http://www.bull.com
GTE http://www.gte.com
Hewlett-Packard http://www.hp.com
Hi/fn http://www.hifn.com
I-Planet http://www.iplanet.com
IBM http://www.ibm.com
ICL http://www.icl.com
Identix http://www.identix.com
IEEE http://www.ieee.org
IETF http://www.ietf.org
Information Resource Engineering http://www.ire.com
Information Technology Industry Council http://www.itic.org
Informix http://www.informix.com
Integralis http://www.integralis.com
Intel http://www.intel.com
Intelitrak http://www.intelitrak.com
Internet Mail Consortium http://www.imc.org
Internet Security Systems http://iss.net/~iss
IriScan http://www.iriscan.com
IRS http://www.irs.gov
ISO http://www.iso.ch
ISOCOR http://www.isocor.com
Junkbusters http://www.junkbusters.com
Key Tronics http://www.keytronics.com
LanOptics http://www.lanoptics.com
Lintel http://www.lintel.com
Livermore Software Laboratories http://www.lsli.com
Lotus http://www.lotus.com
Lucent http://www.lucent.com
Mastiff Electronic Systems http://www.mastiff.co.uk
MCI http://www.mci.com
Microsoft http://www.microsoft.com
Milkyway Networks http://www.milkyway.com
MIT http://www.mit.edu
Motorola http://www.mot.com
Mytec Technologies http://www.mytec.com
NetDox http://www.netdox.com
Netect http://www.netect.com
Netscape http://www.netscape.com
Network Associates http://www.networkassociates.com
Network-1 http://www.network-1.com
NIST http://www.nist.gov/
Nortel http://www.nortel.com
Norton Utilities http://www.norton.com

NTT Electronics Technology Corp. (NEL) http://www.nel.co.jp/index-e.html
OECD http://www.oecd.org
OpenVision http://www.veritas.com
Oracle http://www.oracle.com
PenOp http://www.penop.com
PGPI http://www.pgpi.com
Pilot Network Services http://www.pilot.net
Portcullis http://www.portcullis.co.nz
Privnet http://www.privnet.com
Qualcomm http://www.qualcomm.com
Raptor http://www.raptor.com
RSA Data Security http://www.rsa.com
RSA Data Security Labs http://www.rsa.com/rsalabs
RSA Data Security S/MIME Interoperability Center
http://www.rsa.com/rsa/S-MIME/html/interop_center.html
Schlumberger http://www.schlumberger.com
SCM Microsystems http://www.chips.com.au/products/scmhome.html
Secure Computing http://www.securecomputing.com
Security Dynamics http://www.securid.com
Security Industries Association http://www.sia.com
SGS Thomson http://www.st.com
Siemens Nixdorf http://www.siemens.com
Software.com http://www.software.com
SouceFile http://www.sourcefile.com
Sprint http://www.sprint.com
Spyrus www.spyrus.com
Stanford University http://www.stanford.edu
Sun Microsystems http://www.sun.com
Sybase http://www.sybase.com
Symantec http://www.symantec.com
Tandem http://www.tandem. com
Terisa Systems http://www.terisa.com
Texas Instruments (TI) http://www.ti.com
Thawte Consulting http://www.thawte.com
The Open Group http://www.opengroup.org
Thomson-CSF http://www.tcs.thomson-csf.com
Trusted Information Systems (TIS) http://www.tis.com
Tivoli http://www.tivoli.com
TradeWave http://www.tradewave.com
Trend Micro http://www.trendmicro.com
Unisys http://www.unisys.com
Utimaco http://www.utimaco.com
ValiCert http://www.valicert.com
Vanguard Integrity http://viplink.com
Vasco Data Security http://www.vasco.com
Veridicom http://www.veridicom
VeriFone http://www.verifone.com
VeriSign http://www.verisign
Versit Consortium http://www.versit.com
Viisage http://www.viisage.com
VPNet Technologies http://www.vpnet.com
Whistle Communications http://www.whistle.com
World Trade Organization http://gatekeeper.unicc.org/wto.welcome.html
World Wide Web Consortium (W3C) http://www.w3.org
Zoomit http://www.zoomit.com

6 Software

6.1 Application Development Environments

6.1.1 Executive Summary

In today's highly competitive business environment, organizations in all industries are challenged to accomplish more work with fewer resources. Within information technology (IT) groups, the need for powerful and robust software development tools has never been greater. The graphical user interface (GUI) already has become a necessity for modern applications. New software often is required to support distributed use over fast and reliable local area networks (LANs) as well as over slower and undependable wide area networks (WANs). Most businesses now have some presence on the Internet, many have a private intranet used internally, and a few support an extranet used for communicating with partners and suppliers. Creating applications with sophisticated visual interfaces and supporting robust networking complicates the task of software development.

Development tools are evolving rapidly to meet new business needs and to stay in step with the changes taking place in IT and infrastructure. The computer languages used for software development also are evolving, albeit at a slower pace. During the past several years, many new development tools have emerged, particularly object-oriented programming and Web application tools for client/server and general business applications. The goal of these object-oriented technologies has been to deal with the complexity of modern software development by better supporting large-scale development efforts. These tools are designed to work with various object-oriented languages in meeting developers' needs for robustness, portability, and efficiency. Object-oriented languages now have become the dominant language for new software development.

Java is receiving wide acceptance in the software development community and with tool vendors, particularly because more applications use the Web browser as a portable client front end and the Web server as a form of middleware. Java also is quickly becoming the preferred language for introductory programming courses in academia. Although much of the interest in the developer community is still exploratory, Java is well positioned to displace other object-oriented languages and become dominant for Web applications.

Although new development is based on object-oriented programming, the majority of code in place today was written in older, third-generation computer languages. Furthermore, many corporations will be spending a large part of their software development budgets manage the Year 2000 problem and internally developed applications. Legacy applications that use only a two-digit year field in dates will need to have this field widened to four digits to distinguish the year 1900 from the year 2000. The Gartner Group estimates that 90 percent of all mainframe applications will fail by the millennium without corrective action. Although the Year 2000 problem may boost the demand for special software and mainframe testing tools, it may dampen the market for other software development products. Many organizations may need to defer other projects and the adoption of new technologies to devote greater resources and budgets to resolving the Year 2000 problem.

6.1.2 Computer Languages

To facilitate language portability, the American National Standards Institute (ANSI) and the International Organization for Standardization (ISO) have made efforts to standardize computer languages. In the U.S., the National Institute of Standards and Technology, or NIST (formerly called the National Bureau of Standards), provides validation suites based on similar standards for popular language compilers and interpreters and publishes lists of products that have passed the NIST validation process.

In the past, compilers for a particular computer language and platform (hardware and operating system) were sold as separate products. Today, language compilers are more commonly bundled with a suite of other programming development tools as part of an integrated development environment (IDE) for a particular hardware platform and operating system. Along with the primary language compiler, these suites usually include program editors, source code browsers, source code management tools, a GUI builder, debugging tools, execution profilers, and limited testing tools. The separate tools work with each other in a visual environment (32-bit Windows on the desktop or OSF/Motif on UNIX workstations, for example).

Examples of this approach are IBM's VisualAge (BASIC, C++, COBOL, Java, PL/1, and Smalltalk) on OS/2 and 32-bit Windows, Microsoft's Visual DevStudio (C++) on the 32-bit Windows desktop, and SunSoft's Visual Workshop products (Workshop C++ and Workshop Fortran, for instance) on UNIX. Development tool vendors continue to add new components to these IDEs to compete in the marketplace and add value. Often these enhancements are based on acquiring smaller tools vendors that offer separate products and then incorporating the products under the tool suite umbrella.

Procedural and Functional Languages – the Old Paradigm

Procedural languages often are termed third-generation languages (3GLs), with machine language and assembly language representing the first two generations of computer languages. Procedural and functional languages now represent the older paradigm, with object-oriented languages as the newer one. Although these languages can be used to handle general-purpose problems, each language was developed with features to handle a specific class of problem particularly well.

The two oldest languages in this category, Fortran and COBOL, were first introduced in the 1950s and continue to be used today. Once a language has come into widespread use, the user community pushes to have that language evolve to handle new classes of problems and adopt modern features found in newer computer languages.

Software applications also can have a long lifetime during which maintenance likely will be required to fix problems and make changes. The Year 2000 problem for applications written in COBOL is an example of a situation where the lifetime of the software was longer than its software developers envisioned.

The more popular procedural languages in use today include Fortran, COBOL, LISP, PL/1, RPG, BASIC, Pascal, C, Ada, and Modula-2. Table 6-1 lists the most common use of each of these languages, which are presented in the order they were introduced.

Table 6-1: Popular Older Programming Languages

Language	Year Introduced	Most Common Use
Fortran	1954	Science and engineering problems
COBOL	1959	Business data processing
LISP	1960	Artificial intelligence
RPG	1964	Business applications on AS/400
PL/1	1965	Older applications on mainframes
BASIC	1965	Originally a teaching language; Microsoft's Visual Basic used for client front ends on Windows
Pascal	1970	Originally a teaching language; Borland's Delphi used for database applications on Windows
C	1973	Systems programming, UNIX applications, and PC/Macintosh desktop applications
Ada	1974	Embedded systems and military projects
Modula-2	1979	System programming

■ FORmula TRANslation (Fortran)

Developed in 1954 by IBM, Fortran was the first high-level programming language and compiler. It was designed to handle mathematical problems, although it also was used for business and other applications. The language was standardized in 1966 (Fortran), 1978 (Fortran 77), and 1991 (Fortran 90). (See "Language Standardization" on page 454 for more information about Fortran.) The language is used primarily for science, engineering, and mathematical applications, particularly where computational time is a critical factor. Simulations and other complex scientific problems are its strong suit. Fortran 90 added many object-oriented features and modern constructs to the language. Extensions and dialects for High-Performance Fortran (HPF) also have been developed for use on supercomputers and parallel computer architectures.

■ COmmon Business Oriented Language (COBOL)

Introduced in 1959, COBOL is a high-level language that has been the primary language for developing business data processing applications, particularly on mainframes and minicomputers. By original intent, COBOL is a wordy language as a way to self-document programs. COBOL is widespread in mainframe environments where it is used with IBM's Customer Information Control System (CICS) programming environment for transaction processing applications. The language was standardized in 1968 (ANSI COBOL), 1974 (COBOL-74/68), and 1985 (COBOL-85). The continuing popularity of COBOL over other procedural languages for business applications results partly from its native support of decimal numbers (base 10), so that financial applications produce exact results. Other popular procedural languages have native support for numbers (integer and floating point) represented in a way that occasionally can introduce small errors in financial calculations since some decimal numbers cannot be exactly represented.

■ LISP (LISt Processing)

A high-level language developed in 1960, LISP generally has been used in artificial intelligence (AI) research and programming to handle complex notions (mathematical equations), manipulate large sets of data (lists), process text, and resolve rules. LISP was the primary language used by the MIT AI Lab for several decades. The language can be modified and extended by the programmer with new data types. Both compiled and interpreted versions of LISP are available. LISP also is popular as an embedded macro language for its strength in processing text (the famous GNU EMACS programming editor written by Richard Stallman of the Free Software Foundation is an example). A Common LISP dialect was well documented in books published by Addison-Wesley in 1984 and 1988. Work started on standardizing the Common LISP dialect in the late 1980s has yet to be completed.

■ RPG (Report Program Generator)

Introduced in 1964 by IBM, this language was designed for generating business reports. In 1970, RPG II included enhancements that made it a mainstay programming language for business applications on IBM's System/3x midrange computers. RPG III, which has added more programming structures, including database support, is widely used on the AS/400.

■ PL/1 (Programming Language 1)

A general-purpose programming language introduced in 1965 by IBM on the System/360 mainframe series, PL/1 was designed to combine the number-crunching features of Fortran with the business orientation of COBOL. PL/1 was slow to gain acceptance in its early years because the available PL/1 compilers were inefficient and unreliable. The language was first standardized in 1976 with a subset (Subset G PL/1) standardized in 1981. PL/1 is a large language still used to a limited extent in IBM mainframe environments. Despite size and complexity, PL/1 is available on the desktop and found limited use on UNIX systems.

■ BASIC (Beginners All-purpose Symbolic Instruction Code)

Originally developed in the mid 1960s at Dartmouth College, BASIC was designed as a simple teaching language for interactive use. Starting in the late 1970s and early 1980s, BASIC was popular on small computers and PCs because the language was small (able to run on limited hardware) and it made finding errors easy when run as an interpreted language. Compiled versions of BASIC that mitigate execution deficiencies associated with the interpreted aspect have been available since the early 1980s. Microsoft's Visual Basic, a customized version of BASIC for developing Windows applications, has been very popular and has extended the life of BASIC. Part of the attraction of Visual Basic is the easy-to-use visual GUI builder and other tools bundled with the Microsoft product. The modern dialects of BASIC have not been standardized and are entirely different than the original ANSI BASIC language standard.

■ Pascal

A highly structured programming language developed by Swiss professor Niklaus Wirth in the 1970s, Pascal was designed for teaching programming. Pascal traces its roots to ALGOL, an early computer language used by researchers. The Pascal language was standardized in 1983 (ANSI Pascal). Until recently, Pascal was popular in academia as a modern teaching tool; it was often the first computer language taught to students. Pascal also was popular on the PC as a result of Borland's Turbo Pascal and Object Pascal products. These Borland products are still the fastest language compilers (which translate program source to executable) on the PC desktop. Pascal has been displaced on the PC and other environments by C and C++. Borland's Delphi, which extends Object Pascal, continues to be popular on the PC desktop.

■ C

Developed at AT&T Bell Laboratories in 1973, C has traditionally been associated with the UNIX operating system, which was written primarily in C. This general-purpose language has constructs that closely parallel assembly language and modern computer hardware, particularly 32-bit microprocessors. C has been attractive because of its portability and efficiency for developing operating systems and applications that manipulate text, integers, and structured data. C has been the language of choice for developing applications (other than financial programs) for UNIX workstations and servers. C also is used as a cross-compiler to develop programs for some target hardware (a hand-held device, for example) on a desktop PC or UNIX workstation.

During the 1980s and early 1990s, C became the language of choice for developing commercial application software on the PC and Macintosh. C compilers are available for computers ranging from the largest IBM mainframes to the smallest embedded processors. As a result of C's popularity, C language compilers are available from multiple vendors on most hardware platforms and operating systems. The language was officially standardized in 1989 (ANSI C).

■ Ada

This is a general-purpose language developed in 1974 by the U.S. Department of Defense (DOD) along with the European Economic Community and other industrial organizations. Based on Pascal, Ada was designed as a modern language for large-scale and real-time systems that would promote reliability and simplify maintenance. Ada is a large, complex language with special features targeted toward embedded systems (military aircraft, missiles, and tanks, for example), process control, and logistic applications. Ada supports concurrent processing and constrains overloading data types, features not available with other procedural languages. Since the middle 1980s, the Ada language has been required for use on certain DOD software contracts. Because of the size and complexity of the language and the required DOD certification process, Ada was available from only a limited number of vendors. Outside of the defense establishment, the use of Ada has been limited. The language was standardized in 1980, 1983, and 1993 (Ada 9x).

■ Modula-2

A direct descendant of Pascal, Modula-2 was introduced in 1979 by Niklaus Wirth. Modula-2 is a general-purpose language that allows individual parts of a program to conceal information from other parts, preventing them from interfering with each other. The Modula-2 language allows the program to be divided into several independent modules. A module is a group of related procedures and data collectively known as objects. Each module supports specific public elements, constants, data types, variables, and functions. The remaining elements are private. Modular programming particularly facilitates group programming, allowing different parts of the program to be written separately. The

Modula-2 language can be considered a precursor of object-oriented programming. Modula-2 was designed primarily for writing software systems. Although Modula-2 has a following, the bulk of the developer tools market adopted C++ in the late 1980s and early 1990s.

Object-Oriented Languages – the New Paradigm

Object-oriented languages evolved from research in computer language dating back to the 1960s. SIMULA, a language developed by Norwegian researchers, is considered the parent of modern object-oriented languages. The attraction of object-oriented programming and object-oriented languages is that they allow the developer to combine program algorithms and data structures in objects to hide the internal operations and data representation. This ability allows software to be assembled as a collection of separate objects or components that interact with each other in known ways while hiding internal details. The capabilities of an object also can be inherited in the creation of new objects that extend its features by overriding specific behaviors. A class in object technology is a template for creating an object. (See Table 6-2.) Object technology also allows software development to be divided so that the most talented developers design the core objects and the classes that define these objects (usually the more critical task), while less experienced developers assemble the objects into specific programs.

Table 6-2: Object Technology Definitions

Term	Definition
Object	The primitive element in object-oriented programming. Objects are entities that encapsulate within themselves both the data describing the object and the instructions for operating on the data.
Message	A request sent to an object to change its state or return a value.
Method	The function or procedure that implements the response when a message is sent to an object. Methods determine how objects respond when they receive messages.
Class	The description of a set of objects that share common methods and general characteristics. A class is a template for creating an object.
Instance	The concrete representation of an object after it has been created.
Subclass	The refinement of a class into a more specialized class, sometimes referred to as "derived" or "child" class.
Inheritance	A mechanism for automatically sharing methods and data declarations among classes and subclasses. An new class can be defined such that it inherits characteristics from other classes including multiple differing classes (multiple inheritance). Inheritance allows programmers to program only what is different from previously defined classes.
Polymorphism	The ability of the same message to be interpreted differently by different objects. Polymorphism is implemented through the inheritance mechanism. When a parent object class is specialized into more specific subclasses, it is possible to send generalized messages to an object. Depending upon the type of the object, the behavior of the method may be different.

Given the complexity and size of modern software development, object technology allows large teams of developers to create software more efficiently and with fewer defects. Although there remains scepticism regarding the benefits of object-oriented programming, most new software development efforts have adopted object-oriented languages and object technology. The latest and most powerful development tools are now targeted at object technology. This avail-

Table 6-3: Popular Object-Oriented Programming Languages

Language	When Introduced	Comments
Smalltalk	1970s	A major strength of Smalltalk is its mature class library of reusable components.
C++	1980s	C++ is currently the dominant language used by developers, but it suffers from complexity and a long learning curve.
Objective-C	1980s	Objective-C is the language and development tool supplied on NeXTstep (and now Apple's Rhapsody), bundled with a powerful visual interface builder and a large collection of predefined classes.
Java	1990s	Java is a cross-platform language frequently used for Web programming that leaves out many weaknesses of C++ and is rapidly gaining popularity.

ability of new state-of-the-art tools supporting object-oriented languages makes the move to object technology more attractive. Popular object-oriented languages include Smalltalk, C++, Objective-C, and Java. (See Table 6-3.)

■ Smalltalk

This language was created by scientists and engineers at Xerox Palo Alto Research Center (PARC) in the 1970s. Smalltalk is considered a "pure" object-oriented language because all its facilities are centered on the construction of classes and the creation of objects. It was one of the first languages to support garbage collection – that is, memory management via clean-up of objects no longer in use. Smalltalk was developed at Xerox PARC at the same time as many technologies used in today's PC (a graphical screen displaying windows, and use of a pointing device) were being defined. A significant part of the value of the Smalltalk language is its class library, which is a repository of reusable code that has been refined and cultivated for more than two decades.

The language has two flavors: Smalltalk-80 and Smalltalk/V. The official standardization process for Smalltalk was begun in 1993. With the increasing interest in the use of object technology, Smalltalk also has experienced a renewed surge of interest. Today, there are two major suppliers of the Smalltalk language (IBM and ObjectShare) and a growing number of new companies supplying desktop versions on Windows and Macintosh platforms and supporting tools. In October 1993, IBM put its weight behind the language when it announced support for Smalltalk (IBM's Visual Age for Smalltalk) on OS/2, Windows, AIX, OS/400, and MVS. IBM now offers Visual Age for Smalltalk products on 32-bit Windows (Windows 95 and Windows NT) and on Sun and HP-UX platforms.

■ C++

Designed by Bjarne Stroustrup at AT&T Bell Laboratories in the early 1980s, C++ is a general-purpose language based on extending the C language with native support for objects, classes, and inheritance. Aside from minor details, C++ is a superset of the C language. The key concept in C++ is the class: a user-defined type. Classes offer many programming advantages, especially data hiding (explained later). C++ also includes a useful feature called typesafe linkage that forces separately compiled modules of a program to link together only if the parameters passed to functions are of the identical data type. Typesafe linkage can catch and prevent a certain class of programming errors during development that often would be discovered only during final testing and quality assurance.

The original AT&T version of C++ ran as a preprocessor that translated C++ source code to C source and then compiled the C source to an executable program. As C++ became more popular as a replacement for C, native C++ compilers that generate executable code directly were developed and marketed. Like C, C++ language compilers are ubiquitous and are available from multiple vendors on most platforms.

The C++ language has been going through a long and arduous standardization process over the last 7 years and should finally become standardized in 1998. During this time, C++ has been a moving target for developers and has grown to be a large, complex language. The size of the language and its complexity equate to a long learning curve for new developers. The C++ standard now includes the standard template library, a powerful library of generic container classes that can be used in building applications. Over time, C++ has become the dominant language used for software development.

■ Objective-C

Originally developed by Stepstone in the late 1980s, the Objective-C language was another attempt to extend the C language to support object technology. Objective-C can be considered a cross between C and Smalltalk-80. Objective-C adds class definition and message-passing syntax to ANSI C. The Stepstone implementation of Objective-C operates as a preprocessor that translates the Objective-C source language to standard ANSI C source files that must then be compiled. Like Smalltalk, Stepstone's Objective-C is delivered with a large library of reusable software components. Stepstone also offers a large, separate collection of classes that support the development of graphical interfaces.

In 1989, the Objective-C language was adopted by NeXT as the preferred language for object-oriented development in its NeXTstep operating system. The NeXT version of Objective-C uses a native compiler and is bundled with a powerful visual interface builder and a large collection of predefined classes called the Application Kit. In June 1997, Apple purchased NeXT and will use a revised version of the NeXTstep operating system and development tools called Rhapsody to replace the MacOS on Apple's Power Macintosh computers.

Stepstone currently supports Objective-C on most UNIX platforms, OS/2, MS-DOS, Windows, Windows 95, and Windows NT. The Objective-C language also is supported by the freely available GNU C/C++ compilers (from the Free Software Foundation) on UNIX workstations.

■ Java

Developed by James Gosling at Sun in the early 1990s, this programming language was originally designed for use in embedded consumer-electronic applications, particularly interactive TV set-top boxes. After several years of experience with the language and contributions from other Sun researchers, it was retargeted to the Internet and renamed Java.

As a general-purpose object-oriented programming language, it was designed to have as few implementation dependencies as possible. Java supports concurrency (multiple threads of execution) and is based completely on classes. Java supports a rich library of application programming interfaces (APIs) for network communications, graphics, and window creation. Java allows application developers to write a program once and then be able to run it on any platform incorporating the Java Virtual Machine (JVM). As an object-oriented language, Java was designed to be portable without all the pitfalls and complexity of C++. Like C++ and Objective C, the Java language builds on the popular syntax and control structures of C.

The strength of the Java language is that it leaves out many features of C and C++ that tend to introduce inadvertent errors into programs. In particular, Java eliminates pointers and the constructs in C and C++ for freeing memory by use of pointers; instead it substitutes automatic storage management. By most conservative estimates, errors caused by the improper use of pointers and memory allocation routines (memory leakage, heap, and stack corruption) account for the majority of errors found in most C and C++ programs.

The Java language also defines all its native data types to be identical in size no matter what platform the language resides on. This characteristic differs significantly from C and C++, in which the number of bits in the most basic integer data type ("int") is system-dependent. Java also defines all strings using Unicode, a two-byte encoding that allows easy internationalization (Unicode supports coding in most international languages).

Although the Java language has been strongly associated with Web development and Java applets that run inside a JVM, the Java language can be used as a general-purpose language. Most Java compilers also can create stand-alone applications. Java quickly has become popular in academia as the language adopted for use in teaching modern programming to beginning students.

Many older procedural languages have been extended in recent versions with support added for object-oriented programming, visual interface features, and rapid application development (RAD). With the increasing use of GUIs on client platforms, the ease of developing a visual interface or front end for client applications has become an important consideration in choosing development tools. Microsoft's Visual Basic extends the BASIC language on the desktop with support for objects and wizards to build visual front ends for Windows.

Borland's Delphi extends Pascal in similar ways with additional support for database access. Micro Focus now offers a development environment and class library for the development of GUI applications created with Object COBOL. As these tools and older languages evolve, the distinction between 3GL environments, fourth-generation languages (4GLs) or RAD environments, and object-oriented programming environments is blurring.

Specialized Languages

A number of specialized languages have been used over the years for software development, and recently several newer languages have become prominent. In particular, the use of scripting languages has become common in developing distributed applications for the Web. The advantage that an interpreted scripting language offers is a quick and simple way to prototype and test an application and even get an application up and running. A scripted language also is flexible because the program can be changed and modified easily.

Because most of these languages are interpreted at runtime, the speed of execution is limited compared with that of compiled applications. However, for many distributed applications, network communication or the speed of database access may be the limiting factors, so scripted applications in those cases do not affect performance negatively. This issue is particularly true for applications that communicate with clients over the Internet, where delays caused by routing packets over multiple network hops, network latency, and slow-speed client dial-up connections are variable and can become substantial.

Although a few of these languages operate on the client in the context of a Web browser (JavaScript and VBScript, for example), an equally common use of many of these languages is on the server side. When used on a server, a simple scripted application is often the glue layer between a Web server and other server resources (database servers, for instance). This glue layer often assembles a dynamic response to a client query (such as a custom HTML page) that will ultimately be sent to the client.

Specialized languages include xBase, Perl, JavaScript, and VBScript.

■ xBase

One of the oldest specialized languages is the dBase language and other xBase dialects derived from it. These languages traditionally have been used on the PC for creating custom database applications. Predating the IBM PC, the dBase language was first introduced on the CP/M operating system in about 1980 as part of a database application, also called dBase. The dBase program was later ported to MS-DOS and was purchased first by Ashton-Tate and later by Borland and incorporated into dBase offerings on the desktop. The dBase language proved so popular that several vendors developed competing products based on the same language. The most successful dBase emulators have been from Nantucket Software (Clipper) and from Fox Software (FoxPro), which was purchased by Microsoft. At the time of its acquisition by Microsoft, FoxPro ran on the PC, Macintosh, and SCO Xenix/UNIX.

The xBase language (a way of referring to all the dBase language variants) is a high-level language with modern flow control and I/O with special features that directly support database access and field manipulation in a database. dBase, FoxPro, Clipper, and other compatible products have been used for more than a decade on the desktop and still have a substantial following. Although most xBase dialects support embedded Structured Query Language (SQL) calls, xBase has been used primarily for multi-user databases that depend on sharing the database files for access.

As the software application market has moved to the client/server paradigm, xBase has been displaced by SQL relational database software. Large quantities of data still are stored in the dBase or FoxBase file format on the desktop and PC file servers. The xBase language has never been standardized officially, but until recently most of the versions available have been close enough to allow an application developer some measure of portability between vendors.

■ Perl

Perl is an interpreted language with sophisticated features for manipulating text. Perl provides a mix of power and flexibility that seems to be an excellent match for writing Web scripts. Unlike other approaches (using UNIX shell scripts or PC batch files, for example), Perl also offers an effective mechanism for minimizing the risk that someone will find a way to misuse the script to gain access or do damage to a system. A large archive of useful Perl scripts for Web and database access is available over the Internet.

Because of its features, Perl is currently the favorite scripting language used by Web developers. Perl was developed by Larry Wall on UNIX and is provided as freely available C source code and as ready-to-run interpreters on most UNIX platforms. Versions also are available for Windows 95, Windows NT, OS/2, and the Macintosh.

■ JavaScript

Developed independently by Netscape, JavaScript is a scripting language designed to be incorporated in the company's Web browsers. Originally called LiveScript, it was renamed (with Sun's permission) when Netscape saw how popular Java was becoming. JavaScript code is embedded directly in HTML pages using several extended HTML tags. When a browser downloads an HTML page containing JavaScript, the script is executed. JavaScript is associated with the HTML page in which it was embedded and has limited access to some of the browser's resources (the history list and the Forward and Back buttons, for example); this feature can be disabled by the user for security reasons.

JavaScript can read and set the contents of forms and manipulate windows. The syntax of this scripting language is similar to C's. JavaScript includes a large library of functions for math and text manipulation. One of the most common uses of JavaScript is to perform validation on data entered into forms. JavaScript is supported by Web browsers from Microsoft, Netscape, and Spyglass (Mosaic).

■ VBScript

Developed by Microsoft, VBScript is a scripting language based on a subset of the Microsoft Visual Basic for Applications (VBA) language. The VBScript interpreter is incorporated only into Microsoft's Web browsers and Microsoft's Web servers. Like JavaScript, this feature on the client provides access to Web browser resources and can be used for validating forms and creating new ones. Also like JavaScript, this browser feature can be disabled by the user for security reasons. On the server side, VBScript can be used to access databases using Open Database Connectivity (ODBC) drivers or Object Linking and Embedding (OLE) database providers. The main attraction of VBScript over JavaScript for client-side scripting would be to developers with prior Visual Basic expertise working on intranets where only Microsoft products will be used.

Fourth-Generation Languages (4GLs)

Original computer language development was focused on portability for migrating code and on improving programmer productivity. All commonly used languages (C, C++, COBOL, and Fortran, for example) can be considered somewhat general-purpose in nature. Although Fortran is suited to solving time-consuming engineering problems, it also has been used for commercial data processing.

In the late 1970s and early 1980s, 4GLs designed to make it easier and faster to create transaction-oriented business applications emerged. The term 4GL has no universally accepted meaning but has become associated with high-level languages and products designed to meet these goals.

Most products considered 4GLs share some common characteristics:

- Minimize low-level programming by providing a powerful, high-level language and toolkit

- Provide a fast way of coding a business transaction through event-oriented procedures
- Feature a data dictionary to store definitions of the various application components and encourage reuse

In general, the products considered 4GLs are all proprietary languages. Providing support for business transactions is the major focus. Some of these products and languages are targeted to application development on mainframes and minicomputers, and other products aim more at distributed client/server development using SQL database servers running on UNIX or legacy systems.

Several 4GLs are designed for use on mainframe and midrange systems using a character-oriented terminal interface. These products are marketed by database and other software vendors. Examples of products in this category include these:

- **Cognos' PowerHouse** – PowerHouse was one of the early 4GLs that has been used successfully in a variety of applications. It is a comprehensive, host-based development environment used primarily for developing character-based applications. PowerHouse is popular on HP's HP 3000 platform and Digital's VAX running VMS.
- **Compuware's UNIFACE Seven** – UNIFACE is positioned as a high-end 4GL product targeted at the enterprise-level developer. It can be used to create large, complex applications that are very scalable. UNIFACE supports a broad range of development and deployment environments, making its applications very portable. Architectures supported include host terminal interfaces, two-tier client/server, multi-tier client/server, and Web-enabled interfaces.

  UNIFACE is a graphical, object-based environment that uses data modeling to define data structures, objects, and business rules stored in a repository. Development is based on using components composed of modules that share information from a common repository of application objects, some provided by the vendor. The architecture of the product is based on the same independent layers found in three-tier computing: the application layer, the data management layer, and the presentation layer. Consequently, changes can be made to the underlying database system or presentation interfaces without affecting the application.

  UNIFACE can be integrated with more than 20 CASE tools and works with more than 30 database systems. UNIFACE can generate code for the Macintosh, Windows 3.x, Windows95, Windows NT, OS/2, and OSF/Motif on UNIX as well as character-based interfaces.
- **Information Builders' Focus Release 7.0** – Focus is an application development toolset for creating database applications. This tool is capable of developing complex applications that are portable among a wide range of hardware platforms and relational and non-relational databases. Applications developed with Focus support major databases from Adabas, Computer Associates, IBM, Informix, Oracle, and Sybase.

  With the rise of the client/server model and distributed computing, the focus of new application develop for data processing is now targeted to this architecture. Consequently, most 4GL products originally created for host systems have been enhanced to support client/server development. As well, many products designed solely for developing client/server applications are available from other vendors. Focus also provides reporting tools that allow customers to develop custom reports easily.

- **IBM's VisualAge Generator** – Generator, formerly know as Cross Systems Product (CSP), is a unique tool in that the coding is done in a 4GL, but the output is a 3GL program. Currently, Generator supports C++, COBOL, and Java output.

Products for client/server development have been categorized as first-generation or second-generation tools. The features associated with second-generation tools include the following:

- The ability to create and deploy application logic on the application or database server as well as on the client
- The ability to partition the application logic easily among client and server nodes and often the ability to change the partition through simple recompilation or runtime actions without redesigning the application
- A scalable, RAD methodology
- A development environment that supports automatic code generation and code maintenance
- An object-based or object-oriented architecture
- An architecture supporting component development from class libraries

With the growing interest in using Web browsers as a portable universal client for business applications, many client/server tool vendors have added support for Web-enabled application development. The basic development approach remains similar, with the developer designing GUI screens and generating code. The primary difference is that client business logic now happens behind the Web server, typically in the form of a database server gateway. A few of the more advanced Web-enabled tools support generating Java applets for client code.

Many prominent 4GL products targeted at client/server development are in the marketplace today, including Cactus, Centura, Developer/2000, Forté Application Environment, and Powersoft.

- **Centura Software's Centura** – A graphical application development environment, Centura creates Windows applications that use SQL to access server-based databases, including a broad range of data sources and systems. Centura Team Developer provides connectivity to relational databases via native interfaces and includes both Windows and Web development components. This package is targeted for programmers building both client/server and Web applications. Centura Web Developer is an entry-level package for programmers building only Web applications. It comprises the subset of Team Developer components required to build a Web application that connects to SQL databases. The Centura toolsets are based on graphical environments that include an application debugger.

- **Forté Software's Forté Application Environment** – First introduced in 1994, Forté Application Environment is considered an advanced second-generation client/server development tool. It consists of an integrated set of object-oriented tools for building, deploying, and managing distributed applications. Its focus is on using visual and object-oriented programming facilities to build reusable business objects for distributed applications. The Forté tools emphasize flexible application partitioning and platform independence. Deployment decisions can be made at runtime without making changes to the application logic. The toolset includes an object-oriented 4GL, an application generator, a GUI designer, an HTML page generator (for use with Web-enabled applications), a comprehensive set of class libraries, and a repository to support team development. The product is tightly integrated with the leading relational database management systems and can interface with external systems such as legacy applications and electronic data feeds.

- **Information Builders' Cactus** – A platform-independent, database-independent tool suite designed for building two-tier, three-tier, and Web-enabled applications, Cactus builds highly reusable components for distributed applications through a visual object-based development environment. The Cactus toolset includes Workbench, Application Manager (object repository), Partitioning Manager, Object Browser, and File Painter to build database objects. The powerful Partitioning Manager allows developers to drag locally developed procedures and drop them on icons representing servers anywhere in the enterprise – irrespective of platforms and communication protocols involved – where they can execute. After an application has been logically defined, it can be split into separate partitions that execute on different computers in the network. As requirements change and evolve, the application can later be re-partitioned for other environments without changing the logical definition of the application. Cactus includes support for Microsoft ActiveX controls and bundles an OLE custom control (Cactus OCX) that allows any Cactus procedure to be called by an OLE-enabled third-party application.

- **Oracle's Developer/2000** – A set of development tools tightly integrated with the Oracle relational database management system, Oracle Developer/2000 is a unified client/server development environment. The product features support for Microsoft ActiveX controls, a common repository, flexible modeling and methodology support, and a portable open architecture for adding other tools. When combined with Oracle's Designer/2000 CASE tool, Developer/2000 is one of the industry's most complete client/server design and development environments.

- **Sterling Software's Composer** – Formerly Dynasty Development Environment and Texas Instruments' Information Engineering Facility, Composer is a three-tier, object-oriented toolset based on a repository and oriented to large, enterprise-sized programming projects.

- **Sybase's Powersoft Product Family** – Designed for building business applications for client/server environments and the Internet, PowerBuilder provides a RAD 4GL supporting object-oriented features, compiled code, multi-tier partitioning, and a presentation layer. Powersoft's Internet Developer Toolkit adds components, libraries, and tools for building Web applications and Web-enabling existing PowerBuilder applications. PowerJ is a RAD Java tool for developing Java applications, applets, and JavaBeans components. Power++ is a RAD C++ tool used for creating component-based applications in C++. PowerDesigner provides a comprehensive database design and modeling toolset with integrated modules that provide data discovery, database design, and construction; physical data modeling and application object and data-aware component generation; information sharing and model management for development teams; data warehouse design and implementation; and a viewer for read-only, graphical access to model information.

  Although these tools can be used with SQL databases other than versions of Sybase's SQL Server and Powersoft's SQL Anywhere, the application may be required to connect through a Sybase gateway connector product. PowerBuilder applications can be developed and deployed on Windows, Macintosh, and Sun Solaris platforms.

In addition to the foregoing tools, several products are available that support only the client portion of client/server application development. Tools in this category (Borland's Delphi, Borland's C++ Builder, and Microsoft's Visual Basic, for example) evolved from desktop tools designed for building visual interfaces and modest desktop applications quickly. In the past, these desktop products were used with shared-file databases such as Access, dBase, Foxpro, and Paradox where the database files resided on file servers or the local machine. Today, these products also can build applications that make SQL calls over the network to a database server by use of an ODBC driver or database vendor-specific APIs.

In 1997, Microsoft released beta versions of its OLE Database (OLE DB) software development kit (SDK). OLE DB represents Microsoft's new architecture using OLE and its underlying Component Object Model (COM) for providing access to a variety of data sources from Windows applications. The OLE DB SDK also includes support for Active Data Objects (ADO), a standard set of objects that allow data access from Visual Basic applications. OLE DB and ADO represent a new set of APIs for database access designed to replace ODBC.

Although these PC-centric products can be used for the client portion of client/server development, they lack many of the features provided by the more advanced client/server development tools. In particular, choosing a different underlying database mechanism (changing from ODBC to OLE Database or ADO, for example) or re-partitioning the logic between client and server can be a formidable task with these products. This difficulty arises because the application logic is generally part of the client piece or associated with the SQL server (if it supports triggers and stored procedures). Either way, these products can deliver only two-tier applications, and ultimately, the application logic is mingled with the client code or the server code.

6.1.3 Development Process Support Tools

As software becomes larger and more complex, the size of the team needed to complete the development process increases. On large-scale projects, this team will often consist of application designers, software developers, graphical interface designers, product testers, technical writers, lab-and-build engineers, product support staff, and management. On smaller projects, several of these tasks often are handled by the same team member. When the development team grows past the single programmer, the advantages offered by development tools that facilitate teamwork and collaboration quickly become apparent. With larger teams, tools supporting teamwork and collaboration are a necessity.

Teamwork and management tools used for software development projects include those for analysis, modeling, and design; software configuration; system building; automated software testing and quality assurance; software project management; and specialized reusable class libraries and code repositories.

Analysis, Modeling, and Design

These tools assist in generating requirements for applications, including data definitions, business rules, and programming specifications. Analysis, modeling, and design tools typically use formalized methods and approaches derived from research in computer science and computer assisted software engineering (CASE). These methodologies include functional decomposition, entity relationship modeling, data flow diagramming, event or control flow diagramming, state-transition analysis, and screen prototyping. (Table 6-4 describes how some of these methodologies work.) These tools are used in the analysis and design phases to create models of different levels of detail and depth. Modeling and design data are defined centrally and stored in a shared repository. A good model can be understood and critiqued by application experts who are not programmers.

The primary distinction between older CASE products and newer or updated modeling and design tools is the latter's less-rigid structure, greater process flexibility, and the ability to integrate with other development tools. The most basic feature of these tools is support for diagrams and drawings typically using customizable icons for notations and symbols.

Lightweight modeling tools for database applications have become more prominent with the increase of client/server programming. Many of these tools now directly support object-oriented systems with models and diagrams useful in designing and documenting the object classes and inheritance to be used in a software project. Popular lightweight database modeling tools used for client/server applications include products such as Logic Works' Erwin and Sybase's Powersoft PowerDesigner (formerly S-Designor).

Advanced features considered desirable in these tools are the ability to export the requirements to a skeleton of the source code (by use of code generators) and to import or update the data model from changes made in the actual source code (reverse engineering). At the high end are tools such as Sterling Software's Composer, capable of modeling mainframe and client/server applications.

Table 6-4: Analysis and Design Models and Diagram Types

Model/Diagram Type	Principal Use
Data Processing/ data flow diagrams	Data flow diagrams depict how data is processed at different phases in the system.
Composition/ entity-relation diagrams	Entity-relation diagrams illustrate how some entities are composed of other entities.
Classification/ inheritance diagrams	Object class and inheritance diagrams are used to illustrate how objects have common characteristics.
Stimulus-response/ state-transition diagrams	State-transition diagrams describe how the system reacts to internal and external events.
Process/ process models	Process models depict the primary activities and deliverables involved in completing a process.
Data dictionary	A data dictionary is a catalog containing names (variables, functions, classes, and objects, for example) used by the system with a description of the named entity, the date it was created, the creator, and the representation.

Developed by three prominent developers of object-oriented analysis and design methods (Grady Booch, Ivar Jacobson, and James Rumbaugh, all now with Rational Software), the Unified Modeling Language (UML) is an attempt to standardize a modeling language for design. A modeling language represents the graphical notation and syntax that various methods use to express and document designs. UML does not represent or restrict the methods that can be used for analysis and design. UML Version 1.1 was forwarded to the Object Management Group (OMG) in late 1997 for adoption as an OMG standard. The hope is that a UML standard will benefit developers and eventually allow tools to be developed that provide some level of portability between products from different vendors.

Software Configuration Management

These tools are used by application development teams to provide software revision control, source code versioning, and release management capabilities. Other more advanced features include process management and the ability to track requests for changes and bugs discovered during testing. Development team members are required to check out models, specifications, source code modules, documentation, and other files from a central repository to make changes and then check these files in after changes have been made. The software configuration management tool keeps track of revisions, different versions, and releases and can back out of changes when necessary.

These tools have existed for UNIX and mainframe environments for some time. On UNIX, the classic software configuration management tools have been source code control system (SCCS) developed by AT&T as part of UNIX and the later publicly available revision control system (RCS).

A freely available tool called Concurrent Version System (CVS) extends the capabilities of RCS by creating a software repository rather than operating on individual source-code files and directories. CVS maintains not only the source code for an application, but also the full data related to a given project.

Tools such as these can be categorized by the target market: application developers or technical programmers. Examples of tools targeted more to the former include Computer Associates' Endeavor, Intersolv's PVCS, and Platinum Technology's CCC/Harvest. Products targeted more to the technical community include the RCS/CVS tools available with UNIX, Rational Software's Pure Atria ClearCase, and IBM's TeamConnection. The last two products go a step further than other tools and supports archiving the complete build environment, including the development tools (the language compiler and library files, for example) used in the build process.

Some tools in this category also focus on process control as well as the traditional version control functions. These products offer features for organizing and controlling the processes and procedures used in software development and deployment. (See Table 6-5.) Vendors with a process control focus include Continuus, Platinum Technology, and SQL Software.

Table 6-5: Software Configuration Management Capabilities

Capability	Description
Version identification	Versions and releases can be automatically assigned identifiers when they are added to the system. Some tools support the assignment of attribute values for identification.
Change or version control	Versions of components must be explicitly checked out in order to make changes. The team member making the change is automatically recorded. When the changes are checked in, a new version is created and identified with a tag of team members. The old version is never destroyed or overwritten.
Storage management	Version management tools provide features to reduce the storage space required by the different versions. Most tools try to minimize storage by describing differences from some master version.
Build environment	To replicate a build, versions of development tools and ancillary files (compiler "include" files, for example) that were used in the process also must be managed and archived under change control.
History recording	All changes made to a module or system are recorded and listed.

System Building

Actually running the compilers and other code-generation tools to build any sizable application is a computationally intensive process that can take several hours or even days. Hundreds or even thousands of separate files may be involved, and the possibility of human error is substantial. Modern program development uses tools that automate the build process of an application using the latest source code files that have been approved for check-in. Most of these build tools are based on the "make" utility originally developed on UNIX. The purpose of these tools is to perform the minimum amount of work needed to build a new version of the application (for example, there is no need to recompile a module that has not changed since the last compilation). Integrated development environments generally include a simple "make" facility bundled with the product. System-building tools also are bundled with some software configuration management products. These tools typically use the date and time stamp on source files to determine which modules have been changed and need to be recompiled or rebuilt. These tools also allow a build engineer to start a build process that can be left to run unattended, creating an error log that can be reviewed later for problems.

System-building tools require a structure that documents the dependency between various components, a list of components or modules that are required for a specific build, and information about where these files are stored. Some build tools can create the dependency structure automatically from the list of files. Many modern compilers support an option to generate this file dependency list while parsing the source code. Other tools such as "makdepend," which is bundled with the X Window System on UNIX, can identify these dependencies. Without these tools, this dependency structure must be created manually.

The most recent advance available in some of these tools is a parallel and distributed "make" feature. Parallel "make" takes advantage of multiprocessor machines, spreading the build process across multiple processorss in the same machine by executing parallel compiles of different source files. Distributed "make" spreads the work across multiple computers in a distributed environment. The goal of a parallel or distribute make feature is to shorten the time needed to rebuild the application. SunSoft's Workshop integrated environments are bundled with a parallel make facility. The freely available GNU make utility found on UNIX also supports parallel make.

The last step in building an application is the linking phase. In this step, the numerous modules that constitute an application are brought together and linked into one whole. The link step can be slow and resource-intensive, especially on large projects. The most advanced linkers implement incremental linking, in which only the modified modules are re-linked, rather than relinking all parts of the executable program from scratch every time a single module is changed. This process results in significantly shorter link steps.

On UNIX systems, incremental linking is included in SunSoft's Workshop Products, HP Softbench, and GNU C++. Rational Software's PureLink does the same as a stand-alone product for UNIX platforms. Visual C++/Visual Studio, Watcom C++, and Borland C++ Builder support this capability on the PC. With some products, such as Visual C++, incremental linking is available only when creating debugged builds.

Automated Software Testing and Quality Assurance

These tools represent a wide range of processes and technologies used to ensure that software does not contain bugs. The traditional tools in this category are static analyzers, debuggers, and execution profilers. (See Table 6-6 on page 448.) Static analysis tools, such as the "lint" utility used with C files, are designed to check the source code for potential programming errors. Modern static analysis includes sophisticated checks on control flow, data use, function interfaces and parameters, information flow, and paths of execution.

Another form of static analysis with a different purpose is represented by software metrics. In this case, the tool parses the source code and assigns values based on complexity and other measurements to lines of code, functions, and components to indicate where the more complex, less understandable, and statistically more error-prone programming occurs. Similar metrics for complexity and readability can be applied to documentation files. Based on these metrics, it is possible to focus testing in the areas that have been identified as more error-prone. Tools that determine software metrics include McCabe and Associate's Visual Quality Toolset, which is available on UNIX and Windows and can analyze code in Ada, C, C++, COBOL, Fortran, Pascal, PL/1, Visual Basic, and several other languages.

The newest form of static analysis is represented by tools that parse and check source code against rules that represent poor programming practices in a particular language. If the rule set can be modified by the developer, these tools also can enforce local software coding standards. An example of this category of tool is Parasoft's CodeWizard for C++. This tool checks for rules of poor programming in C++ based originally on the books *Effective C++* and *More Effective C++*.

Profiling tools, bundled with many integrated development environments, report how much processor time is used in executing each line of program code in order to pinpoint areas that should be targeted for optimization and performance improvement efforts. UNIX systems include the "gprof" command line utility for this task. The newer profiling tools incorporate an easy-to-use graphical front end for controlling profiling and reviewing the results. Rational Software's Pure Visual Quantify available on UNIX and TracePoint Technology's HiProf available on Windows 95/Windows NT represent products with graphical interfaces. Similarly, McCabe and Associate's Visual Quality ToolSet uses a graphical interface to present the results of its software metrics analysis.

Software defect testing is usually handled using tools that automate running the application through predefined test suites. Automated defect testing can be applied to unit testing of individual components, module testing of a collection of dependent components, and system testing of the entire application. Automated testing tools generally consist of a test suite manager (often called a "harness") and some mechanism to record and play back tests. Tools targeting client/server applications that automate testing for defects in the GUI and the interaction between the client and server are available. A new category of load and stress testing also can be applied to client/server and Web-based applications to ensure reliability, robustness, and responsiveness. Major vendors with GUI and distributed client/server testing tools are Mercury Interactive, Rational Software (having acquired SQA), and Segue Software. Several other tool vendors also have OEM agreements with companies selling automatic testing software. For example, Intersolv and Parasoft have agreements to resell Segue Software's QualityWorks. Tool vendors that offer automated testing on mainframe and legacy applications include Compuware, Computer Associates, Cyrano (formerly IMM and Performance Software), and IBM. Vendors offering load testing tools include Cyrano and Rational Software.

One of the most popular kinds of products now used for quality assurance is runtime error detection for C and C++. These tools check for memory management problems (memory leaks) and parameter or range errors that occur when calling operating system or library functions. These tools interface with the application and the operating system or libraries to track memory management calls and other system function calls to catch parameter errors dynamically at runtime. Leading runtime error detection tools include NuMega Technologies' Bounds Checker for Windows 95/Windows NT, Parasoft's Insure++, and Rational Software's Pure Atria. One clear advantage of programming in the Java language is that these programming errors cannot occur, although the runtime features needed to confer this benefit impose a performance penalty each time the program is run. Other products that have runtime error checking built in are the development environments from Centerline and SunSoft's Workshop tools.

A code coverage tool measures what portion of the source code for an application actually is executed during automated testing. The goal is to test as high a fraction of the code base as possible. Runtime error detection and code coverage often are included as background processes while automated defect testing is running. Visual tools in this category include Parasoft's TCA (bundled with Insure++) and Rational Software's Pure Coverage on UNIX and TracePoint Technology's Visual Coverage available on Windows 95/Windows NT.

Table 6-6: Software Testing and Quality Assurance Tools

Approach	Description
Static analysis	Analyzes the source code to locate potential defects in programming.
Defect testing	Automates finding areas where the program does not conform to its specifications, usually based on a test manager application using specially created application test suites.
Runtime error detection	Monitors and locates certain classes of runtime errors, particularly memory leaks.
GUI testing	Tests the GUI component of an application by simulating user keystrokes and mouse movements.
Code coverage	Determines how much source code in an application is covered by automated tests and defined test suites.
Load/performance	Stresses the application to determine its load/performance characteristics.
Profilers	Determines how much time is spent during execution in various functions and components to locate the heavily used sections that could be limiting performance.
Software metrics	Analyze the source code or documentation and generate metrics based on complexity.
Quality assurance	An approach usually based on the ISO 9000 standard concerned with defining how an organization intends to achieve quality in its products, including defect tracking.

Software Project Management

Project management involves planning and scheduling software development projects. Managers supervise and monitor progress to ensure that the work complies with requirements and that development is completed on schedule and within budget. The scheduling of staff and resources becomes an important management consideration for larger jobs. The time necessary to complete development on a large software project may be several years. Achievable deliverables and milestones need to be determined and set. Hardware and software resources need to be purchased and delivered on a timely basis to complete development. Given the dependencies and the estimated time needed for specific activities, an activity chart can be generated that shows which activities can be carried out in parallel and which tasks must be executed in sequence.

Several packages can be used to manage staff and resources and to produce time lines, critical-path diagrams, and Project Evaluation and Review Technique (PERT) charts. Another important component of project management is tracking changes in user requirements over the lifetime of a software project.

Reusable Class Libraries

For years, language compilers have included runtime libraries to support functions and operations that are required by the language standards but not directly part of the programming language. For example, Fortran compilers come with a large library of mathematical and other functions that are defined by the Fortran 90 standard but must be implemented outside the language compiler. One enduring strength of C has been the definition of the standard C library, which includes a large collection of portable functions for I/O, character and string handling, mathematics, and other purposes. The draft C++ language standard defines a similarly rich standard C++ library, the cornerstone of which is the standard template library (STL) of generic container classes.

With the migration to object-oriented programming, reusable class libraries have become an important building block for application developers. Depending on the object-oriented language to be used, libraries of reusable classes or components may be available. One key benefit of an object-oriented language (Smalltalk, for example) is its class library, a repository of reusable code that has been refined and cultivated over time. For most of the popular object-oriented languages, libraries of reusable classes and components are available on the Internet.

Several companies now offer mature and tested class libraries designed for a particular set of tasks. This option allows development teams to buy reusable class libraries as an alternative to building the classes themselves. Rogue Wave Software sells a large collection of popular C++ class libraries that are available across platforms (MS-DOS, Windows, Macintosh, OS/2, and most versions of UNIX) and that can be purchased in both source code and object form. These class libraries include basic tools, the STL, mathematics ("linpack" and "lapack"), and database access, networking, and financial functions. Rogue Wave Software's Tools.h++ class library also is bundled with many of the major C++ compilers and development environments. Many C++ class libraries have been rewritten as Java class libraries and are now becoming available. Rational Software also offers C++ and Java class libraries targeted for specific applications. Many vendors now offer software components designed to work with Visual Basic, C++, Borland's Delphi, and Java. The leading vendor of Java components is KL Group.

When purchasing precompiled libraries of functions for C and C++ on the PC, care must be taken to ensure that the object libraries have been built for the specific compiler (and sometimes even the same version of the compiler) that will be used for development. Incompatibilities can result from differences in calling conventions, initialization and startup procedures, internal library function names, and object file formats. This potential problem is less of an issue but still a possibility when purchasing prebuilt Dynamic Link Libraries (DLLs). UNIX systems rarely have these incompatibilities because system functions (the Single UNIX Specification) and file formats are standardized on each platform. These concerns can be eliminated by purchasing the source code to any libraries along with the prebuilt components. Most vendors offer source code to their libraries at additional cost or in some cases include it with the precompiled libraries.

Code Repositories

Large organizations that develop software accumulate millions of lines of code, some of which can be reused on other projects. The difficulty under present conditions is that no systematic way exists to classify and archive code to facilitate reuse. The reuse that does occur is usually limited and done on an ad hoc basis. Software developers may recall some previous work and search through older code they have written for some function name or keyword in order to find the appropriate code to copy and reuse.

Code repositories offer the hope of greater code reuse. The primary problem that a development organization faces is the work involved in creating a useful repository. With millions and millions of lines of existing code, a large organization might have to dedicate 30 or 40 developers for several years, charged with the task of organizing and classifying functions or objects from old code and archiving them in the code repository. Rather than undertake this Herculean effort, organizations often choose to use the repository only for new development. For code repositories to work effectively, software development teams need to be in agreement on processes, classification procedures, and other criteria. Several vendors now offer software tools that facilitate setting up and maintaining code repositories, but these are only as good as the procedures actually employed by software developers. Vendors offering software for creating and maintaining code repositories include Intersolv, Platinum, Rational Software, and Transtar. Some high-end development tools (Sterling Software's Composer, for example) are adding code repository capabilities in their latest releases. This market is still in its infancy.

6.1.4 Year 2000 Analysis and Conversion Tools

As the new millennium approaches, IT departments are struggling to ensure that their applications will still function correctly in the year 2000. Analysts suggest that about 70 percent of all business information still resides on mainframe systems. IBM estimates that there are more than 150 billion lines of code in COBOL alone, representing an investment of more than $5 trillion in software still processing data on mainframes. The Year 2000 problem stems from business applications in which the lifetime of the software was longer than its software developers envisioned. The problem lies in the way dates are represented in the applications companies built during the past 30 years and still rely on to conduct day-to-day business. Most of these applications were written in COBOL, but this problem can exist in business software written in any computer language.

Storage space was relatively expensive in the early days of computing, so programmers often went to great lengths to reduce the data storage required by their programs. Consequently, many business applications were written to use only two digits instead of four for the year portion of date fields, with the first two digits assumed to be 19.

However, this scenario only explains part of the origin of the Year 2000 problem. In many cases, the older mainframe operating systems and library functions supplied with language compilers (COBOL, for example) returned the system date as a 6-byte or 8-byte string (MMDDYY or MM-DD-YY) using only two digits for the year. Therefore, it was natural that programmers would adopt a similar restricted format for storing dates. In addition, the library routines that convert dates to days of the week were not written properly to handle the year 2000, which is a special case and will be a leap year. (Normally, century years are not leap, except when they are evenly divisible by 400.) The libraries supplied with some COBOL compilers available on UNIX (HP/3000 COBOL II, for example) have similar problems because these products attempted to support and emulate the IBM COBOL compiler for compatibility.

Similar Year 2000 problems exist for some desktop applications, although the primary focus by industry has been on business data processing applications for mainframes and minicomputers. The clock chip used in PCs has a two-digit (1-byte) year field and a separate century field. However, the ROM BIOS chips in many older PCs do not update the century field when the year field rolls over from "99" to "00." In fact, the ROM BIOS in these machines considers a "00" year date as invalid and sets the date back to 1980. These problems are remedied easily with some software modifications, although this could affect millions and millions of older PCs. An informal survey performed by the Gartner Group indicates this could be an issue for any PCs purchased from major vendors before 1997. The more difficult Year 2000 problems on the desktop are older decision-support and database applications that, like their mainframe counterparts, use two-digit year fields.

Today, programmers face the monumental task of finding these occurrences and converting these two-digit year representations to four digits so that dates ending in "00" or "01" are not interpreted as 1900 and 1901. This conversion effort involves changes to the data files, databases, on-screen forms, reports, and software, as well as a huge coordination task.

A problem of similar magnitude will exist with the European Monetary Union when a uniform currency is adopted: All fields used to store currency data in business applications will need to be changed and the applications will need to be upgraded to handle both the local currency and the new currency, the Euro. Ultimately, these systems will migrate back to only one currency: the Euro.

A variety of tools are available to help with various facets of the date conversion process. In fact, the number of vendors offering some form of Year 2000 software has exploded during 1997 to reach into the hundreds. Unfortunately for IT professionals and developers, product reviews and information comparing Year 2000 tools in action are almost nonexistent.

The Year 2000 tools available for mainframe software can be separated into several categories: needs assessment, impact analysis, source code conversion, date simulators, and testing tools. To assess problems quickly, date-simulation tools let developers manipulate the apparent computer clock to the year 2000 or later and then run key applications to see how they respond. This is accomplished by trapping all requests for dates and passing the date value according to user-defined criteria (jobs or specific users, for example). These tools are used on MVS where the TIME macro (SVC 11) can be intercepted by vendor software to perform this date manipulation. These tools also support various clock formats. Obviously, this technique cannot be used on mainframes running production business applications using real data. Tools in this category include Advanced Software Product Group's Date/2000, Compuware's Xchange & Xpediter, Isogon's TicToc, ManWare's HourGlass 2000, Prince Software's Simulate 2000, and Viasoft's ValidDate.

The needs assessment usually takes the form of a cost/benefit analysis to assess the extent to which existing systems are susceptible to the Year 2000 problem and to estimate how much it will cost to keep them functioning properly. This process entails inventorying all major components of existing applications and using this information (lines of code and numbers of data files, for example) in an estimating model to determine the resources required to update these software systems. Adpac's SystemVision Year 2000, Computer Associates' CA-Impact/2000, Isogon's SoftAudit/2000, Quintic Systems' Century Source Conversion, and Viasoft's Estimate 2000 can help automate this process.

Impact analysis is the most time-consuming aspect of the code conversion process. Of the available tools, most fall into two basic categories. One technique involves searching source code for text strings that stimulate date or year computations to find suspect fields. The second approach starts with a similar search phase but adds sophisticated parsing tools that trace data flow to identify places in the code affected by date field occurrences. These parsing tools understand the relationships between the various program components and how to identify date fields based on computations, movements, and comparisons of known date variables.

In general, simple search tools are appropriate for small-scale projects (50,000 lines of code or less), but they are not deemed very reliable. Consultants estimate that using a simple search tool will find, with luck, 60 percent of the date-related strings. The better search tools support an iterative approach that allows adding a list of words to be excluded for the string search. For UNIX developers, these products may not offer much over basic utilities such as "grep" for searching text.

Data flow analysis tools, on the other hand, are more expensive and require a large investment in infrastructure. However, once the initial date-related variables are found, developers are reasonably assured of finding other variables that depend on these date fields. For large scale projects, professional consultants usually recommend data flow analysis. These programs often are reengineering tools that have been enhanced to check for Year 2000 date inconsistencies. Products in this class for use with COBOL on mainframe systems include Computer Associates' CA-Impact/2000, Intersolv's Year 2000 Wizard, Micro Focus's Challenge 2000 (an add-on to Revolve), Millennium Dynamics VANTAGE YR2000, Prince Software's Survey 2000, Quintic Systems' Century Code Conversion, and Viasoft's VIA/ValidDate. Viasoft's ESW2000 is an integrated set of tools based on its Existing Software Workbench with additions to handle Year 2000 assessment, analysis, conversion, and testing. The Viasoft product supports five different COBOL dialects and standards, three versions of IBM assembly language, PL/1, and IBM Fortran.

Several tools are available on UNIX that can analyze software written in Ada, C, C++, Fortran, and several COBOL dialects. These products include McCabe and Associates' Visual 2000 Environment and LRDA's Testbed (re-marketed in the U.S. by Eastern Systems). Software Emancipation sells Discover Y2K, an integrated C/C++ development tool enhanced to locate date-related items in the code, for use with UNIX systems.

Once the applications have been inventoried and analyzed, the next phase is making the necessary changes to the source code. Generally, this phase entails expanding the date fields and date variables to accept date values with a four-digit year. Once the date field has been modified, the source code must be recompiled, data files need to be redesigned and converted, reports often need reformatting, and some screens may need to be rearranged. IBM has stated that it will not make its older COBOL compilers (IBM COBOL-II, for example) and assemblers Year 2000-compliant. Therefore, to fix problems with library date routines, companies may have to rebuild applications with newer compilers and development libraries that are Year 2000-compliant. At some sites, this may require purchasing and installing new COBOL or other language compilers and development tools.

Testing is the most expensive part of the conversion process. The Gartner Group estimates that approximately 50 percent of a Year 2000 project will be devoted to testing. The scope of the testing effort is expected to be much larger than normal because 40 to 50 percent of the functions in the application likely will be affected. In addition, the extent of date testing will have to be much more than normal due to the high cost of making an error.

Automated software testing tools can be used to facilitate testing and quality assurance. Some of the Year 2000 software integrated toolsets (Viasoft's ESW2000, for example) include an automatic testing component that can be used in conjunction with a date simulator. Any automatic software testing tool appropriate for the target platform can be used for this phase. The challenge becomes creating the appropriate test cases, both for dates before and after the year 2000.

Although the Year 2000 problem has been discussed and anticipated for more than two decades, many IT organizations kept waiting for technology to provide a silver bullet. However, the quick fix never came. Applications running on standard UNIX systems have a similar destiny with date rollover in 2038, the limit of the signed integer used to store the system date, which is measured in seconds elapsed since January 1, 1970. On the Macintosh, date rollover occurs in 2040.

6.1.5 The Euro Problem

The European Union (EU) long has envisioned adopting a single European currency. The current plans for the European and Monetary Union (EMU) and the adoption of the Euro are still somewhat uncertain. If EMU takes place, Gartner estimates the cost of corporate software changes will exceed $100 billion.

Those countries that are to participate in the EMU will be decided in May 1998 and are most likely to include Austria, Belgium, Finland, France, Germany, Ireland, Luxembourg, Netherlands, Portugal, and Spain. Starting on January 1, 1999, the Euro will become legal tender in these countries. The current plan is to fix the exchange rates between the EMU member countries on this date. National currencies will continue to circulate until they are replaced by Euro coins and bank notes by the middle of 2002.

The adoption of the Euro will have a substantial affect on software used to process financial information for each of the participating national currencies. During the period from 1999 through 2002, affected software applications may need to support parallel transactions in both older national currencies and the Euro. Aside from its effect on data input and output reports, the adoption of the Euro imposes a set of rounding rules, data conversions, decimal data requirements (several European currencies lack decimal units), ranges, and thresholds on financial calculations. The European Commission has adopted a new currenct symbol for the Euro. This currency symbol and a keyboard mapping for it must be added to existing systems. Currently, none of the fonts included with Microsoft Windows, Windows 95, and Windows NT has this symbol. A similar problem exists for most UNIX and mainframe systems. Most of the Macintosh-compatible systems from 1997 onward include the Euro symbol in fonts.

Like the Year 2000 problem, modifying software to be in compliance with the EMU and Euro will involve making changes to data fields, input screens, and output reports as well as making the necessary technical changes to handle the Euro rounding rules. The Euro conversion also may affect the business logic incorporated in some applications, particularly when it is necessary to support dual currencies during the transition period. Desktop PC applications such as spreadsheets also will require modification.

Special tools to facilitate the Euro conversion are not yet available. Potential software tool vendors (those offering Year 2000 tools now, for example) are waiting for the final technical rules to be issued. Some of the same tools used in Year 2000 compliance can be used to determine the scope of the task, search for currency data fields, analyze existing software, and support a reengineering effort. In the U.S., the Euro conversion primarily will affect applications that must deal with currency from the adopting countries; however, the effect on this software, such as for international banking, will be more substantial.

6.1.6 Standards for Interoperability

Several national and international standards for computer languages, APIs, and network protocols have been developed to facilitate portability and simplify program development. When applications are developed in compliance with standards that are widely supported, these applications should be easier to move to other operating systems and hardware platforms. In contrast, vendors of system software or large application suites often encourage developers to use proprietary APIs on a platform to improve performance and take advantage of special features. The downside of a dependency on proprietary APIs is that they lock the developer into a particular platform and make migration much more difficult and expensive.

The application development process also may need to deal with a series of evolving APIs, some of which have achieved the status of a de facto standard. These interfaces can include APIs for windowing, graphics, audio, video, and networking along with accepted file formats for documents, databases, spreadsheets, graphic images, sound, and video.

Language Standardization

One strength of the C language has been portability. Even before the language was standardized officially, the definition of the C language and the standard C library published in 1978 by Brian Kernighan and Dennis Ritchie had become a de facto industry standard. ANSI C became an official standard in 1989 and within a year or two, vendors released C compilers that were in full compliance. Because of the widespread use of C for both application and system development, ANSI C has been the most important language standard adopted in the last decade for the software industry.

No popular object-oriented language has yet been standardized. In an effort that was started in 1990 after adoption of the C standard, C++ has been going through a slow process to achieve stabilization and similar standardization. The proposed C++ language and the standard C++ library have evolved significantly during this time, with new features being added and numerous changes taking place. Some C++ compiler vendors were slower than others to add support for new language and library features (the lack of support for templates and exceptions in Microsoft C++, for example).

In some circumstances, this situation has had a negative impact on software development efforts, preventing programmers from using some of the most powerful features of the language. The software development community has been waiting patiently for a C++ standard to be finalized and adopted. The hope is the C++ standard will approved sometime in 1998. Once adopted, it will likely be 1 to 2 years before compiler vendors have released products that are in full compliance. Therefore, C++ development likely will be plagued with some level of language and library compatibility problems through the end of the century.

COBOL was last standardized in 1985. Vendors have been supplying COBOL compilers compliant with this standard for more than a decade. Although some new extensions have been proposed for an object-oriented version of COBOL, any such changes will take a long time. One major limitation of COBOL arises from the age of the language. When it was conceived, terminals with character-based screens were the norm, not the graphical displays of today. For applications to take advantage of the GUI, developers often are forced to use vendor-specific extensions. No standards that deal with a GUI have been adopted for COBOL library functions. Because of Year 2000 problems, a large amount of reengineering of legacy COBOL applications will be occurring over the next few years. In turn, this revision may prolong the lifetime of some of these applications.

The Fortran 90 standard was officially adopted in 1991, replacing Fortran 77. This new version of the language even changed the naming convention for the language from FORTRAN to Fortran. The Fortran 90 standard added a large number of powerful new features to the language, including object-oriented facilities, pointers, modules, structures, and array syntax. A large library of functions also was standardized and must be supported by compliant systems. Because of the magnitude of the changes, it took some Fortran vendors until the mid-1990s to release compliant software. This modern version of Fortran will more than adequately serve the scientific and engineering community that uses this language well into the new century. A proposed Fortran 95 standard, which will clarify some features and add facilities for high-performance computations spread across multiple processors, is nearing adoption.

Core System Functions

To use advanced features not directly supported by a computer language and its standard libraries, application development requires access to functions provided by the operating system and supported by underlying hardware. These system functions generally deal with processes, threads, memory management, file I/O, access to other input/output devices, networking, and other system-dependent features. On a system where a GUI is integrated tightly with the operating system, such as Microsoft Windows and the MacOS, there are additional programming interfaces for the GUI subsystem. The programmer's ideal would be the adoption of a standardized set of function interfaces defined with various language bindings for core system functions that could be used across a range of hardware platforms and operating systems. Unfortunately, the application development environment strays far from this ideal.

On the PC desktop, Microsoft is the dominant force supplying MS-DOS, Windows 3.x, Windows 95, and Windows NT. MS-DOS, Windows 3.x, and Win32 (Windows 95/Windows NT) each offer different APIs for accessing core system functions. MS-DOS uses software interrupt calls (80x86 supervisor instructions) to access the functions. Windows 3.x and Windows 95/Windows NT use dynamic link libraries (DLLs) that export a large number of API functions by name for this purpose. Although some of these differences can be hidden by wrapper functions included in the libraries supplied with the PC language compilers, the variations between MS-DOS and Windows and the divergences between Windows 3.x (with its Win16 API set) and Windows 95/Windows NT (the Win32 APIs) are dramatic enough that major differences occur in these core APIs. Although considerable overlap exists between the Win16 and Win32 APIs, Windows 3.x lacks support for some significant features, including interprocess communications, threads, access control functions, flat memory management, and other features. Even in 32-bit Windows, several APIs supported by Windows NT are not supported on Windows 95.

Although there have been several proposals to create a public Windows API standard (the European Computer Manufacturers Association proposal, for example), Microsoft has refused to participate and has energetically opposed these efforts, preferring to maintain complete control of its proprietary Windows APIs. The Win32 API continues to change with each new Microsoft operating system release. Some existing functions are enhanced and a large number of new APIs are added in each new release, although many of these functions can be considered higher-level APIs layered on top of various core kernel routines.

Windows NT includes a Portable Operating System Interface for UNIX (POSIX) subsystem, which supports a set of portable system functions that have been standardized. These POSIX functions make up a small part of the Single UNIX Specification (discussed below), providing some portability between UNIX application development and Win32. Unfortunately, applications that run under the Win32 POSIX system do not have access to other Win32 APIs, such as the Windows GUI, and these POSIX applications can create files with case-sensitive names that may not be accessible under the default Windows subsystem. As a result, the POSIX interface to Windows NT is rarely used for software development.

Less widely used, IBM's OS/2 operating system for desktop machines provides a different set of core APIs for application development. Similarly, the core APIs provided by MacOS, the Macintosh operating system, are fundamentally different from the corresponding core functions supplied with OS/2 and Windows.

One of the earliest attempts to define UNIX system APIs came from the Institute of Electrical and Electronic Engineers (IEEE) in the form of the POSIX project. This large standard specifies not only APIs, but also the form of command-line interaction with computers. As mentioned previously, part of the POSIX standard is implemented in Windows NT.

UNIX systems fare much better than their PC counterparts because of work done after the POSIX effort, particularly the X/Open Single UNIX Specification, a common set of APIs supported by the major UNIX system vendors. This specification has enabled a considerable degree of application portability for developers of applications on UNIX, provided they do not use proprietary vendor extensions, a fairly common habit. The Single UNIX Specification combines the most common software interfaces with several international standards. The specification standardizes core APIs, core UNIX commands, and system header files (C/C++ compiler system "include" files).

Adopted in 1994, Version 1 of this specification (originally called Spec 1170) standardized more than 1,100 APIs. Systems that comply with Version 1 of the Single UNIX Specification can carry the UNIX 95 brand. Most major versions of UNIX have been certified to carry this brand. The latest Version 2 specification released in March 1997 increases the number of APIs to more than 1,400. Systems that implement the Version 2 Specification can carry the UNIX 98 brand. In addition, a UNIX 98 Workstation profile adds support for the Common Desktop Environment (CDE) GUI based on OSF/Motif and the X Window System and increases the API count to more than 3,000. Systems in compliance with these additional UNIX 98 Workstation profiles can be branded with the UNIX 98 Workstation logo.

One result of the Single UNIX Specification is that it has reduced the number of incompatibilities between various UNIX implementations. In addition, several non-UNIX systems have added support for these APIs in order to be compatible. IBM's MVS became UNIX 95 branded in 1996, and IBM's OS/400 will likely follow. IBM's OS/390 already provides UNIX interfaces in its Open Edition.

Application Frameworks and the GUI

Several application frameworks, composed of large collections of reusable C++ class libraries, have been developed for specific purposes. The most prominent and widely used application frameworks are those designed for programming the GUI. On the PC desktop, GUI frameworks are available for Windows, OS/2 Presentation Manager, and the MacOS. On UNIX systems, these GUI application frameworks are designed to work with OSF/Motif and the X Window System.

For PC desktop applications, Borland and Microsoft developed competing application frameworks that are bundled with each vendor's C++ integrated development environment. These frameworks make it much easier to implement features such as windows, dialog boxes, menus, toolbars, and other common window interface elements in an application. These frameworks also simplify use of Microsoft's ActiveX, COM, and Distributed COM (DCOM) extensions. The Microsoft Foundation Class (MFC) library is supported by older Microsoft C++ compilers for use in developing 16-bit Windows applications and by its latest C++ compiler for use in developing Windows 95/Windows NT applications. Similarly, Borland's Object Windows Library (OWL) is supported in the latest Borland C++ development environment, targeting both 16-bit and 32-bit Windows applications. Both products include wizards to facilitate constructing the user interface using drag-and-drop tools that generate a skeleton of the code for the application.

Of the two frameworks, MFC has become the most widely used for developing GUIs on Windows. MFC is now licensed and bundled by other desktop C++ vendors (Powersoft/Watcom and Symantec, for example) and can even be used with the Borland C++ products. MFC also has been licensed and ported to UNIX using OSF/Motif by Bristol Technology and MainSoft. Although a port of OWL was announced for UNIX, it has not materialized to date.

Other, more portable frameworks for developing the GUI are available for use on a wider variety of operating systems (Windows, OS/2, Macintosh, and UNIX). The most prominent cross-platform C++ application frameworks are zApp (now a product of Rogue Wave Software), Visix's Visix Galaxy, XVT Software's XVT, and Zinc Software's Zinc Application Framework. (Of these, Visix Galaxy is the most oriented toward enterprise-scale development.) The advantage of using this approach is that the GUI portion of an application can be developed and tested once using these frameworks and then ported to a different platform with a simple rebuild. The disadvantage of this approach is that the ported applications do not always have the correct look and feel of the native platforms.

One attraction of Java is that the Java runtime environment comes with an Abstract Windowing Toolkit (AWT), an application framework for creating graphical interfaces. Although the AWT is not currently as powerful as MFC or OWL for developing the user interface, JavaSoft is adding a Java Foundation Class (JFC) library to the next release of the Java Development Kit (JDK 1.2) that should equal the Windows frameworks in features and flexibility. Because the JFC is provided by Sun and included in Java, these frameworks will be available on all Java implementations. Most vendors that now sell special-purpose C++ class libraries also are selling similar Java class libraries.

Design Patterns

During the 1970s, Christopher Alexander, a professor of architecture at the University of California at Berkeley, was promoting the concept and use of design patterns as a way to create better buildings. A few years later, Alexander and several other collaborators would publish their defining work, *A Pattern Language*, followed by *A Timeless Way of Building*. Design patterns in architecture can be thought of as rules of thumb that can be employed when initially laying out the basic design of a building and structuring the arrangement and characteristics of spaces or buildings. Like a cookbook in the hands of a competent chef, patterns can be used to create better buildings. These ideas held considerable sway for a time until the excitement and general interest in design patterns for architecture eventually fell out of favor.

In the early 1990s, the concept of design patterns was revived by the object-oriented software development community. Researchers and theoreticians noticed that well-structured programs were full of patterns. In contrast with buildings, large software projects often were assembled from many smaller components. The architectural design of C++ classes in larger projects begged for some form of methodology that could offer improvements. Software design patterns as a cookbook for good program design offered this potential.

One of the more difficult aspects associated with C++ and other object-oriented languages is the design and architecture of classes that will be used as the building blocks for an application. Early examples of C++ classes were often merely thin wrappers around the existing C function call interface supported by the operating systems and libraries. A procedural function typically becomes a method or operation within an object-oriented class that may use the object's constructor to perform some modest initialization.

The design patterns movement was established through a series of conferences and books on a Pattern Language of Programs (PLoP). The watershed was the 1994 release of *Design Patterns: Elements of Reusable Object-Oriented Software*. The book resulted from a collaboration of work over several years. The authors,

Erich Gamma, Richard Helm, Ralph Johnson, and John Vlissides, represent a wealth of experience in software design and hailed at the time from Taligent, DMR Group, the University of Illinois at Urbana-Champaign, and IBM, respectively. The authors, or "gang of four" as they often are called, built on the concept of Alexander's design patterns but applied it to object-oriented design. The book consists of a collection of low-level design patterns that describe simple, thoughtful solutions to commonly encountered problems in software design. Many more books and articles have been published since then on design patterns, and these ideas have entered the mainstream for object-oriented programming. Today, most vendors of object-oriented components commonly use these best practices when designing and building class libraries in C++, Java, and Smalltalk.

Networking

The low-level socket and transport library interface (TLI) functions are standardized on UNIX platforms by the Single UNIX Specification. These APIs provide a portable interface for developing network applications. Based on Berkeley sockets, the UNIX socket APIs also were used as a starting point for developing a similar set of APIs for Windows.

Microsoft standardized on the Windows Socket (WinSock) APIs for low-level TCP/IP networking on Windows 3.x and Windows 95/Windows NT. Microsoft also supports more than a dozen other low-level network programming APIs on Windows (NetBIOS, named pipes, and mailslots, for example). Microsoft has developed several high-level API libraries that implement some popular Internet protocols (HTTP and FTP, for example), as well.

One of the more interesting class libraries for networking is the Adaptive Communication Environment (ACE) toolkit originally developed by Doug Schmidt at Washington University. The ACE toolkit is a large, rich framework of high-level C++ classes for writing portable network applications on UNIX and 32-bit Windows. It uses design patterns, a recent advance in software development technology, in the design of its C++ classes. A version of the ACE toolkit implemented in Java also is available. The ACE toolkit is provided as freely available source code and has been ported to most UNIX systems and Windows 95/Windows NT.

Rogue Wave Software sells several C++ class libraries based on design patterns that provide a high-level framework for network applications. As with all the Rogue Wave products, these class libraries are supported on most UNIX systems, Windows 3.x, and Windows 95/Windows NT.

6.1.7 Web-Based Computing

The Internet and the Web have changed the dynamics of the software industry dramatically. Once used primarily with UNIX, the TCP/IP protocols have become the de facto standard for networking, even for once-proprietary architectures. Apple, Microsoft, and Novell all have embraced and adopted TCP/IP as core network protocols to replace proprietary lower-level network transports, such as NetBEUI, IPX/SPX, and AppleTalk.

Virtually all application vendors have been forced to respond to the new computing models ushered in by the popularity of the Web. The first Web applications were designed primarily for information access. Companies now are actively developing

Web-based systems that include transaction processing and communicate with existing applications and database systems. The Web browser has become the ubiquitous client under this popular extension to the client/server model.

The new Web applications are designed to be interactive. These dynamic applications might allow a remote user to order a product, make an airline reservation, or check the status of a claim. Application development tools are changing rapidly to accommodate these shifts in development architecture. For example, a Web-enabled application might use Java applets, distributed objects, and other new technologies that pose new challenges for the software developer.

Several architectures for creating Web applications have emerged. One of the simpler designs uses a middleware layer at the Web server to implement business rules (data validation, for example), communicate with the database system, and send custom-created HTML pages to the Web-based client. With this thin-client approach, the client is just a simple Web browser capable of displaying standard Web pages formatted in HTML. The client starts the application process by accessing a specific Web page on a remote Web server and uses the forms-based capability of the browser to provide input to the application.

Several development tools allow the middleware layer to be created in a variety of languages using 4GLs and RAD techniques. For example, BlueStone's Sapphire/Web, a RAD tool available on UNIX and Windows 95/Windows NT, uses a GUI builder to connect database fields with elements on a Web page. After the layout is complete, code for building the application can be generated automatically in C, C++, or Java. Sapphire/Web can create ActiveX controls and Java applets and includes native support for most of the popular databases. Similar products are available from NetDynamics (formerly Spider Technologies) and Unify.

Java is receiving wide acceptance in the software development community and with tool vendors as a powerful object-oriented language for new application development. Implemented in the Web browser, the JVM provides a sandbox that isolates and limits what resources a downloaded Java applet can access. (See "Java" on page 404 for more information about the Java security model.) Instead of implementing all business logic as server-based middleware, Java allows some of the application to be implemented in a Java applet downloaded from a server. Using the Java Database Connectivity (JDBC) APIs supplied with Java, a Java applet can communicate directly with the database system or alternatively use some middleware layer through the Web server.

Compared with other interpreted languages, Java programs generally will execute faster because the Java language files are precompiled into Java bytecode. The JVM executes the Java bytecode instructions stored in a Java class file, far less work than interpreting programming language constructs on-the-fly. The primary drawback to Java is that as an interpreted language, Java applications execute more slowly than a similar application compiled into native code from C++ or some other programming language. Just-In-Time (JIT) code generation is one of several techniques applied to address these shortcomings. When a Java applet is downloaded into a JVM (or when a Java application is started), a JVM compiles the Java bytecodes directly into native machine instructions so that execution will be faster. Instead of being interpreted at runtime, the Java applet or application using JIT compilation now runs closer to native execution speed. There usually is a short delay when the applet is first downloaded during which the bytecodes are compiled. The JVMs included in both Microsoft's and Netscape's browsers support JIT compilation. Stand-alone JIT compilers are available from Digital, HP, Silicon Graphics, and Sun on their respective UNIX operating systems and from Asymetrix, Microsoft, Sun, Symantec, and Visix on the PC.

As a result of Java's rising popularity, many vendors now offer integrated development environments (IDEs) for Java development. These tools include wizards and code generators for quickly designing the user interface with editors and class browsers, Java bytecode compilers, and a Java runtime with a debugger for testing the applet or Java application. The latest releases of these environments also support creating and using JavaBeans, the Java standard for distributed objects. Vendors offering Java IDEs include Asymetrix, Borland, IBM, Microsoft, SunSoft, Symantec, and Visix.

6.1.8 Distributed Objects

As object technology has become more pervasive for new application development, the need to develop standards for interoperability between objects has become apparent. Although object-oriented programming supports the notion of component-based development, there must be agreement on a standard way of defining the public interfaces in order for component objects to be shared. Two schemes for object interoperability are in direct competition: the Common Object Request Broker Architecture (CORBA), developed by the Object Management Group (OMG), and Microsoft's ActiveX approach. IBM and Apple have discontinued support for a third scheme called OpenDoc and thrown their weight behind JavaBeans. Sun's JavaBeans component architecture currently is receiving substantial support from software developers as a way to enable enterprise component development with Java. (See Chapter 5.2, *Software Component Architectures.*).

6.1.9 Market Overview

IDC estimates that sales of development tools and languages accounted for more than $21.5 billion in revenues worldwide in 1996. Of this amount, only about 5 percent was devoted to object-based tools. The worldwide market for software development tools over the next 5 years is expected to grow at a compound annual growth rate (CAGR) of about 16 percent to $41.5 billion. Total revenues for the object tools market is expected to quadruple. Development tools that support teamwork and collaboration will continue to increase in importance. Automated software testing and software configuration management tools are expected to continue with a CAGR of more than 20 percent.

There is a continuing migration to client/server computing away from older host-based application development. This trend is evident in the projected growth for development tools and languages on Windows 95/Windows NT and UNIX. Development tools for Windows 95/Windows NT are forecast to experience the greatest growth, increasing from $3.5 billion in 1996 to $16 billion by 2001. Revenue from UNIX development tools are predicted to more than double from $6.8 billion in 1996 to more than $17 billion within the same timeframe. In contrast, development tools revenue for host-based mainframes and minicomputers is anticipated to grow at one-fourth this rate.

Database management systems (DBMS) and 4GL/RAD tools historically have been the driving force behind the growth of the overall client/server tools market and accounted for 61 percent of total revenues in 1996. Several development tool categories projected to increase the fastest (object and Web development tools, for example) are often used by the newer Web-enabled style of client/server computing. As a result, some of the revenue spent on Web development tools, object-oriented tools, and component software also may represent an investment in client/server computing but may be executed in a slightly different way from the conventional market tools.

Overall, the worldwide market for development process support tools is expected to increase from $21.5 billion in 1996 to $46.5 in 2001, a CAGR or more than 16 percent. Web professional development tools are forecast to experience an extremely high CAGR of almost 45 percent, indicative of a market in its infancy. Object-oriented tool revenue amounted to about $1.2 billion, representing only 5 percent of the total. However, object tools are one of the fasting-growing market segments.

The revenue for client/server DBMS and 4GL/RAD tools is forecast to grow to 63 percent of the total development tools market by 2001. Of these two categories, DBMS tools account for the lion's share of the increase, growing from $9 billion in 1996 to more than $23 billion in 2001.

Some of the emerging markets in application development tools, especially those that support teamwork and collaboration activities, continue to show strong growth. Automated software quality tools (testing and maintenance) and software configuration management tools are examples of tool categories expected to sustain a better-than-average CAGR or more than 20 percent. For automated testing, some of this growth may be spillover from Year 2000 software investments.

In contrast, sales of 3GL tools will be relatively flat, and the market for analysis, design, and modeling tools is expected to decrease. Part of this market loss stems from continuing fallout from older CASE tools as well as the competition from client/server and 4GL/RAD tools that now incorporate some design and modeling features. Figure 6-1 gives projected revenue figures by development tool category.

Figure 6-1: Worldwide Development Tools Revenue by Category ($thousands)

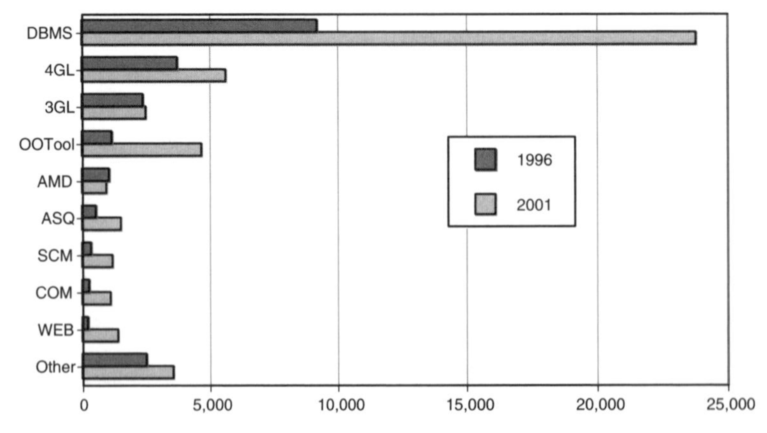

Source: IDC, 1997

A rarely discussed but expanding area for 3GLs and 4GLs is the AS/400 platform from IBM. Traditional leaders among tool vendors on this platform include IBM, Lansa, and Synon. These companies sell 4GL products oriented toward business applications. Recent announcements have signaled a renewed interest in this platform as a Web server, so Web development and Java tools either are promised or are shipping from all vendors.

In addition, traditional PC vendors, such as Borland, have begun promising to port their development products to the AS/400. Borland's tool market share is shown in Figure 6-2 under the heading Host (along with mainframes and some midrange systems).

Figure 6-2: Worldwide Development Tools Revenue by Platform ($millions)

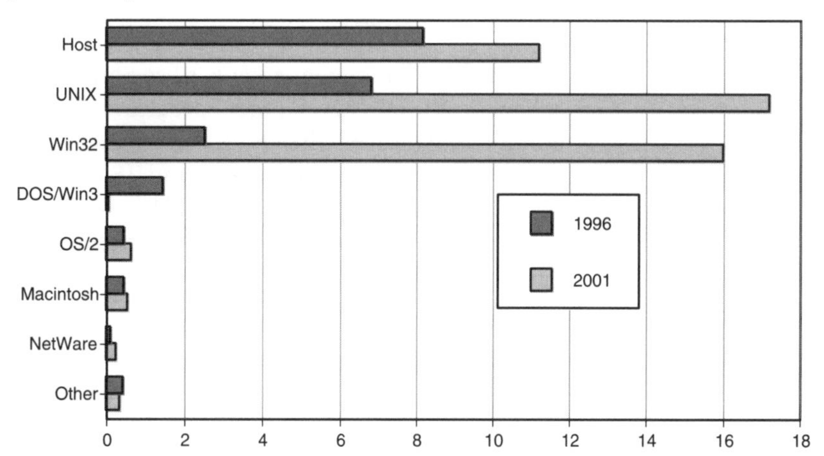

Source: IDC, 1997

Programming Languages

Although purchases of third-generation programming languages account for less than 12 percent of total development tool revenue, the total installed base of these products is considerable. Language products also tend to be priced lower per developer than most tools sold for client/server development. The enormous amount of existing business software written in 3GL languages combined with the value to business represented by the investment in these applications ensures a continuing presence for 3GL tools.

Estimates by Software Productivity Research place the number of programmers in the U.S. at approximately 1.7 million. Of these developers, an estimated 35 percent primarily use COBOL and another 17 percent primarily use C. The proportion of all commercial and business applications written in COBOL is estimated at 80 percent, a consequence of previously developed programs that run on legacy systems. Organizations are estimated to be spending $30 billion annually to maintain these applications, and this number is expected to rise dramatically for the next few years as a result of Year 2000 conversion efforts.

Other popular computer languages including Fortran, C, and RPG have been used to create a large number of applications still in use today. Fortran was the dominant language used to write older applications on Digital VAX and remains the preferred language for scientific computing. C has traditionally been the dominant language used for writing applications on UNIX and the PC, although it is being displaced by C++. C and C++ also have been used to write the operating system software on UNIX and the PC (Windows 3.x, Windows 95, and Windows NT). RPG has been the prevalent language used on IBM minicomputers such as the AS/400.

A variety of newer products, such as Microsoft's Visual Basic and Borland's Delphi, have helped to extend the use of older languages (BASIC and Pascal, for example) into the future. These tools have been very effective at building the client component of client/server applications. In parallel, a market has been created for software components supporting application development efforts using these products.

The use of object-oriented languages, such as C++, Smalltalk, Objective-C, and now Java, is on the increase. This popularity, in turn, has spawned a rise in the use of software components targeted for use with these languages.

Java and ActiveX

The Java programming market has been growing rapidly, and the tools for enhancing the Java development process are evolving in tandem. More than two dozen Java IDEs have already reached market, many of them offered by traditional C/C++ vendors such as Borland, IBM, Microsoft, SunSoft, and Symantec. Revenue generated from these Java IDEs doubled from 1996 levels to $110 million in 1997, with projections to reach more than $350 million by 2001.

The market for distributed object technology is new and untested. The software component market currently provides ActiveX components (ActiveX or OLE controls, for example) commonly used with Visual Basic and C++ that are employed locally, not as distributed objects. Currently, it is difficult to distinguish the market for ActiveX Controls supporting Visual Basic development from ActiveX software components designed as distributed objects. As Java IDE tools that support creating and using JavaBeans are adopted, a corresponding increase in the number of software components available for purchase should develop.

Figure 6-3: Java Integrated Development Environments Revenue

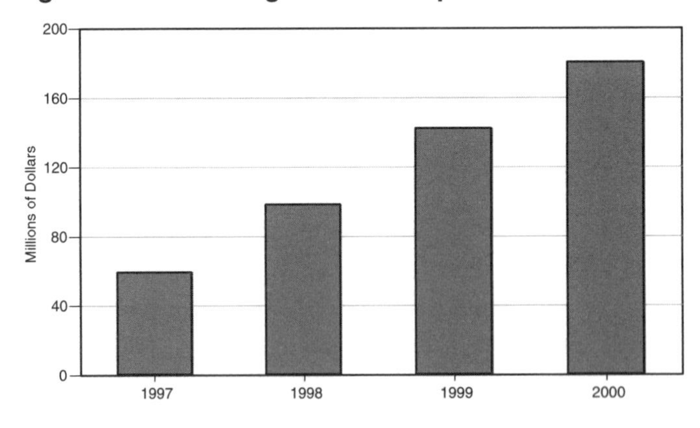

Source: IDC 1997

Web Development Tools

New companies are entering this market at the same time that most of the traditional client/server development tool products are being enhanced to support Web clients. Consolidation is occurring among companies selling client/server development tools. Products that support RAD and can generate code in both Java and C++, such as BlueStone's Sapphire/Web, still have the greatest mind share among developers. Meanwhile, RAD tools for Web computing have been released by Borland, Centura Software, Powersoft, Symantec, and others.

CASE Tools

IDC expects the market for analysis, modeling, and design tools to decrease slightly over the next 5 years. Part of this decline is a continuation of the shift away from older, rigid CASE tools that did not live up to expectations. The market for these tools also is being eroded by 4GL client/server development tool vendors that have added design and modeling to the feature set of their products.

6.1.10 Forecast

Third-Generation Languages (3GLs)

- The older 3GL procedural languages such as COBOL, Fortran, and RPG will continue to have an important role because they are used to support a large base of legacy code that cannot be reengineered easily. Organizations already will be hard-pressed to make the changes necessary to fix the Year 2000 problem in business applications in time. The effort required to replace all these programs would be massive. As a result, 3GLs will continue to dominate in the installed base of business applications into the foreseeable future.
- C will continue to be the language of choice for systems programming and will continue to find use in new software development. Its reputation for source-code portability continues to encourage software developers to choose this language for new applications that will need to be moved across platforms. However, traditional markets will continue to erode due to an increased use of C++. This process will accelerate after the C++ language and standard template library become official standards and vendors release standard-compliant products.
- Object-oriented extensions to 3GLs will tend to blur the distinction between these older languages and newer ones. Fortran already has been modernized to meet the needs of the scientific and engineering community, and its use will continue in this particular market. An object-oriented standard for COBOL is expected in the next few years that will modernize this language. IBM's and Micro Focus' object extensions to COBOL will remove the incentive for COBOL-oriented development organizations to move to a different language.
- Integrated 3GL development environments supporting rapid visual development and software components will continue to grow in popularity. As more features are added to these products, the gap between these visual 3GL tools and 4GL tools will narrow.
- Microsoft's Visual Basic and similar products increasingly will be used on the desktop for quickly generating PC-based applications and visual front ends for client/server applications. Growing use of these products will in turn reinforce the market demand for software components.

Fourth-Generation Languages (4GLs)

- The overall use of 4GL, CASE, and other tools will remain small when compared with the installed base of 3GL and object-oriented languages. However, the market for 4GLs will continue to expand its share of the overall language market. Most of this growth will be at the expense of 3GLs such as COBOL used for host-based application development. 4GL tools will continue to be improved and marketed as second-generation client/server products designed for developing applications that support transaction processing.

- All 4GL tools will be modified to support Internet/intranet computing using Web clients and servers. To remain competitive, these features will be included at no additional cost. As with other Web-oriented products, vendors will begin offering lower-cost Web-only versions of these tools to gain market share.
- The leading 4GL products will provide integrated development environments comprising tools for development (including development process support), deployment, and systems management.
- To enhance their offerings, 4GLs tools will become more powerful by using higher-level constructs and adding more analysis and modeling capabilities for object design. These developments will have a corresponding negative impact on the traditional CASE tool vendors.

Object-Oriented Languages

- The overall use of object-oriented languages will increase, particularly for business applications. This growth will be spurred by greater support from development tool vendors and through the use of integrated development environments supporting visual development.
- C++ will continue as the dominant object-oriented language and will be used increasing as a replacement for C for both systems programming and for developing new applications.
- The use of Java will increase rapidly, particularly for Web applications. Much of the increase in the use of Java will be at the expense of C++.
- Java will play a significant role in the Internet development industry, particularly for vendors who value cross-platform object-code portability. For vendors focusing on the Windows environment, ActiveX will be a strong alternative. Integrated development environments for Java will become more advanced, with better visual development tools, Java compilers, and debuggers.

Software Development Process Management Tools

- The use of software configuration management tools will continue to expand as the complexity of modern software development increases. These tools will be enhanced by including more features that support, control, and archive the build process.
- The market for tools to automate software testing and quality assurance will continue to increase. Ongoing efforts will be made to integrate separate tools under a single visual framework for control and reporting.
- The sales of traditional analysis, modeling, and design tools will continue to lag behind the other categories of development tools and languages.
- CASE tool vendors will continue to refine and expand their products to support reverse engineering and code generation. Support for languages other than COBOL will be added, with C, C++, Java, Smalltalk, and the more popular 4GL products the most likely candidates.
- CASE and 4GL development tools will be integrated to provide a more complete solution for software engineering. This integration will include program management, configuration management, and version control features.
- Organizations doing their development in COBOL will remain the largest single market for CASE tools.

- Code repositories will emerge as a key development tool for improving programmer productivity and code reuse. More high-end development tools will include a code repository or will support integration with other tools that offer this capability.

Year 2000 Tools

- Large organizations with legacy applications will be preoccupied with Year 2000 problem-solving and testing for the next 2 years. Companies doing business in Europe will face the additional challenge of modifying their systems to handle the new EMU currency, which will be phased in beginning on January 1, 1999. Many smaller organizations that have postponed dealing with these problems will be forced to race to meet this deadline.
- Expenditures on Year 2000 repair projects will be large and will have a negative effect on other software development spending for the remaining years of this century.
- The process of addressing the Year 2000 problem by modifying older code will bring costly and sophisticated development tools into many organizations that lacked them. In the process of fixing Year 2000 problems, millions of lines of legacy application code will be inventoried, organized, and processed by these tools, which will have a positive effect on other unrelated software maintenance issues. This process will prolong the use of the legacy applications and facilitate incorporating other software changes.

6.1.11 References

■Articles

Baker, Steven. 1995. Complying with Fortran 90: how does the current crop of Fortran 90 compilers measure up to the standard? *Dr. Dobb's Journal.* January: 68.
Baum, David. 1996. Tool up for 2000. *Datamation.* January 1: 49.
Linthicum, David. 1997. Driving development. *DBMS Magazine.* April: 36.
Lyons, Daniel. 1997. Tools of the trade: which way to go. *Computer Reseller News.* August: 95.
Schwartz, Jeffrey. 1997. Java use grows in enterprise. *CommunicationsWeek.* July: 25.

■Periodicals, Industry and Technical Reports

Hubley, Mary. 1997. *The Open Group single UNIX specification.* Datapro Information Services Group. May.
UNIFACE. September 1997. Datapro Information Services Group.
Heiman, Richard, et al. 1997. *Programmer development tools: 1997, worldwide market and trends.* IDC. July.
Hotle, M. and B. Conway. 1997. *Year 2000 testing crisis: an overview.* Gartner Group.
Jones, N. 1997. *The IT cost of a single European currency.* Gartner Group. May.
O'Reilley & Associates. 1997. *Scripting languages: automating the Web.* World Wide Web Journal. Spring.

■Books

Alexander, Christopher, et al. 1977. *A pattern language.* New York: Oxford University Press.
Alexander, Christopher, et al. 1979. *A timeless way of building.* New York: Oxford University Press.
Bergin, Thomas J. and Richard G. Gibson. 1996. *History of programming languages.* Reading, Mass.: Addison-Wesley Publishing.
Booch, Grady. 1996. *Object solutions: managing the object-oriented project.* Reading, Mass.: Addison-Wesley Publishing.
Fowler, Martin and Kendall Scott. 1997. *UML distilled: applying the standard object modeling language.* Reading, Mass.: Addison-Wesley Publishing.
Gamma, Erich, et al. 1995. *Design patterns.* Reading, Mass.: Addison-Wesley Publishing.
Kernighan, Brian and Dennis Ritchie. 1978. *The C programming language.* Englewood Cliffs, N.J.: Prentice Hall.
Maguire, Steve. 1994. *Debugging the development process.* Redmond, Wash.: Microsoft Press.
Martin, Robert. 1995. *Designing object-oriented C++ applications.* Englewood Cliffs, N.J.: Prentice Hall.
Meyers, Scott. 1992. *Effective C++.* Reading, Mass.: Addison-Wesley Publishing Co.
Meyers, Scott. 1996. *More effective C++.* Reading, Mass.: Addison-Wesley Publishing Co.
Sethi, Ravi. 1996. *Programming languages: concepts and constructs* (second edition). Reading, Mass.: Addison-Wesley Publishing.

Sommerville, Ian. 1996. *Software engineering* (fifth edition). Reading, Mass.: Addison-Wesley Publishing.
Stevens, W. Richard. 1992. *Advanced programming in the UNIX environment.* Reading, Mass.: Addison-Wesley Publishing.
Stroustrup, Bjarne. 1997. *The C++ programming language* (third edition). Reading, Mass.: Addison-Wesley Publishing.
Vlissides, John, et al. 1996. *Pattern languages of program design 2.* Reading, Mass.: Addison-Wesley Publishing.

■URLs of Selected Mentioned Companies

Adaptive Communications Environment (ACE) Toolkit
`http://www.cs.wustl.edu/~schmidt/ACE-overview.html`
Adabas `http://www.adabas.com`
Addison-Wesley `http://www2.awl.com/corp`
Apple Computer `http://www.apple.com`
ASPG `http://www.aspg.com`
Asymetrix `http://www.asymetrix.com`
AT&T Bell Laboratories `http://www.belllabs.com`
Bluestone `http://www.bluestone.com`
Borland `http://www.borland.com`
Bristol Technology `http://www.bristol.com`
Centerline `http://www.centerline.com`
Cognos `http://www.cognos.com`
Computer Associates `http://www.cai.com`
Compuware `http://www.compuware.com`
Continuus `http://www.continuus.com`
Cyrano `http://www.cyrano.com`
Forté Software `http://www.forte.com`
Fox Software `http://www.foxsoftware.com`
Free Software Foundation `http://www.fsf.org`
Hewlett-Packard `http://www.hp.com`
IBM `http://www.ibm.com`
Informix `http://www.informix.com`
Intersolv `http://www.intersolv.com`
Isogon `http://www.isogon.com`
Lansa `http://www.lansa.com`
Logic Works `http://www.logicworks.com`
MainSoft `http://www.mainsoft.com`
McCabe and Associates `http://www.mccabe.com`
Microsoft `http://www.microsoft.com`
NetDynamics `http://www.netdynamics.com`
Netscape `http://www.netscape.com`
NeXT Computer `http://www.next.com`
ObjectShare `http://www.objectshare.com`
Oracle `http://www.oracle.com`
Parasoft `http://www.parasoft.com`
Platinum Technology `http://www.platinum.com`
Powersoft `http://www.sybase.com/products/tools/index.html`
Prince Software `http://www.princesoftware.com`
Quintic Systems `http://www.quintic.com`
Rational Software `http://www.rational.com`
Rogue Wave Software `http://www.roguewave.com`
SCO `http://www.sco.com`
Segue Software `http://www.segue.com`
Software Productivity Research `http://www.spr.com/index.htm`
Spyglass `http://www.spyglass.com`
SQL Software `http://www.sql.com`
Sterling Software `http://www.sterling.com`
Sun `http://www.sun.com`
SunSoft `http://www.sun.com/software`
Sybase `http://www.sybase.com`
Symantec `http://www.symantec.com`
Synon `http://www.synon.com`
TracePoint Technology `http://www.tracepoint.com/frames.html`
Unify `http://www.unify.com`
Viasoft `http://www.viasoft.com/main.htm`
Visix `http://www.visix.com`
XVT Software `http://www.xvt.com`
Zinc Software `http://www.zinc.com`

6.2 Databases

6.2.1 Executive Summary

After a long period of relative stability, the major database vendors have begun to introduce a series of new features in database access. Structured Query Language (SQL) standards (ODBC and SQL-92) provide a large number of new, advanced, relational modifications that are showing up in new database products. Numerous other features also are being deployed, and many of these are likely to be codified in upcoming standards, particularly, SQL3.

Parallel Database Systems (PDBSs) running on platforms such as IBM's SP2 and other UNIX platforms are considered crucial for building new, large-scale decision-support systems (DSS). Most vendors have been marketing integrated PDBS capabilities since 1997, and competition with the more traditional parallel platforms, such as NCR's Teradata, is heating up. PDBSs speed up DSS queries and on-line transaction processing (OLTP) transactions mainly on hardware platforms that have some unused resource capacity. The primary reason to adopt parallelism is not improved response or resource cost savings at peak utilization (there is probably none), but better resiliency to individual processor failures and improved response and resource utilization during slack periods.

Update transactional performance is believed to be near its theoretical limit for current hardware; however, DSS queries, with individual queries often requiring hours of processor and disk access, have not yet been optimized fully. All major database vendors have announced or introduced new advanced indexing capabilities in the last year to improve query price/performance. Price/performance is defined as dollar cost per unit of throughput, and on fully loaded systems, parallelism alone cannot improve price/performance because throughput improvements on such systems are expensive. That is the reason for the recent interest in advanced indexing, which can improve response in a way that parallelism alone cannot. Most vendors have just started to tap this technology but continuing improvement are expected for the next few years in query price/performance.

Object-relational databases (ORDBMSs), with capabilities driven by features of object-oriented database management systems and improved SQL capabilities, are reaching the market. ORDBMSs are suited for multimedia applications as well as data with complex relationships that are difficult to model and process in a relational DBMS. Because any type of data can be stored and operated on within SQL, ORDBMSs allow for fully integrated databases that hold data, text, pictures, voice, and video. ORDBMSs attempt to provide database solutions that can handle complex data types and relationships using development environments based on object-oriented programming languages.

All database vendors that have released object-relational versions of their systems (IBM's DB2 UDB, Informix, Oracle, and Sybase) have relabeled their products "Universal Server," or in IBM's case, "Universal Database." The idea is that the object-relational model is better suited to deal with variant forms of information, such as geographic data and text abstracts, than with the traditional alphanumeric, business data fields found in most applications of the relational model.

Database replication capabilities have been adopted by many companies and continue to expand their role in commercial processing. Replication offers the ability to keep different copies of a database up-to-date by communicating changes made on one or more versions. A few vendors have announced releases to tie replication in with mobile computing when standard facilities will allow laptops to upload data entered by the user automatically and download new information created while the user has been out of contact. The cutting edge in mobile computing is delivering the flexibility to locate the same data or database functionality (rules, stored procedures, and so on) across Java stations, laptops, and desktops.

6.2.2 Database Management Systems (DBMSs)

A DBMS is a specialized piece of software that provides functionality for storing, updating, and retrieving information. Most DBMSs also provide mechanisms for maintaining the integrity of stored information, managing security and user access, recovering information after the system fails, and accessing database functionality from within an application written in a third-generation language (3GL) such as COBOL, C, or JAVA. DBMSs are essential for managing large volumes of data in a secure manner.

Three main aspects define the architecture of a DBMS:

- **Data model** – The way data is conceptually structured. Some examples include relational, object-oriented, object-relational, and hierarchical.
- **Data Manipulation Language (DML)** – The language used to query the contents of the database and to store or update information in the database. SQL provides the DML facilities in relational and object-relational databases.
- **Data Definition Language (DDL)** – The language used to define what types of information will be recorded in the database and how it will be structured. Instructions in the DDL provide the facilities to define the data template or schema into which DML instructions will record data. SQL provides this capability in relational and object-relational databases.

DBMSs are no longer entirely the domain of data-processing departments in large organizations. As desktop applications become more sophisticated, powerful DBMSs are available to personal computer (PC) users. Many issues faced by database developers are now also being faced by a large contingent of desktop developers. Because these problems are now being addressed by more end users, database vendors have made their databases easier to use, relying on visually oriented interfaces for development and data manipulation.

The following DBMS models have emerged over time: hierarchical DBMSs, relational DBMSs, object-relational DBMSs, object-oriented DBMSs, and data warehouses.

Hierarchical Database Systems

DBMSs have been around since the late 1960s. The first DBMSs, such as IBM's Information Management System (IMS), were hierarchical and, according to IDC, IMS still commands a market share of 10 percent of total dollars in the DBMS market. As with all later database systems, the structure of access to data in a hierarchical model was predefined by the database administrator before the programs that access the data were written. Programmers needed to follow the hierarchy established by the data structure, and hierarchical systems offered only a program-oriented navigational interface to data. For example, the program might select a division of a company, then navigate down to a group within the division, then a section within the group, and finally an employee within the section. The original hierarchical systems were extremely efficient for OLTP applications, where quantitative fields of detailed records were modified and hierarchical roll-up aggregations were maintained at higher levels. However, it was not possible with the original hierarchical products to achieve flexible ad hoc query retrieval.

Hierarchical databases provided a major improvement for programmer productivity and data integrity when compared to just storing data in files. The programmer, though needing to be concerned about the organization of the data hierarchy, did not have to know the exact physical layout of the data. Furthermore, many functions related to managing the physical data were performed by the DBMS; therefore, individual programmers did not need to write these functions into the programs.

Hierarchical databases have been enhanced continuously and have a wide range of industrial-strength features. Transactional concurrency and recovery techniques were first developed on hierarchical database systems and reached a high performance level. In addition, many constraints found in relational systems (such as referential integrity, a safeguard that, among other things, ensures records on which other records depend cannot be deleted without resolving the needs of the dependent records) were added to reproduce data-integrity features found in hierarchical systems. For example, in a hierarchical database system, it is not possible to insert a new employee into a department that does not exist.

IBM's IMS now offers a query capability based on SQL. This feature is useful for companies already using IMS, but it is inappropriate for companies planning to acquire a new relational or object-relational system because IBM does not market IMS as a full-featured relational product. IMS is generally considered a legacy database in companies that use it. Although it does the job for which it was designed well, its market share (for support and upgrades) is shrinking.

Relational Database Management Systems (RDBMSs)

Using hierarchical database systems, any change to the database structure required all applications accessing the data to be rewritten. This practice was necessary because hierarchical data is located by navigating the data hierarchy. In contrast, RDBMS products use logical references, such as department ID values, to connect rows that are related in two different tables. This logical view, which avoids knowledge of the particular structure of the data, is known as "program-data independence." With RDBMSs, major modifications of the database rarely require rewriting existing data-access routines.

RDBMSs also separate database design from database implementation better than prior models. Auxiliary data structures such as views can be added to provide additional windows into the data. Indexes can be added or removed to tune database performance, and the application programmer need not be aware of such changes.

When RDBMSs were introduced, they offered other advantages. One of the most important was that new approaches to database design, such as database normalization and entity-relationship diagrams, could be used to design database tables. Normalization identifies certain anomalies that can exist in a database table when too many attributes have relationships between them; it solves the problem by decomposing the table into a set of tables that join to give the same information.

In practice, complete data normalization often results in too many individual tables. This situation may increase update performance, but it also tends to degrade data access performance (because of the need for multiple joins) and complicate the programmer's task. These problems often preclude full normalization, especially in decision-support applications. (See Chapter 7.2, *Data Warehousing.)*

As experience was gained with relational applications, several additional capabilities were added to most RDBMS products:

- Embedded SQL allows application programs to access data through cursors – a SQL select statement that establishes the rows for processing at the point when the cursor is opened. Opening a cursor simply positions the next input/output (I/O) action at a specific record. Successive fetch statements on a cursor relax the set-orientation of interactive SQL for programs that need to access data one row at a time.
- Precompilation of embedded SQL statements to improve runtime efficiency
- Dynamic SQL allows greater program flexibility, permitting the program to select data using criteria that are not entirely known when the program is written.
- Integrity constraints allow business rules to be maintained automatically when the database is updated. These rules can be accessed easily as data from database catalogs.

Today, several newer features are entering the standards. (See "SQL Standards" on page 476.)

- Procedural languages that allow program logic to be performed within the SQL database manager. The first example of this feature was Sybase's stored procedures. (Note that such a feature departs from the original non-procedural concept of RDBMSs.)
- Triggers that run procedures when certain events, such as row updates or deletes, occur in the database
- Improved SQL capabilities to perform more complex transitive queries; for example, searching an organizational chart for all the employees who report to a manager, regardless of how many intermediate managers are in between, as opposed to searching for only those employees who are directly assigned to that manager.
- New cursor capabilities, such as the ability to scroll forward and backward
- Object-relational extensions (discussed in the next section)

Object-Oriented Database Systems (ODBMSs)

An object database management system (abbreviated ODBMS, or sometimes OODBS) attempts to integrate a DBMS with the capabilities of specific object-oriented programming languages such as C++, Smalltalk, or – most important in recent releases – Java. Although an application using a RDBMS must read relational data and then translate it into application-specific data structures, an ODBMS attempts to make object-oriented programming language objects migrate between persistent storage and memory without the necessity for such translation. Of course, some application data structures, such as memory-resident lists of objects, cannot themselves be treated as single objects, so programmatic interpretation of stored data into application data structures is still required.

ODBMSs provide the following characteristics also found in relational databases:

- Ability to provide ACID concurrency and recovery in a multi-user environment. ACID stands for Atomicity, Consistency, Isolation, and Durability. Atomicity means that each transaction executes completely or not at all and leaves no partial results; consistency means that each transaction moves from one consistent state to another consistent state, as regards consistency requirements such as the balance of assets and liabilities; isolation means that each transaction executes as if it were running alone, without concurrent transactions needed to properly utilize the processor resources; durability means that on return from COMMIT, the results that have been committed will not be lost in a failure of memory or disk. (See "Transactional Capabilities" on page 475 for more about the ACID properties.)
- An ad-hoc query facility based on SQL

Meanwhile, ODBMSs uniquely provide the ability to handle complex objects and make changes that are persistent on disk.

Object-oriented database systems originally were developed to fill needs in special application areas, such as computer-aided design (CAD) and computer-aided manufacturing (CAM), where complex structures are most efficiently represented as structured objects. In the relational model, these structures (such as parts in a subassembly) would consist of large sets of relational rows in different tables referencing each other with foreign keys and would require index lookups for navigation. This was an inefficient operation in a relational implementation. The job of accessing such structures is improved, both in terms of programming simplicity and performance, by dealing with single, self-contained objects.

ODBMSs also specialize in memory-resident performance to meet the needs of applications such as CAD/CAM systems. For example, general object pointers between related objects in disk storage can be transformed into memory pointers when referenced objects become memory-resident. (Transforming the object pointer in this way is termed "swizzling" the pointer.) Object-oriented databases do not have tables in the relational sense, so query capabilities are inferior to the SQL standard; most vendors have been trying to address this problem in different ways. Because object-oriented database systems have been finding new applications in traditional businesses, some relational database vendors have begun to show interest in cooperative agreements.

ODBMSs are being adapted to support Internet access to the data. Some businesses also have learned to use these systems for applications integrating multiple data sources and applications.

ODBMSs are particularly well suited for organizations that are committed to development using object-oriented programming languages. Most ODBMS vendors have provided a transparent interface for one or more object programming languages, so that objects not in memory are automatically located and retrieved when needed. Some ODBMSs also will reflect updates to database objects automatically within an application program at the time of a transaction. Most ODBMS vendors have focused on Java recently, but C++ and Smalltalk are alternatives in use.

Object-oriented database systems allow user-defined object classes similar to those provided by object-relational systems, with many products offering additional features such as versioning. Versioning means that a new version of an object can be created while an old version continues to exist and be referenced by other users. This is a very useful feature in CAD environments, one that is not offered in any current object-relational product. Versioning also is important in a Java environment in which dynamically loaded content has to be matched to the right version of the code to manage it.

An ODBMS standard known as ODMG-93 was released in 1993 by the Object Database Management Group. In December 1995, an updated Release 1.2 followed, and Release 2.0 was announced in July 1997. ODMG-93 included definitions for an object model, an object definition language, an object query language (OQL), and bindings to C++ and Smalltalk. The goal of the 1993 and 1995 standards was to allow customers to write portable applications that could run on more than one ODBMS product. A goal for the 2.0 standard was to allow heterogeneous distributed ODBMSs to exist in component form and communicate through the OMG Object Request Broker (ORB). (See "ORBs" on page 338.)

Currently, vendor revenues suggest that the ODBMS model remains in the niche product category. One important motivation for the object-relational (ORDBMS) model (discussed shortly) was to satisfy customer needs not met by the purely relational model, and thus keep customers in the RDBMS vendors' camp (soon to be the ORDBMS vendors' camp). There is good reason for customers to be conservative in adopting a new model, but the final outcome of this contest has not yet been decided. ODBMS products have some special capabilities not yet provided by ORDBMS (such as versioning in some products and the ability to perform data access and update transparently in programming statements), and ODBMS products lack some other capabilities (such as truly integrated SQL Select capabilities in the absence of table structures) found in ORDBMS.

Object-Relational Database Systems (ORDBMSs)

A new category of object-relational database products is emerging to supplement the pure relational model. ORDBMSs have some of the capabilities of ODBMSs and additional unique capabilities, such as the ability to contain sets of record-structured values in a single column of a row. (See "Object-Oriented Database Systems (ODBMSs)" on page 472.) Acceptance of this new model has achieved a critical mass among the database vendors, and in a few years, a pure relational DBMS will be viewed as a legacy product. (The upgrade path from relational to object-relational will be easy to take.) Currently, however, the object-relational and relational databases are separate products, although converging.

The proposed ANSI SQL3 standard, to be released in 1999, will describe an object-relational model in detail. The SQL3 proposal has been under consideration for some time, and most relational vendors have been developing versions of their products that have object-relational extensions and will come to market before the release of standard SQL3. In many cases, the features of these products will probably extend beyond the SQL3 specification.

New features classified as object-relational extensions include the following:

- Columns of tables can contain "composite" types (a record structure with multiple components) and collections of multiple values: sets, multi-sets (no order, but duplicates permitted), and lists (ordered sets). Thus, Codd's First Normal Form rule for the relational model is broken. A row in a PATIENT table (a specific patient) will now be able to contain a "Drugs" column value, which is basically a table in its own right, containing drug names, dosages, and time all doses were administered to the patient. This new capability has an important impact on methods of database design, but many of these implications have not yet been worked out by the experts in the area.

- Columns can be Abstract Data Types (ADTs), which define visible properties of (invisible) stored column data in terms of functions that have been added by application programmers. The data stored for the column could be a bitmap representation of a satellite picture, for example, and the functions provided could search for specific shapes in arbitrary orientation using Fourier transform techniques. SQL then will allow users to retrieve all rows with pictures having shapes that look like a given variable value. Many business applications can use such capabilities, including multimedia, text retrieval, and geographic information retrieval.
- The ability to reference a row of one table from a column of another table in an absolute way will be provided by a new "ref" data type. Refs are something like the reference IDs (RIDs) that many vendors have supplied for some time, but they break an important relational rule: Access rows by content only. Refs are being added for performance reasons; they will make index lookups unnecessary to retrieve rows of different tables that are identified as a single complex object. The unreleased SQL3 draft disallows "sets" of refs, but Informix's Illustra already permits this capability, and it is likely other vendors will follow suit. This is another capability that will have an important impact on logical database design.

Some issues related to database design for ORDBMSs are still unresolved. Even so, users can take a conservative approach and employ older methods of design in cases where the advantages offered by new methods are not compelling. There are a lot of new capabilities in the object-relational model, however, and all ORDBMS products will contain older RDBMS capabilities as a subset. Therefore, the upgrade path is expected to be simple and worthwhile. (See "SQL Standards" on page 476 for more information on SQL3 object-relational plans.) The ability of major RDBMS vendors to deliver increased ORDBMS functionality is threatening to close off growth prospects for pure object-oriented database systems.

6.2.3 Transactional Capabilities

One basic function of a DBMS is to provide transaction management, which allows multiple users to access a database at the same time without loss of data integrity. Without special handling, problems can develop if multiple users are updating the same record and the update from one user overlays an image that was read or updated by another user.

Most SQL-based products provide database transactions (except for some data warehouse products, which are designed for queries and normally do not support concurrent multi-user updates). Database transactions guarantee data integrity to the application programmer through the ACID properties (atomicity, consistency, isolation, and durability) described previously.

To guarantee durability, all database transactional products provide recovery utilities that will recover from various types of hardware and software failures. Recovery must be supported by runtime mechanisms that write out logs to disk. At the same time, isolation must also be supported by runtime locking mechanisms that keep concurrent transactional actions from interfering with one another. These two runtime duties add significant overhead to transactional data processing.

Most database textbooks quote a number of classical results about transactional isolation, a feature that database systems typically have implemented by taking locks on accessed data items, using a protocol known as two-phase locking. This approach also can cause lost productivity when users make requests that might cause a conflict and are forced to wait, making no progress, until the potential for conflict is resolved. Not all applications need a guarantee of perfect isolation; certain types of conflict may prove to be impossible or unimportant for the access patterns involved. To improve performance in such cases, less-restrictive guarantees have been provided by a number of products. These lower-level guarantees are known as isolation levels; they are defined in the SQL-92 standard. This definition has been challenged, however, and it is likely to be modified. (See Berenson, Hal, et. al., "A Critique of ANSI SQL Isolation Levels," in "References" on page 489.)

An example of a lower isolation level is IBM's DB2 Cursor Stability (CS), which allows more concurrency than the perfect isolation of DB2's Repeatable Reads (RR). Some products provide isolation levels based on multi-version concurrency, which guarantees that reads will never have to wait for update locks to be released. An example is Oracle's "Serializable" isolation, which is sufficiently powerful that it can be used in the TPC-C benchmark where completely serializable output is required, yet readers never have to wait for writers and writers never have to wait for readers. (Writers might interfere with each other, however, and force transaction rollback.)

Currently, vendors are looking to introduce less-restrictive isolation levels, but there is little guidance for users on how to recognize an application where these less-restrictive levels of isolation pose no risk of data corruption. An inappropriately low isolation level can result in some changes being overwritten, but it is also common for users to take a large performance loss by using an inappropriately high (overly restrictive) level of isolation. This area requires careful evaluation by application developers and database administrators.

6.2.4 SQL Standards

The SQL language is the feature that most clearly differentiates the relational and emerging object-relational products from object-oriented products. SQL makes the basic assumption that all data is held in relational tables. SQL is a vendor-independent language defined by the long-existing ANSI standards SQL-86 and SQL-89. These have since been replaced by the latest version, SQL-92. Early ANSI SQL standards were minimalist, in the sense that most large vendor database systems were able to meet the standards requirements with a minimum of effort. At the same time, many vendor SQL offerings provided extensions that were not in the standard, and users were forced to deal with several different SQL dialects.

Efforts to extend the standard to encompass these special capabilities seemed relatively fragmented at first. The X/Open standards committee added to the basic capabilities of SQL-89 in an attempt to provide a common language that would be truly interoperable but would allow vendors to add new features incrementally. The ANSI SQL-92 standard was a surprising departure from previous approaches in that it anticipated new features rather than codifying existing ones. In fact, it specified many new features that in 1998 are still not offered by all major vendors. However, SQL-92 has three different levels of compliance that allow vendor SQL offerings to adjust incrementally to the new standard.

The Open Database Connectivity (ODBC) standard originally provided several new SQL capabilities, such as scrolled cursors and a programming interface, which on most database packages is referred to as a call level interface (CLI). Scrollable cursors later appeared in the ANSI SQL-92 standard, and the CLI was written up by the X/Open SQL committee. The ANSI standard currently in development, SQL3 (SQL-92 was originally called SQL2), has adopted a CLI, which has been implemented in IBM's DB2 Version 2.

Although the standards process is moving ahead, the current situation is somewhat confusing. It is appropriate for a corporate IT organization to settle on a particular standard that requires documented vendor compliance and then settle on a specific product with defined compliance that can be used for development. The SQL-92 standard in its various compliance levels should be the best practical choice for organizations to adopt. Unfortunately, most vendor products (including Oracle8.03) comply only to the Entry Level SQL-92 standard, although they implement a number of syntax items from the Intermediate and Advanced compliance levels. Because no product claims to achieve any but the lowest bar of SQL-92, it becomes difficult to judge the power and flexibility of the different products except in terms of individual features. Probably the best approach for a corporation to take is to require that the SQL of a potential database product support the X/Open standard and only those detailed capabilities of SQL-92 and the draft version of SQL3 deemed crucial by the company's database development team.

Following are a few SQL-92 capabilities that are not yet provided in all RDBMS products:

- **Extended set and join operators** – SQL-92 provides for new Boolean operations in the Select statement to combine tables, such as Intersect and Except in addition to Union, as well as several new types of join operations in the From clause. For example, a SQL-89 user would have found it difficult or impossible to join a salesperson table with a table containing total sales by salesperson-ID, where the latter would have no rows for salespeople with zero total sales, and still report names of salespeople who have zero total sales. Because a sales manager would particularly want to see names of salespeople with no sales, this is an unfortunate limitation. Using a SQL-92 outer join, names of salespeople with zero total sales would appear naturally in the result – a useful extension.
- **Scrollable cursors** – Programs should be able to fetch rows in either a backward or forward direction from a selected set of rows as well as move either relative to the current position or to absolute locations (like moving to the 1,000 row). SQL-89 cursors were only able to fetch in a forward direction, so a simple request by a user to scroll backward in a data window required enormous programming attention. In SQL-92, default behavior of changes to data under an existing cursor (cursor sensitivity) has been more carefully defined.
- **Constraints** – New referential integrity actions in SQL-92 include greater control over what was formerly default behavior, such as effects arising from cascaded delete. For example, if a customer row is deleted and an order row in the orders table references that customer, the cascade clause defines what should happen. Two alternatives are that the customer row cannot be deleted while orders refer to it, or when the customer row is deleted, all order rows referring to it are also deleted (which is referred to as a "cascaded delete").

SQL3 is the coming standard that will incorporate the object-relational model. (See "Object-Oriented Database Systems (ODBMSs)" on page 472.) Database vendors are trying to anticipate this ORDBMS standard with earlier product releases (examples are DB2 Universal Database Version 5, Informix Universal Server Version 9.1, and Oracle Release 8). The American National Standards Institute (ANSI) (and the International Organization for Standardization, ISO) have decided to split SQL3 development into a multi-part standard. The final release of the complete SQL3 standard is expected sometime in 1999, though many parts are scheduled for completion earlier.

Following are a few of the more significant parts of SQL3:

- **Foundation** – The core specification. This portion includes the basic syntax and semantics for the object-relational model, including all the new features of advanced data types (ADTs). It also includes several other new standard features, some of which are already released by a few RDBMS vendors. Some of these are explained below, including recursive union queries and triggers.
- **SQL/CLI** – The call level interface capability was previously implemented in ODBC. It is a form of dynamic SQL that works well in a client/server environment by shipping SQL statements in text strings to the server for compilation and execution. The client logic creates "handles," structures available to the client program that maintain the accuracy of information associated with certain operations. Client logic may create multiple connection handles to different databases. Statement handles that can track the status of a cursor supply the means to bind statement parameters and retrieve columns to program variables. Existing CLI implementations, including DB2 Universal Database, support retrieving multiple rows of a cursor into a program array to avoid the overhead of one-row-at-a-time fetch; they also provide several function calls to handle large objects flexibly and efficiently.
- **SQL/PSM** – The Persistent Stored Modules specification provides computational completeness. This capability generalizes a number of current vendor implementations for procedural SQL modules stored on a server (for example, Informix's SPL, Oracle's PL/SQL, and Sybase's Stored Procedures). Procedural SQL supports temporary variables and logic constructs (IF. . .THEN. . .ELSE, WHILE, and UNTIL) so that the SQL language has the power to perform any type of calculation; this is known as computational completeness. This feature has the potential to improve performance in client/server applications by allowing the programmer to perform operations flexibly on the server, reducing the data that must be transmitted back to be handled by the client. Triggers use procedural capabilities such as these, which is why SQL standards had to wait to define triggers until a standard procedural SQL could be agreed upon.

Following are some short descriptions of SQL3 features that are already present in some released products:

- **Recursive union queries** – Using this feature, a select statement is able to handle what is referred to as a "parts explosion" query, retrieving all parts and their subassemblies at all levels of recursive descent (down to a specified limit) that make up a given high-level assembly. DB2 and Oracle both include this capability. This feature could only be achieved using a rather complex Embedded SQL program in SQL-89.
- **Triggers** – Triggers are the mechanism for procedural SQL programs to be activated by certain events in the database, such as rows being inserted, deleted, or modified. When these events occur, the triggered procedure can make other changes to the database, such as creating a new row in a distinct table, that are

required by the application logic. (Database replication is often implemented using triggers.) Triggers can also be used to define constraints that are not provided for in the non-procedural constraints built into the product; this allows greater constraint flexibility. However, unlike declarative constraints, it can become difficult to understand complex triggered procedures that define business rules. Triggers were not defined in SQL-92, although many products supported them at that time, because SQL/PSM had to become part of the standard first. Triggers are now part of the proposed SQL3 standard.

6.2.5 Database Replication

Distributed databases transparently update and retrieve data from multiple databases at different locations. Classical distributed databases are difficult to administer and implement because all databases must synchronize their updates using two-phase commit protocol (both databases have to ready to perform the transaction in tandem for it to be performed at all), and outages at any one site can affect the ability of any transaction to be completed. Distributed queries are difficult to optimize and are usually costly to process. For this reason, distributed databases have not been as successful as originally expected. Many organizations are viewing asynchronous replication services as a more practical alternative to distributing data.

Database replication is the ability to make copies of a database and then communicate transactional updates that take place on one of the database copies to all the others, so that the different copies end up with the same data. These replicated database copies can exist on the same processor, but are more often found on different processors, connected by high-speed local area networks (LANs), slower wide area networks (WANs), or even dial-up telephone lines.

Vendors often mean different things when they refer to replication. The following terminology is used to differentiate types of replication that are available:

- **Synchronous replication** – This type of replication guarantees that different database copies remain in lockstep, so that changes in one database are communicated to all the others in the most timely manner (two-phase commit). This is not the common meaning of replication. The data on the update source remain locked until all target copies have acknowledged they are ready to commit their duplicating updates. The problem with synchronous replication is that update transactions might result in unacceptably long locks on the data, especially if communication is lost to some target processors maintaining copies – in which case, the correctness guarantee might fail. However, this approach normally does guarantee the ability to make consistent updates on multiple copies simultaneously. The approach is often used with very high value transactions, such as electronic funds transfers.
- **Asynchronous replication** – In asynchronous replication, after an update occurs on the source, locks are released immediately and normal processing continues while the message to duplicate the update is communicated to target copies at a slower pace. This is the more common meaning of the term replication. Because data locks are released quickly, it might seem that this approach can accommodate high volumes of transactions. However, many replication products have target databases perform single-threaded recovery of a concurrent transactional source. In this case, update messages can accumulate on disk while the replicated updates fall further and further behind the original transactions. This delay can cause problems for some applications, and anticipated transaction volume should be tested before adoption.

The following terms further differentiate replication approaches:

- **Single-master replication** – With single-master replication, updates can occur on only one of the database copies; all other database copies are passive and can act only as targets of replication messages from the master copy that permits updates. (Of course, remote sites can request an update of records at the master site.) The normal purpose of single-master replication is to allow applications at remote target sites to query local copies of an important central database. The drawback is the possibility the queries might read data that is significantly out of date at the remote target sites because of delayed asynchronous replication.

- **Multi-user replication** – The alternative to single-master replication is multi-user (or peer-to-peer) replication, where multiple copies of the database can be updated concurrently, and update messages may cross between databases to communicate changes. Unfortunately, in the common case of asynchronous replication, the updates that are performed concurrently on different copies can cause conflicts that disobey consistency rules and corrupt the database. For example, constraints that are meant to hold for all of the data may remain true when individual updates are performed on each of two copies of a database, but fail when both updates are applied in common. Or, more simply, the changes applied at one site may be overwritten by changes arriving from another replication. These conflicts will be detected by replication products, but currently, there is no known automatic way to resolve all conflicts. Therefore, this approach is only acceptable when these conflicts can be reasonably managed; for example, where the frequency of updates to an individual record is low. Often, this means data consistency is not absolutely crucial to the enterprise, or that business rules can be formulated to accommodate resolution of conflicts.

The tradeoff between the different replication approaches needs to be understood before any decision is made to adopt one as a business solution. Some products supporting database models with limited transactional capability, such as Lotus Notes, are often believed to have solved the problem of peer-to-peer replication. However, this solution frequently is achieved by requiring applications to follow an insert-only data-update model where certain distributed constraints are not supported.

For example, several knowledge workers might make concurrent modifications to a text document using revision bars. The conflicts resulting in overstruck revisions cannot be resolved automatically, but the two writers involved or someone acting as an editor can resolve the conflicts. More subtle problems, where one writer refers to a concept that another writer has renamed or deleted, are not detected by revision bars.

All major database vendors now provide a replication feature. Sybase's Replication Server handles this task for Sybase, although some vendor replication offerings will now handle replication to and from Sybase as well as Oracle and DB2 databases. Oracle's Symmetric Replication module handles replication to and from Oracle databases only, but it promises some heterogeneity in a forthcoming release. Praxis, which provides a generally available heterogeneous replication product called OmniReplicator that works for all ODBC-compliant databases, offers a separate high-performance product for replicating information between Informix databases. Some of these products support disconnected laptop environments. Microsoft's Access product offers disconnected replication, as does Lotus Notes, but with a more limited set of supported operations.

6.2.6 Database Performance

Depending on the specific application involved, some database products provide better performance than others. The difference can be significant, and it is often not easy to determine the best system because a product that has good performance in one application domain might not do as well in another. With hardware costs dropping over the past decade, there has been some debate about whether database performance should have a significant role in acquisition decisions. Certainly in some applications, performance considerations are entirely secondary because the throughput demands are low and the application logic complex. However, performance is important to a significant number of applications, especially in emerging product categories that promise accelerated response, such as data warehouses and parallel systems. (See "Parallel Database Systems (PDBSs)" on page 482.)

Database Benchmarks

A typical database benchmark measures price/performance and peak throughput on a database platform (hardware and software) for a well-defined domain of applications. Standard generic benchmarks are usually performed by hardware or software vendors to demonstrate the superior performance of their products. The ratings from such benchmarks often appear in advertisements in industry journals. Most users are familiar with the older DebitCredit benchmark, which was codified and expanded by the vendor consortium, the Transaction Processing Performance Council (TPC), under the names TPC-A and TPC-B. These benchmarks were created to measure the performance of database systems that support concurrent order entry by thousands of users in large OLTP application systems.

OLTP is an important application domain, and the TPC benchmarks measure a number of database system features that involve high volumes of update transaction throughput. The features that must be optimized include row locking, logging updates for later recovery, disk I/O buffering in memory, and the ability to support large numbers of user threads efficiently. However, generic benchmarks exist for other important application domains as well. For example, DSS applications generally support a relatively small number of users performing complex queries involving large sets of data (quite unlike the few rows inserted or updated by OLTP transactions). The database system features measured by DSS benchmarks are quite different from the features measured by OLTP benchmarks. This is the reason generic benchmarks are domain-specific.

In making acquisition decisions about database systems, it is important to understand that different systems might perform well in one application domain and poorly in another. Late in 1995, the TPC decommissioned the TPC-A and TPC-B benchmarks (they no longer validate performance ratings) and officially replaced them with the TPC-C benchmark for OLTP. The TPC-C benchmark measures price/performance for complex transaction processing (cost per transaction per minute). A new benchmark, TPC-D, measures decision-support performance on a series of 17 complex queries, with some update activity as well. The OLAP Council also has put out a new OLAP benchmark for comment.

Several results of the TPC-D have now been released and are available on the TPC's Web site (`http://www.tpc.org`). The TPC-D variation between products in price/performance is high, reflecting the fact that query performance is a new application domain for many RDBMS vendors. Analysts expect to see dynamic changes in performance for some time, especially as new methods of indexing are invented and adopted by vendors. More information is available on the OLAP Council's Web site (`http://www.olapcouncil.org`).

In spite of the domain-specific nature of benchmarks, the idea that a single benchmark can predict the performance of all applications, even within a broad domain such as DSS, is questionable. Two different DSS applications easily could achieve their best performance on different database platforms. The best benchmark to predict performance for a specific application is a custom benchmark that has been created with the specific requirements of that application in mind. This is the approach chosen by a large number of businesses making major acquisition decisions. Custom benchmarks can be costly to perform. However, a number of approaches can reduce costs, including asking potential vendors to perform the benchmark and designing a custom benchmark that uses many of the predefined features included in a related standard generic benchmark.

Parallel Database Systems (PDBSs)

A PDBS is one where the work of a single SQL statement can be shared among multiple processors working in parallel. SQL statements that access only a single row through a unique key do not have much to gain from parallelism. In the same way, OLTP systems whose transactions access only a few rows altogether are often programmed to process each transaction on a single processor when possible, avoiding the overhead of combining transactional pieces using two-phase commit. Because many OLTP transactions have short durations, a form of parallelism can still be achieved when hundreds of transactions per second are assigned, seemingly at random, to different processors, keeping them all busy. However, for a complex query involving access to indexes and data from thousands or even millions of rows, limiting execution to a single processor would result in long response times even if other processors had very short queues of tasks to occupy them. The ability to spread out the processing of a single query over multiple processors requires a feature known as intra-query parallelism, which has only recently become common in database system releases.

From the perspective of benefits, database parallelism does not provide improved price/performance for tasks that can be performed using normal processors. (Note that price/performance here refers to dollar cost per unit of throughput.) Although parallelism with multiple processors will increase possible throughput, the user pays more for the added processing power, thereby restoring the price/performance balance. Therefore, parallelism is done as a throughput increase mechanism, rather than as a cost-savings approach. Two measures of parallelism efficiency are outlined in *The Benchmark Handbook*. (See “References” on page 489.) The first, Speed-up, measures the extent to which a task speeds up with added hardware resources. A system provides linear speedup if twice as much hardware can perform the task in half the elapsed time. (This is considered the best case). The second, Scale-up, measures whether a constant response time can be achieved as a workload is increased by adding a proportionate number of processors and disks. If a company has twice as much work in its batch workload and can maintain a constant response time by doubling the computer resources, it is doing as well as can be expected. Linear Speed-up and Scale-up define the best a company can hope for as an improvement from parallelism; in most cases, linear improvement is impossible to achieve. Another advantage of multiprocessor parallelism is improved reliability, because loss of a single processor should not affect the system's ability to continue processing, although at a somewhat reduced level.

Three different types of parallelism are classically defined: shared nothing (in which two nodes provide failover for each other, but otherwise share no resources), shared disk (in which two nodes process transactions separately, but can access the same disks), and shared everything (in which disks and memory are shared, very much like the symmetric multiprocessing (SMP) model). (See "Multiprocessor Architectures" on page 353 for more information on these distinctions.)

- **Shared everything** – This parallel system is basically an SMP machine, in which threads running on different processors can communicate through shared memory and access the same set of disks. Most hardware vendors offer SMP processors, with up to about 16 processors. However, there has been a limit to the number of processors that can be added effectively because of bandwidth limitations and the fact that the different processors must use more and more time keeping track of memory access conflicts (known as a cache coherency problem). The SMP approach has been thought not to scale well to extremely large volumes of throughput or numbers of processors because the overhead increases due to the interprocessor communication required. The Non-Uniform Memory Access (NUMA) memory architecture is challenging the common belief in such limitations.
- **Shared disk** – This parallel system is one where different threads running on different processors communicate through messages on high-speed LAN or other interconnect methods, but share access to the same set of disks. The communicating processors are often called nodes of the parallel system, and each node might itself be an SMP. SMPs and shared disks are becoming more powerful and scalable. Digital's VAX and Alpha clusters take the shared disk approach, and Oracle supports shared disk systems for UNIX platforms.
- **Shared nothing** – This parallel system is one where a number of different nodes communicate by high-speed LAN or other interconnect methods; the nodes take responsibility for disks they own and that are not shared by other nodes in the system. In the shared-nothing approach, disks usually have been accessible between node pairs to allow for continued operation in the event one node crashes. The paired node takes over responsibility for the failed node's data (a process known as failover). A shared-nothing design must partition the database before the data is loaded in such a way that each node takes responsibility for a particular fraction of the data. After the load is complete, transactions must then perform runtime processing on the nodes where relevant data appear. There is no choice about how much effort is to be expended at each node – this is determined by the preexisting partitioning of the data. If the data access is uneven with respect to the partitioning (known as a data skew), it will lead to uneven processing effort on the nodes.

Many parallel database products started by using the shared-nothing approach, including IBM's SP2 database, Informix's OnLine XPS, Tandem's Nonstop SQL, and Teradata. More recently, however, sets of nodes have begun to share disks in clusters, approaching a hybrid architecture between shared disk and shared nothing. The NUMA memory architecture may provide an entirely new form of architecture, with hundreds of nodes potentially sharing a NUMA memory; this approach is not quite SMP, but communications between nodes is much faster than in the shared-nothing model.

Many practitioners believe the shared-nothing architecture is the direction for the future because it scales better than shared disk, and much better than shared everything, for very large processing requirements. The other approaches, particularly shared-disk, have their adherents as well, however, with claims they can handle all practical problems. The shared-nothing argument for scaling is certainly becoming less certain because several 100-GB TPC-D benchmark results appeared that seem to indicate much better performance on SMP than shared-nothing platforms.

6.2.7 Market Overview

For the past several years, worldwide database revenues have been continuing their migration from pre-relational products to those based on the relational model. Pre-relational products include hierarchical database systems, such as IMS, and products built on superannuated models. Market revenue for traditional systems is shrinking, although post-relational revenues are growing, leading to the slow projected growth in non-relational products. (Post-relational refers to ORDBMS and OODBMS.) The relational market's revenue growth has been slowing, but this market still saw healthy revenue increases of more than 18 percent for calendar year 1996. UNIX servers and IBM mainframes continue to dominate the platforms for which database servers are sold, whereas Windows NT has the greatest acceleration of revenues. (See Table 6-7.)

Table 6-7: Worldwide Server Platform DBMS Revenue – 1996

Server Platform	Pre- and Post-Relational		Relational	
	Revenue ($millions)	Growth	Revenue ($millions)	Growth
IBM mainframe	$911	11.4%	$1,151	8.3%
Other mainframe	$426	9.4%	$350	8.3%
UNIX	$196	25.1%	$3,676	19.6%
Windows NT	$40	279.4%	$918	241%
Other (including PC platforms)	$196	-13.4%	$954	-21.9%
Totals	$1,769	10.6%	$7,049	18.5%

Source: IDC 1997

Although pre-relational and relational systems will continue to be used for some time into the future, the relational model will begin to be replaced by the object-relational model during the forecast period. In addition, newer types of database system products are gaining mind share, including object-oriented database systems, OLAP/ROLAP, and data warehouse systems. For example, object-oriented database products have been available for several years, doubling in sales each year until about 1995, when the rate of growth dropped to a more sustainable 45 percent.

Total worldwide database software license revenue in 1996 was $8.08 billion. However, revenue growth is projected to slow, especially for RDBMS engine sales, as these products increasingly are sold as a bundle with SAP R/3, Baan, and other enterprise applications. Such sales represent less revenue for database vendors.

Table 6-8: Worldwide Software Revenue for Database Vendors – 1996

Company	Relational		Pre- and Post Relational		Combined	
	Revenue ($millions)	Share	Revenue ($millions)	Share	Revenue ($millions)	Share
Oracle	$2,372.0	33.7%			$2,372.0	26.9%
IBM	$1,166.7	16.6%	$392.3	22.2%	$1,559.0	17.7%
Sybase	$525.7	7.5%			$525.7	6.0%
Informix	$464.8	6.6%			$464.8	5.3%
Computer Associates	$112.1	1.6%	$327.6	18.5%	$439.7	5.0%
Microsoft	$237.0	3.4%	$3.0	0.2%	$240.0	2.7%
Other	$2,163.6	30.7%	$1,044.1	59.1%	$3207.7	36.4%
Total	**$7,041.9**		**1,767.0**		**$8,808.9**	

Source: IDC, 1998

Competitive Issues for Database Vendors

The primary competitive issues for the database market in 1998 included the following:

- Providing support for high-performance data warehouses
- Providing object management and object-relational SQL, especially for use with distributed computing and Web applications
- Providing support for network computers (NCs)

Support for high-performance data warehouses breaks down into two main capabilities: good query parallelism and advanced indexing capabilities. Informix is generally viewed as having the best parallelism solution, and Sybase IQ seems to have the best indexing.

With Release 8.0, Oracle has provided object-relational capabilities, but Informix, which acquired Illustra some time ago, probably has the preeminent object-relational product at the present time, the Universal Data Server. One problem with Informix, however, is that there are three different product lines: Universal Data Server (UDS), Online Dynamic Server (ODS), and Extended Parallel Server (XPS). This setup means that users wanting to use object-relational capabilities on a parallel server will be unable to do so until the product lines are merged.

NCs are the thin client solution to the high cost of PC support in the office. (See "Network Computers (NCs) and Network PCs (NetPCs)" on page 359.) At a minimum, database vendors must make it possible to use NC applications with their servers; all vendors are adding this capability. Oracle, in particular, is supporting the new application model with its Network Computing Architecture.

Database Vendor Update

The key database vendors include Computer Associates, IBM, Informix, Microsoft, Oracle, and Sybase, which are discussed in this section. Other vendors worth noting include NCR, Software AG, and Tandem.

■ IBM

IBM offers two significant database system products: IMS and DB2. IMS is a hierarchical database system on the MVS operating system that has been IBM's preferred solution for large, mission-critical applications for many years.

DB2 is IBM's relational database product line, with several different implementations on different hardware/operating system platforms. The original flagship DB2 ran on MVS on the 3090. IBM's DB2 Version 2 was released in July 1995, with several new capabilities, including numerous object-relational features. DB2 Version 2 is currently available for Intel- and UNIX-based platforms such as HP-UX, IBM AIX and OS/2, Silicon Graphics IRIX, Sun Solaris, and Windows NT.

IBM's DB2 Universal Database 5.0 (UDB) was released in September 1997 with several capabilities to support multimedia, including an expanded set of object-relational features. For its initial release, DB2 UDB 5.0 was made available in server form on Windows NT and in client form on OS/2, Windows NT, Windows 95, and Windows 3.1.

IBM provides a highly parallel hardware platform known as SP2. In 1995, IBM released DB2 Parallel Edition for SP2 with a comprehensive toolset. Most other relational vendors also offer SP2 products, including Informix, Oracle, and Sybase – creating a good deal of competition on the platform. In 1997, IBM reduced prices for the SP2 platform, clearly responding to some price/performance measures in TPC benchmarks.

■ Informix

Informix is a relational database vendor for UNIX and PC platforms, well-known for providing good technical solutions. Informix has a respected parallel database system architecture, OnLine XPS. In 1997, it released a new indexing capability, known as DSS Indexing, which provides bitmapped indexes and join indexes to accelerate query performance. Informix acquired Stanford Technology Group in 1995 for its MetaCube OLAP product and plans to support OLAP aggregation optimizations in its database engine. In 1996, Informix purchased Illustra for its object-relational solution. This acquisition immediately made Informix the major vendor in terms of object-relational capability, with the ability to build customized DataBlades for application areas such as geographic information, time-series analysis, and Internet multimedia. These capabilities were prominently featured in the Universal Server offering. A large marketing push by Informix to promote these capabilities met with disappointing results because the market for these features turned out to be smaller than expected.

RDBMS revenues for Informix were less than projected in 1997 due to a company-wide focus on the Informix Universal Server, whose acceptance proved to be lower than expected. In addition, Informix accounting errors for 1995 and 1996 have made it difficult to ascertain the actual product-line growth for these years.

As mentioned previously, a growing technical problem with Informix has been the splintered product line that offered different capabilities in different products, but not all capabilities at once. In November 1997, Informix announced that it would be repackaging and offering a single core database product along with a menu of options supporting the other capabilities. At the same time, the company reduced prices. For example, where UDS previously had cost $2,500 per user, these features would now be a $300 option to the core offering, which remained at $1,500 per user.

■ Microsoft

Microsoft has mounted a significant effort to enter the enterprise database market. In 1994, it ended a joint marketing agreement with Sybase and began offering its own version of the SQL Server database system (estimated 1996 revenues of $214 million). This product (SQL Server 6.0) was an older version of Sybase's product, released with a few Microsoft enhancements.

In December 1997, Microsoft released SQL Server 6.5, which featured several improvements, including certified conformance with SQL-92. Microsoft still has not announced SQL Server solutions in some important areas of current concentration, such as object-relational capabilities and advanced indexing for queries; an initial version of intra-query parallelism in a SQL Server code-named Sphinx is due in mid-1998. Microsoft has shown itself to be a low-cost provider, aiming to make its profit through large volume at the low end.

Microsoft's desktop database product, Visual FoxPro (estimated 1996 revenues of $50 million), is known in vertical markets for its good execution performance. Microsoft's Access database product is popular for home users and vertical application developers, with 10 million copies sold and estimated 1995 revenues of $300 million. (More recent figures for Access database sales are unavailable because Access has been bundled into the Microsoft Office suite of applications.) Access has strong mobile database features, which allow disconnected update and replication of updates on docking.

■ Oracle

Oracle was an early entrant into the relational database market, creating an important presence by offering a database system that was portable across all key platforms, from MVS to UNIX and OS/2. Many other platforms have followed, including Windows NT, into which Oracle has put enormous resources. On the technical side, Oracle performs original development of each new release on the Windows NT platform, rather than simply porting products to it. In addition, the Windows NT Oracle product line has a separate marketing organization. This concentration has paid off: Sales of Oracle on the Windows NT platform currently equal or exceed sales of Microsoft SQL Server.

According to IDC, total Oracle license revenue in 1996 from UNIX and Server Operating Environments (SOE) grew 21 percent, compared to average industry growth of 3.1 percent. In the Windows NT market, Oracle's total revenue grew 223 percent, reaching parity with Microsoft SQL Server revenue.

Oracle's new Release 8.0, with object-relational capabilities, was released in 1997. The Oracle object-relational model is powerful, supporting refs and nested tables as column values (nested tables take the place of "sets" or "lists"). Oracle also provides new object types and the capability to Select from an arbitrary Select statement in the From clause.

Oracle provides an OLAP product, Oracle Express, with the largest market share among competing OLAP products. During 1997, Oracle provided the capability for Express to reference relational data within Oracle tables to answer queries that were too detailed for normal preaggregated data internal to Express. This feature is known as ROLAP, while the preaggregated data in Express is multidimensional data, known as MOLAP. Therefore, users can now choose which approach they want for individual queries.

■ Sybase

Sybase was an early relational database player on UNIX platforms and was the first vendor to offer stored procedures in Transactional SQL, an important performance advantage for client/server systems because it allowed complex procedures to be performed on the server without a great deal of communication back and forth with the client. This feature is now part of the future SQL3 standard, appearing under the name Persistent Stored Modules.

In 1995, a problem with Sybase's Symmetric Multiprocessing performance was noted in a number of industry publications, along with the difficulty Sybase was having in making the switch from transactional page locking to row locking. These articles gave the impression that technical problems had escaped Sybase management, translating into poorer performance; as a result, customers began to fall away. Sybase experienced revenue losses in 1995 and 1996. By late 1997, however, Sybase had announced its fourth consecutive quarter of profitability.

On the technical side, Sybase began shipping a new DBMS Engine release in September 1997, Adaptive Server Enterprise 11.5, which it characterizes as the fastest database available. However, this release is still missing important features, especially row locking. To address this deficiency, Sybase announced plans to release a row-locking version of Engine in early 1998. This version will also offer full support for Java. In part because of the delay in shipping these features, recent improvements in sales have been mainly in the area of tools rather than database engines.

Sybase offers a product known as IQ, a data warehouse capability with advanced types of bitmap indexes. IQ was released in mid-1996, and it has demonstrated strong performance recently in benchmarks.

■ Computer Associates

Computer Associates (CA) markets CA-IDMS, a network database product running on IBM mainframes (originally developed at Cullinet Software); Datacom DB, another pre-relational product; and CA-Ingres, a relational product that came out of a public domain database project developed at the University of California, Berkeley. CA currently is in fourth position in total worldwide DBMS engine and tools revenues, and it has achieved good growth. However, CA generally is not considered to be one of the top technology providers in the RDBMS area, so its products may face difficulty matching new features of products in the field.

6.2.8 Forecast

- The object-relational approach will replace pure relational products completely in the next 5 years because the coming SQL3 standard is based on the object-relational model. All major relational vendors will be providing new releases with object-relational extensions.
- The object-relational approach will squeeze pure object-oriented databases into an increasingly smaller niche.

- Universal servers that are based on new DBMS object capabilities promise to provide convenient access to and enable new complex operations on many non-traditional data types. Databases and enabling toolkits will manage new information types such as multimedia, text, time-series, and geographic data. These capabilities will continue to improve as more object-relational capabilities are rolled out with new DBMS releases.
- Due to the widespread migration to object-relational approaches, database design will become a critical part of the system development life cycle. New methodology and retraining will be required.
- Advanced indexing will be needed to improve query performance, especially for data warehouse and other decision-support applications that avoid concurrent update and more easily support complex indexes. Some of these algorithms and techniques are already known, and others remain to be developed. Parallelism continues to be important, but parallelism alone will not improve price/performance. Without advanced indexing, parallelism simply enables more processors to use the same inefficient techniques. All major vendors have released advanced indexing features, but additional improvements can be expected in the future.
- Preaggregation will be made a part of the underlying database engine, with preaggregation comparable to indexing as a technique for speeding up queries.
- Database replication will continue to be an attractive and easier-to-implement alternative to two-phase commit for applications not requiring second-by-second concurrency of data across multiple databases.
- Databases increasingly will include features to support data warehousing, such as bit-mapped indexes, parallel query execution, and multidimensional capabilities.
- IMS will continue to be used widely in existing S/390 sites throughout the forecast period.

6.2.9 References

■Articles

Berenson, Hal, et al. 1995. A critique of ANSI SQL isolation levels. *Proceedings of the 1995 ACM SIGMOD*: 1-10.

Norman, Michael G., et al. 1996. Much ado about shared-nothing. *ACM SIGMOD Record*, Vol. 25, No. 3. September.

O'Neil, Patrick, and Dallan Quass. 1997. Improved query performance with variant indexes. *Proceedings of the 1997 ACM SIGMOD*: 38-49.

■Books

Cattell, R. G. G., et al. (editors) 1997. *The Object Database Standard: ODMG 2.0.* San Mateo, Calif.: Morgan Kaufmann Publishers.

Date, C. J. 1995. *An introduction to database systems.* (6th edition) Reading, Mass.: Addison-Wesley Publishing.

Gray, Jim. (editor) 1993. *The benchmark handbook for database and transaction processing systems.* (2nd edition) San Mateo, Calif.: Morgan Kaufmann Publishers.

Kimball, Ralph. 1996. *The data warehouse toolkit.* New York: John Wiley & Sons.

O'Neil, Patrick. 1996. *Database principles, programming and performance.* San Francisco: Morgan Kaufmann Publishers.

■On-Line Sources

X/Open CAE Specification, Data Management: Structured Query Language (SQL), Version 2. `http://www.XoSpecs@xopen.org`.

■URLs of Selected Mentioned Companies

American National Standards Institute `http://web.ansi.org`
Baan `http://www.baan.com`
Computer Associates `http://www.cai.com`
Digital `http://www.digital.com`
IBM DB2 `http://www.software.ibm.com/data/db2/udb`
Informix `http://www.informix.com`

Microsoft http://www.microsoft.com/products
NCR http://www.ncr.com
Object Database Management Group http://www.odmg.org
OLAP Council http://www.olapcouncil.org
Oracle http://www.oracle.com/
Praxis http://www.praxisint.com
SAP http://www.sap.com
SQL Standards http://www.jcc.com/sql_stnd.html
Sybase http://www.sybase.com/
Transaction Processing Performance Council http://www.tpc.org
X/Open Standards Committee http://www.opengroup.org

6.3 Groupware, Workflow, Document and Knowledge Management

6.3.1 Executive Summary

Over the last 10 to 15 years, office work in the United States and in the industrialized economies around the world has undergone a profound technological transformation. Executives, managers, and clerical workers alike no longer rely on paper-based documents, transmitted via the postal service, for their critical on-the-job information. Rather, they routinely capture information electronically – using keyboards and telephones – and then transmit electronic artifacts over ubiquitous networks at the speed of light. Knowledge workers read, interpret, and act upon these electronic bits and bytes at remote locations – defining them as voice or e-mail messages, attached documents, or facsimile transmissions. More often than not, they also store these "bags of bits" for later retrieval in some kind of on-line repository. There seems to be no end to the number of ways in which people might communicate and send information to one another – leading to a multitude of alternatives competing for an individual's time and attention. However, creating the organizational and technical frameworks through which people can meaningfully share and use electronic transmissions is an entirely different matter.

This chapter describes how people can use the power of work-group computing environments to communicate and share information to enhance their work-a-day activities substantially. Groupware, workflow, and document management technologies are helping corporations, government agencies, and other businesses improve communication among work groups, manage paper documents by routing them electronically, and control computer-generated documents and data through document library functions. Perhaps most important, the collective knowledge of an organization – methodologies, client contacts, success stories, case studies, and more – can now be stored, retrieved, and used economically by anyone authorized to do so. Leading-edge companies are using groupware, workflow, and document management technologies to capitalize on this past knowledge to avoid reinventing the wheel and to make more informed decisions – leading to an additional set of related technologies, which have come to be known as knowledge management.

Individually, or used together, these four kinds of work-group computing technologies provide tremendous benefits that may not otherwise be realized. Groups can collaborate on a project by accessing project documents, schedules, calendars, and other group tools. Workers can either work independently by adding their own documents or comments to the project, or they can work collaboratively on the same documents. People working together at a distance can coordinate the flow of their work activities so that the reviewers of a report might automatically have access to the latest draft, and so that the production specialists might automatically assemble revisions from multiple sources.

Groupware and document management products help knowledge workers make more efficient use of their time and skills. As corporations continue to reengineer their businesses, they become more reliant on technology to replace manual processes. As companies move to electronic systems, more documents are authored and stored on-line, thus improving the efficiency of the organization and reducing the need for physical facilities and resources.

The Internet now provides the underlying network connections to link people and organizations together regardless of location. It serves to fuel the growth in Web-based solutions for groupware, workflow, and document management that allow for widespread distribution of information and expertise. All the major vendors have released new or enhanced products that capitalize on the connectivity of the Internet. A few products, such as Internet-enabled workflow, provide new functionality that would not be possible without the Internet.

Knowledge workers now function in a world of ubiquitous connectivity – where anybody can communicate and share information with anybody else – provided, of course, that they know their address and how to contact them. Organizations large and small are struggling to learn how to leverage this connectivity for competitive advantage – where communications and information-sharing activities are determined by business policies and standard operating procedures, rather than by network topologies or other artificial (technical) barriers. These rapidly changing business needs are leading to a host of new products in the marketplace for work-group computing environments and to important developments in the workplaces of the advanced industrialized economies around the world. With the realization of anytime, anyplace connectivity, business managers and technology leaders alike must now be able to outline, as a core aspect of their business strategies, who needs to exchange information with whom, and for what competitive business purposes.

6.3.2 Groupware

Groupware is a general class of software products that facilitates communication, coordination, and collaboration among people. Groupware is important because it allows work groups – people who need to interact with one another within an organization – to communicate and share information, even when they are working together at a distance. Groupware has come to encompass a variety of applications and functions that use the power of electronic networks and on-line connections to bridge both time and space, enabling people to work together effectively.

When people work in groups, they need to discuss the various issues at hand, share perspectives, develop insights, make decisions, and communicate results. Group work consists of an ongoing series of interpersonal conversations – discussions, debates, fact-finding analyses, exchanges of perspectives, location of previously stored documents, and so on. Exchanging, managing, and directing the flow of information is a crucial aspect of group life. Groupware provides the enabling environments for people to manage their personal needs for interpersonal conversations.

Groupware allows designated work groups to share documents electronically and perform tasks such as writing, editing, and electronic note-taking – supplementing real-time communications with asynchronous, electronic connections. For instance, one person might create a memo and send it to immediate colleagues for review before distributing it to others within the organization. Colleagues might be on the road or working remotely at different locations around the world. Thanks to electronic networks, e-mail, and shared discussion databases, group members have the ability to communicate at any time and from any place. Remote colleagues might suggest revisions within a matter of hours and post their comments in a shared database. Or traveling colleagues could forward their comments automatically to the author and have them collated automatically with the original message. In either case, the author has to look in only one place to find all the comments, can quickly make required modifications, and then can distribute the memo without delay.

Groupware enables people to continue an ongoing set of discussions where one person makes a statement, others reply, and all can read (and then continue to comment upon) the results – as if all are participating in an extended meeting. Groupware includes the capabilities for determining the boundaries of group discussions – whether only a designated group of people can participate, whether many people might read the comments and discussions that only a few can author, or whether anybody who has an interest in a specific topic might join the discussion as a full participant. In addition to providing impromptu meeting capabilities, groupware enables the results of the meeting and subsequent documentation presented to be captured, stored, and later retrieved as part of the project history.

Groupware embodies two interrelated perspectives about how work groups can exploit on-line information technologies:

- **A communications-centric perspective** – Groupware capitalizes on the flow of information across electronic networks – the ability to foster both one-to-many and many-to-many communications. People no longer need to wait for printed materials to arrive.

  E-mail itself represents a simple, yet direct, form of groupware, where one person can send a message to another, or to an entire distribution list of people. Similarly, many people can continue to communicate with one another via e-mail simply by continuing to redistribute their messages via predetermined distribution lists. At minimum, people receive continued notifications of one another's messages as well as their own personal copies in their e-mail in-boxes. Individuals can share information easily with one another by continuing to pass messages through electronic networks – adding their unique perspectives and insights.

  These conversations and communications can expand into sets of "threaded discussions" where groupware applications track who is posting which messages, and then the individual messages are redistributed to other group members. They can be further managed by active agents, capable of automatically directing (or redirecting) the flow of information from one person to another – or from one location to another. Among other things, groupware adds a modicum of structure to unstructured e-mail message exchanges.

- **A database-centric perspective** – Groupware capitalizes on the ability to access shared information repositories. Group members might store all their official memos, formal reports, and informal conversations related to particular projects in a shared, on-line data store, such as the underlying file system or a formally defined database. Then, as individual people need to check on the contents of individual project reports, they might simply access the shared data store to find the information they need. When they want to share information with other group members – such as update one another about the status of their own individual projects or raise concerns – they then add the information to the common information repository. The shared information repository determines group membership – who has the privileges to access, modify, or create new documents within the on-line data store.

Of course, information can be categorized in many different ways. Group members need to understand how to browse through the various information categories. The underlying repositories facilitate browsing depending upon their specific data structures. For instance, a shared filing system (a very simple groupware environment) has a fixed hierarchy of folder names and file names. Alternatively, a formally defined data-store includes capabilities to group and sort individual items based on titles, authors, dates created or last modified, specific keywords, particular action items or status alert flags, or other criteria that the group might find useful.

Groupware's increasing popularity is due to its ability to support the initiatives behind business process reengineering programs easily. Many corporations today realize that groupware has more potential and capability than simply allowing two or more people to share information with one another remotely. Groupware allows geographically dispersed people to collaborate by sharing critical information in a timely manner – electronic networks help to bridge time and space. Corporations can easily build applications that leverage the intellectual assets of individual business teams – providing team members with critical information that helps them do their existing jobs better or work with one another in entirely new ways. When all is said and done, groupware allows firms to redesign work functions to leverage the capabilities of distributed business teams.

Another reason for the increasing popularity of groupware is the need to share information throughout an enterprise. Many different people within an organization can access product information, technical data sheets, technology white papers, and similar information, using one or another groupware environment. Information that was previously located in a departmental file or self-contained database at one location can be easily accessed over the network and replicated, if necessary.

The underlying computer network topology is critical. On self-contained local area networks (LANs) – as have been widely deployed in corporations throughout the 1980s and early 1990s – groupware databases need to be replicated from one physical location to another. Using database replication, remote users access and update information held on a shared LAN server. At periodic intervals (hourly, daily, and so on), these servers check with one another and replicate their contents – either through a dial-up telephone line or some other kind of computer-to-computer network connection. Users connecting to their specific LAN server then have access to the information created by other users and stored on remote (albeit replicated) servers. Replication capabilities take care of synchronizing the information and performing periodic database updates. The replicated data stored at an individual LAN server, however, is only as fresh as the latest replication.

With the advent of enterprise-wide intranets (a private network that uses Internet software and standards and provides employees with easy access to company information), anybody with connections to the network can access the shared information. With greater bandwidth readily available on business networks, from the mid-1990s onward, there has been less need to replicate the data stored in database servers. Remote and mobile users can be located almost anywhere in the world, as long as network connections allow access to a server. Mobile users still need to be able to upload or download the most current data and synchronize database replicas for offline access to the shared file stores.

Groupware on an enterprise level also allows departments to share information that previously did not cross departmental lines except as paper documents. For example, the collaborative writing of a contract could involve a sales representative, a customer service representative, a contracts specialist, and the legal department. Using groupware, each department – although actually located in different buildings and geographical locations – would have on-line access to the contract.

Groupware provides the following productivity benefits to organizations, departments, and business teams:

- **Project management** – Allows controlled management of project documents, schedules, and personnel
- **Location independence** – Allows project personnel to be dispersed geographically without regard to location or time
- **Communication capability** – Provides for communication among project personnel through a variety of means, including e-mail and discussion forums. Even outside the strict confines of a defined project, "birds of a feather" discussions by employees are enabled by groupware.
- **Information availability** – Because project personnel and project information are linked, project information remains current and accessible. The right information gets to the right people at the right time.
- **Business processes** – Provides many different kinds of capabilities to sequence the flow of information from one person to another, and from one work group to another, across the network to expedite work activities

Groupware is providing increasingly important features and functions for leveraging the capabilities of distributed information networks. The major vendors (and products) in the groupware market are IBM/Lotus (Lotus Notes), Microsoft (Exchange), Netscape (SuiteSpot), and Novell (GroupWise).

Lotus Notes

Lotus Notes is a client/server platform for end users and developers to build and deploy groupware applications. (Lotus was acquired by IBM in mid-1995, and is a wholly owned subsidiary of IBM.) Notes is a sophisticated, document-oriented database that allows users to share many different types of unstructured information, including text, graphics, spreadsheets, reports, and word processing documents, regardless of platform or network. The product also offers group calendaring and mail functions. Notes clients run on Apple's Macintosh operating system (MacOS); HP's HP-UX; IBM's OS/2 and AIX; Microsoft's Windows; Novell's NetWare; Sun's Solaris; and The Santa Cruz Operation's SCO UNIX.

Notes Release 4, released in mid-1996, made significant improvements to the user interface. In addition, the programming language – LotusScript – is a powerful programmer productivity tool, with more support for standard application programming interfaces (APIs). Its agent-building function improves the macro capabilities and is intended to allow end users to do more programming themselves. Release 4's replication, which in previous versions of Notes took place on a document level, occurs on a field level, allowing only those parts of a document or record that have been changed to be updated on the server. This feature improves the efficiency of distribution, particularly to mobile remote devices.

Although the underlying architecture for Notes predates the advent of the Web, Lotus has capitalized quickly on the capabilities of Internet connectivity. In December 1996, Lotus released a new product family, Domino, which extends the capabilities of its Notes server and fully embraces the core underlying standards of the Internet and the Web.

Domino adds support for the Web's HTTP protocol to the Notes server, allowing both Web browsers and Notes clients to access server-based Notes documents. Domino allows users to access the same documents through a Web browser, converting documents on-the-fly or through regular Notes applications. Domino allows the creation of Web-based Notes applications that may combine workflow and groupware capabilities. Thus, virtually any current standard Notes application can become an Internet application with full Notes functionality.

At the same time Domino was introduced, Lotus also released Notes 4.5 client software, supporting HTML 3.2, Java applet execution, Netscape plug-ins, the Secure Sockets Layer (SSL), and agent technology that can monitor selected Web pages and notify the user when they have changed. The new client also includes group calendaring and scheduling functions.

The current version of the Web-based environment, Notes/Domino 4.6, adds support for customized applications, including Java applets. Both the standard Notes server and Domino can now be administered via a Web browser. The first "mail-only" version of Domino – the Domino Mail 4.6 server – includes e-mail discussion forums, and calendar/scheduling. Domino Mail now supports POP3, SMTP/MIME, IMAP4, LDAP, NNTP, and full access via HTTP and HTML.

Notes Release 5.0, announced for shipment in mid-1998, offers several significant changes over previous versions. Notes 5.0 will introduce a major user interface change, replacing the familiar Notes "tile" interface in favor of one that more closely resembles a Web page. In addition, Lotus is encapsulating key Notes technologies such as Views, Forms, and Navigator as JavaBeans, enabling applications previously deployed only on Notes clients to be accessible via a Web browser. This component strategy will be a key part of the 1998 next-generation Domino server release (code-named Maui) as well as the Notes 5.0 release.

Microsoft's Exchange

Microsoft's Exchange is primarily an electronic messaging server and also incorporates basic groupware functionality for sharing ad hoc information. It provides e-mail with a universal in-box and supports work group activities with such additional features as interactive scheduling, built-in access to shared bulletin boards, forms design, and access to publicly shared folders on computer networks. It also offers built-in connectivity to the Internet or corporate intranets.

Exchange is based on Microsoft's Mail API (MAPI), which provides application developers with the capabilities to tie electronic messaging capabilities into business applications and support "mail-enabled" applications. Thus, users might query a database, develop a report, and send it (via e-mail) to business team members across the country without having to exit from the business context and invoke a separate messaging application.

Now in its third major release, Exchange 5.5 shipped at the end of 1997. (Microsoft has been shifting version numbers so that all its layered products will correspond somewhat with the underlying versions of Windows NT.) Exchange 5.5 supports relevant Internet standards, including POP3, IMAP4, S/MIME, LDAP-V3, X.509 Version 3, and HTTP/HTML client access to mail, NNTP client access to public folder discussions, and SMTP for native e-mail transport. This server now supports message stores of virtually any size, removing the previously imposed hard limit of 16 GB. Exchange 5.5 includes new collaboration-related features for server-side scripting, allowing developers or administrators to create event- or time-driven collaborative applications using VBScript or JavaScript. Unfortunately, Exchange 5.5 runs only on Windows NT 4.0 and later.

Until recently, Microsoft has had a confusing mail client story. It is now trying to make clear that Outlook, which began shipping with the Exchange 5.0 Server and includes integrated scheduling features, is the preferred mail client. Microsoft continues to ship the Exchange client for those accounts that want to use it for a plain e-mail environment; it is particularly applicable for non-Windows environments. Microsoft also includes a mail client, dubbed Outlook Express, with Internet Explorer 4.0.

Netscape's SuiteSpot Servers

Netscape expects to leverage the capabilities of corporate intranets through its SuiteSpot 3.0 family of messaging, threaded discussion, directory, calendaring, and information sharing servers. Team members might send and receive e-mail using the E-mail Server, read shared documents stored on ordinary Web servers, track threaded discussions on a Collabra Server, create a catalog of shared information with Compass Server, and coordinate schedules via a Calendar Server – all while maintaining a consistent and uniform set of group access rights and privileges via the Directory Server. Thus, SuiteSpot ensures network-wide membership within individual work groups – the Directory Server determines who has access to what kinds of information resources within an enterprise.

The Collabra Server within SuiteSpot is the heart of the groupware aspect of Suite Spot. With Collabra, users now share any HTML document or other kind of digital content accessible over the Web. They can organize ad hoc conferences with descriptive names (rather than the cryptic nomenclature of the Internet), control membership and content through a robust suite of moderator access privileges, and enable full-text on-line searching using the underlying capability of Verity's Topic text retrieval engine. Individual group members can replicate and download the contents of individual conferences using NNTP – replication occurs at the document level and users can decided whether to include both binary attachments and GIF files.

The Compass Server, released at the end of 1997, extends the capabilities of Netscape's original Catalog Server and is designed to facilitate information sharing among work groups within an organization. Compass Server continues to spider remote Web sites to create enterprise-wide indexes of distributed resources. It adds an additional capability – automatic categorization. Thus, a spider that monitors documents on remote sites and finds such terms as "waste water," "recycling," "hazardous waste," "guidelines," and "mandates" might reasonably assume that this site contains information relevant to "environmental policy," and hence, be part of a "government and politics" section of an index catalog. Compass Server includes sets of categorization rules based on such factors as URLs, metatags, or other kinds of relevant information found at remote sites. This kind of capability is useful for knowledge-management solutions (discussed below).

Novell's GroupWise

GroupWise integrates calendaring/scheduling, task management, shared folders, threaded conferencing, workflow, and remote and Internet access into a Universal Mail Box. The Universal Mail Box gives users single-point access to personal calendaring, group scheduling, tasks, voice mail, faxes, and other message types from the desktop, the Internet using GroupWise WebAccess, and even from the telephone (users can call in and have their e-mail and appointments read to them as well as their voice mail). The seamlessly integrated workflow available in GroupWise 5.2 provides a complete environment for designing and executing messaging-based administrative and collaborative processes. All GroupWise items (mail messages, workflow packets, appointments, voice messages, folders, and so on) are managed by an underlying object management system.

GroupWise 5.2 allows companies to leverage their corporate network investments by providing a collaborative solution for the corporate intranet. The product provides an open, extensible architecture, which can be enhanced by customers, value-added resellers (VARs), and third-party developers through the use of industry-standard and GroupWise-specific APIs. GroupWise is scalable from the work group to the enterprise, and it provides a reliable, cost-effective communications infrastructure.

GroupWise 5.2 allows network administrators to manage users, groups, network devices, and applications centrally and hierarchically. Users benefit by easily locating and using network devices, applications, and files in a distributed network environment.

GroupWise 5.2 is platform-independent, supporting MacOS, PowerMac, UNIX, and Windows clients as well as NetWare NLM, OS/2, UNIX, and Windows NT server platforms. In addition, GroupWise also supports industry-standard communication protocols such as NetBEUI, NetBIOS, POP3m, SMTP, TCP/IP, and X.400.

Other vendors beginning to challenge the established players in the groupware space include Oracle, Radnet, SoftArc Inc., and the TeamWare Group.

Oracle's InterOffice

Oracle has redesigned its Oracle Office product line, creating InterOffice, a product line that provides equivalent functionality from a desktop thick client and from a Web browser. The product is being optimized for the Network Computing Architecture that Oracle is promoting, and a Java e-mail client is the first implementation of this architecture in InterOffice.

Oracle is positioning InterOffice as "database messaging." The goal of database messaging is to provide a framework for delivering critical information to users and other applications while leveraging the security, stability, and integrity of the underlying database infrastructure – an area where Oracle has proven leadership.

Oracle InterOffice includes two integrated components: the Messaging Cartridge, which includes e-mail, calendar/scheduling, and directory services; and the Document Cartridge, which includes document management and workflow. Both Cartridges also include the ConText linguistics product, which can summarize any document as well as determine common themes in documents, and the Oracle InterOffice Composer, a rudimentary word processor and presentation graphics package that can be used from any standard Web browser.

Radnet's WebShare

WebShare utilizes the native capabilities of the Web – ordinary Web servers accessed through Web browsers – to support work-group computing applications on corporate intranets and the Internet. WebShare is designed to enable business application designers to create work-group computing applications quickly that share document-oriented information in a structured fashion. As a result, business teams can easily pool knowledge and expertise, track items and issues, communicate across geographic and time boundaries, collaborate to solve problems, or update team members on the status of specific business activities.

Released in September 1997, WebShare 2.1 comprises a middleware layer running on Windows NT platforms, together with a Windows-based development tool for creating (and managing) customized forms and views. It is an entirely "open" environment that utilizes the underlying capabilities of any ODBC-compliant relational database – such as SQL Server or Oracle. It includes full client/server database replication capabilities so that mobiles users can take their laptops on the road and then synchronize their personal data with the shared information stored on a work-group server at periodic intervals.

SoftArc's FirstClass

FirstClass is designed for departmental e-mail, electronic forms processing, group discussions, and real-time chat over both LAN and dial-up connections. FirstClass 3.5, released in late 1996, includes client-side replication for the capability to work offline, and an upgraded Client Extensions Toolkit that lets users access their organization's critical corporate data from where it normally resides. FirstClass provides access to SQL client/server databases through the graphical FirstClass forms engine that then connects directly to the underlying business databases. Originally designed to run on Macintosh servers, FirstClass now also supports Windows NT platforms. It continues to feature easy system administration and very low management overhead and support costs.

FirstClass Intranet Server, released in July 1997, allows core services to be accessed through Internet protocols. It includes calendaring and scheduling features plus improved document-editing tools that support hot links, embedded graphics, and tables.

TeamWare's TeamWare

TeamWare (a division of Fujitsu Ltd. and part of Fujitsu-ICL Systems Inc.) has developed a modular client/server-based groupware product that provides e-mail (with gateways for SMTP, MHS, X.400, and fax), group scheduling, document management, discussion database, workflow, and remote access for mobile workers. It scales from work group to large enterprises, and it can replicate databases and directory information across a WAN. It encompasses an overall framework for developing and managing groupware applications in a consistent manner that adapts to the underlying culture and business requirements of an organization.

6.3.3 Workflow Technologies

At the most basic level, workflow is the automation of business processes, managing the movement of information as it flows through the sequence of steps that make up the work procedure. A workflow application defines all the steps in a process, from start to finish, including all exception conditions, usually based on established business rules. Key to workflow management is tracking process-related information and the status of each instance of the process as it moves through an organization.

Workflow applications fall into four categories:

- **Ad hoc workflow** – Ad hoc workflow is a one-time-only process. It is typically e-mail-based and can be thought of as an intelligent routing slip, where, once one person is done with the information, the document is routed automatically to the next person on the list.
- **Collaborative workflow** – Collaborative workflow coordinates the process of people working together to achieve a goal, providing a management layer to processes that are used regularly in an organization. Collaborative workflow is usually supported via e-mail or document management-based workflow products. An example of a collaborative workflow would be creating an annual report.
- **Administrative workflow** – Internal administrative processes, such as expense reimbursement and purchase order requests, are usually addressed with forms-based workflow products. The intelligent forms can serve as a front end for data from underlying corporate databases, which are then delivered from person to person (or department to department) using the organization's messaging infrastructure.
- **Production workflow** – The premier workflow applications are high-end, mission-critical, strategic, transaction-oriented processes. Often, these applications include image processing requiring specialized hardware and software. They may also include the use of intelligent forms, database access, and ad hoc capabilities. However, in production workflow applications, the database transaction is key. A good way to think about transaction-based workflow is that, in these cases, the business process is the business itself, rather than merely supporting the business. For example, consider the strategic nature of claims processing to insurance companies and loan processing to banks.

It is important to note, however, that the boundaries between the different categories of workflow applications are not solid. The four categories can be used as guidelines to help narrow down the list of appropriate products to consider for implementation. However most processes, especially as they become more complex, span these categories. When defining processes and selecting workflow tools, one needs to consider the entire process cycle. It can then be broken down into smaller applications that are often addressed with different tools. (See Figure 6-4.)

Figure 6-4: The Workflow Continuum

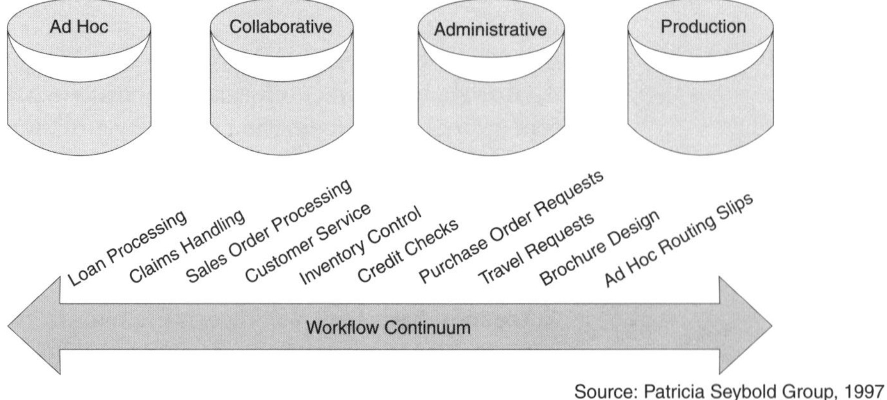

Source: Patricia Seybold Group, 1997

An early step when planning to implement workflow is to analyze and define a process by examining the three "Ps" of workflow: process, policies, and practices.

- **Process** – It is important to determine exactly what each process is that an organization is trying to automate. What are the goals and the business impact of these processes?
- **Policies** – Every business process has official policies – or the rules that govern how work is done and how decisions are made. Policies are reflected in the workflow applications that organizations build, often as business rules.
- **Practices** – Policies are useful, but formal, rules. In practice, many official policies are bypassed or modified for efficiency and effectiveness. When planning an automated workflow, firms and work groups must discover the actual practices followed by the people participating in the process and incorporate those that are valuable in the workflow application design.

Business-process reengineering (BPR) tools are available to help organizations define the business process and perform analysis on the planned process. However, products in this category do not automate the actual workflow process. Rather, they are designed as planning tools. However, many products available today in this space have automatic conversions to commercial workflow engines, automatically generating the core of a workflow application. Some vendors in this market include Delphi Consulting, Holosofx, IDS-Professor Scheer, Meta Software, and Visio (the latter offers BPR-specific templates for the standard drawing package).

Often, these BPR design tools include simulation capabilities so that business teams can test the proposed process and gather vital metrics on its operation. A small number of workflow engines also provide similar simulation capabilities in their process mapping components. Several BPR tools are now integrating with workflow products not only for the creation of the applications, but to establish feedback loops where the actual metrics of running the process can be fed back into the BPR tool for continuous analysis. When defining a workflow application in a workflow product, organizations should take into account the three "Rs" of workflow: routes, rules, and roles.

- **Routes** – The order in which information is flowed through the steps of the work process. Information, often included into a work packet, can be routed serially (on a single path from one task to another); in parallel (routed into multiple paths, each being handled simultaneously and independently of each other, usually resolving into a single work packet at the end of the parallel routes); or conditionally (the routing path is chosen based on a defined business rule). (See Figure 6-5 on page 502.)
- **Rules** – The routing of information often depends on the rules of the business and decisions that must be made in the course of the process. For example, loans exceeding $100,000 may need two levels of approval, whereas smaller loans might need only one. Rules can be defined based on the information being routed (usually based on data fields in routed forms or as metadata for routed documents) or can be based on the explicit selection of authorized options defined in the workflow task. Workflow rules also are used to ensure the proper load balancing of work assignments among process participants.
- **Roles** – The information being routed as part of the process is usually not sent to specific individuals, but rather to the roles that individuals play. Thus, in an expense-reimbursement workflow application, for example, the approval request is sent to the manager, rather than to Sue. Similarly, group roles are often utilized, as in the case of loan officers. Rather than any particular loan officer handling the application, it can be sent to the next available officer.

Figure 6-5: Routing Models for Workflow Processes

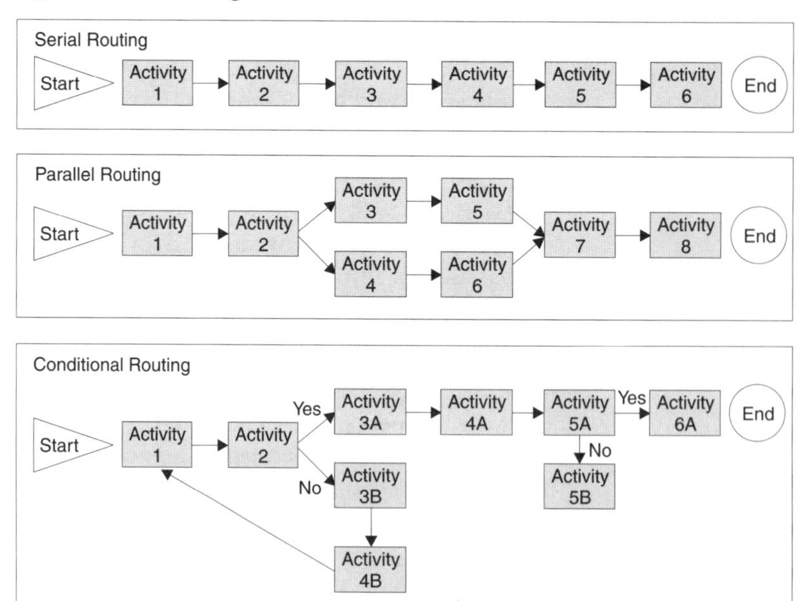

Source: Patricia Seybold Group, 1997

Today's workflow systems are available from independent workflow vendors, PDM vendors, and also from imaging systems vendors, in which case the systems are tightly integrated into the imaging products. In the past few years, workflow has also been tightly integrated into a number of document management products. In 1997, many popular business applications vendors, such as PeopleSoft and SAP, integrated workflow capabilities into their core products. In addition, several makers of object-oriented application development tools, including Forté and Template, added design tools specifically for building workflow applications.

Ad Hoc and Collaborative Workflows

In general, workflow products can be broken down by the category of applications they address as well as the underlying architecture. Ad hoc and collaborative workflows – as well as many administrative workflow applications – are typically supported by messaging-based products that are less structured than products that target transaction-based production processes. In the former category are products from companies such as Action, FileNet, JetForm, KeyFile, Reach Software, and Ultimus.

Action (with its Metro product line), KeyFile, Reach, and Ultimus are all also addressing e-mail-based workflow. Reach and Ultimus are moving quickly to provide intranet and extranet workflow solutions, while KeyFile's KeyFlow product is tightly integrated with Microsoft Exchange, workflow-enabling that environment for document-routing applications.

FileNet is a long-time leader in workflow, primarily through its production imaging product line. For more collaborative workflow applications, the company offers Ensemble, which is designed to complement existing e-mail and groupware software. Ensemble lets users author work processes graphically at their desktops – defining, tasks, establishing deadlines, and attaching documents.

It also provides workflow monitoring capabilities. Ensemble can be linked with FileNet Visual WorkFlo, Microsoft Exchange, and Novell GroupWise. Users receive work items in their standard e-mail in-boxes.

JetForm, the undisputed leader in forms processing (especially since the acquisition of key competitor, Delrina FormFlow) focuses on administrative workflow applications with forms-based processes. Users can workflow-enable both JetForm and FormFlow forms, including linking form fields to underlying data sources and establishing business rules for routing. There is a graphical environment for defining the forms as well as the workflow. JetForm leverages standard e-mail systems to move work from user to user, including Internet mail support.

Production Workflows

The more complex workflow applications are addressed with production-level tools, provided by such vendors as Action, Eastman Software, FileNet, IBM, InConcert, Mosaix, Staffware, and TeamWare.

Action has been a leading voice in workflow for many years, providing a workflow environment with graphical application development tools, a scripting language, and a robust workflow engine.

Eastman Software, formerly Wang, offers OPEN/workflow, a production-oriented workflow environment with graphical development via RouteBuilder and FormBuilder, and a robust set of rule building capabilities. The product supports a variety of workflow process, but it is optimized to address imaging applications. The company is focusing on what it calls work management applications.

FileNet's Visual WorkFlo is an object-oriented production workflow and document/image management system. The product includes Visual WorkFlo/Composer (to design and assemble applications), Visual WorkFlo/Performer (the runtime software for participants), and Visual WorkFlow/Conductor (addressing management and administration). The company has begun delivering industry-specific workflow solutions built in Visual WorkFlo, such as Visual WorkFlo/Payables, a group of modules that automates accounts payable applications.

IBM FlowMark is an object-based production workflow product that supports the graphical design of complex processes. FlowMark components include FlowMark Buildtime (the application development environment), FlowMark Server (the workflow engine), and FlowMark Runtime (for users' access to workflow assignments). Business processes can be initiated from a Web browser using Internet Connection to FlowMark, and the Lotus Notes client for FlowMark provides a Notes front end to FlowMark worklist items.

InConcert provides object-oriented workflow with a graphical workflow builder, a non-graphical screen builder, workflow engine, and runtime client. InConcert excels at providing ad hoc capabilities from within structured and complex workflow processes.

Mosaix ViewStar was the first Windows NT-based workflow product, initially targeted at imaging applications. The product features a graphical development environment, workflow engine, runtime client, and interface builder. Mosaix is now integrating ViewStar with its call center technology to provide complete customer service workflow solutions.

Staffware was an early pioneer in the workflow industry with the first non-imaging-based production workflow solution aimed at financial applications. The Staffware product is available on almost every platform (including Windows NT, various UNIX versions, VMS, and OS/2) and supports multiple clients, including Windows, Windows NT, OS/2, character-based terminals, and the Macintosh. The company was the first to offer full participation in production workflow via a Web browser client. The product line includes a graphical workflow builder, runtime client, forms builder for developing front end forms, and the workflow engine. Staffware is the leading workflow vendor in Europe.

TeamWare's TeamWare Flow combines structured workflow with collaborative capabilities supporting business teams in planning, executing, participating in, tracking, and managing business processes. The product also includes document management capabilities such as version control.

Workflow Standards

One growing concern within a corporation is the proliferation of workflow technologies and lack of standardization across the enterprise. Because workflow technology can be purchased off-the-shelf, work-compatible departments accidentally can purchase and implement incompatible workflow products. The result could be that the purchasing department might not be able to talk to accounts payable, and accounts payable might not be able to talk to shipping and receiving.

To avoid this problem by developing standards for workflow technology, more than 175 companies formed the Workflow Management Coalition in 1993 to establish standards for APIs for workflow products. The goal of the coalition is to bring order to competing technologies while allowing competition to exist. The proposed standards, when completed, will allow different workflow products within a company to communicate with each other. In addition, these standards will also allow workflow products to communicate between companies, for example, enabling a manufacturing company to communicate with a purchasing company via workflow. This type of interoperability between companies and products was demonstrated at the Workflow Canada industry conference in Toronto in June 1996.

The Coalition has proposed a framework for the establishment of workflow standards. This framework includes five categories of interoperability and communication standards that will allow multiple workflow products to coexist and interoperate within a user's environment. Of the five interfaces being defined, as of late 1997, Interface 2 has been published and Interfaces 1, 3, 4, and 5 have been released for comment.

The five key areas being developed by the Coalition include the following:

- **Interface 1: Process Definition** – Defines a standard interface between the process definition tool and the workflow engine
- **Interface 2: Workflow Client Application** – Defines standards for the workflow engine to maintain work items that the workflow client presents to the user
- **Interface 3: Invoked Applications** – A standard interface to allow the workflow engine to invoke a variety of applications
- **Interface 4: Workflow Interoperability** – Defines a variety of interoperability models and the standards applicable to each

- **Interface 5: Administration and Monitoring Tools** – Defines the monitoring and control functions that allow passing status information between workflow systems

6.3.4 Document Management

Much as database managers organize and maintain the integrity of structured data, document management provides the same capabilities for unstructured data in documents. The mechanism for managing documents is to add structure in the form of attributes or metadata, typically stored in an underlying database. Documents can be searched for, retrieved, and managed based on a combination of the metadata fields and full-content indexing.

What exactly is a document? Today, the term document has come to mean almost any file that can be created or stored electronically, including text, graphics, images, spreadsheets, lines of code, even sound and video. Document management systems maintain information on documents of any media type. In addition, many systems allow businesses to capture attributes about offline documents, with one of the attributes being the physical location of the paper file. Document management systems also can organize and maintain compound documents – documents that are built dynamically from several separate components. For example, a book is made up of chapters. Each chapter may be a separately managed document, but the book is also a document – a compound document – that may have its own set of attributes.

Document management systems provide tools for managing the document life cycle:

- **Creation** – The ability to specify different types of document, each with different attributes. For example, an invoice might have an attribute called Customer, whereas a memo would have an attribute called Recipient.
- **Modification** – Maintaining the integrity of new versions as the document is being edited, often by a number of different people. Access restrictions come into play to ensure that only those with editing permission can modify the document. Other users may be able to read the document without changing it. In fact, some users may not even know the document exists because they do not have permission to access it at all.
- **Approval** – Many document management products provide some level of document routing for approval – a basic form of workflow.
- **Distribution (publishing)** – Publishing has changed significantly over the past decade so that, today, documents are distributed as often via electronic media (such as e-mail or published on the Web) as on paper.
- **Archival (storage)** – The retention of documents on-line and the archiving scheme are usually managed by the document management system.

Document management becomes more important as people increasingly rely on electronic documents and on-line connections in their daily work. In collaborative environments – often supported by groupware tools – democratic access to the information in documents requires systems to support the appropriate information sharing. Documents are now being created in any variety of media, and are being maintained in repositories distributed around the world. Systems are needed to ensure secure and timely access to these documents in readable formats.

Document management systems are also being employed to help users filter through the mass of available information on the Internet, organizing it in meaningful ways. In fact, document management is becoming the integration environment for many other technologies such as workflow, imaging, e-mail, groupware, and filtering tools. The major document management companies – Documentum, PC DOCS, and Saros – have integrated workflow and document imaging, with an emphasis on assembly and routing of compound documents that contain imaged documents, text files, spreadsheets, embedded sound, and graphics. (See Figure 6-6.)

Figure 6-6: Document Management as an Integration Environment

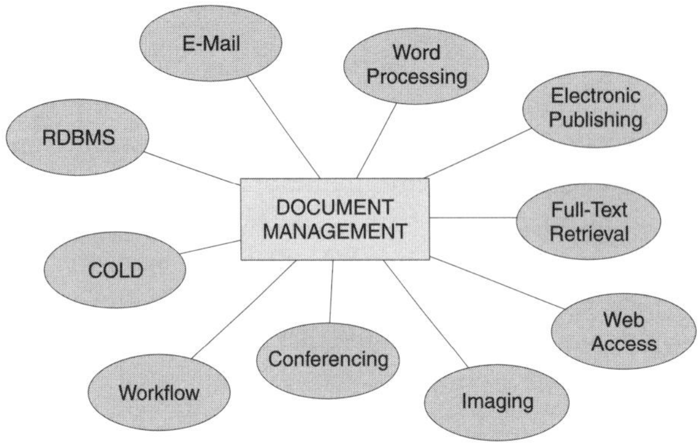

Source: Patricia Seybold Group, 1997

In light of new requirements by collaborative teams as well as the Internet phenomenon, document management is changing rapidly, with many new players entering the market, new technologies evolving, and new alliances being formed.

The most visible players in the document management market are Altris Software, Documentum, Interleaf, PC DOCS, and Saros (FileNet). Several other companies are fast becoming significant players, including NovaSoft and OpenText. All provide complete document life-cycle management and are working to deliver full document management capabilities on thin client architectures.

- **Altris Software's Enterprise Document Management System** (EDMS) – A turnkey system that provides integrated document management for files and images. EDMS offers secure access to all documents, structured and unstructured data; scanning and object capture from applications such as Microsoft Word and Excel; and a complete indexing scheme by application, document, and folder. A modular design allows cross-platform flexibility and expandability. EDMS also provides multimedia, voice, and video plus the ability to manage and add objects as annotations on other documents. EDMS runs on several varieties of UNIX, the MacOS, and Windows NT. In 1997, Altris introduced EB, its next-generation architecture for document management, which is multi-tiered and Web-centric. EB offers full functionality from a Web browser client, including administration.
- **Documentum's EDMS** – A client/server, object-oriented document management system targeted at Fortune 1000 companies with strategic document process requirements. Documentum's EDMS includes the Documentum Server, Documentum Workspace, and several different clients for Web access to documents, depending on functional need – for example, for read-access only. The company also provides a tool called RightSite, which manages the relationship of documents on a Web site. Documentum clients run on HP-UX, IBM's AIX, MacOS, Motif, Solaris, SunOS, and Windows.

Documentum's EDMS can work with non-Documentum clients, such as Notes or Interleaf, and provides replication services for the documents. A focus for Documentum is integrating document management with the document process via workflow technology called "routers." A graphical route builder is provided, similar to those included in most dedicated workflow systems. Routers can be used to send documents in an ad hoc manner or using pre-defined, permanent templates that can support complex, strategic, document-related processes.

- **Interleaf's Intellecte** – A client/server document management application that combines document management with configuration management (establishing relationships among documents), electronic document viewing and distribution, and automated document assembly, all customizable with the same API set. Intellecte includes the RDM Document Manager, the WorldView electronic distribution system, a Production Manager for creating dynamic collections of publishable documents based on rules and queries, and the Liaison API for extending and integrating the tools with other applications. Rudimentary workflow is provided to support the document approval process. Intellecte runs on Macintosh, Windows, and UNIX clients.

  Interleaf is now refocusing on its publishing roots and has introduced a new product, Interleaf Xtreme, which provides easy, low-cost, low-maintenance Web publishing capabilities.

- **NovaSoft's NovaManage** – A document management and workflow system that focuses on managing complex document processes, particularly in the engineering environment.

  In late 1997, the company announced its NovaWeb product line with the first available product, NovaWeb/Approve, for searching and viewing documents and for participating in document-based workflows from a Web browser.

- **OpenText's LiveLink** – Provides Web-based document management and workflow as well as traditional client/server functionality. The product focuses on the easy organization of information combined with visual workflow management and participation capabilities. The company was one of the first to jump onto the Internet with full document management support.

- **PC DOCS' DOCS Open** – A client/server document management system designed for group productivity in document-intensive LAN and WAN environments. DOCS Open provides searching capabilities, administrative features, security, version tracking, and full-text search capabilities in an intuitive interface. The user environment is also customizable by both administrators and individual users. Release 3.5 incorporates DOCS Unplugged document management for mobile users. It offers DOCS Interchange to work with Microsoft Exchange as well as Lotus Notes. Also included is the DOCS Document Sentry Agent, which provides a layer of configurable security for DOCS Open libraries. Basic document routing and image manipulation are available via the DOCS Enterprise Suite. Client platforms include Windows, Windows NT, and UNIX.

  In 1997, PC DOCS released its new architecture for supporting thin clients. The DOCSFusion server, used in conjunction with the CyberDOCS client, provides document access and querying from a Web browser to the identical document repositories used in DOCS Open.

- **Saros's Mezzanine** – A division of FileNet since 1996, Saros provides the Mezzanine document management system with the Saros Document Manager front end client. Web access to documents is provided via the @mezzanine product. The products run on MacOS, Windows 3.1, Windows 95, Windows NT, and UNIX clients.

  With its acquisition by FileNet, the company has begun integration of various FileNet application components. FileNet is a document imaging and workflow company that is rapidly becoming an all-purpose document management company, with additional purchases of Greenbar Software, for its computer output to laser disc (COLD), and Watermark Software, for its document imaging software.

  Document Manager 4.0 has introduced some new capability into the core product set, including compound documents. It also supports Lotus Notes and Microsoft Exchange, serving as the document library and controlling check-in/check-out, revisions, and document versions.

There is a new generation of document management companies springing up rapidly, most of which are focused on Web publishing with rudimentary document management capabilities. Some of the established players are addressing Web publishing, such as Interleaf with Xtreme and Xerox with InterDOC. Other newer participants in this market include companies such as InterNet Solutions and NetRight Technologies. All the leading products in the document management arena offer developer's toolkits and published APIs, allowing third parties and customers to build customized applications using the products' core technologies.

Document Management Standards

Like workflow, document management systems have suffered from a shortage of standards that would allow diverse systems to communicate.

The first standard to be embraced by the industry as a whole is the Open Document Management API (ODMA), a client-based set of APIs allowing document management systems to integrate with any ODMA-compliant editing environment, including word processors, spreadsheets, graphics tools, workflow products, and specialized editors. With ODMA integration, documents can be stored directly into managed repositories from the File menu (or equivalent) on the editor.

The Document Management Alliance (DMA) is addressing a more complex integration problem. Its charter is to deliver industry specifications for universal interoperability among all document management applications, services, and repositories – not just client access. A DMA standard would let users find, retrieve, and share documents from different systems and across different platforms and networks. The DMA is part of the Association for Information and Image Management (AIIM). AIIM originally promoted and advanced the development of systems that store, retrieve, integrate, and manage document images. AIIM's focus now includes imaging, workflow, text management, document exchange, forms processing, optical character recognition/individual character recognition (OCR/ICR), mass storage, multimedia, computer output to microfilm (COM)/COLD, micrographics, databases, networking, electronic data interchange (EDI), voice, and CD-ROM.

The DMA has defined an enterprise-wide document management specification as well as a middleware layer specification to allow access and search for documents between different document management systems, flat file repositories, file servers, and potentially any other defined document management service.

DMA technical specifications define three core elements:

- A common interface for integration of the access and search methods of individual library services
- A uniform API for accessing and searching across diverse document management services
- An object-based data model for standardizing access to enterprise library services. The model allows for modular integration of library services where vendors could support either specific components or implement the complete model.

The DMA completed its multi-vendor trial-use prototype implementation of the DMA specification in October 1997. The specification was scheduled to be published in early 1998.

Future work will include the following areas:

- DMA/ODMA interoperability
- DMA/Workflow Management Coalition interoperability
- DMA/Internet interoperability
- Additional support for compound document types

Eventually, the Workflow Management Coalition and the DMA will work together or possibly merge in order to provide a common set of specifications for documents and workflow.

6.3.5 Knowledge Management

Work group members can readily create and modify reports, memos, or even occasional messages for all to read. As information consumers, team members have the task of finding, reading, understanding, and then acting upon the contents of individual documents. With multiple people writing multiple documents, keeping track of rapidly changing on-line contents and finding the essential information when needed to complete particular tasks is not easy. The problem is compounded when considering that information is also coming into the group from outside the corporation.

Groupware, workflow, and document management currently work best when restricted to narrowly defined business processes rather than open-ended, ad hoc queries. At the enterprise level, problems often arise in finding and processing information due to the sheer volume and diversity of information sources. Standard mechanisms such as keyword-based searches may work to find documents that mention a well-defined topic, but they are not successful when trying to find the specific answer to a question, especially when the question consists mainly of common words. With so much electronic content available with just a few mouse clicks, individuals and business team members want to utilize the capabilities of advanced on-line resources to find essential information that is directly related to their work.

The dilemma of exploiting electronic information to solve hard-to-define business problems seems clear enough. If only business team members could combine the information stored on their computer networks with experiences and insights, then they would have the knowledge necessary to resolve pressing problems. If they could capture the insights of experts and codify their understandings and intuitions into sets of rules to solve specific business problems, then business team members would have an environment for "managing knowledge" across disparate on-line information resources.

The Knowledge Management Continuum

Today, most work-group computing technologies are designed to help individuals deal with the overwhelming amount of on-line information available on their desktops. Relatively few technologies have the capability to manage the knowledge that can be gleaned from this information. Nevertheless, as illustrated in Figure 6-7, managing knowledge on the network is part of an overall continuum for exploiting work-group computing technologies. Data put into context creates information. Information combined with experience creates knowledge.

Figure 6-7: The Knowledge Continuum

DATA + context = INFORMATION + experience = KNOWLEDGE

Source: Patricia Seybold Group, 1997

NRather than simply reporting data in context, people working together on the network can begin to mix information with experience. They can record, track, and share the metadata about connections and relationships – how information is used in real-life situations. With the assistance of software environments that help them manage knowledge on the network, people can begin to locate the essential information related to solving tricky problems just in time.

Capturing Know-How

A knowledge management framework has a problem-solving focus. It includes information related to a specific domain or area of expertise as well as the capabilities to manage an extensive taxonomy of relevant relationships. It incorporates one or more on-line information repositories capable of easily storing and quickly retrieving many different kinds of business information. Various text retrieval engines on the market today, such as Excalibur's RetrievalWare, Fulcrum's SearchServer, and Verity's Information, provide the core capabilities for creating a knowledge management environment – the ability to locate relevant documents must then be tightly integrated into the flow of workday activities. Dataware with its Knowledge Management Suite and Fulcrum with its Knowledge Network are also offering overall suites for knowledge management – products that help organizations structure and organize their interactive content as a way of enhancing core information retrieval technologies.

An individual repository needs to incorporate a comprehensive, extensible set of business objects – such as the people, the tasks, the activities, and the tools required to address specific problems. It needs to capture the business rules – such as defining how specific people are responsible for individual tasks or mapping which tools are needed for individual activities.

Flexibility is key. Users frequently need to create new business objects on-the-fly, to capture the experiences of recent encounters. Often, they do not precisely know what they are looking for, and they want to examine the network of connections in order to spark new insights. A knowledge management framework incorporates various capabilities for recording and exploring the structure of implicit relationships, based on interpreting the meaning of language recorded as unstructured words and phrases in individual documents or database records.

To some extent, users can depend on the core capabilities of text retrieval engines to search explicitly for identified words and phrases – relying upon their own intuitions to find the relevant terms and deduce meaning from the results of individual queries. Many times, however, team members would like some guidance, and seek to benefit from the expertise of subject matter experts who have a detailed understanding about an individual problem domain. Knowledge management frameworks provide insights into a problem-solving process. These frameworks structure the topic area and essential issues that users expect to explore. Interactively, the framework identifies the key concepts and suggests the relevant words and phrases that team members should use to find answers to questions in predefined areas of interest.

Mining and Tracking Unstructured Information

A knowledge management framework can exploit various techniques for "mining" and tracking unstructured information. Beyond trying to match the specific contents of individual queries to the exact words and phrases that occur within individual documents, users might try to find items based on associated meanings, such as thesauri of related terms or additional information about concept hierarchies. Users need not look for documents that only contain specific words and phrases – rather, they can begin to rely on the power of the underlying retrieval capabilities to interpret their queries and expand their searches to locate both more general and more specific items of immediate interest.

Team members might also come to rely on knowledge management frameworks to help stay abreast of rapidly changing developments in their specific fields of interest. They might utilize various kinds of active agents to monitor ongoing information streams and to notify them about new information that meets their specific areas of interest.

A knowledge management framework enables end users to track information on the basis of their underlying concepts or external ratings compiled by subject matter experts. They can launch persistent queries into the network of on-line information resources and then have individual agents monitor different information sources for documents of interest, based on keywords, phrase matching, automatic analyses of content, or analyses by subject matter experts.

Collaborative Filtering

Another approach to knowledge management is based on collaborative filtering – enabling technologies that recognize that prioritization in decision-making is often a social process. People frequently make judgments based on what others say, whether it involves the merits of artistic works such as CDs or movies or what they should read or do next in a business context. For example, there are some music and movie critics who have the power to influence public taste. In organizations, some people similarly function as "gatekeepers," deciding what information is important and who should see it. Collaborative filtering allows teams of people to make such recommendations.

GrapeVINE, from GrapeVINE Technologies, implements collaborative filtering in a way that allows information to be routed to different gatekeepers based on its subject matter, using a structure known as a KnowledgeChart. Gatekeepers review information and can escalate its priority, so that it will be seen by others who have declared that they wish to see only items of a certain priority or higher. Although both technical and political issues arise in the creation and maintenance of KnowledgeCharts, the central idea of filtering by both subject and priority is quite powerful.

Collaborative and other types of filtering will play an increasingly important role in the growth of personalized information access capabilities based on products from such innovative companies as BackWeb Technologies, Charles River Analytics, Firefly Network, Net Perceptions, and PointCast.

Transforming the Groupware Paradigm

One reason for the growth of groupware, workflow, and document management systems is the advances made in key enabling technologies, such as network bandwidth, magnetic storage, optical storage, and desktop computers. Greater network bandwidth, for example, is allowing groups to share image and word documents in real time, and even to change and revise those documents interactively.

Network connectivity, represented by the exponential growth of the Internet and the Web over the past 3 years, is transforming the groupware paradigm for ready access to information and almost instantaneous communications. Many concerns about the reliability of the Internet as a vehicle for business-to-business communications are now being addressed as major telecommunications and computer corporations such as AT&T, GTE, IBM, MCI, Sprint, and WorldComm now compete fiercely for the backbone Internet business – based on both price and quality of service. Now that the Internet is in place, businesses and individuals alike seem to have an insatiable appetite for bandwidth as they seek faster access to more information.

Fueling this demand for network bandwidth is the coming convergence of data and voice communications. Telephony has been a key tool of business communications for almost 100 years. Until relatively recently, however, electronic communications over corporate LANs and the Internet have been entirely separate from telephony. The possibilities for groupware are changing with the realization that real-time voice messages might flow through the packet-switched networks of the Internet at a fraction of the cost of today's circuit-switched telephony networks. (See "Internet Telephony" on page 261.) That is, rather than allocating a fixed connection for an interactive telephone conversation (used by today's telephone systems), a telephone call might consist of a series of real-time voice messages, each of which is broken into individual data packets and instantaneously transmitted through a series of routers to its destination. This new technique, which is still in its infancy, efficiently utilizes the scarce network bandwidth of the Internet and promises to transform work-group communications within the telecommunications industry in the coming decade.

The advent of mobile digital devices, particularly in Europe and the Far East, is an early indicator of the coming capabilities of digital telephony. Lead by such vendors as Ericsson (in Sweden) and Nokia (in Finland), users can send and receive telephone calls from almost anywhere in Europe and major economic centers in Asia using GSM, a common international digital cellular standard. GSM also permits the exchange of digital messages between devices. Currently, these messages are restricted to the 160 characters of the Short Message Standard (SMS), but they are likely to be expanded to encompass a generalized digital data stream over the next few years. It is altogether conceivable that mobile users will be able to surf the Web to find the information they need, interconnect with individual work-group-oriented information stores, read their e-mail, and redirect their long documents to local fax machines using the capabilities of mobile digital devices. Telephony and digital data communications, both essential aspects for groupware, are likely to become one within the next few years.

Converging Technologies

Groupware, workflow, and document management technologies were created independently, and until recently, have evolved separately. The three technologies are now beginning to merge and assume each others' characteristics. For example, Lotus Notes includes imaging and document management software with Release 4.0. This release also incorporates Internet functionality that allows users to share work over an Internet/intranet connection. Similarly, Novell's GroupWise incorporates groupware, workflow, and document management along with Internet capability. In addition, PeopleSoft and SAP have introduced workflow and document management capabilities into their basic corporate application products.

This convergence can be seen as the result of a switch from a tools and technologies focus to a more work-centric approach. A typical department of knowledge workers requires the combination of groupware, workflow, and document management applications in its daily operations. For example, developing a business report using groupware products could result in an end product that contains a mix of imaged documents, spreadsheets, text, graphics, and corporate boilerplates, all of which need to be managed, published, and maintained. Users do not want to keep switching environments, importing and exporting documents as they move through the process just because a different tool is being employed. What users want is a unified framework for collaboration that supports business communications, information sharing, workflow, document management, and knowledge management in an integrated and comprehensive environment.

When the business process takes center stage, true knowledge management can take place. When users can focus on the work rather than on shuffling a lot of dissimilar tools, the work process itself can be examined, improved, and updated more easily. Updating group processes in a collaborative environment is important because the way people work is changing rapidly. With the increased reliance on Web technologies for extranet support – creating on-line collaborative opportunities with our customers and our partners – these processes must be flexible enough to support a mix of corporate policies and cultures.

Groupware, as a product category separate from workflow, document management, and knowledge management, may not continue to exist. Indeed, these technologies are all as integral to supporting collaborative group work as are dedicated "groupware" tools.

6.3.6 Market Overview

Groupware, workflow, and document management technologies will continue to evolve into more unified collaborative environments by adding new capabilities, such as workflow support in document management, and by providing seamless integration among the various products – both via adherence to standards and by direct integration on the part of the vendors.

Groupware

Long the undisputed leader in groupware, Lotus Notes is now feeling heavy competition from a variety of products, most notably Microsoft Exchange. In 1997, according to IDC, new users of Notes products increased by 31 percent, compared with 26 percent growth for Microsoft Exchange. Product revenues in 1997 grew by 37 percent for Notes, compared with 31 percent for Exchange. Although, architecturally, the two products are very different, specifically in the ability to communicate, share, and organize information electronically – Notes operates in continuously connected and intermittently connected environments, and Exchange is designed primarily for continuously connected environments (as Microsoft continues to make headway in basic groupware functionality) – the distinctions between the two will become less apparent.

Netscape SuiteSpot has now moved into third place in the groupware market, measured in terms of revenue growth, and is tied for third place (with Novell's GroupWise) when measured in terms of numbers of new users.

Netscape's SuiteSpot Servers, Radnet's WebShare, and other Web-based solutions will increasingly challenge Notes as the market leader for groupware. These exclusively Web-based solutions promise to capitalize on the flexibility, robustness, and adaptability of corporate intranets – the ability to access shared information repositories over open business networks. Although total cost of ownership studies have not yet been released, these products may deliver the business benefits of groupware with lower initial costs of ownership.

However, Lotus Notes will continue to evolve as a mature product within the groupware market. Notes has, in fact, become a vital legacy system with large repositories of information in most corporations. This information will not be abandoned easily. Through Domino, Lotus has deftly adopted Notes to the challenges of corporate intranets and the Web.

Workflow

Workflow will continue to grow as a technology, with new innovations being added, especially in the areas of providing ad hoc capabilities in the most structured of production environments. Workflow as a distinct technology will become more of a niche (although strategic) market, addressing the high-end, complex, transaction-based applications. The ad hoc and collaborative types of workflow will become integrated into document management systems, e-mail systems, groupware, and client/server applications.

As extranet technology and business practices become more accepted, inter-organizational workflow processes will become a major collaborative technology, allowing partner organizations to work together on shared business processes.

Document Management

Document management will continue to increase in importance, especially in the areas of managing the relationship of documents available on the Web. Technologies for the conversion of multiple formats to Web readable formats will continue to emerge, as will tools to support the business process of doing the conversions. Text retrieval vendors will increasingly emphasize their abilities to manage on-line information. Monitoring and filtering ongoing information streams are integral aspects of competitive text retrieval solutions.

Other technologies will continue to be integrated under the document management umbrella, such as workflow and compound document assembly. Document management will continue to play an important role in the management of knowledge.

According to many industry analysts, documents will become a common interface into information for most users, modeled on the Web paradigm. As a result, documents will no longer be static entities, but will be generated dynamically, combining multiple pieces of information from different sources in different media into document containers or templates for easy access.

Knowledge Management

The ability to create and manage knowledge – adding experience and insight to information – will emerge primarily from enhancements to existing technologies, including document management systems, search engines, and linguistics tools. These all lay the foundation for determining context and meaning out of content.

In addition, new companies are emerging that are addressing knowledge management more directly. These companies provide products that explicitly filter through large volumes of on-line information, allowing users to add insights and opinions and to organize the information in meaningful ways. Other products allow users to relate information from multiple sources such as documents, databases, and Web sites, among others, enabling users to build their own personal and sharable knowledge networks by creating links between information that might not otherwise seem related.

The Internet and the Web

As evidenced in all these areas, the Web will continue to have a profound impact on the groupware marketplace. New developments for the groupware marketplace will begin on the Web and will move to encapsulate functionality of legacy applications (such as Notes and Exchange) insofar as they remain commercially viable.

6.3.7 Forecast

Groupware

- Groupware will continue to play a major role in developing virtual offices that allow workers to work without regard to geographic location, time, and personnel locations.
- Groupware will provide a valuable tool for forming virtual companies and providing the technology needed to make them work.
- As the markets for the enabling technologies mature, groupware will cease to be a self-contained product category and will become an enabling capability for easy access to distributed information within a work group.
- Lotus Notes will continue as the leading high-end groupware product because of its integration of messaging, collaboration, workflow, and application development capabilities.
- New releases of Notes will focus on integrating both client and server further into the Web environment.
- All leading enterprise groupware/messaging products will support Internet mail standards such as SMTP, POP, IMAP, and LDAP.
- Low-end groupware products increasingly will be Web-based and offer real-time chat in addition to Internet mail and threaded discussion capabilities.

Workflow

- Corporate application suites will add workflow into their products. The driving factor behind this development is the need to move information from the application suites to desktops both inside and outside the organization. In some cases, corporate application vendors will develop their own embedded workflow products. In other cases, they will interface with existing workflow products.
- Workflow continues to accelerate its transition onto the Internet, providing companies with worldwide connectivity for moving data and information. Using extranets and the Internet, workflow activities will routinely cross organizational boundaries.
- Workflow vendors are cooperatively developing standards that will allow differing workflow products to talk to each other, making inter-enterprise applications possible. These standards promise to ensure interoperability both between individual clients and between separate workflow engines, providing the potential for seamless, end-to-end business processes.
- Workflow products will use message-oriented middleware (MOM) as a transport mechanism.
- Production workflow applications will address a strategic but niche market; ad hoc and collaborative workflow will become absorbed into collaborative computing and line-of-business applications.

Document Management

- Document management systems will more tightly integrate workflow as part of the core software.

- Web-related technologies will accelerate the deployment of document management solutions throughout organizations. Document management systems will no longer be restricted to writers, editors, and production specialists. Rather, almost everybody within an organization will begin to rely on the underlying capabilities of their networked document repositories to manage, control, and direct access to information as part of their daily work activities.
- Document management systems increasingly will be able to manage all document (data) types, including formatted text, images, graphics, audio, visual, and computer output data, such as accounting reports.
- Document management increasingly will address the needs of remotely located personnel, providing them with easier access to documents located in central office repositories.
- Document management solutions will no longer simply focus on the production of paper-based documents. Rather, these solutions will address the needs of electronic distribution and on-line storage, where explicitly formatted, paper-based documents serve only as one kind of potential output.
- The document publishing life cycle will be shortened dramatically as documents are revised continuously and redistributed electronically.
- Issues of records management and document authentication are a major problem that remains to be addressed.

Knowledge Management

- Active agents – autonomous processes capable of taking independent actions on behalf of individual users and business teams – promise to simplify, personalize, and customize access to on-line information. These tools will help to manage knowledge across business networks.
- Collaborative filtering, both within individual organizations and on the Internet, increasingly will be an important technique for avoiding information overload.
- Firms that develop unique approaches to managing their access to the intellectual resources of their customers, suppliers, products, and business staffs will have a distinct competitive advantage in the marketplace.

6.3.8 References

■Articles

Cole-Gomolski, Barb. 1997. Delays strike Lotus. *Computerworld.* October 13: 2.

Moeller, Michael and Christy Walker. 1997. Lotus turns to JavaBean client components. *PCWeek.* August 4: 6.

■Periodicals, Industry and Technical Reports

Bock, Geoffrey. 1998. *Information retrieval tools for knowledge management.* Workgroup Computing Report (Vol. 21, No. 1). Patricia Seybold Group. January.

Bock, Geoffrey and Lynda Applegate. 1996. *Technology for teams.* (Teaching note #9-196-008). Harvard Business School Press.

Marshak, David. 1997. *Notes as groupware...revisited.* Workgroup Computing Report (Vol. 20, No. 4). Patricia Seybold Group. April.

Marshak, Ronni. 1997. *Document management systems report: managing intellectual assets.* Patricia Seybold Group.

Marshak, Ronni. 1996. *Selecting and implementing workflow.* Patricia Seybold Group.

■Books

Applegate, L. M. 1996. *Managing in an information age.* Boston: Harvard Business School Press.

Bock, Geoffrey and David Marca. 1995. *Designing groupware: a guide book for designers, implementers and users.* New York: McGraw Hill.

Coleman, David (editor). 1997. *Groupware: collaborative strategies for corporate LANs and intranets.* Upper Saddle River, N.J.: Prentice Hall.

Fischer, Layna (editor) 1995. *The workflow paradigm.* (second edition) Lighthouse Point, Fl.: Future Strategies Inc., Book Division.
Kahin, Brian and James Keller (editors). 1997. *Coordination of the Internet.* Cambridge, Mass.: MIT Press.
Marshak, David. 1996. *Mission-critical Lotus Notes.* Upper Saddle River, N.J.: Prentice Hall.

■On-Line Sources

Center for Coordination Science at MIT. `http://www.ccs.mit.edu`

■URLs of Selected Mentioned Companies

Action Technology `http://www.actiontech.com`
Altris Software `http://www.altris.com`
Apple `http://www.novell.com`
AT&T `http://www.att.com`
BackWeb Technologies `http://www.backweb.com`
Center For Coordination Science `http://www.ccs.mit.edu`
Dataware `http://www1.dataware.com`
Delphi Consulting `http://www.delphigroup.com`
Delrina FormFlow `http://www.delrina.com`
Documentum `http://www.documentum.com`
Eastman Software `http://www.eastmansoftware.com`
Ericsson `http://www.ericsson.com`
Excalibur `http://www.excalib.com`
FileNet `http://www.filenet.com`
Firefly Network `http://www.firefly.com`
Forte `http://www.forte.com`
Fujitsu Ltd. `http://www.fujitsu.com`
Fujitsu-ICL Systems Inc. `http://www.fjicl.com`
Fulcrum `http://www.fulcrum.com`
GrapeVINE Technologies `http://www.grapevine.com`
Greenbar Software `http://www.greenbar.com`
GTE `http://www.gte.com`
Holosofx `http://www.holosofx.com`
HP `http://www.hp.com`
IBM `http://www.ibm.com`
IDC `http://www.idc.com`
InConcert `http://www.inconcert.com`
Interleaf `http://www.interleaf.com`
InterNet Solutions `http://www.internetsolutions.com`
JetForm `http://www.jetform.com`
KeyFile `http://www.keyfile.com`
Lotus `http://www.lotus.com`
MCI `http://www.mci.com`
Meta Software `http://www.metasoftware.com`
Microsoft `http://www.microsoft.com`
Mosaix `http://www2.mosaix.com`
Net Perceptions `http://www.netperceptions.com`
NetRight Technologies `http://www.netright.com`
Netscape `http://www.netscape.com`
Nokia `http://www.nokia.com`
NovaSoft `http://www.novastor.com`
Novell `http://www.novell.com`
Open Sesame `http://www.opensesame.com`
OpenText `http://www.opentext.com`
Oracle `http://www.oracle.com`
Patricia Seybold Group `http://www.psgroup.com`
PC DOCS `http://www.pcdocs.com`
PeopleSoft `http://www.peoplesoft.com`
PointCast `http://www.pointcast.com`
Radnet `http://www.radnet.com`
Reach Software `http://www.reachsoft.com`
SAP `http://www.sap.com`
Saros `http://www.filenet.com`
SCO `http://www.sco.com`
SoftArc Inc. `http://www.softarc.com`
Sprint `http://www.sprint.com`
Staffware `http://www.staffware.com`
Sun `http://www.sun.com`
TeamWare Group `http://www.teamware.com`
Template `http://www.template.com`
Ultimus `http://www.ultimus.com`
Verity `http://www.verity.com`
Visio `http://www.visio.com`
Wang `http://www.wang.com`

Watermark Software `http://www.watermark.com`
Workflow Management Coalition `http://www.aiim.org/wfmc`
WorldComm `http://www.worldcomm.com`
Xerox `http://www.xerox.com`

6.4 Systems and Network Management

6.4.1 Executive Summary

Systems and network management traditionally has focused on performance management, fault management, and configuration management. Today, as businesses make the transition to managing a network-based computing environment, the management focus has expanded to encompass all components and resources that make up enterprise networks. Administrators are meeting the needs of this expanded focus with an extensive array of tools for managing desktop systems, business applications, mass storage resources, and security.

The total cost of ownership and management of desktop or client devices has emerged as a highly visible issue. Although most businesses have chosen a personal computer (PC) platform, each PC varies in hardware and software. Fast access to knowledge about each PC in the enterprise and tools to manage tasks (such as software upgrades from a remote network management platform) are critical factors in being able to manage these valuable resources cost-effectively.

Although the desktop system provides the hardware window into the enterprise network, its ultimate value lies in delivering business applications efficiently and reliably to users. The true measure of performance is not the speed of individual systems and network segments, but whether business applications are available to the desktops with the necessary performance to meet relevant business needs. A growing emphasis on ensuring applications are available when and where they are needed in a global business environment has led to the development of new tools that focus on application delivery.

Once applications are running and available, the next step is to ensure the applications have access to necessary data and that new data generated by applications is managed properly. Storage management tools and strategies address the need to protect critical corporate on-line information. Today's challenge is meeting storage management needs across a heterogeneous mix of enterprise systems and data types. Effective storage management systems must provide efficient, reliable, and highly automated backup, restore, and long-term archival of corporate data network-wide without compromising security.

Controlling and monitoring access to data stored in databases, in mass storage systems, transmitted over public networks, and in use through applications are important activities associated with enterprise network management. Greater use of wireless devices, telecommuter PCs, remote office systems, and even other vendors' systems through extranets create specific security and management demands.

Many tools for managing all elements of the enterprise network now take advantage of the increased use of the World Wide Web and Web technologies. Web traffic is increasing as more data on existing systems becomes accessible via a Web browser. The Web browser has become a "universal" client and the interface of choice for many management applications. Browser-managed printers, for example, enable end users to determine easily which printers have the shortest queues or that a particular printer has a paper jam or other problem that requires attention.

When added to traditional management systems, Web browsers facilitate higher levels of service by supporting remote management from Web-enabled systems; administrators may not need to travel to specific sites to provide service. Network users can access network information through Web-enabled management packages, reducing calls to help desks and enabling administrators to spend more time solving problems.

Managing system and network performance will continue to be a challenge. Installed infrastructures typically are a heterogeneous mix of technologies and media that tends to expand on an as-needed basis. Response time can degrade for users if new data- and graphically intensive applications and an increasing number of network and Web users are not anticipated and incorporated properly.

Today's broad management focus extends management capabilities beyond the network infrastructure to the desktops, peripherals, and applications through which users interact with the network. These tools can help users better understand management issues and help administrators better meet user needs.

6.4.2 Systems and Network Management Overview

Systems and network management tools have evolved along different paths that are converging to provide a unified solution to current management needs. Systems management developed from a mainframe perspective; network management responded to the need to manage telecommunications systems and, eventually, data communications systems such as wide area networks (WANs) and local area networks (LANs). An understanding of both areas is necessary for today's combined systems and network management requirements and solutions.

Mainframe Systems Management

Although distributed computing in a client/server environment is now an established element of enterprise computing, mainframe systems play a key role in the information processing landscape. There has been a resurgence of respect for mainframe systems and their ability to meet the scalability, availability, and reliability requirements of critical corporate applications.

Mainframe systems management is an established process, with products available both from mainframe hardware vendors and third-party software providers. Mainframe-based systems management tools continue to manage traditional centralized system configurations as well as combinations of distributed client/server, intranet, and other applications.

Enterprise Management Tools

Enterprise management is the collection of activities that involve configuring, controlling, monitoring, analyzing, diagnosing, repairing, operating, supporting, and securing all components of a networked computing environment. It includes many of the same aspects of traditional mainframe computing such as storage or security management and some additional elements associated with distributed computing.

Enterprise management tools can be proactive or reactive, accessible through the Web or a graphical user interface (GUI), and used by internal information systems departments or third-party management services. Enterprise management tools include those used for monitoring, controlling, and planning. Monitoring provides status information on all elements; controlling facilitates response to information obtained from the monitoring process. Planning uses the information obtained in monitoring and the decisions implemented in controlling to predict future needs and prepare for change requirements.

The following list reflects elements of traditional mainframe and distributed computing enterprise management tools:

- **Accounting** – Computing and network resources represent a significant investment, encouraging companies to track assets and inventory and monitor resource usage. Efficient accounting enables administrators to optimize purchase decisions and to charge other organizations appropriately for network use. Automated data collection tools allow administrators to focus on keeping networks operating smoothly and set aside time for proactive planning.
- **Automated operations tools** – These tools automatically perform activities such as writing a file to a tape system. They are extremely sophisticated and include intelligence that supports actions such as rerunning a task if it is interrupted.
- **Batch processors** – Batch processors collect a series of transactions and process them as a batch. A company may gather payroll transactions during the day and process them all at night when a computer has a lighter workload. Batch processors help maximize system resource use and contribute to a more efficient reporting process.
- **Capacity planning tools** – Capacity planning determines and projects how much computer power a company needs. These tools examine the amount of processor, memory, and disk storage an application will require before it is deployed; help determine whether upgrades are needed to run the application; and facilitate decisions about capacity expansion. Planning involves forecasting loads, adjusting for growth, and incorporating new technologies as necessary. Trending data and data modeling are key components of successful planning. Network and system capacity planning is critical because changes in business requirements must be supported immediately by the corporate network. Accurate planning enables administrators to respond to new demands effectively without burdening existing resources.
- **Configuration management** – Managing a network configuration involves collecting configuration information about all enterprise resources, including systems, servers, network elements, and devices. This information typically is stored in a database, where systems and devices are associated with physical and logical network segments. Configuration changes are logged into this database to facilitate troubleshooting, detect unauthorized changes, and assist in inventory management. Configuration management tools ensure that configuration data is always available and accessible.
- **Disaster recovery** – Disaster recovery comprises the procedures, policies, and products that allow for the timely resumption of the computer-based elements of an enterprise's business processes following a significant large-scale interruption in service. The fundamental assumption behind a disaster recovery plan is that the site where the computers reside either is not usable or not accessible. Disaster recovery is a systems management discipline allied with but separate from security and data availability. Although the consequences of a security breach certainly could be disastrous, security should be treated as a distinct systems management topic of its own. Moreover, although backup and duplicate hardware can improve data availability, the existence of a backup tape is only one part of successful recovery from a disaster.

- **Disk backup systems** – These tools copy information from disks to tapes and other backup systems. They identify where files are located, may compress information so that it is stored efficiently, and may include mechanisms to restore corrupted or deleted files.

- **Event management** – Information supplied by event management systems can include events such as the addition of a new system or router, an early warning of crossing a performance threshold, or notice of a fault such as a device failure. Some problems can be solved with automatic, built-in procedures while others are routed to designated staff. Event management is an important reactive tool for network administrators, and it also provides information useful for future planning. Some network management products integrate both event management and fault management.

- **Fault Management** – Fault management involves recognizing alert conditions, generating alerts, and forwarding alerts to the appropriate management resource. This area may include trouble-ticketing systems, help desk tools, historical problem data, and measurements of problem-solving effectiveness, such as mean time to repair. Efficient fault management ensures rapid problem response and assists administrators in assessing why faults occur and preventing future instances.

- **Job schedulers** – These tools control traffic and determine when specific jobs can use a resource such as an operating system. Job schedulers establish procedures to ensure that high-priority transactions are processed quickly: They can move a mission-critical application, such as an airline reservation or funds transfer, ahead of a less-critical application.

- **Network monitoring** – Network monitoring tracks network statistics at the port, line, and device level. Tools can notify the management application whether workstations and servers are available and supply port and device status, configuration views, trend analysis for problem prevention, alarms on events, bandwidth utilization, error detection, diagnostics, and power supply monitoring. Comprehensive network monitoring helps administrators maintain an overall view of network "health."

- **Performance monitoring tools** – Performance monitoring examines how well a system is operating and determines the location of any bottlenecks. Such tools may direct a company to install more internal memory to ensure, for example, that all transactions will be completed in less than 5 seconds. Enterprise performance management collects performance data on all areas of a network and triggers alarms when thresholds are exceeded. Responses to alarms may include load balancing by reassigning applications to appropriate processors and tracking applications to ensure availability. Historical logging and trend analysis may be used for performance planning. Performance is a key indicator of network and system efficiency and often correlates with customer satisfaction.

- **Print management** – Print management tools route documents to appropriate print resources, manage print queues, and monitor printer functions. Documents can be routed to an alternate printer if a problem arises. In many situations, print servers facilitate printer sharing and maximize print throughput. Print servers can be dedicated hardware connected directly to the network, software resident on a network PC, or a card that plugs into the printer.

- **Security management** – These tools provide remote access monitoring, notification, and management of systems that authenticate users, authorize their requests to use system resources, and audit the use of these assets. Most operating systems include basic features such as password security, but more sophisticated security is available through add-on modules that provide flexible assignment of application privileges to individual users and selected groups. Data collected by security systems includes access records that provide detailed information on log-in attempt locations and times and on modem activity for remote access. Automatic logging and responses to security violations are also essential features of a security solution. (See Chapter 5.4, *Security.*)

- **Storage management** – Storage management enables companies to use available disk, tape, and other storage systems efficiently. These tools enable administrators to view, define, sort, filter, and analyze current storage information. They feature highly available, redundant storage or intelligent caching as well as uniform naming systems so data is stored in a consistent manner that facilitates retrieval. Third-party solutions are now available to manage network data backup and recovery and simplify the administration of enterprise-wide storage management. These systems provide automated backup of network-based corporate data and a range of solutions for data location and retrieval. Using a variety of storage technologies, including traditional disk, magnetic tape, and optical disk, these systems offer capabilities such as tracking data usage and maintaining most-used data on disk and least-used data on tape.

Desktop Systems and Server Management

When the client/server environment first became part of the business computing landscape, management tools for desktop systems and servers were virtually nonexistent. Today, business-critical applications based on client/server architectures and managing the costs of distributed desktop systems and servers have become an integral part of managing the enterprise networks.

With the help of new tools and open standards, system administrators are gaining better control over desktop systems and servers. System and network administrators and some users need advanced management tools for distributed systems. Mainframe and network management platform vendors and LAN management vendors all have been developing tools for managing resources at the desktop level. The shared goal is to bring the power and functionality of centrally oriented mainframe tools to distributed environments.

Due to the diverse nature of personal computing resources, a key issue is monitoring the attributes of each desktop or client system. Software attributes include which applications are installed in the system as well as operating system and applications release levels; hardware attributes include the amount of memory and disk space; and communications attributes might include type and speed of the modem used.

■ Desktop Management Interface (DMI)

Administrators now can monitor desktop system attributes and manage related tasks such as software upgrades with the help of desktop systems and management software products that conform to the Desktop Management Interface (DMI). DMI is a standard created by the Desktop Management Task Force (DMTF), an industry-wide consortium chartered with making PCs easier to use, configure, and manage. (See “Desktop Management Interface (DMI)” on page 533 for more information on DMI.) For example, by querying the DMI database, an administrator can determine whether a user’s desktop system has the resources necessary to load and run a requested program.

An example of DMI-compliant management software is NetFinity from IBM. NetFinity is a Web-enabled, PC management solution that operates across multiple environments, including OS/2, Windows 3.1, Windows 95, Windows NT, and Novell NetWare. Using NetFinity, administrators can manage desktop platforms remotely to perform tasks such as taking a failed disk drive offline and automatically rebuilding the data onto a standby disk or installing new device drivers to access peripherals such as printers.

Many vendors now offer PC products that support DMI, including 3Com, Compaq, Dell, Gateway 2000, HP, IBM, Intel, NEC, Premio, and Toshiba. However, levels of DMI compliance vary among vendors.

Seagate Software's Desktop Management Suite for Windows is an integrated system that combines automated asset management, software distribution, network backup, remote administration, and optional virus control of networked computers systems to control Windows-based desktops from a single console.

In a move to provide better management options for its business PCs and servers, HP purchased Symantec's PC desktop and server administration product line in early 1997. The product line, which includes Norton Administrator for Networks, Expose server management software, and Norton Desktop Administrator, is being integrated into HP's OpenView product line.

■ Reducing Total Cost of Ownership

A key aspect of developing tools for distributed environments is meshing mainframe-class systems management features with the open, standards-based environment of desktop systems and servers. Such tools are needed to improve user and administrator productivity as well as reduce the total cost of ownership of distributed systems. Gartner estimates that an unmanaged PC running Windows 95 has an average annual cost of ownership of $9,784. With today's corporate networks commonly encompassing thousands of such systems, even small reductions in per-system cost of ownership can be significant.

Reduced cost of ownership has been a major marketing emphasis for vendors of network computers (NCs), who essentially remove from the PC the elements that require management. Data, software, software backups, upgrades, and maintenance are received from a network server. NCs are designed to be almost maintenance-free and lack the requirements for disk storage or high-end computing power, making them an economical alternative to PCs. Some major players in the NC arena include IBM, Oracle (which offers an NC product through Network Computing), and Sun.

Although the NC may provide a solution for reducing cost of ownership in the future, the number of network computers that actually have replaced PCs is not great enough to have significant impact on the cost of system ownership in large enterprises. NCs may offer a better replacement for terminals, which are already relatively low-cost. (See Chapter 5.3, *Computing Platforms,* for more information on network computers.)

One effect of the marketing efforts of NC vendors has been an increased awareness of PC cost of ownership. The response from PC hardware and software vendors has been a greater emphasis on reducing cost of ownership and offering "zero administration" desktops. In 1997, Microsoft and Intel announced a diskless PC standard, called the NetPC, that is in between a "thin client" and a PC to help address concerns about high PC ownership costs. Compaq, Dell, and HP collaborated in developing the NetPC specification. Companies will benefit from these initiatives because they serve to lower PC costs over time.

Microsoft's Zero Administration for Windows (ZAW) initiative includes features designed to reduce the cost of managing desktop PCs, including automatic installation of operating system updates when the computer is turned on; automatic installation of applications as they are invoked by the user; storage of user files and profile information on a server rather than on the local hard disk; and the ability for a central administrator to specify system configurations that cannot be modified by the user. Microsoft's Zero Administration Kit (ZAK) was the only component of the company's ZAW initiative available in late 1997 and was developed to complement Microsoft's System Management Server (SMS) software. ZAK provides for the management of users through policies. SMS offers automated software and hardware inventory, software distribution, and remote diagnostics.

During 1997, Intel continued to build on its Wired for Management initiative, which aims to bring manageability features to conventional PCs in a consistent way. HP announced plans in October 1997 to integrate features based on Intel's Wired for Management specification into its line of Vectra PCs and commercial workstations. The specification will give users standard instrumentation, remote boot ROM, and PC wake-up features.

HP works with other systems manufacturers to reduce the cost of networked computing through a program known as HP OpenView-Ready to complement HP's management platform, HP OpenView. Jointly developed by HP and Dell (also the first program participant), the HP OpenView-Ready program includes the HP OpenView-Ready Network Node Manager at no additional cost when customers purchase selected Dell PowerEdge servers running Microsoft's Windows NT or Novell's IntraNetWare. The OpenView-Ready Network Node Manager allows network administrators to monitor the health of their network servers and manage parameters of their servers, desktops, workstations, notebook computers, and network devices. This information is integrated with HP OpenView.

Companies have a stake in making computers more manageable in interconnected enterprise environments, using management standards such as the Simple Network Management Protocol (SNMP), Windows Management Instrumentation (WMI), and Web-Based Enterprise Management (WBEM). With less money required for management, customers' finances are available for investing in new technology and network growth instead of maintaining existing technology.

■ Management Services

An increasing number of vendors have integrated product and service offerings that take advantage of communications over the Internet to help manage desktop and server systems.

In December 1997, IBM Global Services announced Asset Services to help customers track, manage, and maintain control of their multi-vendor desktop computers, software, and of the costs associated with distributed computing. Asset information is managed throughout the life cycle of a PC or software product. Designed to support large groups of users, these desktop management services are priced on a per-seat basis and are delivered via the Internet. Using Lotus Domino, customers can track inventory, request acquisitions, and manage moves, adds, and other changes via an IBM-hosted Internet Web server. Management and coordination of the changes is performed by a centralized group of IBM personnel.

IBM's Asset Services are designed to help lower total cost of ownership and they combine many of the customized capabilities that are part of IBM's Desktop Acquisition and Asset Services announced earlier in 1997. AT&T, Compaq, HP, and other systems and services vendors also have services that help customers manage and monitor desktop systems and servers.

6.4.3 Network Management

Today's corporate networks commonly may encompass thousands of components, including mainframe systems, servers, desktop and notebook computers, wireless devices such as "smart" phones, printers, and networking devices such as routers and switches. End user systems and network devices are connected via LANs that carry information at high speeds over short distances. LANs in turn are connected to high-speed backbone networks that interconnect via WANs.

With the advent of client/server computing in the mid-1980s, the issue of LAN management increased steadily in importance. From an environment with few tools and where LAN users were responsible for backing up their own data to floppy disks, LAN management has become extremely sophisticated. Until recently, however, LAN and distributed systems management tools focused on hardware: monitoring performance, providing status information (such as whether a system or a specific port was available), and implementing device control. LAN management tools now play a significant role in applications management. These tools are capable of performing a wide range of tasks, such as determining whether a particular instance of database software is available at a particular server.

Within this environment, the emphasis is on managing applications end-to-end, which typically encompasses not only LAN traffic but WAN traffic as well. Issues such as software distribution, configuration management, and security control are just as important as traditional operational systems management. In addition, the inherent size and complexity of the corporate network demand expanded capabilities. Tasks such as capacity planning can be far more complex for a heterogeneous, distributed enterprise network because of the potentially large number of factors associated with applications, users, and network media.

With recent advances in standards such as additions to the Remote Network Monitoring specification (RMON2), network management tools are capable of addressing every element of the enterprise network, and they have become robust enough to rival those found in legacy mainframe environments. The tools used to manage both LANs and WANs are based on SNMP and offer similar capabilities.

These tools have helped administrators control the distributed computing architecture of enterprise networks encompassing mainframes and distributed systems in a mixed LAN and WAN environment. However, a new networking paradigm has exploded onto the networking scene: Internet-based Web technology and corporate intranets. (An intranet is an enterprise network based on Internet technology.) Internet technology includes a standards-based network that makes heavy use of Web site and browser technology to provide easy access to information. (See Chapter 4.4, *The Internet and Intranets.)*

The Internet has contributed to the convergence of systems, network, and application management, and it is generating a need for new management tools designed to take advantage of Web technologies. Intranets also have intensified existing networking trends, such as increasing the number of applications on the network, the need for end user access to these applications, and the need for administrator access to management data about each application. Extranets (portions of an intranet accessed by external partners, customers, or suppliers) also pose security, management, and planning challenges.

Self-Healing Networks

Tools are evolving to meet management needs and are moving toward high levels of automation, with an eventual goal of enabling "self-healing" networks. Components of this solution will include higher levels of local intelligence throughout the network in the form of intelligent agents and policy-based network management applications that work proactively to prevent significant network problems. The combination of advanced network monitoring and distributed management intelligence will allow network management applications to take immediate corrective action without network administrator involvement.

Large global networks have many thousands of entry points, and to probe each one would yield an overwhelming amount of information. Tools that help ensure service quality for groups of users mean fewer experts are required to maintain complex internetworking configurations.

Traditional offerings frequently present and track information about service levels. However, many products report performance statistics only or still require a great deal of operator intervention. Others may show an alarm about a violation, but they may not show customers what went wrong. Although routers and switches have become more reliable, recovery processes have become more automated, and new products offer improvements, widespread implementation of technology that promises end-to-end network reliability has not yet occurred.

Network management tools from 3Com and Cisco announced in 1997 illustrate the direction in which the industry is heading. 3Com also plans to guarantee service levels for specific users. 3Com announced plans in 1997 to integrate its Transcend management tools with service-level monitoring software from InfoVista to enforce service policies for traffic moving through its internetworking equipment.

Cisco's Netsys Service-level Management (NSM) automated software offers self-healing capabilities for networks with Cisco equipment. Cisco acquired Netsys Technologies in November 1996 and modified Netsys router modeling software to add features that will let network managers guarantee service quality. NSM software works with Cisco's Catalyst 5000 LAN switches and with routers from Cisco and Bay Networks. (Cisco also plans to support StrataCom's WAN switches.)

NSM is designed to help customers predict where problems will happen and react to failures. NSM verifies whether enough bandwidth is available on network circuits before multimedia traffic is added and analyzes circuit integrity to help minimize connectivity problems. NSM helps monitor network performance, diagnoses problems, and suggests possible solutions. It also helps define service policies and track end-to-end performance network-wide.

Third-Party Network Management

With today's open networks and standards-based management tools, the decision whether to outsource applies as much to network management as any other resource. The Gartner Group terms the market for external delivery of primarily operational network support "managed network services." Outsourcing network management in whole or in part is a viable option today, with a growing number of systems and service vendors offering a wide range of services.

Since its separation from AT&T, Lucent has developed a network management service for data networks known as NetCare. NetCare offers fault management, configuration recovery and management, and performance management services. Lucent also offers a resolution management service that provides a single point of contact for ongoing trouble isolation and problem resolution.

In February 1997, AT&T announced its own network management service by consolidating its managed network services throughout the organization into one offering termed AT&T Managed Network Solutions (MNS). MNS plans to provide planning, design, implementation, operation, and business administration of an enterprise's physical, logical, WAN, and LAN data networks. Both AT&T and Lucent plan to expand their data network management to multimedia applications such as voice and video.

Network and Systems Management Standards

Standards enable the management of large, heterogeneous networks as a single, integrated resource. Networking standards are developed by organizations such as the Institute of Electrical and Electronics Engineers (IEEE) and the Internet Engineering Task Force (IETF). Although numerous standards exist in the networking industry, the most significant contributions to network management have been made by SNMP and RMON/RMON2. Other standards, such as DMI, have been developed but not as widely adopted; still others are in development.

■ Simple Network Management Protocol (SNMP)

The first version of SNMP was developed in 1988 by the U.S. Department of Defense and TCP/IP vendors as a network management architecture for the Internet. It was adopted in 1990 by the IETF as a standard for monitoring network devices over TCP/IP networks. Although SNMP was designed as the TCP network management protocol, it now manages many different types of networks and devices. SNMP is a specification that enables vendors to exchange management information between devices and consoles by providing a single mechanism under which to integrate different types of devices. SNMP provides a means to monitor and set network configuration and runtime parameters. Its use has expanded to include many different types of networks, making it a widely implemented network management standard. (See Figure 6-8.)

Figure 6-8: SNMP Components

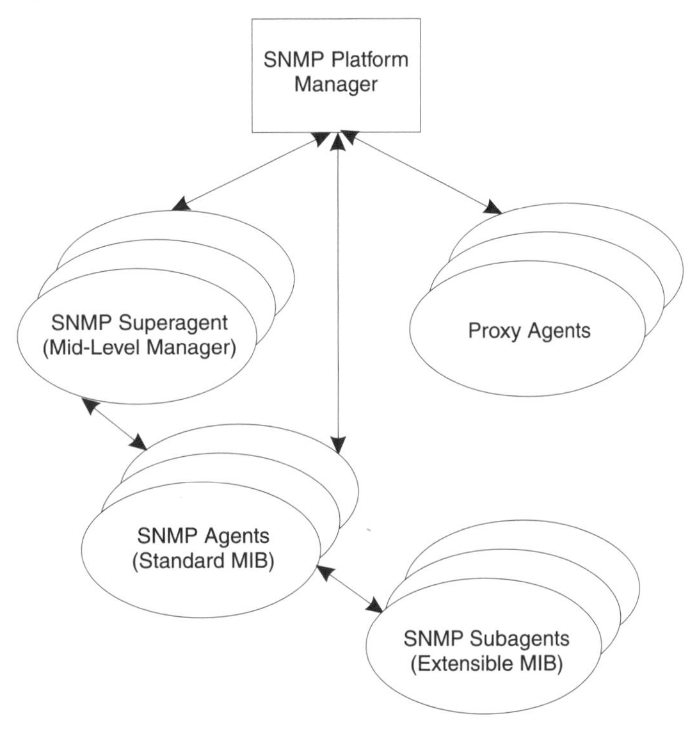

Source: Gartner, 1995

SNMP defines a standard for collecting and storing device data that is easily accessible through a common language for display on a central management station. This standardization paved the way for multi-vendor network management by providing a methodology for SNMP-compliant products to collect Management Information Base (MIB) data.

SNMP comprises three components: the Structure of Management Information (SMI), the MIB, and Protocol Data Units (PDUs). These elements interact to transfer management messages. SNMP-managed entities also must have an agent that implements all the relevant MIB objects and a managing process. SNMP agents store MIB variables and send them to SNMP managing processes (or consoles) when asked to do so. This management process has a list of all MIB objects implemented on entities that it manages.

The MIB is a hierarchical name space, with node registration administered by the Internet Assigned Numbers Authority. The term MIB also refers to collection of objects that can be accessed via a network management protocol. A MIB acts like a directory that lists all managed entities (such as hubs, routers, or other devices) residing on a network. MIB-I was the original collection of objects and attributes. MIB-II, adopted after MIB-I, is an expanded collection of objects and is the current MIB standard.

MIB objects under the MIB-II node include RMON MIB objects and other generic MIB objects; those under the enterprises node include all proprietary MIB objects. Standard MIBs provide network information regarding 165 objects; the remainder is provided through private MIB extensions developed by vendors. There are now many more proprietary MIB objects than generic ones.

MIBs usually are divided into groups of related objects. For example, MIB-II groups of objects include "system" and "interfaces." MIB objects have attributes such as name, object identifier, syntax field (which selects one of several possible data types), a field that specifies access level permitted, a status field, and a text description of the object. MIB objects are compiled to a format that agents and managing processes can load. MIBs exist for bridges, printers, and other entities, and they provide useful common information for management application developers. However, most vendors have defined proprietary objects to optimize their products' capabilities.

The SMI sets the rules for defining MIB objects. The SMI specifies allowable data types in a MIB and defines the rules for naming and identifying MIB components. It also defines a hierarchical naming structure that ensures unique, clearly understood names for managed objects, the components of a MIB. The SMI lets a vendor write an SMI-compliant management object definition, run the text through a standard MIB compiler to create executable code, and install the code in existing agents and in management consoles. These consoles then can generate reports about occurrences of specified events.

SNMP also specifies the packet layout for messages between management agents and managing processes. These elements of different possible management messages are called PDUs. Initially, only five messages were possible. For example, a GetRequest message asks for values associated with one or more object instances, and a GetResponse message returns the object instances with their values. Traps are messages initiated by the SNMP agent and sent to the managing process.

SNMP has some shortcomings in the area of security. Proposals for Version 2 of SNMP (SNMPv2) had enhanced security and improved support for systems and applications management and other extensions. SNMPv2 also enabled manager-to-manager communication, a benefit for distributed management environments. Adopting HTTP to transport management data also has been discussed. Management agents would act like a specialized Web server and managing processes would exist as browsers. However, there were disagreements between two different vendor proposals, and SNMPv2 was not implemented. An IETF work group on SNMPv3 was formed; it posted drafts for different components of the new standard during the last half of 1997. SNMPv3 incorporates some elements from earlier versions of SNMPv2. In 1997, vendors began developing product based on these draft specifications. All SNMPv3 specifications are scheduled to be submitted for consideration as proposed standards by April 1998.

■ Remote Network Monitoring (RMON)

The RMON standard provides detailed information about network devices being managed. RMON provides much of the functionality offered by proprietary network and protocol analyzers and is believed by some to be the most important network monitoring advance of the 1990s. Developed by the IETF to support monitoring and protocol analysis of Ethernet and Token Ring LANs, RMON brought more open, comprehensive fault diagnosis, planning, and performance-tuning features to the market than any monitoring solution previously available. Using RMON, managers can see the traffic on any LAN segment throughout the network and identify trends and bottlenecks from a single management station.

RMON implementations generally are delivered as a two-part client/server solution. The client is the application that runs on the management station and presents the RMON information to the user.

Although RMON is an extension of the MIB, traditional SNMP agents are not capable of capturing most RMON data. Special RMON probe devices or built-in probe functions that can see an entire segment are necessary to collect and forward RMON information.

The servers are the monitoring devices distributed throughout the remote networks that collect the RMON information and analyze network packets. The monitoring device, or probe, runs a software program called a RMON agent. RMON agents are in dedicated devices and also in network infrastructure devices such as hubs and switches. The application and the agent communicate across the network using SNMP. A RMON probe captures and processes more data than ordinary device agents, and it can reduce traffic by storing intermediate results locally and forwarding them to an application on demand.

Virtually all network device vendors now embed RMON agents in their products and offer dedicated RMON probes for more extensive monitoring applications. Large vendors have either acquired RMON probe vendors, as 3Com did in acquiring Axon Networks in 1996, or licensed technology from RMON specialists such as Technically Elite Concepts.

SNMP and RMON focus primarily on monitoring and controlling a specific network device. Although these device-specific tools are important, they do not provide a comprehensive picture of the network with all its devices, servers, applications, and users. RMON2 moves beyond RMON's ability to provide information about an individual network segment. RMON2 extends up to the enterprise network and supplies data for monitoring end-to-end communications. This information, such as the amount of bandwidth used by a particular application, provides crucial insights for deploying and troubleshooting client/server applications. With this detailed knowledge of traffic patterns and client/server application usage, network managers can ensure that resources optimize performance and reduce costs.

The growing number of products that support RMON2 is consistent with the broadening responsibilities of network administrators, who now manage network health as a business and applications resource rather than just a collection of systems and wires.

■ Desktop Management Interface (DMI)

The DMI specification was first released by the Desktop Management Task Force in 1994. DMI standardizes desktop systems management by outlining ways to identify items such as processors, coprocessors, expansion slots, motherboards, operating systems, and power supplies.

DMI defines hardware and software components in a format that can be accessed by PC management applications and technical support personnel using a phone line. DMI permits access to an inventory of hardware and software components, to changes made in parameter settings, and to data views generated by software agents and diagnostics routines.

Technical support and management applications use a management interface (MI) to access data stored in management information files (MIFs). The specification calls for a DMI service provider (also known as a DMI service layer) to handle hardware and software instrumentation through a component interface (CI) and to handle management requests regarding this information through an

MI. DMI can help desktop devices monitor themselves and send information to management software packages, resulting in better information for network and systems managers.

DMI 2.0 was introduced in 1996 and finalized in 1997. By the last half of 1997, a significant number of desktop systems supported DMI. Various levels of DMI implementation are available. Some vendors offer products that are DMI-enabled with product components that meet DMTF guidelines. However, these products may lack critical software for full manageability, such as the DMI service layer. Other vendors offer tools that enhance manageability, such as a MIF browser that can be used view DMI system attributes.

Vendors such as Compaq, HP, and IBM offer functionality that extends beyond the basic DMI specification to include advanced features such as waking systems across a LAN or WAN to perform routine maintenance, diagnostics, or inventory.

Managers will benefit from DMI in the long term. However, in the short term, many organizations still have thousands of older PCs that do not comply with the DMI 2.0 standard and do not work with software intended to manage more sophisticated desktops.

■ Web-Based Enterprise Management (WBEM) Initiative

The WBEM initiative was launched in July 1996 by a consortium made up of BMC Software, Cisco, Compaq, Intel, and Microsoft. By late 1997, WBEM had the support of 70 major vendors, and compliant products are expected to be available in 1998. Though not necessarily limited to use with browsers, WBEM is intended to let enterprises manage any component of their infrastructure and value-added applications by third parties via any Web browser.

WBEM includes three components: the HyperMedia Management Schema (HMMS), a data model for representing the managed environment; the HyperMedia Management Protocol (HMMP), which lets browsers access data from management platforms and devices, runs over the HyperText Transport Protocol (HTTP), and embodies HMMS; and the HyperMedia Object Manager (HMOM), which handles data from applications based on SNMP and DMI. Figure 6-9 illustrates the WBEM architecture.

Figure 6-9: WBEM Architecture

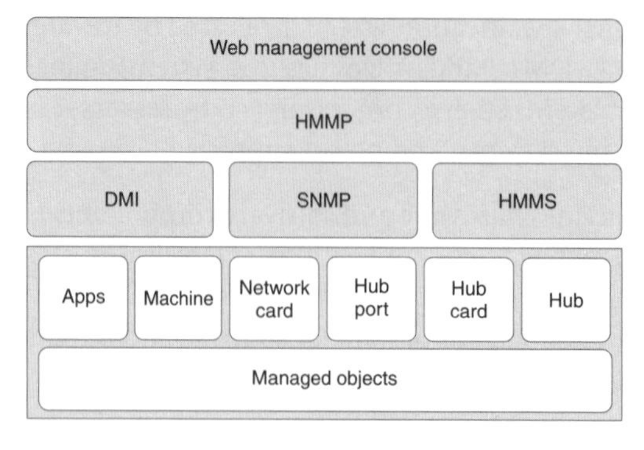

Source: *PCWeek*, 1997

The WBEM draft standard defines a generalized management information model that will work with DMI and many non-DMI proprietary management systems. WBEM is a superset of other standards, encompassing several new protocols and some current Internet standards. It relies heavily on the same Common Information Model (CIM) metadata structure that the DMI 2.0 specification introduced. CIM allows any existing protocol, either standard or proprietary, to provide data for WBEM use. Because it assumes nothing about the object model or protocol used, CIM is platform-independent.

6.4.4 Issues In Systems and Network Management

Just as the client/server paradigm forced a radical change in management systems, Internet and Web technologies are changing the management rules again. At the same time, more proactive approaches to management are being pursued. For example, the network maintains a given performance level by predicting and quickly reacting to anticipated network events accurately. This view is reflected in many vendors' strategies as an eventual goal of "self-healing" networks. Some products to support this goal already are available in the form of policy-based management applications and embedded intelligent agents. However, developers of management tools are just beginning to implement this type of automation. (See "Self-Healing Networks" on page 529.)

Using Web-based management tools can yield operational advantages and potential cost savings. Web-based management products provide a universal client that enables relatively simple, geographically dispersed, multi-user access to management functions and information. These solutions range from simple, read-only access of network monitoring data to interactive tools that support advanced management.

Distributed access to advanced management capabilities offers the freedom to solve network problems from home or while traveling. Technical end users benefit from access to network status information as well, eliminating a call to a busy support desk or planning work to avoid peak use times when the network is congested.

One significant aspect of Web-based management is platform independence. Users with Web access can obtain information with equal ease and without special setup. Development costs and time-to-market are reduced substantially because only a single server-based application accessible from all platforms is required. Platform independence also means Web-based management systems can monitor all elements of the enterprise network, including LANs, WANs, and telecommunication systems. Even peripherals such as printers now offer browser-based access to management information, allowing users to manage their time and productivity better by making use of these resources.

■ Load Balancing and the Web

Load balancing and performance management of Web traffic have become bigger issues as the number of Web users has grown steadily. Traditional methods using a Domain Naming System (DNS) server allow for situations where some servers may be overloaded and others idle. Incoming traffic may be sent to a Web server without regard to that server's existing load.

Offerings such as Cisco's Local Director, HydraWeb's HydraWeb Load Manager, IBM's Network Dispatcher, and RND Networks' Web Server Director Pro help ISPs, corporate intranet managers, and large Web site operators increase reliability and performance of a Web site. These offerings help regulate traffic between the Internet and a group of servers. Proxy server standards are being developed that may help optimize performance as well.

Cisco's Local Director is a high-end hardware and software offering that has several options for distributing traffic. Servers can be brought on-line gradually, and load is based on server response time.

HydraWeb's software and software/hardware products offer load balancing, monitoring, management, and fault-tolerant features. The Load Manager product can send out alerts via e-mail, fax, or pager. Servers also can be managed remotely using token-based authentication.

In September 1997, IBM and its Tivoli Systems subsidiary introduced Java-based tools for managing enterprise hardware, networks, and Web-based business systems. As part of this announcement, IBM's Internet business launched Interactive Network Dispatcher 1.2, a low-end, Java-enabled, load-balancing tool for Web servers.

RND Networks' Web Server Director uses a combination of load balancing techniques (similar to Cisco's Local Director), and provides some security, alert, and redundancy features; however, in its 1997 version, it does not support 100-Mbps Ethernet.

Intermediary or proxy servers also are being used to increase performance. Proxy servers store data that individuals request from the Internet. Requests for this same data later then can be served locally rather than going back over a congested network. With the existing Internet Caching Protocol (ICP), proxy servers communicate with each other to see whether another server already has the requested data before searching the Internet. Using this protocol may be helpful but servers may be storing redundant data. Also, a large group of proxy servers generates more queries between servers, which creates more overhead. A new proposed standard, the Cache Array Routing Protocol (CARP) by Microsoft and Netscape, uses scripts to maximize efficient query routing. Loads can be balanced better among servers through the use of simplified routing and configurable load factors. (See Figure 6-10.)

Figure 6-10: CARP Caching

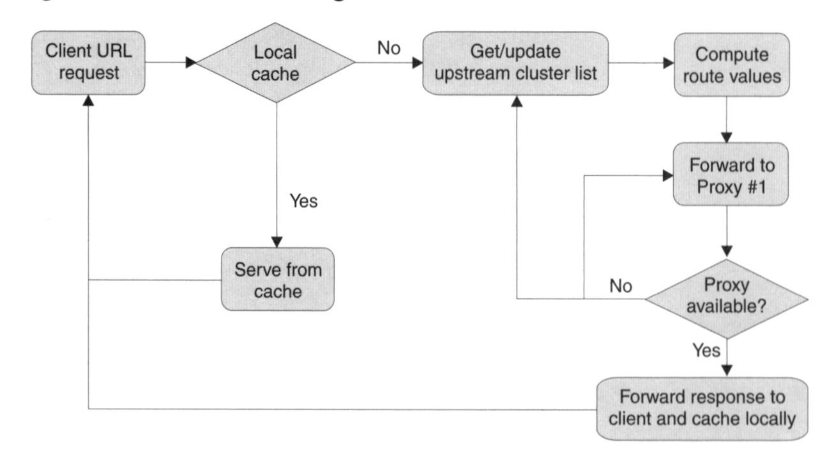

Performance Management

Performance management offerings address performance, monitoring, simulation, management tools, and some resource tracking. These products represent the largest area of systems management in terms of revenue and are still concentrated in the large systems or mainframe area. For example, Computer Associates' products just for IBM mainframes include CA-PMA, CA-LOOK, CA-Explore, CA-TSO/MON, CA-Paramount, and CA-MICS. Other products in this area include Boole & Babbage's MainView product and Candle's OMEGAMON series; vendors including BMC Software, CompuWare, and IBM also offer products.

Change and Configuration Management

One of the most rapidly growing areas of systems management software is change and configuration management. Traditional configuration management dealt with tools to manage large, complex mainframe systems. Change management offerings were designed to make changes in or updates to system configurations without disrupting system operations. Software updates also are included in this category.

The challenges of managing software changes in distributed computing environments stimulated demand for products that could work in distributed environments across multiple platforms. Change and configuration management offerings include CA's CA-Librarian and CA-Panvalet; IBM's ISPF/PDF and IBM/Tivoli Courier; and Intel's LANDesk. Other vendors include HP, Microsoft, Network Associates (McAfee), and Symantec. Initiatives such as Microsoft's ZAW and some of the design criteria behind NCs have attempted to reduce some of the costs and complexity associated with managing software changes.

Storage Management

The number, size, and complexity of databases in enterprise networks is growing, reflected by an increasing need to manage data storage as a corporate resource. Although data is backed up to tape, administrators often place a low priority on the backup, restore, and long-term archival of corporate data. Storage assets are not managed efficiently, high availability or redundancy may not be sufficiently addressed, and backup and store processes often are inefficient.

With businesses now relying on network-based data, the ramifications of data loss are a significant concern. There is a growing emphasis on storage management tools that offer high performance archiving, backup, and recovery of network data. Storage costs are a major part of a server purchase. The percentage of hardware dollars spent on storage exceeded 35 percent worldwide in 1997, according to Dataquest.

Managing data storage tasks across heterogeneous networks and distributed databases is a complex process. Issues include storage backup and recovery performance, maintaining historical data states, and tracking data usage to maintain data on different media according to degree of use, cost, speed, availability, and other considerations.

Basic backup software with Microsoft's Windows NT and Novell's NetWare does not include features such as disaster recovery, flexible scheduling, reporting, or tape rotation. For vendors that do not develop their own storage management systems, third-party solutions are available. Third-party storage management systems commonly work with multiple operating systems, including Apple's MacOS, HP's HP-UX, IBM's AIX and OS/2, Microsoft's Windows and MS-DOS, Novell's NetWare, and Sun's Solaris. As an alternate solution, Novell helps backup system vendors integrate products with NetWare via its Storage Management System. Novell's Storage Management System is a set of interfaces that provides third-party developers with version-independent NetWare file information, reducing development time and ensuring compatibility with future releases of NetWare.

Storage management application packages such as CA's Cheyenne ARCserve and Seagate's BackupExec are priced at approximately $700 for a single server, and they work with more expensive tape drive solutions.

In June 1997, IBM announced the Seascape storage enterprise architecture. The Seascape architecture is based on an open, industry-standard storage server designed to scale in power and performance. Seascape addresses many critical needs of enterprise storage management, including mixed data types, file structures, databases, and operating systems. Software "building blocks" and libraries contribute to functionality that addresses complex issues such as data sharing.

In September 1997, IBM announced the second release of 3466 Network Storage Manager, part of the Seascape family architecture. This hardware and software system provides storage capacity of 1 terabyte or more and allows businesses to backup data in an integrated way automatically. The system includes a processor, disk subsystem, automated tape library storage options, distributed storage manager software, support for 25 operating systems, networking connectivity options, and a Webshell interface that allows use with browsers for backing up and restoring files.

Intelligent caching features also are being integrated into disk drives and storage management solutions. Examples include EMC's drives, Memorex's Scimitar, and Virtual's tape volume manager. Drives often are equipped with onboard buffers with intelligent caching for additional speed; intelligent caching devices throughout a network can help boost performance. (See Chapter 3.2, *Storage and Batteries,* for more information on storage solutions.)

With storage management an increasing concern of enterprise information systems departments, large systems and software vendors are forming partnerships with key storage management vendors. Partnerships also can take the form of acquisitions by large software vendors, such as CA's acquisition of Cheyenne Software and Seagate's purchase of Palindrome.

In July 1997, Oracle and Legato announced a strategic alliance. (Legato supplies enterprise storage management software for heterogeneous client/server environments.) This alliance will add Legato's NetWorker technology to Oracle8, providing users with an integrated base-level backup and recovery solution.

Microsoft also has chosen a partnering strategy to round out the company's management offerings. For example, Microsoft has joined forces with HighGround Systems, which offers solutions in the emerging category of management applications known as storage resource management (SRM). HighGround offers the Storage Resource Manager and Media Series families of SRM applications, which are designed to simplify and centralize storage resource administration.

Other leading storage management vendors include HP with its OmniStorage and OmniBack II; IBM with its Adstar Distributed Storage Manager (ADSM) and Optical Library Dataserver; OpenVision with its AXXION-NetBackup; and Stac with its Replica for NetWare and IntraNetWare for LAN server data management.

Applications Management

As systems management tools evolve to manage in distributed environments, the emphasis on traditional areas such as performance management has expanded to include application performance. The discipline of applications management is based on the concept that the bottom line for business management is not whether the systems and networks are running at high performance levels, but whether the business applications themselves are available with the necessary performance to meet relevant business needs. Ideally, applications management should include administration, availability, life cycle management, performance, process automation, and recovery from unscheduled events.

Traditional systems management metrics such as system latency or the number of network segment packets dropped can provide a misleading perception of application performance efficiency. These indicators measure attributes of the system that delivers services to clients, but they do not indicate the quality of service (QoS) actually realized by the end user. To determine QoS, metrics must focus on meeting user expectations of availability, performance, accuracy, and affordability.

Determining whether applications meet relevant business needs first requires that those needs be quantified by end users and systems administrators. For example, a hospital's accounting department may find a given response time acceptable, but doctors using an on-line diagnostics system might find the same response time unacceptable. Performance numbers no longer can be treated simply as numbers but must be examined with regard to what those numbers represent in terms of appropriate service to the user.

Once appropriate application requirements are determined, administrators must have suitable tools to ensure that those requirements are met. Defining and addressing these needs, categorized as quality of service management, can be addressed by service level agreement (SLA) tools. SLA software helps information systems departments ensure that application performance and availability mesh with business needs.

One popular SLA product is InfoVista from InfoVista. InfoVista addresses both the creation and implementation of service level agreements. Using InfoVista, administrators prepare predefined or customized QoS reports to ensure clients are receiving needed services. The reports are also useful for the information systems departments themselves to ensure they have the resources they need to provide the services and service levels required by end users.

Some application-specific management tools are also available. Tivoli Systems' Application Management for Notes and Domino monitors the health of the Notes/Domino application across a network with up to hundreds of Notes servers and thousands of Notes clients. The application includes systems administration capabilities for UNIX and Windows NT machines and allows security policies to be set based on policy regions and roles for scalability.

BMC Software has taken a different approach with its Patrol Series. Rather than provide extensive management for a single application, the Patrol Series uses "knowledge modules" that focus on the performance and availability of applications running under a variety of operating systems, database management systems, key business applications, and middleware products.

A key technology used in application management is the RMON2 standard. RMON2 provides application-specific monitoring rather than the device port or network segment monitoring provided by RMON. RMON2 provides information from an application perspective, such as which users are using which applications and what percentage of available bandwidth each application uses. With information from RMON2, administrators can manage network resources to ensure that critical applications are performing at necessary performance levels and that people have access to the applications they need.

Security Management

Networks now are accessible from wireless devices, telecommuter PCs, remote office systems, and even other vendors' systems through extranets. Use of the Internet and the Web opens a new realm of security issues for network managers with data transmitted over a public network. Managing sensitive information is especially important, whether that information involves corporate secrets or a personal credit card number. Managing systems properly to protect data and prevent unauthorized access is critical.

An integrated suite of tools will let administrators view systems' problem information as it relates to various departments. For example, administrators may benefit from products that provide detailed information about specific problems associated with remote access across groups over time. Automatic monitoring, logging, and alerts provide valuable information. (See Chapter 5.4, *Security,* for more information on security solutions.)

6.4.5 Network Management Trends

The increase in implementation of switched architectures for data networks is influencing trends in network management. As network bandwidth needs increased and the technology to provide cost-effective data switching became available, switched data networks became the solution of choice for bandwidth-hungry users frustrated by the overburdened shared media technology of the 1980s.

Virtual Networking

Switching technology also supports virtual networking, which allows "logical" network connectivity without physical reconfiguring. With virtual networking, users in different parts of the company can be configured onto a single, logical network segment, providing high-speed connections and improved response times.

Virtual networking eliminates the correspondence of subnets to physical wiring topologies and allows the creation of task-oriented work groups that are not attached to the same network segment. These work groups may be located in different geographic areas, but they require access to the same set of network resources.

Managing virtual networks requires tools that integrate physical network views based on physical networking devices with logical network topology views based on virtual work groups. Although switching has become an established networking technology, it is still a relative newcomer to internetworking. The tools for managing virtual networks within a switched environment are still evolving and will continue to be an important aspect of the overall network management solution in the future.

Virtual network management tools need to provide physical inventory views, monitoring functions, and troubleshooting capabilities as well as ensure that change management is simple and straightforward. Many tools are meeting this need with comprehensive network maps visible through a GUI that meshes physical and logical network topologies. Network managers can switch nodes across the network using drag-and-drop graphics without changing node addresses.

Automation and Intelligent Agents

Today's enterprise networks are too large and complex to be managed manually. Even a task such as setting performance thresholds can be overwhelming when it must be repeated for hundreds of devices. Network management solutions are addressing this issue with the same type of agent used in SNMP, with the addition of task-specific intelligence. Such intelligent agents can monitor network functions, generate reports to the network management station, or even take action automatically when activities warrant, leveraging network administrative time and expertise.

3Com embeds SmartAgent management agent software in devices to process data and take corrective actions independent from the network management console. For example, SmartAgent software agents in some 3Com hubs allow network managers to implement automatic, hub-based network monitoring and self-recovery. SmartAgent capabilities include switching a link to a standby path, reporting if traffic on vital segments drops below minimum usage levels, and carrying out specific actions at different problem severity levels.

Graphical User Interfaces (GUIs)

Well-designed GUIs contribute to product ease of use and can present large amounts of information quickly for analysis. With thousands of network devices and users, the amount of data pouring into management systems can be analyzed more quickly with high-quality color or even 3-D graphics. These features allow more relevant information to be displayed on one screen than via previous text formats. With the help of graphics such as network maps and graphic problem indicators such as flashing lights, administrators can evaluate network situations rapidly and implement necessary actions. Graphics are also an important tool in areas such as configuration management. Using drag-and-drop capabilities, network administrators can reconfigure network management information quickly to reflect adds, moves, and changes.

6.4.6 Market Overview

IDC estimates that the worldwide market for systems management software grew from $6.88 billion in 1995 to $8.29 billion in 1996, an increase of 20.5 percent. When database administrator tools and utilities are added, the market in the same period grew from $7.33 billion to $8.29 billion, an increase of 20.7 percent. Performance management represents the largest market segment with 1996 revenue of $1.8 billion, followed by storage management with $1.71 billion.

Figure 6-11 shows projected growth in worldwide systems management software revenue by market segment from 1996 to 2001.

Figure 6-11: Worldwide Systems Management Software Revenues

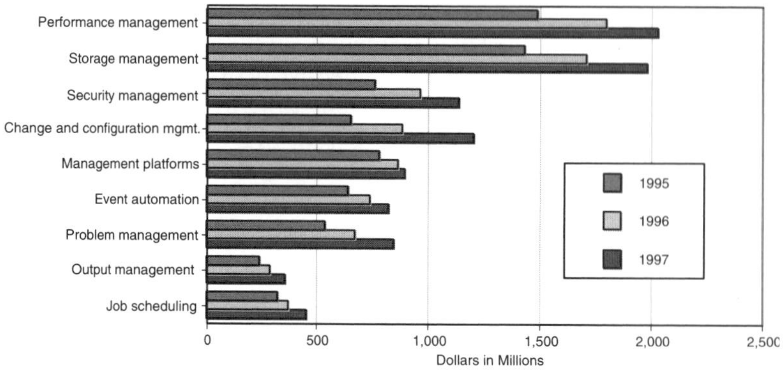

Performance management
Storage management
Security management
Change and configuration mgmt.
Management platforms
Event automation
Problem management
Output management
Job scheduling
1995
1996
1997
0 500 1,000 1,500 2,000 2,500
Dollars in Millions

Source: IDC, 1998

Worldwide systems management software revenue for ISVs in 1995 and 1996 is shown in Table 6-9.

Table 6-9: Worldwide Systems Management Software Revenue for ISVs

ISV	1995 Revenue ($millions)	1996 Revenue ($millions)	1996 Worldwide Market Share
CA	$1,736	$1,906	22%
BMC Software	$363	$488	6%
Candle	$207	$233	3%
Sterling	$169	$206	2%
Platinum Technology	$151	$206	2%
Cheyenne	$140	$202	2%
McAfee (now Network Associates)	$90	$180	2%
Boole & Babbage	$151	$179	2%
Symantec	$110	$173	2%
Seagate	$127	$144	2%
Intel	$52	$113	1%
Remedy	$30	$77	0.9%
CompuWare	$34	$66	0.7%

Source: IDC, 1997

Table 6-10 shows management platforms software revenue for systems vendors, which makes up approximately 85 percent of the market, for 1995 and 1996 and as a percentage of worldwide market share. Management platform revenue for ISVs composes the remaining 15 percent of worldwide management platforms software revenue. The total worldwide management platforms market is expected to grow from more than $942 million in 1997 to more than $1.1 billion in 2001.

Many independent software vendors offer products for event automation, job scheduling, output management, performance management, problem management, change and configuration management, security management, storage management. These "element" products can be mainframe-oriented, open systems network-oriented, or LAN-oriented.

Table 6-10: Management Platforms Software Revenue

Systems Vendor	1995 Revenue ($millions)	1996 Revenue ($millions)	1996 Worldwide Market Share
IBM	$443	$478	55%
HP	$88	$96	11%
Tandem	$63	$66	8%
Sun	$30	$42	5%
Hitachi	$18	$21	3%
Groupe Bull	$9	$19	2%
Amdahl	$6	$9	1%
Unisys	$6	$6	0.7%
Others	$3	$4	0.5%

Source: IDC, 1997

IDC predicts the 1997 network element management software market will surpass $285.1 million in end-user spending, based on 215,000 shipments, with continued steady growth.

In the mainframe arena, the leading independent software vendor in the storage management software market is CA, with 1996 revenues of $475 million and a market share of close to 28 percent from its range of backup and restore, storage migration, and tape management products on a wide array of platforms.

In the systems vendor market, IBM reached software revenue in 1996 of $280 million, resulting in a worldwide market share of more than 16 percent, substantially higher than most other vendors. The second-largest system vendor in this area, Groupe Bull, with $46 million in storage management software revenue, had a worldwide market share of approximately 3 percent.

Table 6-11 shows the anticipated compound annual growth rate (CAGR) in storage management software revenue from a total of $1.7 billion in 1996 to an estimated $4.1 billion in 2001, according to IDC.

Table 6-11: Worldwide Storage Management Software Revenues

Geographic Region	1996 Revenue ($millions)	2001 Revenue ($millions)	1996 - 2001 (CAGR)
United States	$878	$2,068	19%
Western Europe	$557	$1,339	19%
Asia/Pacific	$153	$440	24%
Rest of the world	$119	$246	16%

Source: IDC, 1997

Management Suites

Although the market for many products providing solutions in different areas of management is growing, the fastest area of growth has been in the area of enterprise management frameworks, comprehensive solutions that are multi-discipline products. Gartner estimates that the worldwide network and systems management (NSM) market will grow from $4.5 billion in 1997 to $9 billion in

2000. Gartner says NSM suites represented about 35 percent of NSM revenue in 1997 and predicts they will represent an estimated 55 percent of the market in 2000. CA and IBM/Tivoli are the two largest management suite vendors today.

The network and systems management market emerged from a blending of the network management platform and systems management markets, often through mergers and acquisitions. Consolidation continues among network and systems management companies. Mergers and acquisitions offer the potential benefit of working with a single vendor for management needs, but product portfolio integration will be an ongoing challenge for vendors and users alike.

Management frameworks from systems vendors in 1996 included HP's OpenView, IBM/Tivoli's NetView, and SunSoft's Solstice Enterprise Manager. Network systems vendors offer management applications that are compatible with these accepted frameworks. HP's OpenView Network Node Manager and SunSoft's distributed Enterprise Manager product have the largest number of installed licenses, according to IDC.

Enterprise framework solutions offer the benefits of an integrated suite of capabilities, bundled pricing, a single vendor, and a wide choice of network management products. Because it is difficult for any one product to excel in all areas, customers typically use NSM suites as appropriate and fill in with compatible products from ISVs. Following is an overview of leading framework solutions vendors.

■ Computer Associates

CA is a large mainframe systems management software vendor that moved into the client/server market in 1992 with CA-Unicenter. CA-Unicenter is a comprehensive systems management tool that includes disk backup, help desk, report distribution, security, and tape backup features. In 1994, CA also moved into the applications management area, acquiring and renaming three ICL management tools: CA-Unicenter/Software Delivery, CA-Unicenter/DB Alert, and CA-Unicenter/Systems Alert. CA is working with Candle to link CA-Unicenter to Candle's performance-monitoring tools.

In January 1997, CA shipped a new systems and network management platform known as Unicenter TNG (the next generation). Unicenter TNG provides tools for managing the IT environment and business processes, including security management, event/status/exception management, storage management, workload management, service/help desk, output management, performance and accounting, database and application management, discovery, and network management. Unicenter TNG uses SNMP for network management that can be integrated with HP's OpenView and Sun's NetManager.

In September 1997, CA began beta testing a Java-enabled version of Unicenter TNG. The new capabilities allow network administrators to access Unicenter TNG through any browser and Java-based devices from either a browser or the standard Windows NT interface. Figure 6-12 illustrates CA's Unicenter TNG architecture.

Figure 6-12: CA's Unicenter TNG Architecture

Unicenter TNG
Network management
Workload management
Security management
Storage management
Performance management
Output management

CA Options
Software delivery
Asset management
Advance help desk
SAP R/3
MS-Exchange
System agents
Web management
Single sign-on

3rd. Party
Legato
Insight Manager
Unison Maestro

Customer
In-house built tools

Unicenter TNG SDK

Unicenter TNG Framework
Distributed Services
Auto discovery
Event
Reporting
Real world interface
Scheduling
Virus detection
Object repository

HP NCR Sun Digital IBM Tandem Unisys SGI Sequent Win NT

Source: Computer Associates, 1997

CA's offerings in the area of help desk management are built on the Unicenter TNG Framework. CA's Advanced Help Desk and Paradigm Service Desk provide enterprise solutions for automation and management of service delivery.

In late 1997, CA announced an agreement with Cisco that will allow CA users to manage Cisco routers and networking products. The integration of CiscoWorks for Switched Internetworks (CWSI), CiscoView, and Unicenter TNG allows Unicenter TNG clients to monitor and manage Cisco products.

■ Hewlett-Packard

With more than 100,000 installations worldwide, HP OpenView is the de facto standard for managing integrated networks, systems, applications, and databases in multi-vendor computing environments. More than 290 complementary HP OpenView-based management solutions are available on a variety of operating system platforms, including HP-UX, Microsoft Windows 95 and Windows NT, and Sun Solaris. HP OpenView includes the HP OpenView Professional Suite, the HP OpenView Network Node Manager for Windows NT, and the HP OpenView Network Node Manager for UNIX.

HP OpenView professional suite provides an integrated network, system, and desktop management environment on Windows 95 and Windows NT. The HP OpenView professional suite is an open, standards-based solution that allows network administrators to control and proactively manage heterogeneous PC-LANs. This HP OpenView product helps reduce support costs through the use of industry standards (SNMP and DMI) by using a consistent user interface and infrastructure.

Network Node Manager (NNM) for Windows NT provides concise, in-depth views of network connections in a graphical format. It gives network management staff the ability to evaluate network performance, preempt network disruption, and anticipate network growth or realignment. This product was leveraged from HP's OpenView Network Node Manager for UNIX, which has been available since 1990 and is the established market leader, with more market share than its top three competitors combined. NNM for Windows NT is a native Win32 implementation and carries Microsoft's Designed for BackOffice logo, signifying its compatibility with that integrated family of server software.

HP's OpenView IT/Operations (IT/O) is an enterprise framework that integrates hundreds of systems and network management offerings. HP's OpenView IT/O 4.0 network management platform, released in September 1997, filters event notification information to help administrators troubleshoot problems quickly and efficiently across an enterprise. An integrated event-correlation engine pinpoints important messages by weeding out extraneous or duplicate error messages; it displays only those that identify the root problem. A Web-enabled management console, a management server, and agents for different client platforms also are included.

In July 1997, HP announced that although the company will continue to market HP OpenView, it will integrate a version of CA's Unicenter TNG management platform into the company's HP-UX operating system. HP said it will bundle CA's framework software with all its servers.

■ IBM

IBM continues to be the largest network management and monitoring system vendor with a large number of Systems and Network Architecture (SNA) networks. IBM is the largest provider of storage and event automation management software, and is second only to CA in change and configuration management and job scheduling software revenue worldwide.

IBM is also a force in the LAN management arena, with its Web-enabled, cross-platform PC management NetFinity management solution for Intel-based systems. IBM NetFinity Manager software can manage multiple network segments running different protocols, operated from a desktop or notebook system. NetFinity Manager supports PC operating systems such as Windows NT, Windows 95, Windows 3.1, Novell NetWare, and OS/2; NetFinity Manager also supports network protocols such as NetBIOS, IPX, SNA(LU6.2), and TCP/IP. In addition, NetFinity Manager provides these functions over an Internet connection with a Web browser.

The acquisition of Tivoli in early 1996 added visionary technology – the Tivoli Management Environment (TME) systems management suite – to IBM's strong base of SystemView and NetView products. The combined product based on Tivoli's TME and SystemView (IBM's software for host-based and network management) is known as TME 10 NetView. NetView has been the keystone of IBM's network and systems management products for network operations, error detection, error correction, and management. The first version of NetView inherited the functions of Network Communication Control Facility (NCCF), Network Problem Determination Application (NPDA), and Network Logical Data Management (NLDM). A key feature of NetView Version 2 was a graphical control facility called NetView Graphic Monitor Facility (NGMF).

The SystemView series was unveiled by IBM in May 1995, offering change, configuration management, problem management, and performance management capabilities. SystemView is a framework for planning, coordinating, and operating installations originally based on Open Systems Interconnection (OSI) reference model concepts across IBM's operating systems (MVS, VM, OS/400, AIX, and OS/2). SystemView provides functions for user interfaces, shared data, and enhanced automation.

In September 1997, IBM and its Tivoli subsidiary announced several new products. Tivoli introduced new Java-enabled versions of TME 10 NetView – the SNMP platform previously known as NetView for AIX – and TME 10 NetView for OS/390, which used to be IBM's NetView for System/390. TME 10 NetView Version 5 lets customers access information from the platform via any Web browser, where they can execute all but the most complex NetView commands. IBM's introduction of the new OS/390 system accompanied a new company architecture. Administrators can set up their management tools in a centralized or distributed fashion. A UNIX-based NetView system serves as the central clearinghouse for SNMP, mainframe, and attended or unattended work group management systems. IBM's Network Hardware Division launched Nways Workgroup Manager for Windows NT and Nways Manager for AIX, Java-based tools designed to manage IBM switches and routers and the work groups attached to them.

IBM/Tivoli will continue to add capabilities to the management suite. Acquisitions in the second half of 1997, which should offer additional functionality, included the following:

- **Unison Software** – In September 1997, Tivoli and Unison Software announced that IBM would acquire Unison. The companies will combine Unison's Maestro product with Tivoli's TME 10 products. Maestro offers workload scheduling and other capabilities. Unison already has developed modules for linking Maestro and TME 10.
- **Software Artistry** – In December 1997, IBM announced plans to purchase Software Artistry, which provides two suites of enterprise applications: SA-Expertise for Enterprise Support Management (ESM) and SA-Expertise for Customer Relationship Management (CRM). The ESM suite addresses an organization's service level management requirements through integration of decision support, problem management, asset and change management, and network and systems management. The CRM suite combines aspects of sales and marketing management, customer support, customer self-service, and decision support to support customers.

■ Sun

Sun's Solstice addresses the needs of enterprise management. The Solstice family from SunSoft (a division of Sun) includes the Solstice Enterprise Manager (SEM), Solstice Site Manager, and Solstice Domain Manager. SEM is SunSoft's multi-user, multi-protocol, enterprise management platform designed to manage anything in the network from anywhere on the network. Multi-user support allows customized views as well as access controls for each staff member. An administrator determines exactly what each management staff member sees, the tools that can be used, and the privileges that are allowed.

The major element of SEM is known as the Management Information Server (MIS). A distributed framework supports multiple MISs throughout the enterprise. Each MIS exchanges information, forwards alarms, and ensures that management coverage extends throughout the enterprise.

SEM operates with a distributed, object-based management repository. An object request broker locates objects on any MIS. Object technology also is used for automating discovery; for example, when a new object of a certain class, such as a router, is discovered, default monitoring and thresholds are applied. This process allows administrators to create policies that can be applied to any managed object of the appropriate class.

SEM also supports remote application execution. A Portable Management Interface separates management tools from the management platform, placing the tools where staff are located and allowing tools to be moved through the network, if necessary.

6.4.7 Forecast

- With desktop systems growing in number and customer awareness of manageability issues increasing, customer requests for proposals to vendors increasingly will include manageability requirements such as DMI compliance.
- As the issue of total cost of ownership (TCO) remains a point of contention between large corporate IT users and PC vendors, the PC vendors will continue to find creative methods to reduce TCO dramatically.
- The challenge from low-cost NCs and a growing, less-technical consumer audience for PCs will cause vendors to focus on ease of use as a key asset in the hope of reducing the need for vendor and corporate help desk support.
- As network administrators become familiar with the application-centric view of network management, use of RMON2 to monitor applications will increase in 1998 and will show measurable benefits. Also, RMON2 capabilities now available only in probes will be embedded in devices such as switches.
- Management offerings based on SNMPv3 will continue to become more widely available during the forecast period as the standard is finalized.
- In keeping with the application-centric management trend, new tools will measure application responsiveness over time, so administrators will be able to ensure consistent application response time on a particular workstation from day to day.
- Increased distributed management intelligence throughout networks will reduce overhead traffic by limiting the need for polling and other management-specific communication.
- Web-based user interfaces will continue to become available for a wide variety of management products. Web-based management offerings are available from most of the major vendors, even though they may be limited to a subset of capabilities. Companies will expand their offerings in 1998 with Web access to a broader range of management tools.
- A continued increase in proactive, automated management tools based on distributed intelligence in network devices will play an important role in the eventual development of self-healing networks. Although advances in RMON2, SNMPv2, and software algorithms make self-healing networks closer to reality, it will be past the end of the forecast period before these concepts are widely implemented.
- Although integrated systems and network management suites will expand their functionality, none will provide all the capabilities needed for managing a large enterprise. Products addressing specific functional requirements ("point products") still will be required. Some of these will be designed for transparent integration with systems and network management suites at the console level, and others will require integration by the purchaser.

6.4.8 References

■Articles

Derfler, Frank J., Jr. 1997. TKO your TCO today. *PC Magazine*. October 21: 223.
Dryden, Patrick. 1997. Self-healing networks. *Computerworld*. June 2: 4.

Dryden, Patrick and Bob Wallace. 1996. Taking control. *Computerworld*. December 9: 6.
Boyle, Padraic. 1997. Web site traffic cops. *PC Magazine*. February 18: NE1.
Seachrist, Dave, Russell Kay, and Al Gallant. 1997. Wolfpack howls its arrival. *Byte*. August: 126.
Steinke, Steve. 1997. Lesson 102: Simple Network Management Protocol – SNMP. *LAN Magazine*. February: 21.
Sullivan, Eamonn. 1997. CARP divvies up the duties. *PCWeek*. September 15: 111.

■Periodicals, Industry and Technical Reports

Bair, Traci. 1997. *Lucent expands data network service offerings*. IDC. February.
Goodhue, C. 1997. *PC management functionality and vendor differentiation*. Gartner. July 16.
Keyworth, B., R. Paquet, D. Scott, and I. Stenmark. 1997. *Flexibility vs. stability*. Gartner. April.
Mason, R. Paul. 1997. *Application management defined: examples of breadth versus depth*. IDC. February.
Mason, Paul. 1997. *System management software: 1997 worldwide markets and trends*. IDC. July.
McGee, K. 1997. *AT&T MNS – is it a total managed network service?* Gartner. February 26.
Rainge, Elisabeth. 1997. *Network element management software year-end market review*. IDC. February.
SNMP is still necessary, but no longer sufficient. 1997. Giga. August 25.
Villars, Richard. 1997. *Management and the Internet – turning consoles into conduits*. IDC. March.

■On-Line Sources

Simple-Times. `http://www.fv.com/pub/simpletimes/issues/4-3.html`

■URLs of Selected Mentioned Companies

3Com `http://www.3com.com`
Apple `http://www.apple.com`
AT&T `http://www.att.com`
Bay Networks `http://www.baynetworks.com`
BMC Software `http://www.bmc.com`
Boole & Babbage `http://www.boole.com`
Candle `http://www.candle.com`
Cheyenne `http://www.cheyenne.com`
Cisco `http://www.cisco.com`
Compaq `http://www.compaq.com`
Computer Associates `http://www.cai.com`
CompuWare `http://www.compuware.com`
Dell `http://www.dell.com`
Desktop Management Task Force `http://www.dmtf.org`
Gateway 2000 `http://www.gateway.com`
Hewlett-Packard `http://www.hp.com`
HighGround Systems `http://www.highground.com`
HydraWeb `http://www.hydraweb.com`
IBM `http://www.ibm.com`
ICL `http://www.icl.com`
IEEE `http://www.ieee.org`
IETF `http://www.ietf.org`
InfoVista `http://www.infovista.com`
Intel `http://www.intel.com`
Legato `http://www.legato.com`
Lucent `http://www.lucent.com`
McAfee (now Network Associates) `http://www.networkassociate.com`
Microsoft `http://www.microsoft.com`
NEC `http://www.nec.com`
Netsys Technologies `http://www.netsys.com`
Novell `http://www.novell.com`
Oracle `http://www.oracle.com`
Palindrome `http://www.palindrome.com`
Seagate `http://www.seagate.com`
Stac `http://www.stac.com`
Sun `http://www.sun.com`
SunSoft `http://www.sun.com/software`
Symantec `http://www.symantec.com`
Tivoli Systems `http://www.tivoli.com`
Toshiba `http://www.toshiba.com`
Unison Software `http://www.unison.com`

7 Applications

7.1 Electronic Commerce

7.1.1 Executive Summary

Until recently, electronic commerce (EC) referred primarily to electronic data interchange (EDI) offerings and electronic messaging technologies that facilitated the exchange of information between businesses and organizations. During the past 2 years, the term EC has broadened to encompass business-to-business and business-to-consumer transactions conducted over the Internet and the World Wide Web. The Internet and Web browsers have created an easy-to-use, standardized infrastructure for conducting business and have made new products and ways of conducting business possible. Now, the term electronic commerce is frequently associated with Internet commerce. The use of EC for business transactions is expected to explode over the next several years, fueled by the availability of sophisticated Internet and Web technology and stronger security mechanisms.

The Internet and the Web are changing many aspects of EC. Although the amount of Internet users has grown tremendously, large numbers of people still do not have connections or access. The United States is the largest provider and user of Internet services in the world: 50 percent of U.S. households have a personal computer (PC). Survey results published in *Business Week* in October 1997 found that 19 percent of Internet consumers had used it to purchase a product and the majority rely heavily on the Internet for information about products and services. People have learned to use the Internet for queries and are just beginning to trust it for transactions.

Although consumer-to-business Internet commerce is still in its infancy, business-to-business Internet commerce is moving rapidly through the early adopter stage. Paralleling the historical use of EDI technology in the EC market model, manufacturing companies have taken the lead in employing Internet commerce technology. These early adopters are enjoying the benefits of improved productivity, improved customer service, and reduced operating costs, thereby realizing tangible, competitive advantages.

At the same time, EDI service providers are looking for new strategies to build customer loyalty and satisfy more sophisticated customer needs. They are incorporating the Web and the underlying Internet to accommodate some of these needs. Although EDI is characterized by predetermined and defined relationships between businesses and trading partners, EC over the Internet is providing the opportunity for more ad hoc relationships and transactions.

Traditionally, EC services have used private network systems. The use of a public network-based infrastructure such as the Internet may not offer all the benefits of a private network. However, the Internet can reduce costs and level the playing field for small and large businesses, allowing them to extend their reach to customers globally.

Successful electronic commerce solutions exist today; however, the EC marketplace will continue to remain unsettled because of the continued flow of new products and new Web-enabled variants of existing products into the marketplace. The definition of electronic commerce also will continue to evolve and expand.

7.1.2 Electronic Commerce Overview

Historically, a broad range of technologies has been used to exchange business information electronically between organizations; these include EDI, bar codes, electronic forms, inter-enterprise messaging, and electronic funds transfer (EFT). This information can be transmitted by various electronic means – direct program-to-program, e-mail, or even via fax. A subset of these technologies is used for electronic commerce, that is, to carry out actual transactions. The Web has captured the attention of merchants, consumers, and vendors alike and has caused the scope of electronic commerce to expand to incorporate Internet commerce. With the expanding purview of EC has come confusion due to rapid innovation cycles, resulting in constant changes in terminology.

Definition of Electronic Commerce

Discussions of EC and many market size estimates include not only traditional EDI, but also public e-mail and other forms of Internet-based electronic business activities. This approach results in a large market that in some cases may include the entire spectrum of computer-based external communications.

This chapter focuses more narrowly on technology, applications, and standards for business-to-business and consumer-to-business electronic commerce transactions. Electronic commerce is the use of electronic information technologies to conduct business transactions among buyers, sellers, and other trading partners. EC combines business and electronic infrastructures, allowing traditional business transactions to be conducted electronically. EC also enables the on-line buying and selling of goods and services via the communications capabilities of private and public computer networks, including the Internet.

Electronic commerce is defined in this chapter as possessing all the following attributes:

- Direct electronic interaction between two computer applications (application-to-application) or between a person using a computer (typically a Web browser) and another application (typically a Web server)
- The interaction involves the completion of a specific transaction or part of a transaction
- The transaction crosses enterprise boundaries, either between two businesses (business-to-business) or between a business and a consumer (business-to-consumer)

This chapter limits its discussion to EC as defined above. It does not discuss other ways that the Internet and Web are being used in business today, such as for promotion, advertising, or content distribution, which fall within the larger realm of electronic business. Also excluded are other areas such as e-mail or collaborative information sharing over the Internet that do not involve actual transactions. (See Chapter 6.3, *Groupware, Workflow, Document and Knowledge Management,* for more information on some aspects of the latter topics.)

Types of Electronic Commerce

The two main forms of electronic commerce covered in this chapter are EDI and Internet-based electronic commerce. Internet commerce largely consists of Web-based electronic commerce.

Today, EDI features and technologies differ from those offered by Internet commerce, but these differences will become less pronounced as Internet commerce matures and as traditional EDI utilizes new Internet-based technology. For example, some EDI services now use the Internet, rather than a traditional dedicated network, as a transport mechanism. Meanwhile, Internet service providers (ISPs) increasingly will offer higher-quality services, which today are the province of EDI, to those concerned about reliability and security.

■ EDI

Historically, the main form of electronic commerce has been EDI. EDI is a form of program-to-program communication that lets business applications in different organizations exchange information automatically to process a business transaction. EDI transactions involve predefined relationships between trading partners, suppliers, and customers and typically are carried over specialized networks known as value-added networks (VANs). These relationships and the use of private EDI networks allow EDI service providers to offer a degree of security, performance, and reliability that is more difficult to accomplish with the ad hoc relationships and Internet-based communications that characterize Web-based EC.

EDI typically has the following characteristics:

- Direct application-to-application exchange of information; for example, an auto parts supplier's computer system may generate invoices automatically and submit them to the auto manufacturer's accounts receivable system when parts are shipped
- Well-defined, tightly specified message formats and industry standards
- Store-and-forward messaging to transport messages through an intermediary over a VAN
- Batch-oriented (or "asynchronous") rather than interactive operation; that is, one computer application is sending messages that are queued up for delivery to another computer system over a store-and-forward network
- Business-to-business (not business-to-consumer) interactions
- Interactions based on preexisting contractual relations between the two parties so that EDI is used to carry out transactions that effectuate an existing business relationship, rather than create a new business relationship
- Used primarily within a given industry (or an industry and its trading partners) and characteristically concentrated in specific industries such as manufacturing, health care, and consumer goods retailing
- Often established at the behest of a single company that requires its trading partners to adopt EDI as a condition of doing business

Many established EDI software vendors and VANs are incorporating the Internet as another enabling technology and communications vehicle for their corporate customers to implement their EDI strategies. New technologies and capabilities developed for the Internet are influencing EDI information transport technology and applications. These capabilities sometimes are referred to as EDI Lite or EDI over the Internet. The fact that EDI is evolving from store-and-forward to event-driven and interactive implementation techniques also reflects a shift in the EC business model.

■ Internet Commerce

Internet commerce involves managing and conducting a business transaction using the Internet. Web commerce, a subset of Internet commerce, goes beyond using the Internet as a transport mechanism and presupposes that participants have Web access. (Figure 7-1 illustrates how Web commerce fits into EC.) Typically, the Web browser is used as a software client for interactive access to a Web server implementing electronic commerce. Currently, Web-based EC is the most widely used form of Internet commerce.

Components of the transaction may include catalog display, ordering, order fulfillment, payment processing, and back-end integration. Internet commerce embraces all stages in the trading cycle: from information exchange and relationship building, negotiation, and contract agreements to transactions and fulfillment logistics.

Figure 7-1: Internet Commerce

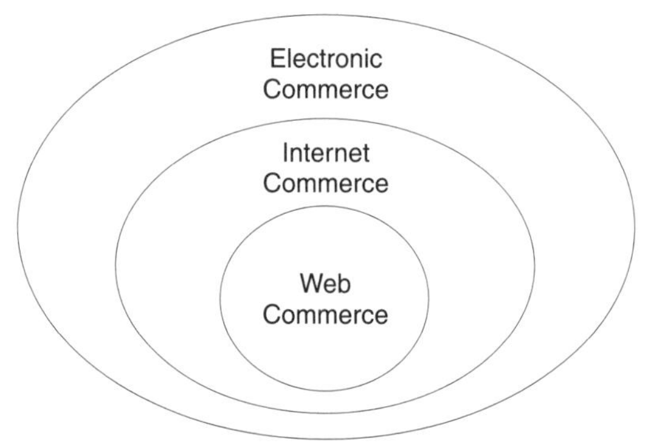

Source: Giga Information Group, 1997

■ The Move to Web-Based EC

The rapid growth in business and consumer use of the Internet and the Web – beginning with the introduction of the first graphical Web browser, Mosaic, in 1992 – and the subsequent elimination of the National Science Foundation's (NSF's) "acceptable-use policy" (which prohibited commercial use of the Internet backbone) created the potential for new forms of EC. These options now are attracting more attention than EDI, are growing much faster, and eventually will be much larger, both in terms of participants and in the value and volume of transactions.

As interest in the Web exploded during the mid-1990s and as the number of consumers with access to the Internet at work or at home grew, companies that originally had established Web sites primarily for marketing purposes – to promote their corporate or brand identity or to provide information about their products – soon became interested in using those sites for sales purposes as well (that is, to take orders). In other cases, the Web was used in support of transactions that already had occurred or were ongoing, for example, in tracking of shipments being handled by the major package delivery services.

The compelling advantage of the Web as an infrastructure for EC is that it provides a universal software client, the Web browser, and a ubiquitous infrastructure, the global TCP/IP network known as the Internet, that can serve as a ready-made platform for EC. This situation vastly reduces the costs of setting up as an EC merchant because it eliminates the need for each vendor to develop, distribute, and support a software client and maintain a dedicated network and dial-in access facilities.

Although business-to-consumer Web-based EC has garnered more attention recently, business-to-business EC will continue to account for the bulk of transaction dollar volume in the next few years, in part due to existing infrastructure and to compelling financial benefits. Barriers to business-to-consumer EC have included concerns about security and poor performance often experienced by consumers from Internet congestion, slow modems, the use of large graphics files, and other factors. At the same time, business-to-business EC has been accelerated because businesses already have the necessary technology infrastructure. The growth of corporate extranets (intranets that have been extended to include business partners and key customers) has fueled growth in Internet-based business-to-business EC.

The two forms of Web-based EC, business-to-business and business-to-consumer, share many common characteristics and technologies. However, they also differ in important ways – in characteristics and technologies as well as in the business drivers for adopting EC.

Common technology characteristics of both business-to-business and business-to-consumer Web-based EC include the following:

- Use of the Web server as a platform and the Web browser as a client
- Interaction that is often real-time and interactive, not store-and-forward (as in the case of EDI)
- Interaction that is often human using computer program (Web browser) to program (Web server), not direct program-to-program (as in the case of EDI)
- Message formats that are not tightly defined or highly standardized (as in the case of EDI). Instead, each Web site has its own structure, content, procedures, and so on. The only way to interact with that site is to navigate through it with a browser and populate forms by typing into them.
- Technology issues involved in linking a company's Web site and electronic commerce server to its back-end systems for functions such as generating the content of an on-line catalog from a product database or passing orders taken over the Web to an order entry and fulfillment system
- The need for authentication and encryption because the transactions are carried out over non-secure networks

There are also important technology differences between the business-to-business and business-to-consumer EC models. For example, the need to process payments via credit card securely has been a major driver for the development of the broad range of security technologies and payment systems discussed in this chapter. However, this need is felt most acutely on business-to-consumer sites because EC merchants expect payment at the time an order is placed, and consumers want to be able to pay on-line to avoid the delay and inconvenience associated with having to mail a check to the merchant. On business-to-business EC sites, this issue is not as pressing because the merchant typically is willing to invoice the buyer and collect payment later, and because business customers are used to ordering via a purchase order rather than paying immediately via credit card. For business-to-business transactions, additional safeguards are built into the existing processes to protect the companies against possible fraud.

Both business-to-business and business-to-consumer Web-based EC also share certain non-technology characteristics, which highlight additional distinctions between Web-based EC and EDI. Figure 7-2 illustrates the differences between EDI and Web EC.

Figure 7-2: EDI and the Web

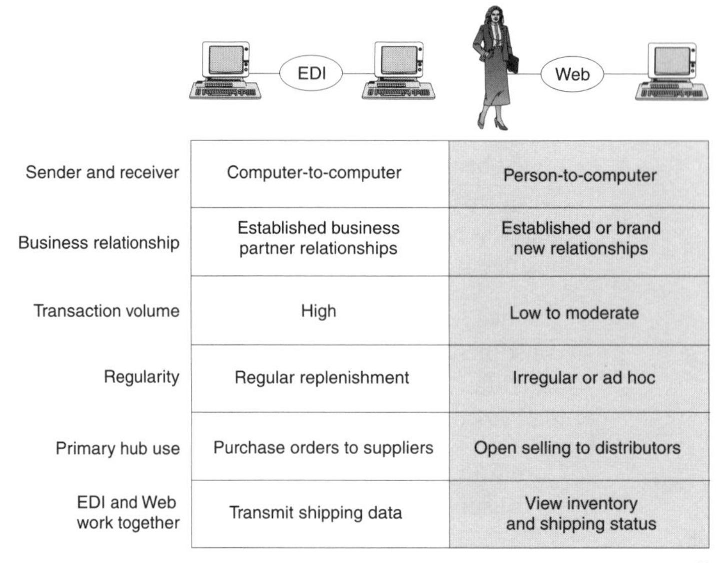

	EDI	Web
Sender and receiver	Computer-to-computer	Person-to-computer
Business relationship	Established business partner relationships	Established or brand new relationships
Transaction volume	High	Low to moderate
Regularity	Regular replenishment	Irregular or ad hoc
Primary hub use	Purchase orders to suppliers	Open selling to distributors
EDI and Web work together	Transmit shipping data	View inventory and shipping status

Source: Forrester, 1997

Non-technology characteristics of Web-based EC include these:

- Not necessarily based on a preexisting business or contractual relationship between the buyer and seller – the buyer can decide to do business with the seller for the first time just by (or as a result of) visiting the Web site
- Not confined to participants within a given industry group and not characteristically associated with any particular industry (today, the highest concentration is probably found in the sale of computer-related products)
- Not imposed by a "hub" company on its trading partners

EC Business Models

Both traditional EDI and newer forms of business-to-business and business-to-consumer EC are growing due to the variety of benefits they offer. The typical drivers for the adoption of EDI are to do business more efficiently or more cost-effectively. This outcome sometimes can be a result of simply reducing the cost of processing the transactions themselves, for example, by eliminating the need to receive invoices in paper form and then manually re-key them into an accounts payable system. It also can allow the underlying business process to function more efficiently and cost-effectively, for example, by eliminating the need to hold excess inventory because EDI is used to arrange delivery of needed parts or merchandise on a "just-in-time" (JIT) basis.

Some examples of the use of EDI to improve business processes include the following:

- **Quick Response (QR)** – Uses EDI, Universal Product Code (UPC), and bar coding for carton marking. Retailers can improve their profitability by increasing the number of stock turns during a season and eliminating end-of-season markdowns. EDI enables market data gathered by point-of-sale (POS) terminals to be delivered from retailers to suppliers more quickly. Retail industry studies estimate that a fully implemented QR system returns about 5 percent of gross sales to the bottom line.

- **Model Stock Replacement** – An approach used by large retailers and key suppliers. A retailer identifies the desired level of inventory, known as "model stock," for each location and provides POS data to suppliers on a daily basis. The suppliers then restock the shelves as needed. The retailer monitors the suppliers' activities while allowing the original model stock level to be adjusted by suppliers as the volume of sales changes over time.
- **Materials Management** – Widely adopted in manufacturing, particularly in the automotive industry. Materials Management uses EDI, materials requirements planning, and JIT manufacturing to reduce the level of parts inventory kept on-site to virtually zero.
- **Efficient Customer Response** (ECR) – Similar to Quick Response and Model Stock Replacement but found in the grocery industry. Sales data is transferred electronically between supplier, distributor, and retail store. The goal is to match product flow to consumption in a seamless, timely, and accurate manner.
- **Evaluated Receipt Settlement** – Eliminates the invoice from the purchase order cycle. The customer authorizes payment to the supplier upon confirmation of the arrival of goods, making the issuance of an invoice unnecessary.
- **Electronic Funds Transfer** (EFT) – The transfer of a value payment electronically from buyer to seller via a financial institution. An EDI/EFT transaction is made through a bank, either by wire transfers or Automated Clearing-House (ACH) transfers.

In the case of Web-based EC, the benefits are more varied and may differ significantly between the business-to-business and business-to-consumer models. In the business-to-consumer EC market, the anticipated benefits to the vendor vary according to the business model for becoming involved with EC initially.

In some cases, merchants have taken a familiar business model, such as catalog shopping, and transported it to the Web as a new medium, using their Web site instead of a paper catalog to provide product information and using on-line ordering to replace calling an operator to place an order. Today, these Web sites may not generate many incremental orders (orders that would not be placed if the company was not on the Web). Rather, merchants are investing in EC to extend their presence to the Web so they will not become "visible by their absence"; to give consumers an additional mechanism through which to do business with them; and to gain experience with Web-based EC so that when (or if) it becomes a larger part of their sales, they are ready to take advantage of it.

Gaining experience with Web-based EC is particularly important so that a company can avoid being outflanked by competitors who take advantage of the Web more quickly and use it to gain market share. In the future, as usage of the Web becomes more widespread and possibly begins to replace (rather than supplement) existing channels, other business drivers may become important as well. For example, expenses of direct mail merchants (such as printing and postage) could be reduced dramatically if Web-based EC replaces traditional catalog sales.

In other cases, entirely new business models are developed around Web-based commerce. These include the on-line retailer, the aggregator, and the direct seller.

The On-Line Retailer

The on-line retailer is a company set up specifically to sell via the Web, such as the Internet bookseller Amazon.Com. Under this model, all the Web-based orders are incremental because these companies would not exist without the Web. They need to take advantage of some characteristic of the Web – such as the ability to have a catalog that is unlimited in size or the ability to provide additional information about the product not available elsewhere (for example, other customers' comments about a particular book). They also may need to provide another benefit over non-Web shopping, such as discounts over retail stores (made possible by their lower cost of doing business), to attract customers away from non-Web shopping and away from their Web-based competitors.

The Aggregator

Another new business model is the aggregator, or the electronic mall. Here, the Web site provides the consumer with a selection of products from multiple vendors, but typically limits the selection to a particular type of product, such as computer accessories. The aggregator may be able to offer a broader selection of products than would normally be found elsewhere. For example, an on-line wine retailer might sell wine from smaller wineries that do not have their own Web sites or that have a limited presence in retail stores. In other cases, the merchant may be taking advantage of the Web's ability to provide more detailed information about products, such as comparative product specifications, than is possible in a traditional retail establishment. Some vendors also provide a forum for their customers to discuss their products, essentially creating a virtual community.

The Direct Seller

EC also has been seen as resulting in disintermediation, where the ultimate supplier eliminates distribution channels and sells directly to the ultimate customer to reduce distribution costs. For example, the airline industry has used EC to reduce the role of travel agents and sell tickets directly to passengers instead. This example also contains elements of the on-line retailer model. The airlines already sell tickets directly over the telephone, but the Web offers additional benefits: the passenger gains the ability to obtain detailed schedule and fare information in a way that might be hard to do in a telephone conversation, and the airline gains a way to reduce costs by replacing telephone reservation agents with direct customer access to its reservation system.

On the other hand, disintermediation is not a one-way process. Although EC may eliminate the need for intermediaries that do not offer valued-added services, the value of many distributors or resellers will increase if they leverage their personal relationships with customers and offer services not available elsewhere. Even successful Internet companies acknowledge the need for third parties to reach the substantial marketplaces between very large enterprises and individual consumers accessing products directly from the Web.

In the case of business-to-business EC, many of these same business models and their related benefits also apply. A well-known vendor may set up a Web site to expedite the order-taking process, resulting in increased convenience and access to more detailed information for its customers. In this case, the benefit is less likely to be incremental orders; however, the savings due to a reduced cost of order processing and the handling of related inquiries can be enormous. In other cases, an existing product distributor may act as an EC aggregator, combining product information from multiple manufacturers with order-taking capability.

Requirements for Electronic Commerce

Significant investments in business process redesign, IT infrastructure enhancement, and marketing reorientation are required to deliver customized, personalized, information-based products and services. IT organizations also are challenged with implementing new applications that must handle increasing volumes of data. In addition, accommodating a growing number of consumer-to-business EC transactions may require an upgraded network and systems infrastructure. (Table 7-1 details Dataquest's view of required EC applications by industry group.)

Table 7-1: EC Applications by Industry Groups

Industry	Applications
Consumer Goods Manufacturing (CGM)	Order processing, accounting, procurement, shipping, delivery, replenishment, inventory management, marketing automation, and supply chain integration
Distribution	EDI: tracking, labeling, routing, shipping, returns, value-added services, direct store delivery (DSD), and efficient consumer response (ECR)
	E-mail: shipping, returns, value-added services
	Intelligent messaging: decision support, tracking, routing, and DSD
	Internet: routing, value-added services, ECR
Utilities	Materials management information systems
	Energy services
Financial Services	Electronic funds transfer

Source: Dataquest, 1997

Consequences of EC

EC facilitates new types of business processes for reaching customers as well as new types of products and selling environments – interactive shopping malls, electronic books, and catalogs. Use of EC technologies can result in improved efficiencies in finding and servicing customers, in communicating with trading partners, and in developing new products and markets.

Customers are learning about products through information on the Web, buying products using electronic cash and secure payment systems, and having information and services delivered in ways not previously possible. Consequently, how customers commit their loyalty to a brand, manufacturer, retailer, or service provider is changing. Given these shifts in purchasing patterns, companies need to adapt to a world where the traditional concepts of product, brand differentiation, quality, content, and distribution may no longer apply.

Small companies are gaining the benefits previously realized only by large corporate and government organizations that depended on fast, economical, computer-to-computer communications to conduct business transactions. Today, small companies faced with the need to compete globally in a cost-effective manner have the opportunity to do so.

IT vendors are developing and marketing products for these new business models, and numerous opportunities exist for them in a variety of EC tools and technologies, including EC Web-site design, commerce servers, security, payment systems, databases, high-speed networks, and integration of new EC systems with existing applications. The EC tools market also is attracting banks, credit card companies, Internet startups, ad agencies, EDI vendors, and systems integrators. Web site infrastructure products are being created by software vendors, systems integrators, and messaging vendors. Infrastructure providers also include database vendors, imaging software publishers, VANs, governments, telecommunications companies, and LAN vendors. Industry changes will continue to result in a flood of new offerings over the next few years.

7.1.3 EDI

Back-office software and services such as EDI, EFT, and VAN services still compose the largest segment by revenue of the EC market. EDI use has continued to grow, but not as fast as Web-based EC. EDI and other back-office activities are expected to shrink as a percentage of the total EC market in the next 4 years.

Traditional EDI Components

EDI implementation requires software that adheres to well-defined standards because moving a business document from one application to another requires interfaces, translators, consistent data, and mapping protocols. It also requires a careful examination of a company's business strategy and focus together with an understanding of the practical implementation of technologies.

Traditionally, EDI required data translation and message transport capabilities to be identified, defined, and addressed. However, significant changes are underway in the transport mechanisms with the increasing use of the Internet as an alternative to VANs.

■ Data Translation

Translation is an integral part of an overall EDI solution. Translators describe the relationship between the data elements in the underlying business application and the EDI standards (ANSI X12, EDIFACT, and others) by interpreting the information and putting it into a standard EDI document format.

Translation can occur at different points. Older corporate applications often were unable to export standards-compliant documents, and translation software emerged to provide this service. These EDI translators often were integrated with communications functionality. Newer applications are incorporating EDI translation as part of their base functionality.

Early adopters of EDI technology frequently had little choice but to develop their own translation software. Generally, these were large companies that were able to enjoy the competitive advantages that EDI brought to their supply chain process, but these advantages also came at a high internal IT cost. This custom software often led to fairly restrictive transaction sets and limited the EDI capability to engagements with trading partners of comparable size and scope. It was technically difficult to create new trading partner profile information. Commercial EDI translators overcame these obstacles, and translator vendors often included many value-added services, such as data mapping, trading partner training, and extended customer support programs.

Today, these software packages, which run as stand-alone applications or are integrated with VAN services, are available from almost all corporate application suppliers for platforms from PCs to mainframes. Despite these developments, traditional EDI is an expensive and time-consuming activity; the Gartner Group estimates initial implementation and ongoing operational costs to be a minimum of $10,000 per trading partner.

A newer generation of corporate applications provides opportunities to leapfrog some of these obstacles. Businesses today can buy ready-made core business applications from vendors such as Oracle, PeopleSoft and SAP. Functionality previously required of separate EDI translator systems is a part of many application programs today.

■ Message Transport

Traditional EDI relied on VANs and private messaging networks (also referred to as point-to-point communications and virtual private networks, or VPNs), both characterized by high costs and limited connectivity. Previously, whether EDI messages could be transported was dependent largely on whether another enterprise or trading partner had a mailbox on the same VAN. These services are and will continue to be satisfactory for the guaranteed movement of EDI-formatted documents such as purchase orders, invoices, and waybills. (VANs are designed with security and redundancy, but liability for delivery failure typically does not extend beyond the actual transmission cost.) However, traditional VAN capabilities are limited when it comes to advertising and interactive functions such as product browsing, serving text and graphics, and ad hoc trading relationships.

Even with the Internet providing a low-cost alternative to VANs and private messaging networks, all EDI traffic will not be moving to the Internet because VANs offer a large array of services beyond simple message transport. These services include any-to-any processing services, which accommodate differences in protocols, speed, media, and content among trading partners. EDI transactions can be converted from or to e-mail messages, facsimiles, telex, or hard-copy for delivery to specific trading partners that may not be able to receive EDI. VANs also offer screening and sorting services that will assign documents to different mailboxes, depending on sender or document type. In addition, VANs can offer transaction management and application management services that extend to the management of e-mail, Lotus Notes, or Web servers. Finally, VANs can track cargo, provide informational databases, and offer implementation services. VANs and other dedicated networks will also continue to be used by closed user communities with specialized needs, for example, for inter-bank fund transfers.

■ Internet-Based EDI

Today, EDI can be transported over the Internet using the Multipurpose Internet Mail Extension (MIME) protocol. EDI and other systems over the Internet can provide a lower-cost alternative for many smaller businesses. In fact, company extranets may use a complementary combination of both EDI and the Web.

For example, Premenos, a supplier of business solutions for EC, provides EDI over TCP/IP networks with its Templar product. Templar uses encryption and authentication for EDI over the Internet or when switching EDI transactions between a VAN and an ISP. (Premenos was acquired by Harbinger in late 1997.)

Transporting traditional EDI over the Internet does not make it interactive – a requirement for catalog browsing, customer service, and other applications. Combining traditional EDI with Web access enables small companies to exchange EC documents with a larger population of organizations using standard messages and translation software. The Web browser becomes a dumb terminal connected to older EDI systems at the other end. Many back-end systems are still store-and-forward messaging systems with batch processing that cannot accommodate the real-time, on-screen bid confirmations associated with interactive Web systems.

EDI Standards

Traditional EDI standards include ANSI X12 and EDIFACT. ANSI X12 is the American National Standards Institute standard that is the dominant U.S. EDI standard. X12 was designed to support the exchange of electronic business transactions across several industries and specifies the terms and format to be used for these transactions. The Electronic Data Interchange For Administration Commerce and Transport (EDIFACT) standard is an International Organization for Standardization (ISO) standard that has been proposed to replace X12 and Tradacoms (a European standard) to become the worldwide standard for EDI.

A movement is underway to supplement public standards for electronic commerce, such as ANSI X12 and EDIFACT for document content, with additional de facto market standards and industry-specific standards.

This trend is reminiscent of the state of EDI in the early 1980s, when influential industry groups successfully developed and deployed EDI standards that specifically addressed their unique needs. In the U.S., there were Voluntary Industry Communications Standards (VICS) for retail, Automotive Industry Action Group (AIAG) for automotive, Uniform Communications Standard (UCS) for grocery, and Transportation Data Coordinating Committee (TDCC) for transportation, among others, prior to full adoption of ANSI. Then, as now, time-to-market is the key. Early availability and early market adoption are often critical for a company's survival, making timely agreement on standards essential.

■ VAN and VPN Standards

Communications and networking systems and protocols in use by VAN providers and in VPNs include the following:

- Asynchronous and synchronous transmissions
- X.25 for transmitting data over a packet-switched wide area network (WAN)
- SNA for peer-to-peer communications among a variety of host systems
- Advanced Peer-to-Peer Networking (APPN)
- ITU-TSS (International Telecommunications Union-Telecommunications Standards Section, formerly CCITT) X.400 and X.435
- EDI-based structured document standards

Senders and receivers must use a compatible format for content and structure for EDI forms. That is, a sender's computer needs to use software that will generate a format the receiver's computer will be able to read.

■ Internet EDI Standards

The Open Buying on the Internet (OBI) standard was developed by the Internet Purchasing Roundtable, a group of Fortune 500 companies, with technical assistance from Actra Business Systems, Intelisys Electronic Commerce, Microsoft, Open Market, Oracle, and others. OBI's aim is to ensure interoperability between Internet commerce systems for business-to-business transactions. OBI specifies an EDI-based standard for the formatting of electronic orders but does not use other transport-related aspects of EDI.

7.1.4 Web-Based EC Technologies

Electronic commerce is enabled by an infrastructure of technologies and services, many of which are also used to support other applications. This infrastructure includes computer systems, operating systems, databases, development tools, networks, Web servers, and Web browsers. In fact, this large existing base of technology has made the rapid growth in Web-based EC possible.

These technologies, when coupled with specialized EC software and standards, are being used to create a new architecture for electronic commerce. (See Figure 7-3.) EC systems are being designed to integrate security, order entry, and payment schemes used for on-line buying and selling.

Figure 7-3: Electronic Commerce Architecture

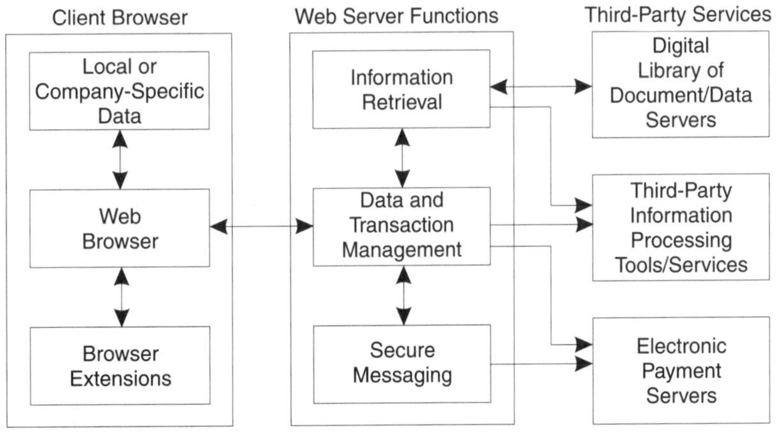

Client Browser
Local or Company-Specific Data
Web Browser
Browser Extensions

Web Server Functions
Information Retrieval
Data and Transaction Management
Secure Messaging

Third-Party Services
Digital Library of Document/Data Servers
Third-Party Information Processing Tools/Services
Electronic Payment Servers

Source: *Frontiers in Electronic Commerce*, 1996

This section discusses the areas of information technology specific to Web-based EC. Much of the technology required for Web-based EC is not unique to EC; instead, it has been developed to support a broader range of Web-based applications. Just as in the case of EDI, early adopters of Web-based EC had to develop their own EC systems, using existing Web servers and development tools.

However, system and software vendors have responded to the growth in EC by developing and marketing EC-specific products. Today, a Web site designed for EC can be created using EC-specific products. These vary widely in the range of their functionality, in particular in the extent to which they can be integrated easily into the merchant's existing business systems. They also span a wide spectrum in terms of the degree to which they offer complete, ready-made solutions, ranging from prepackaged EC sites that only need to be populated with the merchant's product catalog to basic frameworks that provide a platform requiring extensive custom development.

Also unique to EC are the technologies required for secure processing of payments for Web-purchased merchandise. (See "Electronic Payment Systems" on page 572 for a discussion of these technologies.)

The vendor landscape varies widely, ranging from well-established organizations providing EDI, EC, and on-line networking services to startups with a hot new product. In the past 2 years, numerous alliances, partnerships, and joint ventures have been formed that combine the strengths of the mature players with the nimbleness of the smaller companies.

EC Web Site Building Tools

Web site building tools provide merchants with preset and configurable functionality for the creation of EC Web sites, payment systems access, marketing applications, and shopping tools. EC Web sites must support functions such as user authentication, catalog browsing, ordering, payment, and reporting.

Early EC Web sites were created using low-level programming tools such as C++. Several companies now offer products that do not require sophisticated programming skills. BroadVision, Eicom, iCat, and Open Market are examples of small firms that offer such tools, in addition to IBM/Lotus, Microsoft, and Netscape.

For example, BroadVision's One-To-One Commerce offering distributes functionality into logical application services, including database access, content management, transaction services, and user profiling. BroadVision applications can be accessed from most Web servers.

Commerce Servers

A commerce server is a particular set of extensions to a basic HTTP server designed to meet the specific needs of EC. (See "Web Server Software" on page 266 for basic information on HTTP servers.) These commerce servers provide server software for conducting secure EC and communications on the Internet and on other TCP/IP networks.

Commerce server software offerings enable the publishing of hypermedia documents that use the HyperText Markup Language (HTML) for everything from electronic catalogs to Web storefronts. In most cases, the HTML code is generated on-the-fly in response to a user's query, with the underlying data being pulled from a back-end database. Data security is addressed through advanced security features such as server authentication, data encryption, data integrity, and user authorization. Communications are based on open standards such as HTTP, the Common Gateway Interface (CGI), and the Secure Sockets Layer (SSL), which allows for secure communications. (See Chapter 4.4, *The Internet and Intranets,* for more information on Web and Internet protocols.)

Electronic commerce frameworks such as Microsoft's Site Server, Netscape's SuiteSpot, and others provide starting points for system developers or system integrators to build highly tailored Web commerce sites. For example, Netscape's SuiteSpot EC components include a certificate server that can create, sign, and manage certificates for users. The certificate server can configure other servers to accept only authorized certificates. These frameworks can provide Web server capabilities via application program interfaces (APIs).

Financial transaction management products are important for Web merchants accepting credit or debit cards for payment. These products facilitate the exchange of funds and are used primarily by high-volume commerce sites.

VeriFone's Integrated Payment System, for example, provides software that encompasses everything from merchant point-of-sale devices (electronic cash registers), Internet, and smart card applications. IBM sells CommercePoint payment solutions and Open Market offers OM-Transact for similar functionality. Third parties typically provide this functionality for low-volume sites.

■ Server Software Providers

Server software vendors provide server technology as well as software tools and interfaces. EC software is emerging as one of the fastest growing segments of the software industry. In 1997, more than 20 vendors released applications that provide some type of EC service.

The widespread availability of products has helped lower prices. For example, Microsoft's Commerce Server 2.0 could be purchased at the end of 1997 for $5,000, down from its $15,000 price when initially released as Merchant Server 1.0 in 1996. The packaged software marketplace has segmented itself into three general pricing categories. At the high end are Connect's OneServer, Open Market's Transact, and Trade'ex's Distributor, among others. In the mid-tier are shrink-wrapped, business-to-consumer products such as iCat's Electronic Commerce Suite and Microsoft's Commerce Server. Low-end products include on-line-store software from vendors such as AOL, CommerceWave, Viaweb, and Virtual Spin.

Business-to-business EC is generating the majority of the market's attention right now, and it is where the high-end (greater than $250,000) applications are being sold. Products from BroadVision, Connect, One-to-One, Open Market, Sterling Commerce, Trade'ex, and others address the needs of this segment by providing capabilities that tie front-end buying processes to the back-end financial systems of an enterprise.

Open Market's OM-Transact software manages Internet commerce within a corporation's or commerce service provider's operations by addressing order management, on-line customer service, security, authentication, record-keeping, purchasing and payment models, and secure transaction processing, including sales tax and shipping charges. Transact particularly benefits business-to-business manufacturers and distributors of items such as electronic components, computers, office supplies, and industrial supplies by enabling enterprises to move from basic on-line cataloging toward an end-to-end Internet commerce solution.

Microsoft's Commerce Site Server 2.0 runs on top of the Microsoft Internet Information Server and Windows NT operating system. It is designed for establishing business-to-consumer and business-to-business commerce on the Internet. The Enterprise Edition will include Microsoft Wallet component technology that supports several electronic payment methods using security protocols such as SSL and Secure Electronic Transactions (SET).

Microsoft Wallet is available as a free add-on to the Internet Explorer 3.0 browser or as a plug-in for Netscape Navigator 3.0 and 4.0 browsers. Additionally, Microsoft has announced plans to incorporate two electronic wallets – one from Microsoft and one from VeriFone – into its Internet Explorer 4.0 browser. Eventually, it will converge around a single wallet.

iCat Electronic Commerce Suite 3.0 provides tools to set up Web-based storefronts, including sample databases and applications. The program's templates include guides for categorizing entries, registering and tracking users, and creating indexes and searches. Version 3.0.1 includes performance enhancements, security, and support for Informix databases and Silicon Graphics' Irix platform.

Netscape also offers a full line of software to enable EC and secure information exchange on the Internet and private TCP/IP networks. Its software line includes the widely used browser, Netscape Navigator, plus Netscape Servers, Netscape Commerce products and extensions, and Netscape Tools. Netscape has many partnerships and alliances, including one with CyberCash to bundle its CyberCoin (micropayments system) with Netscape LivePayment. Netscape LivePayment provides a financial payment solution that supports real-time credit card processing to banks over the Internet. Netscape's Commerce Platform comprises distinct extensible modules, each addressing a different aspect of EC. Commerce solutions based on this platform include Netscape Merchant System, Netscape Community, and Netscape Publishing System.

Companies such as eFusion are focused on using multimedia capabilities to bring buyer, retailer, and product information together on the Internet. Their concept is to provide Internet technologies that enable the Internet customer and the call center to talk to each other while viewing the same information on the screen, but requiring the use of only one telephone line by the customer. This setup is accomplished by making the split between voice and video at the ISP's point of presence (POP). eFusion's eStream and eBridge offerings make use of basic H.323 communications software available through Intel Phone, Microsoft NetMeeting, and Netscape Communicator.

Another interactive multimedia product under development is from n.ABLE, which focuses on business-to-business applications for EC and interactive commerce. n.ABLE's products combine EC, multimedia, and interactive relationships to facilitate the purchasing process. The design and architecture of the product does not use the Internet's CGI technology but instead uses object-oriented modular services as part of an interactive commerce server, which enables more applications. The server contains all networking and communications, as well as object sharing data and operations.

Intranets and Extranets

Many companies are using Internet technology to transform their local area networks (LANs) into intranets. This process not only simplifies internal information access and communication, but also opens up links to outside organizations or individuals on the Internet.

Extranets – externally focused Web sites with restricted access that provide connections to outside customers, suppliers, partners, and other entities – have emerged out of the original intranet concept.

Using the Internet in the form of extranets, virtual private Internets, or supply chain intranets, suppliers, distributors, and other trading partners are able to exchange information concerning contracts, deliveries, and other matters easily. Some manufacturers already allow their customers to access their networks to place and check the status of orders or receive product information.

For example, General Electric's GE Lighting division saw extranets as a cost-effective way to streamline the order entry and bid process with its suppliers. The Trading Partner Network (TPN) Post product connects GE Lighting's IBM mainframe-based purchasing application to an extranet. When suppliers seek information with their browsers, purchasing data is downloaded from the mainframe to an HP-UX server, then loaded into an Oracle database and uploaded onto a Novell file server. The TPN Post product offering from General Electric Information Services (GEIS), an outgrowth of the original TPN Post project, was expected to generate more than $1 billion in contract opportunities for all GE's worldwide suppliers in 1997.

Extranets facilitate sharing among enterprises data and communications that are necessary for transactions. Extranets will fuel EC spending by giving businesses additional opportunities for market differentiation and competitive advantage. However, extranets likely will face the same challenges with regard to unified industry standards for EC transaction processing that previous efforts have. Standards are likely to develop first within specific industries and tradition communities, and only later will more general standards for use by all industries be developed.

Web-Enabled Corporate Applications

Corporate applications vendors are developing Web front ends to their applications and adopting standards that will enable EC. SAP has joined with Microsoft to create the Business API (BAPI). The companies plan to use Microsoft's Internet Information Server software and ActiveX Internet technologies to open key interfaces in SAP's R/3 enterprise resource planning (ERP) software to allow business transactions to be conducted over the Web. (See Chapter 7.3, *Corporate Applications,* for more information on Web-enabled applications software packages.)

In 1997, SAP and Intel formed a joint venture called Pandesic to sell computers, software, and support services to small and midsized companies wanting to sell products and services on the Internet to consumers. Pandesic takes orders through the Internet, coordinates the orders and shipping, monitors inventory within a business or with its suppliers, and tracks payments. Pandesic addresses and connects each participating company. Merchants also pay a monthly fee, based on Internet commerce transaction revenues. Figure 7-4 illustrates the process used by Pandesic.

Figure 7-4: Pandesic Process

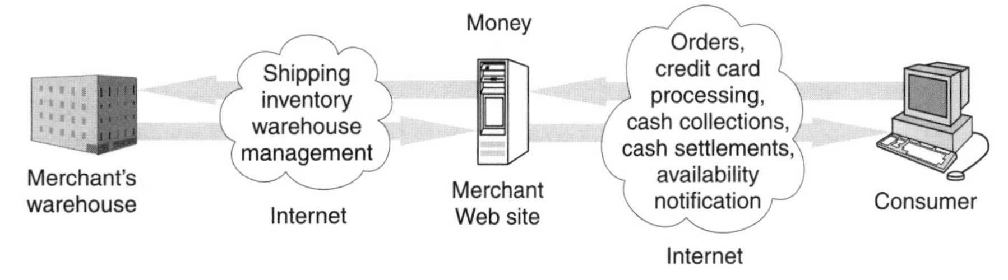

Source: *PCWeek*, 1997

In October 1997, IBM and SAP America began discussions that would allow IBM's commerce Web server, Net.Commerce, to work with enterprise resource planning (ERP) application modules from SAP by making IBM's Net.Commerce APIs and SAP's APIs interoperable. Customers then could link back-office applications to the Internet for a business-to-business commerce solution.

System Vendor Initiatives

Most of the hardware systems and servers being used to run EC applications are not unique to EC and are being used in other segments of the information technology industry. The exceptions include POS devices for scanning credit and debit cards and certain types of smart cards.

POS devices are used at checkout counters for credit and debit transactions. Although they comprise all elements of a computer (processor, keypad, memory, display, and connectivity), they are dedicated rather than general-purpose devices, such as desktop computers. They can be deployed in stand-alone, stand-beside (with LAN connectivity), or integrated (with electronic cash registers) configurations.

Smart cards can be used as stored-value cards and are widely used in Europe. (See Chapter 3.3, *Smart Cards.*)

Although hardware is not EC-specific, the major systems vendors see EC as a key market opportunity and have developed initiatives in this area. In support of these initiatives, platform vendors have formed strategic alliances with multiple software partners and component and service providers.

With HP's acquisition of VeriFone complete and its strategic alliances with AT&T, Microsoft, Netscape, and Oracle in place, the company stands ready to compete in the EC marketplace with end-to-end solutions. EC solutions range from smart card applications for consumers, businesses, and financial services institutions (at the appliance level, by embedding card readers into HP PCs) to Internet commerce applications. A key HP focus is on selling end-to-end solutions to financial institutions based on the SET protocol for payments. The systems, selling for hundreds of thousands of dollars, consist of Microsoft Commerce Server software, VeriFone's vGate and vPOS software, and HP systems.

The HP and AT&T partnership provides an opportunity for merchants using HP-UX or Microsoft NT-based Domain Enterprise Servers to maintain Web sites on their own premises and outsource the back-office functions such as processing secure credit card transactions to AT&T's SecureBuy Service. Users create dynamic Web pages and conduct secure transactions that can access data from a legacy database.

In addition to strategic alliances, EC products are being offered by several IBM divisions. In 1997, IBM began promoting e-Business solutions that target both existing customers with IBM systems and new small businesses. IBM believes that electronic business will alter the way major companies operate and wants to establish a leadership position in helping customers hook up with suppliers, vendors, customers, and employees through the Web. IBM products included in this effort will range from hand-held devices to mainframes.

For example, IBM is developing and marketing Internet solutions designed to run on its AS/400 family of minicomputers. The new AS/400e servers are intended to provide "out-of-box" Web server functionality and are designed to offer Internet transaction security, access control features, faster processing speed, and increased storage capacity. Features include an add-in card for Internet firewall functionality, native support for Java, and a native implementation of Lotus Domino software (beginning in 1998).

IBM has developed a series of EC products and technologies to support secure electronic shopping environments. These products help businesses create virtual storefronts on the Internet where consumers are able to shop, purchase, and pay on-line. The Net.Commerce system is intended for use by merchants and business partners that want to develop highly customized, on-line stores on the Internet. The system has been designed to showcase a business's product line, allow immediate purchases with instant payments through secured credit-card transactions, and integrate with a merchant's existing business systems. Net.Commerce 1.0 includes a database server, Web server, and EC components.

Electronic Commerce Standards

The Internet is an environment where rapidly adopted, popular technologies became standards. Interoperability will continue to be an issue as many of the Internet standards continue to evolve and as manufacturers seek to extend and add value to these standards. Lack of consistent standards could hinder the growth of EC solutions in some sectors. (See "Internet Standards and Protocols" on page 241.)

EC is a way of doing business that employs many technologies, each with its own standards, protocols, and varying degrees of industry acceptance and use. Therefore, adherence to a true EC standard in the next few years is highly doubtful because it has been difficult for the groups involved to reach consensus. In fact, the trend is toward more standards rather than fewer standards.

Historically, the formal standards process has been a lengthy one. ANSI X12 and the ITU-TSS are examples of standards published by formal organizations. Typically, it has taken years for a formal standard to be established, for example, for an EDI invoice. Vendors then have needed up to 2 years to develop and release products supporting the standard. With the market moving to the commercial use of the Internet, the process may be compressed to as little as 18 months for market mind share to be achieved, for a de facto standard to be established, and for products to be developed.

Today, although organizations such as the Internet Engineering Task Force (IETF) and the World Wide Web Consortium (W3C) provide some formal processes for Internet and Web standards, the process is generally shorter. Or, a company or vendor may produce a "reference" (standard) document, post it on the Internet where it is accessed and tested by users and other vendors, and get feedback via the Internet, eventually resulting in a de facto standard.

Standards for security are a particular area for concern in EC. Although many security technologies are being incorporated into secure EC, one of the most important is public-key cryptography. Public-key encryption is used for peer-to-peer secure transactions, authentication, and digital signatures. (See Chapter 5.4, *Security,* for a complete discussion of security approaches and issues.)

7.1.5 Electronic Payment Systems

A variety of Internet payment systems are starting to emerge, and the roles of existing payment systems and service providers are evolving rapidly. In addition to terms such as digital currency and digital cash, the term electronic cash, or e-cash, is being used to refer to a broad range of payment models, including cards, wallets, tokens, and so on. Table 7-2 summarizes some of the different types of electronic payment systems.

Table 7-2: Different Types of Electronic Payment Systems

Payment Type	Description
Open exchange	A simple, widely used payment method that extends a mail-order merchandising model to the Internet. Credit card information is provided to a merchant via e-mail or Web page response instead of by a telephone call. Security may involve the content of the current message only. Traditional clearing procedures are used to process payments. U.S. regulations protect consumers against fraudulent credit card losses of more than $50.
Secured linkage or presentation	Encrypted transmission of information between buyers and sellers over the Internet, providing authentication by means of common software (such as security-enhanced Web browsers using SSL).
Trusted third party	A trusted intermediary service that provides credibility for both buyer and seller, manages Internet-based transaction accounts, and provides for the transfer of funds between accounts or for the authentication of the transaction with banks or credit card clearing facilities. This third party defines service terms and signs up buyers and merchants.
Digital check	A network-based replacement for the paper check that uses secure encrypted data interchange and existing inter-bank clearing systems.
Digital cash	The use of serial-numbered, digitally signed tokens that represent secure currency (preventing double spending) and allow direct buyer-to-seller transfer of value.

Payments increasingly will be made by a variety of consumer-driven devices such as PCs, laptops, and other portable devices. Financial institutions have a strong interest in accelerating the growth of EC. Although electronic shopping and ordering do not require electronic payment, a much higher percentage of these transactions use payment card products instead of cash or checks both in the consumer and commercial marketplaces. In fact, many on-line vendors cannot accept cash and do not accept checks.

Micropayments

Microtransactions are used for sales of goods or services ranging from $0.25 to less than $10 and typically include sales of newspaper articles, on-line research sources, on-line gaming or on-line chat, and software distribution. Microtransactions or micropayments could be widely implemented within 2 years if vendors can agree on one or two easy-to-learn, easy-to-use payment technologies and standards. Credit card transaction fees make it costly for a merchant to allow credit card purchases of $10 or less. Visa estimates that each year, $1.8 trillion is spent in transactions of $10 or less around the world.

To date, the two primary electronic cash (e-cash) models for microtransaction payments are smart cards and electronic wallets. The development of low-cost, low-maintenance technology will be a key factor in successful adoption of these systems. In addition to implementing one or both of these models, most vendors and merchants are likely to adopt the SET protocol or other systems to safeguard larger financial transactions.

■ Smart Cards

Smart cards, in particular stored-value cards, are a well-established technology and are widely available in Europe. Austria, France, and Germany have smart payment cards that can be used for a variety of consumer purchases. Smart cards allow customers to purchase goods and services by debiting the value that has been previously transferred to the card. In addition to being used for stored value, smart cards also can perform the encryption and authentication (digital signature) functions needed for EC. (See Chapter 3.3, *Smart Cards.)*

Because of the versatility of smart cards, banks, insurance companies, car rental companies, and other businesses appear willing to help subsidize the cost of smart card networks, thereby lowering the fees charged to consumers. Both disposable and reloadable cards are being tested by card providers.

Mondex was established as an independent payments company in July 1996, with ownership by 17 major banks and organizations from several different countries. MasterCard acquired 51 percent of the company in February 1997. Mondex's payment offering uses a smart card to store electronic cash and is being tested for payment of goods and services on the Internet. Trials have been conducted in Australia, Canada, China, New Zealand, the U.S., and the U.K. Direct consumer-to-merchant card transactions can save banks the costs involved with having to establish an independent clearing-house or settlement process. Consumers can use modified ATMs and PCs to reload their cards. In September 1997, Mondex announced an agreement with Cellnet to develop a service that will allow digital mobile phone owners to withdraw, deposit, or transfer electronic cash with their Mondex smart cards.

During 1997, Mondex began testing the AT&T/Mondex system. It is a small transaction payments system that will be integrated into AT&T's SecureBuy service. AT&T-hosted Web sites that use the SecureBuy plan will be eligible to use the Mondex system.

Visa also has a similar smart card offering. Visa Cash cards are available in more than 15 major cities, and approximately 250,000 Visa smart cards were given out as part of a public trial at the 1996 Summer Olympic games in Atlanta, Georgia. In May 1997, several hundred Visa and Bank of America employees began a 6-month pilot program to use Visa Cash cards for purchases on the Internet.

■ Electronic Wallets

CyberCash, through its CyberCoin system, will charge merchants between $0.08 and $0.30 to process transaction amounts ranging from $0.25 to $10. This "software wallet" employs a browser plug-in as the foundation of its payment system. To buy something using CyberCoin, both the consumer and the merchant need to have CyberCash software on their computers. To use the system, a consumer must have either a credit card (American Express, Discover, MasterCard, and Visa) or a checking account with a participating bank. Money can be transferred from the account to a CyberCoin "wallet" in increments of $20. Banks supporting the wallet technology hold the money transferred into the wallet until it is used. To make a purchase, a consumer clicks on a Web page and is queried about the purchase and payment amount; when the consumer approves the purchase, money is then transferred from the wallet to the merchant. As with other systems, consumers incur no charges for using the wallet but can purchase goods or services only from subscribing merchants.

Digital's Millicent product, which allows for transactions of a $0.01 or less, became available for public trial during the summer of 1997. Millicent's client component sits as a peer to the Web browser on the desktop. As with other products, an electronic wallet is used in conjunction with a credit card or bank account managed by a broker that the user has a financial relationship with. The Broker server (banks or service providers) collects the real money and in exchange gives the user SCRIP– electronic coupons – deposited into the wallet. The Vendor server accepts payments as part of a URL request and gives back coupons and creates change in accordance with the user's purchases.

SCRIP is an electronic coupon with transaction integrity that is encrypted. The SCRIP does not work if it is detached from the URL request. It links a single user to a single subscription. Only one person at a time can use it, unlike password technology. According to IDC, Digital expects SCRIP to be of interest to IT organizations that need a method to meter usage and charge user organizations.

Processing microtransactions is a good task for a Java applet. As a result, Sun plans to release a Java Wallet that will organize all a customer's credit cards and e-cash accounts.

Secure Electronic Transactions

The majority of payments on the Internet are made with credit cards. A certain degree of security is achieved when SSL encryption techniques are employed; the credit card number is transmitted and other information about the transaction is secured against eavesdropping of data packets on the Internet. However, many consumers remain unwilling to use credit cards on the Internet.

Even with the use of SSL, potential problems include that merchants have no way of determining whether the user of a credit card is in fact the owner of the card; the same problem occurs when retailers accept catalog orders over the telephone. In addition, if the merchant retains the customer's card number electronically, it could be accessible for subsequent fraudulent use. Also, Web sites may be set up to look like legitimate businesses, but in fact they may be set up solely to collect credit cards numbers for fraudulent use. Furthermore, the implementation of SSL available for export outside the U.S. is limited to 40-bit key size, which provides relatively weak encryption.

Solutions to these issues are needed to create better security around credit card transactions over the Internet and, equally important, to create the appearance of greater security.

The SET protocol specifies how all members in an electronic transaction – credit card holders, merchants, issuing banks, acquiring banks, and processors – interact to ensure secure payment processing over the Internet. The SET system prevents access to sensitive credit card information because it substitutes an X.509 certificate for the credit card number. Participants in SET development, led by MasterCard and Visa, include CertCo, GlobeSet, GTE, IBM, Microsoft, Netscape, RSADSI, SAIC, Spyrus, Terisa Systems, VeriFone, and VeriSign.

A typical SET payment authorization follows the process as illustrated in Figure 7-5:

1. The consumer and merchant receive digital certificates from a certificate authority prior to ordering.
2. The consumer then verifies the legitimacy of a merchant and sends order and payment information.
3. The merchant authenticates that the customer is a legitimate credit card holder and sends back a message confirming the order.
4. The merchant sends an encrypted payment authorization request to a "payment gateway" (operated by the merchant's bank or third-party processor).
5. The gateway decrypts the message and verifies the merchant, order, and payment information.
6. The gateway then sends a traditional authorization request over the existing credit card network to the customer's bank. The customer's bank then sends its response back across the credit card network to the merchant's bank or third-party processor.
7. The bank or processor sends the approval (or denial) to the payment gateway, which encrypts the response and sends it back to the merchant via the Internet. The merchant lets the customer know whether or not the purchase has been approved.

Figure 7-5: SET Payment Authorization

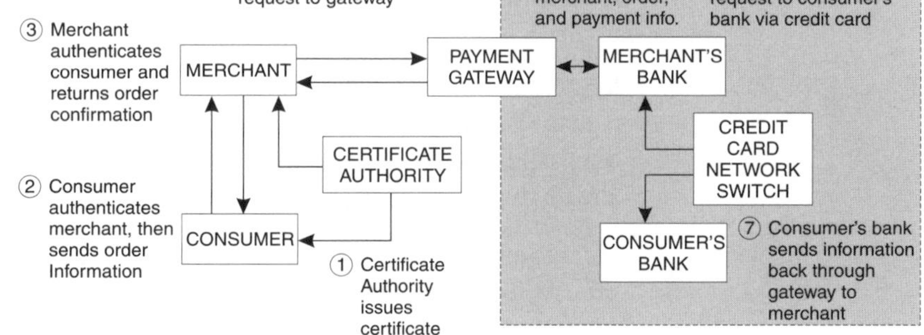

Source: Piper Jaffray, 1997

Transactions become more affordable for merchants because they can purchase a packaged software security solution, rather than develop their own. According to IDC, integrating SET into the transaction and avoiding the cost of a transaction server with a dedicated telephone line to a clearing bank's credit card processing center can save the merchant tens of thousands of dollars. Adoption of SET also reduces the merchant's risks. The major card issuers have announced plans to indemnify merchants against claims that a purchase was unauthorized, if it was carried out using SET.

■ SET Implementation Update

In January 1997, Visa successfully issued the first branded SET digital certificates over the Internet using software from multiple vendors, including IBM and VeriSign. The SET 1.0 specification was released in June 1997 (with a final version released in September). Software based on SET should become generally available from major vendors during the first half of 1998. For example, IBM will be shipping SET-enabled software as part of its CommercePoint payment product suite.

The SET 1.0 protocol is scheduled to be used in more than 70 pilots involving more than 150 banks in 28 countries. Previously announced SET pilots are moving forward in Europe, Japan, Singapore, and Taiwan. Six banks are participating in pilot SET studies in North America, and bank's fees are reportedly being waved by Visa to promote SET's use.

SET 1.0 currently is based on RSA encryption. Development is already underway on the SET 2.0 specification, expected in 1998, which will include extensions for international use, smart cards, debit cards with PIN numbers, and an encryption-neutral architecture that is algorithm independent. The shift toward independence from RSA was prompted by MasterCard's announcement in April 1997 that RSA's key length and the computing power required to implement it might be excessive. Instead, MasterCard suggested that newer technologies, such as Elliptic Curve Cryptosystem (ECC), might make a better core because they are smaller and faster than RSA. Some pilot projects already are testing ECC implementations, and that technology is being advocated by Certicom and Apple (due to its acquisition of NeXT's ECC technology). Certicom has developed an Elliptic Curve-enabled Secure Electronic Transaction (ECSET) specification, which details the adaptations necessary to incorporate ECC into the SET specification. (See "Encryption Techniques" on page 385 for a detailed discussion of encryption algorithms.)

■ Remaining SET Challenges

Concerns about SET include its ability to support true interoperability and the length of time for a SET transaction. One key limitation of the SET specification is that SET addresses front-end systems for processing credit card transactions only. Although the goal is to have this standard complement existing systems already in place, interoperability testing is still being resolved.

Serious questions about SET remain as yet unresolved, including who pays for the process, how the digital certificates will be distributed and protected, and who will be responsible for fraudulent transactions – Visa and MasterCard (as with current credit card exchanges) or the merchant? Although Visa and MasterCard both have policies stating that customers will be held unaccountable in case of fraud, neither has stated publicly whether they will pass these costs along to the merchant.

Additional problems are linked to the SET approach itself. For example, because SET does not use credit card numbers, processes such as customer profiling, disputes, and chargebacks (usually linked to card numbers) become difficult or impossible to manage. Also, SET does not guarantee nonrepudiation. In addition, it allows considerable latitude in how it deals with credit card numbers: banks can give the numbers back to merchants after the transaction, and the numbers can then be stored (possibly insecurely) on the merchant's system. However, this appears to be a policy, not a technology, issue.

Whether these issues are fully resolved or not, the support of the major credit card companies is a significant advantage for this standard. Alternative encryption methods from companies such as JCP Computer Services are being tested by European banks, service providers, and merchants. Credit card companies and new entrants into the banking sector are more likely to support SET, whereas traditional banks also are evaluating other security and encryption methods.

■ Alternatives to SET

The proposed SET standard is not the only way to handle Internet credit card transactions. Two French consortia are supporting two competing smart card-based electronic payment implementations: one is compliant with SET (e-COMM), and one is not (Europay C-SET). e-COMM began to test Internet payments using SET and smart cards in late 1997; C-SET began an implementation rollout in April 1997.

- **e-COMM consortium** – Formed in July 1996; includes three French banks, France Telecom, Gemplus, and Visa
- **Europay France consortium** – Formed in January 1996; includes four French banks, the Post, and Europay International

Table 7-3: Alternative Electronic Payment Standards

Technology Differences	Europay C-SET	e-COMM
SET compliance	Requires software at merchant site to convert SET transactions to C-SET, allowing interoperability but adding cost and time to the transaction	Supports SET but has no provision to convert transactions to C-SET
Portability	Certified identity and address of C-SET's translator is stored on chip card, allowing for portability	Currently not portable; plans are to support future Europay, MasterCard, and Visa standard compliant chip card with cryptographic functions including SET, enabling portability
Hardware/software requirements	Requires a personal reader (developed by Groupe Bull), attached to a PC into which the Europay smart card is inserted when customer makes purchase. Merchant sites that are C-SET-equipped allow customers to use magnetic-strip credit cards for purchases.	Requires a PC, the e-COMM software, a card reader (developed by Gemplus), and browser-enabled access to the Internet

Source: Gartner, 1997

These consortia were prompted to develop alternatives to SET because of concerns about two fundamental security weaknesses in that protocol: authentication (which is still not regarded as secure enough for on-line credit transactions in France) and nonrepudiation (which is not strongly implemented in SET, according to consortia members).

C-SET, for example, addresses both security concerns. It moves the authentication API logic to a tamper-evident card reader (with a keyboard for entering PIN codes); offline computation prevents interception of keyboard input. C-SET also uses its smart card to physically store audit trails of transactions, prohibiting users from claiming card number abuse, thus strengthening nonrepudiation.

SET will possibly never become an industry-standard method for secure electronic payments. Visa and MasterCard must continue to cooperate rather than look for competitive advantage in a non-SET mechanism. SET alternatives have advantages, especially nonrepudiation. However, at least in those parts of the world where smart cards have limited distribution, including the U.S., there are few alternatives to SET, and it will likely see a slow but steady progression toward widespread adoption.

To help consumers feel more secure when conducting business over the Internet in the U.S., some Web site operators with electronic stores or merchant links, such as Excite, Netscape, and Yahoo!, offer to limit cardholder liability in the event of unauthorized credit card use following an on-line transaction. In the U.S., credit card companies by law must limit a cardholder's liability to the first $50 of unauthorized charges. Netscape and other vendors are offering to cover any remaining liability, up to the $50 limit, for shoppers who make on-line purchases through their sites.

Other Secure Payment Standards

ANSI X12.4-#820 is the Payment/Remittance Advice document used by a corporation to initiate payments to another corporation via its bank. The National Automated Clearing-House Association's (NACHA's) CCD+ and CTX are used for bank-to-bank corporate trade payments.

In April 1996, the W3C and CommerceNet announced the formation of the Joint Electronic Payments Initiative (JEPI) to create standard protocols and a uniform API set to help manage the selection of different electronic payment schemes for Internet commerce. The group estimates that 35 incompatible electronic payment schemes currently are in existence. JEPI has proposed a protocol to define negotiation between a browser, server, and merchant involved in an Internet commerce transaction, helping to establish a common payment scheme between buyer and seller. JEPI protocols will work with other protocols such as SET. The first phase of JEPI specifications and runtime scenarios was completed in October 1996, and protocols have been submitted to the IETF for approval. A second phase of JEPI is being discussed and may occur in 1998, but plans had not been completed by late 1997.

Checkfree, Intuit, and Microsoft are promoting a single specification known as Open Financial Exchange (OFX) for on-line banking and financial transactions. OFX is a back-end format that provides a secure way for banks and customers to share information via the Internet. OFX supports banking, paying bills (from customers to businesses), and investing in stocks, bonds, and mutual funds. Support for bill presentment (from businesses to customers) and insurance purchase transactions is planned.

In March 1997, Integrion Financial Network unveiled its Gold protocol for requesting and receiving on-line financial data. Integrion's Gold is comparable to the OFX protocol from Microsoft. Unlike OFX, Gold will handle data carried between the bank and the processor as well as between the consumer and the bank. However, the initial version will not be able to handle investment positions like the recent OFX release. Integrion and Microsoft have stated they plan to merge their separate "standards" in 1998.

Microsoft announced the Windows Distributed interNet Architecture for Financial Services (DNA FS) in December 1997 with delivery expected in 1999. DNA FS is a framework for financial industry software. The DNA FS specification goes further than OFX's simple financial record format to standardize financial industry software components for routine transactions. Components (small software modules designed to work together) adhering to the DNA FS specification will form the basis for retail and back-office financial service applications. Companies such as Corillian, Diebold, and Edify have announced support of DNA FS. These vendors will write (and sell) reusable components for basic transactions such as funds transfers and balance inquiries, and will use these components to build applications that run across multiple delivery channels, including automatic teller machines, bank branches, and the Internet.

7.1.6 EC Implementation Requirements

The success and continued growth of EC is predicated on many factors, the most important being that EC must offer some compelling benefit to the participants, such as increased convenience or reduced cost. Beyond that, however, the success of EC depends on the resolution of a number of key technology and infrastructure issues that could otherwise limit its adoption. These include integrating new EC systems with existing business applications; providing the consumer with fast, reliable access to the Internet from widely available computing devices; and satisfying the consumer that EC is secure. EC's success also depends on the adoption of appropriate government policies and regulations that foster, or at least do not hinder, its growth.

Integration Issues

Companies need to modify back-office applications to align data with their trading partners. EDI addressed this requirement with the implementation of ANSI X12 and EDIFACT standard messages. As more companies begin to use the Internet as a transport mechanism for conducting business electronically in new ways, data integration issues continue to expand.

High-end, transaction-enabled Web server software is already available and adaptable to most existing presentation, database, transaction processing, order entry, and payment systems. Unfortunately, message and data exchange standards have not yet been established, and many of the available EC offerings are not compatible.

Many companies still are using a batch-oriented, mainframe application model using formal, structured EDI transactions. Large companies will continue to protect their investment in their legacy applications until it becomes feasible and cost-effective to migrate to new client/server-based applications. The pace at which these companies move to adopt new EC technology will correlate with perceived return on investment.

Broadening Internet Access

Before Web-based EC can be broadly used, a greater fraction of the population must have access to the Internet. In an effort to build critical mass for consumer-to-business EC by providing inexpensive Internet access products to the 60 percent of U.S. homes that do not have a PC, Microsoft and Oracle are competing in the network computer marketplace. With simpler, easier-to-use, lower-cost access devices, more people will have access to and will feel comfortable using EC sites on the Internet and making on-line purchases. (See "Network Computers (NCs) and Network PCs (NetPCs)" on page 359.)

Public Network Congestion and Reliability

Internet congestion will continue to be a barrier to Internet commerce. For many potential consumers, low-speed access to an ISP vendor and limited bandwidth in local connections often present the biggest problems. Until significant investments are made in the local access networks, a major constraint will remain. (See Chapter 4.3, *Telecommunications Services.)*

Delays throughout the Internet backbone in most countries will continue for several years. For network backbone congestion, differing levels of service quality and priority, with extra charges for business users requiring guaranteed delivery or response time, are likely. The business-to-business Internet commerce market is looking for a cost-per-transaction billing system, which carriers such as ISPs are not able to accommodate yet.

The Internet is a network of networks with no centralized management to guarantee message and transaction integrity, so EC Web site personnel must be experienced in handling network and systems reliability issues. An EC Web site requires distributed applications that may involve multiple vendors and perform multiple transactions against one or many databases. Data integrity issues arise when a transaction is terminated prematurely or when it is only partially completed. The risk of a premature termination is higher on the Internet than on a LAN.

Security

Because the Internet is a public network, concerns often are raised about how commercial transactions over the Internet can be made secure. In fact, security is the most frequently cited reason for delay and reevaluation of Internet commerce implementation decisions. Several tools are available to protect information and systems against compromise, intrusion, or misuse. Each needs to be weighed in terms of cost and appropriateness to the specific EC application. Systems security measures include firewalls, encryption, and message integrity, customer and merchant authentication, and nonreputability of transactions. (See Chapter 5.4, *Security,* for more information about security protocols and alternatives, including firewalls, encryption, and message authentication.)

Also closely related to security are the issues of privacy and trust. Electronic commerce initiatives in the U.S. addressing these issues include these:

- **TRUSTe Initiative** – A global consumer privacy initiative (formerly eTRUST), introduced in June 1997. TRUSTe will work in conjunction with existing standards such as SET to increase the trust between buyers and sellers engaged in electronic commerce through issuing Trustmarks, which certify compliance with standards on information privacy. (See "Invasion of Privacy" on page 407 for more information.)
- **WebTrust** – A certification criteria developed by the American Institute of Certified Public Accountants and the Canadian Institute of Chartered Accountants that allows accounting firms to provide a certification service for business-to-consumer EC Web pages. Web pages that are certified will receive a WebTrust seal from VeriSign to place on the page. The criteria involves business practices disclosure, transaction integrity controls, and information protection controls. The aim of WebTrust is to provide a visible seal and corresponding attest report so that consumers may have confidence in the security of an EC Web page.

Tax and Regulatory Issues

Extended intranets now include communications and business transactions between organizations and their customers and suppliers via the Web. As more organizations conduct product sales, customer support, and distribution electronically with their customers, they are grappling with international sales tax and other issues, such as how local sales offices will be notified about account activity when it occurs.

Many merchants and businesses feel that governments need to adopt a nonregulatory, market-oriented approach that supports global Internet commerce of all types. Governments play a role in taxes, duties, licensing requirements, standards development, and restrictions on certain types of content on the Internet.

The Clinton administration is recommending that any taxes on Internet sales should be simple and transparent, be consistent with principles of international taxation, and neither hinder nor distort commerce.

In Europe, the Organization for Economic Cooperation and Development (OECD) and the European Union (EU) take the view that national governments should define Internet commerce policies for the EU and their own countries, leaving industries free to implement Internet commerce. The EU's view is that the Internet will not be a duty-free zone but taxes should be nondiscriminatory.

Joint initiatives have been set up by the EU and the U.S. for Internet commerce solutions. Internet codes of conduct are emerging for countries such as France, Germany, and the U.K. A current OECD study is evaluating member states' laws, regulations, and policies affecting the Internet.

7.1.7 EC Business Scenarios

Entrepreneurs have been creating new business models for retailing, publishing, and advertising that are working and generating profits. It has been only 3 years since companies began transacting commerce on the Internet, and the majority of the more than 250,000 commercial Web sites have little more than a year's experience. Table 7-4 shows three industry groups at the forefront of this growth: manufacturers, intermediaries, and service and utilities.

Table 7-4: Early Adopters of Internet Commerce

Manufacturers	Durable goods manufacturers, chiefly electronics and airplane parts, will total $3 billion,or 38%, of all Internet commerce business in 1997. Examples: Boeing, Cisco, and Dell Nondurable goods, in comparison, will generate only $182 million.
Intermediaries	Middlemen will thrive by realizing $2 billion in Internet-based resales in 1997. Examples: MicroAge and TechData leverage Dell's and 3Com's strengths; BT and Boise Cascade brought office supplies to the Internet. Categories of intermediaries include electronic clearing-houses, on-line support and services companies, information brokers, multimedia content delivery, on-line software delivery, transaction brokers, market-place concentrators, and virtual trading communities.
Service and Utilities	Utilities, services, and transportation will achieve more than $3 billion in 1997. Utilities - $2 billion Services (including software) - $1 billion Transportation - $30 million

High growth in commerce between companies should continue as suppliers, distributors, and customers come on-line. Today, the market is dominated by early adopter manufacturers striving to be Internet leaders.

Not surprisingly, Web sites for business-to-consumer EC are often those belonging to high-technology companies. As early adopters of the technologies, these companies offered an excellent environment in which to create, test, change, and "mature" their Web sites to satisfy customer needs. In addition to computers and related products, best-selling products on the Web continue to be in the areas of consumer entertainment and leisure products, such as music CDs, books, and airline tickets. Table 7-5 shows *PCWeek OnLine*'s selection of the top ten Web sites (in order of popularity).

Table 7-5: *PCWeek*'s Top Ten Web sites

Web Site	Purpose	URL
Microsoft's Expedia	Travel Research Reservations Purchasing	http://www.expedia.com
E* Trade	Investing	http://www.etrade.com
GE Trading Process Network (TPN)	Commerce Negotiation Ordering Web services	http://www.tpn.geis.com
Amazon.com	Bookstore	http://www.amazon.com
Manheim Online	Used automobile supply chain	http://www.manheim.com
PhotoDisc	Digitized stock photography	http://www.photodisc.com
Internet Liquidators	Auctions	http://www.internetliquidators.com
CD Universe	Music CDs	http://www.cduniverse.com
Virtual Vineyards	Wine and food	http://www.virtualvin.com
Cisco Connection Online	Product ordering	http://www.cisco.com

Dell and other computer and software vendors already are selling large volumes of product over the Web. In November 1997, Dell announced that the company's Internet Web site was generating sales of more than $3 million per day, equivalent to an annual run rate of approximately $1 billion. Compaq offers home computer users the opportunity to buy computer products through its Online Store.

The steps from viewing advertising and marketing to conducting retail transactions have become compressed as the Internet-based Web becomes a channel of distribution to speed up the sales process and capture more orders. Consumers can view an ad and using the Web to navigate electronically, can view merchandise and place electronic orders, send e-mail, and even communicate via Internet phone software during transactions.

Typical EC Applications: Business-to-Consumer

The general category of business and commerce on-line includes a wide variety of traditional and alternative retail options as well as financial, travel, and other services. The following is a list of some industries where business-to-consumer electronic commerce is being conducted.

■ Retail

On-line purchases via the Internet are growing, offering a low-cost method of distribution to many businesses. Computer and software vendors are doing substantial business over the Web, and growth is expected in other industries as well. The size of the potential market has attracted large companies such as television's Home Shopping Network, which owns the Internet Shopping Network, the largest Internet mall for computer products and peripherals. Purchases may be made via the Web using a credit card or by calling a toll-free number. A registration process to record the credit card number that will be used for on-line purchases is required.

Amazon.Com's Web site offers customers a huge selection of books, including hard-to-find or out-of-print books, and uses Web technology to provide services that traditional book retailers cannot. For example, when a customer selects a book, Amazon.Com also displays a list of related books. Customers can list their favorite authors and topics to receive an e-mail with recommendations or ask to be notified (also by e-mail) when a particular book comes out in paperback.

■ Auctions

On-line auctions have grown more popular and offer possible alternatives to traditional retail business models. Onsale brings computer buyers and sellers together in an on-line silent auction that offers a game-like experience. Onsale capitalizes on the Internet's strengths of real-time communication and community interaction because it lets users see and respond to one another's bids seconds after they are placed. Onsale specializes in refurbished and close-out software and electronics such as computers, printers, consumer electronics products, and related accessories. In addition to consumers who are placing bids, the site attracts many people who just watch the auctions. Onsale's weekly sales topped $1 million in 1996. Since the site's inception in 1995, it has hosted 75,000 unique purchasers. The site attracted an estimated 750,000 visitors each week in the first half of 1997.

Electronic Travel Auction's Web site lets consumers bid on hotel rooms, airplane seats, and other travel-related merchandise. The company partnered with eBay, which specializes in developing auction technology for Web sites.

■ Shipping

Consumers can purchase services and check on orders through shipping company Web sites. Federal Express uses its Web site to reduce its customer service costs by allowing users to arrange package pickup, track delivery status, and order merchandise on-line as well as request free shipping and tracking software. United Parcel Service has a similar Web site, which also allows users to ship and track merchandise, request a pickup, and download its software. These sites are examples of extended intranets, or extranets, because they allow a customer with a waybill number to track package status via the site. Both the FedEx and UPS sites offer links to the respective company's international facilities.

■ Banking

An increasing number of banks are offering customers on-line services. Wells Fargo Bank enables customers to use the Internet to view account balances and history and transfer funds between accounts. Through Wells Fargo's bill payment service, consumers can set up the automatic payment of recurring, fixed-amount bills (such as mortgage or cable TV); make one-time or regular payments of varying amounts (such as the phone bill or plumbing bill); send money to any individual, anywhere in the U.S.; and review upcoming scheduled payments and make changes or cancel them.

Security First Network Bank began its Internet banking service with individual checking accounts, then added joint checking accounts and the ability to purchase certificates of deposit and open money market accounts.

Banks are focusing on initiatives that will add convenience and help attract and retain customers. Banks continue to work on electronic versions of checks. New bill payment and funds transfer options over the Internet will become more widespread over the next few years.

Jupiter projects that the total number of U.S. on-line banking households will increase to 4.5 million households by the end of 1997. Forrester predicts that the total on-line banking market will grow to more than 9 million households in 2001.

■ Investment Services

Web sites maintained by E*Trade Securities, Charles Schwab, Lombard Institutional Brokerage, and others provide access to stock, bond, and mutual funds by individual investors. The three sites provide similar services, including stock quotes and buying and selling, at prices similar to those charged by discount brokers. In addition, Schwab's e.Schwab site provides the ability to download information into a spreadsheet and cash management and tax applications as well as the ability to analyze and compare performance. Forrester predicts the on-line brokerage market will grow from 2.77 million accounts in 1997 to 10.05 million accounts in 2001.

The Investor's SuperSite On The Web provides stock and mutual fund information for investors, including analysis by industry group. It includes an interactive, customizable stock ticker written in Java. Users enter the symbols for the stocks they want to track, and a Java applet then provides real-time quotes from the various exchanges in an on-screen ticker.

■ Travel

Consumers are using the Web for on-line airline reservations and ticket sales. Customers can compare prices, check flight schedules and availability, and book their own reservations. Some airline customers can use electronic ticketing to avoid the need for paper tickets. Forrester predicts that the on-line travel market will grow from $827 million in 1997 to more than $8 billion in 2002.

The Internet Travel Network allows users to select destinations around the world and book airline flights, cars, and hotels through the site. Users can build a travel profile (destination, activity, and desired price for travel) and send it to ITN, which then sends back an e-mail with information that meets these requirements. The Sabre Group's Travelocity site and Microsoft's Expedia site are two other popular travel sites.

■ News and Information

A wide variety of newspapers and news bulletins are available via a Web browser. Most are related to hard copy versions of publications. Search engine sites such as Yahoo! offer personalized news in an advertising model that resembles television more closely than traditional EC.

The Wall Street Journal and many other publications are also available on the Web for a fee. These on-line versions usually are offered at a discount or at no cost to subscribers of the hard copy version. Local newspapers are coming on-line as well, primarily because the Internet provides another outlet for their classified ads.

Jupiter Communications estimates that by the year 2000, only 40 percent of Web users will pay for subscription services. Subscriptions for information over the Internet are not anticipated to be a major source of revenue – reaching only $966 million by the year 2000.

Charging subscription fees generally has proven difficult for most general-interest, content-based Web sites. Microsoft's on-line magazine, *Slate*, for example, delayed charging for a subscription and ultimately abandoned plans to move to a subscription model. However, on-line communities are succeeding. The WELL, for instance, charges its 11,000 members $10 per month for access to 260 ongoing conferences. Internet communities can reshape the way buyers and sellers conduct EC. Consumers naturally will gravitate to communities offering a wide choice of information or product suppliers.

■ Government

Government programs also can use the Internet and electronic means to gather information and distribute funds and benefits. Governments are using EDI for purchasing goods and services and EFT for payments. The U.S. government spends about $200 billion per year with suppliers of goods and services, arguably making it the largest EDI "hub" in the world. Smart cards are being used to access government benefits, and in the future, recipients could have Medicare, Social Security, or welfare information (or all three) on their cards.

In addition to government-to-citizen transactions, the Internet also is being used for government-to-business applications. Beginning at the end of 1997, the U.S. government started a year-long market test of its Web server software to pay government contractors over the Internet. The test is sponsored by the Financial Services Technology Consortium, a nonprofit group composed of banks, other financial service companies, high-technology businesses, universities, and government agencies. The two banks involved will deposit the electronic checks into vendor's checking accounts, clearing them through the Federal Reserve Bank electronically.

The Government Information Technology Services Board (GITSB), an inter-agency group set up by the Clinton administration, has been working on a plan for setting up network connections among federal agencies as well as state, local, and international governments. The group is focused on improvements in the electronic delivery of government benefits, on setting up an inter-governmental wireless law enforcement network, and on ensuring that these networks are inter-preted. The GITSB also will look at new ways for citizens to access electronic services and electronic tools for some businesses.

Typical EC Applications: Business-to-Business

Cisco's corporate Web site, known as Cisco Connection Online (CCO), is an example of a full-service, business-to-business EC site. Cisco, the leading supplier of routers and other internetworking products, is not using its Web site primarily to generate incremental sales. Instead, benefits accrue in the form of improved efficiency and reduced costs of handling orders and supporting its customers.

One way Cisco has improved efficiency is to create Web-based applications that allow its customers to locate, configure, and order exactly the right products, all without Cisco incurring the overhead of traditional sales and marketing expenses. In fact, Cisco provides software agents on its Web site that directly parallel the roles and responsibilities typically present in a nonelectronic purchase. These software agents include the following:

- **Configuration Agent** – Allows customers to search for Cisco products that are configurable, choose a particular model, and create a configuration on-line
- **Pricing Agent** – Allows customers to access Cisco's on-line price list. Customers can search for prices based on product family, product description, or product number. An additional feature enables customers to download the entire price list to their own computers.
- **Lead Times Agent** – Provides information about current availability and delivery lead times for Cisco products
- **Status Agent** – Provides quick and easy access to current information on existing orders, allowing customers to track the progress and status of orders 24 hours a day, 365 days a year
- **Invoice Agent** – Lets customers view Cisco invoices on-line
- **Service Order Agent** – Lets customers access information on service orders
- **Contract Agent** – Lets customers access information on service contracts
- **Upgrade Agent** – Allows customers to request software upgrades, hardware upgrades, or documentation free of charge, under the provisions of their Cisco service contract
- **Notification Agent** – Allows customers to specify action criteria to send notifications automatically on Cisco's order status and pricing changes
- **IPC Extract Agent** – Allows users to view orders submitted to the Internetworking Product Center (IPC) and download copies to their desktops for integration into local reporting or purchasing systems

Cisco estimates that its Web site saved the company $268 million during its most recent fiscal year. Specific savings included $85 million and $40 million that would have been spent providing customers with software updates and hard-copy documentation, respectively. Meanwhile, Cisco is using its Web site to process orders at the rate of almost $10 million per day.

Figure 7-6 illustrates the architecture of CCO.

Figure 7-6: Cisco Connection Online

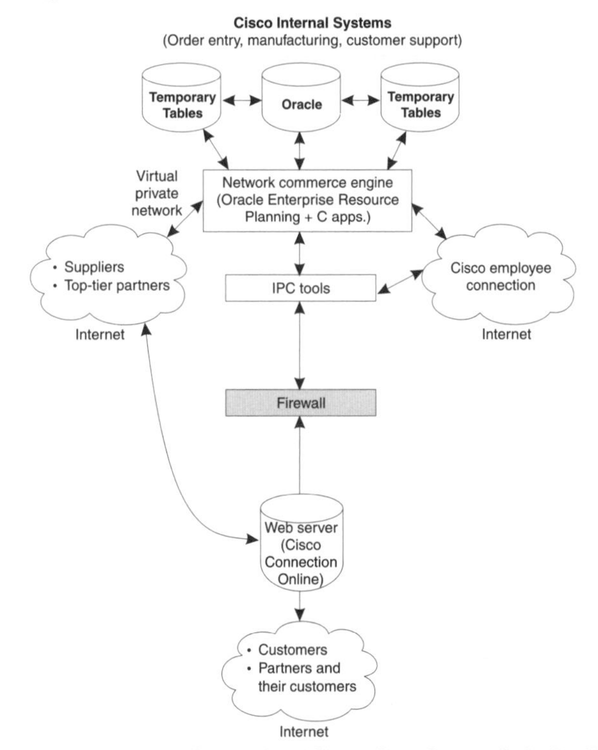

Source: David Clark, "Cisco Connect Online" in *IEEE Internet Computing*, © 1997 IEEE

7.1.8 EC Service Providers

EC service providers provide infrastructure and management services. They facilitate one-stop shopping by providing the software, network, platform, and components such as payment, billing, and security. Service providers can be VANs, network providers, ISPs, or on-line services. Traditional EDI made use of VANs or VPNs. However, just as traditional EDI has evolved, most VANs are adding to their service offerings, incorporating new technologies, and becoming network providers in a broader sense.

EDI Service Providers

This market segment also contains a mix of traditional VAN, VPN, and other service providers and newer offerings. This area will continue to be a dynamic mix over the next few years, one that will include telecommunications carriers as well. EDI and VAN providers offering new services and EC application outsourcing and integration should enjoy excellent growth prospects.

EDI service providers include GEIS, IBM Global Network, and Sterling. GEIS, although second in market share by revenue to the IBM Global Network, has the largest customer base, with more than 44,000 subscribers. EDI VAN Harbinger acquired EC software vendor Premenos in 1997. The combined company will likely focus on affordable Internet EDI for midsized companies. Telecommunications companies such as AT&T, MCI, and Sprint also are EDI service providers.

During the next few years, the average cost for services will be driven down, and the number of subscribers will increase. Customers from new industries, expansion into international markets, and approval of security standards will drive much of the subscriber growth. Providers' pricing will reflect service value rather than network transport. Charges also will be based on several levels of service rather than a flat fee pricing strategy, which often does not reflect service levels.

In a traditional EDI business model, hub vendors (typically larger, more centralized companies that drive transactions and may convince trading partners to sign up for a given service) may be 15 to 20 percent of a service provider's subscribers but may account for 60 percent of provider revenues. Trading partners sell products and services. Overall numbers for both hub and trading partner companies on service provider networks are growing, and an increasing number of these companies will likely be both hubs and trading partners. IDC sees the average yearly spending ratio shifting over the next few years, with trading partner spending surpassing hub-based spending. By 2001, IDC predicts that 28 percent of all subscribers will act as hubs, and 94 percent will act as trading partners. Only 1 percent of subscribers acted in dual roles in 1996; this number is expected to grow to 22 percent in 2001. More peer-to-peer transactions can occur as companies take on dual roles.

Financial processors will work with commerce service providers and intermediary service players to incorporate merchant server software offerings where it makes sense. Although VANs are in a good position to take advantage of developments in EC, they will have to evolve to add new value to Web services and transaction processing over the Internet.

Web-Based EC Service Providers

GEIS provides EDI networking, software, and integration services to more than 44,000 companies. In 1996, GEIS partnered with UUNet (a commercial ISP owned by WorldCom) to add Internet services to the information services already provided to its GE BusinessPro service customers. GE's TradeWeb is an Internet-based system for EDI that is targeted at small businesses, many of whom do not participate in EC today. Early in 1997, a second generation of TradeWeb was released. It lets companies that do not have conventional EDI software use Web browsers to link to companies that do. GEIS began operating its Trading Process Network (TPN) in January 1997 as a Web-based service linking corporate buyers and sellers. In its first 3 months, TPN has logged $350 million of industrial products purchased electronically by GE divisions. The GE Capital Services unit created a consulting practice that will provide EC services to GE Capital units and, eventually, outside customers. The consulting group is tapping talent from GE Information Technology Solutions and from the acquisition of AmeriData.

A joint venture between GEIS and Netscape, called Actra Business Systems, will provide Internet-based business-to-business solutions. Actra's CrossCommerce consists of five products: ECXpert, OrderXpert Seller, OrderXpert Buyer, MerchantXpert, and PublishingXpert. CrossCommerce does not require businesses to install special purpose-desktop clients; it relies on the core capabilities of the Internet and the Web. Actra assumes that companies are linked with industrial strength Internet connections and that TCP/IP networks are almost as ubiquitous, reliable, and secure as telephone lines or proprietary EDI VANs. ECXpert supports the exchange of structured business documents among heterogeneous mission-critical business applications over the Internet. It supports both event-driven and store-and-forward EDI solutions, uses the Web, and incorporates capabilities for any-to-any data and object transformation. Authentication, encryption, and nonrepudiation of electronic document exchanges are also supported.

IBM's Global Network is linked to a variety of services. For example, the Automotive Network eXchange (ANX)-based virtual network, EnterpriseXspan, helps customers in the automotive industry to establish a virtual supply chain using the Internet or other networks.

As part of the IBM program, companies receive firewalls, proxy servers, or Web servers to conduct business on the Internet. Users can exchange EDI documents, computer-aided design (CAD) drawings, and other data and can access folders, discussions, and newsgroup information that might relate to a given part of the design or manufacturing process. Suppliers also can participate in the review or design process through access to certain folders. EnterpriseXspan can help reduce development time at manufacturing companies because engineers can collaborate in near real-time, in addition to or instead of exchanging documents.

Pilot projects for EnterpriseXspan will run through early 1998. Components will include Lotus's Domino Web server, electronic folders, directory services for ANX or other IP networks, connections to the IBM Global Network, dial-up leased lines, and consulting services.

The EC Company has created its own VPN for managing EDI business-to-business commercial transactions. The 4-year-old company is introducing a product targeted at the needs of midsized companies and positioned as a solution for back-office EC. Hosted on a series of Windows NT servers running SQL Server databases, the hub system processes EDI transactions among registered trading partners and their banks. The EC Company has contracted with UUNet for commercial Internet services nationwide, enabling all transactions to occur within a single network provider's TCP/IP network. Customers are required to access the system via a proprietary Windows-based client. The EC Company has applied for two patents to protect its technology for conducting EDI transactions over public networks, which includes the following characteristics:

- Secures the contents of EDI transactions through a two-stage encryption and data compression process
- Claims a fully extensible common data dictionary of individual business documents and their respective terms

Internet Service Providers

Commercial subscribers typically access the Internet through one of the larger ISPs. Web access and site hosting became attractive service opportunities that required relatively little initial capital and so attracted many entrants into the ISP marketplace in the early to mid-1990s. Strong competition among ISPs from carriers and other service providers, spurred by the U.S. Telecommunications Act of 1996 and by low, flat-rate pricing schemes is driving consolidation of providers in this market.

The growth in the number of independent ISPs in the early 1990s paralleled the growing popularity of the Internet and the Web. Based on information from American Business Information's database, there were 2,298 Internet service firms at the end of 1996, an increase of more than 9,000 percent from the beginning of that year.

ISPs can vary from large companies such as UUNet (the first commercial ISP, founded in 1987, which became a WorldCom subsidiary in December 1996) to regional service providers that include university-affiliated enterprises and local service providers that include small businesses serving between 10 and 100 customers.

In addition to an increasing number of EDI service providers incorporating the Internet, ISPs will offer services in the future that will more closely match the capabilities of some VANs, and VANs also will partner with ISPs. ISPs and on-line service providers include telecom companies, cable companies, and on-line service firms such as AOL and CompuServe.

Commercial on-line services offer relatively secure environments, reliable access, simplified user interfaces, interest groups, and marketplaces. Merchants gain a relatively controlled environment to host their applications, an authenticated customer base, and a mechanism to charge for access to content. As the Web user community has shifted from its traditional base of government and educational users to significant numbers of commercial and consumer users, on-line services and other providers must continue to provide greater ease of use, security, and functionality. The resulting improvements will be part of and also help drive EC opportunities.

7.1.9 Market Overview

The EC market, including EDI over the Internet and Internet commerce segments, is experiencing high rates of growth and a high degree of interest as well. Market watchers and industry consultants such as Forrester, Gartner, and IDC assert that EC is the fastest growing application area in the computer industry.

During the last few months of 1996 and the beginning of 1997, a noticeable shift occurred from the EDI-centric, business-to-consumer and business-to-business EC market to a focus on business-to-business Internet commerce. Companies using these public network-type technologies are doing so primarily because they perceive value in cost savings, reduced order processing time, and better information flow. Quantifiable benefits are being realized by taking orders over the Internet rather than through telephone-based salespeople; extranet systems have reduced the time between order placement and delivery often from months to weeks; and order volumes are increasing due to easier access to a larger pool of customers and suppliers.

Fortune 1000 companies typically spend more than 5 percent of revenue on purchases of nonproduction goods. This business-to-business exchange of goods comprises a significant market, exceeding $250 billion per year. In general, it costs the average company between $0.01 and $0.02 to purchase $1 of supplies, which creates an enormous clerical and administrative expense and effort. Companies such as CommerceOne are trying to take advantage of this market with multi-vendor catalog solutions.

It is expected that business-to-business Internet commerce will predominate over the next 2 years in both U.S. and European markets. Figure 7-7 compares the growth of business versus consumer Internet commerce.

Figure 7-7: Worldwide Internet Commerce Market

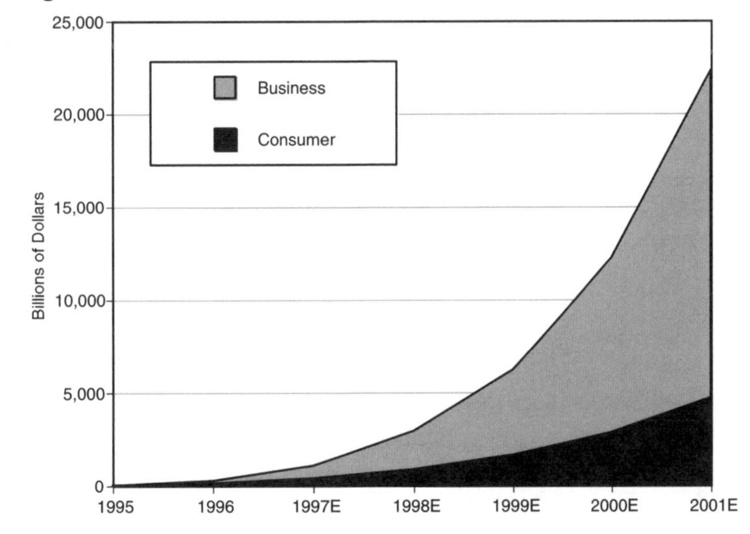

Source: IDC, 1997

Gartner estimates that by 1998, approximately 100,000 enterprises will be using EC for business transactions and that this figure will increase to nearly 300,000 in 1999, doubling to 600,000 in the year 2000.

From a worldwide perspective, the majority of the traffic on the Internet and the Web clearly is generated in the U.S. However, with 734,400 Internet host computers and with an estimated 6.7 million on-line users in 1997, Japan is firmly in second place. One-fourth of all European companies have Web access, 20 percent of employees have Web access, and 13 percent have a home page.

Forrester Research's Business Trade & Technology Strategies group predicts that the total value of goods and services traded between companies in the U.S. over the Internet will jump to $327 billion by the year 2002 from $8 billion in 1997, realizing a 40-fold increase. The 1997 figure for business-to-business Internet trade represents a 1,000 percent increase from 1996.

The on-line population represents a growing group of potential EC consumers. In 1997, approximately 40 million people were on the Web in the U.S., up from nearly 20 million in 1996 and 1 million in December 1994. U.S. women, who often are responsible for major family purchases, composed 41 percent of the population on the Web in 1997, compared with 23 percent in September 1995. An estimated 67 percent of on-line users are now 30 years and older, with 19 percent older than 50. Almost one-fourth of on-line users, or 10 million people, has purchased something on-line. Adult Internet users generally are more affluent and better educated than the U.S. population as a whole.

EC market growth can be viewed from three perspectives:

- EDI
- Internet
- Web products and services

EDI

Market leaders have changed little in 1997, but the landscape for EC is changing rapidly. Three U.S.-based service providers have enjoyed dominant EDI market share positions, by revenue, for the past 3 consecutive years: IBM Global Network (IGN) with 29.1 percent, GEIS with 28.1 percent, and Sterling Commerce with 16.4 percent of total 1996 revenue of $664.2 million.

In 1996, U.S.-based providers' worldwide EDI services revenue represented an increase of 41.1 percent compared with 1995 figures, according to IDC. The Business Research Group predicts that the EDI market will nearly double from 1996 to 2000, reaching a worldwide value of almost $2 billion. The following traditional forces will continue to fuel this market's growth:

- Smaller trading partners being added into the trading community
- Large hub companies demanding that their trading counter parts become EDI enabled
- Services being expanded into new international markets and vertical industries

However, the market is expected to slow briefly between 1998 and 2000 as companies expand their trading communities and initiate peer-to-peer trading; both market revenues and subscriber base will slow for a short while. There will be increased instances in which an EDI subscriber will be a transaction driver for one transaction and the seller for another.

Only a small percent of the potential EDI market has been penetrated because of the complexity and high cost of setting up EDI systems. Although EDI standards and technology have been available for more than 25 years, EDI's implementation continues to be concentrated among larger corporations and the top 20 percent of their trading partners. Of the 2 million U.S. companies with 10 or more employees, approximately 100,000 choose to deploy today's EDI, according to Forrester – the other 1.9 million small to midsized companies are potential participants.

These small and midsized companies present a big opportunity to providers of EDI services over the Internet who can deliver cost-effective, flexible solutions. In 1996, approximately 7 percent of companies using EDI exchanged transaction sets over the Internet. This number has been growing slowly over the last few years, with users moving with some caution into use of the Internet as a transport medium. More than 15 percent of U.S. business traffic may move from VANs and leased lines to the Internet by the year 2000. According to a 1997 Forrester survey, more than 40 percent of those interviewed believed that at least some EDI over the Internet and Web front ends will take place in companies by the year 2000.

In 1996, EDI with Web access was estimated at $100,000 worldwide, and Web-based EC generated no measurable services revenue, according to Forrester. However, Forrester expects EDI with Web access to increase to $500 million and EC on the Web to reach $300 million by the year 2000.

U.S.-based providers' worldwide EDI services revenue are projected by IDC to grow from $889 million in 1997 to more than $2.5 billion in 2001. In 1996, 73 percent of the revenue generated by EDI service providers originated within the U.S., according to IDC. During the next 5 years, as providers expand their market reach, the distribution of revenue origination will begin to change. By 2001, IDC estimates that the U.S. will represent 66 percent of this market, the Latin American market will increase its share to nearly 5 percent, the Asian/Pacific market will increase to more than 10 percent, and Europe will maintain its position at approximately 17 percent. GEIS spent 1996 developing new markets – such as the Asia/Pacific market to create CHINA EDI – rolling out Internet services, and targeting the small business market.

EC Server and Software Offerings

As of late 1996, Gartner estimated the North American EC software and services market to be between $2 to $3 billion, growing annually at 25 percent per year through 2002. This market is categorized by frequent technology changes, new startup ventures, consolidations, and changing vendor business models.

The largest segment of the EC software and services marketplace consists of back-office software from EDI, EFT, and VAN vendors. Gartner estimates 1997 revenues for vendors in this space to be between $800 million and $1.2 billion. Revenue for this segment, according to Gartner, is expected to remain flat through the year 2002.

Web-site building tools, EC marketing applications, electronic catalogs, payment system access, and shopping tools are all products in what Gartner calls the Web Front Office. Currently, there is turmoil in this segment, and a consolidation is expected during the next 5 years. However, revenue is expected to increase from $400 to $600 million in 1997 to $1 to $1.2 billion by 2002.

Vendors that have garnered market share in this area include traditional players such as GEIS, HP, IBM, Open Market, and Premenos and newer entrants such as Microsoft, Netscape, and Oracle.

Systems vendors are bundling increasing amounts of software into hardware system offerings in an attempt to gain a foothold in this competitive horizontal market. For example, Apple's alliance with Microsoft, announced in August 1997, gives Microsoft a 7 percent stake in Apple, and Microsoft's Internet Explorer software becomes the easiest browser to choose from a Macintosh computer. Additionally, Apple and Microsoft are collaborating on their own version of the Java programming language and computer platform – a move that could give them further command of the desktop Internet market and potential end points for transactions.

By 1998, many early Internet commerce systems will need to be updated or even replaced, spawning a nearly $3.2 billion commerce software market by the year 2000, according to Forrester.

Table 7-6 lists some of the vendors involved with each of the various parts of the infrastructure required for EC.

Table 7-6: Key Electronic Commerce Infrastructure Providers

Sector	Vendors
Electronic Market Managers	AT&T, CompuServe, GEIS, IBM, MCI
Platform/Service Providers	Digital, HP, IBM, Silicon Graphics, Sun
Service/Software Developers	BroadVision, Microsoft, Netscape, Open Market, Spyglass
Component Providers	CheckFree, CyberCash, DigiCash, Digital, First Virtual Holdings, Harbinger, Integrion, MasterCard, Premenos, Sterling Commerce, Visa

Source: Gartner, 1996

EC Consultants and Systems Integrators

The consultants and systems integrators that make up this market will thrive. Most enterprise applications are still not fully prepared to handle EC functionality, and it is estimated by Gartner that it will be 5 to 7 years before truly packaged EC solutions are available. Also, in-house IT organizations' resources may be dedicated to solving Year 2000 problems for the next few years, preventing them from focusing their attention on EC. Forrester Research estimates revenue from e-commerce and Internet consulting and systems integration will be $8.8 billion in 2002, with a compound annual growth rate of 88 percent between 1996 and 2002.

7.1.10 Forecast

- The value of goods and services traded between companies over the Internet will grow substantially during the forecast period.
- Business-to-business Internet commerce will explode in the next 2 years. This segment of EC is largely being driven by manufacturing companies and intermediary service providers utilizing the Internet and extranets.
- The fastest-growing volume of transactions in business-to-business EC will come from people accessing on-line catalogs and other Web sites, not from direct application-to-application communications.
- VAN vendors based in the U.S. will continue to dominate the VAN market for the next 5 years, although their market share (based on revenue) will decrease over the next few years.
- Web advertising revenue will continue to grow rapidly during the forecast period and will successfully subsidize a growing number of Web sites.
- The growth of EC will be influenced by the availability of better payment systems. Payment by credit card will continue to be the most popular method of paying for goods and services on the Web in 1998 and probably beyond, despite the proliferation of new payment systems. Credit card usage will get a boost from more advanced, security agreements by card issuers and merchants to indemnify consumers in the case of fraud and from consumer familiarity.
- Digital cash will become a standard micropayment mechanism for paying small amounts (less than $1) for on-line information within the forecast period.
- Smart cards will become more widely used throughout the world, including the U.S., for business-to-consumer electronic commerce payments. This trend will be driven by improved card technology, lower card costs, and ease of use. PC manufacturers are contributing by adding card readers as standard (embedded) features of their hardware.
- The number of businesses of all sizes involved in EC will grow substantially, aided by hosting and other services that can take care of network, system, and transaction management as well as other tasks.
- The use of EC in particular business segments will be pioneered by innovative startups pursuing new business models, but they will be followed closely by established vendors moving into EC to defend their market position.
- To appeal to a broader audience, the use of EC will need to offer some compelling advantage to the buyer over traditional purchase methods. The fastest growth of Web-based EC will occur where new business models are developed.

- The growth of extranets will facilitate electronic commerce, initially with computer and software vendors but increasingly with a variety of businesses as Internet and intranet usage becomes more common.
- EC will allow many service industries to sell directly to their end customers, avoiding the need for intermediaries.
- Integrating Web-based EC systems with existing corporate applications will be a major challenge for companies implementing EC. This task will be easier for companies who take advantage of the new EC front ends being offered by the major packaged application vendors as part of their product suites.
- The growth of business-to-consumer EC will be limited by Internet access problems, including both limited local loop bandwidth and congestion in the network backbone, that prevent high-speed access to EC Web sites. These problems will not be addressed meaningfully for most consumers until well into the forecast period.
- Implementations of Version 1 of the SET protocol will be deployed and used by companies that backed the standard such as GTE, IBM, MasterCard, Microsoft, Netscape, and Visa.
- Two issues of concern to the public – how demographic and personal information is gathered and used, and the risks associated with sending confidential information such as credit card numbers over the Internet – will remain the focus of attention during the forecast period.
- By the year 2000, comprehensive Internet/Web system security will be readily available and affordable at the desktop, helping to boost consumer confidence and Internet commerce transactions.

7.1.11 References

■Articles

Carr, Jim. 1997. Users wade through electronic-commerce market. *InfoWorld*. June 23: 75.
Fabris, Peter. 1997. Electronic commerce EC riders. *CIO*. June 15.
Gylinsky, Gene. 1997. Sales are clicking on manufacturing's Internet mart. *Fortune*. July 7.
Halper, Mark. 1997. Meet the new middlemen. *Computerworld Emmerce*. April 28.
Himelstein, Linda, Ellen Neuborne, and Paul M. Eng. 1997. Web advertising starts to click. *Business Week*. October 6: 128.
Hoard, Bruce. 1997. Enterprise infrastructure for electronic commerce. *EC.COM*. June.
Hof, Robert D., Seanne Browder, and Peter Elstrom. 1997. Internet communities. *Business Week*. May 5: 65 -80.
Hoffman, Thomas. 1997. Inside extranets. *Computerworld*. June.
Lohr, Steve. 1997. Beyond consumers, companies pursue business-to-business Net commerce. *The New York Times*. April 28: D1.
Strom, David. 1997. Setting up shop on the Web. Windows Sources. June, V5, No. 6, page 167.
Suited, surfing, and shopping. 1997. *The Economist*. January 25: 59.
Survey finds Internet services growing fast. 1997. Reuters. February 10.
Top 10 e-commerce sites. 1997. *PCWeek*. January 6.
Voight, Joan. 1997. The advertising world is scrambling for new ways to pull dollars out of Web sites. *Wired*. December.
Wilder, Clinton and Marianne McGee Kolbasuk. 1997. The Net pays off. *InformationWeek*. January 27.

■Periodicals, Industry and Technical Reports

Bock, Geoffrey. 1997. *Creating an end-to-end solution for Internet commerce*. Patricia Seybold. June
Burnham, Bill. 1997. *The electronic commerce report*. Piper Jaffray. August.
Burton, B. 1997. *Tracking missing Internet transactions*. Gartner. June 27.
Dolberg, Stan, Donald A. DePalma, Michael Mavretic, and Jesse Johnson. 1997. *Internet commerce software*. Forrester. February.
Elliott, Tom. 1997. *Measuring the payback of Internet commerce*. Giga Information Group. June.
Erwin, Blane, Mary A. Modahl, and Jesse Johnson. 1997. *Sizing intercompany commerce*. Forrester. July.

Guptill, B. 1997. *Defining EC, EMs and other terms.* Gartner. May 23.
Mesher, A. 1997. *EDI and the Internet: with free EDI, who will do the work?* Gartner. June27.
Moore, Cynthia. 1997. Electronic commerce opportunities in vertical markets. Dataquest. May 12.
Natis, Y. 1997. *The reality of Web transaction processing.* Gartner. June 27.
Reilly, B. 1997. *Chaos in the EC software and services market.* Gartner. July 27.
Robertson, Caroline. 1997. *Electronic commerce services: 1996 - 2001.* IDC. February.
Special issue on E-commerce. 1997. *IEEE Internet Computing.* December.
Spieler, G. 1997. *Delivering an electronic commerce product in a competitive market.* Gartner. June 1.
Taylor, D. 1997. *Extranets: their impact on the EC market.* Gartner. March 28.
Taylor, D. 1997. *Kleline: international electronic-payment service.* Gartner. June 9.

■Books

Colberg, Thomas P. 1995. *The Price Waterhouse EDI handbook.* New York: John Wiley & Sons.
Kalakota, Ravi and Andrew B. Whinston. 1996. *Frontiers of electronic commerce.* Menlo Park, Calif.: Addison-Wesley Publishing.
Loshin, Pete. 1995. *Electronic commerce: on-line ordering and digital money.* Rockland, Mass.: Charles River Media.

■Conferences

Kalin, Sari. 1997. "Business-to-business e-commerce standard aims to ensure interoperability." Netscape Developer Conference. June 4.

■URLs of Selected Mentioned Companies

@plan `http://www.@plan.com`
3Com `http://www.3com.com`
AICPA `http://www.aicpa.org`
Amazon.com `http://www.amazon.com`
American Express `http://www.americanexpress.com`
AmeriData `http://www.ameridata.com`
ANSI `http://www.ansi.org/default.htm`
AOL `http://www.aol.com`
Apple `http://www.apple.com`
Bank of America `http://www.bofa.com`
BroadVision `http://www.broadvision.com`
BT `http://www.bt.com`
Cellnet `http://www.cellnet.com`
CertCo `http://www.certco.com`
CertiCom `http://www.certicom.com`
Charles Schwab `http://www.eschwab.com`
CheckFree `http://www.checkfree.com`
Chicago Industrial Communications Association (CICA) `http://www.cica.org`
Cisco (CCO) `http://www.cisco.com`
CommerceWave `http://www.commercewave.com`
Computer Associates `http://www.cai.com`
CyberCash `http://www.cybercash.com`
Dell `http://www.dell.com`
DigiCash `http://www.digicash.com`
Digital `http://www.digital.com`
E*Trade Securities `http://www.etrade.com`
eBay `http://cayman.ebay.com`
eFusion `http://www.efusion.com`
Eicom `http://www.eicom.com`
Expedia `http://www.expedia.com`
FedEx `http://www.fedex.com`
Financial Services Technology Consortium `http://www.fstc.org`
First Virtual Holdings `http://www.firstvirtual.com`
GEIS `http://www.geis.com`
General Electric `http://www.ge.com`
GlobeSet `http://www.globeset.com`
GTE `http://www.gte.com`
Harbinger `http://www.harbinger.com`
Hewlett-Packard `http://www.hp.com`
IBM `http://www.ibm.com`
iCat `http://www.icat.com`
IETF `http://www.ietf.org`
Informix `http://www.informix.com/infmx-cgi/webdriver`
Intel `http://www.intel.com`
Internet Shopping Network `http://www.isn.com`
Internet Travel Network `http://www.itn.com`
Intuit `http://www.intuit.com`
Investor's SuperSite on the Web `http://www.wallstreetcity.com`
Lombard Institutional Brokerage `http://www.lombard.com`

Lotus http://www.lotus.com
MasterCard http://www.mastercard.com
MicroAge http://www.microage.com
Microsoft http://www.microsoft.com/microsoft/htm
Mondex http://www.mondex.com
NACHA http://www.nacha.org
National Science Foundation http://www.nsf.org
Netscape http://www.netscape.com
NeXT http://www.next.com
Novell http://www.novell.com
OECD http://www.oecd.com
One-to-One http://www.one-to-one.com
Onsale http://www.onsale.com
Open Buying on the Internet (OBI) http://www.supplyworks.com/obi
Open Financial Exchange (OFE)
http://www.microsoft.com/industry/finserv/m_online_inv.htm).
Open Market http://www.openmarket.com
Oracle http://www.oracle.com
PeopleSoft http://www.peoplesoft.com
Premenos http://www.premenos.com
RSADSI http://www.rsa.com
SAIC http://www.saic.com
SAP http://www.sap.com
Security First Network Bank http://www.sfnb.com
Silicon Graphics http://www.sgi.com
Slate http://www.slate.com
Sprint http://www.sprint.com
Spyrus http://www.spyrus.com
Sterling Commerce http://www.sterlingcommerce.com
Sybase http://www.sybase.com
TechData http://www.techdata.com
Terisa Systems http://www.terisa.com
The EC Company http://www.eccompany.com
The Wall Street Journal http://www.wsj.com
The WELL http://www.thewell.com
Trade'ex http://www.tradeex.com
Travelocity http://www.travelocity.com
TRUSTe http://www.etrust.org
UPS http://www.ups.com
UUNet http://www.uunet.com
VeriFone http://www.verifone.com
VeriSign http://www.verisign.com
Viaweb http://www.viaweb.com
Virtual Spin http://www.virtualspin.com
Visa http://www.visa.com
Well Fargo Bank http://www.wellsfargo.com
World Wide Web Consortium (W3C) http://www.w3.org

7.2 Data Warehousing

7.2.1 Executive Summary

Access to the accurate and timely information needed for the management of daily operations and long-term strategic planning has become increasingly important in today's global marketplace. Unfortunately, it has not always been easy to identify, access, and retrieve the required information.

In recent years, companies have focused on a series of concepts, with the intent of making data easier to access. The key concepts include Information Centers, decision-support systems, and Executive Information Systems. These concepts have been implemented using a variety of tools and technologies from mainframe file management systems to relational databases on client/server platforms to spreadsheets on PC file servers and individual desktops.

The data warehousing concept aims to integrate the best of the previous concepts and make a quantum leap forward by creating an integrated enterprise architecture based on the need for what used to be called reporting and analysis and is now called on-line analytical processing (OLAP). The recognition of the need for a completely separate architecture to the traditional on-line transaction processing (OLTP) framework (with reporting and analysis created as part of system maintenance and support function) is what makes the data warehousing approach different and more effective than past efforts in reporting and analysis.

Conceptually, the data warehousing approach creates a clearinghouse in which to gather and organize critical business knowledge and information. Companies are combining data warehousing techniques and technologies with classic business analytics (from sales reporting to statistical sampling and other deep, quantitative analyses) to streamline and automate processes that support fact-based decision-making in the same manner that enterprise resource planning (ERP) has transformed the transaction processing capabilities. Companies are leveraging data warehousing concepts, techniques, and technology to facilitate strategic decision-making across the supply chain; from early identification of sales trends to early identification of product failures, from understanding the causes of customer behavior to understanding patterns in vendor activity and competitive analysis.

From a management perspective, data warehousing implies the existence of an oversight program that links development and implementation to the enterprise business strategy and to the enterprise technology strategy. From a development perspective, data warehousing implies an underlying discipline, structure, and organization that ensures people will be able to access and apply business information on a timely, need-to-know basis.

Data warehousing on an enterprise scale is one of the most demanding systems integration and change integration challenges facing organizations today. There are several technology components required for data warehousing (in addition to the technology components upon which data warehousing is dependent). The components required for data warehousing generally are categorized as warehouse generation (getting the data in), data management (storing the data), and information access (getting the data out). Data warehouse generation and management use tools for designing and populating the data warehouse. Data management encompasses the multi-user database engines necessary to store the data. Information access tools help users access and analyze the data in the data warehouse.

Operationally, data warehousing can allow organizations to spend significantly less time and fewer resources on finding, cleaning, and accumulating data to build reports that reconstruct a view of the past. More important, data warehousing delivers strategic business value when it enables management and other workers that require business information to make faster and more effective decisions. The implementation of data warehousing can also facilitate synergistic thinking and establish fact-based dialog about business performance across the enterprise. As a result, organizations put more focus on management and control of key business performance indicators, business processes, business events, and trends.

7.2.2 Rationale for the Data Warehouse

There are several ways that data warehousing can be used successfully by companies:

- To provide business users with a "customer-centric" view of the company's heterogeneous data by helping to integrate data from sales, service, manufacturing and distribution, and other customer-related business systems
- To provide added value to the company's customers by allowing them to access better information when data warehousing is coupled with Internet technology
- To consolidate data about individual customers and provide a repository of all customer contacts for segmentation modeling, customer retention planning, and cross-sales analysis
- To break down barriers between business departments and functions by offering a way to consolidate and reconcile views from multiple "islands of automation," thus providing a look at activities that cross functional and departmental lines
- To provide "macro-level" information views of the company's critical success factors from multiple data sources such as shareholder value drivers and performance measures
- To enhance traditional financial reporting by providing more timely and detailed access to information
- To report in global trends on group results across multidivisional, multinational operating units, including trends or relationships in areas such as merchandising, production planning, and so on

IDC estimates that data warehousing is currently in use at 20 percent of U.S. companies, and that another 33 percent will implement data warehousing by the year 2000. Also, as users within those companies that have data warehouses begin to realize the benefits, data warehouse usage will increase, and companies will demand access to more data. As a result, Sentry Market Research estimates that the raw data in data warehouses will triple in the next 2 years. As company data increases in both volume and complexity, the data warehouse will become an even more critical repository of timely and accurate data for decision-making.

7.2.3 Data Warehousing Overview

For many years, companies have been working to improve access to data for decision-making and analysis. To date, most of these efforts have focused on enhancing or replacing the transaction processing systems that are the entry point of most company data. Now that these operational systems contain vast amounts of data, companies are refocusing their attention on making this information available to end users through efficient organization and access management. The focus and technology has shifted from data input to information availability. Data warehousing is a comprehensive, systematic approach that addresses this challenge for improved information access.

In a data warehouse, a company transforms the data into a more useful resource by summarizing it (grouping it more conveniently for end users), transforming it (putting it into more usable formats), categorizing it (thus providing the ability to "slice and dice" information), and distributing it (dispersing it to appropriate groups to increase its availability and accessibility). Data warehouses always have an explicit time dimension – that is, any data value stored is associated with a particular time. This feature enables historical information to be maintained in the data warehouse. A key factor in designing a successful data warehouse is utilizing a high-level, enterprise-wide data working model.

An enterprise-wide data model serves as the foundation of a company-wide data warehouse by providing the framework for analyzing the performance measures that underlie the business decision-making processes of the enterprise. A concerted effort is needed at the front end of a data warehousing initiative to identify these performance measures as well as the data elements and algorithms needed for support. Then, these requirements must be mapped onto the enterprise-wide data model.

The data is then organized for easy access into categories (such as by subject area). The information in these subject-area warehouses (or data marts) is integrated through metadata. The data warehouse is the mechanism necessary to store and deliver the required data to a company's business users, and it cannot be an effective mechanism without metadata.

Data cleansing is particularly important because the results obtained from a data warehouse are only as good as the data entered in the first place. Many organizations focus closely on their data warehouse implementation projects but fail to give sufficient attention to data cleansing. In addition to affecting the quality of results fundamentally, data cleansing is an ongoing activity throughout the life of the data warehouse and therefore represents both a significant cost factor and a critical success factor. Therefore, organizations should consider appointing a data quality manager to oversee this area of warehouse operation.

Data Warehousing versus Operational Databases

One way to develop a better understanding of data warehousing is to contrast it with the databases that hold most of the data from current business applications. (See Table 7-7.) OLTP describes systems that support day-to-day business operations. They are transactional in nature, with many users concurrently accessing a database to insert, retrieve, update, and delete individual records. In contrast, OLAP describes analytical systems that support business decision-making. OLAP may be ad hoc in nature and require access to data representing the company's performance over an extended period of time. OLAP emerged due to the inability of OLTP systems to deliver ad hoc query support.

OLAP data is lightly to highly summarized, unchanging, and accessed as read-only. OLTP data is at the most detailed level. A query executed twice in an OLAP system will return the same results each time, but an OLTP system may return different results because the underlying data can change.

Another way to contrast the two processing types (and associated databases) is that OLAP helps a company decide what to do, and OLTP helps a company do it. OLTP focuses on efficient operational management, while OLAP can be operational and focuses on value-added activities (speedy decisions, customer service improvements, and revenue increases). OLTP systems capture and store the data needed to run the business on a daily basis, which also can be used to create OLAP systems. The resulting OLAP systems are aggregations, transformations, integrations, and historical collections of OLTP data from one or more systems.

Table 7-7: Contrasts Between OLAP and OLTP

Characteristic	OLAP	OLTP
DATA		
Organization	Subject-oriented	Process-oriented
Scope	System-wide	Single-function
Time horizon	Near current and historical data	Current data only
Breadth	Broad	Narrow
Type of data	Summary and derived data	Detailed operational data
Accuracy	Snapshot of data at time *x*	Accurate at time of access
Volatility	Cannot be updated directly	Can be updated directly
Sources	Internal and external	Internal
Design	Denormalized	Normalized
USERS		
Number	Fewer concurrent users than OLTP	Many concurrent users
Type	Managers and analysts	Operational and administrative staff
USAGE		
Processing	Low volume of large queries	High volume of small transactions
Access	Unstructured; ad hoc	Structured;repetitive
Usage focus	Data-analysis-driven	Transaction-driven
Typical operation	Reports and analyzes	Updates
Response time	Performance is less sensitive	Very performance-sensitive
FOCUS		
Operational focus	Flexibility	Performance and reliability
Business focus	Tactical and strategic; supports managerial needs	Operational and tactical; supports day-to-day operations
Managerial focus	Effectiveness	Efficiency

■ Operational Data Store

One component of the data warehouse process has emerged recently that combines some characteristics of OLAP systems with OLTP systems. This component is generally referred to as an operational data store (ODS).

The ODS involves combining data from one or more operational databases into a repository from which users can access granular information. As such, the ODS is similar in many ways to an OLAP system. However, an ODS also shares several characteristics of an OLTP system: it contains detailed operational data, it contains current data, and it can be accessed by many concurrent users.

The ODS is particularly popular as a means of integrating various OLTP systems and supporting operational querying and reporting while minimizing the impact on the performance of the OLTP system. Most ODSs are built to support business operations such as sales and marketing, customer service, or report generation from transactional systems.

As a component of the data warehousing process, the ODS is being integrated into the data warehouse as a repository for atomic-level detail generated by the extraction process. As such, companies are finding the ODS provides the granular data source and the technology to support data mining applications. In addition, if an ODS is developed, it can be used as the source from which the data warehouse is populated and regenerated over time.

Enabling Technologies

Data warehousing is the latest generation in decision-support systems. It is closely related to and builds upon previous technological approaches such as the Information Center, the Executive Information System (EIS), and improvised spreadsheets.

Many technologies associated with data warehousing are not new. These technologies have been evolving continually and now have reached the point where data warehousing has become a feasible and cost-effective solution for achieving the goals of earlier decision-support systems. Some of these key supporting technologies are desktop computer systems, relational database management systems (RDBMSs), Internet/intranets, extraction/transformation products, and metadata repository tools. Another important development has been the technological advancement (scalability) of important components such as disk storage, hardware, networking technologies, relational databases, and so on.

The low cost of terabyte disk arrays as well as the availability of powerful, easy-to-use, inexpensive desktop computers and software (such as information access tools) are key enablers for the improved data access provided by data warehousing. The graphical user interface (GUI) and functionality associated with today's desktop computers have made them much easier to use. In turn, they can be connected via powerful networks to the company's data warehouse, making the data more accessible than was possible with previous decision-support systems.

The introduction of the RDBMS and its widespread adoption is another key enabling technology for data warehousing. The RDBMS introduced the concept of data independence and standardized data access via Structured Query Language (SQL). Subsequent improvements in the areas of database replication, OLAP capabilities, improved query performance through parallelism and partitioning, and greater standardization and interoperability are also key contributors to data warehousing.

Advances in extraction/transformation and data cleansing products are also assisting with the development and ongoing management and administration of data warehouses. They simplify and automate the enormous task of extracting and formatting the data from operational databases.

■ Multidimensional Databases

Multidimensional database management systems (MDBMSs) are a more recent innovation for data warehousing. Based on relational technology, they store data in "cells" organized by the main dimensions of the data – time, product line, town, and customer name, for example. Unlike a traditional RDBMS, which stores data in two-dimensional (2-D) tables, an MDBMS maintains links between all the dimensions. A three-dimensional (3-D) database can be viewed as a cube, with each edge being a dimension. A four-dimensional (4-D) database adds a measurement of some value (often revenue) for a particular product at a particular customer located in a given town at a specified point in time.

A 4-D (or larger) data structure of this type is known as a "hypercube." Many older products require that the dimensions be identified in advance, and the entire hypercube pre-computed. Access to the data is then very fast, allowing highly interactive analysis by end users. New data is generally uploaded overnight and the hypercube re-computed. Because updates and changes to fixed hypercube databases require re-computation of the entire data set, they are not suited to an environment where the primary dimensions change. Because everything must be computed in advance, storage must be allocated for each cell, whether used or not, and therefore, storage requirements for hypercubes become very large very quickly.

Consider the storage demands of a 4-D example: storing a 4-byte floating-point number (such as revenue) with 12 time intervals (months), 100 products, 200 customers, and 50 towns would require 12 * 100 * 200 * 50 * 4 or 48 MB of storage for the hypercube. Adding a fifth dimension (such as supplier) with 100 values would increase the storage requirement to 4.8 GB. In general, hypercube storage requirements increase as *o(x**n)*, where *n* is the number of dimensions; therefore, large numbers of dimensions are not feasible.

To overcome this problem, a technique known as the "sparse hypercube" was developed. Here, only those regions of the total hypercube that actually contain data are computed. This approach significantly reduces storage requirements, but at the expense of complexity, because the database manager must know which areas contain values, which do not, and how they are related. This technique is common where the hypercube is stored on a server (so-called "server-based OLAP") and is used, for example, in the latest release of Oracle's data warehouse.

Some recent products, however, employ a technique known as the "dynamic hypercube" (also known as a "micro-hypercube"). In this scenario, only those portions of the data that are actually being used are turned into hypercube form. Because the active data set is generally small, this computation can be performed on-the-fly without performance degradation. This technique can be used not only on the server, but also on the client. Data values can be retrieved as simple textual lists using a SQL query and downloaded to a client PC. Analysis of the retrieved data can then be performed locally on a desktop (or laptop) machine without requiring further access to the database. This process not only reduces server load, but allows users to disconnect their machines from the network and still perform full function analyses – an advantage for telecommuting and mobile applications as well as for places where the number of data warehouse users is large. This way of working is often called ROLAP (relational OLAP), and an example of a product that uses this technique is Brio Technology's BrioQuery.

■ MDBMS versus RDBMS

The internal design of a true MDBMS differs from that of a conventional RDBMS. The former is better suited to data warehousing applications and generally offers better performance – sometimes as much as a five- to eight-fold increase.

A standard RDBMS joins tables two at a time (a "pair-wise join"). However, joining three tables (A, B, and C) requires two steps – join A and B to produce an intermediate table T, and then join T and C to produce the result R. Extending this concept shows that joining *n* tables requires (*n-1*) pair-wise joins. However, a true MDBMS can perform what is known as a "star join," typically joining up to eight tables at once, with related performance increases.

In the same way, a conventional RDBMS typically uses a 2-D index structure (normally a B-tree). An MDBMS offers this type of index but also provides star and bitstream indexes, both of which allow direct indexing of multidimensional data. Because the MDBMS uses these indexes to store complex data relationships, the path length between related data values (cells) is shorter, and retrieving values requires fewer table joins, thus provides better performance. The more recent MDBMS systems can determine which type of index is optimal for a given group of data values and create it automatically; the underlying index structure is transparent to the user.

An MDBMS is optimized for data retrieval, and an RDBMS is optimized for fast data updates. Because warehouse data values are read-only, they are not updated, and the requirement for features such as journaling, sophisticated locking, and two-phase commit is eliminated. Removing these features results in significant performance increases.

An MDBMS bulk data loader typically runs at up to 1 to 1.5 million rows per minute, which is considerably faster than an RDBMS loader. Achieving this speed requires optimizations. For example, in most warehouse load cases, referential integrity checking can be safely omitted, and data format conversions are unnecessary because these operations have been performed during the preceding data extraction and cleanup steps.

Although RDBMS data spaces and tables tend to be single entities with fixed size limits, MDBMS data spaces (and often individual tables) can be segmented. This feature offers significant advantages in two areas. First, the size of the database is essentially unlimited (databases in excess of 100 TB are not uncommon) because all that is needed to add more data is the creation of new segments. Second, individual segments can be stored on offline media such as tape or optical storage, significantly reducing disk requirements and therefore overall system cost. Hierarchical storage management techniques are used to manage the data; if an offline segment is required by an application, it is retrieved automatically from tape and placed on-line.

The more ad hoc queries an application performs, the greater the advantage of an MDBMS over a conventional RDBMS. Applications with predominantly fixed queries (such as report generation) typically see little performance gain, although there may well be gains in other areas, such as data loading.

Unfortunately, the product market is confused because all the database vendors active in the data warehouse market claim to have multidimensional databases. Although all indeed can store multidimensional data, most are closely derived from conventional RDBMS products such as IBM's DB2 and Oracle. Very few MDBMS products have been designed as such from the ground up; an example of one that has been is the Red Brick database (`http://www.redbrick.com`).

7.2.4 Data Warehousing Approaches

Data warehousing approaches can range from simple to complex, with many variations in between. There is no one-size-fits-all solution. However, to be successful, a company's data warehousing strategy must be directly aligned with its overall business strategy. Once the enterprise data warehousing strategy has been agreed to, a company can choose a tactical plan that uses one or more of the basic approaches that fit best with its current applications, data, and technology architectures.

There are two major approaches to consider within the overall architecture: the data mart and the enterprise data warehouse. These approaches differ greatly in scale and complexity.

Data Marts

A data mart is a scaled-down version of a data warehouse that focuses on a particular subject area. The data mart is usually designed to support the unique business requirements of a specific department or business process. A company can have many data marts, each focused on a subset of the company. The trend toward greater use of the data mart strategy often is driven by end users and departments that want to build local data marts focused on their specific needs instead of waiting for a company-wide enterprise data warehouse.

There are two major types of data marts: dependent and independent.

- **Dependent data mart** – Contains a subset of the data from the data warehouse, which acts as its authoritative source. The data is fed from the central data warehouse (the hub) to the data marts (the spokes), which are a means of accessing the data – sometimes referred to as "data retailing" or the "hub and spoke" model. The dependent data mart is basically a distribution mechanism for data, and as such, is tightly integrated into the enterprise data warehouse.
- **Independent data mart** – Derives its data directly from transaction-processing systems and operates autonomously. An independent data mart focuses on one subject or problem, and it may not be as easy to integrate with a data warehouse, particularly if an enterprise has multiple, independent data marts.

The movement to data marts is also being accelerated by the availability of lower-priced software and hardware as well as packaged data mart applications. For example, Informatica's PowerMart and Sagent Technology's Data Mart Builder are suites of tools to help a company design and build a data mart.

Because of its reduced scope, a data mart takes less time to build, costs less, and is less complex than an enterprise data warehouse; it is easier to agree on common data definitions for a single subject area than for an entire company. Therefore, a data mart is appropriate when a company needs to improve data access in a targeted area, such as the marketing department. However, the indiscriminate introduction of multiple data marts with no linkage to each other or to an enterprise data warehouse will cause longer-term problems. Because these data marts proliferate quickly, they may give rise to the same "islands of information" problem that plagued earlier initiatives. Isolated data marts are undesirable and should be developed and deployed within the overall data warehousing strategy so that eventually they can evolve into an integrated decision-support environment if business circumstances require.

Data Warehouses

The enterprise data warehouse provides an enterprise-wide view and reconciles the various departmental perspectives into a single, integrated corporate perspective. (Sometimes, the data warehouse can be the hub that feeds data to data marts.) This strategy is the most complex because a data warehouse is generally company-wide in scope, and therefore requires the company to establish a centralized, structured view of all its data.

A data warehouse can be physical or virtual. A physical data warehouse is a central repository where all data is staged after it has been gathered and transformed. A virtual data warehouse contains pointers to the data that is stored in the various business-area data marts, creating a logical view of all the company's data.

The enterprise data warehouse provides a consistent, comprehensive view of the company, with business users employing common terminology and data standards throughout the enterprise. When put together, these two strategies can represent stages in the evolution of a data warehouse. (See Table 7-8.)

Table 7-8: Data Mart versus Data Warehouse

Typical Characteristic	Data Mart	Enterprise Data Warehouse
EFFORT		
Scope	A single subject area	Many subject areas
Time to build	Months	Years
Cost to build	$hundreds of thousands	$millions
Complexity to build	Low to medium	High
DATA		
Requirements for sharing	Shared (within a business area)	Shared (across the company)
Sources	Few operational and external systems; a data warehouse	Multiple operational and external systems
Size	Megabytes to low gigabytes	Gigabytes to up to 50 terabytes
Time horizon	Near-current and historical data	Historical data (usually more than 5 years' worth of data)
Frequency of update	Daily, weekly	Daily, weekly, monthly
Number of instances	Multiple data marts distributed across the company	Single data warehouse
TECHNOLOGY		
Hardware	PCs; work-group servers; small to midsized Windows NT or UNIX servers	Midsized to large UNIX servers; mainframes
Typical operating systems	Windows NT, UNIX, AS/400	UNIX, MVS
Database	Workgroup database	Large database

Table 7-8: Data Mart versus Data Warehouse (Continued)

Typical Characteristic	Data Mart	Enterprise Data Warehouse
USAGE		
Number of concurrent users	Tens	Hundreds or thousands
Type of users	Business area analysts and managers	Knowledge workers, corporate analysts, and senior executives
Business focus	Optimizing activities within the business area	Cross-functional optimization and decision-making

Table 7-8 above shows the main characteristics of and differences between data marts and data warehouses. In practice, few organizations begin by implementing a data warehouse because the cost and time frame are prohibitive. Instead, an organization often ends up with a cooperating collection of data marts forming a virtual data warehouse. Organizations often find that a single large data warehouse is possible but not desirable for a number of reasons, including geography and access to data, differing requirements, scale and scope of the organization's activities, and so on.

Therefore, data marts are implemented most often, sometimes to test the waters and prove the usefulness of technologies and business projects. Such activities are highly beneficial, but it is extremely important for an organization to develop an overall coordinated strategy and policy to prevent the creation of multiple, incompatible data marts.

Whatever route is taken, it is critical to ensure that development, implementation, and deployment are performed iteratively. Each step should be of a short, fixed duration and offer real (measurable) benefits.

7.2.5 Data Warehouse Technologies

Data from operational systems and other external systems are periodically extracted and transformed (cleansed and summarized) by data warehouse generation tools, and then loaded into a data warehouse. The data warehouse is managed through the use of data management tools, and is accessed using information access tools. (See Figure 7-8.)

Data Warehouse Generation

Data warehouse generation is the complex process of extracting data from the operational databases, transforming (and cleansing) the data, moving the data to the server on which the data warehouse is located, loading the data into the data warehouse, and managing the data warehouse. With most data warehouses available for use 16 to 22 hours per day in a read-only mode, there is usually only a 2- to 8-hour period in which to complete these processes (often in the early morning) if daily updates are needed.

Figure 7-8: Creating the Data Warehouse

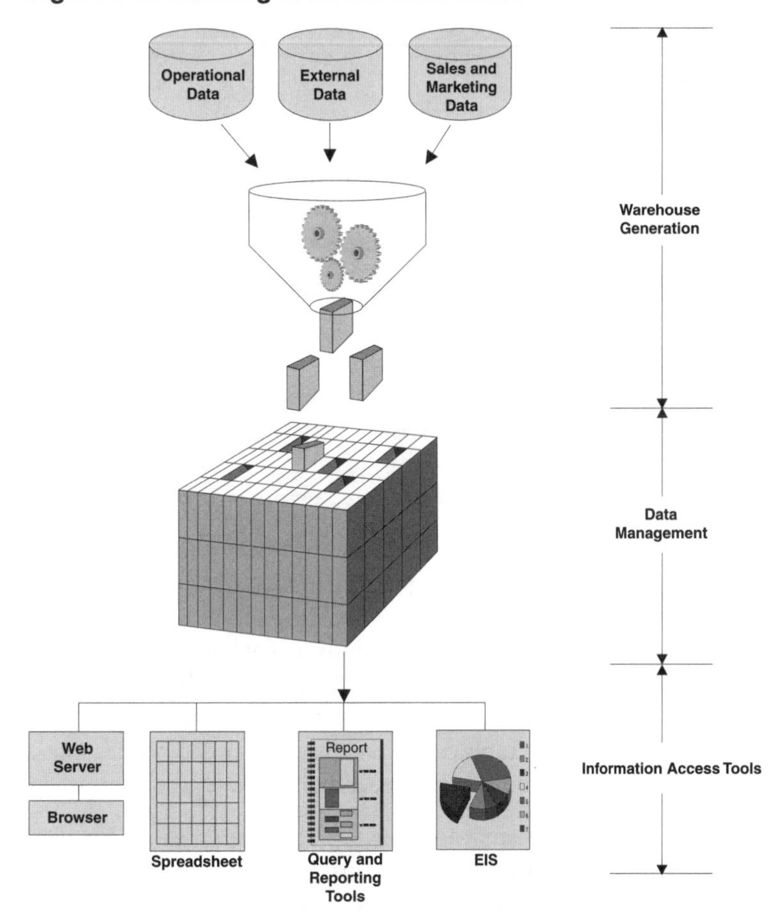

■ Preparing Data for the Data Warehouse

Preparing data to be loaded into a data warehouse involves extraction, transportation, and transformation. Extraction programs are run periodically to extract and collect the relevant data from the operational systems (reading the transactional system data formats and identifying the changed records). Customized programs often perform the data extracts, but shrink-wrapped software tools are now becoming available to assist in this process. These programs specify the source operational system or external system and the extraction criteria.

Database gateways that access different database and file formats and database replication are often used in conjunction with other software to extract data from operational systems. The software transports the extracted data to an intermediate file.

After the data has been transferred to the intermediate file (which can be an ODS), transformation programs or utilities are run to prepare the data for the data warehouse. The transformation process includes the following steps:

- **Consolidating** the data from multiple sources
- **Filtering** the data to eliminate unnecessary details or fields
- **Cleansing** the data to eliminate incorrect or duplicate data
- **Converting and translating** the extracted operational data
- **Aggregating** the data

Consolidating the data involves integrating data from multiple sources (such as internal OLTP applications or external information providers) to create a consolidated view of the data. The records from these various data sources are combined into one file, which is further processed prior to being loaded into the data warehouse.

Filtering the data in the consolidated file involves selecting only the data that the data warehouse needs and identifying any unnecessary data. Unnecessary records are removed from the consolidated file. The remaining records are filtered to remove unnecessary or duplicate attributes.

Cleansing the data is a particularly important and difficult process. Data such as customer or mailing information must be accurate, but it is often difficult to recognize that multiple entries representing the same entity exist in data sources (such as Bill Jones – 123 Willow Street, William Jones – 123 Willow St., and W. Jones – 123 Willow Street). Stale, redundant, or poor quality data must be identified and corrected before it is introduced into the data warehouse.

Sophisticated data-scrubbing tools have emerged to help improve the quality of data before it is put into a data warehouse. For example, Vality Technology's Integrity Data Reengineering Environment and Apertus' Enterprise/Integrator tools use pattern analysis, fuzzy logic, lexical analysis, and statistical matching to identify and consolidate logically redundant data such as names and addresses, resolve value conflicts, and validate the integrated result against business rules. These tools can also help uncover data values that are outside the normal range and can standardize the different legacy file formats and data structures.

Converting and translating the source data involves mapping the source data to the target data structure and converting the source data to the database format. The mapping process invokes rules that convert the values of the data used in the source application into the values used in the decision-support environment. The mapping process accomplishes the following:

- Defines the logical association of data from the source systems to the target database (for example, CUST_N field with the format FIRSTNAME MI LASTNAME in the source data to CUSTOMER field with the format LASTNAME FIRSTNAME MI in the target database)
- Translates column, field, and table names (which are often cryptic technology names) into the business terminology of the target database
- Translates data codes (such as 1 for Male) to more meaningful codes (such as M for Male) or descriptions

The availability of data models (description of a database's organization, often created as entity relationship diagrams) is a key contributor to success in the complex and critical mapping process. Converting the data involves converting the data layout (data type, length, format mask, and so on) of the source system to the target database format. Data conversion engines typically accept SQL data or flat or sequential files as input, and they output data in the appropriate file format for loading into the target database.

Aggregating the data involves sorting and summarizing the data, and creating generalized keys for the aggregates so that they can be easily queried. Data can be aggregated along any dimension, independent of any other dimension. For example, if atomic data exists at the product, store, or day level, users can aggregate it by product category, region, or month. In general, it is possible to create more levels of aggregates than are practical, given storage constraints and the time required to compute these aggregates. The challenge is to determine where aggregates are used frequently or where they are useful to a significant percentage of the user queries. Incorrect summarizations can yield incorrect or misleading results, or impair the efficiency of the data warehouse.

■ Loading Data into the Data Warehouse

After the data has been transformed into the appropriate format, it is loaded into the data warehouse, which includes preparing the data for access. The loading process can be a bulk load (to establish the data warehouse), a trickle load (to replicate changes as they occur in the operational system), or a periodic incremental load (to refresh the database with new snapshots of the operational data on a regular basis). Errors and exceptions from the loading process must also be identified and resolved.

Some representative companies in the data warehouse generation tools market and their products are shown in Table 7-9.

■ Modeling the Data in the Data Warehouse

The data must also be organized properly for easy access. This process includes summarizing the data within hierarchies; in addition, it may include denormalizing the data, time-stamping the data, dimensioning the data, building appropriate indexes, and updating models of the data. The terms multidimensional, star schema, and snowflake modeling are used to describe this modeling technique.

Besides being highly summarized, OLAP data is also denormalized (by including descriptive information in the summary table, typically resulting in fewer but larger tables). This process reduces the number of joins necessary when querying the data to improve performance; in contrast, OLTP data is normalized to improve data integrity for updates. Another major difference in data organization is the use of data partitioning: some OLAP applications use data partitioning (breaking up a big table into several smaller tables) to speed data loading and improve data access; however, OLTP applications rarely use this form of partitioning because of data integrity concerns associated with updating the data.

Data Warehouse Management

Data warehouse management tools help manage operations of the data warehouse throughout its operational life cycle. These operations include managing the metadata, designing the data warehouse, ensuring data quality, and managing the systems.

■ Metadata

A key enabler of effective data warehouse management is metadata. Metadata is known as "data about data." It is a central point of reference for designing, constructing, retrieving, and controlling the warehouse data. The metadata of the data warehouse is like the floor plan for a regular warehouse that identifies how the warehouse is organized, where the goods are kept, and how much is in inventory. Similarly, data warehouse users need to know what data is available, what its source is, where it is, and how to access it. To identify what they want from the data warehouse, users may browse through a data catalog in the same way customers might browse through a store catalog from a retailer such as Land's End.

Table 7-9: Representative Warehouse Generation Vendors

Company	Warehouse Generation Product(s)
Apertus Carleton	MetaCenter is a suite of tools that includes Apertus' Enterprise/Integrator data cleansing and integration software, Carleton's Passport data extraction and transformation product, Intellidex's Warehouse Control Center end user information management, and Software AG's SourcePoint high-performance scheduled data movement.
D2K	D2K introduced its data migration and transformation product, Tapestry, in April 1997. Tapestry uses BEA Systems' Tuxedo messaging middleware to handle the scheduling and execution of processes such as the extraction of data from operational databases and the movement of that data into a warehouse or mart. D2K says that an upgrade, Tapestry 2.0, will begin shipping in early 1998. Improvements in version 2.0 include support for more source databases (including Teradata, Adabase, Rdb, and AS/400 databases) and modules to move data from PeopleSoft or SAP applications into a data warehouse.
Evolutionary Technologies International (ETI)	ETI-EXTRACT Tool Suite automates data collection, conversions, and migrations, and provides metadata management. Master Set and Data Conversion Tool components interactively convert selected data from multiple sources in any file or database format to any other format.
IBM	Visual Warehouse, an application for the quick development of data marts, has been enhanced with data-loading agents that eliminate size constraints and allow it to build and manage terabyte-sized data warehouses. New management capabilities allow users to generate code, set up automated scripts, and monitor the extraction of data from sources to the warehouse. The foundation of Visual Warehouse is Business Views, which defines the mapping of source data to target data and the scheduling of extraction and load processing. The Visual Warehouse solution package includes DataGuide, a metadata tool, Lotus' Approach for analyzing data from the warehouse, IBM Net.Data for integrating data warehouses and the Web, and DB2.
Informatica	PowerMart Suite assists users with designing, deploying, and managing enterprise scalable data marts. It consists of PowerMart Designer, which visually defines data maps and transformations; Repository, which stores map, transformation, and data mart definitions; Server Manager, which configures and monitors the warehousing process; PowerMart Server, which maps and transforms data from the operational system and loads it into target data marts; and PowerCapture, which allows a data mart to be refreshed incrementally, as changes occur in the operational system.
Information Builders	SmartMart is a turnkey solution that integrates tools for designing data marts and data warehouses, transforming and cleaning data, and querying, analyzing, and managing data. In addition to extraction and transformation support, SmartMart components include data modeling, directories, analysis, OLAP, data mining, distribution, and publishing as well as a multidimensional database. Java and OLE DB support have been being added to boost browser access and improve metadata management.
Platinum Technology	InfoRefiner automates the extraction, cleansing, and movement of legacy and mainframe data to client/server databases. InfoHub is an application development and database access tool that provides direct SQL access to non-relational mainframe data from client/server applications. InfoPump provides bidirectional replication of data between heterogeneous databases.
Praxis International	OmniEnterprise is a suite of enterprise information movement and management products that enables companies to move, synchronize, manage, and maintain updated database information locally, remotely, and via the Internet. It includes OmniDirector, for modeling distributed information flow and monitoring production information movement across an enterprise; OmniReplicator, for replicating data on a transaction-by-transaction basis; OmniCopy, which combines cross-table mapping and filtering with networked-routing and automated scheduling; and OmniLoader, for extraction and load synchronization.

Table 7-9: Representative Warehouse Generation Vendors (Continued)

Company	Warehouse Generation Product(s)
Price Waterhouse	Geneva V/T exploits the high data bandwidth, parallel processing, and shared memory/disk architecture of the S/390 platform and MVS operating system to extract, cleanse, and reformat data from operational systems to feed lightly and heavily summarized data warehouses. It also feeds specialized data mining processes from multiple sources.
Prism Solutions	Prism Warehouse Executive provides integrated capabilities for design, construction, and maintenance of data warehouses and data marts. It generates code to extract, integrate, and transform data between a wide variety of source and target databases on mainframe and client/server platforms. The metadata is stored in the Prism Warehouse Directory companion software.
Sagent Technology	The Sagent Data Mart Solution is a family of integrated products for populating, managing, and accessing data marts. It includes Sagent Data Mart Server, Sagent Admin, Sagent Information Studio, Sagent Design Studio, Sagent WebLink, Sagent Analysis, and Sagent Reports.
Vality Technology	Integrity Data Reengineering Environment is a data cleansing environment that reconditions and consolidates large volumes of legacy data through pattern processing, lexical analysis, and statistical matching. It can also uncover data values that stray from their meta labels and business rules and standardize different legacy file formats, data structures, and character and value representations that vary from system to system.

The creation of quality metadata is a key success criterion for data warehousing because it represents the shared understanding of the company's data. Metadata is critical for both business users and technical users because it provides guidance through the data warehouse. (See Figure 7-9 on page 614.)

Technical metadata identifies the following:

- **Where the data comes from** – internal and external data sources, system of record
- **How the data was changed** – data mapping, consolidation procedures, transformation rules, and aggregation policies
- **How the data is organized** – logical structure and content of the data warehouse
- **How the data is stored** – physical structure and content of the data warehouse
- **How the data is mapped** – to a multidimensional database format from a relational database format
- **Data ownership/stewardship** – who owns the data, who is responsible for the data, what is the contact information
- **Security** – who can access what data
- **System information** – date of last update, extract history, archiving criteria, and data usage statistics
- **Scheduling information** – delivery schedules, scheduling details for data extracts, data transformations, and data loading

Business metadata describes the following:

- **What data is available** – catalog of available data
- **Where the data is** – location, access instructions

- **What the data means** – description of the data contents, units of measure, definitions and aliases for the data, details of how the data was derived or calculated
- **How to access the data** – business model of the data, indexes, keys
- **Predefined reports and queries** – for getting the data
- **How current the data is** – when the data was last refreshed

Figure 7-9: Technical and Business Metadata

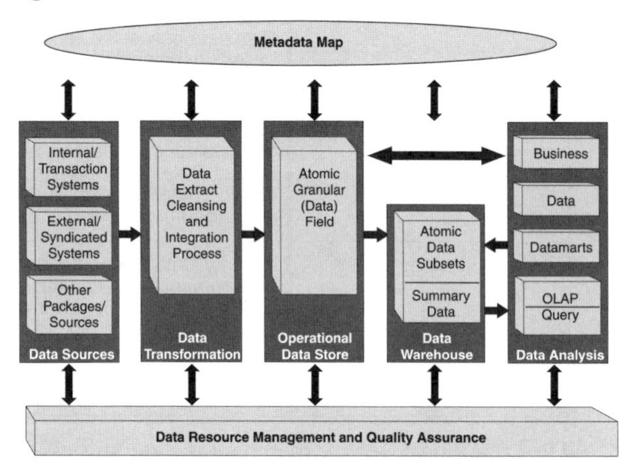

Metadata management tools assist in automatically synchronizing the metadata used by the various components of the data warehouse (warehouse generation, data management, and data access tools). Unfortunately, metadata standards are not yet widely supported, and synchronizing metadata is often restricted to exporting and importing metadata (requiring manual synchronization) between the various components. (See "Data Warehousing Standards" on page 627 for more information about metadata standards.)

Metadata management tools also support versioning, which is important for accessing historical data that may differ in format from the current data format. The repository stores the formats of the data at the time the data was archived.

■ Design Tools

The design process matches the needs of business users to the available data by defining the form and structure needed to create the data warehouse. These tools are used by information technology professionals to define user requirements; identify and qualify source data; design data warehouse data structures and translate them into a physical design; establish data extraction, transformation, and integration policies and procedures; and define mappings between data sources and targets. Examples of design tools include Prism's Warehouse Executive and Xpert Corp.'s PlanXpert for Data Warehouse.

■ Quality Assurance

Data quality and consistency are key issues because the data warehouse is only as good as the data in it. Programming and operational errors can introduce data reliability problems in any of the various steps from extracting data from source systems to producing final results through information access tools. Therefore, a framework for error identification, error correction, and reconciliation must be in place and operational when the data warehouse is created. Data validation and testing tools help monitor and resolve data quality problems by identifying and correcting invalid and inconsistent data.

The data load results must be reviewed to ensure that loads are successful and that the quality of the data meets at least the minimum quality requirements. Often data is evaluated in two stages: quick reviews and in-depth validations.

- **Quick reviews** – Performed soon after the data is loaded to ensure that the correct amount (for example, number of records) of data was loaded and that a few high-level record value totals (for example, total dollar amount) match with the originating applications. These reviews are performed quickly to determine whether data was extracted and loaded properly. If any problems are detected, typically the data needs to be reloaded and possibly re-extracted from the originating application. An error in the source application could also have introduced inconsistencies in the originating data.
- **In-depth validation** – Increasingly done as part of a formal data stewardship program that assigns ownership and accountability for the data. In-depth validation assesses whether the data is complete and valid; whether the relationships between the data elements are correct; whether business rules are followed; whether the transformation process (consolidation, filtering, cleaning, converting and translating, and aggregating) is done correctly; and whether data loads correctly. This process often evaluates specific data element values, identifies data that is outside tolerance levels (where the data values fall outside allowed ranges or do not conform to allowed values for the associated domains), and reviews data that was assigned default values. When errors are identified, they can be fixed manually or automatically before proceeding. The root of these errors should be investigated so that the appropriate changes can be made in the source data systems, the transformation process, or the business procedures.

Often, major benefits of the data warehousing process are achieved in the area of quality of data. Because the data warehouse consolidates and rationalizes data from many sources, it is only after the data is collected from these sources and analyzed that it becomes evident the data content of the source systems are inconsistent or incorrect. Even though it is often painful and costly to rectify these conditions, the end result is more reliable, high-quality information for the enterprise.

■ Systems Management

Systems management includes scheduling warehouse operations (such as warehouse generation and backups), managing performance, ensuring security, and monitoring the data warehouse.

The scheduler program initiates and monitors the processes that extract, transfer, transform, and load the data at scheduled times (usually overnight or over a weekend). Specialized tools to automate more of the process are beginning to appear, including HP's Open Warehouse Management Suite, version 2.0 of Pine Cone Systems' integrated suite of data warehouse and data mart management software solutions, or Software AG's SourcePoint.

For example, HP is repositioning its HP OpenView network-management platform as a console for managing HP-based data warehouses. The HP Open Warehouse Management Suite is a set of tools for managing data warehouse query response times, warehouse availability, and currency of data. Tools include the IT/Operations central management console; PerfView, which examines characteristics such as OLAP and Web-server performance; and MeasureWare, agent software that funnels alarms to the IT/Operations console when performance goals are not met. HP's management platform can monitor databases such as those from Informix, Microsoft, Oracle, and Sybase.

Security is also an important consideration. The data warehouse management tool must protect the data in the warehouse from unauthorized access. Such protection is especially important because a data warehouse by its design makes data easier to locate and access without automatically imposing the access control restrictions found in the original application. Besides security, safeguarding the data also requires facilities for data backup and recovery, and for archiving little-used data.

Performance management tools are important for maintaining high throughput. Such tools help monitor the data warehouse by gathering metrics on workload and identifying bottlenecks. When the use patterns of the data warehouse change, the physical organization and indexing of the data need to be adjusted to keep performance reasonable. Using pre-stored summaries (aggregates) and creating indexes for frequently used data elements are two effective techniques for improving performance; however, aggregation and indexing take time and space, so a constant tradeoff must be made between query performance and maintenance overhead.

Data warehouse monitoring and management tools track and report on data warehouse usage. They help monitor the following:

- How well the data warehouse is functioning (generally through response time)
- Who is using the data warehouse (and who is not), and how often they are using it, to help assess whether the data warehouse is an effective business mechanism
- What data is being accessed (and what data is not), and how often is the data being accessed
- How much it costs to run the data warehouse and how to allocate the costs to users

Data Management

The data warehouse database stores, manages, and stages the data for end-user access. It is the core component of a data warehouse.

A decision-support database can be visualized as a multidimensional data cube. Dimensions are the edges of the cube. They represent the primary "views" of the business data. For example, sales data could be represented as a three-dimensional cube with the dimensions of product, geography (market), and time. (See Figure 7-10.)

Figure 7-10: OLAP as a Multidimensional Data Cube

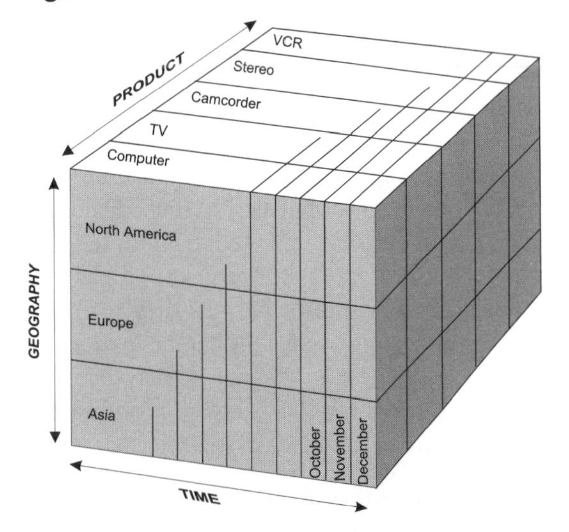

Like a spreadsheet, a multidimensional database stores related data in blocks or cells. In a traditional 2-D spreadsheet, rows might represent products, and the columns might represent time. A 3-D spreadsheet is created by stacking additional sheets for different geographical markets. These additional sheets also have product rows and time columns. A specific block (at the intersection of all the dimensions) in the 3-D spreadsheet (a cube) would be defined as the sales of a specific product, to customers in a specific market, on a certain date (time).

Although visualized as a multidimensional cube, the underlying architecture in an OLAP database is the star schema. As the name suggests, a star schema looks like a star with one large central table and a set of smaller tables linked to it. (See Figure 7-11.)

Figure 7-11: A Typical Dimensional Model

Sales Fact
time_key
product_key
store_key
dollars_sold
units_sold
dollars_cost

Time Dimension
time_key
day_of_week
month
quarter
year
holiday_flag

Product Dimension
product_key
description
brand
category

Store Dimension
store_key
store_name
address
floor_plan_type

The central table is called the "fact" table, and the other tables are called the "dimensional" tables. The fact table is the only table in the schema with multiple connections to other tables. The dimensional tables all have only a single connection attaching them to the central table.

The fact table is where the business measures, events, or transactions such as sales, costs, and similar data are stored. Each measurement is taken at the intersection of a specific value for each of the dimensions. (The foregoing assumes that the data has already been highly summarized. Alternatively, the database might consist of detailed transaction records, resulting in a set of transaction records characterized by the "intersection of the dimensions" instead of "measurement.") In Figure 7-11, the numerical measurements are the number of dollar sales, the number of units sold, and the extended cost.

Dimensional tables contain key values that act as indexes for the fact table and text attributes that define these dimensions. Each attribute helps describe a member of the respective dimension; for example, each record in the product dimension represents a specific product. There are usually many attributes (fields) in a dimensional table that contain information about the dimension, such as a description, brand, or category for the product dimension. Text attributes are most common because they are most useful in describing one of the items in a dimension. Also, they can be used as constraints and row headers by users when querying the database. When these dimensional tables are normalized, the dimensional model becomes a variant of the star schema known as the snowflake schema.

When a multidimensional model is translated into an OLAP database, it is known as a multidimensional database (also known as MOLAP, for Multidimensional OLAP) or an analytical engine (with an associated multidimensional model) that front ends a relational database (commonly referred to as ROLAP, for Relational OLAP). Many traditional EIS vendors, such as Arbor and Pilot Software, are MOLAP vendors; however, most traditional RDBMS vendors have implemented ROLAP or partnered with vendors who do.

MOLAP architectures traditionally have excelled in calculations across rows of data. These calculations are delivered extremely fast, often as a result of precalculation of different aggregations across the cube, which tends to come at the cost of longer load cycle. MOLAP calculations can be complex (for example, matrix arithmetic) and support operations not found in traditional RDBMS products (such as ranking). The variety of proprietary storage structures in the different MOLAP tools are designed to support speed of calculation versus speed of update, sparse or dense data. The better tools provide multiple types of storage structures for these different requirements, including RAM-based and various disk-based cubes. Increasingly, MOLAP vendors are seeking to drill through into RDBMS structures as well for what is called Hybrid OLAP (HOLAP), described later.

ROLAP architectures build multidimensional views dynamically from data stored in RDBMS-based data warehouses. To achieve this goal, ROLAP vendors employ sophisticated SQL generation engines and multidimensionally oriented metadata. The multidimensional processing can be done within the RDBMS, on an application server, or on the client. ROLAP products often are implemented as three-tier, client/server architectures with a database layer, an application logic layer, and a presentation layer distributed across multiple systems interconnected across the network. Three-tier ROLAP products isolate cache results on the middle tier, improving flexibility and scalability.

ROLAP products were originally developed by innovative startups such as Information Advantage, MicroStrategy, and Stanford Technology Group (STG). Increasingly, however, ROLAP capabilities are being added to RDBMS products by the major RDBMS vendors. These vendors are forming partnerships and, in some cases, acquiring MOLAP vendors (such as Informix's acquisition of STG and Oracle's acquisition of IRI Software for its Express family of OLAP products), and are integrating these capabilities into their own products. ROLAP products are often used for more sophisticated data warehousing applications and are predicted by some analysts to become the preferred approach for most data warehousing applications during the next few years due to the enhanced scalability of RDBMS servers compared with pure MOLAP databases. As a result, RDBMSs are now becoming the hub of most data warehousing solutions.

In 1997, another form of OLAP has emerged to bring the ROLAP and MOLAP approaches closer together. Hybrid OLAP is viewed as a way to reduce MOLAP storage requirements as well as the need for multiple ROLAP components to work together. HOLAP products allow direct access to relational data for multidimensional processing and in addition provide their own optimized multidimensional database (for frequently accessed data). An example of a HOLAP product is IBM's DB2 OLAP server, which IBM created by integrating DB2 with Arbor Software's Essbase. Essbase communicates directly with a star schema stored in DB2 to access DB2 data as well as the data in the Essbase multidimensional database.

The key attraction of an OLAP database product is its design. In addition to its support for intuitive navigation of the data, other key benefits of OLAP database products include these:

- The ability to define aggregation hierarchies and interrogate all aggregation levels at any dimensional intersection
- Built-in analytical and computational features such as roll-up (the aggregation hierarchy) and drill-down capabilities
- Improved query performance optimized for complex ad hoc queries due to data pre-aggregation, pre-calculation, and defragmentation
- Direct support for the concept of time through the provision of logic to support aggregations by different time frames (hour, day, week, month, year)

Some representative OLAP data management companies and their products are shown in Table 7-10.

Information Access Tools

Information access tools access and analyze the data in the data warehouse. As the size and amount of data increases, these tools have become more essential. The ease-of-use and capabilities of these tools are key determinants of the user's perception of the value and success of the data warehouse. A good user interface minimizes the number of user actions required to obtain the desired result, and is based on recognizing and pointing, not remembering and typing.

These information access tools can support predefined and ad hoc data access, data analysis, data visualization, and, increasingly, data mining.

■ Predefined Access

Most predefined access is provided through applications that have been written to present views of the data stored in a data warehouse. Fourth-generation languages can be used to develop applications that perform more complex analyses of the data and present the results to analysts and knowledge workers. These applications generally are targeted to the specific needs of individuals or small work groups.

Table 7-10: Representative OLAP Data Management Vendors

Company	Product(s)
Arbor	Essbase Server is a multidimensional database that lets companies simultaneously share, analyze, and update data using an unlimited number of dimensions, such as time, account, region, channel, or product. It includes a set of graphical tools for database management as well as security and administration for deploying OLAP applications.
Gentia	Gentia is a Java-enabled decision-support system that incorporates a distributed networking infrastructure, integrated application development tools, and intelligent agents designed to let end users issue event-driven queries from outside applications such as multi-vendor spreadsheets and decision-support tools. Gentia 3.5, released in October 1997, supports access to data in the Gentia OLAP database from Excel spreadsheets plus drill-down capabilities on Java-based data charts.
IBM	DB2 Universal Database combines OLTP performance, object-relational extensibility, and the optimizer and relational feature set of DB2 Common Server with parallel processing and clustering, query performance, replication support, data warehousing, and very large database support. IBM will add a new OLAP component to DB2 Universal Database in early 1998.
Information Advantage	DecisionSuite Server is an ad hoc ROLAP engine. The multidimensional analysis engine formulates intelligent SQL queries and performs computations on the fly in response to user requests issued from information access tools.
Informix	Informix-MetaCube Analysis Engine is a ROLAP engine that takes advantage of the relational Informix Universal Server, eliminating the need to manage and maintain a separate multidimensional database.
Microsoft	Microsoft SQL Server is a scalable DBMS that features built-in data replication, management tools, and Internet integration. New OLAP capabilities help summarize and aggregate data, making SQL Server 6.5 a suitable platform for data warehousing and query-intensive applications. SQL Server 6.5 also can replicate information to databases including IBM's DB2, Microsoft Access, Oracle, and Sybase.
MicroStrategy	DSS Server provides the core ROLAP engine and multithreaded server functionality necessary to perform sophisticated analyses directly against an RDBMS. It consists of a SQL Query Engine (which translates multidimensional requests into the SQL necessary to retrieve the appropriate data from a relational database) and an OLAP Engine (which stores and manages data requests).
NCR	NCR Teradata is a large-scale, decision-support parallel relational database. It provides a high-performance, scalable database with seamless connectivity and low administration requirements. It employs a patented, shared-nothing massively parallel processing (MPP) architecture that can be deployed on SMP and MPP hardware platforms.
Oracle	Oracle Express Server is an OLAP server based on a multidimensional data model and optimized for query and analysis of company data. It is a key component of the Oracle Universal Server.
Pilot	Decision Support Suite (DSS) 6.0 is a data analysis tool that delivers OLAP and data mining in a hybrid database environment. DSS allows users simultaneous access to both multidimensional and relational databases (such as Oracle7 and Microsoft SQL Server) from DSS applications. Support for other databases will be added in the future.
Platinum Technology	InfoBeacon leverages an advanced three-tier architecture to provide ROLAP solutions. It provides OLAP capabilities for RDBMSs by performing the analytical processing functions and interacting directly with the RDBMS.

Table 7-10: Representative OLAP Data Management Vendors (Continued)

Company	Product(s)
Red Brick Systems	Red Brick Warehouse is an RDBMS designed specifically for data warehouse, data mart, data mining, OLAP, and database marketing applications. Red Brick Warehouse supports very large data warehouse operations.
SAS	SAS Warehouse Administrator is a central console that tracks the movement of data from a database to a data warehouse. Interfaces are available for exception handling, copying, and deleting data, job status viewing, scheduling, and browsing of technical metadata. Support for the creation of warehouse exploitation tools such as DSS, OLAP, and data mining is provided through access to the information catalog, a centralized directory of metadata.
Seagate	Crystal Info 6.0, code-named "Black Widow," was released at the end of 1997. It represents the integration of two existing products – Crystal Info 5.0 and Holos-OLAP – along with other new features. Crystal Info provides the infrastructure for reporting (with features such as page-on-demand processing), and Holos provides multidimensional and relational OLAP processing.
Sybase	Sybase IQ database access software is the centerpiece of the company's overall data warehousing solution. Sybase IQ is a DBMS server designed specifically for decision-support databases and complex data analysis queries. Sybase IQ was designed for interactive data warehousing. Architected around Sybase's patented Bit-Wise query-processing technology, it uses advanced performance-optimized algorithms for complex ad hoc queries. Version 11.2 of Sybase IQ, released in August 1997, includes support for SQL views, an administrative GUI, and additional back-up functions.

Complex and sophisticated queries and reports are often developed by IT professionals and set up as predefined reports that can be run on demand by business users. Increasingly, predefined queries and reports are being published on a company's intranet or via e-mail or groupware packages so that end users can run pre-defined queries or view reports more easily. (See "Trends" on page 630 for more information.)

■ Ad Hoc Access

There are several categories of ad hoc access tools, each of which is suited to different types of end users.

- **Tools for end users requiring occasional access** – Personal productivity software such as the Microsoft Office Suite (Excel, MS Query, and Access) can be used to access the data from a data warehouse. Internet browsers are becoming another popular means of accessing data as information access tools become more Web-enabled.
- **OLAP tools** – These are designed to allow users to navigate through a multi-dimensional environment, typically small cubes generated from large databases. This approach has emerged as a powerful and intuitive way for end users to select, analyze, and present data from a business perspective. Three key operations are aggregation, analysis, and view organization.
  - **Aggregation** – Product, market, time, and other dimensions can be organized by hierarchies. These dimensions consist of units that can be broken down into smaller units (by drilling down) or grouped into large ones (by rolling up). A typical hierarchy for time might be month, quarter, year. Typically, users want to break down a particular dimension into its components (such as sales by region) and explore the data. A business user can move up (roll up) or down (drill down) the hierarchy.

- **Analysis** – OLAP tools are a popular means of analyzing data in a data warehouse. It is especially important for an OLAP tool to have comprehensive arithmetic and statistical capabilities.
- **View organization** – This multidimensional data cube can also be manipulated. Grouping items resembles slicing the cube, which allows a user to view and analyze the cube one slice at a time. The user can expand the analysis by rotating or pivoting the cube to look at the problem or trend from a slightly different perspective or dimension. These capabilities give a user the ability to slice-and-dice along each of the dimensions of the data cube (re-creating various 2-D views of the data) and drill-down through several layers of consolidated data.

- **Tools for end users who access data frequently** – Query and reporting tools (also known as managed query tools or managed query environments, MQE) allow technical users to define and run more sophisticated ad hoc custom reports interactively. These users typically account for 20 percent or less of the population accessing a data warehouse, but may account for up to 80 percent of the queries.
- **Advanced data access and analysis software** – These advanced tools include statistical analysis tools (which are undergoing a rebirth in popularity), geographic information systems (which take data and create a graphical map), and intelligent agents (which gather data according to a profile established and maintained by the user). Users of this type of tool typically account for only a small proportion of the population accessing a data warehouse. This functionality also is being built into many existing tools.

Some representative data access tool companies and their products are listed in Table 7-11.

Table 7-11: Representative Data Access Tool Vendors

Company	Product(s)
Andyne	Andyne GQL (Graphical Query Language) allows users to query, report, export data, and automate activities. GQL Reports allows users to create reports that contain multiple data sources and presentation styles. Andyne Pablo is an end-user OLAP client tool for multidimensional data sources.
Brio	BrioQuery integrates ad hoc query, analysis, 3-D charting, and reporting with dynamic metadata integration and a portable OLAP engine. Its complementary Web-based products include Brio.Insight, which enables interactive OLAP, charting, and reporting within a Web browser; Brio.Quickview, which presents fully formatted BrioQuery reports within a browser; and BrioQuery Server, which processes queries and distributes reports with or without encapsulated data sets to printers, file servers, e-mail, or across the Web.
Business Objects	BusinessObjects combines querying, reporting, and OLAP capabilities in a single package. Intelligent agents inform users about changes in data based on business rules and conditions. A design module provides a graphical, automated way to set up the software mapping that provides end users with a business representation of data stored in corporate databases and data warehouses.
Cognos	Impromptu is a Windows-based SQL query and reporting tool that lets users perform complex queries without detailed knowledge of SQL or database structure. PowerPlay is an OLAP tool that provides corporate business performance data access. Scenario is a decision-support tool that features a neural network-based engine and automates the discovery and ranking of critical factors affecting a business. Scenario, which integrates with Impromptu and PowerPlay, also can find statistically relevant relationships and trends within large business data repositories.

Table 7-11: Representative Data Access Tool Vendors (Continued)

Company	Product(s)
Constellar Hub (formerly The SQL Group)	Constellar Hub (formerly Information Junction) is designed for data warehousing, migrating data between disparate legacy and client/server platforms, and facilitating the flow of data between operational systems in the enterprise, such as between HR and payroll systems. Constellar Hub allows data to be transformed, consolidated, and moved between systems without the need for custom interfaces between data types. Users can maintain a central repository with business rules, metadata, and data transformations.
IBM	The Intelligent Miner toolkit is a comprehensive set of data mining tools for classification, association, sequence discovery, time series, clustering, and regression. IBM supplies multiple technologies for both classification (decision tree and neural net) and clustering (demographic and neural net); most of the algorithms have been parallelized for scalability. Intelligent Miner is packaged with DB2 and operates in DB2, Oracle, Sybase, and flat files.
Integral Solutions	The Clementine data mining toolkit allows users to access data in Informix, Oracle, Sybase, and other databases; display data in a variety of visual, interactive formats; extract decision-making knowledge automatically; and visually program the data.
IQ Software	IQ (Intelligent Query) is a database-independent query and reporting tool that lets business users quickly create columnar, freeform, and matrix reports; mailing labels; and graphs.
Microsoft	MS Query, Access, and Excel are software for PC-literate users to query and report on the data in a data warehouse.
Platinum Technology	Forest & Trees is a decision-support tool featuring built-in charting, which coverts numerical data into business graphics to facilitate trend analysis; cross-tabulation tables; simultaneous access to multiple data sources; and an HTML interface that enables the Web to be used as a data source.
SAS	Enterprise Miner (announced for availability in early 1998) lets users choose from a variety of data mining algorithms. It features statistical, data preparation, and advanced data visualization tools. Enterprise Reporter (also available in early 1998) is a desktop ad hoc query and report tool for SAS data warehouses. System for Information Delivery is an integrated suite of business intelligence tools, including EIS, OLAP, applications development and decision support, and applied analysis, including technical data analysis, data mining, and neural networks.
Seagate – Information Management Group	Crystal Info is a reporting and analysis system that includes an OLAP module; Web components for interactive analysis across Internet and intranets; secure, distributed processing; comprehensive scheduling capabilities; and easy administration and scalability.
Silicon Graphics	MineSet is a set of data mining tools that combines classification and association algorithms with visualization. It integrates data mining analytic tools with high-end visualization tools for user exploration and navigation of data sets and mining results.
Visible Decisions	In3D is a comprehensive toolset for constructing interactive 3-D applications. It is composed of a high-level, cross-platform C++ library and an interactive builder. Intuitive navigation paradigms allow the user to rotate, zoom, pan, and explore information using the mouse and keyboard. Applications built using In3D can be deployed on the desktop, in a Web browser, or on a server.

■ Data Visualization

Recently, data visualization, a technique long used by scientists and statisticians to represent masses of data visually, has begun to be applied to corporate data. Visualization tools have been used to develop applications that graphically display data, highlight relationships between data, and give an overall qualitative feel of the data. Data visualization involves representing data in graphic forms to uncover new insights, as suggested by the expression, "a picture is worth a thousand words."

Data visualization provides an intuitive way to comprehend large quantities of complex information quickly and easily. Data visualization involves creating 2-D or 3-D pictures of the data. Approaches can range from simple graphical presentations to sophisticated, 3-D renderings with interactive animation. By interactively changing the underlying data using a 2-D or 3-D interface and then observing the corresponding changes in the visual representation, users can gain a new understanding of the data.

Data visualization tools are available from a wide range of vendors, who can be separated into three groups: those offering tools that create static charts (such as Microsoft and Seagate); those offering interaction and playback capabilities (such as Advanced Visual Systems, IBM, and Visual Numerics); and those offering interactive animation via a live data feed with continuously changing data (such as Silicon Graphics and VDI).

For example, Silicon Graphics' MineSet 2.0 (introduced in August 1997) is an integrated suite of high-end analytical and visual data mining tools that includes a facility for launching visualization from Web browsers. VDI builds customized visualization applications from its Discovery for Developers toolkit.

Other offerings in this area include Business Objects' BusinessMiner 4.0 (introduced in February 1997), which features a large selection of data visualization graphs; Cognos' Scenario (introduced in April 1997), which offers "at-a-glance" data visualization using an innovative interface; and Speedware's Esperant 4.0 (introduced in June 1997), a query and reporting tool that includes functions such as data pivoting and data visualization.

■ Data Mining

Data mining provides a means of extracting previously unknown, actionable information from the growing base of accessible data in data warehouses. Data mining users typically are few in number because of the degree of sophistication and knowledge that is required and the relatively specialized nature of this work.

Data mining tools use sophisticated, automated algorithms to discover hidden patterns, correlations, and relationships among the millions of data elements collected by an organization. There are currently four general classes of data mining algorithms:

- **Neural networks** – Designed to take a pattern of data and generalize from it. A neural network improves its performance on a particular task by trial and error (it "learns").
- **Decision tree analysis** – A graphical representation of the various parameters or alternatives involved in a decision-making process. Unlike neural networks, which do not explain how data is categorized, decision trees provide rules that define the relationships discovered in the data.

- **Clustering analysis** – An analytical technique often used in data mining that identifies "clusters" or groups of closely related data. For example, a clustering analysis may identify 20 clusters, with each cluster described by hundreds of different attributes. Using information visualization, these clusters and their properties can be visualized to enhance interpretation of the data.

- **Association rules analysis** – Automatically analyzes data, revealing the nature and frequency of relationships between data types. The resulting associations define a predictability model based on observed behavior that can be used to understand trends and forecast future results. Analyzing these association rules provides greater insight into the nature of a particular data set and reveals the quantity and relative strengths of relationships between elements.

Some tools offer one or two data mining algorithms, and others offer a full suite of data mining capabilities. Classical statistical methods such as regression models also are widely used.

Although the mathematical algorithms underlying most data mining tools have been in use for more than a decade, it is only in the past few years that these algorithms have been packaged in user-friendly tools. Advances in technology and more accessible data (through relational databases and data warehousing) have encouraged the development of these products.

More powerful hardware such as symmetric multiprocessing servers and massively parallel processors as well as advances in artificial intelligence (AI) and neural network technology are making data mining a valuable extension to many data warehousing initiatives. Data mining is an area where active research is ongoing to bring together statistical analysis and AI techniques. Many AI software products, both old and new, are marketed as data mining tools. (AI refers to the ability of a program to make judgments based on incomplete data and experience.) Machine learning – that is, learning from experience and generalizing based on the data presented to the computer – is also central to data mining.

Although predefined and ad hoc access tools provide top-down, query-driven data analysis, data mining provides bottom-up, discovery-driven data analysis (also known as "knowledge discovery"). The predefined and ad hoc access tools allow users to test their theories or hypotheses by exploring the data and allow routine operational data to be collected from the data warehouse.

In contrast, data mining identifies facts or conclusions based on sifting through the data to discover patterns or anomalies. Data mining tools typically access more granular data than other information access tools.

The Gartner Group defines five main functions that data mining can perform: classification, clustering, association, sequencing, and forecasting.

- **Classification** – Infers the defining characteristics of a certain group (for example, customers who have been lost to competitors)

- **Clustering** – Identifies groups of items that share a particular characteristic. Differs from classification in that no pre-defining characteristic is given.

- **Association** – Identifies relationships between events that occur at one time (for example, the contents of a shopping basket)

- **Sequencing** – Similar to association, except that the relationship exists over a period of time (for example, repeat visits to a supermarket or use of a financial planning product)

- **Forecasting** – Estimates future values based on patterns within large sets of data (for example, demand forecasting)

Data mining complements predefined and ad hoc access tools by enabling users to discover new relationships they may have overlooked previously. However, a user's "gut feeling" is often not a reliable indicator of where to look for patterns.

Some representative data mining tool vendors and their products are listed in Table 7-12.

Table 7-12: Representative Data Mining Tool Vendors

Company	Product(s)
Angoss	Version 5 of KnowledgeSeeker features support for DCOM, enabling users to move into distributed computing environments, and is accessible from a Web browser via a plug-in. It will also accept data from all major databases as well as from popular query and reporting, spreadsheet, statistical, and OLAP and ROLAP tools. Angoss' KnowledgeStudio data mining software allows users to mine large databases in place without reading records out using ODBC or other protocols by using a more efficient "query wave" technique to extract desired information. KnowledgeStudio also can be embedded in other applications and can work with programming languages such as Visual Basic.
Business Objects	Based on intuitive decision tree technology, BusinessMiner provides hierarchies that graphically depict relationships in data. BusinessMiner provides all the classical data mining functions, including modeling, discovery, visualization, what-if, and segmentation.
DataMind	Version 2.0 of the DataCruncher data mining tool gives users direct access to databases from Oracle and Informix and includes up to four business templates for use in up to 75 data mining tasks. The templates have an underlying engine that guides users through data mining analysis for specific business situations.
IBM	The Intelligent Miner toolkit is a comprehensive set of data mining tools for classification, association, and sequence discovery, time series, clustering, and regression. IBM supplies multiple technologies for both classification (decision tree and neural net) and clustering (demographic and neural net); most of the algorithms have been parallelized for scalability. Intelligent Miner is packaged with DB2, and can import data from DB2, Oracle, and Sybase databases as well as flat files.
Integral Solutions	The Clementine data mining toolkit allows users to access data in Informix, Oracle, Sybase, and other databases; display data in a variety of visual, interactive formats; extract decision-making knowledge automatically; and visually program the data.
Magnify	Magnify's Pattern data mining system integrates a data warehouse, a toolkit for data mining, a toolkit for predictive modeling, and industry-specific vertical applications to support the entire data mining process from cleansing and ingesting data to decision-making.
Pilot Software	Part of Pilot's Decision Support Suite, Pilot Discovery Server is a data mining product that uses customer metrics such as profitability, lifetime value, or new product return on investment to drive a focused market segmentation and proactive analysis of customer behavior. Working directly with and residing in a relational data warehouse, Pilot DiscoveryServer issues standard SQL queries for decision-tree analysis.

Table 7-12: Representative Data Mining Tool Vendors (Continued)

Company	Product(s)
SAS	Enterprise Miner (announced for availability in early 1998) lets users choose from a variety of data mining algorithms. It features statistical, data preparation, and advanced data visualization tools. Enterprise Reporter (also available in early 1998) is a desktop ad hoc query and report tool for SAS data warehouses. System for Information Delivery is an integrated suite of business intelligence tools (including EIS, OLAP, applications development and decision support) and applied analysis (including technical data analysis, data mining, and neural networks).
Silicon Graphics	MineSet is a set of data mining tools that combines classification and association algorithms with visualization. It integrates data mining analytic tools with high-end visualization tools for user exploration and navigation of data sets and mining results.
Thinking Machines	Darwin 2.0 lets users export models in Java, C, and C++, enabling them to do data mining over the Web, integrate data mining results in custom applications, and embed them in SQL queries. Rather than creating statistical printouts of the results, Darwin offers on-line visualization for end users. Features can include the ability to display n-dimensional graphics, dynamically rotate graphs so they can be viewed from many different perspectives, and show related 3-D graphs on screens.

7.2.6 Data Warehousing Standards

Today, there is a great deal of inconsistency in defining, storing, and managing metadata. Each data warehousing product has its own metadata approach, so companies must develop their own metadata strategies to ensure effective integration. Standards are needed to reduce the complexity, give companies more choice, improve access to data, and lower data warehousing deployment costs. Because many groups are advocating a broad selection of emerging standards to the vendor and user communities, companies should be careful when deciding which standards to adopt for their data warehousing solutions.

Metadata Standards

Because a widely accepted metadata standard does not yet exist, every data warehouse tool vendor handles metadata differently. A metadata standard will enable metadata to be shared among various warehouse generation, data management, and information access tools from different vendors so that products can be combined more readily into a data warehouse solution. For example, metadata standards would enable information access tools to learn about the structure and meaning of the data in an OLAP database automatically by reading the metadata, instead of requiring configuration.

■ Metadata Interchange Specification

One current leader in the metadata standards development process is the Metadata Coalition, a group of more than 50 data warehousing vendors and users that includes companies such as Business Objects, ETI, IBM, Platinum Technology, Price Waterhouse, Prism Solutions, R&O, and SAS. The Coalition made some early progress in facilitating the exchanging of metadata between the various data warehouse tool vendors with its Metadata Interchange Specification (MDIS).

MDIS Version 1.0 was approved by the Coalition and released in July 1996. MDIS 1.0 provides a model to address the main types of commonly shared metadata objects and a standard import/export mechanism that enables the exchange of these metadata objects between tools. The metadata objects include schema, record, data element, dimension, level, and the relationship between a given object and any other(s). Each object type has a standard naming syntax that creates a unique identifier for each individual metadata object. Any proprietary or tool-specific metadata is treated as an attribute that is carried along with the metadata object to which it belongs. To comply with MDIS 1.0, a tool's metadata import and export mechanisms must conform to a standardized read and write procedure that uses this standard naming syntax and carries along the proprietary metadata as required.

As of July 1997, Apertus Carleton, ETI, Intellidex, IBM, and R&O Inc. had announced their intent to support MDIS 1.0; no products have been announced to date.

■ Microsoft's Open Information Model

The relational database vendors are also attempting to establish de facto standards for metadata. Microsoft is positioning its Repository as a focal point for its data warehousing initiative. The Microsoft Repository consists of the Open Information Model (OIM), a set of published Common Object Model (COM) interfaces, and a metadata repository built on top of SQL Server. Any third-party tool that writes to the OIM and COM interfaces will be able to input metadata into the Repository or export from it.

Although its Repository strategy is centered around Windows NT, Microsoft has also developed a partnership with Platinum to broaden the applicability of this standard. Platinum will port the Microsoft Repository to non-Microsoft environments, such as UNIX and IBM's OS/400 and MVS operating systems, and provide support for other database products.

The Metadata Coalition is now working cooperatively with Microsoft to review extensions to the Microsoft Repository to extend component interoperability. The Coalition announced in October 1997 that it also will provide a translator for conversions of metadata between the Coalition's MDIS and the Microsoft Repository.

■ Metadata Exchange Architecture

A proposed metadata standard, called the Metadata Exchange (MX) Architecture, could provide needed standardization. Developed by Informatica, MX lets vendors create links between their data access and query and reporting tools and Informatica's data warehouse. MX will let users choose the front end they want to use to tap into the information in a data mart.

Informatica claims that the MX architecture is dynamic, rather than "programmatic," making it easier to implement changes. MX also offers access to metadata via a visual Web browser such as Netscape Navigator or Internet Explorer. Informatica published the Metadata Exchange Architecture API in February 1997, enabling third-party developers of querying and reporting tools to build in compatibility. MX has already been implemented by suppliers of OLAP software such as Brio, Business Objects, and MicroStrategy. Andyne, Cognos, Information Advantage, Infospace, and IQ Software have committed to completing implementations by the end of 1997.

■ Oracle's Warehouse Technology Initiative

Although most of the relational database vendors do not have as much influence as Microsoft, Oracle is also a contender to establish a de facto metadata standard. Oracle established the Warehouse Technology Initiative (WTI) partnership program (part of its Oracle Warehouse strategy). The initiative helps data warehouse product vendors integrate their technologies with Oracle's database to offer business tools that can design, develop, deploy, and manage data marts and data warehouses. The relationship also enables warehouse generation and information access tools to exchange metadata with Oracle's data management solutions such as Oracle's Datamart Suites.

Initiative participants include Apertus Carleton, Business Objects, Comshare, Evolutionary Technologies, Information Builders, Information Discovery, IQ Software, KPMG, Price Waterhouse, Prism, Saxe, SPSS, Thinking Machines, Virtual Integration Technology, and VMark Software.

■ Other Metadata Standards Initiatives

Some vendors, such as Apertus Carleton and Prism, are publishing APIs to promote third-party development using their products. Apertus Carleton's MetaCenter is an integrated data warehousing toolset (actually, a metadata repository) developed with vendors such as Intellidex and Software AG. IBM is packaging Passport, a component of the MetaCenter, with the AS/400 to enhance its data warehousing capabilities. Prism introduced its Open Metadata Architecture in 1996, and many of the same vendors now partnering with Informatica have already built interfaces to Prism's metadata.

Other vendors are forming joint ventures to ensure a seamless exchange of metadata. For example, to increase interoperability, IBM and Evolutionary Technologies have agreed to modify their Data Guide metadata browser and extraction/transformation tool suite, respectively.

OLAP Data Access Standards

An OLAP data access standard promotes interoperability through conformance to a specifically designed front-end architecture. The ultimate goal is transparent access between any client and any server through an OLAP database API. Thus, information access tools and warehouse generation tools would be able to access and put data into any OLAP data source easily.

The cross-industry standards initiative in the OLAP area is facing competition from vendor-specific approaches. The current leaders in the OLAP data access standards development process are Microsoft and the OLAP Council.

■ Microsoft's OLE DB for OLAP

Microsoft has developed and published a set of APIs to access and manipulate OLAP data stored in a database. These APIs are called OLE DB for OLAP. Any OLAP client product that uses OLE DB for OLAP can access any OLAP data source that supports these APIs. OLE DB is part of Microsoft's universal data access strategy and provides a single method for accessing a wide range of data sources. Given Microsoft's influence in the marketplace and its earlier success with the Open Database Connectivity (ODBC) standard, OLE DB for OLAP is likely to have a significant influence on the data warehousing marketplace.

Microsoft formed the Microsoft Alliance for Data Warehousing in late 1996 with eight other vendors – Business Objects, Execusoft, Informatica, Pilot, Platinum, Praxis, SAP, and Teradata/NCR – to make it easier for customers to build Windows NT-based data marts and data warehouses and to get input and support for its OLE DB for OLAP standard from other vendors. To integrate tools and products from alliance members, Microsoft created the Active Framework for Data Warehousing, which provides a common architecture and interfaces for third parties. The Active Framework is a set of COM objects and products that provide for metadata storage, data transformation, data movement, and a common administration tool.

■ OLAP Council's Multidimensional API

The OLAP Council's Multidimensional API (MD-API) was published in late 1996. Although some vendors have implemented this standard, most have adopted a wait-and-see attitude. This hesitation has provided an opening for Microsoft and others to exploit, further complicating and fragmenting the standards efforts in this area. The founding members of the OLAP Council include Arbor Software, Comshare, IRI Software, Oracle, and Pilot Software. General Council members include Business Objects, Cognos, IBM, IQ Software, Kenan Technologies, Management Science Associates, Planning Sciences International, Ltd., Platinum Technology, Inc., and TM1 Software Corp.

In August 1997, Infospace released SpaceOLAP 1.0, an OLAP application for the Web and corporate intranets that features native access to Arbor's Essbase and Oracle's Express as well as to Gentia Software's Gentia through the first implementation of the OLAP Council's MD-API.

7.2.7 Trends

Recent major trends in data warehousing from the customer's perspective include linking the data warehouse to the Internet/corporate intranets and buying data warehousing solutions as opposed to building them. Trends in the data warehousing vendor community include vendor consolidation and product integration and the incorporation of data warehousing into knowledge management.

Data Warehousing and the Internet/Intranets

From an IT perspective, the Internet provides new ways of sourcing and delivering information – both internally and externally, to the company and throughout the world. The Web (internally in the form of intranets, and externally in the form of extranets or the Internet itself) is a key way to raise the visibility of a data warehouse and to extend its benefits beyond the company to customers, vendors, and so on. Using standard Web browsers and an Internet connection, IT can roll out a global, distributed data warehousing solution to users inside or outside the company through the existing network infrastructure without installing specialized client software or significantly increasing support and licensing costs.

The impact of the Internet is currently most pronounced on information access tools. Information access tools are being linked to the Web to various degrees – from making them Web-accessible, to Web-enabled, to Web-exploited.

■ Web-Accessible Tools

Making tools Web-accessible means using a Web server to provide basic file distribution services. (See Figure 7-12.) Companies save or convert reports and documents as HTML files and store them on the Web server. Users then employ Web browsers to access these files. These Web-accessible tools are ideal for delivering predefined information to internal and external users. The main drawback to their use is the lack of flexibility to customize reports or perform ad hoc queries.

Figure 7-12: Web-Accessible Architecture

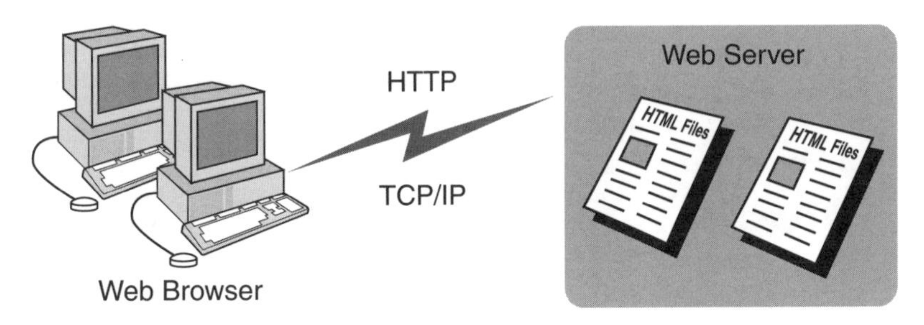

Source: Patricia Seybold Group, 1997

■ Web-Enabled Tools

Making access tools Web-enabled means supporting dynamic HTML publishing so that HTML documents can be created on-the-fly in response to user requests. Web-enabled information access tools let end users interact with the database by filling out HTML forms on a Web browser and submitting them to the Web server to be run by the information access tool as queries against the data warehouse. In this approach, the access tool is converted into an application server, which is accessed via the Web.

Web-enabled information access tools implement a four-tier architecture consisting of Web browsers, Web servers, applications servers, and databases. (See Figure 7-13.) Web servers transfer user requests to an application server via a gateway. The application server translates HTML requests into calls or SQL statements and connects to the database where requests are processed. The application server then packages up the result sets (including performing any necessary calculations) and returns them to the Web server in HTML format. The Web server forwards the results to the clients.

Figure 7-13: Web-Enabled Architecture

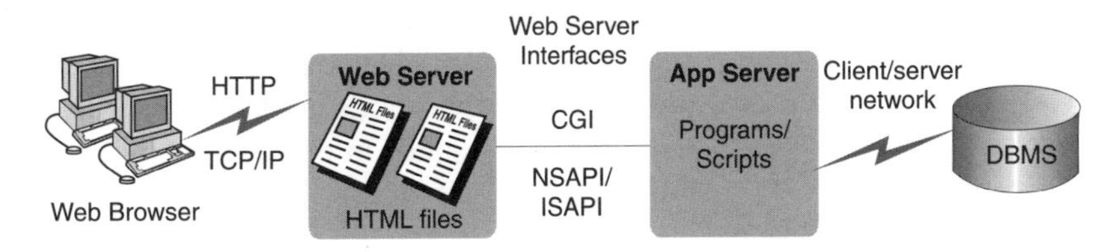

Source: Patricia Seybold Group, 1997

This approach gives end users some flexibility to submit ad hoc queries, but as yet cannot be used to handle advanced queries or create polished reports effectively. In general, these tools do not provide the enhanced flexibility or features that most mature decision-support products currently offer.

- **Web-Exploited Tools**

The most advanced forms of information access tools are Web-exploited. These architectures are designed, maintained, and executed entirely on the Web using Java code. (See Figure 7-14.) The tools employ a three-tier architecture in which processing can be partitioned among Java applets downloaded to the client, a Java applications server, and a database. In this approach, the information access tool itself is split between a Java applet (running on the client) and the server (unlike the previous Web-enabled approach where the tool is completely server-based).

Figure 7-14: Web-Exploited Architecture

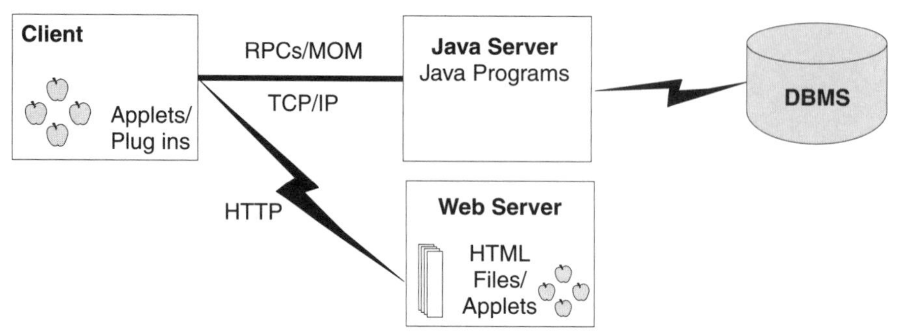

Source: Patricia Seybold Group, 1997

The Web server stores and downloads Java applets to the client (functioning much like a file server). The Java applets and Java server communicate directly using remote procedure calls (RPCs) or message-oriented middleware (MOM). Unlike Web-accessible tools, which are constrained by the HTTP protocol, these RPCs or middleware are geared to support long-running, repeated, or complex queries or transactions. The Java server then interacts with the database using native database drivers with Java wrappers or the Java Database Connectivity (JDBC) interface (a version of Microsoft's ODBC interface).

The Java server typically outputs data in an application-specific format for display within a Java window in the client browser. It could also output the data in HTML or other formats.

Web-exploited information access tools are just emerging and will continue to become easier to use, more extensible, and more flexible, rivaling the look and feel and range of functionality of client/server products available today.

Packaged Data Warehousing Applications

Although the focus to date in the data warehousing market has been on development tools, the transition to packaged data warehousing applications is beginning. IDC predicts that 1998 will see the emergence of the packaged data warehousing applications market. This would be a continuation of the general trend that has resulted in most companies buying rather than building their OLTP applications today.

The trend toward packaged data warehousing applications is gathering momentum. Companies are seeking to jump-start their data warehousing efforts by buying data warehousing applications to reduce effort and costs.

Warehouse capabilities are also being embedded in cross-industry applications and vertical industry packaged solutions. Cross-industry packaged applications are incorporating packaged data marts or data warehouses that generally need only limited customization.

Vertical industry packaged data warehouse solutions have emerged from several vendors, including Comshare (retail), Platinum (insurance), and SAS (pharmaceuticals). These solutions usually are based on prebuilt data models for a specific industry and are bundled with consulting to tailor the solution to a company's specific needs. The availability of these prebuilt, vertical-specific data models eliminates many of the problems associated with data preparation and data quality.

Integration of Data Warehousing and Enterprise Resource Planning

A Forrester Research survey of 50 Fortune 1000 companies confirmed the trend toward integrating data warehousing with enterprise resource planning (ERP) applications: almost 75 percent of the executives interviewed put data warehousing with ERP corporate applications high on their priority list for the next 3 years.

These prebuilt OLAP applications sit on top of OLTP applications. Who better to understand and package the data than the vendors who designed and built the applications? (See Chapter 7.3, *Corporate Applications.*) For example, SAP is positioning itself to deliver a data warehousing solution integrated with its ERP software. SAP is evolving its Open Information Warehouse (an integration framework) into the Business Information Warehouse (a data warehouse solution) by embedding data warehousing capabilities into its R/3 business process software suite. (See Figure 7-15.) Beta testing began in October 1997. SAP expects to release BIW in early 1998 with R/3 4.0.

Figure 7-15: SAP Business Information Warehouse

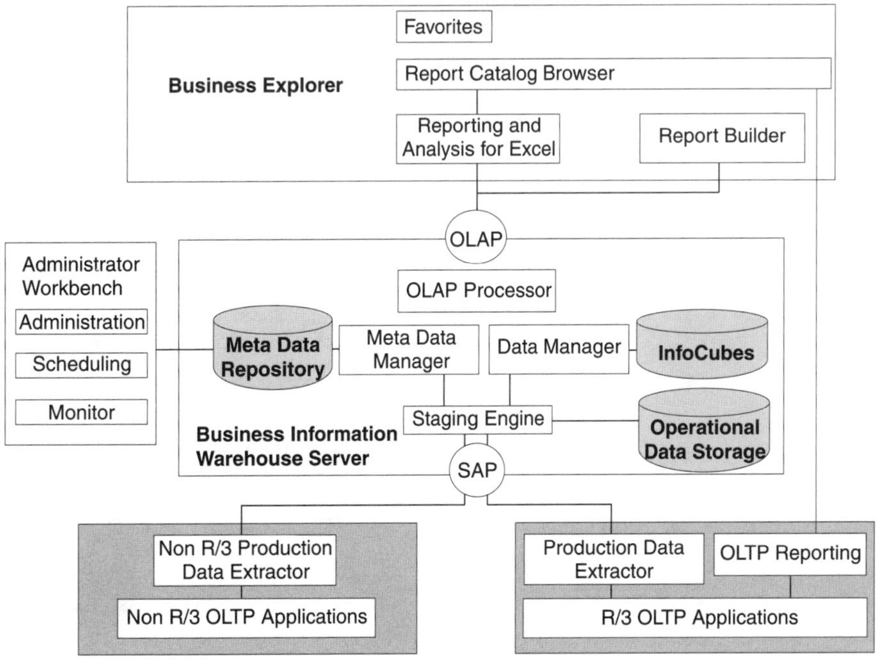

Source: SAP, 1997

The SAP solution includes the following components:

- **The Business Information Warehouse Server** – An on-line OLAP engine and a metadata repository, both of which are pre-configured with business content.
- **The Business Explorer** – A 3-D-oriented access interface. The Business Explorer's navigation capability allows companies to build a customized catalog of reports easily for ongoing or recurring queries and reports, displaying data using Microsoft Excel.
- **Automated data extraction and loading capabilities** – Supplies the Business Information Warehouse Server with data from R/3 applications, R/2 applications, and any other data source. This feature is the major advantage of the ERP with integrated data warehousing approach.
- **The Administrator Workbench** – Provides a single point of control for creating, monitoring, and easily maintaining the complete data warehouse environment.
- **Open BAPIs** – Link external data sources and applications such as third-party information access tools.

Some highlights of SAP's data warehouse include these:

- Built as a separate component linked to an SAP OLTP application by SAP's Application Link Enabling (ALE) technology
- Pre-configured with business processes included in R/3
- Can be deployed as a central repository or at multiple locations
- Contains a staging layer for automated data extraction and loading into the data warehouse from R/3 or other applications

SAP estimates that this embedded functionality will cut the time needed to deploy a data warehouse by 60 percent or more. Other benefits of a packaged data warehousing solution include avoiding the need to build custom interfaces between the repository and the applications, and less exposure to changes in underlying application data structures.

There is a downside to selecting a packaged data warehousing solution, however. Application-specific warehousing may prove somewhat inflexible, rather than providing the ability to analyze data from a variety of production systems. On the other hand, the tight link between the data warehouse and the packaged application (such as SAP R/3) is precisely what makes this choice attractive to some users.

Some application package vendors generally support the data warehouse generation process through their own data extraction, transformation, and transport technology, taking advantage of the tight links to business rules, metadata, and transaction events, and their detailed knowledge of the application data. These offerings are then coupled with leading data management and information access tools through partnerships.

Application package vendors such as Geac (with its Enterprise Warehouse), Lawson (which incorporates Arbor's Essbase multidimensional server into its applications software), Oracle (with its Application Data Warehouse), and PeopleSoft (partnering with Arbor and Cognos) are also integrating data warehouse functionality into add-on packages for their core products.

Vendor Consolidation and Product Integration

Vendor consolidation in the data warehousing market is occurring as established vendors in other market segments enter the data warehousing market (such as Microsoft through its purchase of Israel-based Panorama Software) or acquire new products (such as Informix's purchase of MetaCube and Oracle's purchase of IRI's Express product). Within the information access tools market, examples of vendor consolidation include Cognos acquiring Right Information Systems (with its neural-network technology) and Interweave (a developer of Web-based information access solutions), and Speedware acquiring the Esperant Division of Software AG.

One of the most active areas of vendor consolidation is warehouse generation. Platinum Technology has been particularly aggressive. Among its acquisitions in the last few years are Paradigm Systems (modeling tools), RelTech Group (repository), Software Interfaces (data conversion utilities), and Trinzic (data movement, data analysis, and transformation).

Niche vendors within the data warehousing market are actively forming alliances with the major database providers (for example, Sagent's alliance with Oracle) or with other leading data warehousing vendors to remain competitive. However, these niche vendors must also focus on a market niche and on meeting specific needs, such as those of a particular industry, or focus on a specific data warehousing market segment, such as data marts. Niche vendors must provide unique functionality because the established vendors continue to extend their range of products within the data warehousing market; for example, Oracle recently introduced Discoverer, an information access tool, and Sybase has introduced dbQueue software to query databases and integrate applications over an intranet or the Internet.

■ Category Integration

Integration of various data warehousing products is another trend as products within each category (warehouse generation, data management, and information access tools) are integrated and often combined with products in other categories.

More comprehensive warehouse generation tools are starting to emerge that cover more than one area (metadata management, warehouse monitoring and management, design tools, and so on). As discussed earlier, the relational database management systems are integrating OLAP capabilities into the database engine.

The trend toward product integration is most advanced in the information access tools market. Many vendors have merged their query tools and report writer tools by adding report writing and distribution features to the ad hoc tools. In addition, some query and reporting tool vendors have begun to blend OLAP viewer functionality with traditional query tools into a single environment (for example, Cognos' Impromptu and PowerPlay).

Other trends include the integration of data mining capabilities with OLAP/query/reporting/Web information access tools (such as BusinessMiner from Business Objects) and the integration of data mining with data visualization tools.

■ Cross-Category Integration

Integration across categories is another trend that is gaining momentum as companies increasingly look to vendors and systems integrators to provide one-stop shopping. Currently, products from multiple vendors are required for a data warehouse. These components must be made to work together by the purchaser, greatly increasing the complexity and cost of building a data warehouse and making it a significant systems integration challenge.

The ultimate integration goal is to avoid the need to integrate these products and have them all work together seamlessly from a single metadata source, or have one vendor deliver a complete data warehousing solution. No vendor currently offers an end-to-end data warehousing solution. However, vendors such as IBM (with its VisualWarehouse), Platinum, SAS, and some of the relational database vendors (such as Oracle with its Data Mart Suite and Sybase with its Warehouse-NOW) are starting to make progress toward this objective. In the interim, hardware vendors, software vendors, and system integrators are creating programs where they select and certify best-of-breed products and provide a framework for them to work together (for example, HP with its OpenWarehouse). These vendors pre-integrate products, which simplifies product selection and reduces the time needed to enable various products to work together.

7.2.8 Market Overview

IDC expects the data warehousing market to grow significantly, from $16.8 billion in 1996 to $40.5 billion worldwide in 2001. The software component of the data warehousing market represented $1.9 billion in revenues worldwide in 1996. Warehouse generation accounted for 17 percent of this total ($327 million), data management for 48 percent ($933 million), and information access tools for 35 percent ($664 million). (See Figure 7-16.)

Figure 7-16: Worldwide Data Warehouse Market

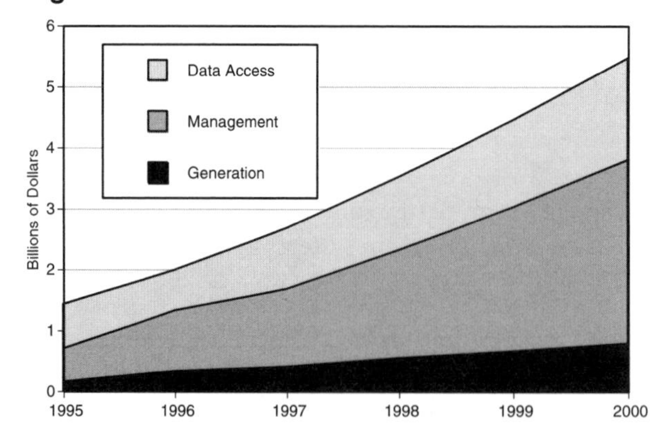

Source: IDC, 1997

The software component of the worldwide data warehouse market will continue to undergo rapid growth. IDC estimates that this market will grow to $7.2 billion by the year 2001 – a compound annual growth rate (CAGR) of more than 30 percent. Dataquest provides a similar estimate of $6 billion by the year 2000.

Data management products account for the largest section of the data warehouse market. IDC expects the data management market to increase as a percentage of the total market as it experiences the fastest growth (37 percent per year to the year 2001, resulting in a $4.4 billion market). The four leading RDBMS vendors – IBM, Informix, Oracle, and Sybase – will retain the largest market share, according to IDC.

The warehouse generation market will also experience strong growth (28 percent CAGR), resulting in revenues of $1.1 billion by the year 2001. This market is an area of intense activity: it has almost doubled in size since 1995, with new products steadily emerging and rapid innovation occurring in existing products. The data cleansing tools market is expected to become one of the fastest-growing segments of the warehouse generation market.

The information access tools market is undergoing more modest growth (21 percent CAGR, resulting in a $2.3 billion market in 2001). This market is showing signs of reaching maturity, and with the large number of vendors in this segment, further consolidation is likely. Within this market, however, niches such as data mining, data visualization, and multidimensional analysis will realize high growth over the next several years.

When other software-related costs such as system-level software ($899 million) and applications ($610 million) are factored in, the overall 1996 software market for data warehousing was $3.4 billion. Warehouse generation, data management, and information access tools accounted for the bulk of this total, at $1.9 billion.

Putting these costs into context, IDC estimates that the overall market for data warehousing was $16.8 billion in 1996, with software accounting for $3.4 billion, hardware $9.4 billion, and services $4 billion. The total market is expected to grow to $40.5 billion by the year 2001 (a 19 percent CAGR), with software growing the fastest (29 percent CAGR), followed by services (18 percent CAGR) and hardware (15 percent CAGR).

7.2.9 Forecast

Data Warehousing

- The number of companies using data warehouses will increase significantly during the forecast period.
- Companies will see the size and complexity of their existing data warehouses grow considerably.
- Data mining will become a more integral part of OLAP environments.
- Web-enabled data warehouses will be open to more users and therefore increase the importance of scalability and performance issues.
- The market for application-specific data warehouse products will emerge while ERP vendors such as Oracle and SAP develop data warehousing solutions that sit on top of their core applications.
- Vendor consolidation will occur within the data warehousing market.
- Companies will increase their use of consulting partners when implementing data warehousing solutions. These consultants and system integrators will differentiate themselves by bringing vertical expertise and industry-specific data models to their data warehousing engagements.

Warehouse Generation

- Better data warehouse generation tools are essential to the long-term success of data warehousing. The warehouse generation market will undergo rapid growth as the price of these tools declines significantly (through being bundled with data warehouse suites and the increasing number of vendors competing in this segment) and as increased functionality of the tools makes it more attractive for companies to buy them rather than to build their own.

- Warehouse generation tools will have much greater functionality as strong advances in metadata management capabilities, integrated data transformation tools (with AI capabilities to enhance data cleansing), and more intelligent data extraction tools (with more automated data transfer and extraction capabilities) are incorporated into these products. Advances will be made in tools used to design the data warehouse and initially populate it as well as in tools that automatically update the contents.

- Specialized audit software and a renewed information resource management function will emerge as companies decide that it is better to be vigilant about data integrity rather than deal with the consequences of inconsistent and inaccurate data.

- Data and process modeling that is focused on developing performance measures will become a key technique whereby a common understanding of business decision-making processes and their associated data can be developed. This understanding will be recognized as being essential to the process of integrating data from different business areas and meeting the needs of company resource allocators and decision-makers. Increasingly, business models for specific functional areas and industries will become available from business modeling product vendors, systems integrators, professional services organizations, and industry associations.

- MDBMSs will become more popular. There is already a trend toward a dual-system strategy: RDBMS for operational databases, and MDBMS for the data warehouse.

- The trend is toward three-tier client/server solutions for the data warehouse, often with multiple application servers performing different functions (OLAP, Web servers, and so on) as front ends to the underlying database server.

Data Management

- Rapid growth in the number of businesses implementing data warehouses and the average data warehouse size will be fueled by the improving OLAP capabilities in leading RDBMS products.

- Thanks to partnerships, acquisitions, and substantial R&D expenditures over the past few years, IBM, Informix, Microsoft, Oracle, and Sybase will continue to make strong advances during the forecast period to support larger and more sophisticated data warehouses.

- More widespread access to data warehouses (including access via the Web) will place a heavy burden on data management servers and drive the need for continuous improvements in data management products.

- Data management vendors will respond by improving their support for very large databases through improved performance and scalability (via enhanced parallel processing, bit-mapped indexes, improved query algorithms, preaggregation, and other techniques).

- More companies will introduce operational data stores into their architectures to support the need for more detailed information.

Information Access Tools

- Data mining will be a high growth niche within the data access market, especially as these products become simpler and less expensive, and are bundled with other information access products. OLAP tools will continue to incorporate more elaborate data mining functionality until the distinction between the two becomes virtually nonexistent. A single toolset will be able to satisfy all the functions required of a particular business solution.
- Data visualization will emerge as a means of exploring large volumes of complex data. It also will be used to enhance the effectiveness of data mining tools.
- Web browsers will replace low-end query and report writing tools as as interface to data warehouses as information access tools are rewritten to exploit the Web.
- Different types of information access tools will converge as query and report writing tools, OLAP tools, data visualization tools, and data mining tools increasingly are bundled.

Standards

- There will continue to be a lack of integration among metadata tools. Integrated metadata will not be readily available until at least 1999, when repository-based solutions should begin to emerge.

7.2.10 References

■Articles

Brooks, Peter L. 1997. Visualizing data. *DBMS*. August: 38.
Darling, Charles B. 1997. Ease implementation woes with packaged datamarts. *Datamation*. March: 94.
Dyck, Timothy. 1997. Tapping a new vein in data mining. *PCWeek*. May 5: 61.
Eckerson, Wayne W. 1997. Web-based query tools and architectures. *Journal of Data Warehousing*. Vol. 2, No 2. April.
Edelstein, Herb. 1997. Mining for gold. *InformationWeek*. April 21: 53-70.
Finkelstein, Richard. 1995. MDD: database reaches the next dimension. *Database Programming & Design*. April: 27-38.
Gardner, Stephen R. 1997. The quest for standardized metadata. *Byte*. November: 47.
Hurwitz, Judith. 1997. The evolution of metadata. *DBMS*. July: 12.
Madsen, Mark. 1997. Warehousing meets the Web. *Database Programming & Design*. August: 37-45.
Marshall, Martin. 1997. Metadata standardization debuts. *CommunicationsWeek*. February 24: 21.
Nadile, Lisa. 1997. Links for database tools. *InformationWeek*. February 24: 76.
Perez, Juan Carlos. 1997. Taming the metadata monster. *PCWeek*. February 17: 8.
Rettig, Hillary. 1997. 3D business data visualization. *VARBusiness*. May 1: 112-117.
Watterson, Karen. 1997. Attention, data-mart shoppers. *Byte*. July: 73-77.
Wright, William. 1997. Information visualization: the fourth dimension. *Database Programming & Design*. April: 50-55.

■Periodicals, Industry and Technical Reports

Brethenoux, E. and K. Strange. 1997. *Data warehousing and data mining: separating the two*. Gartner. April 29.
Cameron, Bobby, Stuart D. Woodring, Mamie H. Hsien, and Ketty C. Lieu. 1997. *Warehousing apps data*. Forrester.
Hendrick, Stephen D. 1997. *The data warehouse market: perceptions and realities*. IDC.
Morris, Henry. 1997. *Packaging the data mart: the application-centered data warehouse emerges*. IDC.
Open Computing & Server Strategies. 1997. *The evolving data warehouse blueprint*. META.
Open Computing & Server Strategies. 1997. *2nd quarter 1997 trend teleconference: information mover infrastructure*. META.
Strange, K. 1997. *Data warehouses: clarifying the hype and confusion*. Gartner. January 22.

van den Hoven, John and Bill McKenna. 1996. *Data warehousing – inventorying the company's data.* White paper. Noranda.

Wayson, Dennis. 1997. *Data warehousing goes mainstream.* Dataquest.

Wingfield, Teresa. 1997. *Data mart: departmental vs. enterprise trade-offs.* Giga (PA-I-96-00053).

■Books

Inmon, W. and R. Hackathorn. 1994. *Using the data warehouse.* New York: John Wiley & Sons.

Inmon, W., John A. Zachman, and Jonathan G. Geiger. 1997. *Data stores, data warehousing, and the Zachman framework.* New York: McGraw-Hill.

Kimball, Ralph. 1996. *The data warehouse toolkit: practical techniques for building dimensional data warehouses.* New York: John Wiley & Sons.

■On-Line Sources

The Data Warehousing Information Center `http://pwp.starnetinc.com/larryg`
The Data Warehousing Institute `http://www.dw-institute.com`
The OLAP Council `http://www.olapcouncil.org`
The OLAP Report `http://www.olapreport.com`

■URLs of Selected Mentioned Companies

Advanced Visual Systems `http://www.avs.com`
Andyne Computing (acquired by Hummingbird) `http://www.andyne.com`
Angoss Software Corp. `http://www.angoss.com`
Apertus (now Apertus Carleton) `http://www.apertus.com`
Apertus Carleton `http://www.carleton.com`
Arbor Software `http://www.arbor.com`
BEA Systems `http://www.beasys.com`
Brio Technology `http://www.brio.com`
Business Objects `http://www.businessobjects.com`
Cognos Corp. `http://www.cognos.com`
Comshare `http://www.comshare.com`
Constellar Hub (formerly The SQL Group) `http://www.constellar.com`
D2K Inc. `http://www.d2k.com`
DataMind `http://www.datamindcorp.com`
Evolutionary Technologies Inc. (ETI) `http://www.evtech.com`
Execusoft `http://www.execusoft.com`
Geac `http://www.geac.com`
Gentia Software `http://www.gentia.com`
HP `http://www.hp.com`
IBM `http://www.ibm.com`
Informatica `http://www.informatica.com`
Information Advantage `http://www.infoadvan.com`
Information Builders `http://www.informationbuilders.com`
Information Discovery `http://www.datamining.com`
Informix `http://www.informix.com`
Infospace `http://www.infospace.com`
Integral Solutions `http://www.integralsolutions.com`
Intellidex `http://www.intellidex.com`
Interweave (acquired by Cognos) `http://www.iweave.com`
IQ Software `http://www.iqsoftware.com`
IRI Software `http://www.infores.com`
Kenan Technologies `http://www.kenan.com`
Land's End `http://www.landsend.com`
Lawson Software `http://www.lawson.com`
Lotus `http://www.lotus.com`
Magnify Inc. `http://www.magnify.com`
Management Science Associates `http://www.msa.com`
Microsoft `http://www.microsoft.com`
MicroStrategy `http://www.microstrategy.com`
NCR `http://www.ncr.com`
NeoVista Solutions `http://www.neovista.com`
Oracle `http://www.oracle.com`
Panorama Software `http://www.panorama.com`
Paradigm Systems `http://www.paradigmsystems.com`
PeopleSoft `http://www.peoplesoft.com`
Pilot Software `http://www.pilotsoftware.com`
Platinum Technology `http://www.platinum.com`
Praxis `http://www.praxis.com`
Price Waterhouse `http://www.pw.com`
Prism Solutions `http://www.prismsolutions.com`
Red Brick `http://www.redbrick.com`
Sagent Technology `http://www.sagenttech.com`
SAP `http://www.sap.com`
SAS Institute `http://www.sas.com`

Seagate Software `http://www.seagate.com`
Silicon Graphics `http://www.sgi.com`
Software AG `http://www.softwareag.com`
Speedware `http://www.speedware.com`
SPSS `http://www.spss.com`
Sybase `http://www.sybase.com`
Vality Technology `http://www.vality.com`
Virtual Integration Technology `http://www.vit.com`
Visible Decisions (VDI) `http://www.vdi.com`
Visual Numerics `http://www.vni.com`
VMark Software `http://www.vmark.com`
Xpert `http://www.xpert.com`

7.3 Corporate Applications

7.3.1 Executive Summary

Global markets, electronic commerce, and the Internet continue to have a significant impact on corporate applications. With fast technological and business changes now a way of life, customers want functionally superior applications that make key features easily available. Therefore, customers' requirements for flexibility and ease of implementation will continue to increase.

Unless there are significant competitive advantages, building custom applications is too expensive, time-consuming, and risky. Instead, customers continue to move to packaged corporate applications, requiring the ability to create or modify applications quickly according to changing business needs. In particular, companies lacking the resources of large corporate enterprises share these requirements, indicating that the market for corporate applications is changing. Those changes require that enterprise corporate applications change as well. Because growth is slowing due to product maturation and near-saturation at the high end of the market, all the major vendors – Baan, Computer Associates (CA), Geac, J.D. Edwards, Oracle, PeopleSoft, and SAP – are targeting smaller companies with a variety of initiatives and strategies that involve easier and less-expensive turnkey implementations, a focus on vertical and midsized markets, and shifting distribution channels.

With about 60 percent of large companies already running some form of corporate application, the growth is clearly in the midsized market. Analysts define a midsized business as having annual revenues between $50 million and $250 million. They also point to a market of between 30,000 and 70,000 midsized companies in North America that have no corporate applications technology. Such companies could account for billions of dollars in potential software licenses.

Customers increasingly require non-uniform corporate applications. They want individual components – configurable software modules that collaborate via standard interfaces – that can be combined as necessary to meet changing business needs. Vendors are responding to this demand by separating their integrated application modules into smaller components.

As the scope of implementation projects becomes more process-focused, it also begins to extend across the border of the enterprise, bringing partners', suppliers', and customers' needs into the integrated solution.

Supply chain management has become critical to global companies seeking to link with suppliers, distributors, business partners, and customers. As a result, all major corporate application vendors are integrating this functionality into their product lines. In fact, all major corporate applications vendors believe companies will use the Internet to operate their entire supply chain – from obtaining quotations from suppliers, to ordering goods or services electronically. The belief is that the Internet will facilitate global commerce by eliminating geographical and financial boundaries, and increasingly, companies are relying on corporate applications for this functionality.

Mergers, acquisitions, and joint development alliances among major corporate applications vendors and smaller supply chain management software providers continue to be major trends. For example, PeopleSoft ships software that it acquired when it purchased Red Pepper in 1996. Baan also has used acquisitions strategically to purchase critical functionality such as sales force automation (with its acquisition of Aurum), sales configuration systems (with its acquisition of Antalys), and supply chain planning (with its acquisition of the Berclain Group).

By using object technology, corporate application developers are able to encapsulate and reuse business knowledge, rules, and software code for faster and easier development. For example, SAP officials expect SAP Business Objects to be critical to SAP's architecture. (See "Business Objects" on page 655.) Several of SAP's primary competitors – including Baan and Oracle – are also developing object-oriented versions of their client/server enterprise software. Smaller vendors such as Infinium (formerly Software 2000), J.D. Edwards, Marcam, and SSA already market object-based versions of their applications.

7.3.2 Business and Technology Trends

As businesses continue to look for more value from their information technology (IT) operations, over-stretched IT departments consider several options, including integrated corporate applications (rather than individual best-of-breed products) and outsourcing development and support. Some companies do not want to deal with integrating best-of-breed corporate packaged applications from multiple vendors, however. Although they may believe that best-of-breed applications offer better functionality and flexibility, the cost and effort involved in integration are too onerous for them to undertake. To minimize both, many corporations also purchase systems integration and consulting services.

Complex, expensive, monolithic packaged applications represent another set of challenges. However, corporate applications vendors are heeding their customers' demands for less-expensive software that is based on industry standards, interoperates with other vendors' products both internally and external to their enterprises, and is easier to configure and install. All of the major vendors – Baan, Computer Associates, Geac, J.D. Edwards, Oracle, PeopleSoft, and SAP – are developing standards-based software that will make implementation by IT personnel easier. SAP's component-based software modules are a direct response to its customers' requirements for faster, less-expensive deployment of corporate applications.

Of additional concern to corporations are the imminent changes they face from Year 2000 and European Monetary Union (EMU) issues. In fact, the Year 2000 problem poses one of the most critical and expensive business IT challenges ever. The Gartner Group, for example, predicts companies worldwide may spend $600 billion to deal with the problem.

Analysts believe industries that rely heavily on mainframe computers face the most daunting Year 2000 challenges. One alternative for such companies is to forgo fixing the problem in their mainframe code and replace it with packaged corporate applications. All the major corporate applications vendors promise Year 2000 compliance. Baan, CA, Geac, J.D. Edwards, PeopleSoft, and SAP applications claim Year 2000 compliance. Although Oracle was the last major corporation applications vendors to deliver Year 2000-compliant versions of its products with Release 10.7 NCA (Network Computing Architecture), prior releases are not fully Year 2000-compliant. Oracle will stop supporting the earlier releases by the end of 1998.

Despite possible benefits from migrating legacy-based systems to packaged applications, the mainframe remains a critical link in many corporate technology infrastructures because it continues to support mission-critical applications. One of the biggest challenges facing IT departments is the lack of strategies and tools for migrating legacy applications and data to new packaged applications and platforms. Therefore, instead of implementing an integrated corporate application, many companies have been forced to adopt a piecemeal approach, delaying benefits and increasing implementation costs. In fact, Forrester projects that the mainframe will continue to play a major role as a platform for delivering applications in corporate and Internet computing. Increasingly, for example, companies are integrating corporate applications with mainframe data and transaction engines.

Vendors and users also are moving to three-tier and multi-tier architectures. The three-tier architecture allows the presentation logic on the client layer, business process logic on the application layer, and data management logic on the database layer. Multi-tier architectures provide additional layers to handle Internet logic, data warehousing logic, and other special-purpose applications such as real-time, available-to-promise, and supply-chain optimization. These architectures are better-suited than two-tier, fat-client/thin-server configurations for the current crop of corporate applications, and pave the way for Internet computing.

Companies also must deal with the EMU issue, which is arriving faster than the Year 2000 problem. Starting on January 1, 1999, the 15 member countries of the European Community will begin using a new, common currency (the Euro) in addition to their respective national currencies. During the 3 transitional years (to a mandatory December 31, 2002 deadline), companies inside and outside Europe will have to account in multiple currencies, with multi-currency system capabilities being a prerequisite for completing business processes during the transitional phase. After the transition period, the Euro will be the only legal currency in circulation.

All organizations that use electronic commerce and the Internet as well as financial institutions that deal with European-based businesses will face the Euro currency problem. At the heart of the problem is the need to bill European companies in national currencies and Euros after January 1, 1999; they can then send back payment in either, and the recipient has to account for gain and losses in both currencies. The process is so effort-intensive that some analysts have predicted the cost for organizations to become Euro-compliant could outweigh that related to the Year 2000 issue by a factor of five or more.

In response, corporate application vendors are introducing features to help customers doing business in Europe. For example, Oracle is adding new capabilities to Oracle Financials to enable European-based and multinational companies to operate smoothly as they migrate to a united currency environment. (See "Financials" on page 650.) Likewise, SAP has introduced SAP Euro, a software package that includes conversion tools and special functions for the dual-currency phase. (See "Euro Component" on page 651.)

7.3.3 Major Corporate Applications Vendors

Enterprise resource planning (ERP) software touches all areas of an organization, automating tasks such as supply-chain management, inventory replenishment, ordering, and logistics coordination between a company, its suppliers, and its customers – often by using the Internet to connect disparate corporate sites or remote users. Traditional ERP includes manufacturing resource planning, human resources, and accounting.

Until a few years ago, there was a clear segmentation in the ERP software market between high-end vendors (such as Baan, Oracle, PeopleSoft, and SAP) that served large, multi-billion dollar firms and the middle-market vendors (such as CA, Geac, and J.D. Edwards) that offered applications for smaller firms.

All the major vendors are focusing on ways to boost market share and revenue. Today, with the increasing saturation of the high-end business and the decentralization of multi-billion dollar firms into smaller, independent businesses, both high-end and middle-market suppliers are targeting customers in the $100 million to $1 billion range. Many vendors are narrowing their focus still further to those companies in the $50 million to $250 million range that need full-featured ERP solutions that can be implemented quickly and cost-effectively.

The drive into this market encompasses several approaches, including "componentizing" application implementations to reduce costs and effort, targeting specific vertical markets, and shifting channel distribution strategies.

Forrester predicts that high-end vendors will have a hard time breaking into the middle market because of the perception that their solutions are too expensive, complex, and inflexible. On the other hand, middle-market vendors with component-based architectures, such as Geac and J.D. Edwards, feature components that can be moved around, distributed, and easily reconfigured or integrated with other systems.

The Internet represents an enormously lucrative battleground for all corporate applications vendors. In fact, Paul Wahl, CEO of SAP America, probably summed it up best when he told a trade publication that SAP's objective is a complete electronic-commerce system that connects every Internet-based order to the supply chain, distribution partners, third-party freight carriers, banks, and the manufacturing and transportation units of an enterprise – all in real time.

SAP will enable third-party supply-chain vendor software to integrate with R/3; it has also delivered its own supply-chain initiative as an alternative to other products. Baan and Oracle are working on their own software as well as partnering with third-party supply-chain vendors.

However, SAP is not the only corporate applications vendor targeting the Web. J.D. Edward's Configurable Network Architecture accommodates the Internet, intranets, and extranets. Oracle Applications for the Web seek to extend access to corporate data to employees, customers, and suppliers. Baan and PeopleSoft have also delivered significant Web enhancements to their product suites. Geac began to ship its Web-enabled SmartStream products in October 1996.

Packaged Applications versus Best-of-Breed

The trend is to buy prepackaged software for enterprise use rather than construct it in-house, with the goal of replacing legacy systems or solving specific problems (such as Year 2000 and EMU compliance). Many offerings still need to be customized to the particular business, however, and that requirement, coupled with potential integration or migration issues, could slow deployment. As companies continue to link electronically with suppliers, customers, and outsourcing partners, "internally focused" enterprise applications could prove insufficient. When faced with projects that require years to install, companies will also turn to smaller, more "agile" applications that focus on one or two particular business problems and provide short-term paybacks.

Because of the cost, complexity, and time needed to implement enterprise-wide corporate applications, many companies are considering a variety of alternatives to comprehensive packaged applications. One popular solution is to purchase only the specific application required, such as manufacturing, financials, or sales force automation. Automating such departmental functions has proved successful for smaller vendors such as Siebel Systems (sales force automation) as well as market leaders such as SAP (manufacturing) and PeopleSoft (human resources). The newest entry in this field is automated purchasing, with offerings from a half-dozen vendors, including Ariba Technologies. Ariba's software can be integrated with other corporate applications, such as Oracle Financials, and is written in Java, allowing it to be used on virtually any desktop.

Vendors that provide only part of a complete enterprise solution may face an uncertain future, however. The larger ERP vendors are already entering the market: Oracle is developing a sales force automation product, SAP has released a customer-service module, and PeopleSoft partnered with Vantive to provide a front office solution rather than enter the market directly. SAP has also announced plans to release a sales force automation module, although the company has not yet stated whether the module will be developed in-house or whether SAP will acquire or resell third-party technology. In response to such developments, specialized providers are rushing to expand their product lines; for example, Siebel, a sales force automation vendor, has introduced new customer-support products.

Vertical Markets

Because the leading ERP vendors all provide robust functionality, they are all seeking to differentiate themselves by offering industry-specific solutions targeted toward a particular vertical market. SAP delivers the widest range of industry-specific offerings, with 13 industry-specific units, including high tech and electronics, chemicals, oil and gas, utilities, and the public sector. In December 1997, SAP announced the formation of a subsidiary targeting the public sector. SAP America Public Sector Inc. will focus on the vertical market composed of schools, nonprofit organizations, and state, federal, and local governments.

Other vendors are following suit. PeopleSoft, for example, reorganized into nine industry-specific business units, including retail, higher education, financial services, federal government, manufacturing, and health care. Under PeopleSoft's plan, each vertical unit has its own sales and support organizations and industry-specific product development team. Each unit will develop or purchase its own industry-specific functionality and be responsible for sales and service. Core development for HR and financials modules, however, will rest with a central development center.

Specific applications in Geac's SmartStream product suite focus on the vertical industries being targeted by the company. For example, SmartStream Funds Control, introduced in April 1997, provides specific functionality tailored to the health care and public sector markets.

J.D. Edwards is focusing on 11 vertical market segments, with a new push into areas where it has traditionally done well but has not led – consumer packaged goods, fabricated metals, and electronics. The company has led the U.S. market in architecture, engineering, and construction, and has been strong in oil and gas.

Oracle's vertical program is based on partnerships with smaller vendors that specialize in particular industries. To target the consumer packaged goods (CPG) market, for example, Oracle teamed up with IMI, Manugistics, and TRW. Oracle's CPG solution incorporates Manugistics' full suite of supply-chain management solutions, including Demand Planning, Supply Planning, Manufacturing Scheduling, and Transportation Planning. This integration allows customers to manage operations across the entire business, from procurement of raw materials to driving consumer demand. In December 1997, Oracle also broadened its partnership with i2 by agreeing to integrate i2's Rhythm supply and demand planning solutions into its Applications. The products (including Factory Planner, Sequencer, Supply Chain Planner, and Demand Planner) will be aimed at the automotive, aerospace, and high-tech industries. In addition to these markets, Oracle offers solutions for the banking, utilities, and oil and gas industries; solutions for transportation and retail also are being planned.

Baan, on the other hand, seeks to assemble its own vertical market functionality, and currently offers solutions for five industries – aerospace and defense, automotive, electronics, heavy equipment and contract manufacturing, and process industries.

In a July 1997 report on vertical industry solutions, the Aberdeen Group said that Baan counts aerospace and defense and automotive as leading verticals; Oracle is broadly used in the energy and telecommunications markets; and SAP has demonstrated well in the electronics and chemical industries.

In addition to the applications discussed previously, numerous companies provide similar enterprise packages in narrow niches. A prime example is the Pinpoint Energy System from Price Waterhouse, which is sold exclusively to the petroleum and gas industry. This package comprises two parts: Premas Plus and Stars. Premas Plus is an application that handles data processing for the production of crude oil, natural gas, and liquid natural gas; Stars handles the supply, market, and transport processing. Both of these packages generally are installed by an integrator, who matches them with vertical components from other vendors.

In the public utilities area, Price Waterhouse sells Service 2000, a similar client/server package that handles most accounting functions and specialized vertical needs, such as the management of gas-meter readings, customer-premise information, and local tax ordinances.

The following ERP vendors are discussed in order of their prominence in the marketplace as discussed in the Market Overview. (See Figure 7-18 on page 680.)

7.3.4 SAP

The SAP R/3 client/server enterprise computing package is a comprehensive business solution that includes accounting, sales and distribution, human resources, and computer-integrated manufacturing offerings. SAP supports multi-site, multi-currency operations and delivers a wide functional reach across the enterprise. R/3 succeeds SAP's earlier, mainframe-based R/2 line and runs on mainframes, UNIX, Windows NT, and IBM's AS/400 servers.

SAP (`http://www.sap.com`), which offers functionality to a wide set of vertical markets – aerospace, automotive, chemical/process, consumer products, finance/insurance, health care, high technology, manufacturing, oil and gas, pharmaceutical, public sector, retail, telecommunication, and utilities – continues to command almost 30 percent of the client/server enterprise applications market. SAP has achieved this market position because companies believe that R/3 offers the best overall system with which to provide cross-functional integration and support geographically dispersed operations.

In addition to its historical concentration on large enterprises, SAP now targets midsized companies and markets. The company has also broadened its focus beyond traditional ERP into areas such as supply-chain management and sales force automation, and vertical industry applications such as hotel reservation systems.

In an effort to make R/3 more open, SAP makes it easier for third-party software vendors to link directly to R/3 through its business application programming interfaces (BAPIs). More than 100 technology partners exhibited at SAP's user conference in August 1997, demonstrating a wide range of add-on functionality. SAP partners include Acquion, which enables R/3 users to access Web-based catalogs, and Metasys, which introduced a gateway product that lets R/3 users plan and execute product shipments.

Other key initiatives unveiled at the user conference to reduce the time and expense of installing SAP include the following:

- **Ready-to-Run R/3 (RRR)** – A program under which SAP delivers a pre-configured system with a pre-loaded operating system and database to let users begin implementation as fast and inexpensively as possible.

  RRR bundles server and network hardware systems with a pre-installed, pre-configured base R/3 system; operating system; database; and setup for daily R/3 operation, including batch processes and regular database saves. As a complement to SAP's AcceleratedSAP program (SAP's rapid implementation methodology), RRR should speed up R/3 installation efforts by some 20 to 25 person-days. Available for some time, RRR now supports AS/400 hardware as well as a variety of other hardware platforms and database systems.

- **SAPNet** – An on-line repository of SAP training materials and documentation so customers have ready access to other customers' knowledge. SAPNet has been rolled out in Europe and is just beginning to roll out in North America.

  SAPNet is a Web-based service built on the R/3 Information Database. Customers can use SAPNet, which is accessible from any standard Web browser, to obtain customized data on SAP products and solutions, register for courses and perform other self-service functions, or participate in on-line discussion forums. SAPNet also includes intelligent agent information wizards to direct users to specific data and advise them of updates to data they have accessed.

SAP Applications

The strength of SAP's R/3 flagship product has been its high degree of integration, mostly for large global corporate enterprises. This integration also has served as a target for much criticism: its enormous scope and wide range of functionality have made R/3 complex, time-consuming, and expensive to install and integrate with other systems.

SAP has answered many of those criticisms through major changes to its business strategies along with significant technological modifications to R/3 itself. The most notable alteration is the delivery of R/3 Release 4.0 (scheduled for initial shipment to customers by year-end 1997 and general availability by mid-1998), an object-oriented, component-based architecture that breaks away from R/3's formerly monolithic architecture. SAP decoupled its applications into separate components, each with its own database, but maintained the underlying Application Link Enabling (ALE) middleware to tie them and third-party components together.

A component strategy offers several benefits. SAP and its customers will be able to respond faster to market demands because they will not have to wait for major system releases to upgrade some functionality. Combined with several other changes aimed at frustrated customers, components will enable customers to build smaller, simpler, and less expensive R/3 systems. According to Giga Information Group, customers will have access to unbundled product offerings, easier upgrades, industry-standard interfaces (where defined), more installation options, and less lock-in.

SAP R/3 Release 4.0 also includes intranet capabilities for self-service access to human resources, customer service, and purchasing functions; a business information warehouse; a sales configurator and pricing engine; native Java applications; and interoperability with object technologies from IBM, Microsoft, Sun, and applications based on the industry-standard Common Object Request Broker Architecture (CORBA).

Financials

Release 4.0 enhances the core financial functionality with improved handling of networked environments with transfer prices, group calculation, cross-enterprise and multi-industry Business Information Warehouse (BIW), dynamic investment programs, proactive rolling forecasting, risk management, and activity-based costing.

Release 4.0 improves R/3's core financial applications with the following:

- **Transfer pricing** – Supports the valuation of material stocks and goods movements between groups in business units, divisions, organizational units or profit centers.
- **Activity-based costing** – Reduces maintenance needed for such calculations. Using cost-driver data from R/3, this functionality provides a realistic view of the cost structures of products and customers.
- **Financial application integration via BAPIs** – Based on the Business Framework architecture. The financial application BAPIs provide event-based, publish-and-subscribe capabilities, which automate the exchange of information between R/3 and third-party applications.
- **New financial components** – The five new components are as follows:
    - **Treasury** – Includes applications for managing money market investments, foreign exchange derivatives, securities, loans, and market risk management.
    - **Investment Controlling** – Supports global capital investment management. It helps organizations manage the corporate approval process from start to finish, letting users rate requests by their viability and priority.

- **Self-Audit** – Enhances audit quality. The Audit Report Tree, at the core of this component, provides a broad array of auditing functions and default configurations that include auditing procedures, documentation, auditing evaluations, and downloading audit data.
- **Joint Venture Accounting** – Lets several companies combine their resources for projects. It distributes revenues and expenses among partners according to their working interests, transfers material and assets between ventures, and enables expense and revenue netting by equity partners.
- **Consolidation** – Links to the centralized R/3 Executive Information System; aggregates data from all parts of the R/3 system, at the company, business, or profit-center level for financial reporting purposes.

- **Business Information Warehouse** – Provides companies a comprehensive view of data across the enterprise. (See "Business Information Warehouse" on page 656 for more information.)

Euro Component

SAP's Euro solution helps customers convert SAP data to the new European currency that takes effect beginning January 1, 1999. After that date (and continuing through December 31, 2002), software must be able to handle individual business transactions in native currencies as well as the common Euro currency; after the transition period has ended, all financial processes must be handled in Euro only.

SAP Euro also includes the following business application functions to help customers process business transactions during the dual-currency phase:

- **Accounting** – Currency checking, conversion functions for VAT reports, conversion programs for converting national currencies to the Euro, and retroactive accounting
- **Human Resources** – Simultaneous payroll processing in Euro and local currency and payroll accounting conversion
- **Logistics** – Tools to change currencies for business partners on request, conversion functions for reports and pricing, and conversion of other reports based on legal requirements

Logistics

Recent enhancements include new sales and distribution functionality, distribution resource planning, available-to-promise server, flow manufacturing, and sequencing. These enhancements rely in part on some of these new capabilities:

- **External catalog integration** – Provides access to external electronic catalogs for procurement
- **Vendor consignment** – Extends handling of vendor consignment stocks
- **Value contracts** – Handles validity period, agreed total value, and pricing for sales and purchases
- **Free goods** – Automates promotion management
- **Automatic settlement** – Based on purchase orders for services performed, allows for forwarding costs from service purchase order to accounting
- **Letters of credit** – Provide confirmation that the quantity, quality, and nature of the goods delivered are the same as that determined in the contract

- **Automated export system** – Collects and processes export information
- **Warehouse management enhancements** – Support load distribution and warehouse operations
- **Sample management** – Enables flexible planning of actual samples to be drawn, identification of actual samples, administration of sample data in a sample record
- **Calibration inspection** – Supports calibration inspections for test and measurement equipment

New business components that complement R/3 Release 4.0 logistics functionality include the following:

- **Sales Configuration Engine (SCE)** – SCE captures product data and dependencies stored in different databases, which lets customers accurately configure complex products such as electronics. This capability allows salespeople to configure sales proposals and orders offline, while taking advantage of reusable configuration functionality and master data information from their organizations' R/3 system.

  SCE is written in Java and mapped directly into R/3. SCE is compatible with R/3 Release 3.x and 4.0 and is announced for availability in the first half of 1998.

- **Product Data Management (PDM)** – Integrates design and engineering into the supply-chain process from point of design to point of consumption, and lets customers create a product-development environment and manage product-related data across product life cycles. Using PDM, customers can control product data from all parts of an organization's logistics process.

  SAP plans initial customer shipments of PDM for year-end 1997, with general availability expected by mid-1998.

- **Web-based catalog and purchase requisition system** – Directly integrates with third-party catalog systems to provide streamlined procurement capabilities for customers buying goods from approved catalog suppliers. The result is that users can access third-party component catalogs directly from within their R/3 purchase requisition screens.

  The Web-based purchase requisition, which works with SAP Employee Self Service applications, lets companies implement strategic maintenance, repair, and operations (MRO) procurement. This procedure can result in significant savings by standardizing on approved suppliers with volume discounts, as well as simplify the process for employees. This new component requires only a Web browser and uses a standard HyperText Transfer Protocol (HTTP) interface for open connectivity to multiple catalog vendors. The catalog can be inside or outside an organization's firewall. First customer shipments of the Web-based catalog and purchase requisition system, available as a separate business component, are scheduled for late 1997; it supports R/3 Release 3.0 and 4.0.

Human Resources

As result of the major componentization effort of SAP in R/3 Release 4.0, Human Resources (HR) and Travel Management were split off from R/3 as separate components with a dedicated database and an independent release cycle. The HR and Travel Management components include a number of workflow, Internet, and intranet enhancements as well as new country-specific versions. R/3 Human Resources 4.0 provides these enhancements:

- **Globalization** – Personnel Administration, Payroll, and Trip Cost Accounting now addresses more than 30 countries' legal requirements and business procedures.
- **Employee Self Service Solutions** – Lets employees view and maintain their own HR data in R/3 easily via the Internet or corporate intranet using a Web browser.
- **Outsourcing** – The Interfaces Toolbox helps create interfaces needed for outsourcing. An interface to ADP PC/Payroll for Windows via a master extract means customers do not have to configure gross payroll schemes and rules.
- **Compensation Management** – Controls and administers an organization's remuneration policies.
- **Benefits Administration** – Medical insurance continuation enrollment, claim processing within flexible spending account administration, and evidence of insurability processing.
- **HR Reporting** – Includes the HR ad-hoc query, a tool for selecting and processing personnel information based on infotypes within Personnel Administration as well as standard HR reporting tools, which perform simple evaluations for selected payroll results.
- **Training and Event Management** – Includes functions for booking and canceling attendance at training and events, employee participation information and enrollment in benefits, and the use of form-based workflow in SAP Business Workflow.
- **Career and Succession Planning** – Enables personnel departments to create potential career paths within a company, show qualified employees available options, and provide information about employees who currently fulfill (or who will fulfill in the future) the requirements of particular positions.
- **Time Management** – Enhanced functions in the Shift Planning component enable companies to define requirements and find matchups on an hourly basis, if necessary, and specify qualifications as part of the requirements definition.
- **Form-based SAP Business Workflow in HR** – Allows individuals to create simple forms using tools such as Visual Basic or JetForms and integrate them with SAP Business Workflow to create self-service HR applications.
- **Euro functionality, including payroll** – The ability to handle financial payments and accounting in national and Euro currencies.

Supply-Chain Management

SAP's supply-chain management module (SCOPE) is based on the Advanced Planner and Optimizer (APO). The APO component's four main elements are these:

- **Supply-Chain Cockpit** – Provides a graphical interface to the supply chain that synthesizes collaborative business and decision data into actionable information. It includes an Internet-enabled user interface to provide remote viewing and control over the entire supply chain.
- **Available to Promise (ATP)** – Performs fast, high-volume, multilevel, simultaneous, rules-based checks on product and resource availability and allocation of goods between trading partners. The checks offer more accurate promised delivery dates, reservation of required material and product resources, and scheduling of planned orders to meet customer commitments.

- **Advanced Planning and Scheduling** (APS) – Capabilities include rapid and intelligent material requirements planning (MRP), distribution resources planning (DRP), master production scheduling (MPS), and deployment planning and scheduling (DPS).
- **Forecasting** – Lets customers perform high-volume, Internet-enabled collaborative forecasting. Customers can simulate and execute comprehensive sales plans that incorporate data from point of sale, collaborative forecasts, and other sources, using APO's graphical capabilities.

SAP's APO reduces R/3 planning weaknesses – speed, constraint handling, and supply-chain span. However, APO does not provide sophisticated scheduling and optimization algorithms for industry-specific planning and scheduling problems that are available from best-of-breed vendors. Midsized companies will find SAP's supply chain initiative attractive because it is a simpler, built-in solution, says Forrester.

SAP Architecture

R/3 has a two- or three-tier client/server architecture compatible with most major operating systems and database servers from Digital, HP, IBM, Informix, Microsoft, Oracle, and Sun. The three-tier structure provides more efficient transaction processing and increased scalability than two-tier. It also provides an architecture well-suited to Internet computing. A three-tier architecture has distinct layers for presentation, application, and database. The database layer runs database processing, the application layer operates applications logic that implements business processes, and the presentation layer handles client and local preprocessing and offers a graphical user interface. R/3's classic multi-tier architecture, which includes executable code downloaded from the database layer, evolves directly into a platform for Internet computing. (See Figure 7-17.)

Figure 7-17: SAP R/3 Architecture

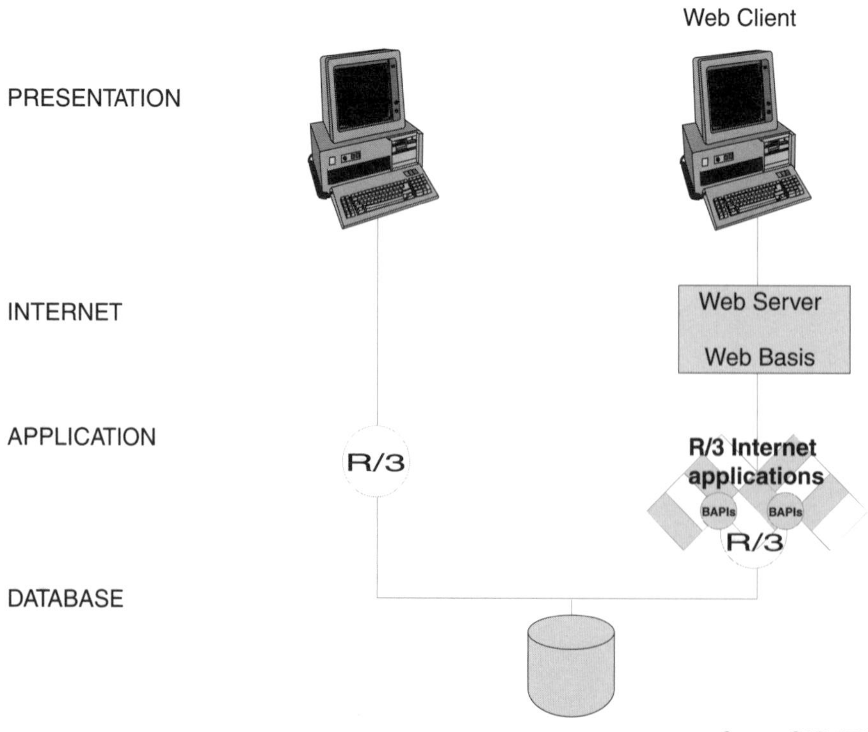

Source: SAP, 1997

- **Two-tier R/3 configurations** – These are implemented with clients that are responsible for the graphical user interfaces (GUIs), generating what users actually see on screen and all other functionality handled by a single server.
- **Three-tier R/3 configurations** – In these setups, the application logic is further separated from the underlying database. Depending upon customer requirements, users can opt to connect to one or multiple application servers. Separating the database and application servers enables support for a larger number of users, which is critical for large enterprises and Internet computing.

With R/3, clients send transactions to application servers, which process the business functions. These servers interact with the database, which has typically resided on a UNIX server or an S/390 mainframe. SAP also supports databases on IBM AS/400s and Windows NT.

R/3 uses remote procedure calls (RPCs) for communications between clients and applications servers and among applications servers. R/3 components communicate over major networking protocols such as TCP/IP, IBM's SNA, and Novell's IPX/SPX.

Windows NT is the fastest-growing R/3 server platform. According to Microsoft, 42 percent of new SAP customers, most of which are midsized companies, run Windows NT. Windows NT is generally easier to use than UNIX, which makes it a suitable platform for small- and midsized companies. UNIX, however, has a richer set of features and supports more simultaneous users than Windows NT. UNIX is regarded as a more mature server platform for enterprise computing.

SAP has built a complete development environment that facilitates the maintenance and development of SAP R/3 applications. The ABAP/4 Development Workbench offers access to SAP development tools that cover all phases of the overall software development life cycle. Tools are available for writing business applications in ABAP/4, for accessing databases, for network communications, and for implementing GUIs. The ABAP/4 Development Workbench includes development tools, such as Screen Painter, Menu Painter, Editor, and Interactive Debugger; tools for performance measurement; and Computer-Aided Test Tools. It also includes the ABAP/4 Repository, the ABAP/4 Repository Information System, the Active ABAP/4 Dictionary, the Enterprise Data Model, and the ABAP/4 Development Organizer.

Customers have the flexibility of using SAP's ABAP/4 fourth-generation language either to develop their own customized applications or to enhance SAP-supplied application logic. All applications developed with the ABAP/4 Development Workbench are platform- and database-independent.

■ Business Objects

SAP has rewritten R/3 Release 4.0 to support components or business objects. Today, SAP provides interoperability of Microsoft's Distributed Component Object Model (DCOM) objects with SAP Business Objects as well as support for CORBA standards; future support for Object Linking and Embedding (OLE) has been announced. SAP is also developing object-oriented extensions to the ABAP/4 programming language, with the objective of making the SAP Business Objects the core of future business applications.

■ Business Framework

SAP's Business Framework is an integrated, open, component-based product architecture that encompasses SAP R/3 enterprise applications and third-party products and technologies. Business Framework provides customers with simplified systems upgrade and maintenance; increased interoperability between R/3, legacy systems, customer-specific, and third-party solutions; and a more flexible platform, enabling continual change.

For example, SAP's Business Framework lets customers install new R/3 applications (business components) without having to upgrade the entire R/3 system. Therefore, an R/3 customer now using Release 3.0 can immediately implement the Internet application components shipped with Release 3.1 as well as several new business components without upgrading to Release 4.0.

R/3 users traditionally have integrated R/3 applications by tying them to a single database. With Business Framework, SAP encapsulates individual applications as business components that can run on their own dedicated databases, involve their own programming languages or operating systems, and be distributed to multiple locations. To ensure tight integration among components, Business Framework includes a series of SAP and third-party integration, interface, and communication technologies that link distributed components and custom applications across wide area networks and the Internet.

SAP's BAPIs, critical elements in the Business Framework, provide stable, reusable interfaces to and within R/3. BAPIs are the key customers and third-party developers using Java, Visual Basic, and other languages to deploy new application components rapidly for the R/3 environment. SAP is also making all BAPIs Java-enabled and publicly available over the Internet.

SAP R/3 Release 3.1 introduced Business Framework, which added the ability to use a loosely coupled integration of the core system, largely for supporting geographically remote sites. Release 4.0 delivers tightly coupled core applications and the introduction of HR as a business component. HR, when used separately with its own database, can retain tight integration with other core applications through the use of ALE distribution scenarios between HR and Financials and Logistics. As announced, R/3 Release 4.0 will mark its evolution into a product line of tightly integrated but separate and upgradable business components, each on its own release cycle.

■ Business Information Warehouse

The BIW, an independent component of the SAP Business Framework architecture, goes beyond SAP's earlier data warehouse effort, Open Information Warehouse. BIW takes a relational on-line analytical processing (ROLAP) approach and lets customers deploy data analysis functionality (or components) as necessary, without disrupting the existing R/3 environment. Compatible with R/3 Release 3.0 or later, BIW is designed as an open solution and delivers new BAPIs for non-SAP data integration and complementary third-party analysis tools. BIW's three main elements – Business Information Warehouse Server, Business Explorer, and Administrator Workbench – assist users in obtaining and analyzing data.

BIW includes all components necessary for installation, use, ongoing changes, and maintenance of a data warehouse. BIW Server includes an OLAP engine and metadata repository, both pre-configured with business content. It provides fast retrieval, interpretation, and preparation of the information stored in the data warehouse. Automated data extraction and loading capabilities supply the BIW server with data from R/3 and R/2 applications, as well as from any other data source. Business Explorer, which provides a graphical Web-access interface, lets customers personalize desktops and display data using the Microsoft Excel spreadsheet application. Administrator Workbench provides a single point of control for creating, monitoring, and maintaining the complete data warehouse environment. Open BAPIs link external data sources and applications.

BIW, scheduled for shipment in early 1998, is announced to be compatible with R/3 Release 3.0 and later. (See "Packaged Data Warehousing Applications" on page 632 for information about other vendor's offerings.)

■ Business Workflow

SAP Business Workflow for Release 4.0 utilizes Workflow Wizards to provide expanded access to business workflow functionality and process information, and to support easier development of complete workflows and workflow components.

SAP Business Workflow 4.0 provides integration with Microsoft Exchange and Outlook and other MAPI-compliant clients, extended support for a variety of electronic forms software, and integration with BIW for analysis and optimization of business processes. Integration with Exchange provides Exchange Server communications options, including SMTP-based e-mail, direct faxing, and the ability to use Outlook and Exchange forms to trigger on-line and offline R/3 workflows. Evaluation of workflow-related process information through BIW enables customers to detect business process bottlenecks and predict future business process responses based on real data. With Release 4.0, users can trigger Web-based workflows through automatically generated HyperText Markup Language (HTML) forms.

SAP Business Workflow 4.0 speeds up and simplifies workflow implementation by generating complete and executable workflows as well as single tasks within a workflow. An example of the former would be the total approval process of a travel expense report; an example of the latter would be sending a status report via e-mail about the travel expense report's approval to the employee who initiated the workflow from a self-service application.

SAP released SAP Business Workflow for Release 4.0 for first customer shipments with R/3 Release 4.0; it is announced for general availability by mid-1998.

■ Business Engineer

With R/3 Release 4.0, SAP has enhanced R/3 Business Engineer, the core enabling tool of AcceleratedSAP, SAP's rapid implementation methodology. R/3 Business Engineer for Release 4.0 lets customers configure R/3 faster and more easily through the use of a knowledge-based interactive system of questions, answers, and built-in validation checks that ensure compatibility of configuration decisions with applicable business rules and R/3 functionality.

In addition to accelerated and simplified implementation, R/3 Business Engineer for Release 4.0 supports continuous business improvement (integrating traditional enterprise modeling with the live R/3 solution by tying business models directly to deployment), delivers pre-configured R/3 systems for vertical industries, and offers a flexible, open configuration environment.

Pre-configured industry systems combine pre-selected business scenarios, processes, functions, and system settings for specific vertical industries – initially for the chemical, consumer products, and steel industries. One hundred business scenarios aggregate the 1,000 business processes in R/3 into manageable views of best practices, which provides a top-down implementation approach in enterprise process areas such as production, procurement, and sales and distribution.

Internet Capabilities

In its efforts to target the Web and midsized companies, SAP formed Pandesic, a joint venture with Intel to market a turnkey electronic commerce package for consumer-to-business and business-to-business markets. Pandesic is an end-to-end system that handles all marketing, order processing and fulfillment, inventory pricing, materials management, tax processing, payment processing, shipping and handling logistics, financial reporting, and vendor payment processes associated with electronic commerce transactions. Customers and vendors will be able to check order status, review inventory levels, requisition balance sheet information, check "available to promise" information, and automatically manage financial transactions. In addition, the product includes R/3 logistical, financial, shipping, and general ledger modules; Microsoft's Windows NT 4.0 Server software; and Intel-based hardware (HP NetServer and Compaq ProLiant servers).

The business model for acquiring Pandesic solutions is different. Users will pay a per-transaction fee of 1 to 6 percent of revenue generated, depending on transaction volume, much as they would under a traditional credit card system. Customized applications can cost two to three times more than what it will cost customers for the Pandesic turnkey solution. Pricing varies, depending on the hardware and software configuration purchased, transaction volumes, and applications used. Also included will be USWeb's support and system integration services, Taxware International's tax-processing software, and CyberCash's payment processing. Additional partners include Citicorp for banking, UPS for logistics, and Yahoo! for advertising.

Pandesic promises a pre-configured product that, according to SAP officials, can be deployed in a matter of weeks for a startup cost of less than $100,000. SAP expects to attract midsized companies with the Pandesic solution. In addition, Pandesic promises ongoing software and hardware updates at no extra charge.

SAP and Intel delivered the Pandesic Internet business solution Release 1.0 during late 1997, with initial availability through Inacom and USWeb. Pandesic initially will operate in the U.S., with plans to conduct business internationally by mid-1998.

7.3.5 Oracle

Oracle has 4,000 customers in 59 countries. Oracle Applications provides support for more than 25 languages and includes more than 35 products for financial management, supply-chain management, manufacturing, HR, and sales force automation. Oracle Applications Release 10.7 NCA is fully Year 2000-compliant; prior releases are not. Because of the expense involved in bringing prior releases up to Year 2000 compliance, Oracle (`http://www.oracle.com`) announced that it intends to stop supporting pre-10.7 releases as of December 31, 1998.

Oracle Applications were introduced in the late 1980s and are based on the Oracle7 and Oracle8 RDBMSs. Oracle Applications Release 10.7 NCA, which shipped in early 1997, takes advantage of several new features of Oracle8, such as advanced queues, objects, and partitioning. Oracle8 server's "connection pooling," which supports more users simultaneously, is particularly useful for users of Oracle Applications for the Web (Oracle Web Employees, Oracle Web Customers, and Oracle Web Suppliers).

Oracle Applications run only on the Oracle database. The company believes that this strategy is one of its chief assets: It can use all the power of its database and tools in its applications. Because the Oracle database runs on more than 80 hardware platforms and over every major network protocol, the approach provides a high degree of platform independence.

Along with many of its competitors, Oracle targets vertical markets and enlists vertical market resellers, integrators, and consultants to sell into those markets. Oracle expanded its line of industry-specific solutions when it announced products in September 1997 for several additional markets, including energy, communications, consumer packaged goods, industrial sector, financial services, public sector, and healthcare. Oracle also added an Industrial Sector Vertical, with complete business solutions for industrial manufacturing customers and suppliers in engineer-to-order/aerospace and defense (ETO/A&D), consumer packaged goods, automotive, and high-tech industries. Each business solution includes industry-specific core Oracle Applications and partner products, consulting and implementation services, and customer support and education. More than 700 customers worldwide use industry-specific versions of Oracle Applications.

Specific vertical offerings include the following:

- **Aerospace and Defense** – Includes capabilities to help companies manage complex, global manufacturing enterprises, synchronize and optimize global supply-chain functions, and manage after-market operations
- **Oracle Project Manufacturing** – Provides a fully integrated solution (for project-based, ETO, and A&D companies) for planning, tracking, and managing the project definition, management, execution, and analysis activities of complex manufacturing environments
- **Oracle Service Resource Planning (SRP)** – For after-market service industry information management needs, SRP supports the entire maintenance, repair, and overhaul process. Initially, Oracle Maintenance, Repair, and Overhaul (MRO) will provide real-time decision support for the entire repair process; slated for availability in early 1998, with future offerings planned for mid-1998.
- **Public Sector** – Public Sector Budgeting and Public Sector Grants Management, announced in December 1997, are integrated with Oracle's Public Sector General Ledger, providing customized financial applications and project tracking/monitoring

Oracle's FastForward program is designed to get 100-user groups in midsized companies up and running on its financial or manufacturing applications within 3 to 8 months. Similar to SAP's AcceleratedSAP program, FastForward combines version 10.7 of Oracle Applications with Oracle's database, consulting, training, and a year of support.

Oracle has also enhanced its distribution channel to focus on midsized companies in the general business market. Oracle believes more than 200,000 public and private businesses make up this market. Under the Oracle Authorized Reseller program, qualified resellers are eligible for expanded training and support from Oracle.

Gartner believes that Oracle's strategy is to shift, through such partnerships, from being exclusively a database, technology-based product company to being an enterprise solutions provider. Forrester reports that many users often choose Oracle's applications because of the company's core technology strengths, database, and tools. Other customers worry that Oracle's focus on its database core may limit its commitment to the corporate packaged applications business.

Oracle Applications

Oracle Applications Release 10.7 NCA, scheduled to ship at the beginning of 1998, is the world's first suite of enterprise business applications based on Java. With Release 10.7 NCA, corporations will be able to implement enterprise applications that have all the benefits of graphical, client/server software without the cost and complexity of installing client software on each user's desktop. Instead, Oracle's Network Computing Architecture moves complex applications and data onto intelligent networks and network-based servers.

Oracle Applications comprises the following seven product families:

■ Financials

Oracle Financials is a suite of integrated financial applications designed for large, multinational organizations that are geographically dispersed. The Financial product suite provides functionality for asset management, financial planning, financial analysis, financial consolidation, expenditure management, billing and cash collection, and cash management.

■ Human Resources

Oracle Human Resources provides proactive management of a full range of HR functions, including recruitment, staffing, training, compensation management, payroll, and planning. Combined with Oracle Workflow, Oracle Human Resources Management facilitates recruitment processes. Oracle Human Resources 10.7 offers 35 applications for HR and finance department needs. Many are Web-enabled. Managers can access hiring, relocation, or assessment information; employees can access their own files.

■ Manufacturing

Oracle Manufacturing is an integrated enterprise resource planning and control solution for a variety of types of manufacturing processes, including discrete, repetitive, assemble-to-order, project-based, just-in-time, cell-oriented, KANBAN, and flow-line. This set of modules provides functionality for manufacturing planning, execution, and complete supply-chain management. Specific capabilities include workflow, purchasing, inventory management, material and capacity planning, production control, cost management, order administration, and quality management.

In early 1997, Oracle acquired Datalogix, primarily for its Global Enterprise Manufacturing Management System (GEMMS). GEMMS is a suite of client/server applications for managing manufacturing operations of multinational manufacturers. Although the GEMMS suite has been integrated with Oracle's enterprise applications for 2 years, Oracle is hoping to cull more from GEMMS into its other manufacturing applications.

■ Projects

Oracle Projects provides organizations with a comprehensive system to track costs. Functionality includes project tracking, cost collection, revenue accrual and billing, capitalization of project costs, and cross-project analysis. Oracle Projects is integrated with Oracle's core Financial Applications, and Oracle Projects' Activity Management Gateway provides integration with a company's preferred project planning and scheduling systems.

■ Sales Force Automation

Oracle Sales Force Automation delivers market and compensation analysis, compensation plan and sales quota modeling, marketing management, sales management, market expansion, and operational support.

■ Supply Chain Management

Oracle Supply Chain Management provides an integrated manufacturing and distribution planning solution, along with an electronic communications framework for internal and external procurement activities. Functionality includes supply chain planning from manufacturing through distribution, replenishment planning and release, materials management, sales order management, post-sales customer service, and quality management.

■ Sales, Service, and Support

Oracle's new Sales, Service, and Support application suite (released in September 1997) is targeted at the sales force automation market. This application suite includes Oracle Sales and Marketing, Oracle Sales Compensation, and Oracle Service. The suite is fully integrated with Oracle ERP, Financials, and HR applications to support sales across multiple organizations and currencies. It provides tight coordination of buying, selling, and manufacturing products and services.

Oracle Applications Release 11

The following features and functionality are expected in Release 11, which Oracle plans to deliver in early 1998:

- **Globalization support** – With the addition of functionality for Eastern Europe and Latin America, Release 11 will support 28 languages and more than 40 countries with specific local features. Release 11 will also include support for Oracle Human Resource Management Systems in Japan.
- **Third-generation self-service capabilities** – Including more than 20 new business flows (for a total of more than 70). Oracle Web Customers offers an improved Internet storefront; Oracle Web Employees includes new time- and cost-savings features such as a new laptop expense reporting system.
- **Best practices application of workflow** – To optimize transaction flows and business procedures

- **Support for networked supply-chain operations** – To allow planning beyond the enterprise
- **Extended multimode manufacturing** – Provides both project and flow manufacturing for customers in the A&D, construction, and other industries with make-to-order projects. Release 11 will also offer flow manufacturing for customers in work-orderless, repetitive environments. Advanced shipment and departure planning capabilities will help customers manage fulfillment.

Oracle Architecture

Oracle's Network Computing Architecture (NCA) is a cross-platform environment for developing and deploying network-centric applications for the Internet and the corporate enterprise. NCA provides a unifying framework that combines the robustness of the client/server environment with the ease of use and universal access of the Internet. NCA focuses on providing easy, component-based development with a high-level of flexibility and extensibility in a network-centric computing environment, tied together using industry-standard protocols such as CORBA, HTML, HTTP, Internet Inter-ORB Protocol (IIOP), and Java.

NCA consists of three tiers:

- **Universal Data Server** – Oracle8, the universal data server, provides an advanced product for data management.
- **Application Server** – Oracle Web Application Server 3.0 provides a scalable platform for delivering transaction-based applications for the Web.
- **Universal Client** – Any client device can be used to access applications and information within NCA, including traditional PCs, Java- or browser-based clients, mobile devices, and network computers.

Oracle's Developer/2000 and Designer/2000 for the Web are the main development tools for NCA. Developer/2000 allows Oracle client/server applications to be moved to an intranet or the Web by recompiling them. The development software has added a native Java applet client that lets its applications run from any Java-enabled browser and any operating system over the Internet, intranets, or virtual private networks. Designer/2000 supports the modeling of complex systems with business process reengineering, analysis, and design diagrammers.

Developer/2000 version 2.0, which shipped in November 1997, features a reusable component Object Library, object partitioning, and open connectivity to all major databases. A companion upgrade of the Designer/2000 database modeler is scheduled for early 1998.

Some companies are using a third-party product, Citrix's WinFrame (a multi-user Windows NT-based applications sharing system) to transform Oracle's architecture into a three-tier approach. WinFrame is a client/server network application that allows remote, server-based execution of any Windows or DOS application from a variety of clients, including PCs or X Window System terminals or network computers. With this approach, the Oracle client runs on the WinFrame server but can be accessed across the network, providing the ability to support thin clients or less-powerful PCs.

Internet Capabilities

Oracle Applications for the Web are designed for secure, self-service business transactions across the Internet and corporate intranets. Because security is a primary concern when using the Web for electronic commerce, Oracle Applications for the Web provides complete, end-to-end security – from the Web browser, to the Web server, and through the corporate firewall to the Oracle Applications server. Communications between the Web browser and Web server are protected by Secure Sockets Layer (SSL) data encryption, while those between the Web server and Applications server use advanced firewall security.

Oracle Applications for the Web is made up of three components that extend corporate information access to three new audiences: employees, customers, and suppliers. Each component allows its particular audience to use a standard Web browser to perform common business functions.

Key features of the three components include the following:

- **Oracle Web Employees** – Allows authorized employees to create and view requisitions, specify default requisition templates for each user, view supplier catalogs and expense reports, drill down to third-party Web sites, and review workflow notifications. This component is tightly integrated with Oracle Workflow, which automatically routes requisitions for approval, lets managers approve them on-line, and then sends confirmations to requesters.

  In September 1997, Oracle added disconnected expense reporting to the module. It will work with Oracle Payables to let mobile employees enter and track expenses offline using a spreadsheet, upload the reports with a standard Web browser, and route them for approval using Oracle Workflow. This module has translation, multi-currency, and multi-language capabilities

- **Oracle Web Customers** – Allows authorized customers to place and view sales orders as well as view invoices, payments, debit and credit memos, item forecasts, and an event calendar.

- **Oracle Web Suppliers** – Allows authorized suppliers to view supplier agreements, purchase orders, supplier items, invoices and payments, delivery schedules, receipts, and returns.

7.3.6 Geac

Having acquired Dun & Bradstreet Software (DBS) in late 1996, Geac promptly reorganized into two separate operating divisions: Geac SmartStream, marketing the SmartStream suite of client/server software applications, and Geac Enterprise Server, marketing the Expert and Millennium suite of enterprise server (mainframe) applications. The SmartStream division will target customers in the mid- to top-tier of global organizations that require the accessibility and flexibility of distributed systems. The Enterprise Server division will target large, global organizations that want to maximize the value of their existing systems or whose transaction volumes are beyond the capability of any client/server system. Geac also acquired Libra Corp.'s Signature Division and Pro-Mation in 1996, and is bolstering its infrastructure to accommodate the new acquisitions.

Geac experienced rapid growth following the acquisition of DBS as total corporate revenue jumped from $139 million to $500 million and employees increased from 1,200 to 3,000. The company's business strategy also changed following the acquisition. At its April 1997 user conference, Geac announced plans to focus on key vertical markets – financial services, health care, insurance, and public sector – as well as core human resources. The company also dropped a manufacturing module from its SmartStream offering. Geac has stated that it no longer considers itself an ERP vendor competing with high-end suppliers such as SAP and Oracle, but instead will concentrate on the specific markets listed above.

Geac Applications

The cornerstone of Geac's product line is its SmartStream enterprise client/server applications suite, originally developed by DBS.

■ SmartStream

SmartStream applications form an integrated solution based on enterprise-wide business processes. The SmartStream Builder toolset allows users to create links that enable the suite to work seamlessly with other applications on client/server and host-based platforms.

SmartStream consists of the following components: Financials, Budget, Decision Support, Human Resources, and Procurement.

- **SmartStream Financials** – This module is made up of six applications:
  - ◆ **Ledger** – Features include user-defined account key, integrated drill-back from ledger balances to supporting documents, multiple-level consolidations, and flexible spreadsheet integration.
  - ◆ **Receivables** – Features include management of receivables and customer remittances via cash, bank drafts, or electronic transactions; on-line histories of receivable and remittance activity, and automated credit line and collections facilities.
  - ◆ **Asset Management** – Features include various depreciation methods, country-specific tax reporting, and workflow-managed processes.
  - ◆ **Payables** – Features include automatic matching of invoices to purchase orders and receiving documents; resolution of matching exceptions via workflow; and flexible payment options (check, bank draft, or electronic transactions) in generic and country-specific formats.
  - ◆ **Allocations** – Features include a range of calculation methods, step-down allocations, and choice of outputs, including automatic creation of journal entries.
  - ◆ **Funds Control** – Targeted to the health care and public sector markets, features include originating a controlling budget for monitoring expenditures; creating the level at which the budget is established and funds checking occurs; real-time checking to ensure budgeted funds are available when processing invoices; preventing processing of documents if funds are not available; and providing query and reporting of funds availability status.
- **SmartStream Budget** – This module consists of three applications:
  - ◆ **Budget Management** – Features include the ability to manage multiple budget versions and create rolling budgets, to define the distribution of budget data throughout the organization, to define class types, and to load data from current financial systems into Budget Management.
  - ◆ **Budget Workbench** – Features include tools to adjust balances, update multiple amount classes or periods, view variances, create backup versions for distribution firmwide, and create reports from Workbench data.
  - ◆ **Budget Analysis** – Takes advantage of Decision Support (details below) for reporting and analysis during the budget process.
- **SmartStream Decision Support** – This module consists of five applications:
  - ◆ **Analyzer** – Features include ranking, sorting, and exception highlighting; drill-down using SmartStream Structures (details below); the ability to create new categories of data as needed; and electronic distribution of Analyzer views to other users.

- **Management Reporter** – Features include reusable report components for consistency and easy maintenance, automatic generation and distribution of reports, and presentation-quality formatting tools.
- **Query & Reporter** – Features include predefined parameters, a wide range of report design capabilities, the ability to export selected data to desktop applications, and electronic distribution of output via e-mail.
- **Connect** – Features include the ability to extract and transfer data to the Decision Support server from other databases and host sources, upload data to the host for processing, and to populate SmartStream's structured query language (SQL) databases automatically.
- **Decision Support Viewer** – Features include the ability to view and print reports; drill down through data; add comments, calculations, data fields and exception highlights to reports; forward modified reports to others; and copy information to desktop applications.

- **SmartStream Human Resources** – This module consists of three applications:
  - **Personnel** – Features include separate management of job and position information; worksheets to manage all steps of recruiting process; automated approval routing for changes; a skills inventory to conduct internal searches; and an integrated system to manage training classes, schedules, registrations, approvals, and costs.
  - **Payroll** – Features include flexible calculation of multiple levels of taxation; internal and third-party payroll deductions; accruals and advances for time off; generation of checks or direct deposit transaction files; processing and audit reports; and automatic general ledger entries.
  - **Benefits** – Features include management of plan parameters; integration with accounting systems; a variety of benefits reporting formats to meet employee and regulatory requirements; and easy changes to employee status.
- **SmartStream Procurement** – SmartStream Procurement includes comprehensive receiving processes and automated matching of invoices to purchase orders and receiving documents. The module consists of two applications:
  - **Purchasing** – Features include processes for repeat purchases (from on-line catalogs or pre-completed requisition forms); routing for pricing and specification requests; workflow routing of requisitions for approval; and automatic generation of purchase orders via fax or electronic transaction.
  - **Payables** – Features include on-line or electronic invoice entry; flexible payment options in generic and country-specific formats; invoice tax processing; and payment scheduling.

SmartStream 5.0

In December 1997, Geac announced SmartStream version 5.0, which includes two new applications, Projects and Supplies Management, both of which are tightly integrated with the existing SmartStream suite of financial and procurement applications.

- **SmartStream Projects** – Features include easy access to information via integration with SmartStream applications such as Ledger, Asset Management, and Purchasing; advance notice of potential budget overruns via warnings or error messages when predetermined tolerance limits are reached; the ability to customize Projects to meet specific needs, decentralize tasks, and distribute activities among users and locations; templates to streamline project tracking; and multinational project management. SmartStream Structures lets users define the structure of a project using a GUI as well as the relationships between project components.

- **SmartStream Supplies Management** – Designed for businesses that consume inventory internally (such as those in the health care, utilities, insurance, and banking industries), this solution enables management of the entire inventory business process, including ordering, receiving, put-away, picking, transfers, and replenishment across multiple sites and stockrooms. Integration with SmartStream Financials lets users view results graphically.

Other new features in SmartStream 5.0 include a new System Verification Tool, which allows system administrators to detect and isolate any SmartStream environmental problems that might delay implementation.

■ Vision 360°

In March 1997, Geac SmartStream announced a new strategy to help secure its place in the global applications software market: Vision 360°. The goal of Geac's Vision 360° solutions is to help companies manage competitive business operations and address issues such as customer service, ensuring employee productivity, leveraging market intelligence, and utilizing strategic alliances. Components of Vision 360° are scheduled to roll out through the end of 1998.

Vision 360° is made up of seven elements:

- **DecisionStream** – Adds decision-support functionality for specific vertical industries via analytic applications that work with subject-specific data marts. Examples of solutions include product line profitability, supplier analysis, customer portfolio analysis, and trading partner relationship information.
- **IntelliScopes** – Gather information from disparate sources and integrate them into a tab-like GUI on the client side
- **IntelliMaps** – Graphically chart workflow
- **Decision 360°** – Analyzes and views information and deploys it to users
- **CyberStream** – Allows users to access SmartStream solutions via the corporate intranet with a Java-based Web front end
- **Operational and administrative systems** – Continuation of SmartStream applications in areas such as financials
- **SmartStream Connections** – A set of JavaScript and ActiveX APIs that link legacy systems to SmartStream Vision 360° solutions. Connections includes two components:
  - **SmartPath** – Facilitates migration from host solutions to SmartStream
  - **SmartStream Connect** – Facilitates data transport between a host and SmartStream

Geac Architecture and Internet Capabilities

Geac has begun to rearchitect and migrate its SmartStream suite to the Internet by using Java and the CORBA Internet Inter-ORB Protocol (IIOP) standard for integrating objects with the Internet. The first application to take advantage of Geac's next-generation architecture is CyberStream Employee (introduced in April 1997), a corporate intranet application that allows direct user access to personnel information. Using any Java-based browser from a PC, an employee can perform self-service, routine HR activities such as name and address changes and benefit inquiries. CyberStream Requestor for purchase requisitions is also available.

Future CyberStream offerings in development (as of November 1997) include the following:

- **CyberStream Manager** – Will enable management of recruiting activities, approval processes, and direct report administration
- **CyberStream Open Enrollment** – Will allow employees to change benefit status
- **CyberStream Time Management** – Will address leave management, vacation scheduling, and labor distribution.

7.3.7 Computer Associates

A developer of client/server solutions, CA offers more than 500 products for heterogeneous, distributed systems. These include visual application development solutions; multiple-access database management systems; customizable, distributed business applications; and end-to-end, distributed enterprise management solutions. This discussion will only cover CA's core business applications.

CA's Prestige Software International division (formerly the Enterprise Financial Management Group) is dedicated to delivering Web-enabled enterprise financial management and HR management software for a range of platforms and operating environments.

CA Applications

The CA-Masterpiece financial management suite, introduced in 1976, has more than 2,000 customers worldwide. CA-Masterpiece links tightly to CA-Unicenter enterprise management software, allowing companies to reach across the Web or a wide area network to suppliers and customers. CA-Masterpiece also uses CA-Unicenter's agent technology to alert users proactively to business conditions that require attention.

CA-Masterpiece provides comprehensive support for multinational requirements, offering multiple currencies, regulatory support (including EMU requirements), multiple languages, and multiple financial consolidations. The CA-Masterpiece suite includes ten modules: consolidation, general ledger, accounts payable/receivable, fixed assets, fund accounting, inventory control, job cost, labor distribution, and purchasing as well as query, reporting, and analysis tools.

■ CA-Masterpiece/ICE

In March 1997, CA announced the release of CA-Masterpiece/ICE, a multinational, multi-organizational enterprise financial management solution that provides Web-enabled access to financial data. The software package covers the same areas as CA-Masterpiece, and lets Masterpiece users build and manage business applications on the Internet. CA-Masterpiece/ICE offers networked multimedia financial management capabilities for businesses to perform accounting functions and financial reporting on the Internet and corporate intranets.

■ CA-Masterpiece/Net

In June 1997, CA announced CA-Masterpiece/Net (New Enterprise Technologies), a "visual" financial management approach to enterprise management, global business communications, and all the CA-Masterpiece functions using any

Java-based browser. CA-Masterpiece/Net financial software supports the same computing platforms as CA-Masterpiece/ICE, and permits access from any browser with Java support.

CA-Masterpiece/Net Consol consolidates and reports on decentralized, disparate, and multi-currency financial information. It enables complex, multinational consolidations independently of the accounting cycle of the general ledger.

■ CA-HRISMA

Introduced in 1993, CA-HRISMA is multi-platform client/server HR management software that provides payroll, personnel, and benefits administration:

- **Payroll** – Provides tax calculations, defined contributions, retroactive pay, time and attendance tracking, direct deposit, on-line manual checks, an integrated tax reporting system, and financial system interface
- **Personnel** – Provides user-defined histories, position control, applicant and requisition tracking, career development, employee profiles, and compensation administration
- **Benefits** – Provides plan administration, employee enrollment, pension, automatic sick leave and vacation accrual, benefits statements, and deferred compensation

CA Architecture and Internet Capability

CA-Masterpiece was one of the first solutions to address the security and management issues raised by the integration of business systems via the Internet. CA-Masterpiece/Net uses browser technology and integrates the client with the Web server to provide significant cost savings.

7.3.8 PeopleSoft

PeopleSoft, with more than 1,700 customers around the world, provides a suite of enterprise solutions for finance, materials management, distribution, supply chain planning, manufacturing, and human resources. In addition, PeopleSoft provides industry-specific enterprise solutions to customers in select markets, including automotive, communications, financial services, healthcare, manufacturing, higher education, pharmaceutical, public sector, retail, services, telecom, transportation, U.S. government, and utilities.

PeopleSoft is expanding aggressively beyond its original HR expertise into the overall packaged applications market. The company's 1996 acquisition of Red Pepper, a leader in developing manufacturing and supply chain optimization software for the enterprise, helped open the manufacturing systems market to PeopleSoft's existing HR, payroll, and financial applications.

PeopleSoft Applications

PeopleSoft 7, introduced in April 1997, debuts the company's move to a three-tier architecture, using a transaction middle layer employing BEA's Tuxedo transaction manager. PeopleSoft 7 supports Windows NT, UNIX, or MVS mainframe database server and a range of RDBMSs, including IBM's DB2, Oracle, and Sybase. In addition, the release adds OLAP integration and Java-based Web applications, and includes new manufacturing modules, including integrated engineering and product configuration. PeopleSoft 7 also delivers Cube Manager, a data integration tool that lets users define their own OLAP dimensions and analysis models.

PeopleSoft partners with developers for key elements, instead of building its own. PeopleSoft also integrates several other vendors' capabilities for "resume" scanning, interactive voice response, expatriate taxes, bar coding and scanning, and sales tax compliance. For example, PeopleSoft 7 integrates Cognos' PowerPlay and Arbor's Essbase OLAP tools, enabling users to analyze multidimensional data stored in various locations. To ensure open applications, PeopleSoft 7 includes an Open Query API, based on the data access standard ODBC API, to let third-party reporting tools access data from PeopleSoft applications.

PeopleSoft 7 offers the following functionality:

- **PeopleSoft HRMS** – Provides personnel administration, recruitment, position management, salary administration, training and development, health and safety, skills inventory, and career planning. Fully integrated modules include PeopleSoft Human Resources, Benefits Administration, Flexible Spending Accounts (FSA) Administration, Payroll, Payroll Interface, Time and Labor, and Pension Administration. HRMS also accounts for all regulatory and legal requirements, both within the U.S. and internationally, and is available in Dutch, English, French, German, Portuguese, and Spanish.

- **PeopleSoft Inventory** – Supports inventory management, including the ability to store and issue stock in response to changing demands, track movement of stock in real time, and automatically replenish stock as needed

- **PeopleSoft Order Management** – Supports real-time order processing and customer service for multinational and multi-site businesses. It applies electronic commerce, workflow, and multimedia technologies to improve the quality of data held in sales orders, and automates the sales order cycle.

- **PeopleSoft Purchasing** – Supports management of requisitions, vendor contracts, purchase orders, and receiving on-line

- **PeopleSoft Manufacturing** – Completely integrated with PeopleSoft Distribution, Financials, and HRMS, PeopleSoft Manufacturing applications include PeopleSoft Engineering, Bills and Routings, Production Planning, Production Management, and Cost Management, as well as the Red Pepper line of supply chain optimization software solutions (details below).

- **PeopleSoft Distribution** – Provides logistics and materials management. Modules include Enterprise Planning and Product Configurator.

- **PeopleSoft Financials** – An integrated suite of financial applications, including PeopleSoft General Ledger, Receivables, Payables, Billing, Asset Management, Budgets, Projects, Treasury, and Expenses.

- **PeopleSoft Universal Applications** – Java-based, cross-platform applications built with an interactive interface based on a standard Web browser are designed for self-service administrative tasks. Universal Applications are linked to PeopleSoft's core Financials, HRMS, and Distribution applications.

New features in the Red Pepper supply chain optimization solutions include these:

- **Sales order prioritization** – Lets users dynamically prioritize sales order shipments by customer, product, margin, revenues, promise and request dates, and sales territory.

- **Equipment preference** – Lets production planners specify alternate equipment preferences, using a weighting factor, if the primary piece of equipment is unavailable because of capacity constraints.
- **Enhanced sourcing logic** – Considers materials at multiple levels, aggregate capacity, or both, to enable users to balance inventory and capacity across the enterprise.

PeopleSoft Select

Like its competitors, PeopleSoft is targeting midsized companies. In a new offering, announced in October 1997, PeopleSoft is partnering with Microsoft to market Microsoft BackOffice as the software platform of choice for PeopleSoft Select, the company's solution for the midsized-market.

PeopleSoft Select is a total software, hardware, and services offering that includes a choice of PeopleSoft applications plus Intel architecture-based hardware from Compaq or HP; the Windows NT network operating system; and Microsoft SQL Server database as the platform of choice. PeopleSoft will deliver PeopleSoft Select through a new division that includes a dedicated sales, service, and development unit, and partnerships with hardware, software, and implementation providers.

PeopleSoft Select, which should enable companies to be up and running within 3 to 6 months, bundles the following hardware, software, and services:

- **PeopleSoft Financials** – General ledger, receivables, payables, asset management, budgets, and projects
- **PeopleSoft Distribution** – Purchasing, billing, order management, and inventory
- **PeopleSoft HRMS** – Human Resources and payroll interface
- **PeopleSoft Manufacturing** – Scheduled to be available with PeopleSoft SelectSet in 1998
- **Microsoft BackOffice:** Microsoft SQL Server database and Microsoft Windows NT
- **Intel-based hardware from Compaq or HP**
- **PeopleSoft implementation and support services**

The full PeopleSoft 7 functionality is available with PeopleSoft Select, including PeopleTools development environment, reporting and analysis tools, OLAP integration, workflow, and Web client options. PeopleSoft's direct sales approach differs from competing vendors, which are shifting to an indirect mode to target smaller companies.

PeopleSoft SelectPath

SelectPath is PeopleSoft's new rapid implementation methodology that provides midsized companies with a low-cost, fixed-bid implementation within 3 to 6 months. SelectPath provides a set of specific steps and activities based on a customer's business requirements. PeopleSoft is developing a configuration tool to automate the SelectPath methodology, and will deliver implementation services through PeopleSoft Select Implementation Centers. The centers will provide training and knowledge transfer; business process deployment; general system setup, build, and reporting interface planning and development, and data conversion. As of December 1997, PeopleSoft Select was available in Boston, Chicago, Los Angeles, New York, San Francisco, and Toronto, Canada; additional locations will be added throughout 1998.

■ PeopleSoft PeopleTools

PeopleTools is an object-based application development, implementation, globalization, and customization environment bundled with all PeopleSoft products. PeopleSoft claims that PeopleTools provides developers with predefined business functionality to simplify design and cut maintenance costs.

Development tools for systems analysts include tools to design records, panels, and menus; operator and object security capabilities; built-in application globalization facilities; an on-line help designer; process scheduling tools; and workflow design tools and capabilities. Productivity tools for IT include automated application installation and Windows-based application upgrades. Similar tools for users include an interactive reporting and analysis tool with drill-down capability, the ability to manage and change enterprise-wide hierarchical structures, and a database query tool with a point-and-click interface.

■ PeopleSoft 7.5

Scheduled for release in mid-1998, PeopleSoft 7.5 will include more functionality, including global distribution and manufacturing and demand planning improvements to supplement the Red Pepper planning engine; industry-specific functionality; several new applications; a new implementation toolset; and global payroll for France, Germany, and the U.K.

PeopleSoft 7.5 will include the following enhancements:

- **Global Distribution and Manufacturing** – Will support the full range of languages and currencies supported in PeopleSoft's global Financials and HRMS application suites.

- **European Monetary Union (EMU) support** – Will enhance existing accounting capabilities with additional EMU functionality, including multi-book capabilities in the Payables and Receivables applications for supporting the automatic conversion of transactions. With the release of PeopleSoft 7.5, PeopleSoft Purchasing will support multi-currency processing, and PeopleSoft Expenses will support the same level of multi-currency functionality as other PeopleSoft Financials applications.

- **European Payroll** (codeveloped with Automatic Data Processing) – Will include implementations for all countries supported by PeopleSoft HRMS. It will let users control all aspects of payroll operation in a multinational environment, including multi-language and multi-currency support for France, Germany, and the U.K.

- **Performance Measurement** – Will let organizations track key performance indicators by providing activity-based costing, economic value added, and multi-dimensional revenue and profit analysis capabilities that are integrated with PeopleSoft General Ledger.

- **PeopleSoft Quality** – Will allow quality engineers to identify trends and address manufacturing process problems by gathering real-time data from the shop floor.

- **Demand Planning** – Designed to integrate the forecasting process with supply planning, the application will support collaboration across multiple departments within the enterprise as well as among supply chain partners.
- **Treasury** – Consists of these applications:
  - **Cash Management** – To help cash managers monitor and forecast cash needs, maintain bank relationship information, and perform bank reconciliation
  - **Deal Management** – Will address the needs of a treasury organization that transacts in a broad range of financial markets by streamlining deal initiation, administration, settlement, accounting, and position monitoring
  - **Risk Management** – Will support the valuation and analysis of a treasury's portfolio of transactions and exposures
- **Expenses** – Will let multinational organizations exercise control over expense management processing while enabling timely and efficient employee reimbursement.
- **Stock Administration** – Will let organizations administer and maintain employee stock purchase plans and incentive and non-qualified stock option grants.
- **Total Compensation** (in the HRMS application suite) – Will allow users to consolidate factors in a cube, including functions from the PeopleSoft Payroll, Human Resources, and Benefits applications as well as third-party compensation components.
- **Global Human Resources support** – Additional support for Belgium, Japan, and the Netherlands as well as expanded global and local functionality of the HR application for companies doing business in Canada, France, Germany, and the U.K. New functionality will include diversity tracking, preferred language tracking, and multilingual reporting capabilities.
- **Financial Services** – In addition to industry-specific functionality, PeopleSoft 7.5 will deliver new applications for the financial services industry.
- **Service Industries** – Enhancements to PeopleSoft Projects will meet the specific needs of service organizations, including complex project-based billings and revenue recognition as well as enhanced optimized resource planning and deployment through integration with Primavera P3 and Microsoft Project desktop applications. New global enhancements to the HRMS application suite will also help address the specific challenges of service organizations in recruiting, deploying, and compensating a multinational work force.
- **New Implementation Tools** – To support low-cost, rapid software implementations, PeopleSoft will make DirectPath, its implementation methodology and toolset, available to all customers. DirectPath is an enhanced version of PeopleSoft's current implementation approach, consisting of a structured implementation method, including a sequence of predefined deliverables and templates, and a software toolset. General availability of the DirectPath methodology and toolset is scheduled for mid-1998.

PeopleSoft applies a two-pronged approach to developing industry-specific solutions. First, it partners with industry-leading vendors that specialize in each particular market. And second, the company uses its toolset to build functionality specific to each industry.

- **Industry-Specific Solutions** – PeopleSoft has announced enhanced functionality and tailored solutions for the following industries:

- **Manufacturing** – New functionality will include date collection and bar code enhancements to streamline reporting of production and allow for more efficient inventory management throughout the enterprise.
- **Retail** – PeopleSoft and Intrepid Systems have partnered to develop a joint solution for retailers that integrates Intrepid's merchandising management software with PeopleSoft's enterprise applications.
- **Higher Education** – PeopleSoft has developed the Student Administration application suite to complete its enterprise solution for the higher education community, including Financials for Public Sector and HRMS for Public Sector. In addition, two new applications will be included with PeopleSoft 7.5 – PeopleSoft Grants to support research proposal development and award management, and PeopleSoft Advancement to support fundraising, donor, and alumni relationship management for universities. PeopleSoft plans to make Student Administration available by late 1998.
- **Federal Government** – Availability of PeopleSoft's HRMS for Federal Government and Financials for Federal Government is planned for late 1998.
- **Public Sector** – Availability of PeopleSoft's HRMS for Public Sector and Financials for Public Sector application suites is planned for late 1998.
- **Healthcare** – Enhancements to PeopleSoft's Materials Management solution (Purchasing and Inventory) will provide the healthcare industry with improved cost management capabilities, including functionality for stockless part inventories, procurement cards, electronic catalogs, and bar code entry.
- **Communications, Transportation, and Utilities** – PeopleSoft Projects will contain improved integration with PeopleSoft Asset Management for capitalizing asset-intensive projects. PeopleSoft's Asset Management application will add expanded support for operating leases to its financial and tax accounting for fixed assets.

■ PeopleSoft 8

PeopleSoft has announced that PeopleSoft 8, scheduled for release in 1999, will move toward a component-based architecture using object technology and will include new applications such as labor scheduling and continuous administration.

PeopleSoft Architecture

With PeopleSoft 7, the company shifts away from its original fat-client architecture and provides Web support by delivering three-tier processing and a Web client with self-service applications. By integrating BEA's Tuxedo middleware into its application software, PeopleSoft has moved toward a three-tier architecture. The three-tier architecture lets users move to a distributed computing environment by moving the processing layer of PeopleSoft applications from desktops to servers.

Companies can continue to use PeopleSoft's traditional two-tiered, fat-client approach that keeps business and presentation logic on the client, or separates business logic and database access onto an applications server managed by Tuxedo. PeopleSoft claims that the second option cuts network traffic and boosts the average response time, both by more than 50 percent.

Internet Capabilities

Although PeopleSoft has lagged behind in Web development, PeopleSoft 7 provides Web interfaces for 15 self-service applications, based on three downloadable Java applets and BEA's Jolt middleware, known as Universal Applications. Rather than place all its software modules on the Web, PeopleSoft will support universal access to the applications users require most, such as employee data updates, benefits checking, and office supply ordering.

Combined with a Java-based Web client for browser access to business transactions, queries, and workflow, users can access those applications from any standard Web browser. Universal Applications provide occasional users with the ability to perform Internet-based self-service tasks associated with HR, payroll, benefits, accounting, and distribution.

PeopleSoft formed a partnership with Netscape to develop the Java-based administrative applications that integrate Netscape's messaging, security, and directory services with PeopleSoft's corporate manufacturing, distribution, financial, and human resources solutions.

7.3.9 J.D. Edwards

J.D. Edwards is one of only a few companies that has developed and marketed its own ERP solution. After having been in business for more than 20 years, J.D. Edwards held an initial public stock offering in September 1997.

To enable continued growth, J.D. Edwards has increased the available platforms for its offerings from primarily AS/400 and other IBM midrange systems to leading UNIX and Windows NT servers through Windows and Internet browser-enabled desktop clients.

J.D. Edwards Applications

J.D. Edwards's products support 20 languages and feature multi-currency functionality; preliminary work has begun on EMU compatibility. J.D. Edwards offers four product suites that are differentiated primarily by technology:

- **WorldSoftware**

WorldSoftware is a procedural, host-centric architecture running on the AS/400 platform. It is also capable of operating with OneWorld in a unified, enterprise-wide environment. J.D. Edwards's coexistence strategy allows both solutions to reside on the same AS/400 platform using the same database. This setup allows users to switch easily between the two, using new modules as they become available, and provides an easy migration path for users moving from WorldSoftware to OneWorld.

- **WorldVision**

WorldVision is a thin-client bridge that provides the GUI look and feel common to the PC while enabling users to make a safe move to client/server that leverages existing host-centric WorldSoftware applications.

The next release of WorldVision, scheduled for late 1997, is Java-enabled, allowing customers to have graphical access to the software through the use of common browser technology or a network computer. For example, users can have WorldVision as a Windows 95- or Windows NT-style GUI for a PC and as a Java-based interface for use across the Internet, intranets, or extranets. The newest release also includes enhancements in repetitive manufacturing, quality management, user-defined depreciation for fixed assets, customer information systems billing for utilities, advanced maintenance management, multi-site consolidation, intercompany settlements, and advanced payment terms.

■ OneWorld

OneWorld's network-oriented environment introduces J.D. Edwards's Configurable Network Computing (CNC), a distributed object architecture. Released in November 1996, OneWorld runs on UNIX and Windows NT as well as AS/400 platforms by using a proprietary object request broker. In July 1997, J.D. Edwards enhanced OneWorld to provide greater Internet and Java functionality. It now features an improved GUI and real-time access to enterprise applications. Enhancements are upgrade-protected or flagged if J.D. Edwards changes that functionality in the next release; and panel displays can be synchronized, displayed simultaneously, and updated dynamically, allowing users to create an event-based interface. The GUI can incorporate other desktop applications via OLE.

The OneWorld suite consists of four modules:

- **Human Resources** – Newest functionality includes supplemental employee data and mass change processing
- **Finance** – New features in the accounts receivable application include simplified transaction processing, the ability to conduct direct audit trails, and database information sharing
- **Manufacturing** – Planned enhancements (in 1998) include repetitive manufacturing, electronic KANBAN, vendor release schedules, procurement-evaluated receipts, and pay-on-consumption
- **Sales and Distribution** – Future enhancements (in 1998) include pricing for setup and maintenance charges, material handling unit bar codes, forecasting for consumer packaged goods, and warehousing

J.D. Edwards also provides WorldSoftware and OneWorld toolsets to enable rapid implementation, customization, and modification. OneWorld's toolset extends the applications with workflow, data warehousing, reporting, and application development and deployment capabilities.

■ Genesis

Genesis is a prepackaged enterprise solution for smaller private and public sector organizations (with annual revenues of less than $100 million). Based on World-Vision technology, Genesis also includes on-line education and implementation tools called Teacher and Coach to help speed implementations. At press time, Genesis was available only in the U.S.

J.D. Edwards is also developing an automated configuration tool called Composer for OneWorld, which will use the same top-down series of questions and answers employed by its Genesis product. Composer will allow users to configure business processes, the GUI, and workflow. Release is scheduled for mid-1998, with fully functional industry templates to follow throughout 1999.

J.D. Edwards Architecture

OneWorld is a component-based application built on J.D. Edwards' proprietary development middleware and tools, which shield OneWorld from platform-related issues.

J.D. Edwards' CNC network-centric architecture allows users to change network configuration at runtime to support changing processes, organizational structures, and technologies. CNC allows users to operate OneWorld as a single, logical view in a heterogeneous platform (hardware, operating system, and database) environment. According to J.D. Edwards, the goal of CNC is "to bring stability, simplicity, and flexibility to the world of open systems."

CNC adopts a modular approach to the computing process in which the architecture comprises five functional elements: presentation, logic, database, data warehouse, and reporting/batch processes. In contrast to a more linear or procedural approach to computing, CNC does not dictate when, how, or where each component process takes place, allowing users to balance processing loads across the network as applications and needs change.

■ Internet Capability

When J.D. Edwards decided to add Java into OneWorld applications, it added Java into the toolset used to build the applications and started recompiling them. This approach allowed the applications to inherit Java capabilities. The process also gave J.D. Edwards two ways to deliver OneWorld applications: by a specific OneWorld client, or via a browser, allowing access to an intranet, an extranet, or the Internet. Applications can be accessed over the network from anything that runs a Java Virtual Machine, such as a browser, allowing almost any device to function as a OneWorld client.

J.D. Edwards also plans to Java-enable WorldVision.

7.3.10 Baan

Baan provides a scalable suite of enterprise business applications that address the complete value chain, from ERP and supply chain management, to customer interaction software from its subsidiary, Aurum Software. To reduce software complexity, Baan employs an open architecture and uses a Dynamic Enterprise Modeling (DEM) approach that puts business requirements at the heart of the implementation process.

As with other major vendors, alliances and acquisitions are key strategies for Baan, which acquired Aurum for sales force automation, Antalys for sales configuration systems, and Berclain for supply chain planning. However, Baan has yet to integrate the companies and their applications into a united business model. Its partnership with Baan will likely cause Aurum's other partners, such as Oracle, to switch to different sales force automation companies such as Scopus or Siebel. Baan also partners with leading software suppliers such as HP, IBM, Informix, Microsoft, and Sybase.

As part of its thrust into the midsized market, Baan was the first ERP vendor to support Microsoft BackOffice, shipping its Baan IV BackOffice software in May 1997. Baan IV BackOffice features a low cost of entry and quick installation on top of its DEM architecture, which supports rapid application implementation and pre-configured, out-of-the-box application solutions to customer's business processes.

The company sells into six vertical industries: high technology/electronics, process industries, automotive, project industries, aerospace and defense, and hybrid manufacturing. Baan has six major applications: Manufacturing, Finance, Project, Service, Transportation, and Distribution. With no HR applications, Baan remains confined primarily to the manufacturing space. However, Baan has entered into a cross-market and joint-development agreement with Hyperion for financial management software.

Baan Applications

Baan IV is an integrated family of business management software consisting of five main components: Baan Orgware, Baan Desktop, Baan Internet, Baan Applications, and Baan Technology. Baan IV was released in April 1996 and enhanced in October 1996. Baan IV features graphical representation of business process flows and system customization, workflow support, and permissions. Baan's products support 30 languages and are Year 2000-compliant.

■ Baan IV

Baan IV is a multi-language, multi-currency business management application that supports various types of manufacturing processes, including make-to-stock, assemble-to-order, make-to-order, engineer-to-order, repetitive and batch as well as process verticals, and transportation environments. Baan IV includes a project control/planning system, supplier scheduling, financials, logistics, process control, shop-floor monitoring, EDI, a product configurator, an integrated service and maintenance system, and distribution and simulation capabilities.

Compared to Baan's earlier Triton applications suite, Baan IV offers expanded supply chain management support, "configure-to-order" ERP, shorter configuration and installation times (Baan claims that most of its software can be implemented in 3 to 12 months), and Internet and intranet capabilities under Baan's DEM. Using DEM, Baan IV matches a company's business processes with the organization, computing environment, desktop, and Internet connectivity functionality of other Baan applications, and then configures the product automatically. Baan's automatic configuration capability stands out, compared with other leading ERP vendors.

In late 1996, Baan enhanced Baan IV with the following:

- **Enterprise Reference Models** – Provide industry-specific templates of standard business practices for quick-start business process models:
  - **Project Industries** – For companies involved in building or planning complex, multi-phase projects with engineering, procurement, and construction phases
  - **Manufacturing Assemble-to-Order** – For discrete manufacturing companies that execute assemble activities to customer order and manufacturing components or subassemblies
  - **Financial Management and Reporting**

In October 1996, Baan also introduced wizard support and functionality for its enterprise reference models. Wizards automate complex application implementations. Baan also introduced Orgware enterprise wizards for customers and service providers to install and implement Baan IV BackOffice applications quickly and easily on Windows NT. Baan expects to deliver similar capabilities for UNIX platforms by the end of 1997.

■ Baan Orgware

Baan Orgware is a dynamic process modeling tool that allows organizations to speed deployment of, or changes to, the applications. Baan's Orgware application modeling suite consists of these components:

- **Enterprise Modeler** – Tools for graphically modeling an organizational structure, business processes, and people

- **Enterprise Performance Manager** – Tools for tracking business performance benchmarks
- **Enterprise Implementer** – Tools and methodology for managing the implementation process, including project management and consulting services

■ Baan IVc

Baan IVc, released in October 1997, runs on Windows NT, HP-UX, Solaris, Digital UNIX and AIX as well as other flavors of UNIX. Baan IVc includes the following features:

- **Internet connectivity across the supply chain** – Including enhanced multi-currency finance support, advanced data search tools, and integration with partner and affiliate products that allow customers to take advantage of a broad range of integration with complementary software products.
- **Enhanced financials support** – Including enhanced multi-currency reporting support, cost control systems, and connections with Hyperion's advanced financial solutions, including on-line analytical processing and sophisticated financial consolidation and budgeting. To address the EMU mandate, Baan extends its multi-currency capabilities to execute financial accounting in up to three currencies simultaneously.
- **Baan Product Data Management** – A product from B.A. Intelligence Networks that cuts time to market, delivers faster product innovation, and improves the ability to adopt concurrent engineering practices.
- **Baan Synchronization** – A constraint-based scheduling and execution management system based on Berclain technology that enables coordination of plant-level production decisions with market demand and enterprise goals.
- **Baan Controlling** – A project management tool for planning and controlling costs. Baan IVc enhancements involve new planning, budgeting, and allocation functionality, which let customers worldwide implement flexible budget accounting systems.

■ Baan Sync

In November 1997, Baan announced a new family of products for global supply chain planning, leveraging the expertise of Berclain Software, which Baan acquired in early 1997. The new Baan Sync series will bridge the gap between transaction-based ERP and planning engines that use complex calculations to manage resource optimization. The Baan Sync series is made up of three integrated applications for constraint-based planning, scheduling, and shop floor execution. Baan says that the new Sync series will be available with Baan V, scheduled for release by mid-1998.

Baan Architecture

Baan bases its Baan IV applications on a three-tier client/server architecture that separates user presentation, applications logic, and data management functions. Baan IV runs on UNIX clients and servers as well as Microsoft Windows NT, Windows 95 (client only), and BackOffice, and is available for Oracle, Informix, DB2/6000, and Microsoft SQL Server databases.

Baan V will start Baan's move to a component-based architecture. It will be Internet-enabled via a separate Java-based user interface, and include new Internet-based workflow capabilities that will focus on push technology. Baan will also drop support for the X Window System and dumb terminals in favor of Internet browsers and network computers. Baan expects the first Web-enabled product to ship will be supply chain applications from its Antalys acquisition. Future Web-enabled supply chain applications will use push technology to manage a complete Web-based order entry process within a single application. Baan V also should improve financial functionality, including EMU multi-currency support.

Baan's component strategy includes componentization of the Baan application family as well as Baan's Business Object Interfaces (BOIs) and its eXtended Middleware Architecture (XMA), which will form the framework for linking both Baan components and third-party solutions.

Internet Capability

In May 1997, Baan introduced its Internet strategy and first Web-enabled modules. Support for Java query and reporting clients will be part of the Baan IVc release; full Java-based access to Baan transactions and Baan Web-specific applications will be featured in the Baan V release.

Baan's Internet strategy includes three parts:

- Development of a Java-based Web user interface for existing Baan applications
- Addition of new workflow-based capabilities that will focus on push technology. Future Web-enabled supply chain applications will use push technology to manage a Web-based order entry process within a single application.
- Creation of a suite of stand-alone Web solutions that run against any ERP suite

In addition to Web-enabling its application suite through a Java-based front end, Baan will create new workflow-based applications for Internet, intranet, and extranet transactions.

7.3.11 Standards Efforts

Users of corporate applications want multi-vendor standards so they can mix and match applications seamlessly, at less cost, and with less effort. The Open Applications Group (OAG) is the primary organization pushing to mandate standard interfaces between packaged corporate applications.

Microsoft and SAP are working together on what both hope will become an industry-standard, administrative-application architecture. For now, Microsoft and SAP recognize that a partnership between the two dominant players can have a far greater impact on the industry than each can have alone. In the long term, however, these two software vendors may have to overcome the historical reality that mega-company joint ventures often fail.

Open Applications Group

OAG is a consortium of vendors formed in 1995 to establish specifications for interoperability between packaged corporate applications from different vendors. Corporate members include many of the world's largest corporate applications vendors, although each has a unique way to link external applications to its own. OAG's most recent specification – OAGIS Release 5 – provides functionality for the most-needed ERP integration functions, including core financial, human resources, manufacturing, and logistics applications.

Although both Microsoft and SAP are improving their relationships with official standards bodies such as OAG, the clout of such organizations cannot match that of the two industry giants. In particular, OAG's interface standards initiative appears to be stalled. Today, no one uses OAG's interfaces, and OAG's primary impact has been for SAP to open up its architecture by sharing data and code with BAPIs to clients, systems integrators, and other software vendors.

SAP has announced even closer collaboration with OAG to speed up the specification and adoption of enterprise application integration standards that enable interoperability between back-end business applications. As part of that relationship, SAP committed to making all relevant BAPIs compliant with the OAG Integration Specification (OAGIS).

Forrester believes that the OAG will "run its course" in the next few years without achieving its vision, and the major corporate applications vendors will, in the end, establish their own integration rules, with SAP's BAPIs and Oracle APIs likely to become de facto interface standards in the near future.

7.3.12 Market Overview

Of the major vendors in the worldwide client/server enterprise applications market, SAP remains the clear leader. With 1996 revenues of $1.7 billion, SAP claims the largest single share of the market, which market research firm International Data Corp. placed at approximately $9.6 billion for 1996. Oracle is a distant second, followed by Geac, CA, PeopleSoft, J. D. Edwards, and Baan. Figure 7-18 shows the revenues of the leading corporate application vendors in 1996.

Figure 7-18: Worldwide Corporate Application Vendor Revenues – 1996

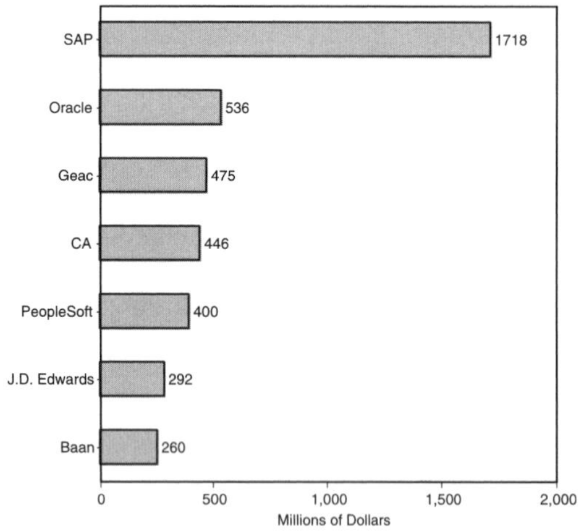

Source: IDC, 1997

Revenue for the top vendors is still growing, albeit at a much slower rate than in previous years. IDC believes this less-robust growth signals market maturation and saturation of the high-end market, which has led the top-tier players to target midsized companies and vertical markets aggressively. Oracle, PeopleSoft, and SAP are all supporting industries outside their traditional strength areas.

Despite a slowdown, the market will continue to grow, driven largely by Year 2000 conversions. Companies are moving away from outmoded legacy systems to Year 2000-compliant packaged software suites. Analysts estimate that businesses will invest up to $600 billion in Year 2000 conversions.

The midsized-market also represents an attractive growth area for all major vendors in this space. Analysts predict that such companies – which generally have between $50 million and $250 million in annual revenues – could represent billions of dollars in sales.

Although SAP's 53 percent growth in 1996 does not come close to its 88 percent growth in 1995, the company continues to dominate each of its corporate enterprise applications segments: accounting, manufacturing, distribution, and human resources. SAP's strategies include targeting vertical markets and smaller companies plus offering additional product functionality.

It is unclear whether SAP wants to remain a solo applications developer. In the past, for example, SAP formed agreements with third-party vendors such as i2 and Numetrix for supply chain management enhancements. However, SAP recently chose to implement its own internally developed module SCOPE (for supply-chain optimization, planning, and execution) rather than adopting a solution from i2 or Numetrix. Likewise, with R/3 4.0, SAP released products that resulted from joint development relationships with Acquion and Aspect Development. What will eventually happen to these partnerships is also unclear, however.

In December 1997, SAP announced that it would acquire Kiefer & Veittinger, a sales force automation vendor. SAP says that it will integrate K&V's technology with its own development efforts, embedding sales force automation functionality in future releases of R/3. This acquisition represents yet another change in strategy for SAP, which typically develops applications from scratch.

Oracle's 71 percent growth echoes 1995's growth of 70 percent. Like SAP, Oracle's plans for continued growth include targeting new vertical markets as well as smaller companies. Sales of its application programs rose 96 percent from the same period in 1996. Revenue from applications and services together represent about one-third of Oracle's business.

Geac's impressive growth of more than 500 percent in 1996 is due primarily to its acquisition of Dun & Bradstreet Software, its reorganization into two divisions, and its successful assimilation of the SmartStream product suite. As discussed earlier, Geac's new strategy calls for it to pull back from the manufacturing area and focus on financial, HR, and procurement products, well-matched to its target industries of financial services, health care, and business services.

CA grew 230 percent in 1996. During 1996, CA spun off some of its product lines into wholly owned divisions, including Acacia (targeting AS/400 manufacturing and distribution applications), MK Group (UNIX and Windows NT manufacturing applications), Prestige Software International (financial applications), and ACCPACC International (accounting solutions for small and midsized companies).

PeopleSoft experienced 90 percent growth in 1996, comparable to its 94 percent growth in 1995. PeopleSoft increased its average contract from $300,000 a few years ago to more than $800,000 in 1996, due largely to a broader product line and a robust vertical market strategy. PeopleSoft had healthy sales across its entire product line, including HR, its most mature segment. In fact, the company's HR product leads the U.S. market, and IDC expects that it will challenge SAP for the top position in 1997 in the corporate applications market.

J.D. Edwards and Baan both experienced strong growth in 1996. J.D. Edwards was not even among the top 15 vendors in 1995 (thus, no growth figures are available), whereas Baan experienced 108 percent growth in 1996. As two of the fastest-growing ERP vendors, both offer hybrid products (process and discrete) with strong overall functionality and state-of-the-art technology. J.D. Edwards has more than 4,000 customers with sites in more than 90 countries, and supports its products worldwide through 46 offices and more than 150 third-party business partnerships. With 2,400 customers and 4,000 installations worldwide, Baan is making its mark in the industry with Baan IV Orgware.

The META Group predicts strong demand for process manufacturing enterprise solutions during the forecast period, and says that Baan and J.D. Edwards are poised to win greater market share. IDC predicts that J.D. Edwards, traditionally strong in the middle market, will be among the vendors to move "upmarket" via new programs with the Big Six consultants.

Because of J.D. Edwards's "do it all" ("do it alone") philosophy, its products lag the competition in defining APIs, making integration more complex and expensive. In addition, J.D. Edwards has not yet established an enterprise functional framework that provides a road map for use of third-party applications to supplement its products. The lack of an architecture for inclusion of best-of-breed partners may put J.D. Edwards at a disadvantage in a marketplace that now requires such functionality. Nevertheless, its earnings of $292 million in 1996, jumping it ahead of Baan, indicate that J.D. Edwards will continue to be considered a force to reckon with in the market.

7.3.13 Forecast

- Corporate applications vendors will adopt component architectures for their own efficiency as well as to ease installation and lower costs.
- Large-scale ERP packages business will become easier to implement and use. Vendors increasingly will offer industry- and vertical market-specific and pre-packaged versions of their software. Customers will use tools such as Baan's Orgware and SAP's Business Engineering Workbench for quick implementations.
- To maintain growth, the high-end packaged application vendors will need to expand their services within global corporations, begin to focus on midsized, high-growth companies, and move into new vertical markets.
- The retailing industry will become a huge market for corporate application suite vendors in the next 3 to 5 years (reaching more than $1 billion, according to META), as the 1,000 largest retailers worldwide seek to emulate Wal-Mart's successful use of information technology. PeopleSoft and SAP already have products for this market, and others will follow soon.
- The trend toward Web-based self-service access to packaged applications will continue to grow, as will support for Web-based electronic commerce.
- The market is moving back to the use of mainframe-class servers and more centralized styles of computing to support enterprise corporate applications.
- Current bundled pricing will come under great pressure as companies look to implement modules at different paces. There will be additional pressure on the vendors to adopt more flexible licensing arrangements that will enable implementation of strategies such as shared services contracts and business process outsourcing.
- Data warehousing will become an integrated function of corporate applications.

- The growing modularity of corporate application packages and the new vendor-provided APIs will allow businesses to pursue something closer to a best-of-breed approach by installing third-party applications for specialized functions while still gaining the benefits of an integrated package.
- The IT architecture of many business organizations increasingly will be defined by the architecture of the major enterprise package they select and implement. However, this precept will not hold true for businesses that choose to undertake significant amounts of custom software development.
- Packaged applications increasing will improve their built-in workflow capabilities.
- Packaged application vendors will continue to form alliances with or acquire software vendors offering specialized functionality not included in their core packages, including sales force automation and supply chain management.

7.3.14 References

■Articles

Angus, Jeff and Tom Stein. 1997. Document management for ERP software. *InformationWeek*. June 30.
Callaway, Erin and Jeff Moad. 1997. SAP thinks small, too. *PCWeek*. August 25: 1.
Fontana, John. 1997. Apps wedded to the Web. *CommunicationsWeek*. August 25: 10.
Fontana, John. 1997. Automating sales caught in Web. *CommunicationsWeek*. August 4: 19.
Higgins, Kelly Jackson. 1997. Dare to Webify your back office. *InternetWeek*. September 1: 83.
Moad, Jeff. 1997. In mint condition. *PCWeek*. September 1: 85.
Stein, Tom. 1997. Fast deployment. *InformationWeek*. February 3: 60.
Stein, Tom. 1997. Industry-specific. *InformationWeek*. April 14: 79.
Stein, Tom. 1997. Market in the middle. *InformationWeek*. March 10: 82.
Stein, Tom, Marianne K. McGee, and Stuart J. Johnston. 1997. Services come in-house. *InformationWeek*. February 17.
Sustar, Lee. 1997. IT moves into mainstream. *CommunicationsWeek*. January 27: 64.
Walters, Glenn and Lynda Taskett. 1997. Your financial future. *InformationWeek*. April 21: 505.

■Periodicals, Industry and Technical Reports

Bond, B. 1996. *ERP market trends: vendors struggle to stay competitive*. Gartner. November 18.
Bond, B., A. Dailey, C. Jones, K. Pond, and J. Block. 1997. *ERP vendor guide 1997: the Baan Company – Baan IV*. Gartner. August 5.
Cameron, Bobby, Stuart D. Woodring, Mamie H. Chen, and Matthew M. Nordan. 1997. *The apps market resets*. Forrester. August.
Doormat for electronic commerce. 1997. META. August 11.
Gillan, Clare M. and Sandra Rogers. 1997. *Client/server "enterprise" applications: leading vendors and market performance 1996*. IDC. May.
Kugel, Robert D. 1997. *ERP vendors now targeting retail applications*. META. July 2.
OAG's strongest impact may be indirect. 1997. Forrester. August.

■On-Line Sources

Aberdeen Group. 1997. *Vertical industry solutions: Baan leads in innovation*. July 1. `http://www.aberdeen.com/secure/viewpnts/1997/v10n8/v10n8.htm`
Stein, Tom. 1997. *SAP buying half of German sales force automation vendor*. InformationWeek Online. December 15. `http://techweb.cmp.com/iw/newsflash/nf661/1215_st4.htm`
Sweat, Jeff. 1997. *SAP goes after PeopleSoft's public-sector stronghold*. InformationWeek Online. December 17.

■URLs of Selected Mentioned Companies

Acquion `http://www.acquion.com`
Ariba Technologies `http://www.ariba.com`
Aspect Development `http://www.aspectdv.com`
Aurum `http://www.aurum.com`
Baan `http://www.baan.com`
BEA `http://www.beasys.com`
Berclain `http://www.berclain.com`
Citicorp `http://www.citibank.com`
Citrix `http://www.citrix.com`
Cognos `http://www.cognos.com`
Compaq `http://www.compaq.com`

Computer Associates http://www.cai.com
CyberCash http://www.cybercash.com
Digital http://www.digital.com
Geac http://www.geac.com
HP http://www.hp.com
Hyperion http://www.hyperion.com
i2 Technologies http://www.i2.com
IBM http://www.ibm.com/
IMI Systems http://www.imisys.com
Inacom http://www.inacom.com
Infinium http://www.infinium.com
Informix http://www.informix.com
Intel http://www.intel.com
Intrepid Systems http://www.intrepidsys.com
J.D. Edwards http://www.jdedwards.com
Manugistics http://www.manugistics.com
Marcam http://www.marcam.com
Metasys http://www.metasys.com
Microsoft http://www.microsoft.com
MK Group http://www.mkgroup.com
Netscape http://www.netscape.com
Novell http://www.novell.com
Numetrix http://www.numetrix.com
Oracle http://www.oracle.com
Pandesic http://www.pandesic.com
PeopleSoft http://www.peoplesoft.com
Primavera http://www.primavera.com
SAP http://www.sap.com
Scopus http://www.scopus.com
Siebel http://www.siebel.com
SSA http://www.ssax.com
Sun http://www.sun.com
Sybase http://www.sybase.com
Taxware International http://www.taxware.com
TRW http://www.trw.com
UPS http://www.ups.com
USWeb http://www.usweb.com
Vantive http://www.vantive.com
Wal-Mart http://www.wal-mart.com
Yahoo! http://www.yahoo.com

8 Back of Envelope Numbers

Introduction

All of us in the technology business periodically need rules of thumb to help us quantify some proposed application. Some typical questions might include these:

- Just what is a T3 circuit? What is its capacity?
- How long would it take to transport an x-ray image on this particular line?

When needed, these numbers often seem elusive. We remember seeing them, but can never remember where. In this vein, we are including the following set of numbers in this appendix to the *Technology Forecast*. These items are a modest beginning. We solicit contributions from readers so that we can share them with others in future editions.

Processors and Systems

This section provides some rules of thumb for commonly available systems. We have included estimates of price changes over time, of important microprocessor statistics, and overall system prices. They are provided for perspective on the building blocks available for system design.

Table 8-1: Historical Component Price/Performance Improvement

Component	Price Improvement per Year
Microprocessor	50%
Memory (RAM)	50%
Disk	25%
Modem	25%
Display	15%
Printer	5%

Source: Disk/Trend, IDC, Dataquest

Table 8-2: Some Microprocessor Architectures

Manufacturer	Processor	Internal	Data Bus	Address Bus	Maximum RAM	MHz
Intel	8088	16-bit	8-bit	20-bit	1 MB	5
	8086	16-bit	16-bit	20-bit	1 MB	10
	80286	16-bit	16-bit	24-bit	16 MB	12
	80386SX	32-bit	16-bit	32-bit	4 GB	33
	80386DX	32-bit	32-bit	32-bit	4 GB	40
	80486	32-bit	32-bit	32-bit	4 GB	50
	Pentium	64-bit	64-bit	32-bit	4 GB	200
	Pentium Pro/II	64-bit	64-bit	36-bit	64 GB	200+
Motorola	68000	32-bit	16-bit	24-bit	16 MB	25
	68020	32-bit	32-bit	32-bit	4 GB	33
	68030	32-bit	32-bit	32-bit	4 GB	40
	68040	32-bit	32-bit	32-bit	4 GB	66
	PowerPC 620	64-bit	64-bit	32-bit	4 GB	133

Source: Intel, 1995; *Microprocessor Report*, 1995

Storage

As systems developers are aware, calculating storage requirements from record sizes and record counts was easy when the records were just characters. In what follows, we have provided some of the sizes of common storage devices and some helpful estimates of what the newer object types for audio and graphical images require. Multimedia applications can absorb a sizable amount of space and must be carefully designed.

Table 8-3: Document Storage Requirements

Document	Bytes
Page	2,400
Report	7×10^{4}
Book	7×10^{5}
Dictionary	6×10^{7}
Encyclopedia	1×10^{8}
Local Library	7×10^{10}
College Library	7×10^{11}
Library of Congress	1.8×10^{13}

Table 8-4: Common Storage Media

Storage Medium	Capacity
5.25" Floppy Disk	1.2 MB
3.5" Floppy Disk	1.4 MB
5.25" Rewritable Optical Disc	128 MB - 256 MB
CD-ROM	650 MB
Magneto-Optical Drives	2 GB
DVD (digital video disc or digital versatile disc)	4.7 GB - 17 GB
Disk Drive (IBM 3390 / 2 HDA)	3.78 GB
3.5" Disk Drive (IBM Ultrastar 2)	10.8 GB
Optical Disk Jukeboxes	2 GB - 3 TB
Cartridge Tape Drive (3490 x 3 compaction)	2.4 GB
Digital Audio Tape	30 GB - 100 GB
Tape Library	9.6 TB

Source: IBM

Table 8-5: Audio Storage Requirements

Audio Level	Resolution	Sampling Rate	Bandwidth	Concurrent Channels	Storage for 1 Minute
CD-DA	16-bit	44.1 KHz	20 KHz	2 stereo	10.09 MB
CD-I Level A	8-bit	7.8 KHz	17 KHz	2 stereo, 4 mono	4.33 MB
CD-I Level B	4-bit	37.8 KHz	17 KHz	4 stereo, 8 mono	2.16 MB
CD-I Level C	4-bit	18.9 KHz	8.5 KHz	8 stereo, 16 mono	1.08 MB

Source: *The Desktop Multimedia Bible*, 1993

Table 8-6: Video Storage Requirements

38 seconds of video with audio requires 1 GB without compression MPEG

Video Type	Resolution	Frames per Sec	Color	Compression	Storage for 1 Minute
NTSC	640 x 480	30	24-bit	uncompressed	1,600 MB
8-bit NTSC	640 x 480	30	8-bit	uncompressed	530 MB
MPEG II	702 x 480 (medium)	30		compressed video	45 MB
MPEG I	352 x 240	24 - 30		compressed video	11 MB

Source: *The Desktop Multimedia Bible*, 1993

PC Video Standards

Special names given to current PC video standards:

VGA: Video graphics array is 640 by 480 pixels at 4 bits of color (16 colors)

SVGA: Super VGA is 800 x 600 pixels at 8 bits of color (256 colors)

XGA: Extended graphics array is 640 x 460 pixels at 16 bits (65,536 colors)

IBM 8514/A: IBM's standard is 1,024 x 768 pixels at 4 bits (256 colors)

High Color: indicates 24-bit graphics ("millions of colors")

True Color: indicates 32-bit graphics ("billions of colors")

Table 8-7: Memory Requirements for Video Resolution Levels

Horizontal (pixels)	Vertical (pixels)	Bits of Color	Number of Colors	Memory Required
640	480	4	16	0.147 MB
640	480	8	256	0.293 MB
640	480	16	65,536	0.586 MB
640	480	24	16,777,216	0.879 MB
640	480	32	4,294,967,296	1.172 MB
800	600	8	256	0.458 MB

Table 8-7: Memory Requirements for Video Resolution Levels (Continued)

Horizontal (pixels)	Vertical (pixels)	Bits of Color	Number of Colors	Memory Required
800	600	16	65,536	0.916 MB
800	600	24	16,777,216	1.373 MB
800	600	32	4,294,967,296	1.831 MB
1,024	768	8	256	0.750 MB
1,024	768	16	65,536	1.500 MB
1,024	768	24	16,777,216	2.250 MB
1,024	768	32	4,294,967,296	3.000 MB
1,280	1,024	8	256	1.250 MB
1,280	1,024	16	65,536	2.500 MB
1,280	1,024	24	16,777,216	3.750 MB
1,280	1,024	32	4,294,967,296	5.000 MB

Telecommunications

The variety of communications offerings is expanding and changing all the time. The tables below offer a handy reference for the most common speeds of services, and some rules of thumb for transmission time.

Figure 8-1: Rate Ranges for Service Types

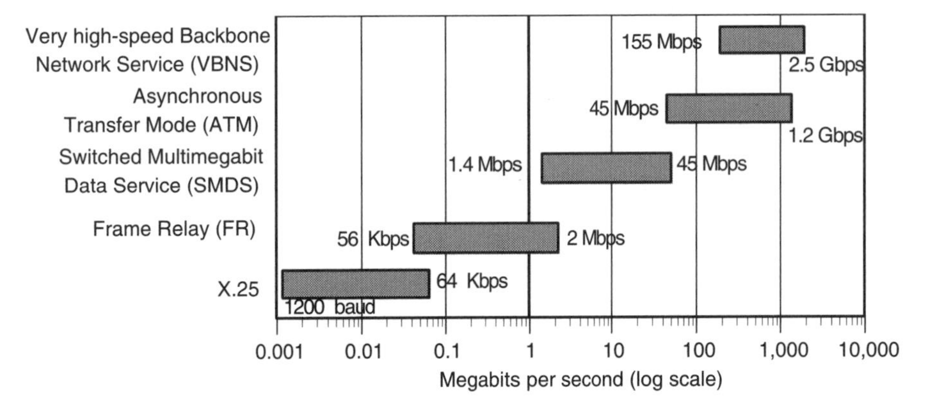

Table 8-8: Telecommunication Line Rates

DS - digital signal, OC - optical carrier, STM - synchronous transport module

Service	Bit/Line Rate
DS0	64 Kbps
ISDN (Basic Rate Interface)	128 Kbps
LocalTalk	230.4 Kbps
DS2 [T1 (North America)]	1.544 Mbps
ISDN (Primary Rate Interface)	1.544 Mbps
E1 (Europe)	1.92 Mbps
DS-2	6.312 Mbps
Ethernet	10 Mbps
Token Ring	4 Mbps or 16 Mbps
DS3 [T3]	44.736 Mbps
OC 1	51.840 Mbps
Fast Ethernet	100 Mbps
FDDI	100 Mbps
CDDI	100 Mbps
OC 3 [STM-1]	155.520 Mbps
OC 9	466.560 Mbps
OC 12 [STM-4]	622.080 Mbps
OC 18	933.120 Mbps
OC 24 [STM-8]	1.244 Gbps
OC 36 [STM-12]	1.866 Gbps
OC 48 [STM-16]	2.488 Gbps
OC 96	4.900 Gbps
OC 192	9.950 Gbps

Table 8-9: Approximate Transmission Times for Typical Documents

Rate	Fax (1 page)	X-Ray	100,000 Pages
9.6 Kbps	1 minute	12 minutes	20 hours
28.8 Kbps	20 seconds	4 minutes	7 hours
64 Kbps	12 seconds	2 minutes	3 hours
1.5 Mbps	0.5 seconds	6 seconds	8 minutes
50 Mbps	.02 seconds	0.2 seconds	16 seconds

Source: Dataquest, 1994

Table 8-10: Approximate Transmission Times for Files

Rate	10 KB	100 KB	1 MB	10 MB
2400 baud	41.7 seconds	6.9 minutes	1.2 hours	11.6 hours
9600 baud	10.4 seconds	1.7 minutes	17.4 minutes	2.8 hours
28.8 Kbps	3.4 seconds	34 seconds	5.8 minutes	1 hour
38.4 Kbps	2.6 seconds	26 seconds	4.3 minutes	43 minutes
230.4 Kbps	.4 seconds	4.3 seconds	43 seconds	7.2 minutes
10 Mbps	.01 seconds	0.1 seconds	1 second	10 seconds
50 Mbps	.002 seconds	.02 seconds	.2 seconds	2 seconds

Table 8-11: Common Contemporary Satellite Uses

Service	Band	Frequencies Used (typical)	Antenna Size (typical)	Notes
Television feeds between companies and distribution to cable operators	C-band	6 GHz (uplink) 4 GHz (downlink)	8 - 20 meters	Increasing use of encryption
Television news gathering	Ku-band	11-12 GHz (uplink) 14 GHz (downlink)	1.5 - 2 meters	
Direct TV broadcast (DBS)	Ku-band Ka-band	17 - 31 GHz	< ½ meter	Increasing use of digitalization, compression, and encryption
Direct audio broadcast (DAB)	L-band	0.5 - 1.7 GHz		
Telephony and data	C-Band Ku-band			
Very small aperture terminals (VSAT)	Ku-band Ka-band		< 2.5 meters	
Personal communications services (PCS)	L-band		Hand-held	
Global positioning system (GPS)	L-band			Increasing use in automobiles

Source: *Communication Technology Update,* 1996

Frequency Allocation

The first band shows the ranges for all electromagnetic waves. The second two bands show more detail on radio and television.

Figure 8-2: Radio Frequency Spectrum

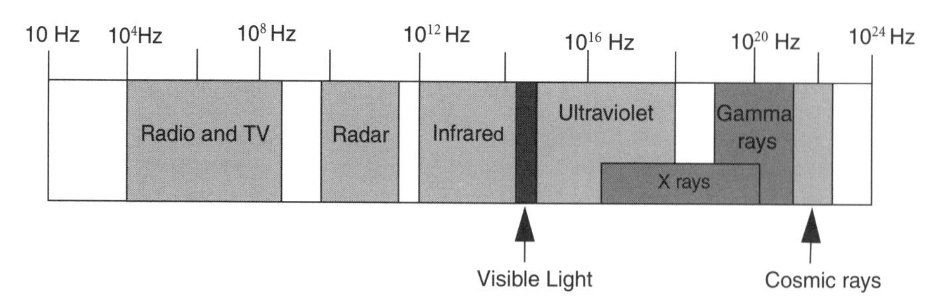

Source: *IEEE Spectrum*, 1994

Figure 8-3: Radio and Television Radio Frequencies (Part 1)

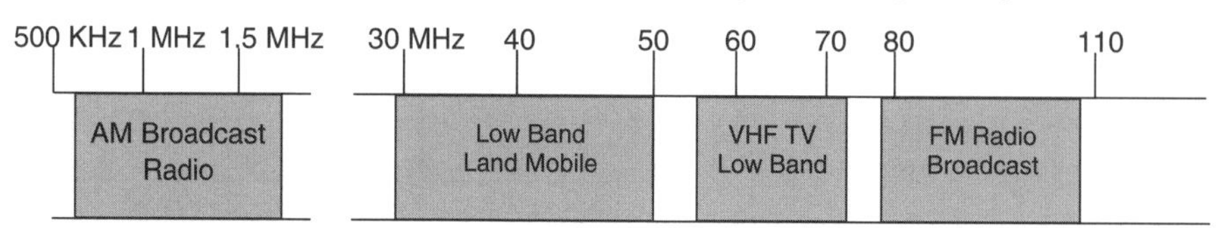

Source: Motorola

Figure 8-4: Radio and Television Radio Frequencies (Part 2)

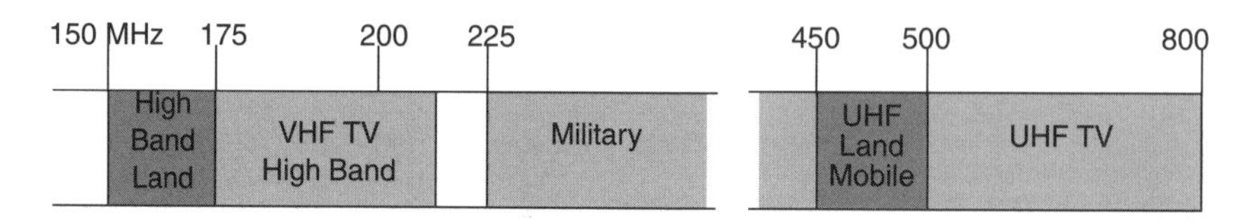

Source: Motorola

Figure 8-5: U.S. Allocated Personal Communications Services (PCS)

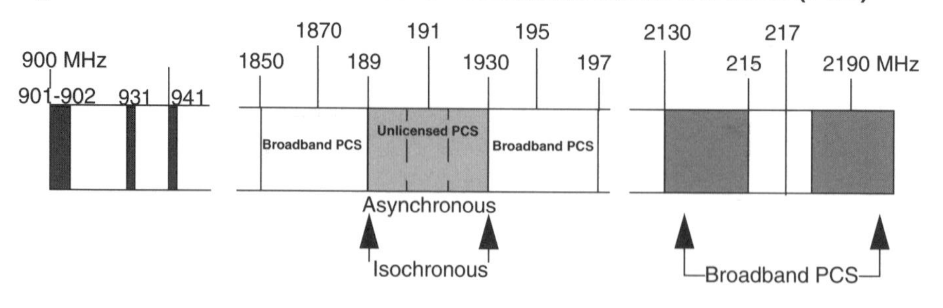

Source: *IEEE Spectrum*, 1994

Budget

It is useful to know how resources are allocated for information technology (IT). The information below shows the average percentage of the total revenue allocated to the IT budget for the identified industries in the U.S. and Canada.

Table 8-12: Information Services Budget as Percent of Total Revenue

Industry	1995
Banking, Finance, Insurance	2.8%
Manufacturing	2.2%
Computing, Communications, Electronic Equipment	3.7%
Aerospace, Defense Contractor	1.7%
Communications (Telcos)	1.9%
Utility	1.9%
Hospitals and Health Care	2.6%
Software Development	5.1%
All Industries	2.1%

Source: Gartner, 1996

The U.S. spends more on IT equipment than any other country. Morgan Stanley reports that in 1994, almost 3 percent of the U.S. gross domestic product was spent on computer and telecommunications hardware. IT spending continues to grow rapidly in many countries. For example, Swiss IT spending grew almost 11 percent annually between 1987 and 1994 spurred by technology investments made by Swiss banks.

Figure 8-6: Computer and Telecommunications Hardware Spending

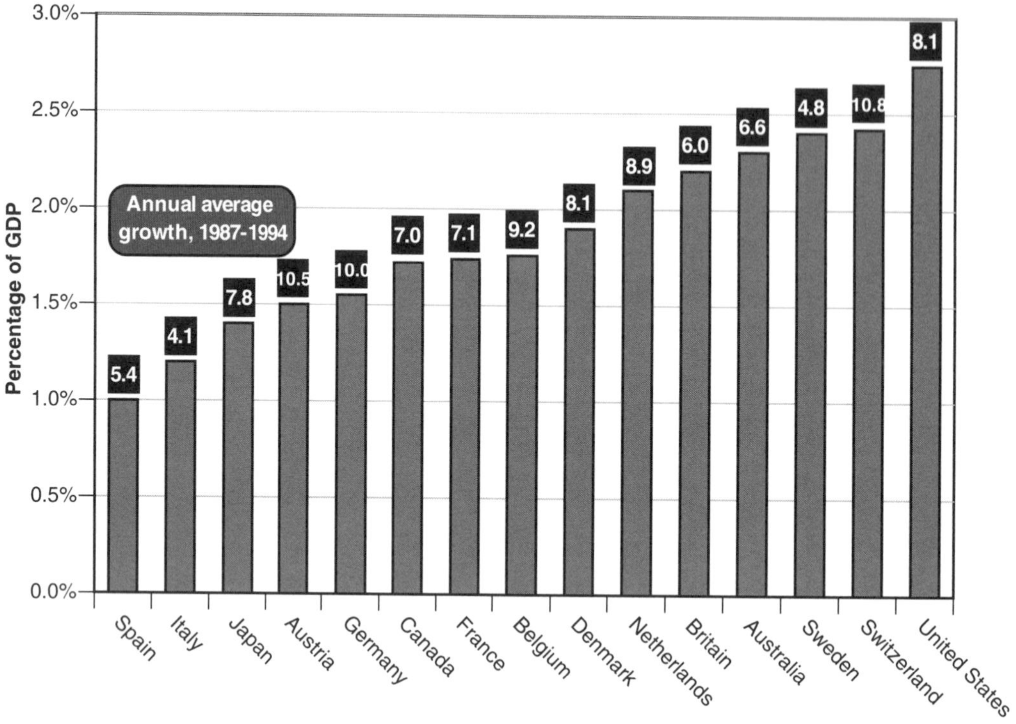

Source: Morgan Stanley, 1996; OECD, 1996

Table 8-13: Prefix Reference

Prefix	Symbol	Power
exa	E	10^{18}
peta	P	10^{15}
tera	T	10^{12}
giga	G	10^{9}
mega	M	10^{6}
kilo[1]	k	10^{3}
hecto	h	10^{2}
deka	da	10^{1}
deci	d	10^{-1}
centi	c	10^{-2}
milli	m	10^{-3}
micro	m	10^{-6}
nano	n	10^{-9}
pico	p	10^{-12}
femto	f	10^{-15}
atto	a	10^{-18}

[1]Kilo with a capital K is used to represent 1,024

Table 8-14: Common Paper and Envelope Sizes

ISO Names	Dimensions
Letter ("A")	8/5" x 11" (216 x 279 mm)
Legal	8.5" x 14" (216 x 356 mm)
Ledger (US "B")	11" x 17" (279 x 432 mm)
Executive	7.25" x 10.5" (184 x 267 mm)
Universal	11.7" x 17.7" (297 x 450 mm)
US "C"	18" x 24" (457 x 610 mm)
US "D"	24" x 36" (610 x 914 mm)
US "E"	34" x 44" (864 x 1118 mm)
ISO A3	11.7" x 16.54" (297 x 420 mm)
ISO B3	13.9" x 19.7" (353 x 500 mm)
ISO A4	11.7" x 8.23" (210 x 297 mm)
ISO B4	10.12" x 14.33" (250 x 353 mm)
ISO A5	5.85" x 8.23" (148 x 210 mm)
ISO B5	.17" x 10.12" (176 x 250 mm)
Common Envelope Sizes	
#10	4.125" x 9.5" (105 x 241 mm)
Monarch	3.875" x 7.5" (98 x 191 mm)
DL	4.3" x 8.6" (109 x 218 mm)
ISO C4	9.02" x 12.76" (229 × 324 mm)
ISO C5	6.37" x 9.015" (162 x 229 mm)

IBM Technology History

Table 8-15: Significant Technology Introductions on IBM Computers

Year Announced	Machine	Technology	Purpose
1945	407	Electro-mechanical	Accounting
1948	604	Vacuum Tube/ Electro-mechanical	Calculating
1952	701	Vacuum Tube	Scientific
1959	7090	Transistor	General-purpose
1960	1401	Transistor	Business
1964	System/360	Solid Logic Technology	Broad Product Line
1970	System/370	Monolithic Circuitry	Upgrade 360
1977	303X	Bi-polar ECL	High-end Mainframe
1979	4300	Multi-layer Ceramic Packaging	Midsized Mainframe
1980	308X	Thermal Conduction Modules	High-end Mainframe
1981	PC	Intel CMOS	Personal Computer
1988	AS/400	BiCMOS	Midrange
1991	RS/6000	RISC Architecture	UNIX Workstation
1994	S/390 Parallel Query Processor	CMOS Air-cooled	Database
1995	S/390 Parallel Enterprise Server	CMOS Air-cooled	New mainframe technology
1995	AS/400 Advanced Series	Power PC 620, integrated RAID Level 5	Uses new CMOS chips in existing midrange
1996	S/390 G3	CMOS Air-cooled	Mainframe
1997	S/390 G4	CMOS Air-cooled	Mainframe
1997	AS/400e	PowerPC AS	Midrange

Source: *IBM's 360 and Early 370 Systems*, IBM

Historical Events

This timeline of events in the history of information technology shows just how far our industry has evolved, especially in the last 30 years.

Table 8-16: Historical Events

500 BC:	Bead and wire abacus originates in Egypt.
1833:	Charles Babbage designs the analytical engine. It is the first general-purpose computer.
1842:	Lady Ada Byron, Countess of Lovelace and daughter of Lord Byron, the poet, documents Babbage's work and writes programs for Babbage.
1876:	Telephone is invented by Alexander Graham Bell.
1944:	Mark I (IBM ASCC) is completed, based on the work of Professor Howard H. Aiken at Harvard and IBM. It is a relay-based computer.
1946:	ENIAC (electronic numerical integrator and computer), with 18,000 vacuum tubes, is dedicated. It measured 8 feet by 100 feet and weighed 80 tons.
1948:	Transistor is invented by William Bradford Shockley with John Bardeen and Walter H. Brattain.
1952:	Nixdorf Computer is founded in Germany.
1953:	IBM ships its first stored-program computer, the 701. It is a vacuum tube, or first-generation, computer.
1954:	FORTRAN is created by John Backus at IBM.
1956:	Government antitrust suit against IBM is settled; consent decree required IBM to sell as well as lease machines.
1959:	COBOL is defined by the Conference on Data System Languages (CODASYL), based on Grace Hopper's Flow-Matic. IBM ships its first transistorized, or second-generation, computers, the 1620 and 1790.
1960:	Digital ships the first small computer, the PDP-1.
1964:	IBM announces the System 360, with integrated, third-generation technology.
1965:	Digital ships the first PDP-8 minicomputer.
1968:	Integrated Electronics (Intel) Corp. is founded by Gordon Moore and Robert Noyce.
1969:	PASCAL compiler is written by Nicklaus Wirth and installed on the CDC 6400.
1970:	IBM announces and ships its first System 370, a fourth-generation computer.
1971:	Intel announces the first microprocessor, the Intel 4004, developed by a team headed by Marcian E. Hoff.
1972:	First electronic pocket calculator is developed by Jack Kilby, Jerry Merryman, and Jim Van Tassel of Texas Instruments.
1973:	C language developed by Dennis Ritchie at Bell Labs. PROLOG language is developed by Alain Comerauer at the University of Marseilles-Luminy, France.
1974:	Intel introduces the 8080, an 8-bit microprocessor that will be used in numerous personal computers.
1975:	MITS introduces the Altair personal computer, named after a *Star Trek* episode, *A Voyage to Altair.* Microsoft is founded, after Bill Gates and Paul Allen adapt and sell BASIC to MITS to be used on the Altair PC.
1977:	Apple Computer is founded and introduces the Apple II personal computer. First ComputerLand franchise store opens in Morristown, N.J. under the name Computer Shack.
1978:	The first COMDEX trade show is held.
1979:	VisiCalc, the first electronic spreadsheet software, is shown at the West Coast Computer Faire.
1979:	The Source and CompuServe Information Services go on-line.
1981:	IBM enters the personal computer market, creating a de facto standard.

Table 8-16: Historical Events (Continued)

1982:	Sun Microsystems is founded. U.S. Justice Department drops IBM antitrust suit begun in 1969.
1983:	Compaq ships its first computer in January and completes the year with $111 million in revenue, the greatest first-year sales in the history of U.S. business.
1984:	Apple introduces the Macintosh computer.
1986:	Compaq makes the Fortune 500 list and introduces the first Intel 80386-based PC.
1987:	IBM announces the PS/2 family of personal computers and OS/2 operating systems.
1988:	IBM, Digital, HP, Apollo, and other major computer companies form the Open Software Foundation to develop a new version of UNIX.
1989:	The battery-powered notebook computer becomes a full-function computer, including hard and floppy disk, with the arrival of Compaq's LTE and LTE/286. The first 80486-based computers are introduced.
1990:	Microsoft introduces Windows 3.0. Microsoft's fiscal year revenue ending June 1990 exceeds $1 billion.
1991:	The first general-purpose, pen-based notebook computers are introduced.
1992:	IBM releases OS/2 Version 2.0 and ships more than 1 million units. Microsoft introduces Windows 3.1 and ships nearly 10 million units. Novell acquires UNIX Systems Laboratory, including Univel, from AT&T for $350 million.
1993:	Lotus releases Notes 3.0. Microsoft ships Windows NT.
1994:	Novell assigns UNIX trademark to X/Open for its branding program. A few CMOS chips become the basis for IBM's newest mainframe computer, eliminating water cooling in new products such as the Parallel Query Server. The Internet is dubbed the "information highway" as it achieves 20 million users. Microsoft settles an antitrust suit with the U.S. Justice Department.
1995:	Microsoft ships Windows 95. Government auctions PCS airwaves. AT&T announces plans to break into three companies. Netscape Communications goes public for more than $2 billion.
1996	The Telecommunications Act of 1996 is signed into law, deregulating communications carriers in the U.S. Worldwide deregulation goes into high gear. Merger wave in the U.S. telecomm industry results. Globally, BT announces plans to acquire MCI. Vendor coalition led by Oracle announces Network Computer Reference Platform specification. First-generation products include the IBM Network Station and Sun JavaStation. Consumer set-top devices for Web access, such as WebTV, become available. Microsoft and Intel announce the NetPC management initiative in response to network computers. Convergence of computing, communications, and consumer electronics results in new standards. The computer and entertainment industries agree on a standard for a high-capacity 5-inch CD-ROM (DVD) large enough to hold a full-feature movie on a single side. Computer and television vendors agree on a digital TV standard for the U.S.
1997	Intel buys Digital Equipment Alpha foundry. Compaq buys Tandem. Apple discontinues Macintosh clone program. IBM announces copper breakthrough in microprocessor fabrication.

Source: *Computer Industry Almanac,* 1993; IBM, Microsoft, Digital

Index

Symbols

$/tpmC
  for cost per transaction per minute
@Home Network 152, 225

Numerics

100Base-T 167
100Base-T4 149
10Base-T 167
3Com Corp. 165, 167, 168, 172, 176, 177, 183, 185, 193, 581
  Axon Networks 533
  CoreBuilder 3500 168
  CoreBuilder 9000 Enterprise Switch 168
  DMI 526
  Fast IP 181
  MPOA 170
  Palm Pilot 360
  SmartAgent 541
  Transcend 529
  U.S. Robotics 183, 185, 196
  X2 183
3-D graphics 356
  controllers 60, 66
  VRML 254
3GLs 430, 431
  Ada 433
  BASIC 432
  C 433
  COBOL 431
  forecast 465
  Fortran 431
  LISP 432
  market 462
  Modula-2 433
  object technology extensions 437
  Pascal 432
  PL/1 432
  RPG 432
4GLs 437, 439, 460
  client/server 441
  client/server vendors 441
  data warehousing 619
  definition 439
  features 439, 441
  forecast 465, 466
  mainframes 440
  market 461

A

Abdullah, Dr. Othman Yeop 33–41
Accent Software International Ltd. 241
access tools
  see data, data mining, data warehousing, groupware, information retrieval, Internet, network management, security, SQL, systems management, World Wide Web
accountability 379
ACD 225
  for automatic call distribution
Acer America Corp. 48, 364
ACH
  for Automated Clearing House
ACID 473
  for Atomicity, Consistency, Isolation, and Durability
  definition 473
ACL 379
  for access control list
  SDSI 392
Acquion Inc. 649, 681
ACR 149
  for attenuation-to-crosstalk ratio
Action Technologies Inc. 502, 503
  Metro 502
Active Directory 333
ActiveX 317, 325, 332
  components 405
  forecast 340
  market 464
  security 405
  see Microsoft
Ada 431, 433
Adabas 440
Adaptec Inc. 79
  AIC-8420 79
  FCL 80
Adesa
  Hawk 390
Adesso Inc. 47
administration 379
ADO 443
Adobe Systems Inc.
  Acrobat Reader 246
  PageMill 256
  Premiere 259
ADPAC Corp.
  SystemVision Year 2000 452
ADSL 161, 162, 186, 212
  for asymmetric DSL
  RADSL 212
  technologies 212
  VH-ADSL 213
Advanced Recognition Technologies Inc.
  smARTwriter 378
Advanced Visual Systems Inc. 624
advertising
  cookie technology 255
  electronic commerce 581
  push technology 248, 249
  retail transactions 582
  search engines 276
  smart cards 128
  streaming video 257
  Usenet groups 252
  VANs 563
  World Wide Web 231, 249, 273, 585, 594
    revenue 253
AER Energy Resources Inc. 98
AES 389
  for Advanced Encryption Standard
AFC 335
  for Application Foundation Class
AGIS 410
  for Apex Global Information Services Inc.
AGP 358
  for accelerated graphics port
AI
  for artificial intelligence
AIAG
  for Automotive Industry Action Group
AIIM 508
  for Association for Information and Image Management
Airbus Industries 8
AIT 72, 73
  for Advanced Intelligent Tape
  capacity 73
  compatibility 72
Alcatel Data Networks 13
ALDC 73
  for Adaptive Lossless Data Compression

data compression ratio 73
Alis Technologies Inc. 241
Alps Electric USA Inc. 47
Alteon Networks Inc. 182
Altris Software Inc.
EB 506
EDMS 506
ALU
for arithmetic logic unit
Amazon.Com Inc. 560, 583
Web site 582
Amdahl Corp. 83, 305, 349, 361
Logical Volume Series 85
Spectris 85
technologies 85
Transparent Data Migration Facility 85
see also Fujitsu Ltd.
America Online Inc.
see AOL
American Airlines Inc. 127
American Express Co. 123, 127, 573
American Institute of Certified Public Accountants 385
Americas
deregulation 209
Ameritech Corp. 208, 222
Ampex Corp.
DST 81x series 71
AMPS 217
for Advanced Mobile Phone System
D-AMPS 215
technologies 214
Andyne Computing Ltd.
Andyne GQL 622
MX 628
Angoss Software Corp.
KnowledgeSeeker 626
KnowledgeStudio 626
Anixter Inc. 149
Level 7 150
ANS Communications Inc. 233
ANS SecureConnection 375
ANSI 169, 389, 430
X9.63 389
X9.F1 389
ANSI SQL
see SQL
antennas
adaptive array 218
PCS/PCN 217
technologies 217, 218
anycast 192
AOL 232, 233, 252, 400, 567
for America Online Inc.
CompuServe 233, 276, 590, 593
spam 410
Apertus Carleton Corp.
Enterprise/Integrator 610, 612
MDIS 628
MetaCenter 612, 629
Passport 612
WTI 629
APIs
for application programming interfaces
core system functions 456
definition 287
GUIs 287
interoperability 454
Java 330
JDBC 460
NetBIOS 459
network programming 459
operating systems 287
TCP/IP 459
Win32 456
Windows 456
X/Open 456
APM
for Advanced Power Management
Apple 46, 357, 459, 593, 698, 699
for Apple Computer Inc.
Apple Open Transport 336
AppleTalk 178, 299, 459
ECC 576
Macintosh 79, 261, 291, 357, 364, 387, 433, 495, 537
MacOS 48, 288, 299, 336
MacOS 8 45, 246, 288, 309, 313
MacOS versus Windows 313
Newton 359, 360
Newton 2100 360
NeXT Software Inc. 286, 300
NeXTstep 436
PowerPC 604e 345
QuickTime 257, 258
QuickTime for Windows 258
QuickTime Media Layer family 300
Rhapsody 286, 300, 313, 436
trackball 47
trends 366
Year 2040 453
applets 460, 632
definition 112
hostile 402, 404
HTML 242
JavaCard 113
monitoring 417
security 373
see also Java
application development
browsers as front ends 429
Year 2000 issues 429
Application Programming and Development Inc.
True Site 385
applications
see corporate applications, Java, object technology, programming
APPN 564
for Advanced Peer-to-Peer Networking
APU
for Authenticated Payload URL
Arbor Software Corp.
Essbase 619, 630, 634, 669
Essbase Server 620
MOLAP 618
OLAP 630
Ariba Technologies Inc. 647
Ariel Communications Corp.
Rascal RS1000 Model 4802 186
Ascend Communications Inc. 172, 178, 185, 193, 196
Pipeline 75 186
Secure Access Firewall 186
Secure Access Manager 186
Whitetree Inc. 171
ASCII 291
for American Standard Code for Information Interchange
Ashton-Tate Corp. 438
Asia/Pacific
deregulation 208
ASICs 60, 176, 177
for application specific integrated circuits
ASMO 94
for advanced storage magneto-optical
Aspect Development Inc. 681
ASPG
for Advanced Software Products Group Inc.
Date/2000 452
AST Research Inc. 364
Asymetrix Corp. 461
Neuron 246
AT&T Bell Laboratories Inc.
see Lucent Technologies Inc.
AT&T Corp. 159, 160, 197, 222, 263, 276, 392, 512, 528, 530, 570, 587, 593, 699
C 433
Managed Network Solutions 530
RCS 445
SecureBuy Service 570, 573
Teleport Communications Group 206

Atalla Corp. 387, 414
WebSafe 2/PCI Internet security processor 414
Atari Corp. 46
ATI Technologies Inc. 60
ATM 137, 149, 166, 169, 203
for asynchronous transfer mode
25-Mbps ATM 170
622-Mbps ATM 149
advantages 169
background 222
capacity 169
cell networking topology 169
costs 169, 171, 196
definition 169, 222
forecast 199
international 222
interoperability 189
LANE 170
market 196, 227
MPOA 170
QoS 169, 222
requirements 169
scalability 222
server bottlenecks 195
server clustering 172
speed 222
technologies 222
trends 227
versus Ethernet 170
versus FDDI 170
versus Gigabit Ethernet 168, 169
versus SMDS 226
VPNs 184
ATM for um 169, 189
MPOA 181
Atmel Corp. 129
ATMI 337
for Application-to-Transaction Manager Interface
Attachmate Corp. 276
audio
data compression 260
Internet 274
real-time 260
streaming 257, 258
auditing 379
authentication 112, 375, 376
bar codes 117
biometrics 375
capture/reply 376
challenges 376
client 414
digital signatures 394
digital watermark 117
distributed systems 381
document management 517
electronic commerce 566
FIPS 392
forecast 422
Internet 375
inter-realm 381
IPsec 184
Kerberos 376, 381
LDAP 391
market 420
methods 375, 377
one-session password 376
operating systems 290
PEM 411
PIN 376
security tokens 376
SET 577
smart cards 114, 120, 390
solutions 420
two-factor 375, 382
umbrella API 376
X.509 391
Authority for Information Technology in the Civil Service 407
authorization 375, 379
ACLs 379
Autodesk Inc. 356
Autonomy Inc. 276
Avantel 209
AVI 257
Avici Systems Inc. 177
AWT
for Abstract Windowing Toolkit
Axent Technologies Inc. 418
Axon Networks Inc.
see 3Com Corp.

B

Baan Co. 484, 643, 644, 646, 648, 676–679
Antalys Inc. 644, 676, 679
architecture 678
Aurum Software Inc. 644, 676
Baan IV 678
Baan IV BackOffice 676, 677
Baan IV Orgware 677, 682
Baan IVc 679
features 678
Baan Sync 678
Baan V 678, 679
Berclain Group Inc. 644, 676
Berclain Software Ltd. 678
Business Object Interfaces 679
DEM 677
Dynamic Enterprise Modeling 676
EMU 678, 679
Enterprise Implementer 678
Enterprise Modeler 677
Enterprise Performance Manager 678
Enterprise Reference Models 677
ERP 677, 679
eXtended Middleware Architecture 679
Internet 678, 679
Java-based Web user interface 679
market 680, 681, 682
push technology 679
trends 679
Triton 677
vertical markets 676
Year 2000 644, 677
backbone
bandwidth requirements 195
definition 195
WAN 195
backup
storage management 537
BackWeb Technologies Inc. 248, 512
Baltimore Technologies Ltd. 412
LJ/CRYPTO 419
bandwidth 81, 149, 203
ATM 169, 222
backbone network 163
backbones 195
cable modems 188
clustering 354
committed information rate 222
frequency allocation 692
Gigabit Ethernet 167
groupware 512
H.323 260
increasing 139
Internet 235, 238, 261, 274
mainframes 349
MPP 355
network bottlenecks 194
PVCs 172
QoS 169
RSVP 190
SDSL 212
servers 195
SMP 354
SONET 142
SVCs 172
switched data networks 540
switches 175, 182
TDM 143
telecommunications 152
video 169
virtual circuits 183
VLANs 182
VPNs 184
WDM 160
wiring 150
Bangemann Report 7

Bangemann, Martin 7–14
Bank of America 573
banking
C-SET 124
digital check 572
EFT 559
electronic commerce 562, 572, 583, 588
electronic wallets 573
hybrid smart cards 125
Internet
services 8
JavaCards 112
OFX 578
optical laser cards 118
Payment/Remittance Advice 578
SCRIP 574
server operating systems 285
SET 574, 575
smart cards 108, 110, 121, 573
standards 578
stored value cards 121
see also electronic commerce
Banque Paribas
Kleline 124
Banyan Systems Inc.
StreetTalk for Windows NT 305
VINES 178, 189, 304, 310
BAPI
for Business API
bar codes 117
2-D 117
uses of 118
Barco Inc.
BarcoVision 9200 58
BASIC 431, 432
batteries
disposable 94
forecast 102, 103
Li-ion 96, 102
lithium 94
lithium-ion polymer 97
lithium-polymer 97, 102, 103
market 101
NiCd 94, 95
NiMH 94, 95
overview 94
rechargeable 94
issues 95
SBS 99
smart 98
technologies 69, 94
trends 70, 94, 98
zinc-air 97, 98, 104
Bay Networks Inc. 165, 167, 168, 172, 176, 183, 193, 196, 529
BayStream Dial VPN Service 191
Rapid City Communications 181
BBN Corp. 392
SitePatrol 375
BEA Systems Inc.
ATMI 337
Distributed Application Framework 337
Jolt 673
Tuxedo 336, 337, 612, 668, 673
Universal Applications 673
Beijing Marine Communications and Navigation Co. 157
Bell Atlantic Corp. 153, 205, 223, 226
Bell Cablemedia PLC 205
Bell Canada 123
Benchmarq Microelectronics Inc. 99
Berkeley Networks Inc. 177
Berkeley Systems Inc.
After Dark 4.0 249
Bernstein vs. the U.S. Department of State 401
BGP 179
for Border Gateway Protocol
BGP4 179
for Border Gateway Protocol Version 4
Bharti Enterprises 207
BioData Inc. 396
Biometric Consortium 378
biometrics 376
challenges 378
devices 378
face recognition 377
fingerprints 377, 420
forecast 422
one-to-one 378
personal privacy 378
retina scans 377
revenues 420
smart cards 114
techniques 377, 420
technologies 377
testing 378
uses 377
vendors 377
voice verification 420
BISDN 223
for Broadband ISDN
definition 221
BLOB 71
for large binary object
BlueStone Software Inc.
Sapphire/Web 460, 464
BMC Software Inc. 534, 537
Patrol Knowledge Module for MQSeries 336
Patrol Series 539
Boeing Co. 581
Boise Cascade Office Products Corp. 581
Boole & Babbage Inc.
Command MQ 336
MainView 537
Border Technologies Inc. 418
Borland International Inc. 438, 457, 461, 464
C++ Builder 442, 446
Delphi 433, 437, 442, 449, 464
Object Windows Library 457
Paradox 443
Turbo Pascal and Object Pascal 433
Visigenic 339
BOT 71
for beginning of tape
bpi
for bits per inch
BPR 501
for business process reengineering
design tools 501
BRI
for basic-rate interface
definition 211
usage 227
bridges 174
advantages 175
definition 175
types of 175
versus routers 176
versus switches 180
Brio Technology Inc.
BrioQuery 604, 622
MX 628
Bristol Technology Inc.
OSF/Motif 457
British Telecommunications PLC 205, 206, 210, 581
BT Internet ISDN LAN 210
MCI 263
SMDS 226
broadcast storms 175, 176, 180
BroadVision Inc. 566, 593
One-to-One 567
One-to-One Commerce 566
Brocade Communications Systems Inc. 80
Brooktrout Technology Inc.
IP/FaxRouter 264
Brother International Corp.
MFC-7550MC 65
browsers
adult content 252
as front ends 429
authoring tools 256, 279
background 234
caching 239, 246

capabilities 244
challenges 246
cookie technology 255, 409
data warehousing 621, 639
definition 234, 244
desktop computing 246
electronic commerce 553, 556, 557
enhanced bookmarks 247
extensions 256
forecast 278, 279
HTTP 241, 266
input devices 46
intranets 264, 275, 276
JavaScript 439
JVM 460
legacy systems 165
Lynx 234
management applications interface 521
management information file 534
market 244, 274, 275
Mosaic 234
offline 253
operating systems 312
overview 231, 244
personalization 253
pre-fetching 246
push technology 247
remote management 522
revenue 274
scripting languages 438
security 402, 403, 410, 411, 414
technologies 245
trends 246, 275
vendors 245
VRML 254
WBEM 534
Web-accessible tools 631
Web-enabled tools 631
see also electronic commerce, Internet, intranets, World Wide Web
BTG Inc. 377
business applications
see corporate applications, groupware, workflow
Business Objects Inc.
BusinessMiner 624, 626, 635
BusinessObjects 622
Metadata Coalition 627
Microsoft Alliance for Data Warehousing 630
MX 628
OLAP 630
WTI 629
Business Software Alliance 386

C

C programming language 431, 433
advantages 454
C++ 435, 466
forecast 465
market 463
Objective-C 436
runtime error detection tools 448
C++ 435
typesafe linkage 435
versus Java 437
CA 311, 440, 448, 452, 485, 543, 544–545, 643, 646, 667–668
for Computer Associates International Inc.
Acacia 681
ACCPACC International 681
Advanced Help Desk 545
architecture 668
CA-Explore 537
CA-HRISMA 668
CA-IDMS 488
CA-Impact/2000 452
CA-Ingres 337, 488
CA-Librarian 537
CA-LOOK 537
CA-Masterpiece 667
CA-Masterpiece/ICE 667
CA-Masterpiece/Net 667
CA-MICS 537
CA-Panvalet 537
CA-Paramount 537
CA-PMA 537
CA-TSO/MON 537
CA-Unicenter 63, 544, 667
CA-Unicenter TNG 379, 544
CA-Unicenter TNG Framework 545
Cheyenne ARCserve 538
Cheyenne Software Inc. 538
Datacom DB 488
EMU 667
Endeavor 445
Internet 668
market 680, 681
MK Group 681
Prestige Software International 667, 681
Year 2000 644
cable
see wiring
Cable & Wireless PLC 205
cable companies 210
services offered 204, 206
Cable Deregulation Act of 1993 213
cable modems
market 225
pricing 188, 225
services provided 188
speed 188
technologies 188
CableComms PLC 205
Cabletron Network Express Inc. 196
Cabletron Systems Inc. 168, 172, 176, 177, 196
NICs 168
Spectrum 63
CAD 47, 49, 54, 55, 473, 474
for computer-aided design
Caere Corp.
OmniPage 50
Caldera Inc.
Linux 300
OpenDOS 293
Caligari Corp.
Fountain 255
CAM 473
for computer-aided manufacturing
Canadian Institute of Chartered Accountants 385
Candescent Technologies Corp.
ThinCRT 57
Candle Corp. 544
Candle Command Center for MQSeries 336
OMEGAMON 537
Canon Inc. 119
CAPs 206, 217
for competitive access providers
Capstone Software 398
Caribbean
deregulation 210
CaroNet 239
CARP 536
for Cache Array Routing Protocol
Carte Bleu 120
Cartes Bancaires 124
Cascade Communications Corp. 196
CASE 443
for computer-aided software engineering
forecast 465, 466
market 465
Casio Computer Co. 298, 360
Castelle Inc.
FaxPress 264
CATV 188
for cable television
forecast 229
Internet 224
services 223
telephony 224
wireless 213
wiring 160

CBD
for component-based development
CBS Corp. 259
CC 383
for Common Criteria
CCD 45, 46, 50
for charge-coupled device
costs 51
definition 51
technology 51
CCITT
see also ITU-TSS
CD Universe 582
CDE 457
CDIF
for CASE Data Interchange Format
CDMA 13, 216, 217, 229
for Code Division Multiple Access
capacity 216
features 216
technologies 215
versus TDMA 216
CDMA Development Group 216
CDPD 216
for Cellular Digital Packet Data
CD-R 89
for CD-Recordable
versus DVD 90
CDR 327
for Common Data Representation
CD-ROMs 89
for compact disk read-only memory
CD-R 89
CD-RW 90
market 100
recordable 89
technology 89
versus DVD 90
versus NFR 78
WORM 89
CD-RW 90
CDSA 400
for Common Data Security Architecture
CDSL 212, 213
for consumer DSL
CEC 207
for Commission of European Communities
Cegetel 208
Cellnet 573
Cellular One Inc. 204
cellular technology 24, 203
AMPS 214
analog to digital 220
antennas 217
CDMA 215
CDPD 216
cell transmitters 214
definition 214
EPS 219
ESMR 219
European 12
forecast 29
global roaming 229
groupware 513
GSM 215
JavaCards 112
Li-ion 96
MSS 158
paging systems 218
PCS 217
PCS/PCN 217
smart cards 127, 573
standards 214, 229
TDMA 215
telephone networks 214
wireless telephones 225
see also antennas, mobile computing, PCN, PCS, satellites, telecommunications
CellularVision Inc. 213
Centerline Software Inc. 448
Centura Software Corp. 464
Centura 441
CEPT
for Council of European PTTs
CERN 233
for European Laboratory for Particle Physics
CertCo LLC 574
Certicom Corp. 576
CE2 Security Builder 389
certificate authority 387, 393
CRLs 393
forecast 422
key pairs 387
validation 393
CFAR
for collaborative forecasting and replenishment
CGI 566
for Common Gateway Interface
definition 267
languages 267
script 267
technologies 267
Charles River Analytics Inc.
Open Sesame 512
Charles Schwab Corp.
e.Schwab 584
Check Point Software Technologies Inc. 185
FireWall-1 417
OPSEC 417
SYNDefender 406
CheckFree Corp. 578, 593
Cheyenne Software Inc. 418
InocuLAN/AntiVirus 403
see also CA
CICS 322, 431
for Customer Information Control System
CIDR 179
for Classless Inter-Domain Routing
Ciena Corp.
MultiWave Firefly 144
MultiWave Metro 144
MultiWave Sentry DWDM 144
CII-Honeywell Bull 108
CIM
for Common Information Model
Circuit City Stores Inc. 91
Cirque 48
CISC 344
for Complex Instruction Set Computing
technologies 344
versus RISC 344
Cisco Systems Inc. 165, 167, 168, 172, 176, 178, 193, 385, 534, 581
196
1988 IGRP 178
7500 routers 177
7x00 177
Catalyst 5000 LAN switches 529
Cisco Connection Online 582, 585, 586
CiscoView 545
CiscoWorks for Switched Internetworks 545
GSR 176, 177
Internet2 239
Internetwork Operating System 190
L2F 186
LANE 170
LightStream 177
Local Director 535, 536
Netsys Service-level Management 529
Pix Firewall 417, 418
RSVP 193
StrataCom 177, 529
tag switching 181
Web Cache 239
Citicorp 658

Citrix Systems Inc.
Picasso 299
WinFrame 299, 662
CKS North Amercia
MyNet 380
Claris Corp.
Emailer 387
class 434
Clear Communications 209
CLEC
for Competitive Local Exchange Carrier
definition 160
CLECs 206
for competitive LECs
technologies 223
CLI 478
for call level interface
client/server computing 632
4GLs 441
administration 379
application server 319
automated testing tools 447
background 318
browsers 278
client devices 355
client forecast 368
client market 362
clustering 289, 354
commerce server 566
corporate applications 648
cost reductions 526
data warehousing 604
document management 506, 507
electronic commerce 557
failover operation 353
Fibre Channel Interconnect 172
forecast 278
groupware 495
hypercube 604
Kerberos 381
LDAP 391
management 525, 527, 528
market 362, 461, 464
monitoring software attributes 525
MPP 354
multiprocessor architectures 353
NCs and NetPCs 355, 359
NetPCs 342
NetWare 304
network bottlenecks 195
network performance 165
NUMA 354
operating system forecast 313
operating system market 310
operating system vendors 301
operating systems 285, 286, 287, 291
OS/2 Warp Server 304
OS/400 305
revenue 361, 362
RMON 532
RSVP 190
security 374, 414, 422
server forecast 367
server market 362
servers 351
shared-disk model 353
shared-nothing model 353
SMP 353
SQL3 478
storage management 538
three-tier 319
transitions 319
trends 342, 348
two-tier 318
UNIX 301
UNIX and NT-based 352
versus mainframes 348, 350
VINES 304
Web servers 266
forecast 278
Web servers as middleware 429
Web-based computing 462
Windows NT Server 302
Wolfpack 353
Clipper 398
CMOS 349, 367
for Complementary Metal Oxide Semiconductor
Advanced CMOS-ECL 351
CMS
for Conversational Monitor Systems
CNC 675
for Configurable Network Computing
COBOL 431
forecast 465, 466
market 463
object-oriented COBOL 455
versus GUIs 455
Year 2000 451, 452, 455
Codabar 118
Cognos Corp. 634
Impromptu 622
Impromptu and Powerplay 635
MX 628
OLAP 630
PowerHouse 440
PowerPlay 669
Right Infor mation Systems 635
Scenario 624
color gas-plasma 56
COM 331, 405, 443, 628, 630
for Common Object Model
COM automation 332
components 332
definition 332
OLE automation 332
COM+ 331, 334
for Component Object Model Services
Comcast Corp. 225
CommerceNet 408, 578
CommerceOne 590
CommerceWave Inc. 567
Commercial Biometrics Developer Consortium 378
Commercial Building Telecommunications Cabling Standard 146
Commercial Electronic Mail Choice Act 409
Commodore Business Machines 46
Compaq Computer Co. 48, 167, 172, 196, 260, 286, 298, 341, 352, 356, 364, 368, 526, 534, 670, 699
NetPC 527
ProLiant 353, 658
SuperDrive 87
Tandem Computers Inc. 276, 311, 352, 353
Tandem Nonstop SQL 483
Tandem Pathway 336
Tandem Safeguard 379
Tandem ServerNet 354
Comparex 361
Compatible Systems Corp. 177, 184
compilers
C 433, 454
C++ 436, 455
COBOL 455
file dependency list 446
Java 437
MIB 532
PC 456
Year 2000 451, 453
CompuServe Inc. 232
computational completeness 478
computer displays
see displays
Computer Oracle and Password System 380
Computer Reseller News 275
Computer Security Institute 374
Compuware Corp. 448, 537
UNIFACE Seven 440
Xchange & Xpedite 452
Comshare Inc. 633
OLAP 630
WTI 629
Connect Inc. 567

OneServer 567
Connectix Corp. 51
QuickCam 51, 52
Constellar Corp.
Constellar Hub 623
contact cards
definition 109
physical dimensions 109
contactless cards
active cards 115
passive cards 115
RF-IDs 118
types of 115
Continuus Software Corp. 445
cookie technology
advantages 255
challenges 255
definition 255
HTTP 266
cookies 402, 404, 409
definition 408
products 408
CORBA 269, 291, 317, 325, 326, 328, 333, 461, 650, 655, 666
for Common Object Request Broker Architecture
CDR 327
dynamic invocation 326
ESIOP 327
IIOP 327, 331
object adapter 327
Corillian 578
Corning Inc. 153
corporate applications
architectures 645
component architectures 682
componentizing 646
cost reductions 526
customer demands 644
EDI data translation 562, 563
electronic commerce 646
EMU 644, 645
enterprise support 682
ERP 645
forecast 682, 683
groupware 494
Internet 646
intranets 264
mainframes 348, 350
management 539
market 643, 680
midsized business 643
midsized market 681
multi-tier architecture 645
multi-vendor standards 679
non-monolithic 643
OAG 679
overview 643
packaged applications 643, 644
process manufacturing enterprise solutions 682
push technology 249
RAID 82
retailing 682
revenue 680
security 374, 385
specific applications 647
standards 679
storage management systems 521
supply chain management 643
systems management 522
three-tier architecture 645
trends 644, 647
vendor targets 646
vendors 645
vertical markets 647
Web front ends 569
workflow 501, 516
Year 2000 644, 680
see also document management, groupware, knowledge management, network management, storage devices, systems management, workflow
COS 113
for card operating system
security 113
COST 397
Council of European PTTs 215
Council of Registrars 238
Cox Cable Communications 225
CPA Web Trust 385
CPG 648
for consumer packaged goods
CPU
for central processing unit
CRL 393
for certificate revocation list
CRT 45, 53
for cathode-ray tube
advances 55
aperture grille 54
dot pitch 54
EnergyStar 55
forecast 66
graphics market 55
interlaced versus non-interlaced 55
setup programs 55
shadow mask 54
sizes 54
stripe pitch 54
technology 54
terminology 54
triads 54
versus FED 57
see also displays
Crumbler 97 409
cryptography 385, 394
certificate authority 393
cryptographic processor 111
DES 386
Diffe-Hellman 388
ECC 389
forecast 422
government regulations 396
law enforcement 398
market 419
PEM 411
PKI 387
smart cards 107, 114, 124, 390
standards 389
C-SET 124
for Card-Secure Electronic Transactions
CSMA/CD
for Carrier Sense Multiple Access with Collision Detection
CSSs 242
for Cascading Style Sheets
CSS Positioning 243
definition 242
CTCPEC 383
CTI 224
for computer telephony integration
Curie point 93
CVP 418
for Content Vectoring Protocol
Cyber Promotions Inc. 410
CyberCash Inc. 593, 658
CyberCoin 568, 573
CyberGuard Corp. 418
CyberMedia Inc.
Guard Dog 404
cybernotaries 393
cyberspace
see Internet, World Wide Web
Cygnus Solutions
KerbNet 381
Cylink Corp. 388, 389, 392, 399
Cyrano 448

D

D2K Inc.
Tapestry 612
Dai Nippon 129
D-AMPS 217
for digital-AMPS
advantages 215
DARPA
for Defense Advanced Research Projects Agency

DAT 70
  for digital audio tape
  MLR 75
data 611
  aggregating 609, 610
  archiving 616
  backup and recovery 616
  cleaning 603, 608, 609, 610
  compression 69, 73, 260
  consolidating 609, 610
  converting 610
  converting and translating 609, 610
  data mart versus data warehouse 607
  data warehousing technologies 608
  denormalizing 611
  evaluation 615
  filtering 609, 610
  gateway 631
  hierarchical storage management 605
  high-value 610
  inconsistencies 615
  in-depth validation 615
  information access 599
  layout 610
  loading 611, 615
  management 599, 616, 636
  mapping 610
  marts 606
  metadata 611
  mining 511
  models 610
  monitoring 616
  normalization 471
  ODS 603
  OLAP 602
  OLAP versus OLTP 602
  OLTP 602
  organizing 611
  partitioning 611
  quick reviews 615
  scrubbing tools 610
  storage management 537
  transfor mation 615
  versioning 614
  visualization 624, 635
  warehouse generation 599
  warehousing 599
  see also databases, data mining, data warehousing, information retrieval, security
data communications
  ATM 222
  bridges 175
  desktop computing 285
  devices 174
  digital 513
  hubs 175
  ISDN 211
  layerless 182
  Layers 1-7 173
  network management 522
  OSI 173
  overview 137
  routers 176
  switches 180
  VLANs 182
  see also ATM, Ethernet, frame relay, network management, networking systems, protocols, systems management, telecommunications
data compression 69
  ratio 73
Data General Corp. 101
  Clariion disk arrays 86
data marts 633
  advantages 606, 608
  and data warehouses 607
  and virtual data warehouses 607
  data retailing 606
  definition 601, 606
  dependent 606
  disadvantages 606
  hub and spoke 606
  independent 606
  technologies 607
  trends 606
  types 606
  versus data warehouses 607
  versus data warehousing 606, 607
  virtual data warehousing 608
data mining 624
  algorithms 624
  artificial intelligence 625
  association rules analysis 625
  clustering analysis 625
  decision tree analysis 624
  forecast 639
  functions 625
  knowledge discovery 625
  machine learning 625
  neural networks 624
  ODS 603
  trends 625, 635
  users 624
  vendors 622, 623, 626
Data Protection Act 407
data warehousing 485
  ad-hoc access 621, 622
  advantages 599
  aggregates 616
  aggregation 621, 622
  application-specific 637
  approaches 606
  availability 608
  available tools 603
  BIW 656
  bulk load 611
  categorizing 601
  category integration 635
  concepts 599
  consolidating and reconciling views 600
  creating 609
  cross-category integration 636
  current usage 600
  customer-centric 600
  data catalog 611
  data cleaning 603
  data evaluation 615
  data extraction 609
  data marts 606, 630
  data marts versus data warehouses 607
  data mining 355, 511, 603, 624
  data modeling 611
    terms 611
  data quality 615
  data scrubbing 610
  data scrubbing tools 610
  data transformation 609
  data visualization 624, 639
  data warehouse 607
  data warehouse scope 607
  decision-support systems 603
  definition 599, 601
  design tools 614
  designing 601
  distributing 601
  enterprise data warehouse 607
  enterprise storage 85
  enterprise-wide data model 601
  extraction/transformation 603, 609
  financial reporting 600
  forecast 637
  generation 599, 608, 634, 635, 636, 637
  generation forecast 638
  global trends 600
  hardware market 637
  HOLAP 618, 619
  indexing 489
  information access 599, 600, 603, 608, 619, 625, 630, 635, 636, 637
  information access forecast 639
  Internet 630

loading data 611
macro-level viewing 600
management 599, 608, 611, 616, 635, 636
market 484, 632, 636
MDBMS 604
MDBMS versus RDBMS 605
metadata 611
metadata standards 627
monitoring 616
normalization 472
ODS 603
organizing data 601
overview 599, 600
packaged applications 632, 634
performance 481
performance management 616
periodic incremental load 611
physical versus virtual 607
predefined access 619
preparing data 609
quality assurance 614
RDBMS 603, 619
ROLAP 619
scheduler program 615
software market 637
solutions 636
standards 627
storage capacity trends 83
summarizing 601
systems management 615
tape drive formats 71
technologies 599, 603, 607, 608
transforming 601
trends 630, 632
trickle load 611
using successfully 600
vendors 612, 613, 635
versus data marts 606, 607
versus operational databases 601
VLDBs 85
Web-accessible tools 631
Web-enabled tools 631
Web-enabling tools 637
Web-exploited tools 632
with ERP 633
World Wide Web 630
see also data, databases, metadata
databases
3-D 604
4-D 604
storage demands 604
access tools 631
ACID 473, 475
acquisition decisions 481
benchmark 481
application domains 481
custom 482
DebitCredit 481
TPC-A 481
TPC-B 481
TPC-C 481
TPC-D 481
competitive issues 485
concurrent multi-user updates 475
data mart versus data warehouses 607
data models 470, 610
DBMS 470
DDL 470
distributed 479
DML 470
DSS 482
electronic commerce 565
embedded SQL 472
forecast 488, 489
gateways 609
hierarchical 470, 471, 484
hypercube 604
dynamic hypercube 604
micro-hypercube 604
sparse hypercube 604
storage requirements 604
indexing 469, 485, 489
intranets 265
intra-query parallelism 482
isolation levels 476
linear speedup 482
market 484
MDBMS 604
MDBMS versus RDBMS 605
multi-version concurrency 476
NCs 485
ODBMS 472, 484
ODS 603
OLAP 617
operational 603
operational versus data warehousing 601
ORDBMS 469, 474, 484, 488
overview 469
parallelism 469, 482, 489
types of 483
parts explosion 478
PDBSs 469, 482
performance 481
preaggregation 489
RDBMS 471
recovery 471, 475
ref data type 475
referential integrity 471
relational 473, 484, 603
relational to object-relational 474
relational versus object 469
replication 469, 479, 494, 609
asynchronous 479
definition 479
multi-user 480
peer-to-peer 480
single-master 480
synchronous 479
tradeoffs 480
types 479
revenues 484, 485
runtime locking mechanisms 475
scale-up 482
search engines 250
shared disk 483
shared everything 483
shared nothing 483, 484
sizes 605
software components 319
SQL3 478
SQL-92 477
storage 605
storage management 537
transactional capabilities 475
transactional concurrency 471
transactional isolation 476
triggers 478
types of 618
Universal Server 469
VANs 563
vendors 480, 483, 485
versioning 473
VLDBs 85
xBase 438
see also OLAP, OLTP
Datalogix International Inc.
see Oracle Corp.
DataMind Corp.
DataCruncher 626
Dataware Technologies Inc.
Knowledge Management Suite 510
DBMS 461
for database management system
architecture 470
definition 470
forecast 489
hierarchical 470, 471
market 462
models 470
revenues 484
transaction management 475
see also databases
DBS 159
for direct broadcast satellite
DCE 321, 380, 382, 383
for Distributed Computing Environment
definition 321
features 382

HTTP 266
MS-RPC 333
RPC 332, 333
versus Kerberos 382
DCOM 268, 331, 333, 405
for Distributed COM
architecture 333
definition 332
forecast 339
DDE
for dynamic data exchange
DDI 209
DDL 470
for Data Definition Language
DDS 73
for Digital Data Storage
advantages 73
debit cards
transitioning to smart cards 125
decision-support systems
data warehousing 603
see data warehousing
DECT 215
for digital European cordless telecommunications
Dell Computer Corp. 48, 353, 368, 581
DMI 526
NetPC 527
PowerEdge 527
Delphi Consulting 501
Delta Three Inc. 263
DEM 676
for Dynamic Enterprise Modeling
Deming Internet Security 412
Demon Internet 239
Department of Telecommunications 233
DES 113, 386, 388, 396, 414, 571
for Data Encryption Standard
cracking 386, 388
exports 399
key for mula 386
PEM 411
Triple DES 386, 388
desktop computing
25-Mbps ATM 170
application frameworks 457
ATM 169
browsers 246
business applications 521
client devices 355
competing technologies 360
cost reductions 526
data warehousing 603
DBMS 470
delivery tools 521
DMI 525, 533, 534, 548
forecast 368
graphics controller 54
graphics workstations 356
hubs 174
JavaOS 301
keyboards 47
MacOS 299
management 525, 527
market 363, 364
mergers and acquisitions 364
microprocessors 345
mobile computing 358
monitoring software attributes 525
MS-DOS 293
multimedia 357
multiprocessing 288
multitasking 288
mutlithreading 288
network performance 194
operating system forecast 312
operating system vendors 293
operating systems 285, 286, 287
operating systems market 308
OS/2 299
PC 98 357
pointing devices 48
pricing 341
printers 60, 62
process management 288
revenue 361
Rhapsody 300
TCP/IP 189
typical configuration 368
UNIX 300
UNIX/RISC workstations 356
Windows 293
Windows NT 356
Windows systems features 298
workstation market 362
workstations 356
Year 2000 451
zero administration 526
desktop publishing
CD-ROMs 89
displays 54
electronic commerce 581
HTML 234
printers 60, 61
scanners 50
SCSI 79
World Wide Web 265
authoring tools 255
vendors 258
DeTeMobil of Germany 157
Deutsche Telekom AG 119, 123, 207, 208, 263
development tools
3GLs 430, 431
4GLs 439, 465
ACE toolkit 459
analysis and design models 444
analysis, modeling, and design 443, 444
application delivery 521
application frameworks 457
automated testing 447
by category 462
by platform 463
CASE 443, 465
code coverage 448
code repositories 450, 467
compilers 430
configuration management 445, 466
CVS 445
defect testing 447
distributed make 446
electronic commerce 565
forecast 465, 466
gprof command-line utility 447
incremental linking 446
integrated development environment 430
interoperability 454
Java 112, 461
languages market 463
linking 446
lint utility 447
mainframes 461
market 461, 462
middleware 460
modeling 444
object languages 434, 435
object technology 462
process control 445
process support 443
profiling 447
project management 448
project teams 443
quality assurance 448
reusable class libraries 449
reverse engineering 444
runtime error detection 448
scripting languages 437
software metrics 447
specialized languages 437
static analysis 447
system building 446
testing 448
testing and quality assurance 447, 466
trends 429
UNIX 461
Web-based computing 460, 462
Windows 95/Windows NT 461
World Wide Web 464

Year 2000 450, 461
see object technology, programming
DHCP 187, 241
for Dynamic Host Configuration Protocol
DHTML 243
for Dynamic HTML
advantages 243
definition 243
Diebold 578
Diffie-Hellman 388
DigiCash 593
Digipass S.A. 418
digital
broadcast channels 253
certificates 575
communications 220
data communications 513
data services 220
effects on telecommunications 9
European technologies 7
telephony 513
digital audio
Divx 91
digital cameras 45, 46, 66
storage devices 88
digital certificates 305, 393, 399
market 419
X.509 391
Digital Equipment Corp. 76, 168, 172, 176, 181, 286, 352, 356, 363, 368, 399, 593, 654
21164 345
Alpha 307, 351
AlphaServer 353
Alta Vista 250, 276
DECnet 178, 188
Digital Alpha 291, 301, 311
Digital StrongARM 301
Digital UNIX 286, 301, 307
DLT 74
Galaxy 315
GIGAswitch/Ethernet 168
historical events 698
Memory Channel 354
MilliCent 574
Object Broker 327
OpenVMS 286, 307, 310, 315
Secure Enterprise Web 383
StorageWorks 173
VAX 74, 307, 351, 440
Windows NT Server 307
Digital Pathways Inc. 418
digital signatures 112, 113, 373, 385, 389, 394
forecast 422
international developments 395
ISO 14888-3 389
legality 394
PEM 411
PKI 392
RSA 388
S/MIME 412
smart cards 114
worldwide status 394
X.509 391
Digital Video Express 91
digital watermark 117
Discover 573
disk array 81
disk controllers
disk array 81
EIDE 78
FC-AL 80
FCL 80
RAID 81
SCSI 78
SSA 79
switching fabric 80
disk drive interface 78
disk drives
capacity 69, 99, 103
forecast 102
jukeboxes 69
market 99
optical 88
removable 100
see also floppy disks, hard disk drives, storage devices
disk mirroring
RAID 1 81
displays 53
color gas-plasma 56
CRT 53, 54
display-projection systems 54
DLP 59
factors 54
FED 57
forecast 66
LCD 53, 55
light-valve projectors 58
sizes 54
three-beam CRT projectors 58
video projection systems 58
distributed computing
client stubs 326
components 324, 325
CORBA 325
definition 319
dynamic invocation 326
forecast 339
IDL 326
JavaBeans 328
overview 318
static invocation 326
technologies 323
distributed objects 319
Divx
for Digital Video Express
movie rentals 91
DLLs 324, 456
for dynamic link libraries
definition 324
DLP 59
for digital light processing
forecast 66
DLT 74
for digital linear tape
availability 74
DLT-7000 74
DMA 508
for Document Management Alliance
compound document types 509
core elements 509
future work 509
DMD 59
for Digital Micromirror Device
adding color 59
technology 59
DME
for Distributed Management Environment
DMI 525, 534
for Desktop Management Interface
and WBEM 534
CIM 535
implementation 534
overview 533
technologies 533
vendors 534
DML 470
for Data Manipulation Language
DMTF 525, 534
for Desktop Management Taskforce
DNS 235, 264, 292, 333, 535
for domain name system
document management
advantages 491
AIIM 509
bar codes 117
converging technologies 513
definition 505
DMA 508
document life cycle 505
exploiting electronic information 510
forecast 516, 517
groupware 493
integration environment 506
Internet 506

market 515
ODMA 508
overview 491
publishing life cycle 517
revision bars 480
standards 508
technologies 505, 506, 515
trends 515
vendors 506, 508
workflow 500
World Wide Web 517
Documentum Inc. 506
EDMS 507
Enterprise Document Management System 506
RightSite 506
DOD 118, 433
for Department of Defense
Dolby Laboratories Inc.
AC-3 91
DOM
for document object model
definition 244
Dr. Solomon Software Inc. 421
Dragon Systems
NaturallySpeaking 53
DRAM 342
for dynamic RAM
performance 343
RDRAM 344
SDRAM 343
SLDRAM 344
technologies 343
versus SRAM 343
DreamWorks SKG
Divx 92
Drexler Technology Corp.
LaserCard Systems Corp. 119
DS-3 177
for Digital Service Level 3
DSL 152, 212
for Digital Subscriber Line
ADSL 212
CDSL 213
distance limitations 212
DSL Access Multiplexer 212
HDSL 212
SDSL 212
technologies 212
DSOM
for Distributed SOM
DSP 53
for digital signal processor
DSS 388, 389, 469
for decision-support systems
DTC 334
for Distributed Transaction Coordinator
Dun & Bradstreet Software
see Geac
Duracell Batteries Ltd. 97, 99
DVD 90
for Digital Video Disc or Digital Versatile Disc
capacity 91
Divx 91
DVD+RW 92
DVD-RAM 92
DVD-ROMs 91
forecast 103
formats 91
rewritable 91
technology 93
versus CD-RW 90
versus NFR 78
DVD Consortium 91
DVD Forum 91, 92
DVD+RW 92
DVD-RAM 92
DVD-ROMs 91
market 100
rewritable 91
trends 91
DVMRP 192
for Distance Vector Multicast Routing Protocol
DWDM 139, 143, 144
for Dense Wavelength Division Multiplexing
forecast 163
market 159
vendors 144
Dynacard 119
Dynapoint Inc. 47
DynaSoft 419

E

E*Trade Securities Inc. 582
Eastern Systems Inc.
Testbed 452
Eastman Kodak Co. 254
Eastman Software Inc. 503
OPEN/workflow 503
EBT
for Electronic Benefits Transfer
EC
see electronic commerce
EC Co. 589
EC Key Establishment 389
ECAPMO
for Electronic Commerce Acquisition Program Management Office
ECC 389, 419, 576
for Elliptic Curve Cryptosystem
issues 389
EchoStar II 159
ECI Telematics 185
ECMA 92, 456
for European Computer Manufacturers Association
e-COMM 577
E-Corp
eMail '97 410
ECR 559
for Efficient Customer Response
ECSET 576
for Elliptic Curve Enabled Secure Electronic Transaction
EDI 411, 553, 554, 558, 562, 565, 585, 587, 589, 677
for electronic data interchange
application-to-application exchange 555
asynchronous 555
business model 588
business practices 558
business processes 558
businesses 561
business-to-business 555
challenges 579
characteristics 555
data translation 562
definition 555
EDI Lite 555
EDIFACT 579
event-driven 555
forms 564
fueling growth 592
interactive 555
Internet 555, 564
invoice 571
market 562, 590, 592
penetration 592
message transport 563
peer-to-peer trading 592
products 562
revenue 592
security 555
slowing growth 592
standards 564
technologies 555, 562
Tradacoms 564
VANs 563
vendors 592
versus Web-based electronic commerce 558
see also electronic commerce
EDIFACT 564
for Electronic Data Interchange for Administration Commerce and Transport
Edify 578

EDO
 for Extended Data Out
EDP
 for Electronic Data Processing
EEC
 for European Economic Commission
EEPROM 109, 111
 for electrically erasable programmable ROM
EFT 554, 562, 585
 for electronic funds transfer
 definition 559
eFusion Inc. 568
 eBridge 568
 eStream 568
EGP 178
 for Exterior Gateway Protocol
EIA 141, 146
 for Electronic Industries Association
Eicom Corp. 566
EIDE 77, 78
 for Enhanced IDE
EIGRP 178
 for Enhanced Interior Gateway Routing Protocol
EIS
 for Executive Information System
EJB 329
 for Enterprise JavaBeans
Elcotel Inc. 123
Electric Mail Co.
 E-mmunity 403
electronic commerce
 aggregator 560
 architecture 565
 attributes 554
 auctions 583
 back-office 566, 579, 593
 banking 583
 business models 558
 business practices 558
 business-to-business 557, 558, 560, 567, 580, 585, 588, 590, 591, 594
 business-to-business versus business-to-consumer 557
 business-to-consumer 557, 559, 567, 579, 581, 582, 595
 consequences 561
 Consumer Goods Manufacturing 561
 corporate applications 643, 646
 definition 554
 digital cash 572, 594
 digital check 572
 direct seller 560
 disintermediation 560
 distribution 561
 e-COMM 577
 EDI 553, 554, 555
 EDI over the Internet 555
 EDI versus Web-based 558, 562, 563
 Efficient Customer Response 559
 EFT 559, 561
 electronic mall 560
 electronic wallets 572, 573
 e-mail 554
 Europay C-SET 577
 European 395
 European currency 645
 Euro 645
 Evaluated Receipt Settlement 559
 extranets 595
 forecast 594, 595
 global 8
 government uses 585
 hardware 570
 infrastructure 562, 565
 integration 579
 Intenet
  market 581
 interactive 555
 Internet 236, 553, 568
  market 590
 Internet commerce 556
 intranets 557
 investment services 584
 mainframes 349
 market 273, 567, 581, 582, 587, 590, 591, 594
  challenges 580
 Materials Management 559
 microtransactions 572
 Model Stock Replacement 559
 multimedia 36
 network congestion 579
 networking systems 165
 on-line retailer 560
 on-line services 590
 open exchange 572
 overview 553
 Pandesic 658
 payment standards 577, 578
 payment systems 572, 594
 POS 570
 products 562
 public versus private network systems 553
 Quick Response 558
 requirements 561
 retail 582
 SCRIP 574
 secure transactions 574
 secured linkage or presentation 572
 security 373, 374, 387, 389, 393, 396, 411, 555, 557, 580
 servers 567
 SET 411, 576
  payment authorization 574
 shipping 583
 smart cards 110, 119, 124, 128, 573
 software 567
  market 593
 standards 564, 571, 581
 success factors 36
 system designs 565
 technologies 554
 tools 562
 transaction management 566
 travel 584
 trusted third party 572
 types of 554
 utilities 561
 vendors 563, 565, 567, 570, 593
 Web site building tools 566
 Web-based 556, 557, 558, 559, 563, 580, 592
 Web-based technologies 565
 see also Internet, security, Web-based computing, World Wide Web
Electronic Commerce Enhancement Act of 1997 395
Electronic Computing Information Systems 397
Electronic Data Security Act of 1997 401
Electronic Engineering Times 97
Electronic Financial Services Efficiency Act of 1997 395
Electronic Frontier Foundation 408
electronic mail
 see e-mail
Electronic Mailbox Protection Act 409
Electronic Travel Auction 583
EliaShim Ltd.
 Protect 404
Elo TouchSystems Inc. 50
ELSE 356
e-mail 234, 273
 bombs 406
 businesses 561
 data warehousing 621
 document management 506
 EDI 563
 electronic commerce 582
 encryption 401, 411
 gateways 403
 groupware 492, 493

market 419
paging 226
PEM 411
PEM-MIME 413
private system 411
S/MIME 412
security 373, 387, 411
smart cards 391
spam 373, 402, 409, 410
trends 23, 374
users 272
VANs 563
viruses 402
workflow 500
X.500 391
EMC Corp. 76, 83, 101, 538
AS/400 Harmonix 83
ESP 86
Intelligent Storage Architecture 83
Mosaic 2000 83
Symmetrix Enterprise Storage Platform 83, 84
technologies 83, 84
Emergent Corp. 238
EMI 154, 155
for ElectroMagnetic Interference
EMILM 144
for electroabsorption-modulated isolated laser module
Empresa Guatemalteca de Telecomunicaciones 210
EMU 644, 646
for European Monetary Union
EMV 124, 131
for Europay/MasterCard/Visa
encryption 18, 373
AES 389
algorithm 385
bar codes 118
CDPD 217
cryptography 385
DES 386
digital signatures 385
DSS 388
electronic commerce 236, 396, 566, 574
e-mail 411
European legislation 402
European legislative update 401
export 388, 396, 422
export exemptions 397
export regulations 373
firewalls 418
forecast 422
foreign 388
government regulation 396
hardware 420
IDEA 386
initiatives 398
IPsec 184
isolating 398
JCP Computer Services 576
key 385
key encapsulation 399
key escrow 398
key length 388
key recovery 398, 401
law enforcement 399
market 419
PGP 128-bit 397
PKI 390
products 387, 388
public two-key 386
public-key algorithms 387
recent developments 399
RPK 389
RSA 387, 576
S/MIME 412
SCRIP 574
SET 576
shared single-key 386
single-key algorithms 387
smart cards 107, 112, 113, 114, 116
software 420
SSL 413
symmetric-key algorithms 387
techniques 388
trends 388
Triple-DES 91
U.S. legislative update 401
unique shortest vector 390
VPIs 184
Enigma Logic Inc. 418
Entegrity Solutions Corp. 397
Enterprise Integration Technologies
S-HTTP 415
enterprise management 522
accounting 523
automated operations 523
batch processors 523
capacity planning 523
configuration management 523
disaster recovery 523
disk backup systems 524
event management 524
fault management 524
forecast 548
job schedulers 524
market 544
network monitoring 524
performance monitoring 524
platform independence 535
print management 524
security management 524
storage 83, 85, 537
cost 101
market 101
subsystem market 83
storage management 525, 538
tools 522
WBEM 534
see also mainframes, network management, systems management
Enterprise Security Initiative 385
Entrust Technologies Ltd. 412, 419
Public-Key Infrastructure 382
EPIC 344
for Explicitly Parallel Instruction Computing
technologies 345
EPROM
for erasable programmable ROM
EPS 219
for enhanced paging services
Epson America Inc. 86
Stylus Color 3000 62
e-purse 121
for electronic purse
reloadable 122
Ericsson Inc. 13, 208, 218, 513
ERP 599, 633
for enterprise resource planning
Open Information Warehouse 633
services 682
vendors 647, 648
eSafe Technologies Inc.
Protect 404
eShop Inc.
see Microsoft Corp.
ESMR 219
for Enhanced Specialized Mobile Radio
market 226
technologies 219
ESMTP 241
for Extended Simple Mail Transport Protocol
Ethernet 148, 166
10Base2 151
10Base5 151
costs 194, 196
Fast Ethernet 145, 167, 195
Fast Ethernet versus ATM 171
forecast 199
Gigabit Ethernet 137, 145, 149, 167, 190, 195
Gigabit Ethernet routers 177
Gigabit Ethernet versus ATM 169
LANE 170
RMON 532

switched 167
switched 100-Mbps 167
switches 180
Thick Ethernet 151
Thin Ethernet 151
transparent bridges 175
versus ATM 170
ETI Solutions Inc.
MDIS 628
Metadata Coalition 627
Etisalat 157
for Emirates Telecommunications Corp.
ETO 659
for engineer-to-order
ETSI 13
for European Telecommunications Standards Institute
EU
for European Union
Bangemann, Martin 7
convergence 8
WOLF 10
EU Data Protection Directive of October 1995 407
Europay France 577
Europay International 124
Europe
deregulation 10, 207
European Bangemann Challenge 7, 13
global 14
European Commission 124, 395
encryption 402
Esprit 10
European Economic Community 433
European Monetary Union
unified currency implications 12
Year 2000 451
European Research in Marketing 396
Evolutionary Technologies International
ETI-EXTRACT 612
Evolutionary Technologies International Inc.
WTI 629
Evolutionary Technologies International Inc.
Data Guide 629
Exabyte Corp. 72
Mammoth 74
Mammoth drive 72
Excalibur Technologies Corp.
RetrievalWare 510
Excite Inc. 252, 276
eXcite 250
NewsTracker 253
Personal Access List 246
Execusoft Systems Inc.
Microsoft Alliance for Data Warehousing 630
ExperSoft Corp.
COBRAplus 338
extranets 514
definition 265
distributed computing 323
EDI 563
electronic commerce 568, 569, 590
forecast 280, 595
market 277
overview 231
security 521
technologies 265
tools 265
uses of 569
workflow 502, 515
Extreme Systems Services 176
EyeDentify Inc.
Retina Biometric Reader 378

F

fax transmission
Internet 264
vendors 264
FBI 374, 396, 401
for Federal Bureau of Investigation
FC-AL 80
for Fibre Channel Arbitrated Loop
advantages 80
switched FC technology 80
FCC 210
for Federal Communications Commission
CATV 223
competing local service providers 205
digital systems 214
U.S. Telecommunications Act of 1996 204
wireless spectrum division 158
FCL 80
for Fibre Channel Loop
challenges 80
FDDI 145, 166, 172, 175
for Fiber Distributed Data Interface
server bottlenecks 195
FED 56
for field emission display technology 57
Federal Express Corp. 118, 583
Federal Reserve Bank 585
FEXT 154
for far-end crosstalk
fiber optics 137, 210
advances 143
advantages 137, 139, 153, 154
bandwidth 144
cable configuration 155
cable TV 160
capacity 139, 140
cladding 139
connections 140
core 139
costs 138, 145, 153, 154
crosstalk 154
dark fiber 159
definition 138
DWDM 137, 139, 144
FC-AL 80
Fibre Channel Interconnect 145, 172
forecast 162
glass 139
hot standby circuits 160
hybrid fiber-coax 211, 213
installation 141, 155
interference 154
layers 139
LEDs versus lasers 141
loss 140
manufacturing and packaging techniques 145
market 159
multi-mode 141
NEXT 154
optical electronics 144
overview 137, 138
photon transmission 140
safety 155
security 154
single-mode 141
SONET 141
specifications 142
speed 140
technologies 138, 139, 140
testing standards 141
total reflection 139
trends 137
types of 141
ultra-wideband optical-fiber amplifier 143
vBNS 239
versus copper 139, 153, 154, 155
WDM 139, 143
Fibre Channel Interconnect 145, 172
capacity 172
forecast 103
server clustering 172
versus Ultra SCSI 79
file management
see corporate applications, data-

bases, information retrieval, intranets
FileNet Corp. 502, 503
Document Manager 508
Ensemble 502
Saros 506
Saros Mezzanine 508
Visual WorkFlo 503
Watermark Software 508
Financial Services Technology Consortium 373, 585
Finjan Inc.
Java Security Alliance 417
SurfinGate 417
FIPS 392
for Federal Information Processing Standard
Firefly Network Inc. 407, 512
Web site 254
Firewall Tool Kit 416
firewalls 112, 306, 373
access control 416
applet monitoring 417
appliances 423
attacks 407
CVP 418
encryption 418
forecast 423
interoperability 184
IP address spoofing 417
IPsec 184
market 421
MISS 374
packet-filtering gateways 416
PPTP 185
proxy 417
routers 176
security 403, 406, 411
stateful inspection 417
technology 416
trends 417
virus scanning 417
FireWire 295, 358
FIRST 191
for Fully Integrated Routing Switch Technology
First Virtual Corp. 183
First Virtual Holdings Inc. 593
Fischer International Systems Corp. 126
Smarty 376
Watchdog 380
flag cards 390
flash memory 111
floppy disk drives 86
formats 86
HiFD 87
high-density 86
market 100
multi-function 87
SuperDrive 87
trends 86
versus CD-ROMs 89
FlowWise Networks Inc. 182
Fore Systems Inc. 176
ForeFront Group Inc.
WebWhacker 253
Forté Software Inc. 502
Forté Application Environment 441
Fortran 431
Fortran 90 455
Fortran 95 455
HPF 431
market 463
Fortune 23
Foundry Networks Inc. 176
Fox 259
frame relay 171, 177, 203
advantages 171
backbones 195
definition 221
pricing 222
revenue 227
speed 222
SVCs 171
technologies 222
versus SMDS 226
Frame Relay Forum 171
France Telecom 119, 207, 208
Free Software Foundation 432, 436
freeware 380
FTP 234, 414, 416
for file transfer protocol
FTP Software Inc. 193
Secure Client 191
Fuji Photo Film Pte. Ltd. 87
Fujitsu Ltd. 56, 86, 124, 276, 349, 355, 362, 364
Fujitsu-ICL Systems Inc.
TeamWare 499
TeamWare Flow 504
Fulcrum Technologies Inc. 276
Knowledge Network 510
SearchServer 510
FutureTense Inc.
Texture 256

G

Gateway 2000 Corp. 364, 526
GCA 123
for Global Chipcard Alliance
GCB 108
for Le Groupement des Cartes Bancaires
GE Trading Process Network 582
Geac Computer Corp. Ltd. 643, 644, 646, 663–667
architecture 666
CyberStream 666
CyberStream Employee 666
CyberStream Manager 667
CyberStream Open Enrollment 667
CyberStream Requestor 666
CyberStream Time Management 667
DBS 663
Decision 360° 666
DecisionStream 666
Dun & Bradstreet Software 663
Enterprise Server 663
Enterprise Warehouse 634
Expert 663
IntelliMaps 666
IntelliScopes 666
Internet 666
market 680, 681
Millennium 663
operational and administrative systems 666
Pro-Mation 663
SmartStream 646, 647, 663, 664, 681
SmartStream 5.0 665
SmartStream Budget 664
SmartStream Connections 666
SmartStream Decision Support 664
SmartStream Financials 664
SmartStream Human Resources 665
SmartStream Procurement 665
SmartStream Projects 665
Vision 360° 666
Year 2000 644
GEIS 569, 587, 588, 592, 593
for General Electric Information Services
Actra Business Systems 588
AmeriData Technologies Inc. 588
CHINA EDI 593
CrossCommerce 588
GE BusinessPro 588
TradeWeb 588
Trading Process Network 588
GEMMS 661
for Global Enterprise Manufacturing Management System
Gemplus 123, 126, 129, 400
flag card 390
GemXpresso 125
General DataComm Inc. 183

General Electric Corp.
    GE Lighting 569
General Signal Networks 195
Generale Des Eaux 208
Gentia Software
    Gentia 620, 630
GEOs 156
    for geosynchronous earth-orbit (satellites)
    DBS 159
    services provided 157
    technologies 156, 157
    TSATs 157
    VSATs 157
GGP 178
    for Gateway-to-Gateway Protocol
Giesecke & Devrient 129
GIF 256
    for Graphics Interface Format
Gigabit Ethernet Alliance 167
GigaLabs Inc. 176
GigaPacket Networks Inc. 177
Gips, Donald 15–21
GITSB 585
    for Government Information Technology Services Board
Global Exchange Carrier 263
GlobeSet Inc. 574
GM Hughes Electronics Corp.
    DirecTV 159
GMR 77
    for giant magnetoresistive
GNU
    C++ 446
Gradient Technologies Inc. 381
    WebCrusader V3 382
GrapeVINE Technologies LLC
    GrapeVINE 512
graphics
    2-D controllers 60
    3-D controllers 60
    3-D coprocessors 60
    3-D renderings 624
    data visualization 624
    graphics accelerator 59
    graphics controllers 59, 66
    graphics processor 60
    multimedia accelerators 60
Graphics, Visualization and Usability Center 272
Groupe Bull S.A. 362, 543
    Bull PTS 126
    CP8 Transpac 129
    DPS9000/900 349
    Integrated System Management AccessMaster 382
groupware
    benefits 491, 495
    business process reengineering 494
    business processes 495
    communication capability 495
    communications-centric perspective 493
    converging technologies 513
    data and voice communications convergence 512
    data warehousing 621
    database-centric perspective 493
    data-store 494
    definition 492
    document management 505
    e-mail 492, 493
    forecast 516
    information availability 495
    Information Lens system 23
    Internet 512
    intranets 264, 265
    key enabling technologies 512
    LANs 494
    location independence 495
    market 514
    overview 491
    project management 495
    security 403
    shared filing system 494
    technologies 492, 493, 494, 495
    telephony 512, 513
    threaded discussions 493
    transforming 512
    trends 23, 514
    vendors 495
    World Wide Web 515
Grupo Financiero Banamex-Accival 209
GSM 13, 114, 215, 217, 229, 513
    for Global System for Mobile Communications
    DCS 1800 215
    PCS-1900 215
    trends 513
GSR
    for Gigabit Switch Router
GTE Corp. 123, 205, 390, 393, 512, 574
    BBN Planet 235
    CyberTrust 399
GTE Laboratories Inc.
    VH-ADSL 213
GTI Government Systems 419
gTLD 238
    for Generic Top-Level Domain
gTLD-MoU 238
    for Generic Top-Level Domain Memorandum of Understanding
Guangdong Mobile Communications 208
Guidelines on Cryptography 401
GUIs 48, 297, 603
    for graphical user interfaces
    APIs 287
    application frameworks 457
    MFC 457
    Motif 290
    network management 541
    PCs 290

H

H&R Block Inc. 233
HAL 297
    for hardware abstraction layer
hand-held computers
    storage devices 88
Harbinger Corp. 587, 593
    Premenos Corp. 563
hard disk drives
    caching 538
    capacity 77
    coating compositions 77
    cost 77
    disk array 81
    disk drive interface 78
    EIDE 78
    FC-AL 80
    forecast 103
    heads 76
    market 99
    MO 77, 89
    new technologies 77
    NFR 77
    on-the-fly self-calibration 77
    PMRL 77
    RAID 81
    removable 87
    SCSI 78
    shipments by capacity 100
    sizes 76
    speed 76
    SSA 79
    technologies 75
    types of 75
    Ultra SCSI 79
    Ultra-2 SCSI 79
    Winchester drives 75
Harris & Jeffries 172
Haystack Labs Inc.
    WebStalker 406
HDDT 72
    for high-density data tapes
    capacity 72

HDSL 212
for High-speed Digital Subscriber Line
Hellman-Merkle 388
Hercules Computer Technology Inc. 60
Hewlett-Packard Co.
see HP
HFC
for hybrid fiber-coax
definition 152
technology 152
Hi/fn Inc. 390
HICOM Bhd 35
HiFD 87
for high-capacity floppy disk
high packing density 73
HighGround Systems Inc.
Storage Resource Manager and Media Series 538
Hilton Hotels International 127
HiPPI 172
for High Performance Parallel Interface
Hitachi Data Systems Corp. 83, 92, 101, 349
5700 82
7700 Scalable Array 85
mainframe technology 351
Multiplatform Data Exchange 85
Multiplatform Resource Sharing 85
Skyline Series 351, 361
tape drives 70
technologies 85
Voyager Family 351
Hitachi Ltd. 92, 125, 129, 176, 181, 305, 364
VisionBook Pro 7560, 56
Hitachi Maxell 94, 96
HOLAP 618, 619
for Hybrid OLAP
Holosofx Inc. 501
Home Shopping Network
Internet Shopping Network 582
Hong Kong Telecom IMS 119
Digital Smart Box 119
House International Relations Subcommittee on International Economic Policy and Trade 401
HP 50, 74, 92, 97, 126, 196, 218, 286, 298, 338, 352, 356, 360, 363, 368, 534, 545–546, 569, 593, 654, 670, 676
for Hewlett-Packard Co.
Color LaserJet 5M 61
DDS 73
DeskJet 720C, 722C, and 890C 62
FC-AL 80
Fibre Channel Interconnect 172
flag card 390
Hewlett-Packard Laboratories 389
HP 3000 440
HP 9000 302, 353
HP Open Warehouse Management Suite 615
HP OpenView 615
HP SoftBench 446
HP-PA RISC 301
HP-UX 83, 300, 302, 495, 537, 545
ICF architecture 390
ImagineCard 376
International Cryptography Framework 400
JetAdmin 63
LaserJet 4V/MV 63
LaserJet 5Si Mopier 65
Merced 301
Net Vectra PC 376
NetPC 527
NetServer 658
Norton Administrator for Networks 526
Norton Desktop Administrator 526
OfficeJet 65
Omnibook 800 98
OmniStorage and OmniBack II 538
OpenView 63, 526, 544, 545, 546
OpenView-Ready 527
OpenView-Ready Network Node Manager 527
OpenWarehouse 636
ORB Plus 327
PA-8200 345
PA-RISC 302, 344
Photo REt II 62
Praesidium Authorization Server 379
Secure Enterprise Web 383
SureStore CD-Writer Plus 90
Tachyon Fibre Channel 172
VCSEL 145
Vectra PC 527
VeriFone Inc. 126, 570, 574
vGate 570
vPOS 570
Web JetAdmin 65
HPCs 298, 309
for hand-held PCs
HPF 431
for High-Performance Fortran
HSM 88
for hierarchical storage management
HSTRA 168
for High-Speed Token Ring Alliance
HTML 234, 439, 460
for HyperText Markup Language
authoring tools 256
CSS 242
CSS versus GIF 242
definition 242
DHTML 243
DOM 244
electronic commerce 566
forecast 279
intranets 265
layers 244
meta-information 242
technologies 242
type=hidden attribute 266
versus XML 243
HTTP 234, 408, 413, 416, 534, 652
for HyperText Transport Protocol
definition 241
technologies 241, 266
hubs 174, 175
bridges 175
Ethernet 167
intelligent agents 541
load balancing 183
MAEs 235
NAPs 235
PoPs 235
Hughes Communications Inc.
Spaceway 157
Hughes Network Systems Inc. 172
Hutchison Max Telecom 162
HydraWeb Technologies Inc.
HydraWeb 536
HydraWeb Load Manager 535
Hyperion Software Corp. 676

I

I/O technologies
for input/output
CCD 45
digital camaras 45
displays 45, 53
forecast 66
input devices 45, 46
keyboards 46
LCD 55
overview 45
pointing devices 48
printers 45, 60
scanners 45, 50

speech recognition 45, 52
touchscreens 45, 49
video cameras 45, 51
i2 Technologies Inc. 681
Ibex Technologies Inc.
Fax-From-Web 264
IBM 15, 76, 78, 83, 101, 126, 168, 181, 183, 196, 218, 242, 265, 338, 341, 352, 359, 361, 363, 364, 368, 440, 448, 461, 484, 485, 503, 512, 526, 534, 537, 543, 546–547, 566, 570, 574, 575, 593, 624, 637, 638, 650, 654, 676
for International Business Machines Corp.
128-track Magstar MP 71
3270 246
3466 Network Storage Manager 538
3480 drive 70
3490 tape cartridge 70
3590 Magstar Tape Subsystem 71
3830 64
3900 Advanced Function Wide Duplex Printing System 65
4381 306
701 Defense Calculator 70
726 Tape Unit 70
9370 306
9729 WDM system 143
acquisitions 547
Adaptive Lossless Data Compression 73
Adstar Distributed Storage Manager 84, 538
AIX 300, 302, 311, 351, 495, 537
Application Management for Notes and Domino 539
AS/400 84, 305, 351, 432, 463, 570, 629, 648, 655
AS/400 Advanced Entry 351
AS/400e series 351
AS/400e servers 570
Asset Services 527
Automotive Network eXchange 589
cc:Mail 412
CICS 305, 322, 336, 338, 431
clustering technology 350
CMOS 7S 341
CMOS mainframe technology 350
CommercePoint 567, 575
Data Guide 629
DB2 337, 338, 480, 486, 606
DB2 cursor stability 476
DB2 OLAP server 619
DB2 Parallel Edition for SP2 486
DB2 Repeatable Reads 476
DB2 Universal Database 469, 620
DB2 Universal Database 5.0 478, 486
DB2 Version 2 477, 486
DB2/6000 678
DCE Security Server 379
Deskstar 14GXP 77
Deskstar 16GP 77
Directory and Security Services 304
DMI 526
Domino Go Webserver 306
e-Business 570
electronic commerce 570
Encina 338
encryption 390
enterprise operating systems 305–307
enterprise storage systems 84
EnterpriseXspan 589
ES/9000 302
FCL 80
Firewall for AS/400 305
flat-panel display 56
floppy disks 86
FlowMark 503
Fortran 431
G3 350
G4 350
G5 350
Global Network 587, 589, 592
Global Sign-On 379, 380
GMR 77
historical events 698
IBM Registry 394
IC Phone 262
Immune System for Cyberspace 403
IMS 470, 471, 486
Intelligent Miner 623, 626
Interactive Network Dispatcher 536
Internet Connection Secure Server for AS/400 305
Internet Emergency Response Service 385
Internet2 239
ISPF/PDF 537
Java 299, 312
Java for OS/390 306
Kerberos 382
Lotus 276
Lotus Domino 265, 267, 305, 311, 495, 496, 514, 527, 571, 589
Lotus InterNotes Web Navigator 265
Lotus Notes 265, 385, 412, 480, 495, 503, 513, 514, 516, 699
Lotus Notes Release 4 495
Lotus Notes Release 5.0 496
Lotus Notes/Domino 4.6 496
LotusScript 495
LU 6.2 337
Maestro 547
Magstar MP 3575 Tape Library DataServer 84
MDIS 628
Metadata Coalition 627
MPOA 170
MQSeries 322, 335
Multiprotocol Switched Services Layer 2/Layer 3 170
MVS 305, 314, 457, 628
Net.Commerce 305, 306, 570, 571
NetFinity 526
NetFinity Manager 546
NetView 544, 546
NetView for System/390 547
Network Dispatcher 535
Network Storage Manager 84
Nways Workgroup Manager 547
OLAP 630
OpenDoc 317
operating systems 546
Optical Library Dataserver 538
OS/2 189, 312, 313, 456, 495, 537
OS/2 Presentation Manager 291
OS/2 product family 299
OS/2 Warp 45, 299, 308, 309
OS/2 Warp Server 285, 286, 304, 311
OS/390 286, 305, 310, 312, 314, 457
OS/390 Open Edition 349
OS/400 286, 305, 310, 312, 314, 351, 352, 457, 628
P2SC 345
Parallel Enterprise Servers 350
Parallel Query Servers 350
Parallel Sysplex technology 350
Parallel Transaction Servers 350
Passport 629
PC 293
PC-DOS 293
PCI cryptographic co-processor/adapter card 420
Personal Computer 46
PL/1 432
POWER 344

PowerPC 301
PowerPC AS 352
PowerPC-based AS/400 Advanced Series 351
RACF 379
RAMAC 3 Array 84
RPG 432
RS/6000 64, 84, 302
S/390 165, 189, 349, 350, 655
S/390 Parallel Enterprise Server – Generation 4 350
S/MIME 412
SA-Expertise for Customer Relationship Management 547
SA-Expertise for Enterprise Support Management 547
Seascape 84, 101, 538
Secure Enterprise Web 383
SecureWay 375
SecureWay Key Management Framework 400
security 376, 400
Selectric 46
SGML 242
smart cards 123
SNA 188, 190, 546, 655
Software Artistry 547
SP2 469, 483, 486
SSA 79, 84
storage barrier 78
System Object Models 291
SystemView 546
technology history 697
Technology-Independent Machine Interface 352
ThinkPad 48
ThinkPad 76 56
ThinkPad 770 56
Tivoli Courier 537
Tivoli TME 63
Tivoli TME 10 547
Tivoli TME 10 NetView 546, 547
Tivoli TME 10 Security Management and RACF, 379
TLI 337
touchscreens 50
TrackPoint 47
trends 366
Unison Software Inc. 547
Universal Database 469
UNIX 285
Versatile Storage Server 84
ViaVoice 53
ViaVoice Gold 53
VisualAge 430
VisualAge 2000 306
VisualAge for Smalltalk 435
VisualAge Generator 441
VisualWarehouse 612, 636
VM 286, 305, 307, 314
VM/ESA 307
VoiceType 299
VSE 286, 306, 314
VSE/ESA 306
Winchester drives 75
WorkSpace On-Demand 304
Year 2000 450, 453
ICAST Communications 193
iCat Corp. 566
Electronic Commerce Suite 567, 568
ICC 124
for integrated circuit card
ICF 400
for International Cryptography Framework
ICL Inc. 362
Access Manager 382
ICO Global Communications 157
ICO Korea Co. Ltd. 157
ICP 536
for Internet Caching Protocol
IDE
for Integrated Drive Electronics
IDEA 386, 387
for International Data Encryption Algorithm
Identification Technologies International Inc.
One-on-One 377
Ideo 240
IDL 326
for Interface Definition Language
IDRP 178
for Inter-Domain Routing Protocol
IDSL 212
for ISDN DSL
IDS-Professor Scheer 501
IEEE 145, 389, 530
for Institute of Electrical and Electronics Engineers
802.3z Task Force 145
IETF 181, 236, 263, 530, 571
for Internet Engineering Task Force
Audio/Video Transport 190
HTTP 241
LDAP 391
OpenPGP 413
RC2 412
IGMP 192, 194
for Internet Group Management Protocol
IGP 178
for Interior Gateway Protocol
IIOP 267, 327, 666
for Internet Inter-ORB Protocol
ImagineCard 390
Imation Corp. 74
SuperDrive 87
IMI Systems Inc. 648
IMS 471
for information management system
Inacom Corp. 658
InConcert Inc. 503
Individual Inc. 276
Indosat 209
Inference Corp. 252
Infinium Software Inc. 644
Informatica Corp. 628
Microsoft Alliance for Data Warehousing 630
PowerMart 606
PowerMart Suite 612
Information Advantage Inc. 618
DecisionSuite Server 620
MX 628
Information Builders Inc.
Cactus 442
Focus 440
SmartMart 612
WTI 629
Information Discovery Inc.
WTI 629
Information Resource Engineering Inc. 375
information retrieval 511
challenges 509
continuum 510
database replication 494
document management 505
groupware 493
intranets 494
leveraging the Internet 492
overview 491
pull technology 253
search engines 250
SQL 476
tape drive for mats 71
technologies 247
trends 515
vendors 510
see also databases, data warehousing, document management, knowledge management, search engines, SQL
Information Society
building 10
definition 9

European 9
Information Technology Industry Council 373
InformationWeek 374
Informix Software Inc. 265, 311, 400, 440, 469, 480, 485, 637, 638, 654, 676, 678
flag card 390
Illustra 475, 486
Informix-Metacube Analysis Engine 620
MetaCube 635
MetaCube OLAP 486
OnLine XPS 483, 486
ROLAP 618
SPL 478
Stanford Technology Group Inc. 618
Universal Data Server 485
Universal Server 486
Universal Server Version 9.1 478
Infoseek Corp. 250, 276
Infospace Inc.
MX 628
SpaceOLAP 630
InfoVista Corp. 529
InfoVista 539
inheritance 434
Inktomi Corp. 250
Traffic Server 239
Inmarsat 157
input devices 46
digital video cameras 66
forecast 66
interactive voice response 52
keyboards 46
pointing devices 47, 48
scanners 50
speech recognition 52
touchscreens 49
verbal command 52
video cameras 51
Inso Corp.
Word Viewer 246
instance 434
Institute for the Management of Information Systems 401
instruction set 344
Integral Solutions Ltd.
Clementine 623, 626
Integralis Inc. 418
Integrated Network Corp. 196
Integrion 593
Integrion Financial Network
Gold 578
Intel Corp. 51, 167, 263, 342, 352, 359, 362, 400, 534
8051 111
80x86 processor 293
Chips and Technologies 366
CISC 344
Covington 346
Deschutes 346
Digital Equipment Corp. 366
DMI 526
forecast 369
historical events 698
IA-64 347, 348
Internet Video Phone 260, 262
Katmai 346
LANDesk 537
Mendocino 346
Merced 301, 348, 370
MMX 347
NetPC 526
Pandesic 569, 658
PC 98 Hardware Design Guidelines 357
Pentium 45
Pentium family 345, 346
Pentium II 345
Pentium Pro 356, 363, 368
Phone 568
RSVP 193
SanDisk 341
StrataFlash 341
Tillamook 346
Ultra DMA-33 78
Willamette 346
Wired for Management 527
workstation sales 363
x86 and RISC processors 291
x86 architecture 345
Intelisys Electronic Commerce LLC 565
Intelitrak Technologies Inc.
Citadel GateKeeper 420
intellectual property 38
Intellidex Systems LLC 629
MDIS 628
Warehouse Control Center 612
Interface Mechanics 118
Interface Repository 326
Intergraph Corp. 356
Interleaf Inc.
Intellecte 507
Xtreme 507
Internal Revenue Service 378
International Ad Hoc Committee 237
International Multimedia Teleconferencing Consortium
Voice Over Internet Protocol 261
International Solid-State Circuits Conference 366
International Wireless Inc.
Prodigy Internet 233
Internet 24
access 233
advertising 21
agent technology 252
aggregation points 274
antagonistic attacks 409
applications 231
ATM 222
auctions 583
audio and video
real-time 260
streaming 257
technologies 260
transmission 257
backbone 235, 580
providers 235
background 232
bandwidth 238
BAPIs 656
BGP4 179
business-to-business 553
business-to-business transactions 274
cable companies 204
cable modems 188
CATV 223, 224
challenges 235, 238
client devices 355
congestion factors 238
consumer trust 236
consumer-to-business 554
cookies 408
corporate applications 643, 645, 646
data integration 579
data warehousing 603, 621, 630, 638
definition 234
delays 438
denial-of-service attacks 406
development tools 429
distributed computing 317, 323
DMA 509
document management 506, 507
domain names 236, 238
EDI 555, 563, 564, 592
electronic commerce 235, 274, 553, 556, 561, 562, 578, 581, 582
electronic payment 577
electronic payment standards 577
e-mail 234, 411
enterprise server operating systems 314
e-purse 128
European currency 645
Euros 645
expansion 238

extranets 568
fax transmission 264
vendors 264
filters 252
firewalls 176
forecast 277, 280
global commerce 643
global effects 20
global issues 18
global markets 240, 581
groupware 512
gTLDs 238
H.323 260
high-speed routers 177
hostile applets 404
HTTP 242
hubs 235
IETF 236
IIOP 666
information access tools 630
information management 492
information retrieval 247
Information Society 9
interactive technologies 254
international growth 240
Internet commerce 554, 593
industry groups 581
market 581, 590, 591
Internet2 239
Internet-enabled workflow 492
intranets 165, 264
IP multicasting 192
IPOC 238
IPsec 184
IPv6 189
ISAKMP 185
ISDN 211
ISPs 232
Java 436
L2F 186
language content 240
LDAP 391
licensing agreements 263
MacOS 299
mainframes 349
management issues 535
management services 527
market 224, 272, 273, 274, 276
forecast 278
international 274
segments 273
MBone 192, 193, 260
metasearch engines 253
MISS 374
mobility support in IPv6 191
multicasting 165
multi-culturalism 240
multimedia 257
market 277
NCs 240
network caching 239
network congestion 579
network management 528, 540
network performance 165
networking systems 166
OBI 565
ODBMS 473
OECD 581
OFX 578
OLSs 232
OLSs versus ISPs 233
operating systems 285, 290
OSPF 179
overview 231
packet-switched networks 512
paging 226
payment systems 572
PCs 357
PEM 411
perfor mance 236
Perl 438
phone packages 262
phone-to-phone 261, 263
printers 65
protocols 192, 235, 241
proxy servers 536
publishing 234, 273
push technology 247
QoS 235
regulation 236
remote log-in 375
revenue 273
routers 176, 181
RTCP 191
RTP 191
S/MIME 412
search engines 250, 276
security 373, 374, 387, 402, 411, 578, 580
service providers 274
services 210, 229
SET 575
SKIP 185
smart cards 119, 124, 127, 128, 391, 573
SNMP 530
societal effects 20
software components 325
spam 410
ST2 191
standards 241, 278, 571, 580
streaming 257
subscription services 585
TCP/IP 189, 241
technologies 231, 235, 236, 274
technology deployment 274
technology providers 273
telecommunications 203
teleconferencing 263
telephony 231, 258, 261, 262
directory access 263
forecast 280
service providers 263
vendors 262
television-based devices 240
thin-client computing 342
third parties 560
traffic 235, 238
transactions 124
trends 25, 374
TRUSTe 580
tunneling 186
users 272, 273, 278, 553, 591
vendors 239, 258, 260
video telephony 51
viruses 402
VPIs 184
VPN 373
Web-accessible tools 631
Web-based computing 459
Web-enabled tools 631
Web-exploited tools 632
white boarding 260
workflow 516
see also browsers, electronic commerce, extranets, intranets, network management, telephony, telecommunications, World Wide Web
Internet Architecture Board 179
Internet Assigned Numbers Authority 237, 531
Internet Liquidators 582
Internet Mail Consortium 407, 412
Internet Purchasing Roundtable 565
Internet Security Systems Inc.
Internet Scanner 379
RealSecure 406
InterNet Solutions Inc. 508
Internet television
see network computers
Internet Travel Network
Web site 584
internetworking
see bridges, hubs, routers, switches
Intersolv Inc. 447, 450
PVCS 445
Year 2000 Wizard 452
Interweave Software Inc. 635
intranets 568
advantages 231
authoring tools 265

browsers 265, 275
challenges 265
corporate applications 646
data warehousing 603, 621, 630
databases 265
definition 264, 494
development tools 429
electronic commerce 557, 580
forecast 280
groupware 265, 494
mainframes 349
market 277
MOM 321
network management 528
network performance 165
overview 231
printing 65
push technology 248
search engines 276
security 373, 374, 385
smart cards 127
technologies 264
uses of 165
versus extranets 265
VPIs 184
workflow 502
Intuit Inc. 578
Investor's SuperSite On The Web 584
Iomega Corp. 75
Clik! 88
Jaz 78, 88, 100
Zip 86, 87, 89
Iona Technologies Ltd.
Orbix 338
IP 176
for Internet Protocol
address spoofing 417
classes of service 189
LDAP 392
MBone 192, 193
multicast
vendors 193
multicast tunnel 192
multicasting protocols 192
switching 181
advantages 181
IPI 172
for Intelligent Peripheral Interface
I-Planet Inc. 421
IPS-SMG 421
IPOC 238
for Interim Policy Oversight Committee
IPsec 418
interoperability 184
vendors 184
versus L2TP 186
versus PPTP 185
Ipsilon Networks Inc. 181
IPX
for Internetwork Packet eXchange
IPX/SPX 176, 459
for Internetwork Packet eXchange/Sequence Packet eXchange
IQ Software Corp.
IQ 623
MX 628
OLAP 630
WTI 629
IRI Software
see Oracle Corp.
IriScan Inc.
IrisCode 378
ISA Corp. 189
ISAKMP 185
for Internet Secure Association Key Management Protocol
ISAM 323
for Indexed Sequential Access Method
ISAPI
for Internet Server API
ISDN 51, 186, 210
for Integrated Services Digital Network
B-channel 211
BRI ISDN 211
Broadband ISDN 221
channels 211
D-channel 211
definition 211
Euro-ISDN 221
international 211
market 227
pricing 221
revenue 228
speed 211
technologies 211, 221
IS-IS 179
for Intermediate System-to-Intermediate System
ISO 242
for International Organization for Standardization
OSI 173
Isocor 391
Isogon Corp.
SoftAudit/2000 452
TICTOC 452
ISPs 210, 587
for Internet service providers
background 232
definition 235
electronic commerce 589
European 233
fax transmission 264
forecast 229
growth 589
hubs 235
market 587
MISS 374
pricing 587
private peering agreements 235
QoS 235
smart cards 127
versus OLSs 233
IT
for information technology
administration 379
auditors 379
budget 693
challenges 645
collaborative computing 23
convergence 7
corporate applications 644
development tools 429
electronic commerce 561
European trends 7, 8
forecast 17
global effects 18
Internet 319
Internet effects 166
network performance 165
network response time and availability 165
security 374, 375
server clustering 172
Year 2000 450, 451, 644
see also data warehousing, document management, groupware, information retrieval, knowledge management, network management, systems management, workflow
ITJ 209
for International Telecom Japan
ITSEC 383, 384, 423
for Information Technology Security Evaluation and Certification
ITU-TSS 169, 187, 564, 571
for International Telecommunications Union-Telecommunications Standards Section
H.323 260
IXCs 204, 206, 210, 221, 229
for interexchange carriers

CAPs 223
forecast 228
frame relays 226

J

J.D. Edwards and Co. 305, 643, 644, 646, 648, 674–676
architecture 675
CNC 675
Composer 675
Configurable Network Architecture 646
Configurable Network Computing 675
disadvantages 682
EMU 674
ERP 674
Finance 675
Genesis 675
Human Resources 675
Internet 676
Manufacturing 675
market 680, 681, 682
OneWorld 675
platforms 674
Sales and Distribution 675
WorldSoftware 674
WorldVision 674, 676
Year 2000 644
Japan Telecom 209
Java 279, 312, 335, 435, 436, 593, 647, 676
Abstract Windowing Toolkit 458
advantages 437, 448
APIs 330
applet execution 496
applet monitoring 417
applets 236, 460, 632
BAPIs 656
bytecode verifier 404
class loader 404
Enterprise Java Platform 330
Enterprise JavaBeans 330
EJB Container 329
forecast 466
interactive Web pages 255
Java Development Kit (JDK) 330
Java Foundation Class 458
JavaBeans 317, 325, 327, 339, 418, 461
beanInfo classes 328
beans 329
component architecture 328
Enterprise Beans 329
introspection 328
technologies 328, 330
JavaCard 111, 112, 113, 132
JavaScript 244, 250, 438, 439
JDBC 331, 460
JFC 335
JIDL 331
JNDI 330, 331
JTS 330
JVM 460
market 464
ODBMS 473
revenue 464
RMI 331
sandbox 404
security 373, 404, 460
security components 404
security manager 404
smart cards 113
software development 429
technology overview 436
tools 461
UNIX 310
versus ActiveX 423
wallets 574
Web animation 254
Web-based computing 460
wrappers 632
see also Sun Microsystems Inc.
JavaBeans
see Java
JavaCard Forum 112
JavaSoft
see Sun Microsystems Inc.
JDBC 330, 460
for Java Database Connectivity
JEPI 578
for Joint Electronic Payments Initiative
Jetform Corp. 502, 503
JFax Communications
Fax/Voice 264
JFC 335
for Java Foundation Classes
JIDL 330, 331
for Java Interface Definition Language
JIT 558
for just-in-time
JNDI 330, 331
for Java Naming and Directory Interface
JPEG 256
for Joint Photographic Experts Group
JRMI 331
for Java RMI
JRMP 331
for Java Remote Method Protocol
JTS 330
for Java Transaction Service
JTS Corp. 76
jukeboxes 69
Juniper Networks Inc. 177
Junkbusters Corp.
Internet Junkbuster 409
Web site 409
JVM 125, 285, 299, 301, 328, 436, 437
for Java Virtual Machine

K

KANBAN 660, 675
Kaypro Corp. 46
Kenan Systems Corp. 630
Kensington Microware Ltd. 47, 48
Kerberos 380, 381, 391, 415
credential-for warding 381
features 381
interoperability 382
inter-realm authentication 381
Kerberos 5 381
products 382
purpose 381
SESAME 382
technology 381
versus DCE 382
Key Recovery Alliance
purposes 400
keyboards 46
application-specific keys 47
chord 47
chord keyboard gloves 47
costs 47
designs 47
ergonomic 47, 66
forecast 66
pointing device integration 47
QWERTY 47
split 46
KeyFile Corp. 502
KeyFlow 502
Keytronics Corp. 376
Keyware Technologies Inc.
Face Guardian 377
Voice Guardian 378
KL Group Inc. 449
knowledge management 27
active agents 517
capabilities 511
collaborative filtering 512
continuum 510
converging technologies 513
definition 491, 515
forecast 517
frameworks 510, 511
gatekeepers 512
market 515
mining 511
overview 491
repositories 511

technologies 509, 510, 511
text retrieval engines 511
tracking unstructured information 511
vendors 510, 512
Kobixx Systems LLC
EZine Publisher 256
Kokusai Denshin Denwa 209
Korea Mobile Telecom 216
KPMG Peat Marwick LLP
WTI 629
KRM 398, 399
for key recovery management
Kurzweil Inc. 53

L

L2F 184, 186
for Layer 2 Forwarding
L2TP 184, 186
for Layer 2 Tunneling Protocol
Landis & Gyr Communications Corp. 123
LANE 170
for LAN Emulation
languages
see 3GL, 4GL, C++, C programming language, Cobol, development tools, Fortran, Java, object technology, programming
LanOptics Ltd.
Guardian 406
LANs 165, 189, 522, 528
for local area networks
100VG-AnyLAN 149
ATM 169
backbones 195
bridges 175
cabling 162
composite 175
database replication 479
devices 174
electronic commerce 562
Fast Etherent 167
frame relay 171
groupware 494
hubs 174
intranets 568
IPX/SPX 189
LAN-to-LAN connections 222
network bottlenecks 194
network management 528
RMON 532
segmentation 194
shared to switched 194
smart cards 127
standards 151
storage devices 74
switches 175, 180
virtual 182
VLAN obstacles 183
VLANs 182
VPIs 184
wireless 217
wiring 137
LaserCard Systems Corp.
see Drexler Technology Corp.
lasers 140, 141
optical laser cards 118
VCSEL 145
WDM 143
Law 975/96 407
Lawson Software Inc. 634
LCD 45, 53, 55
for liquid crystal display
active-matrix 55
active-matrix versus FED 57
costs 53
dual-scan 55
fabrication 56
flat-panel displays 56
forecast 66
ghosts 56
light-valve projector 58
sizes 56
types of 55
LCI International 172
LDAP 263, 333, 390, 391
for Lightweight Directory Access Protocol
issues 392
LDDS WorldCom 160, 205, 239, 512, 590
ANS Communications Inc. 235
UUNet 233, 588
LECs 144, 160, 204, 206, 210, 217, 223, 229
for local exchange carriers
CLEC 160, 206
frame relays 226
versus CAPs 223
Legato Systems Inc. 538
NetWorker 538
Lemcon Systems Inc. 418
LEOs 156, 157
for low-earth-orbit (satellites)
European trends 12
forecast 229
MSS 158
technologies 158
UTMS 12
Lexis-Nexis
see Reed Elsevier PLC
Lexmark International Inc. 47, 63
Optra S series 63
LG Electronics Inc. 94
Libra Corp. 663
Lightscape Technologies Inc. 356
Li-ion 96
for lithium-ion
capacity 96
forecast 103
market 96, 102
LIMDOW 93
for Light Intensity Modulation Direct Overwrite
Lintel Security S.A. 418
Linux 310
LISP 431, 432
for LISt Processing
Common LISP 432
Lithium Technology Corp. 97
Litronic Inc. 126
Livermore Software Laboratories
N.O.A.H. 406
Protus 406
LMDS 213
for local multi-point distribution service
Logic Works Inc.
Erwin 444
Logitech Inc. 48
PageScan Color 50
Lotus Development Corp.
see IBM
LRDA
Testbed 452
LSI Logic Inc. 167
Lucent Technologies Inc. 177
EMILM 144
NetCare 530
Optical Line Systems 143
Personalized Web Assistant 408
UNIX 301
lumen
definition 58
Lycos Inc. 250, 276

M

MAC 173, 413
for Media Access Control
MacOS
for Macintosh operating system
see Apple
Macromedia Inc.
Director 254
Shockwave 254
Madge Networks Inc. 168
MAEs 235
for Metropolitan Area Exchanges
magnetic stripe cards 117, 120, 129
and ATM machines 120
features 117

watermark tape 117
magnetic tapes
advantages 70
DAT 70
forecast 102
mainframe formats 70
market 70
overview 70
silos 71
Travan technology 75
uses for 70
versus CD-ROMs 89
see also tape drives
Magnify Inc.
Pattern 626
mainframes 648
4GLs 440
and client/server applications 645
clustering 289
CMOS 342, 349, 361
COBOL 431
corporate applications 645
critical role 350
disk storage 69
electronic commerce 579
Fibre Channel 145
forecast 367
groupware 494
integrated-logic CMOS 349
market 361
MIPS 361
MVS 305
OpenVMS 307
operating system forecast 314
operating system market 312
operating system vendors 305
operating systems 286, 287, 291
OS/390 305
pricing 349, 361
printers 45, 64
reinventing 350
removable disk drives 87
revenue 361
rightsizing 350
single-tier computing 318
software components 318
storage devices 76, 101
storage management 543
systems management 522
tape drive for mats 70, 71
technology 342
TP monitors 322
trends 342, 348, 350
types of 349
vendors 362
versus client/server 348, 350
VSE 306
wide area networking 350
Year 2000 429, 451, 645
see also systems management
MainWare Inc.
HourGlass 2000 452
Malone, Thomas W. 23–32
managed network services 530
Management Science Assoc. 630
Manheim Online 582
Mannesmann Arcor 208
Manugistics Group Inc. 648
MAOS 107
for multi-application operating system
Marcam Corp. 644
Marimba Inc. 248
Castanet 248, 250, 276
Marubeni Corp. 162
MasterCard International Inc. 123, 390, 573, 574, 576, 593
Mondex 122
MultOS 113, 125
smart cards 121, 123
Mastiff Electronic Systems
Scentinel 377
Matsushita Electronic Industrial Co. Ltd. 92
Divx 92
SuperDrive 87
Maxtor Corp. 78
MBone 192, 193, 259, 261
for Multicast Backbone
McCabe and Assoc. Inc.
Visual 2000 Environment 452
Visual Quality Toolset 447
MCI Communications Corp. 160, 172, 205, 375
vBNS 239
see also British Telecomunications PLC
MDBMS
for multidimensional database management system
ad hoc queries 605
advantages 605
bitstream index 605
bulk data loader 605
data spaces 605
definition 604
star join 605
versus RDBMS 605
MDIS 628
for Metadata Interchange Specification
Memorex Telex Corp.
Scimitar 538
memory
cache 343
cache consistency 289
chips 342
size 343
trends 366
flash 342
Level 1 cache 343
Level 2 cache 343
main 343
management 289
performance 343
protection 289
smart cards 107
virtual memory 289
memory cards 109, 119
contactless 116
protected 110
simple 109
uses of 109
MEOs 156
for medium-earth-orbit (satellites)
advantages 157
shadowing 157
technologies 157, 158
Mercury Communications PLC 205
Mercury Interactive Corp. 447
Mergent International Inc.
Domain/DACS 379
PC/DACS 379
Meta Software Corp. 501
metadata 326, 611, 634, 638
availability 613
BIW 656
business data 613
data warehousing 603
interchange specification 627
management 635
management tools 614
MDIS 627
MOLAP 618
MX 628
objects 628
standards 614, 627, 629
technical data 613, 614
versioning 614
WTI 629
Metadata Coalition 627, 628
Metasys Inc. 649
method 434
MFC 457
for Microsoft Foundation Class
MFS Communications Co. Inc.
see LDDS WorldCom
MIB 531
for Management Information Base
MIB-II 531
objects 532

MIC 73
  for Memory In Cassette
mice
  scrolling wheels 46
Micro Focus Inc. 437
  Challenge 2000 452
MicroAge Inc. 581
Microcom Corp. 196
microcomponents
  see microprocessors
Micropolis PTE Ltd. 79
microprocessor cards 110
microprocessors
  architecture 344, 345, 686
  C programming 433
  CMOS 349
  forecast 369
  trends 342
  vendors 345
Microsoft Corp. 46, 48, 123, 185, 193, 242, 249, 252, 276, 278, 317, 335, 342, 352, 353, 359, 400, 439, 457, 461, 485, 487, 534, 538, 565, 566, 570, 574, 578, 579, 593, 624, 638, 650, 654, 676, 698, 699
  16-bit applications 288
  Access 265, 443, 480, 487, 623
  Active Desktop 243, 246
  Active Directory 292, 303
  Active Framework for Data Warehousing 630
  Active Pages 335
  Active Platform 246, 249
  Active Server 335
  Active Streaming Format 258
  ActiveMovie 258
  ActiveX 249, 268, 269, 279, 291, 339, 373, 404, 405, 442, 457, 461, 569
  AFC 335
  Authenticode 405
  BackOffice 670, 676, 678
  BackOffice Small Business Server 303
  BAPIs 569
  browser market share 275
  C++ 455
  CARP 536
  Channel Definition Format 244
  COM 291, 331, 457
  COM+ 291
  COM/DCOM 331, 334
  Commerce Server 570
  Commerce Server 2.0 567
  Commerce Site Server 2.0 567
  CryptoAPI 388
  DCOM 328, 332, 339, 457, 655
  DocObjects 331
  Domain Enterprise Servers 570
  eShop Inc. 268
  Excel 506, 623, 634, 657
  Exchange 387, 412, 496, 497, 502, 514
  Exchange and Outlook 657
  Expedia 582, 584
  FileMaker Pro 265
  FoxPro 438, 443
  FrontPage 257, 268
  Hydra 299
  Intellimirror 303
  Internet Explorer 236, 245, 295, 388, 405, 411, 628
  Internet Explorer 4.0 243, 244, 249, 252, 262, 274, 275, 276, 332, 414, 567, 593
  Internet Information Server 268, 567, 569
  Internet Services Platform 268
  Kerberos 382
  Mail API 496
  Merced 312
  Merchant 268
  MFC 457
  Microsoft Alliance for Data Warehousing 630
  MIDL 332
  MS Query 623
  MS-DOS 293, 456, 537
  MS-Mail 412
  MSMQ 333
  MSN 233
  MSNBC 253
  MTS 333
  natural keyboard 47
  NetMeeting 260, 262, 568
  NetPC 526
  NetShow 258
  OAG 680
  ODBC 632
  Office 331, 487
  Office 97 244
  Office Suite 621
  OLAP standards 629
  OLE DB 443, 629
  Open Information Model 628
  Open Software Description 250
  Panorama Software Systems Inc. 635
  PC 98 Hardware Design Guidelines 357
  PCT 398, 413
  Personal Delivery Service 253
  Proxy Server 406
  Repository 628
  S/MIME 412
  Secure Enterprise Web 383
  Site Server 566
  Slate 585
  Smart Card Software Development Kit 125
  SQL Server 320, 334, 337, 620, 670, 678
  SQL Server 6.0 487
  SQL Server 6.5 487
  standards 679
  Studio 268
  System Management Server 527
  Transaction Service 332
  VBScript 439
  Visual Basic 268, 332, 410, 432, 437, 442, 464, 465, 656
  Visual Basic for Applications 439
  Visual C++ 446
  Visual DevStudio 430
  Visual FoxPro 487
  VXtreme Inc. 258
  Wallet 567
  WebTV 119, 244
  Win32 API 456
  Win32 Driver Model 296
  Windows 291, 495, 537
  Windows 3.1 288
  Windows 3.1 and 3.11 308
  Windows 3.1 and Windows for Workgroups 3.11 293
  Windows 95 45, 189, 293, 294, 295, 297, 308, 387, 678
  Windows 98 244, 276, 295, 296, 297, 308, 312
  Windows APIs 456
  Windows CE 298, 309, 312, 313, 360
  Windows desktop family 293–299
    market 308
  Windows Distributed interNet Architecture for Financial Services 578
  Windows NT 45, 189, 291, 310, 312, 334, 352, 353, 356, 387, 484, 496, 628, 630, 648, 655, 678
  Windows NT 4.0 Server 658
  Windows NT 5.0 382
  Windows NT New Technology File System 292
  Windows NT Server 285, 286, 302–304, 311, 314
  Windows NT Server 4.0 Enterprise Edition 303
  Windows NT Server 5.0 303
  Windows NT Server Enterprise Edition 290, 311

Windows NT Workstation 294, 296, 309, 312, 313
Windows NT Workstation 4.0 308
Windows NT Workstation 5.0 296, 297, 308
Windows plans 298
Windows system features 298
Windows WAV 244
WinSock 459
Word 127, 506
Zero Administration for Windows 303, 527, 537
Zero Administration Kit 527
MicroStrategy Inc.
DSS Server 620
MX 628
ROLAP 618
Microsystems Software Inc.
CyberPatrol 252
Microtek Lab Inc.
ScanMaker III 50
ScanMaker V300 50
MicroTouch Systems Inc.
ThruGlass 50
microwave transmission
market 162
technologies 159
middleware 331, 336, 632
background 319
definition 320
disadvantages 323
generic 320
messaging 319
MOM 321, 632
overview 317
publish-and-subscribe 336
pure 320
RPCs 321
standards 322
technologies 320
TP monitors 320
MIDL 332
for Microsoft Interface Definition Language
Milkyway Networks 417
Millennium Dynamics
FirstStep SSO 380
VANTAGE YR2000 452
MIME 412, 563
for Multipurpose Internet Mail Extensions
Ministry of Posts and Telecommunications 162, 208
Minitel 206, 240
MIPS
for millions of instructions per second
Miros Inc.
TrueFace CyberWatch 377
MISS 374
for managed Internet security service
MIT
for Massachusetts Institute of Technology
PGP 387
MIT Media Lab 48
Mitsubishi Electronics America Corp. 35, 50, 61, 97, 129, 353
Panther 97
Verbatim Corp. 90
Mitsubishi Motors Corp. 35
Mitsumi Electronics Inc. 87
MLR 75
for multi-linear recording
MMDS 213
for multi-channel, multi-point distribution service
MMX 295
for multimedia extensions
forecast 369
technologies 347
MO
for magneto-optical
ASMO 94
capacity 93
definition 93
LIMDOW 93
MO7 93
technologies 93
WORMs 90
mobile computing
batteries 102
client devices 355
data access 521
data warehousing 604
database replication 470
EPS 219
Finland 9
forecast 368
groupware 492, 513
HPCs 298
Internet 240
intranets 494
market 358
Mobile IP 191
MOM 322
network management 535
PCMCIA cards 360
PCS 226
PDAs 359, 365
products 218
shipments 365
technologies 358
telecommuting 492, 517
Windows CE 298
see also telecommunications, transmission media
modems 186
56-Kbps 187
cable modems 186, 188, 212
CATV and the Internet 224
CDPD 217
PC 98 358
speed 211
technologies 187
V series modulation protocols 187
V.32bis 187
V.32terbo 187
V.34 187
V.42 187
V.42bis 187
wireless 216
World Wide Web access 239
Modula-2 431, 433
MOLAP 488, 618
for multi-dimensional OLAP
calculations 618
vendor aquisitions 618
vendors 618
MOM 317, 321, 322, 632
for message-oriented middleware
callback 322
definition 321
forecast 339, 516
market 335
message queueing 322
multiple threads 322
publish and subscribe 321
Momentum Software Inc.
XIPC 336
Mondex International Ltd. 114, 121, 573
monitors
see displays
Moore's Law
definition 78
Mosaic 556
Mosaix Inc. 503
ViewStar 503
MOSS 411, 413
for MIME Object Security Standard
Motorola Inc. 60, 129, 218, 360, 390, 392
6800 111
68040 processors 300
PowerPC 301
Voice-over-Internet Protocol 263
MPEG 258, 259
for Motion Pictures Experts Group

MPOA 170, 181
for Multi-Protocol Over ATM
MPP 353, 355
for massively parallel processor
challenges 355
definition 354
middleware 355
MRO 652
for maintenance, repair, and operations
MSC 33
for Multimedia Super Corridor
definition 33
technologies 35
MSMQ 333, 334
for Microsoft Message Queue Server
definition 334
MSN
for Microsoft Network
MSNBC 246
MSS 158
for mobile satellite services
MTBF 73
for mean time between failure
MTI Technology Corp.
Gladiator 86
MTS 333
for Microsoft Transaction Server
architecture 334
multicasting 165
definition 192
multimedia
ATM 222
databases 489
electronic commerce 36
forecast 280
Internet 257
market 277
MBone 193
ORDBMS 469
Web servers 266
World Wide Web 254
authoring tools 255
Multimedia Cable Network System 188
Multimedia Development Corp. 33
MultOS 125
for Multi-application Operating System
MX 628
for Metadata Exchange
Mytec Technologies Inc.
Touchstone 378

N

n.ABLE Inc. 568
NACHA 578
for National Automated Clearing House Association
Nantucket Software
Clipper 438
NAPs 235
for network access points
NasTel Technologies Inc.
MQControl and Visual MQ 336
National Association of Certification Authorities 395
National Association of Software and Service Companies 233
National Association of Video Distributors 92
National Center for Supercomputing Applications 234
National Computer Security Center 383
National Electrical Code 151
National Registry
NRIdentity Pass for Portables 420
National Science Foundation 235, 556
National Semiconductor Corp. 99
Cyrix 366
NBase 182
NCA 644
for Network Computing Architecture
NCR Corp. 83, 352, 362
Microsoft Alliance for Data Warehousing 630
NCR Teradata 620
Teradata 337, 469
Top End 336, 337, 338
NCs 355, 359, 526
for network computers
definition 240, 342, 359, 485
disadvantages 359
forecast 368, 548
Internet 240
market 359, 363, 365
NCSA 266, 378, 384
for National Computer Security Association
ftp site 403
Mosaic 274
NDS
for Novell Directory Services
NEC Corp. 92, 162, 176, 181, 260, 349, 355, 362, 364
DMI 526
PC-98 system 364
Neo Networks Inc.
StreamProcessor 182
NeoNet LLC. 177
Net Nanny Ltd.
Web site 252
Net Perceptions 512
NetBEUI 459
for NetBios Extended User Interface
NetCentric Corp.
FaxStorm 264
NetCom Communications Inc. 232
NetDox Inc.
ePackage 399
NetDynamics Inc.
Spider Technologies Inc. 460
Netect Ltd.
Netection 421
Netizens Protection Act 410
NetLink Inc. 196
NetManage Inc. 193
NetObjects Inc.
Fusion 246
NetPCs 342, 355, 359
for network PCs
definition 342, 359
disadvantages 359
forecast 368
market 363, 365
standards 355
NetRight Technologies Inc. 508
Netrix Corp. 172
Netscape Communications Corp. 242, 252, 265, 268, 274, 275, 276, 279, 310, 376, 566, 568, 570, 574, 593, 699
Actra Business Systems 588
browser market share 275
CARP 536
Commerce Platfor m 268
Communicator Deluxe Edition 246
Communicator Internet Access Edition 246
Communicator Professional Edition 246
Communicator Suite 245, 276, 568
Constellation 248
CoolTalk 263
CrossCommerce 588
Enterprise Server 269
FastTrack Server 270
Internet Foundation Classes 269
Java security model 404
JavaScript 439
Lava 392
LAYER support 244
Navigator 236, 243, 245, 250, 252, 295, 414, 568
Navigator Gold and Composer 257

Netcaster 249
Netscape Commerce 568
Netscape LivePayment 568
Netscape Navigator 127, 411, 628
Netscape Servers 568
Netscape Tools 568
NSAPI 269
Open Network Environment 269
plug-ins 496
Publishing Suite 246
Resource Description Format 244
S/MIME 412
SSL-enabled Web browser 415
SuiteSpot 269, 270, 314, 391, 397, 497, 514, 566
SuiteSpot Standard Edition 270
NetSpeak Corp.
WebPhone 262
Network Appliance Inc.
NetCache 239
Network Assoc. Inc. 418, 421, 537
GroupShield 403
WebScanX 404
WebShield 403
network computers 679
cost reductions 526
electronic commerce 579
security 374
smart cards 127, 131
versus HPCs 309
see also NCs
Network Computing Devices Inc. 526
network management
automated management tools 548
backbones 195
background 522, 528
bottlenecks 194
bridges 175
broadcast storms 175
browsers 522
computing platforms 348
configuration management 541
cost reductions 526
desktop computing 525
desktop PCs 357
devices 174
DMI 525, 533
extranets 529
file management 292
firewalls 417
focus 522
forecast 548
global networks 529
GUIs 541
hubs 175
intelligent agents 541
Internet 529
intranets 529
issues 528
knowledge management 510
load balancing 183, 535
managing applications 528
managing systems 173
market 544
MIB 532
mobile computing 494
network computers 526
network traffic patterns 183
operating systems 289, 290, 292
overview 521, 522, 528
performance 522
proxy servers 536
remote access 540
remote access devices 186
RMON 532
RMON2 533
routers 176
security 540
self-healing 529, 535
server bottlenecks 195
SMI 532
SNMP 530
SNMP-2 532
standards 528, 530
storage revenues 543
switched architecture 540
switches 175, 180
third-party 530
transmission media 529
trends 540
vendors 535
virtual networking 540
VLANs 182
Web issues 535
see also networking systems, operating systems, systems management, transmission media, wiring
Network News 276
Network Solutions Inc. 237
networking systems
advances 165
ATM 169
backbones 195
background 318
BGP4 180
bottlenecks 194
bridges 175
broadband networking 180
broadcast storms 175
connections 166
data warehousing 603
database replication 479
devices 174
Ethernet 166
Fast Ethernet 167
Fibre Channel Interconnect 172
file management 292
forecast 199
frame relay 171
Gigabit Ethernet 167
groupware 494
hubs 175
incompatibility 178
installed infrastructure 166
Internet traffic 238
internetworks 176
load balancing 183, 535
management 173, 529
market 196
MPOA 170
operating systems 290, 292
overview 165
packet-switched networks 176
peer-to-peer 318
printers 45, 65, 67
programming APIs and libraries 459
protocols 564
remote access devices 186, 198
response time 166
revenues 196
routers 176
security 379, 421
server bottlenecks 195
smart cards 127
software development 429
storage devices 74
switches 180
Token Ring 168
trends 165
vendor consolidation 196, 200
vendors 177, 186
VLANs 182
VPNs 183
Nevada Bell 204
New Media 258
Newbridge Networks Inc. 172, 176, 239
News Corp. 259
NEXT 148
for near-end crosstalk
NeXT Software Inc.
see Apple
Nextel Communications Inc. 219
NFR 77
for near-field recording
solid immersion lens 77
NFSBUF 380

NIA 183
for Network Interoperability Alliance
NiCds 95
for nickel-cadmium
advantages 95
capacity 95
drawbacks 95
NICs 167, 174, 197
for network interface cards
NiMH 94, 95
for nickel-metal hydride
advantages 95
capacity 95
forecast 103
market 101, 102
Nippon Telegraph and Telephone Corp. 208
Nissan Motor Corp. USA 254
NIST 383, 389, 392, 430
for National Institute of Standards and Technology
NIU 152
for network interface unit
NLSP 179
for NetWare Link Services Protocol
NNTP 416
for Network News Transport Protocol
Noblenet Inc.
RPC 3.0 321
Nokia Mobile Phones Inc. 13, 208, 218, 513
Nokia 9000 240
Nomai USA
750.c 100
Nortel 123, 172, 213
Entrust 390, 391, 394
notebook computers
batteries 97, 99
disk drives 99
Novacom 168
NovaSoft Systems Inc.
NovaManage 507
NovaWeb 507
Novell Inc. 242, 276, 337, 400, 459, 699
GroupWise 497, 498, 503, 513, 514
IntraNetWare 304, 527
IPX 185
IPX/SPX 655
Moab 189, 304, 311
NetWare 168, 178, 189, 285, 286, 288, 304, 310, 311, 314, 495, 526, 537
NLSP 179
Novell Directory Services 292
S/MIME 412
Storage Management System 537
Novonyx Inc. 314
NSA 383
for National Security Administration
NTSC 51
for National Television Standard Committee
NTT Electronics Technology Corp. 412
NUMA 353, 483
for Non-Uniform Memory Access
definition 354
versus SMP 354
Numega Technologies Inc.
Bounds Checker for Windows 95/NT 448
Numetrix Ltd. 681
NVOD 71
for near-video-on-demand
NVRAM
for non-volatile RAM
definition 111
NYNEX Corp. 205, 222

O

O.R. Technology Inc.
LS-120 87
SuperDrive 87
OAG 679
for Open Applications Group
definition 679
OAGIS 679
OAGIS 680
for OAG Integration Specification
OBI 565
for Open Buying on the Internet
object
definition 434
object technology 319
advantages 434, 644
beans 328
C++ 435
components 324
data hiding 324
databases 469, 485, 488
development tools 462
distributed computing 317, 319
distributed objects 461
DOM 244
EJB 329
encapsulation 324
forecast 465, 466
Fortran 90 431
HTML 242
Java 436
JIDL 331
languages 434, 435
market 464
MultOS 125
object orientation 324
Objective-C 436
objects 332
ODBMS 472
ORBs 324
ORDBMS 474
overview 434
reusable class libraries 449
Smalltalk 435
smart cards 113, 131
standards 455
terminology 434
trends 429
Web-based computing 460
Objective-C 435, 436
ObjectShare Inc. 435
OC 142
for Optical Carrier
Océ Technologies B.V.
PageStream 440 64
OCR 50
for optical character recognition
OCR/ICR 508
for optical character recognition/individual character recognition
ODBC 276, 439, 443, 480, 629
for Open Database Connectivity
CLI 478
ODBMS 472
for object database management system
definition 472
features 473
Internet 473
market 484
memory-resident performance 473
ODMG-93 474
programming 473
revenues 474
swizzling 473
versioning 473
versus ORDBMS 474
ODL 332
for Object Definition Library
ODMA 508
for Open Document Management API
ODMG 474
for Object Database Management Group

ODS 602
  for operational data store
  business operations supported 603
  data warehousing 609
  definition 603
  usage 603
OECD 401, 581
  for Organization for Economic Cooperation and Development
OFX 578
  for Open Financial Exchange
OIM 628
  for Open Information Model
OKI America Inc. 129
OLAP 481, 602, 603
  for on-line analytical processing
  applications 633
  benefits 619
  data access standards 629
  data warehousing 611, 621
  definition 599, 617
  design 619
  dimensional table 618
  drill-down 619
  fact table 617
  HOLAP 618, 619
  hypercube storage 604
  model 617
  MOLAP 618
  multidimensional model 618
  ROLAP 618
  roll-up 619
  server-based 604
  snowflake schema 618
  star schema 617
  tables 617
  vendors 620
  versus OLTP 602
OLAP Council 629
  MD-API 630
  Web site 482
OLE 331, 655, 675
  for Object Linking and Embedding
  see also Microsoft ActiveX
OLE DB 443
Olicom USA Inc. 168
Olivetti Office USA 364
OLS
  for on-line service
  versus ISPs 233
OLTP 320, 323, 336, 350, 481, 602
  for on-line transaction processing
  and ODS 603
  applications 633
  benchmarks 481
  data consolidation 610
  data warehousing 611
  databases 470
  definition 599
  parallelism 482
  trends 632
  versus OLAP 602
Olympus Optical 94
  SYS.230 89
OMG 325, 461
  for Object Management Group
OneComm 226
Onsale Inc. 583
ONU 153
  for optical network unit
OO 324
  for object orientation
  definition 324
Open Group 321, 323, 337
  DCE 381, 382
Open Market Inc. 565, 566, 567, 593
  OM-Transact 567
Open Port Technology Inc.
  Open Port Harmony 264
Open Text Corp. 250, 276
  Open Text 251
OpenDoc 461
OpenSoft Corp. 412
OpenText Corp.
  LiveLink 507
OpenVision Technologies Inc.
  AXXion 379
  AXXion-NetBackup 538
operating systems 45, 456
  APIs 287, 456
  ASCII 291
  availability/fault tolerance 290
  backup 537
  browsers 234, 246, 279, 285
  C 433
  character sets 291
  characteristics 287
  client/server 285, 286, 287, 291, 351, 352
  client/server forecast 313
  client/server market 310
  client/server vendors 301
  clustering 289, 314
  core system functions 456
  COS 113
  data mart versus data warehousing 607
  definition 285
  desktop 285, 286, 288, 356
  desktop forecast 312
  desktop market 308
  desktop technology 285
  desktop vendors 293
  directory services 292
  dynamic application load balancing 290
  electronic commerce 313, 565
  Ethernet 166
  features 290
  file management 289, 292
  forecast 312
  functionality 287
  Gigabit Ethernet 168
  I/O 289
  interapplication communication technologies 291
  interoperability 454
  Java 312
  JavaOS 301
  job management 291
  JVM 125, 285
  MacOS 287, 299
  mainframe 287, 291
  mainframe forecast 314
  mainframe market 312
  mainframe vendors 305
  mainframes 286, 287
  market 285, 308
  memory cache 289
  memory management 289
  MS-DOS 287, 293
  multicast 193
  multiprocessing 288
    asymmetric 288
    symmetric 288
  multitasking 288
    cooperating 288
    preemptive 288
  multithreading 288
  MultOS 125
  NetWare 287
  network file access 292
  networking 290, 292
  OS/2 299
  OS/2 Warp 287
  OS/390 287
  OS/400 287
  overview 285
  passive application load balancing 290
  PCs 356
  portability 291
  process management 288
  remote job transfer and manipulation 292
  remote log-in 292
  Rhapsody 287
  scheduling 292
  security 290, 313, 373
  server revenue 310
  smart cards 110, 113, 123, 131

storage management 292
task management 292
TCP/IP 285
technology 286
trends 285
Unicode 291
UNIX 285, 287, 300, 301
virtual memory 289
Web servers 266
windowing systems 290
Windows 293
Windows 3.x 287
Windows 9x 287
Windows NT 287
Windows systems features 298
OPS 407
for Open Profiling Standard
OPSEC 385
for Open Platform for Secure Enterprise Connectivity
optical laser cards 118
optical storage media 88
capacity 88
CD-ROMs 89
Divx 91
DVD 90
DVD+RW 92
DVD-RAM 92
HSM 88
Optus Communications 209
OQL 474
for object query language
Oracle Corp. 123, 265, 310, 337, 342, 359, 376, 440, 469, 485, 487, 526, 538, 563, 565, 569, 570, 579, 593, 604, 606, 635, 637, 638, 643, 644, 646, 647, 648, 654, 659–663, 676, 678
Aerospace and Defense 659
API 680
Application Data Warehouse 634
Application Server 270, 271
Applications for the Web 646
architecture 662
Authorized Reseller program 660
challenges 660
CPG 648
Data Mart Suite 636
Datalogix International Inc. 661
Datamart Suites 629
Designer/2000 662
Developer/2000 442, 662
Dialogix 661
Discoverer 635
electronic commerce 662
ERP 637
Express 618, 630, 635
FastForward 659
Financials 647
GEMMS 661
Industrial Sector Vertical 659
industry-specific versions 659
Internet 662
InterOffice 498
Java applet client 662
market 680, 681
NCA 271, 485, 644, 662
tiers 662
OLAP 630
Oracle Applications 659, 660
Oracle Applications for the Web 662
Oracle Applications Release 11 features 661
Oracle ERP 661
Oracle Express 487
Oracle Express Server 620
Oracle Financials 645, 660
Oracle Human Resources 660
Oracle Maintenance, Repair, and Overhaul 659
Oracle Manufacturing 660
Oracle Payables 663
Oracle Project Manufacturing 659
Oracle Projects 661
Oracle Salesforce Automation 661
Oracle Sales, Service, and Support 661
Oracle Service Resource Planning 659
Oracle Supply Chain Management 661
Oracle Web Customers 659, 663
Oracle Web Employees 659, 663
Oracle Web Suppliers 659, 663
Oracle Workflow 660, 663
Oracle7 659
Oracle8 538, 659
Parallel Server 337
PL/SQL 478
Public Sector 659
Release 8.0 478, 485, 487
Serializable isolation 476
SQL 477
strategy 660
Symmetric Replication 480
TouchNet 378
UNIX 487
vertical markets 659
Warehouse Technology Initiative 629
Web Request Broker 270
Web site 658
Year 2000 644, 658
Orange Book 383
ORBs 324, 331
for object request brokers
definition 326
market 338
ORDBMS 469
for object-relational database management system
advantages 469
composite types 474
definition 474
features 474
forecast 488
market 474, 484
SQL3 478
trends 475
versus ODBMS 474
ORGA Card Systems 129
Organization for Economic Cooperation and Development 395
ORPC 333
for Object Remote Procedure Call
OSI 173
Inter-Domain Policy Routing 180
OSPF 179, 190
for Open Shortest Path First
OTE of Greece 157
OTMs 339
for object transaction monitors
OTS 330
for Object Transaction Services

P

P3 407
for Platform for Privacy Preferences
PACE Switch 193
Pacific Bell 204, 222
Pacific Telesis Group 204
Packet Engines Inc. 167
packets
broadcast 175
flow 181
IPX/SPX 176
multicasting 179
tag 181
paging systems
e-mail 226
EPS 219
technologies 218
PAL 51
for Phased Alternate Line
Panasonic Co.
Divx 92
Panorama Software Systems Inc.
see Microsoft Corp.

ParaGraph International Inc.
Home Space Builder 255
Web site 255
Paramount Pictures Inc.
Divx 92
Parasoft Corp. 447
CodeWizard for C++ 447
Insure++ 448
TCA 448
ParkerVision Inc. 51
autoTRACK 52
CameraMan 51
Pascal 431, 432
PBS Online 259
PBXs 224
for Private Branch Exchanges
ACD 225
wireless 225
PC 98 357
PC Concepts 47
PC DOCS 506
CyberDOCS 507
DOCS Fusion 507
DOCS Open 507
PC Magazine 402
PC/SC Workgroup 125
PCI 357
for Peripheral Component Interconnect
PCN 214
for personal communications network
technologies 217
PCS 15, 203, 214
for personal communications services
antennas 218
EPS 219
forecast 229
market 226
narrowband 219
technologies 217
trends 16
PCs 275
for personal computers
background 318
BASIC 432
C 433
client devices 355
consumer PCs 358
cost reduction 526, 548
databases 484
DBMS 470
DVD-ROMs 91
GUIs 290
hand-held 88
market 363
NCs 342
NetPCs 342
notebooks 364
Office PCs 358
OnNow 357
operating systems 312, 356
Pascal 433
PCI 357
pricing 341
revenue 361, 364
ROM BIOS 451
SCSI 79
storage devices 77
tape formats 72
TCO 359
technologies 356, 357
trends 341
two-tier computing 319
USB 358
versus network computers 526
workstations 356
Year 2000 451
see also desktop computing
PCT 398
for Private Communications Technology
PDAs 359
for personal digital assistants
capacity 360
client devices 355
enhancements 360
forecast 369
handwriting-recognition software 360
market 365
mobile computing 218
prices 360
uses of 360
vendors 360
PDBSs 469
for Parallel Database Systems
advantages 469, 482
definition 482
technologies 482
PDU 531, 532
for Protocol Data Unit
PeakSoft Corp.
PeakJet 246
Web site 246
PeerLogic Inc.
Pipes 336
PEM 411
for Privacy-Enhanced Mail
PEM-MIME 411, 413
PenOp Inc.
PenOp 378
Penril Datacomm Networks Inc. 196
PeopleSoft Inc. 337, 502, 513, 563, 634, 643, 644, 646, 647, 668–674
architecture 673
Cash Management 672
Cube Manager 669
Deal Management 672
Demand Planning 672
DirectPath 672
EMU 671
European payroll 671
expenses 672
financial services 672
forecast 682
global distribution and manufacturing 671
Global Human Resources support 672
industry-specific solutions 672
Internet 673
market 680, 681
ODBC API 669
PeopleSoft 7 668, 670, 673
PeopleSoft 7.5 671
PeopleSoft 8 673
PeopleSoft Distribution 669
PeopleSoft Financials 669
PeopleSoft General Ledger 671
PeopleSoft HRMS 669
PeopleSoft Inventory 669
PeopleSoft Manufacturing 669
PeopleSoft Order Management 669
PeopleSoft PeopleTools 671
PeopleSoft Projects 672
PeopleSoft Purchasing 669
PeopleSoft Quality 671
PeopleSoft Select 670
PeopleSoft SelectPath 670
PeopleSoft Universal Applications 669
performance measurement 671
Red Pepper 643, 668, 669
Risk Management 672
Stock Administration 672
Total Compensation 672
Year 2000 644
Perl 438
Perspecta Inc. 276
Perusahaan Otomobil National Bhd 35
PGP 387, 411
for Pretty Good Privacy
PGP 5.0 387
PGP/MIME 413
PGPI 397, 413, 418
for Pretty Good Privacy Inc.
OpenPGP 413
PGPcookie.cutter 409
PGPdisk file encryption software 420

Philips Electronics Ltd. 89, 90, 91, 92, 108, 129, 360
Philips/Mikron 118, 129
PhotoDisc Inc. 582
Photonic Power Systems Inc. 145
PICS 252
  for Platform for Internet Content Selection
PictureTel Corp. 52
  LimeLight 52
Pilot Network Services Inc. 375
Pilot Software Inc.
  Decision Support Suite 620, 626
  Microsoft Alliance for Data Warehousing 630
  MOLAP 618
  OLAP 630
  see also Dunn & Bradstreet Corp.
PIM 192
  for Protocol-Independent Multicast
PIN 114, 376
  for personal identification number
  master PIN 114
Pine Cone Systems 615
Ping 406
  for Packet INternet Groper
Pirelli S.P.A.
  WaveMux 3200 143
PixTech Inc.
  FED 57
PKI 390
  for Public-Key Infrastructure
  proposals 392
  supported application 390
PL/1 431, 432
  for Programming Language One
Planning Sciences International Ltd. 630
Platform-Independent Cryptography API 400
Platinum Technology Inc. 445, 450, 633, 636
  CCC/Harvest 445
  Forest & Trees 623
  InfoBeacon 620
  InfoRefiner 612
  Metadata Coalition 627
  Microsoft Alliance for Data Warehousing 630
  OLAP 630
  Paradigm Systems 635
  RelTech Group 635
  Repository 628
  Software Interfaces 635
  Trinzic Inc. 635
Pluris Inc. 177
PMD 142
  for polarization mode dispersion
PointCast Inc. 246, 248, 276, 512
  PointCast I-Server 248
  PointCast Network 247
  SmartScreen 247
pointing devices 48
  cost 48
  digitizing tablets 49
  forecast 66
  Intellimouse 48
  mice 48
  touchpad 48
  trackball 48
  TrackPoint 48
polymorphism 434
PoPs 235, 239
  for points of presence
portable computing
  HPCs 298
  LCDs 55
  pointing devices 48
  Windows CE 298
Portcullis Computer Security Ltd. 396
Portugal Telecom 208
POS
  for point of sale
  definition 570
POSIX 307
  for Portable Operating System Interface
  definition 456
Powersoft Corp. 457, 464
  Power product family 442
  PowerDesigner 444
  SQL Anywhere 442
pps
  for packets per second
PPTP 185
  for Point-to-Point Tunneling Protocol
  versus IPsec 185
Praxis International Inc.
  Microsoft Alliance for Data Warehousing 630
  OmniEnterprise 612
  OmniReplicator 480
prefix references 695
Premenos Corp.
  see Harbinger Corp.
Premiere Technologies Inc.
  Orchestrate 262
Premio Computer Inc. 526
Pretty Good Privacy Inc.
  see PGPI
Preview Travel 259
PRI
  for Primary Rate Interface
  definition 211
Price Waterhouse LLP
  Geneva V/T 613
  Metadata Coalition 627
  Pinpoint Energy System 648
  security 400
  Service 2000 648
  WTI 629
Prince Software Inc.
  Simulate 2000 452
  Survey 2000 452
printers 60
  advances 63
  clusters 64
  color 61
  dye-sublimation 62
  forecast 66
  high-volume 64
  ink-jet 61
  Internet 65
  laser 61, 62, 63, 64, 65, 66
  media handling 63
  mid-volume 62
  Mopier 65
  multi-function 65
  overview 60
  paper handling 65
  paper supply 64
  personal 62
  solid ink 62
  technologies 61, 62, 63
  thermal transfer 61
  thermal wax 61
  toner supply 64
  trends 65
  two-up printing 64
  variable-dot-size 61
Prism Solutions Inc.
  Metadata Coalition 627
  Open Metadata Architecture 629
  Prisim Warehouse Executive 613
  Warehouse Executive 614
  WTI 629
privacy
  anonymizing proxies 408
  cookie technology 408
Privnet Inc. 418
PRML 77
  for partial response, maximum likelihood
Prodea Software Corp.
  see Platinum Technology Inc.
Prodigy Service Co. 232
programming
  3GLs 430, 431
  4GLs 439
  Ada 433
  application frameworks 457

BASIC 432
C 433
C++ 435
clustering 354
COBOL 430, 431
compilers 430
component architectures 317
configuration management 445
cookie technology 255
core system functions 455
databases 475
dBase 438
DBMS 470
errors 437
Fortran 430
hierarchical databases 471
HTML 234
IDL 326
instruction sets 344, 345
integrated development environment 430
Java 429, 436
Java versus C++ 437
JavaBeans 317, 328
JavaScript 439
language standardization 430
language standards 454
language trends 429
languages 495
LISP 432
market 463
Modula-2 433
modular 433
MOM 322
MultOS Executable Language 125
object languages 434, 435
Objective-C 436
ODBMS 472, 473
Pascal 432
Perl 438
PL/1 432
reusable class libraries 449
RPG 432
scripting languages 437
Smalltalk 435
smart cards 131
specialized languages 437
tools 429
Unicode 437
VBScript 439
VRML 254
Web-based computing 459
Web-enabled application development 441
World Wide Web 251
xBase 438
Year 2000 429, 430, 450
see also APIs, client/server computing, desktop computing, Internet, object technology, World Wide Web
PROM
for programmable read-only memory
protocols
Applications Layer 173
ATM 176
BGP4 179
CARP 536
Data Link Layer 173
DHCP 187, 241
distance vector 178
DVMRP 192
Dynamic Host Configuration Protocol 182
EGP 178
EIGRP 178
ESMTP 241
forecast 200
GGP 178
HTTP 234, 241, 266
IDRP 178
IGMP 192
IGP 178
Internet 192, 231, 241
IP 189
IP classes 189
IPsec 184
IPv4 185, 189
IPv6 185, 189, 199
IPX/SPX 189
ISAKMP 185
IS-IS 179
JRMP 331
L2F 186
L2TP 186, 191
layers 1-7 173
LDAP 263
linkstate 179
MIME 563
Mobile IP 191
modulation protocols 187
NetBEUI 185
Network Layer 173
NLSP 179
OPSF 179
OSI 173
overview 188, 189
path vector 179
Physical Layer 173
PIM 192
PPTP 184, 185
Presentation Layer 173
RIP 178
RIP2 178
routers 178
routing 178, 179
RSVP 190, 193, 194
RTCP 191
RTP 190, 193
SDRP 178
Session Layer 173
SET 573, 574
SKIP 185
SMDS 184
SNA 188
SNMP 188, 530
TCP/IP 189, 459
Transport Layer 173
tunneling 186
two-phase locking 476
VANs 564
VPNs 564
see also standards
Psion PLC 360
PSM 478
for Persistent Stored Modules
PSTN 264
for Public Switched Telephone Network
PTT 206
for Post, Telegraph, and Telephone
PTT Telecom Netherlands/Unisource 123
Puerto Rico Telephone Authority 210
Puerto Rico Telephone Co. 210
pull technology 273
collaborative filtering 254
definition 247
offline browsers 253
punched key cards 117
push technology 295, 679
advertising 249
challenges 249
corporate 249
definition 247
integration 248
market 249, 276
networking systems 165
self-updating 248
software distribution 249
tuner 248
vendors 247, 248
World Wide Web authoring tools 255
XML 243
PVCs 172
for permanent virtual circuits
Pyramid Technologies Inc. 352

Q

QIC 74
  for quarter-inch cartridges
QoS 222
  for Quality of Service
  definition 200
  forecast 200
Qualcomm Inc. 216
  Eudora 387
  Q Phone 216
  S/MIME 412
Quantum Corp. 70, 85
  Atlas III 79
  DLT 74
  DLT-4000 72
  DLT-7000 74
  FC-AL 80
  Flagship 74
  NFR 78
  Symmetric Phase Recording 74
  Ultra DMA-33 78
  Valueline 74
  Viking II 79
Questal Online 276
Quintic Systems Inc.
  Century Code Conversion 452
  Century Source Conversion 452
Quza
  QuzaSafe 375

R

R&O Inc.
  MDIS 628
  Metadata Coalition 627
RAB 82
  for RAID Advisory Board
  disk classification 82
RACF
  for Resource Access Control Facility
Radiance Software Inc.
  Ez3d 255
  Web site 255
Radnet Inc.
  WebShare 498, 499, 514
RADSL 212
  for rate-adaptive ADSL
RAID 80, 362
  for redundant arrays of independent disks
  A Case for Redundant Arrays of Inexpensive Disks 81
  capacity 81
  definition 81
  failure-tolerance classification 82
  forecast 103
  market 100
  new solutions 82
  overview 81
  RAB 82
  RAID 1 81
  RAID 5 81
  subsystems 81
RAM 110
  for random access memory
  NVRAM 111
RandomNoise Inc.
  Coda 257
Rapid City Communications Inc. 196
  FIRST 191
Raptor Systems Inc. 185, 417, 418, 421
Rational Software Corp. 447, 450
  Pure Atria 448
  Pure Atria ClearCase 445
  Pure Coverage 448
  Pure Link 446
  Pure Visual Quantify 447
RBOCs 160, 204, 206
  for Regional Bell Operating Companies
  ATM 222
  copper cabling 161
  U.S. Telecommunications Act of 1996 205
  WDM 160
RCS 445
  for revision control system
RDBMS 471, 603
  for relational database management system
  advantages 605
  bulk data loader 605
  capabilities 472
  Codd's First Normal for m rule 474
  data spaces 605
  definition 471
  intranets 265
  market 484
  MOLAP 618
  pair-wise join 605
  revenues 486
  SQL3 478
  technologies 471
  TPC-D 481
  vendors 618
  versus MDBMS 604, 605
  versus MOLAP 618
RDRAM
  for Rambus DRAM
Reach Software Corp. 502
RealNetworks Inc. 258
  RealAudio 257, 259
  RealPlayer 259
  RealVideo 258
Recognition International Inc.
  ID-3D Handkey 378
Red Brick Systems Inc.
  Red Brick Warehouse 621
  Web site 606
Red Hat Software 300
Red Pepper Software Company
  see also PeopleSoft Inc.
Reed Elsevier PLC 232
relational database management system
  see RDBMS 641
Reuters
  TIBCO 336
RFI 154
  for Radio Frequency Interference
RF-IDs 118
  for Radio Frequency IDs
  advantages 118
  challenges 118
  SuperTag 118
Ricoh Corp. 90
RIP 178
  for Routing Infor mation Protocol
RISC 182, 344
  for Reduced Instruction Set Computing
  forecast 369
  technologies 344
  trends 363
  vendors 369
  versus CISC 344
RMI
  for Remote Method Invocation
RMON 531, 532, 533
  for Remote Network Monitoring
  RMON2 533, 540
    forecast 548
RND Networks Ltd.
  Web Server Director 536
  Web Server Director Pro 535
Rockwell Semiconductor Systems 213
  K56flex 187
Rogue Wave Software Inc. 449, 459
  Tools++ 449
  zApp 458
ROLAP 488, 618
  for Relational OLAP
  data warehousing 604
  definition 618
  vendors 618
ROM 110, 111, 114, 342
  for read only memory
routers 174
  and ATM 180
  ASICs 177
  BGP4 179
  capacity 177

challenges 176
CIDR 179
definition 175
EGP 178
EIGRP 178
forecast 199
GGP 178
gigabit 176
high-speed 176
IDRP 178
IGP 178
Layer 3 switches/routers 180
MBone 192
multicast 193
network management 529
OPSF 179
overlays 176
protocols 178
QoS 176, 177
RIP 178
RSVP 190
SDRP 178
security 176
switching-routers 176
trends 176
vendors 176, 177
versus bridges 176
versus switches 180
RPCs 321, 382, 632, 655
for remote procedure calls
challenges 321
definition 321
disadvantages 321
HTTP 267
RPG 431, 432
for Report Program Generator
RPG III 432
RPK 389, 419
RSA 113, 387, 388, 390, 415, 419
for Rivest, Shamir, and Adleman encryption algorithm
recommended key sizes 387
versus ECC 389
Web site 387
RSA Data Security Inc.
see RSADSI
RSADSI 387, 390, 399, 400, 412, 418, 574
BSafe 389
ECC 389
forecast 423
J/SAFE 419
RC2 412
RC4 414
RSA SecurPC 399
S/MIME Interoperability Center Web site 412
RSVP 189, 190, 193
for Resource Reservation Protocol
Gigabit Ethernet 190
prioritizing 190
QoS 190
scalability 190
RTCP 191
for Real-Time Control Protocol
RTP 190, 193
for Real-Time Transport Protocol

S

S/MIME 411, 412
for Secure/Multipurpose Internet Mail Extensions
vendors 412
S/WAN 184
for Secure Wide-Area Network
DES 184
Sabre Group
Travelocity 584
SAFE Act 401
Sagent Technology Inc. 635
Data Mart Builder 606
Sagent Data Mart Solution 613
SAIC 574
for Science Applications International Corp.
Samsung Electronics Co. Ltd. 364
Santa Cruz Operation
OpenDesktop 300
OpenServer 302
OpenServer and UnixWare 285
SCO UNIX 495
UnixWare 7 300, 302
Sanyo Electric Co. Ltd. 94
SAP AG 502, 513, 563, 633–634, 637, 643, 644, 646, 647, 648, 648–658, 681
for Systeme Anwendung Produkte AG
ABAP/4 Development Workbench 655
AcceleratedSAP 649, 659
activity-based costing 650
Administrator Workbench 634
Advanced Planner and Optimizer 653
Application Link Enabling 634, 650
applications 649
architecture 654
BAPIs 569, 649, 656, 680
Business Engineering Workbench 682
Business Explorer 634
Business Framework 656
Business Information Warehouse 633, 650, 651, 656, 657
Business Information Warehouse Server 634
Business Objects 644, 655
Business Workflow 657
data warehousing 633, 634
Employee Self Service applications 652
enhancements 651
ERP 633
Euro functions 651
Euro solution 651
financial application integration via BAPIs 650
financials 650
Internet capabilities 658
Kiefer & Veittinger 681
market 680
Microsoft Alliance for Data Warehousing 630
new financial components 650
new logistics functionality 652
OAG 680
OLAP 656
Open BAPIs 634, 657
Open Information Warehouse 656
Pandesic 569, 658
partners 649
Product Data Management 652
R/2 634
R/3 484, 569, 633, 634, 646, 648, 649, 650, 652, 656, 681
R/3 Business Engineer 657
R/3 Human Resources 652
Ready-to-Run R/3 649
ROLAP 656
Sales Configuration Engine 652
SAP Euro 645
SAPNet 649
SCOPE 653, 681
standards 679
strategies 681
three-tier R/3 configurations 655
transfer pricing 650
two-tier R/3 configurations 655
Web site 649
Web-based catalog and purchase requisition system 652
Year 2000 644
Saros Corp.
see FileNet Corp.
SAS Institute Inc. 633, 636
Enterprise Miner 623, 627
Metadata Coalition 627
SAS Warehouse Administrator 621

System for Information Delivery 623, 627
Satellite Phone Japan 157
satellites 137
basic types of 156
challenges 162
considerations 156
DBS 159
GEOs 156, 157
LEOs 156, 158
market 161, 162
MEOs 156, 157
microwaves 159
services provided 156
smart cards 127
technologies 156
types of 156
UTMS 12
versus CATV 223
VSATs 157
Saxe Inc.
WTI 629
SBC Communications Inc. 205
SBS 99
for smart battery systems
scanners 46, 50
convenience 50
flatbed 50
forecast 66
OCR-enhanced business-card 50
sheet-fed 50
specialized 50
Schlumberger/Sligos 108, 123, 126, 129, 397
Scientific American 23
SCM Microsystems Inc.
SwapSmart 376
SCO
see Santa Cruz Operation
Scopus Technology Inc. 676
Scottish Telecom 207
SCSDK 125
for Smart Card Software Development Kit
SCSI 50, 78, 172
for Small Computer Systems Interface
capacity 79
FC-AL compatability 80
Ultra SCSI 79
SDH
for Synchronous Digital Hierarchy
versions 142
SDRAM
for synchronous DRAM
SDRP 178
for Source Demand Routing Protocol
SDSI 390, 392
for Simple Distributed Security Infrastructure
SDSL 212
for symmetric DSL
SDV 153
for switched digital video
Seagate Software Infor mation Management Group
Desktop Management Suite 526
Palindrome 538
Seagate Technology Inc. 74, 76, 78, 624
BackupExec 538
Barracuda 79, 80
Cheetah 4LP 4 GB hard disk 76
Crystal Info 621, 623
FC-AL 80
FCL 80
SSA 79
search engines 264
alliances 276
complex queries 251
content selection 251
databases 250
definition 250
directory-oriented 276
filters 252
forecast 279
intranets 276
keywords 251
market 276
metasearch engines 253
Web sites 253
precision ratio 251
query and results 251
registration 251
relevance 251
guidelines 252
Soundex 251
vendors 276
web crawlers 251
Web sites 250
Secure Computing Corp. 418
SafeWord 420
Secure Enterprise Web 383
Secure Public Networks Act 401
Securities Industries Association 380
security 297
access 112
accountability 379
administration 379
alternatives 374
antagonistic attacks 409
anti-virus 373
auditing 379
authentication 375
authorization 379
bar codes 117
biometrics 376
browsers 414
business environment 373
certificate authorities 376, 393
commerce servers 566
components 375
computer system resources 379
cookie technology 408
cryptography 385
data warehousing 616
DCE 382, 383
denial of service 402, 406
digital certificates 575
digital signatures 394
disaster recovery 374
distributed environments 380
DSS 388
electronic commerce 236, 411, 553, 555, 574, 580
electronic payment systems 572
encryption 388
enterprise management 524
environments 422
evaluation criteria 383, 384
extranets 521
firewalls 186
flag cards 390
forecast 422, 595
freeware 380
hostile applets 404
Internet 128, 278, 411, 580
Internet and smart cards 124
Internet outsourcing 374
invasion of privacy 407
IPsec 184
Java 404
Kerberos 380, 381
key length 388
L2F 186
law enforcement 399
magnetic stripe cards 117
market 374, 418
mergers and acquisitions 418
MISS 374
monitoring 379
multi-platform 422
network management 540
new vulnerabilities 374
OpenCard 376
operating systems 290
overview 373
PIN 376
Ping of Death attacks 406
PKI 390
PPTP 185
products 373
projected growth 419

protection profiles 384
protocols 185
real-time vulnerability monitoring 374
remote logging in 375
resource grabbers 406
RF-IDs 118
risks 402
routers 176
RTP 191
S/MIME 412
SDSI 392
service providers 375
services 374
SESAME 382
SET 575
smart cards 107, 109, 110, 113, 114, 116, 120, 131, 390
SNMP-2 532
SSO 380
standards 373, 383, 385, 423
switches 180
SYN flooding 406
technologies 580
TLS 415
tokens 375, 376, 382
trends 373
Triple-DES 91
virus scanning 374
virus signature 403
viruses 402
virus-prevention programs 403
voiceprint 377
VPIs 184
Web site and extranet hosting 374
World Wide Web 384
Security Administrator Tool for Analyzing Networks 380
Security Alliance 400
Security and Freedom through Encryption Act of 1997 401
Security Dynamics Technologies Inc. 418
SecurID 376
SoftID 376, 420
Security First Network Bank 584
Security Tools for the Financial Industry 389
Segue Software Inc. 447
QualityWorks 447
semiconductor industry
CMOS 10, 349
memory chips 342
smart cards 107
Sentient Networks Inc.
Ultimate 1200 187
Sequent Computer Systems Inc. 352
NUMA 354
NUMA-Q 2000 172
servers
DCE 383
disk drives 69
security 402
storage requirements 69
tape formats 72
Web 266
Web architectures 267
SESAME 380, 382
for Secure European System for Application in a Multi-vendor Environment
features 382
SET 124, 411, 567, 570, 573, 577, 578
for Secure Electronic Transactions
challenges 576
definition 574
digital certificates 575
e-COMM 577
ECSET 576
forecast 595
nonrepudiation 577
payment authorization process 575
pilots 576
technologies 574
SFTP
for Simple File Transfer Protocol
Category 6 147
SGML
for Standard Generalized Markup Language
SGS-Thomson Microelectronics 92, 129, 420
Shared Registration System 238
Sharp Electronics Corp. 94
shielded twisted pair
S-HTTP 415
for Secure HyperText Transport Protocol
SIA
for Semiconductor Industry Association
Siebel Systems Inc. 647, 676
Siemens AG 13
Siemens Nixdorf Information Systems Inc. 126, 129, 349, 362
TrustedWeb 382
Silicon Graphics Inc. 352, 356, 363, 368, 593, 624
Cosmo Worlds and WebSpace Author 255
Cray Computers 354
Cray T90 series 355
Gigaswitch technology 354
Irix 302, 568
MineSet 623, 624, 627
MIPS 301
MIPS R10000 345
Web site 255
SIMD
for Single Instruction, Multiple Data
SIMs 112, 128
for subscriber identification modules
Singapore Telecom 157, 208
SIPC
for Simply Interactive PC
SKIP 185
for Simple Key Management for IP
SLA 539
for service level agreement
SLDRAM
for SyncLink DRAM
Smalltalk 435, 449
forecast 466
Smalltalk-80 436
smart 123
Smart Battery Systems Implementers' Forum 99
smart cards 373, 375, 376, 382, 585
32-bit RISC processor 111
access 112
advantages 116
advertising 128
alternative technologies 116
applications 119
authorizations 115
banking 121
bar codes 107
barriers to the acceptance 108
biometric technology 114
card acceptance device 107
cardholder verification 114
card-to-card transactions 122
challenges 126
closed systems environments 131
combicard 116
communications 113
contact cards 109
contactless cards 115
contro of cardholder information 126
COS 113
cryptography 114
C-SET 124
data integrity 114
debit cards 120
definition 107
desktop computing 127
disposable 109

distance-systems 116
EEPROM 111
electronic commerce 128, 570, 572, 573
EMV 124, 131
encryption 113
e-purse 121, 122, 128, 129
Europay C-SET 577
European use 573
evolution 123
file management 113
flag cards 390
float 119, 128
forecast 131, 132, 594
fraud 120
growth 130
GSM 128
health cards 128
history 108
hybrid cards 125
hybrids 107
in circulation 130
international markets 129
international use 122
Internet 128
Internet-accessing 125
interoperability 123, 130, 131
issuer spending 130
JavaCard 112, 113, 125
lifestyle 127
logical access keys 127
loyalty programs 127, 128
magnetic stripe 107
manufacturing keys 114
market 127
market share 131
mask 111
memory cards 107, 109, 116
microprocessor 107, 110, 111
mini-card 107
multi-application 110, 112, 131
multipurpose 37
MultOS 125
new markets 128
object technology 113, 131
obstacles 126
offline debit transactions 120
operating systems 123, 126, 131
optical laser cards 118
overview 107
packaging 107
PC Cards 128
physical dimensions 109
PINs 114
PKI 390
protected memory cards 110
RAM 110
reducing fraud 121
reloadable e-purse 110
reloading 116, 128
revenue 111, 126, 127, 128
RF 107
RF-IDs 118
ROM 110
runtime operations 113
secret keys 110
security 113, 116, 131, 390
SET 576
simple memory cards 109
simplification 131
SIMs 128
single-application 131
smart payment cards 573
standards 123
stored-value 115, 121
stored-value applications 107, 111
storing information 128
technologies 116
token cards 119, 130
transportation uses 115
types of 108
U.S. forecast 131
usage 130
uses of 107
vendors 129
versus other media 117
versus paper-based stamp cards 127
VOD 119

SMDS 184, 203
for Switched Multimegabit Data Service
definition 223
revenue 226
technologies 226

SMI 531, 532
for Structure of Management Information

SMP 288, 351, 352, 353
for symmetric multiprocessing
technology 353
versus NUMA 354
versus shared nothing databases 484

SMTP 416
for Simple Mail Transfer Protocol

SNA 292
for Systems Network Architecture
see also IBM

SNMP 530, 532, 533, 534
for Simple Network Management Protocol
components 531
definition 531
forecast 548
versus RMON 533

Social Security Administration 378

SoftArc Inc.
FirstClass 499

SoftDD
Complete Cleanup 408

SoftQuad International Inc. 242

Software AG 629, 635
SourcePoint 615

Software and Systems Engineering Ltd. 412

Software and Technologies Inc. 417

software components
background 318, 319
forecast 339
market 335
messaging middleware 319
overview 317
standards 317

software development
3GLs 430, 431
4GLs 439
application frameworks 457
C++ 436
code repositories 450
compilers 430
configuration management 445
design patterns 459
forecast 465
GUIs 429
integrated development environment 430
interoperability 454
Java 429
language standards 455
market 461
object languages 434, 435
object technology 429
overview 429
project management 448
project team 443
scripting languages 437
specialized tools 437
system building 446
tools 429

Software Emancipation Technology Inc.
Discover Y2K 452

Software.com
Post.Office 410

solid immersion lens 77

Solid Oak Software Inc.
CyberSitter 252

SOM
for System Object Model

SONET 141, 143, 162, 177
for Synchronous Optical Network

microwaves 159
technology 142
Sony Corp. of America 72, 75, 87, 89, 90, 91, 92, 129, 218, 360
Advanced Intelligent Tape 72
Advanced Metal Evaporated 72
DDS 73
Divx 92
QIC-Wide 75
SDX series 72
SDX-300c drive 73
tape drives 70
Sony Electronics Inc.
SMO-F541-DW 93
SourceFile 398
specifications
see standards
SpectraLink 216
speech recognition 52
advances 53
challenges 52
feature extraction 53
forecast 66
Hidden Markov Models 53
law and medicine 52
technology 53
Speedware Corp. Inc. 635
Esperant 624
Spider Technologies Inc.
see NetDynamics Inc.
Sportsline USA 259
spreadsheets
2-D 617
3-D 617
Sprint Corp. 160, 172, 222, 235, 375, 512, 587
SPSS Inc.
WTI 629
SPT Telecom 123
SPX/IPX
see IPX/SPX
Spyglass Inc. 242, 593
Mosaic 439
SurfWatch 252
Spyrus Inc. 418, 574
SQL 471, 603
for Structured Query Language
capabilities 472, 477
cascaded delete 477
compliance 477
constraints 477
data conversion 610
DDL 470
definition 476
dynamic SQL 472
embedded SQL 472
entry-level SQL-92 477
extended set and join operators 477
features 472
MOLAP 618
ODBMS 473
ORDBMS 475
overview 476
Persistent Stored Modules 488
precompilation 472
procedural 478
recursive union queries 478
scrollable cursors 477
Select 474
SQL3 474, 477, 478, 488
SQL-92 476
standards 469, 476
Transactional SQL 488
triggers 478
versus xBase 438
X/Open 477
SQL Software 445
SRAM 342
for static RAM
technologies 343
versus DRAM 343
SSA 79, 644
for Serial Storage Architecture
advantages 79
capacity 79
PKI 392
SSA Industry Association 79
SSL 236, 382, 398, 413, 414, 496, 566, 567, 663
for Secure Sockets Layer
secure hash functions 414
SSO 380, 381
for single sign-on
advantages 380
benefits 380
products 380
ST2 191
for Stream Protocol version 2
Stac Electronics Inc. 390
Replica 538
Staffware Corp. 503, 504
standards
100Base-T 167
802.1Q 183
Ada 9x 433
ANSI BASIC 432
ANSI C 433, 436, 454
ANSI Pascal 432
ANSI SQL3 474
ANSI X.12 562, 564
ANSI X12.4-#820 578
ANSI X3T1 145
ANSI X3T9.3 145
ANSI/EIA/TIA-568-1991 146
ANSI/TIA/EIA-568-A 147
APPN 564
ATM 169
BISDN 221
C++ 436, 455
CARP 536
CCD+ 578
CDMA 13
CGI 267
COBOL-85 431
COM 628
CORBA 325
corporate applications 679
C-SET 124
CTX 578
CVP 418
data warehousing 639
DCOM 296
DES 386
DHCP 241
DHTML 243
DMA 509
DMI 525, 530, 533
document management 508
DVD 91, 92
ECC 389
e-COMM 577
ECSET 576
EDIFACT 564
electronic commerce 564
EMV 124, 131
ESMTP 241
Europay C-SET 577
FCL 80
Fibre Channel I/O 172
Fibre Channel Interconnect 172
FIPS 392
Fortran 90 431, 455
Fortran 95 455
FOTP-171 141
GSM 215
H.323 259, 260, 262, 568
H.324 259
HTML 234, 242
HTML 3.2 496
HTTP 234, 241, 496
ICC 124
IEEE 1394 358
IEEE 802.14 188
IEEE 802.3 167
IEEE 802.5 168
interoperability 454
IPsec 184
IPv4 241
IPv6 192, 241
ISO 14888-3 389
ISO 7816 109
ISO-3166 TLD 237
ITU-TSS 391
JavaBeans 461

L2F 184
L2TP 184
LANE 170
LDAP 263
MD-API 630
MDIS 627
MIB-II 531
Mobile IP 191
modems 187
MPOA 170
OAGIS 679
OBI 565
objects 461
ODBC 469, 477, 629
ODMA 508, 509
ODMG-93 474
OFSTP-14 141
OFX 578
OpenPGP 413
OPS 407
OPSEC 385
P1363 389
PC 98 358
PEM Internet 411
PKCS-11 376
Platform for Privacy Preferences 407
POSIX 456
PPTP 184
privacy 407
programming languages 430
QIC 75
RFC 1112 193
RFC 2208 194
RFC 2209 194
RMON 531, 532
RMON2 528, 530, 533
RSVP 190
RTP 193
S/MIME 412
S/WAN 184
SDH 142
SDSI 392
security 423
SET 411, 574, 576
S-HTTP 184
Single UNIX Specification 310
smart cards 123
SNA 564
SNMP 527, 528, 530
SNMPv2 532
SNMPv3 548
SQL 476
SQL3 469, 478
SQL-92 469, 476, 477
SSL 184
Subset G PL/1 432
T.120 260
TCP/IP 241
TDMA 13
Tradacoms 564
TRUSTe 408, 580
TSB-67 148, 149
UMTS 12
vCard 407
VLANs 183
VPNs 184
WBEM 527, 534
wiring 150
WMI 527
workflow 504
X.25 564
X.500 292, 391, 411
X.509 411, 414, 574
X/Open Single UNIX Specification 456
XATMI 337
XML 243
Stanford Technology Group Inc.
see Informix Software Inc.
Stanford University 388
StarBurst Communications Corp. 193
Stepstone Corp.
Objective-C 436
Sterling Commerce Inc. 567, 587, 592, 593
Sterling Software Inc.
Composer 442, 444, 450
STM 144
for Synchronous Transport Module
storage devices
8-mm tape 72
AIT 72
ALDC 73
architecture 80
audio requirements 688
bar codes 117
CD-ROMs 70, 89
DAT 73
DDS 73
disk drive interface 78
diskettes 86
Divx 91
DLT 74
DVD 90
DVD-RAM 92
enterprise management 525
enterprise storage 83, 85
file management 289
floppy disks 86
forecast 102, 103
hard disk drives 75
HiFD 86, 87
HSM 88
increased storage density 78
intelligence 83
jukeboxes 69, 88
LIMDOW 93
magnetic tape 70
management 537
management tools 521
market 99
measuring capacity 70
mediums 687
MIC 73
MLR 75
MO 93
multipurpose systems 71
optical technology 88, 89
overview 69
PC 98 358
RAID 81
removable disk drives 87
requirements 687
silos 71
subsystems 80
tape drives 69, 71
third-party management 537
Travan technology 75
trends 80
video requirements 688
storage media
capacity 341
databases 605
Fibre Channel Interconnect 172
Storage Technology Corp.
Powderhorn 71
tape drives 70
stored procedures 472
stored-value cards 130
small-value items 128
small-value transactions 122
types of 121
see also smart cards
STP 161
StrataCom Inc.
see Cisco Systems Inc.
subclass 434
Sun Microsystems Inc. 83, 85, 126, 167, 242, 338, 342, 352, 359, 363, 368, 526, 547, 593, 650, 654, 699
Convex 355
Encore Computer Corp. 86
Enduser SKIP 397
Infinity SP 86
Java 285, 312, 436
Java Platform for the Enterprise 330
Java Stations 298
JavaBeans 328, 461
JavaCard 112, 113, 125
JavaCard API 376

JavaOS 301, 312
JavaWallet 574
JVM 125
servers 129
SKIP 379
Solaris 168, 302, 310, 495, 537
Solstice 547
Solstice Security Manager 379
SPARC 301
Sunscreen SPF 100 379
Ultra-2 345
UltraHPC 355
VCSEL 145
SunSoft Inc. 461
Solaris 286, 300
Solstice Enterprise Manager 544
Solstice Network Client 292
Visual Workshop 430
Workshop 446, 448
see also Sun Microsystems Inc.
Sunsoft Inc.
Workshop development tools 446
supercomputers 355
market 355
vendors 355
SVCs 171
for switched virtual circuits
switches 174, 175, 177, 180
advantages 180
architecture alternatives 180
ATM 176
capacity 180
costs 168
Ethernet 167
Fibre Channel 173
flow-switching 181
forecast 200
IP switch 181
LANs 194
Layer 2 switching 180
Layer 3 switching/routing 180, 199
Layer 4 switching 182
layerless 182
multicast 193
network management 529
OSI 182
packet-switching 220
speeds 143
switched routing 181
VLANs 182
switching fabric 80
Sybase Inc. 265, 310, 337, 400, 440, 485, 488, 637, 638, 676
Adaptive Server Enterprise 11.5 488
dbQueue 635
FormulaOne/NET 246
IQ 485, 488
Replication Server 480
SQL Server 442
Stored Procedures 478
Sybase IQ 621
Symmetric Multiprocessing 488
WarehouseNOW 636
Symantec Corp. 457, 461, 464, 526
Norton Antivirus Internet Scanner 246
Norton Utilities 2.0's System Genie 410
Norton's Antivirus for Internet E-mail Gateways 403
Symbios Logic Inc. 85
SYM8951U 79
Symbol Technologies Inc.
PDF417 117, 118
SyQuest Technology Inc.
135 MB EZ 86
EZFlyer 230 86, 89
Quest 4.7-GB 88
Rocket 88
SparQ 87, 100
systems management
applications management 539, 548
background 522
change and configuration management 537
core system functions 455
cost reduction 526
data warehousing 615
desktop computing 525
DMI 533
file management 292
market 541, 544
operating systems 289
overview 521, 522
performance 522, 539
performance management 536
performance market 541
revenue 542, 543
standards 530
vendors 542
see also enterprise management

T

T/R Systems Inc.
Multiple PrintStation 024 Performance 64
T11 Committee 80
Tadiran Electronic Industries Inc. 97
Talarian Corp.
SmartSocket 336
Taligent Inc. 317
Tally Printers Corp.
SpectraStar T8050 61
Tandberg Data ASA 75
Tandy Corp. 46
tape drives
8-mm 72
AIT 72
ALDC 73
D2 tape technology 71
advantages 71
DAT 73
DDS 73
DLT 74
HDDT 72
market 70
measuring capacity 70
MIC 73
MLR 75
QIC 74
silos 71
Travan technology 75
trends 69
using effectively 74
see also magnetic tapes, storage devices
TAPI
for Telephone API
TASC Inc.
see Primark Corp.
Taxware International Inc. 658
TCI 225
for Tele-Communications Inc.
TCO
for total cost of ownership
TCP/IP 176, 178, 184, 186, 189, 292, 413, 655
Internet 241
MPOA 170
overview 189, 241
SNMP 530
switches 182
TCSEC 383, 384
for Trusted Computer System Evaluation Criteria
TDCC 564
for Transportation Data Coordinating Committee
TDM 143, 163
for Time Division Multiplexing
TDMA 13, 215, 217, 229
for Time Division Multiple Access
TeamWare Group
see Fujitsu Ltd.
Tech Data Corp. 581
Technical Committee for Device Level Interfaces 80
Technically Elite Inc. 533
Technology Resources Industries Bhd. 208

Tele Danmark 207
Telebit Corp. 196
Telecom Malaysia 123
telecommunications
affordability 10
AMPS 214
ATM 169, 203, 222
bridges 175
cabling 146
CATV 223
circuit switching versus packet switching 220
CLECs 223
competition 206
convergence 7, 8, 9, 17
copper cabling 161
deregulation 203, 228
devices 174
digital services 220
document transmission times 690
DSL 212
electronic commerce 562
European deregulation 7
European market 10
forecast 228
frame relay 171
groupware 512, 513
HFC 152
International Charter 11
international deregulation 207
Internet 210, 260, 274
ISDN 221
ISPs 229
joint ventures 208
line rates 690
LMDS 213
market 206, 229
mergers 204, 228
microwave transmission 159
mixing voice and data 203
modems 187
network management 522
overview 203
power lines 150
pricing 41, 229
private line 220
private line costs 221
QoS 222
rate ranges 689
regulation 11
remote access devices 186
RSVP 190
RTP 191
satellites 156
server-to-server 261
service providers 233
services 219
smart cards 119, 121, 127, 129
standards 229
switches 180
telephony 224
transmission times for files 691
trends 137, 219
types of satellites 156
U.S. market 160
U.S. Telecommunications Act of 1996 203, 204
versus broadcasting 9
VPNs 184
wireless telephones 213
wireless terminals 218
wireless vendors 218
wiring standards 146
see also cellular technology, data communications, Internet, mobile computing, transmission media, wireless connections, World Wide Web
Telecommunications Act of 1996
see U.S. Telecommunications Act of 1996
telecommuting
data access 521
forecast 228
TeleDenmark
Danmønt 122
Teledesic Corp. 158, 213
Telefonica Communicaciones Personales 128
Telefonos de Mexico 209
Teleglobe Inc.
Odyssey 157
Telenor 208
telephones
cellular 112, 203
CTI 224
HFC 152
interactive voice response 52
PBXs 224, 225
smart cards 119, 128
switching facilities 224
trends 512
wireless 225
telephony 210
AMPS 214
ATM 222
background 224
business-oriented 262
cable 224
computer devices 260
computer-to-telephone calls 260
CTI 224
digital 513
full-duplex 262
groupware 512
H.323 261, 262
half-duplex 262
Internet 231, 258, 274
directory access 263
forecast 280
licensing agreements 263
phone packages 262
service providers 263
technologies 261, 262
teleconferencing 263
vendors 262
LDAP 263
market 277
metadirectory 264
mobile 12
paging 219
PCS/PCN 217
phone-to-phone 261, 263
RTCP 191
technologies 213, 224
trends 263
Teleservices Data Protection Act 407
television
cable 160
interactive TV 119
pay-per-view 119
Teleway Japan 209
Telnet 414, 416
Telstra Corp. 123, 209
Template 502
Teradata Corp.
see NCR Corp.
TeraStor Corp.
NFR 77, 78
Terisa Systems Inc. 418, 574
SecureWeb Client 2.0 and Server Toolkit 2.0 415
SecureWeb Toolkit 414
testing
basic link 148
channel configuration 148
specifications 148
standards 148
Texas Instruments Inc. 48, 118, 129, 364, 417
DLP 59
Information Engineering Facility 442
Thailand
overseas competition 209
Thawte Consulting 393
The Benchmark Handbook 482
Thinking Machines Corp.
Darwin 627
WTI 629
Thomson-CSF Inc.
FingerChip 378

Thorn-EMI 72
TIA 146
for Telecommunications Industry Association
TIBCO 336
TIBCO Inc. 335, 336
Information Bus 336
Tiger Technologies 380
Time Warner Inc. 91, 239
TimeStep Corp. 184
TIS 185, 384, 398, 399, 418
for Trusted Information Systems Inc.
Gauntlet 416, 421
RecoverKey Cryptographic Services 399
Tivoli Systems Inc.
see IBM
TLS 415
for Transport Layer Security
TM1 Software Corp. 630
Token Ring 148, 166, 168
challenges 168
encapsulation 168
gigabit 168
LANE 170
RMON 532
source routing bridges 175
Torrent Networking Technologies 177, 182
Toshiba Corp. 91, 92, 126, 260, 526
Divx 92
flat-panel display 56
T3100-series and T5200-series 56
Tecra 740 and 750 56
touchscreens 49
capacitive sensing 49
definition 49
forecast 66
guided acoustic wave 49
technologies 49
types of 49
vendors 50
TP
for transaction processing
TP monitors 330, 338
asynchronous communications 323
CICS 322, 323
disadvantages 323
forecast 339
ISAM 323
market 336
OTMs 339
overview 317
security 323
synchronous communication 323
technologies 322, 323
TPC 481
for Transaction Processing Performance Council
Web site 481
TracePoint Technology Inc.
HiProf 447
Visual Coverage 448
Trade'ex Electronic Commerce Systems Inc. 567
Distributor 567
TradeWave Corp. 418
Trading Partner Network
Post 569
transmission media
ADSL 212
advances 149
alternatives 138
ANSI/EIA/TIA Category Specification 147
backbones 195
bridges 175
cable modems 188
certification 149
choosing media 138
coaxial cable 150
digital transmission 211
fiber optics 138
forecast 162, 228
fractional T1 220
frame relay 203
high speed 145
hubs 175
Internet 239
lasers 141
Layer 3 routing and switching 166
market 159
MIB 531
microwaves 159
modems 187
network management 529
NEXT 149
NICs 167
optical multiplexing 161
overview 137
packet-switching 220
parallel transmissions 149
PC 98 358
power lines 150
Power Sum 149
remote access devices 198
RMON 533
routers 176
satellites 137, 156
SDV 153
speed 147, 149, 341
switches 180
T1 circuits 220
technologies 138
telecommunications 203, 210
testing 148
trends 137
twisted-pair cable 138
types of 138
types of satellites 156
virtual circuits 220
wireless 155
wireline 138
see also bridges, hubs, packets, routers, switches, wiring
Transtar Software Corp. 450
Travan technology
capacity 75
Traveling Software Inc.
WebEx 253
Trend Micro Inc. 404
InterScan VirusWall 406
Interscan VirusWall 403
MacroTrap 403
Web Protect 406
Trillium Digital Systems Inc. 172
Trinzic Inc.
see Platinum Technology Inc.
Triple-DES 414
see also encryption, security
TriStrata Security Systems 396
Trust Technology Assessment Program 383
TRUSTe
types of privacy 408
TRW Inc. 648
Odyssey 157, 158
TSAPI
for Telephony Server API
Tseng Labs 60
TSI International Software Ltd. 400
Mercator 322
Tut Communications 149

U

U.K. Department of Trade and Industry 401
U.S. Defense Advanced Research Projects Agency 232
U.S. Department of Commerce 398
Triple-DES 399
U.S. Department of Defense 218, 355, 383, 433
U.S. Department of Energy
EnergyStar program 55
U.S. Department of Justice 278
U.S. District Court of Appeals 205
U.S. Immigration and Naturalization Service 378

U.S. National Science Foundation 237
U.S. Post Office 47
U.S. Postal Service 118
U.S. Robotics
see 3Com Corp.
U.S. Supreme Court 205
U.S. Telecommunications Act of 1996 17, 18, 160, 203, 206
CLEC 160
ISPs 589
issues 205
mergers 204
results of 204
U.S. Telephone Consumer Protection Act of 1991 410
UB Networks Inc. 167
UCS 564
for Uniform Communications Standard
UDP 193
for User Datagram Protocol
Ultimus LLC 502
Ultralife Batteries Inc. 97
Umax Data Systems Inc.
PowerLook II 50
UML
for Unified Modeling Language
defined 444
UMTS 12
for Universal Mobile Telecommunications System
UN/ECE
for United Nations Economic Commission for Europe
Underwriters Laboratories Inc. 151
unicast 192
Unicode 240, 274, 291, 437
XML 243
Unify Corp. 460
Unispan 172
Unisys Corp. 349, 352, 362
thermogram technology 378
United Nations Commission on International Trade Law 395
Universal Studios Inc.
Divx 92
University of Michigan 391
UNIX 168, 300, 301, 628, 648, 655
C 433
databases 484
Digital UNIX 301
displays 54
features 301
forecast 313
HP-UX 302
IBM AIX 302
IRIX 302
issues 310
Java 310
Linux 300
market 301, 309, 310
Motif 290
OpenServer 302
operating systems 285
portability 291
revenue 310
RISC 300
SCSI 79
security 380, 387
Single UNIX Specification 310
Solaris 302
System V Version 3.2 302
UNIX 95 457
UNIX 98 457
UnixWare 7 302
versions of 301
versus Windows NT 655
workstation market 362
workstation sales 363
workstation vendors 368
workstations 356
Year 2000 tools 452
Year 2038 453
Unwired Planet 216
UPS 583, 658
for United Parcel Service of America Inc.
URLs 234
for Uniform Resource Locators
US West 123, 204, 226
USA Global Link 263
USB 51, 295
for Universal Serial Bus
Usenet
definition 252
USWeb Corp. 658
Utimaco Mergent UK 419
UTP 146, 161
for unshielded twisted pair
Enhanced Category 5 147
Standard Category 5 147
UUNet Technologies Inc. 177, 232, 233, 235, 410, 588

V

Valicert Inc.
ValiCert Toolkit 393
Vality Technology Inc.
Integrity Data Reengineering Environment 610, 613
Vanguard Integrity Professionals Inc.
RACF Administrator 379
VANs 562, 563, 587
for value-added networks
and the Internet 588, 592
EDI 563
market 587
services 563
standards 564
technologies 563
versus ISPs 590
Vantive Corp. 647
Vasco Data Security Inc. 418
vBNS 239
for very high-speed Backbone Network System
VBScript 438, 439
VCSEL 145
for vertical-cavity surface-emitting lasers
VDI 624
Discovery for Developers toolkit 624
VDOnet Corp.
VDOLive 259
VDSL 212
for very high-data-rate DSL
Verbex Voice Systems Inc. 53
Veridicom Inc. 421
Veridicom 378
VeriFone Inc. 400
Integrated Payment System 567
see also HP
Veri-Q Inc.
Vcom 336
VeriSign Inc. 385, 390, 392, 393, 405, 419, 574, 575
certificate classes 393
S/MIME 412
Veritel Corp. of America
VoiceCrypt 378
Verity Inc. 276
Information 510
Versit Consortium 407
VH-ADSL 213
for Very High Speed Asymmetrical DSL
VI 352
for Virtual Interface
Viag AG 208
Viasoft Inc.
Estimate 2000 452
ESW2000 452, 453
Existing Software Workbench 452
ValidDate 452
VIA/ValidDate 452
Viaweb Inc. 567
VICS 564
for Voluntary Industry Communications Standards
video 203
ATM 169

cameras 51, 66
CCD cameras 51
data compression 260
digital video cameras 52
dynamic speaker location 52
frame relay 171
HDDT 72
Internet 274
light-valve projection systems 58
network connections 166
NVOD 71
projection systems 58
real-time 260
RTP 190
storage and retrieval 80
streaming 257
videoconferencing 46, 51, 162, 165, 169, 192, 210, 211, 257, 261
standards 259
VOD 119
Videologic Ltd. 60
Videotron Holdings PLC 205
Videsh Sanchar Nigam 233
Vienna Systems Corp.
Vienna.way 262
Viewsonic Corp.
29GA 54
Viisage Technology Inc.
Viisage Gallery 377
virtual circuits 222
Virtual Integration Technology
WTI 629
virtual networking 540
management 540
Virtual Spin 567
Virtual Technology Corp. 538
Virtual Vineyards
Web site 582
Virtus
Web site 255
Virtus Corp.
3-D Website Builder and Walk-Through Pro 255
Virus Bulletin 402
viruses 402
anti-virus forecast 423
anti-virus market 421
anti-virus software 403
hostile applets 404
types of 402
vendors 421
Visa International Inc. 123, 390, 572, 573, 574, 575, 576, 593
JavaCard 125
SET 124
smart cards 121, 122, 123
Visa Cash 573
Visa Integrated Circuit Card Specification 124
Visible Decisions Inc.
In 3D 623
Visigenic Software Inc.
VisiBroker 339
Visio Corp. 501
Visionics Corp.
FaceIt 377
Visix Software Inc. 461
Visix Galaxy 458
Visual Numerics Inc. 624
Vivo Software Inc.
Events 259
Web site 259
Vixel Corp. 145
VLANs 182
for virtual LANs
VLDBs 85
for very large databases
VLM 352
for very large memory
VLSI 145
for very large-scale integration
VLSI Technology Inc. 167
VM
for Virtual Machine
VMark Software Inc.
WTI 629
VocalTec Inc. 263
VOD 119
for video-on-demand
voice recognition 46
Vosaic Corp.
Audio for Java 259
Java TV Station 259
Web site 259
Voxware Inc.
MetaVoice 263
VPIs 184
for virtual private Internets
VPNet Technologies Inc. 184
VSU-1000X 399
VPNs 183, 191, 373, 385, 587
for virtual private networks
definition 183
exports 399
forecast 199
IPsec 184
L2F 186
L2TP 186
market 587
PPTP 185
standards 184, 564
VRML
for Virtual Reality Markup Language
definition 254
market 255
VSAM 323
for Virtual Storage Access Method
VSATs 157
for very small aperture terminals
VSNL of India 157

W

W3C 236, 266, 407, 571, 578
for World Wide Web Consortium
CSS1 243
HTML 242
HTTP 241
Wahl, Paul 646
WAIS 234
for Wide Area Information Service
Wall Street Journal, The 23, 253, 585
Wal-Mart Stores Inc. 682
Walt Disney Co. 252
Divx 92
WANs 165, 169, 522, 528
for wide area networks
ATM 169
backbones 195
database replication 479
frame relay 171
network management 528
switches 180
switches versus routers 180
Watcom Corp
C++ 446
Watcom Corp. 457
Watermark Software Inc.
see FileNet Corp.
Wayfarer Communications Inc.
Incisa 249
WBEM 527
for Web-Based Enterprise Management
CIM 535
components 534
WDM 139, 141, 143
for Wavelength Division Multiplexing
costs 143
forecast 163
market 160, 161
optical multiplexing 161
products 143
technology 143
Web
see World Wide Web
Web-based computing 459, 460
CGI 267

electronic commerce 556, 557, 558, 559, 565, 566, 592
forecast 548
HTTP 266
JVM 460
management tools 535
market 464
middleware 460
platform independence 535
server architectures 267
servers 266
Web server definition 266
WebCrawler 250
WebTV Networks Inc.
see Microsoft Corp.
WELL 585
Western Digital Corp. 76, 78
WFW
for Windows for Workgroups
Whistle Communications Corp.
InterJet 421
White Pine Software Inc.
Enhanced CU-SeeMe 261
Reflector 261
Wild List, The 403
Winchester drives
see hard disk drives
Windows
see Microsoft Corp.
Windows NT
see Microsoft Corp.
WIPO 11, 238
for World Intellectual Property Organization
Wired 253
wireless connections 137
CATV 213
data access 521
forecast 163, 228, 229
GSM 219
major vendors 218
market 208, 226
microwave transmission 159
MMDS 213
smart cards 122
technologies 155
telecommuting 165
telephones 213, 225
trends 16, 137
see also cellular technology, satellite communications, transmission media
wireline connections
see wiring
wiring
ANSI/EIA/TIA Category Specifications 147
backbone cabling 141
Baseband Transmission 151
Broadband Transmission 151
Category 5 162
Category 5 UTP 145
Category 6 162
certification 149
coaxial cable 138, 150
Commercial Building Telecommunications Cabling Standard 147
copper 137, 146, 149, 162, 210
market 161
copper versus fiber 153
costs 148, 153
European market 161
European types of 161
FC-AL 80
fiber optics 138, 210
fiber optics versus copper 139, 155
foil twisted-pair 146, 161
forecast 162
HFC 152
High-End Category 5 150
horizontal cabling 141
hybrid fiber-coax 211
interference frequencies 155
market 159
parallel transmissions 149
plenum 151
premises 163
RG 58 A/U 151
RG 58 C/U 151
screened twisted-pair 146
SDV 153
standards 146, 149
STP 161
T1 212
Teflon 151
testing 148
twisted-pair 138, 170, 174, 212, 220
twisted-pair cable 146, 147, 149
twisted-pair cables 147
twisted-pair copper cable 138
UTP 161
WiseWire Corp. 276
WOLF 10
for World Wide Web Opportunities in Less Favored regions
WordPerfect Corp.
see Corel Systems Inc.
workflow
ad hoc 500, 515
administrative 500, 502, 503
BPR 501
categories 499
collaborative 500, 515
continuum 500
converging technologies 513
definition 499
document management 506
e-mail-based 502
forecast 516
implementation 500
Interface 1
Process Definition 504
Interface 2
Workflow Client Application 504
Interface 3
Invoked Applications 504
Interface 4
Workflow Interoperability 504
Interface 5
Administration and Monitoring Tools 505
interfaces 504
Internet 516
market 514
overview 491
policies 501
practices 501
process 501
production 500, 503, 516
roles 501
routers 507
routes 501
routing models 502
rules 501
standards 504, 516
technologies 499, 500, 501
transaction-based 500
vendors 501, 502, 503
Workflow Canada 504
Workflow Management Coalition 504, 509
workflows
ad hoc 502
collaborative 502
WorkLink
BAT Personal Keyboard 47
workstation
see desktop computing
World Administrative Radio Conference of 1992 158
World Radiocommunication Conference 158
World Wide Web
agent technology 252
aggregators 276
audio and video
real-time 260
technologies 260

authoring tools 265
cookie technology 255
CSSs 242
market 273
overview 255, 256
vendors 256
background 233
caching 239
clipping services 253
computing 459
content aggregators 276
cookie technology 255, 408
data warehousing 630, 638
definition 234
development tools 462, 464
document management 508, 515, 517
domain names 237
country suffixes 237
second-level 237
suffixes 237
top-level 238
electronic commerce 553, 559, 579
market 581
electronic mall 560
EPS 219
filtering
collaborative 254
forecast 277
front office products 593
groupware 515
growth 272
information retrieval 247
interactive technologies 254
intranets 165
Java-based applications 429
JavaBeans 327
keywords 251
management issues 535
market 272
metasearch engines 253
modems 239
Mosaic 234
multi-culturalism 240
navigation tools 231
network management 521, 528, 540
news services 584
operating systems 312
overview 231
PCS 218
Perl 439
personalization 253
PICS 252
pointing devices 48
publishing 234
market 273
pull technology 253
revenues 273
RTP 191
scripting languages 437
search engines 250, 276
security 231, 373, 382, 383, 384, 413, 423
smart cards 128
streaming 257
subscription fees 585
technologies 236
television-based devices 240
top ten sites 582
travel 584
trends 23
users 591
vendors 254, 258, 260
VRML 254
W3C 236
WBEM 534
Web commerce 556
Web site
cost 232
development tools 265
hosting 210
large-scale 243
uses 231
Web site building tools 566
Web-accessible tools 631
Web-based management tools 535
Web-enabled application development 441
Web-enabled tools 631
Web-exploited tools 632
see also browsers, electronic commerce, Internet, intranets, search engines, Web-based computing, World Wide Web Consortium
WorldCom Inc. 223
WORM 89, 109, 118
for write-once read-many
ablative 90
continuous composite 90
WTI 629
for Warehouse Technology Initiative
WTO 15, 401
for World Trade Organization
WWW
see World Wide Web
WWW Worm 250

X

X.500
security 391
X.509 413
PEM 411
security 391
X/Open Company Limited
see Open Group
XATMI 337
xBase 438
Xerox Corp.
DocuPrint 184 64
Docuprint 4517 63
InterDOC 508
TextBridge 50
WorkCenter 250 65
Xerox PARC 17, 23, 435
for Palo Alto Research Center
IP multicast 192
XML 250
for Extensible Markup Language
definition 243
DOM 244
forecast 278
uses of 243
versus HTML 243
Web site 243
Xpert Corp.
PlanXpert for Data Warehouse 614
XVT Software
XVT 458
Xylan 168, 170, 176, 183

Y

Yago Systems 177, 182
Yahoo! Inc. 276, 584, 658
Yahoo! Web site 250
Yamaha Corp.
CRW2260 90
CRW4260t 90
CRW4260tx 90
Year 2000 306, 342, 429, 430, 594, 644, 646, 680
background 450
data flow analysis tools 452
date-simulation 451
Euro currency 12
forecast 467
impact analysis 452
market 462, 463
needs assessment 452
reengineering 452
small-scale projects 452
source code changes 453
testing 453
TIME macro 451
tools 450, 451

Z

Zenith Electronics Corp.
Divx 92

Zinc Software Inc.
    Zinc Application Framework 458
Zoomit Corp. 418

PRICE WATERHOUSE *TECHNOLOGY FORECAST: 1998* CLIENT SURVEY

We are interested in your feedback on the Price Waterhouse *Technology Forecast*. We would appreciate a few minutes of your time to fill out this questionnaire after you have had a chance to review the *Forecast*. Please mail to the address on the back or fax to +1-650-321-5543.

1. For what purposes have you used the *Forecast*?

__ to prepare for a presentation or client meeting
__ for strategic technology advice
__ for marketing/competitive products information
__ as a general technology reference
__ as a technology implementation guide
__ other: ....

2. Does the *Forecast* meet your needs? (circle one)

Yes, always   Most of the time   Sometimes yes, sometimes no   Usually not   Never

3. Can you easily find the topics you are looking for?

Yes, always   Most of the time   Sometimes yes, sometimes no   Usually not   Never

4. Is the depth of coverage appropriate for the topics?

More information than I need   Just right   Less information than I need   Too superficial

5. How do you rate the organization of the individual sections (Executive Summary, Technology, Market Overview, Forecast, References)?

Excellent   Good   OK   Confusing

6. How do you rate the overall organization of the *Forecast*?

Excellent   Good   OK   Confusing

7. Which of the Notes databases supporting the *Forecast* have you used?

Technology Forecast (Notes version)   Technology Update Service   These databases are not available to me

8. Chapter names or numbers you particularly liked: ....

9. Chapter names or numbers you particularly disliked: ....

10. Are there additional topics that should be covered in the *Forecast*?

11. Other suggestions for improving the *Forecast*:

Optional: Name ....

Company ....

Address ....

....

Phone ....

Please contact me about the Price Waterhouse Technology Update Service ❑

If you would like someone from the Technology Industry Group to contact you, please check here ❑

Printing 10 9 8 7 6 5 4 3 2 1

TECHNOLOGY FORECAST EDITOR
PRICE WATERHOUSE TECHNOLOGY CENTRE
68 WILLOW ROAD
MENLO PARK, CALIFORNIA 94025-3669
U.S.A.